Algorithms for Intelligent Systems

This book series publishes research on the analysis and development of algorithms for intelligent systems with their applications to various real world problems. It covers research related to autonomous agents, multi-agent systems, behavioral modeling, reinforcement learning, game theory, mechanism design, machine learning, meta-heuristic search, optimization, planning and scheduling, artificial neural networks, evolutionary computation, swarm intelligence and other algorithms for intelligent systems.

The book series includes recent advancements, modification and applications of the artificial neural networks, evolutionary computation, swarm intelligence, artificial immune systems, fuzzy system, autonomous and multi agent systems, machine learning and other intelligent systems related areas. The material will be beneficial for the graduate students, post-graduate students as well as the researchers who want a broader view of advances in algorithms for intelligent systems. The contents will also be useful to the researchers from other fields who have no knowledge of the power of intelligent systems, e.g. the researchers in the field of bioinformatics, biochemists, mechanical and chemical engineers, economists, musicians and medical practitioners.

The series publishes monographs, edited volumes, advanced textbooks and selected proceedings.

More information about this series at http://www.springer.com/series/16171

I. Jeena Jacob ·
Selvanayaki Kolandapalayam Shanmugam ·
Selwyn Piramuthu ·
Przemyslaw Falkowski-Gilski
Editors

Data Intelligence and Cognitive Informatics

Proceedings of ICDICI 2020

Springer

Editors
I. Jeena Jacob
GITAM University
Bengaluru, India

Selwyn Piramuthu
University of Florida
Gainesville, FL, USA

Selvanayaki Kolandapalayam Shanmugam
Department of Mathematics and Computer Science
Concordia University Chicago
River Forest, IL, USA

Przemyslaw Falkowski-Gilski
Gdańsk University of Technology
Gdańsk, Poland

ISSN 2524-7565 ISSN 2524-7573 (electronic)
Algorithms for Intelligent Systems
ISBN 978-981-15-8529-6 ISBN 978-981-15-8530-2 (eBook)
https://doi.org/10.1007/978-981-15-8530-2

This Springer imprint is published by the registered company Springer Nature Singapore Pte Ltd.
The registered company address is: 152 Beach Road, #21-01/04 Gateway East, Singapore 189721, Singapore

The ICDICI 2020 conference is solely dedicated to all the editors, reviewers, and authors of the conference event.

Foreword

The International Conference on Data Intelligence and Cognitive Informatics (ICDICI 2020) was held in Tirunelveli, India, during July 8–9, 2020, at SCAD College of Engineering and Technology. The proceedings of ICDICI 2020 conference is presented here with the aim to share and exchange the state-of-the-art research ideas about the different aspects of data and informatics research with a special attention to the encountered practical challenges and the potential solutions adopted to overcome it.

We strongly believe that the research articles of ICIDICI 2020 will give you a technically rewarding experience by providing more research information on the current issues of informatics and general data science interest. IDICI received 247 papers across the country and also from overseas by representing government, industries, and academia out of which 74 papers are duly selected for publication.

ICDICI 2020 promises to be more informative and research simulating with a magnificent array of keynote speakers across the globe. The program consists of invited sessions, presentations, and technical discussion with the most eminent and proficient speaker **Dr. Joy Chen**, Professor, Department of Electrical Engineering, Dayeh University, Taiwan, and session chairs by covering a wide range of topics in data intelligence. Also the conference delegates had a wide range of sessions in different domains of data science, informatics, and cognitive intelligence.

We humbly wish to thank the guest editors **Dr. I. Jeena Jacob, Dr. Selvanayaki Kolandapalayam Shanmugam, Prof. Selwyn Piramuthu, and Dr. Przemyslaw Falkowski-Gilski**, organization staff, technical program committee, and reviewers of the conference for their valuable suggestions and timely response to the authors of

ICDICI 2020. We also extend our gratitude to the authors and conference participants for contributing their novel research results to the conference. Special thanks to Springer publications.

Dr. P. Ebby Darney
Conference Chair
Principal
SCAD College of Engineering
and Technology
Tirunelveli, India

Preface

With a deep satisfaction, I write this Preface to welcome you all to the International Conference on Data Intelligence and Cognitive Informatics (ICDICI 2020) held in Tirunelveli, India, during July 8–9, 2020.

The theme of ICDICI 2020 is *Data Intelligence*, a research topic that is gaining quick research traction in both industries and academia due to its research relevance to the emerging societal and economic issues in the areas like health care, transportation, industries, education, and so on. The well-established research track records on intelligent information systems along with the integration of artificial intelligence techniques and processes make ICDICI an excellent venue for exploring the cognitive foundations in the emerging information systems.

With respect to the potential hardwork of the ICDICI 2020 conference committee, I would like to express my appreciation and gratitude to all the technical program committee members, international and national advisory board members, and review committee members, who have made this conference a successful and possible one.

Finally, I would like to extend my warm thanks to all the keynote speakers, session chairs, and fellow researchers, who have willingly shared their research experience and knowledge to all the readers of this extended conference proceedings.

I hope that this proceedings of ICDICI 2020 will further stimulate research in data mining and intelligent systems and provide practitioners with advanced algorithms, techniques, and tools for deployment. I feel honored and privileged to serve the significant recent developments in the field of intelligent systems and data intelligence to you through this exciting program.

Bengaluru, India

Dr. I. Jeena Jacob
ICDICI-2020

Acknowledgements

We are deeply obliged to all the contributors of this ICDICI 2020 conference and all the reviewers for their informative, cogent, and timely reviews of papers submitted to the conference, and also the SCAD College of Engineering and Technology staffs and international technical reviewer's community with their professional and thoughtful expertise to shape this conference event.

Overall thanks go to all the conference programme and local organizing committee members, who have gone out in their way to make this as a successful event.

Finally, we offer our sincere thanks to conference chair, co-conference chair, and organizing secretary for their continuous effort in organization, preparation, and handling of ICDICI conference administration. Further appreciation is also due to the editors of Springer publications, high standards of editorial productions of conference proceedings.

Contents

About the Editors

I. Jeena Jacob is working as a Professor in Computer Science and Engineering department at GITAM University, Bangalore, India. She actively participates on the development of the research field by conducting international conferences, workshops and seminars. She has published many articles in refereed journals. She has guest edited an issue for the International Journal of Mobile Learning and Organisation. Her research interests include mobile learning and computing.

Selvanayaki Kolandapalayam Shanmugam is currently working as an Associate Professor in Computer Science, Concordia University, Chicago, USA. She had over all 15+ years of lecturing sessions for theoretical subjects, experimental and instructional procedure for laboratory subjects. She presented more research articles in the national & international conferences and journals. Her research interest includes image processing, video processing, soft computing techniques, intelligent computing, web application development, object-oriented programming like C++, and Java, scripting languages like VBscript and Javascript, data science, algorithms, data warehousing and data mining, neural networks, genetic algorithms, software engineering, software project management software quality assurance, enterprise resource planning, information systems and database management systems.

Selwyn Piramuthu is Professor of Information Systems at the University of Florida. He received his B.Tech., M.S. and Ph.D., respectively, from IIT Madras, University of Arizona and University of Illinois at Urbana–Champaign. His research interests include machine learning and cryptography with applications in medical informatics, supply chain management, financial credit risk scoring, IoT, among others. His book, co-authored with Wei Zhou titled, "RFID and Sensor Network Automation in the Food Industry," was published by Wiley in 2016.

Przemyslaw Falkowski-Gilski is a graduate of the Faculty of ETI, Gdansk University of Technology. He graduated 1st degree B.Sc. studies (in Polish) and 2nd degree M.Sc. studies (in English) in 2012 and 2013, respectively. He pursued his Ph.D. studies in the field of electronic media, particularly digital broadcasting systems and quality of networks and services. In 2018, he received the title of Doctor of Technical Sciences with distinction, discipline Telecommunications, specialty in Radio communication. Currently, he works as an academic.

Chapter 1
Text Generation and Enhanced Evaluation of Metric for Machine Translation

Sujit S. Amin and Lata Ragha

1 Introduction

Recurrent neural networks (RNNs) have the capability to learn from the given text and generate new text or translate the given text. They are powerful since they have a lot of hidden states with nonlinear dynamics that permit them to process them previous information. Further, gradients of RNN are cheap to compute. Even with all these features, RNNs are difficult to train properly. The cause is the “vanishing gradient problem” [1]. Vanishing gradients problem occurs when neural networks have many layers in neural network gradients of loss functions approach to zero, making it hard to train. Because of this, there have been limited successful applications using RNNs [2].

Gated recurrent units help to fine-tune neural networks input weights to resolve vanishing gradient problem. Gated recurrent unit (GRU) is a variant of long short-term memory (LSTM) because both have a similar design; GRUs are an improved version of the recurrent neural network [3]. Gated recurrent unit is known to have good text generation in many languages [4].

Our model uses GRU architecture. Our model uses a sequence memorizer [5], which is a hierarchical nonparametric Bayesian method. The memorizer brings dependencies between its predictions by making similar predictions at related contexts.

S. S. Amin (✉) · L. Ragha
Department of Computer Engineering, Fr. C. Rodrigues Institute of Technology, Vashi, Navi Mumbai, India
e-mail: mastersujitamin@gmail.com

L. Ragha
e-mail: lata.ragha@gmail.com

I. Jeena Jacob et al. (eds.), *Data Intelligence and Cognitive Informatics*, Algorithms for Intelligent Systems,
https://doi.org/10.1007/978-981-15-8530-2_1

The problem is identified that the text generated by RNN might not be grammatically correct, GRU, although learns patterns after training for a large number of times. At times, it does make grammatical mistakes. To combat that problem, the grammatical correction model is used using natural language processing (NLP) at the end of text generation. Grammar model uses language modelling, word tagging to check grammar for generated text sentence by sentence. To achieve this, natural language toolkit (NLTK) is used. From framework, it uses part-of-speech tagging, noun phrase extraction, spelling corrections, tokenization and sentiment analysis to achieve grammar auto-corrections.

Text generated after grammar corrections was a significantly high level of linguistic structure and has a considerable amount of grammatical structure in it.

Next component is machine translation (MT), and MT is a field that uses machines to translate text from language to language. During translation, the meaning of the text must be similar in the goal language. The translation is not a replacement of word for the word problem.

In [6], the authors talk about the neural machine translation (NMT) system, which is applied in English to German. Authors have achieved a state-of-the-art model for NMT. In [7], the authors achieved multiple translations of the same paragraphs. Authors have achieved high-quality translations. Translation system uses LSTM networks.

In MT, the sentence-level score is calculated by the BLEU metric. The bilingual evaluation understudy (BLEU) metric is a metric developed by IBM, and it is a way to measure translation quality [8]. Closer the translation to an expert human translation, the superior it is [8]. But BLEU metrics have drawbacks when evaluating Indian languages. So, this drawback aimed to remove using modified BLEU metric for translation. Modified BLEU uses synonym replacing algorithm and after that shallow parsing algorithm was used to get better results (Fig. 1).

This way, two goals are accomplished: one is grammatically correct generated English text from text generator and improved BLEU score for translation.

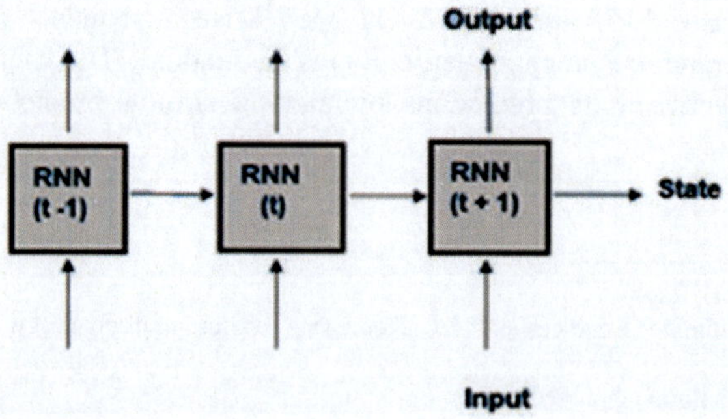

Fig. 1 RNN is a very deep feed-forward network which uses nonlinear activation function

2 System Overview

The system's goal is to generate English text from a given input and then translate text to Hindi with high translation score (BLEU score). As shown in Fig. 2, the system first takes the input and then generates text in utf-8 format. UTF-8 format is chosen because it is difficult to convert text into Hindi with another encoding format. Input can be a word, sentence or paragraph. Automatic text generation module contains a recurrent neural network (RNN) for creating text from the input text. Since automatic text generation module output can be grammatically incorrect. So, the grammar correction module contains NLTK is used for correcting the grammar text sentence by sentence. Now, got our first output English text. This output is given to the English to Hindi translation module. English to Hindi translation module contains RNN for translation. Now a BLEU score is got which measures the translation accuracy. The BLEU score has known to be less for English to Hindi translation since Hindi is a morphologically rich language [9]. So, the modified BLEU score module is used, which uses synonym replacement and shallow parsing to overcome this issue.

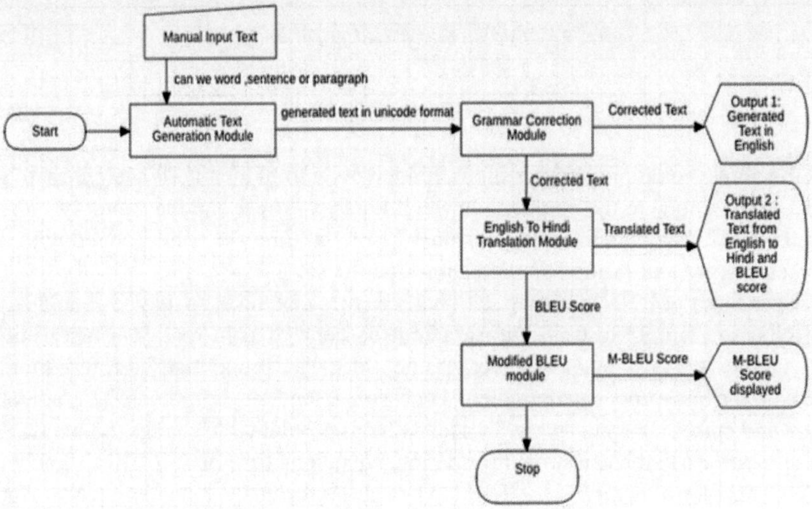

Fig. 2 Flow chart for the overall working of text generation and translation system

3 Neural Networks

3.1 Recurrent Neural Networks

A recurrent neural network is a feed-forward neural network that allows to model sequential data [4]. At each time step, RNN loads an input, revises its hidden state and makes a prediction (Fig. 1). The RNN has a high-dimensional network and has nonlinear activation function over many time steps that enable them to have accurate predictions.

Activation functions are computational equations, which dictate a neural network's output. The function is linked to each neuron in the network and decides if it should be triggered ("fired") or not, based on whether the input of each neuron is essential to the evaluation of the model. Activation functions also help to normalize each neuron's output to a range of around 1 and 0 or -1 to 1.

The standard RNN is written as follows: given a series of input vectors $(x_1, \ldots, x_T)$, the RNN calculates the hidden state sequence $(h_1, \ldots, h_T)$ and the output sequence $(o_1, \ldots, o_T)$ by iterating the following equations for $t = 1$ to T [4].

$$h_t = \tanh(W_{hx}x_t + W_{hh}h_{t-1} + \mathrm{b_h}) \tag{1}$$

$$o_t = W_{oh}h_t + b_o \tag{2}$$

In the above equations, W_{hx} is input, hidden node to hidden node weight matrix is W_{hh}, hidden node to the output node weight matrix is W_{oh} and the vectors b_o and b_h are bias [4]. The undefined expression $W_{hh}h_{t-1}$ at time $t = 1$ is changed with a bias vector $hinit$ and $tanh$ is applied coordinate-wise.

The gradients of RNN are computationally easy to calculate, so maybe concluded that RNNs are simple to train with gradient descent [10]. In reality, relationship and dynamics are unstable. Moreover, gradient decays exponentially through time. Moreover, the backpropagated gradient's occasional tendency improves the gradient descent and makes it very unstable. To deal with this problem is to change the standard RNN structure so that it can store information over a long time period. This is known as "long short-term memory" (LSTM) [11]. It outperformed the standard RNN. But LSTM is computational and memory intensive. LSTM is also less efficient. A better approach is gated recurrent unit (GRU) [3]. GRU is a simpler version that uses two gates, easier to modify and does not need memory units. Moreover, GRU is quicker to train than LSTM. LSTM is good for a machine translation system.

3.2 Gated Recurrent Unit

GRU is a recently proposed method in [3] which allows acquiring dependencies of various timescales. LSTM is similar to GRU. GRU has several gating units that allow the flow between data. GRU is good as a generative model. GRU performs better than LSTM [3].

4 Various Models of RNN

4.1 RNN as Generative Model

The goal of text generative model is predicting the upcoming word in the sequence. GRU is used for the generative model since it is computational efficient and uses less memory. GRU has two gates. GRU does not have internal memory. Also, the nonlinearity model did not apply and clearly, much better than LSTM. Standard RNN uses replace activation with a new value calculated from current input and hidden state [3]. GRU keeps the available content and adds unique content on top of it. This enables conditional distribution in a produced string to get the next character and provide the next input to RNN. Thus, the model is directed non-Markov model; it resembles the sequence memorizers [4].

4.2 Natural Language Processing for Grammar Correction

The goal of grammar correction is to understand if the grammatical error is present and auto-corrects the sentence [12]. This uses NLP to accomplish the objective. This is used since generated text may not be grammatically correct. NLP tool uses tokenization to convert a sentence into tokens. Then language is detected. Once tokenization and language check are done, spelling check is done. This is done to avoid spelling errors. Finally, the model checks for any semantics error in every sentence and checks for what sentence contains (numbers, alphabet and punctuation). After checking if incorrect text (numbers, alphabet, punctuation) is present, then it is replaced with the correct word or remove text.

4.3 RNN as Translation Model

Translation can be done using neural machine translation (NMT) [7]. NMT has produced a state-of-the-art outcome for English to French, English to German and English to Czech [7]. The goal of NMT is to model directly source sentence (English), to a target sentence (Hindi). It accomplishes by encoder–decoder framework [7]. For each source sentence, the encoder calculates a depiction [13]. The decoder produces an output, one word at a time, based on that depiction of the source (see Fig. 3). By taking a weighted mean of hidden states, the context vector is determined. Figure 4 describes the attention mechanism used in our translation system. Translation model uses LSTM networks.

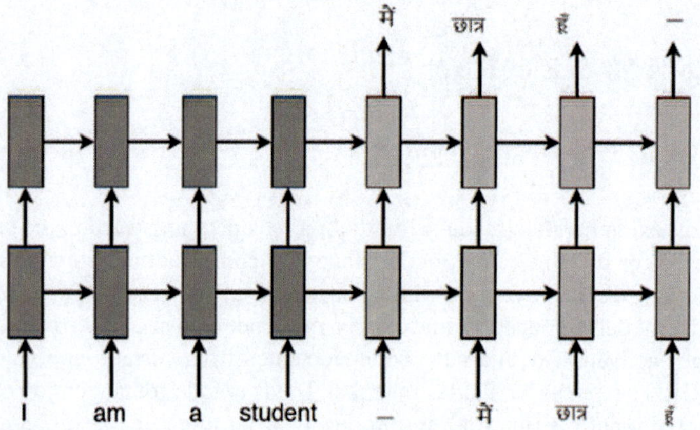

Fig. 3 Neural machine translation—here is an example of RNN for translating English ("I am a student") to ("मैं छात्र हूँ") Hindi [7]

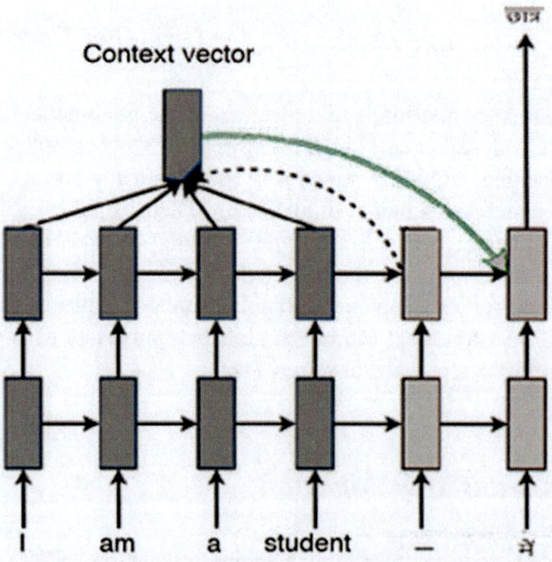

Fig. 4 Attention mechanism—mechanism has two steps: firstly, context vector is calculated by using hidden state; secondly, context vector is used to calculate next hidden state [2]

5 Translation Assessment

Translation assessment answers the accuracy of the translation. Initially, MT assessment was done by humans. It used to be subject from person to person and time-consuming. Assessment varies from one evaluator to another. Hence, the need for automatic translation system is developed. BLEU is the most popular technique developed for the automatic translation system. Human assessment cannot be used every time due to several reasons. On a broader perspective, it has a human assessment and automatic assessment.

5.1 Human Assessment

For human assessment, a speaker of the translated language (e.g. Hindi) and source language (e.g. English) [14] is needed. Human evaluator's translation quality is influenced by various factors such as fluency and adequacy while assessing translation. Moreover, the process can take weeks to months to complete. Monitoring cost also adds up to the total cost. Due to the above variables, there is a need for an automatic assessment scheme

5.2 Automatic Assessment

If the assessment of text is done by machines, then it is automatic machine translation assessment. There are two levels of evaluation. First is sentenced level, and an algorithm is run here on a bunch of sentences then compared against human judgement. Secondly, having a corpus level, the score is calculated by assembling the score of both human and metric judgements. Finally, the aggregate score is calculated. The benefit of automatic machine assessment is that it is much faster.

6 Measuring of Automatic Assessment

The metrics for the MT system are the accuracy of the translated output. However, the quality is subjective. So, metric responsibility is to get the score for translated output text. The score should match with human judgement. Therefore, the metric should give a high score for translations where human will give a high score and low score to translations where humans will give low scores. The BLEU metric has been used for our system.

6.1 BLEU Metric

The bilingual evaluation understudy (BLEU) metric is a metric developed by IBM [8]. The concept of BLEU is nearer a translation output to a translation done by a human, superior the output. Scores are computed for unique segments that are translated. This is usually done by phrasing and comparing translated segments with a set of reference translations of good quality. Those scores are then weighted around the entire corpus to get an estimation of the perceived quality of the translation. No account is taken of the fluency or grammatical correctness. The output of BLEU is always a number between 0 and 1. *N*-gram comparison is used for checking the translation accuracy.

N-gram concept is used in the BLEU metric [8]. The difference is a check between both translations. The score is calculated by the formula:

$$\text{BLEU} = \text{BP} * \exp\left[\sum_{n=1}^{N}\left\{\left(\frac{1}{N}\right) * \log(P_n)\right\}\right] \quad (3)$$

Here N is max n-gram size. n varies from 1 to N [8].

7 Issues with BLEU

BLEU is the most popular method for evaluation of the translated text. But while evaluating English to Indian dialects, it has many disadvantages. [9]. The significant BLEU disadvantages in automatic English-Indian language evaluation are only discussed.

- Poor correlation with judgements of humans.
- The same term in two different forms is considered to be distinct.
- BLEU considers similar meaning words or about the same meaning words as different words.
- The better score does not indicate better translation.
- Many references are required for evaluation purpose.

8 M-BLEU Module

As shown in Fig. 5, the grammatically correct generated text is fed to the machine translation engine. MT engine is two layers deep RNN network for translation. From that, English to Hindi is translated. In the root word extraction part, morphological analysis is done, i.e. is to split the sentence in morphemes. The lexicon contains the stock of words. Lexicon is like a dictionary. Synonyms are grouped. To make correct sentences after synonym replacement process, shallow parsing is used. Then

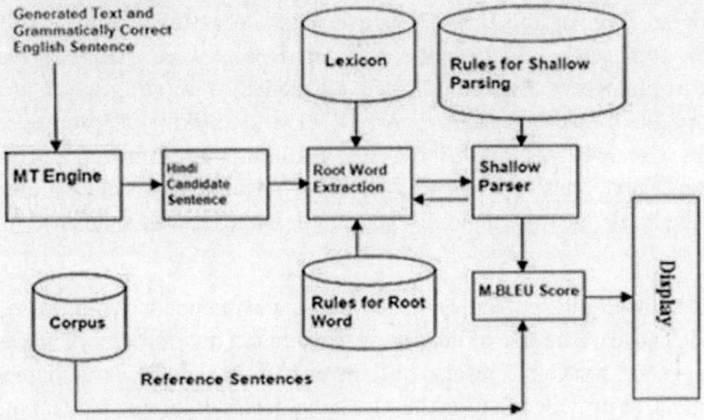

Fig. 5 Modified BLEU module overall working

reference sentences are checked and generated sentences for BLEU score. Shallow parsing and synonym replacement are done several times for every sentence. Until the best BLEU score is reached. Finally, M-BLEU score is calculated. For the entire text, M-BLEU score is calculated by taking the mean of M-BLEU score.

9 Experiments

The goal of our experiments is to demonstrate first that grammatically correct text can be generated from RNN when the gated recurrent unit is used. Second, generated text can be converted to Hindi text with better BLEU score by removing drawbacks.

9.1 Text Generation and Grammar Correction of Generated Text

Aim of this part is to generate text using RNN using the gated recurrent unit. This is demonstrated by comparing LSTM and GRU on English language data sets. Owing to nonparametric nature, LSTM used a large amount of memory while training. GRU, on the other hand, used considerably less memory. The GRU and LSTM are trained on the same size of the data set. Finally, the grammar correction part is applied which gave good correction output text on both LSTM and GRU.

Data sets
The data set is now described. The data set contains text with 80 different types of utf-8 characters which include punctuations, digits and special characters. English text here used is American English. The books have been excluded from other English,

such as Indian English and British English. Data set is made from a collection of books. Ten books are used to make data set. Books were of pdf format. It was converted to plain text. Every book was cleaned by removing any page number, unnecessary line breaks and headers. Any XML and markup were removed to clean the data set. Greek letters, Latin letters and formula were removed since its output is the English text. Punctuations were retained without punctuations output looks obscure. The book's text is randomly permuted. Data set was about 140 MB.

Details

To process the training set (Eq. 2), the entire data set is needed to process, which is infeasible because of the size of the data set system ran out of memory while training. Training the LSTM or GRU can also be done on shorter sequences of character which is just as effective provided they are hundred characters or longer. Advantage of using relatively short sequence is that it is easy to parallelize. Three parallel high-end GPUs and 6 GB of RAM each are used.

Firstly, our model reads the text file and then splits the content character by character. Then characters are sorted. Count of unique characters is also taken. Mapping is done from a unique character to indices. Another index to character mapping is done. Both are used while training. Sequence length is kept to optimal value so that text generated is effective. Data set is taken in slices and run parallel on different GPUs. By looking at the present character, the model predicts the next character. The batch size was 64. The model has 1200 hidden nodes. Small buffer is maintained so that shuffling can be done. It does not shuffle the entire sequence in memory. Model is sequential.

The model has been implemented under the Python programming language. Keras and TensorFlow Library have been used. Keras is the API for neural networks. TensorFlow is a digital computing library. It is devoted to the study of machine learning.

Grammar correction uses natural language toolkit (NLTK). Grammar correction is used since the text generated from the GRU and LSTM recurrent model might not be grammatically correct to sound. Grammar correction uses NLP tools. First, tokens are generated from generated text. Tokens are generated from word to word format. A character-level token does not fit for grammar correction. Then language is detected using spaCy. spaCy is a framework used for NLP. spaCy makes it simple to detect languages. Next, NLTK is used for spelling checks. Spelling checks remove any spelling errors in the text. Next part, check for semantic error; this is done sentence by sentence. Semantic is checked sentence by sentence. Semantic information of text is checked for, i.e. what the text contains (numbers, alphabet, punctuation). After understanding the semantics, if the incorrect word (numbers, alphabet, punctuation) is present, then replaced with the correct word or remove a word.

To validate our approach of generated text, two cases are there by our proposed method.

- LSTM applied to the text: standard LSTM architecture is used on the text.
- GRU applied to the text: as originality of the paper, GRU is used with grammar correction.

Experiment condition is same three parallel high-end GPU, 6 GB of RAM each. A utf-8 encoding has been used. The model took 2 weeks to train properly. The multinomial distribution is used to predict word returned by model. As training progresses, training loss was reduced. Every 100 epochs are gathered in the checkpoint file. Checkpoint file is maintained so that it does not lose model if, for some reason, the machine stops processing. The best set weights are used (least loss) to our generative model.

9.2 Text Translation and Modified BLEU Score for Translation Accuracy

Aim of this module is to translate text from the output of text generation module to Hindi language and generate better BLEU score. The NMT has been used for machine translation. Basically, RNN has been used for the translation since RNN has shown to be very useful in text translation [6].

9.3 Training Details

The data set is now described. Data set is a collection of sentence pairs in English and Hindi which is also made by a collection of books contained both Hindi and English versions. Then the text was carefully placed in pairs. UTF-8 encoding was used. The data set contains about 3 million sentence pairs. Vocabulary is made limited to 50 thousand frequent words. Other words are represented as <unk>. 20% of the data was for validation and test set.

Two layers of LSTM are used with 1000-dimensional embedding. Other methods were applied like attention mechanism and source reversing [15]. The model was trained on the GPU. BLEU score is generated for checking accuracy for translation.

The original BLEU scores were low. As Hindi is a morphologically rich language, so, the modified BLEU (M-BLEU) score is created. M-BLEU module contains shallow parsing and synonym replacement on a different set of sentences. This resulted in better BLEU score as compared to old BLEU score. Synonym data is created from available Hindi data set of synonym. Synonym replacement contains synonyms for every Hindi word, and synonyms of every word are replaced.

At the same time, wrong semantics might be encountered for every sentence where synonym is replaced, and shallow parsing is used. Shallow parsing is basically checked for semantics after replacement if it is correct or not. If incorrect, the sentence is reframed. Once reframing is done, M-BLEU score is checked for.

This works because synonym is considered a different word while calculating BLEU; whereas, while calculating modified BLEU, synonyms are replaced by several times and checked for BLEU score. One with high BLEU score is taken. This is done for every sentence in the translated text. Finally, the overall BLEU score is generated.

10 Results

Every module has been explained output separately. Starting from text generation module, the system is given an input text and output text is displayed. Then after correcting grammar, output of grammar correction module is displayed. The entire generated text is translated; translated text is displayed with the BLEU score. Finally, M-BLEU correct score of BLEU score and M-BLEU score are displayed for the entire translated text. Time taken for entire text generation to translation is 3–5 s on Windows 10 system with i5 processor and 4 GB of RAM.

10.1 Result of Text Generation Module

Here, results of text generation are presented in the form of graphs and tables.

Figures 6 and 7 illustrate the loss function of the first 300 iterations of GRU model and LSTM model applied on a data set. The curve keeps the same shape in both LSTM and GRU model (i.e. both tending to zero).

As you can see from Table 2, standard LSTM takes about 6 times more time to run an epoch. This is significantly higher. The real difference is text generation output. The output of the GRU model learnt much more than LSTM model. LSTM model failed to learn balancing brackets over long distance, and ease of reading is better

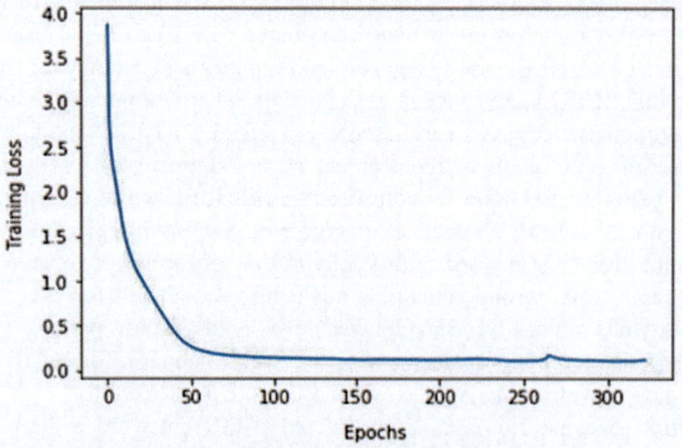

Fig. 6 Shape of the loss function curve for GRU model—it takes a curve path

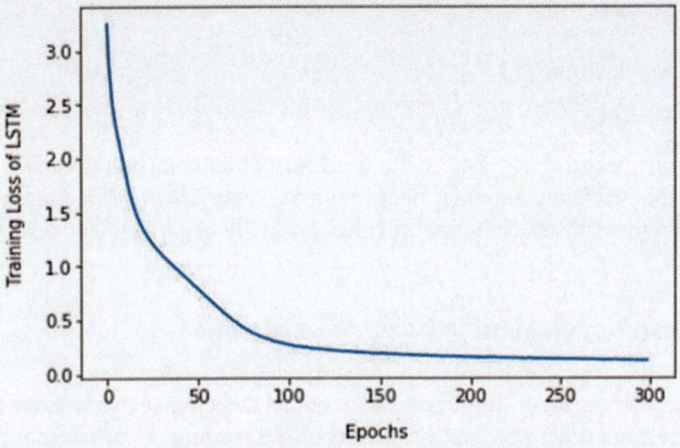

Fig. 7 Shape of loss function for the LSTM model—it takes a curve path

in GRU model to see Table 3. Hence, GRU is a much better approach and also loss function value is less comparatively (Table 1).

Table 1 Loss function value of last epoch

RNN model type	Loss value (last epoch)
Standard LSTM	0.1243
GRU	0.0980

Table 2 Time taken per step in epoch

RNN model type	Time taken per step in epoch
Standard LSTM	741 ms
GRU	136 ms

Table 3 Output by GRU model and LSTM model for same input "supervised learning"

GRU model output for the same input	Standard LSTM model output for the same input
Supervised learning algorithms are used for outlier detection (extreme value or values that deviate from other observations). Supervised learning maps input to output in a data set. Solution provided is usually better than humans	Supervised learning algorithms are used for outer detection. Backpropagation long distance error and the decision boundary, the first training examples in many that are logs to solve the probability density function (of) of the model fit the training data

Input of GRU

Supervised learning

Output of GRU

Supervised learning algorithms are used for outlier detection (extreme value or values that deviate from other observations). Supervised learning maps input to output in the data set. Solution provided is usually better than humans.

10.2 Result of Grammar Correction Module

Supervised learning algorithms are used for outlier detection (extreme value or values that deviate from other observations). Supervised learning maps input to output in the data set. Solution provided is usually better than humans.

Here, "data set" is changed to "the data set" and 'an' is removed. Grammar correction module corrects mostly simple grammatical error. This is because NLTK is used for general applications, not for grammar corrections.

10.3 Result of the Translation Module

Input for Translation Module Supervised learning algorithms are used for outlier detection (extreme value or values that deviate from other observations).

Output
सुपरवाइज्ड लर्निंग एल्गोरिदम का उपयोग बाह्य पहचान (चरम मान या मान जो अन्य टिप्पणियों से विचलित होता है) के लिए किया जाता है। डेटासेट में आउटपुट के लिए पर्यवेक्षित शिक्षण मानचित्र इनपुट। प्रदान किया गया समाधान आमतौर पर मनुष्यों से बेहतर होता है।

BLEU score: 0.588

10.4 Result of M-BLEU Module

After applying shallow parsing and synonym replacement evaluation, results are shown below.

10.5 Result of Translations System

Sentence: Supervised learning algorithms are used for outlier detection (extreme value or values that deviate from other observations).

Translation Text: सुपरवाइज्ड लर्निंग एल्गोरिदम का उपयोग बाह्य पहचान (चरम मान या मान जो अन्य टिप्पणियों से विचलित होता है) के लिए किया जाता है।

Translated Version 1 by M-BLEU Module: सुपरवाइज्ड लर्निंग एल्गोरिदम का इस्तेमाल बाह्य पहचान (चरम मान या मान जो अन्य टिप्पणियों से विचलित होता है) के लिए किया जाता है।

Translated Version 2 by M-BLEU Module: सुपरवाइज्ड लर्निंग एल्गोरिदम का इस्तेमाल बाह्य पहचान (शिखर मान या मान जो अन्य टिप्पणियों से विचलित होता है) के लिए किया जाता है।

Translated Version 3 by M-BLEU Module: बाह्य पहचान (चरम मान या मान जो अन्य टिप्पणियों से विचलित होते हैं) के लिए, पर्यवेक्षित शिक्षण एल्गोरिदम का इस्तेमाल किया जाता है।

BLEU score = 0.7510
M-BLEU score = 0.970 (new score)
Human score = 1.0000

Here, translated versions 1, 2 and 3 are generated for modified BLEU score. One of these is chosen by M-BLEU module for with the highest BLEU score. M-BLEU module algorithm works sentence by sentence rather than the whole corpus so the result is shown for one sentence

As you can clearly see in Fig. 8, M-BLEU score is much better than the BLEU score (Table 4).

Results of translation scores in plot format

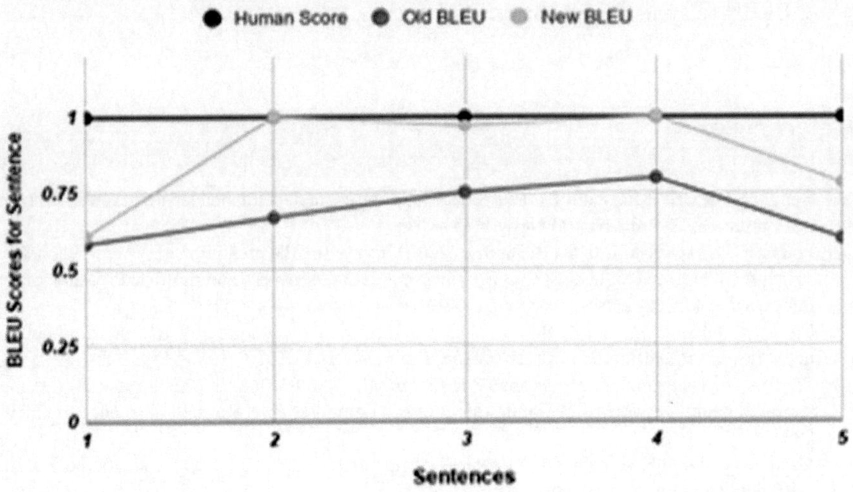

Fig. 8 Line chart showing the comparison of old BLEU score, human score and new BLEU score

Table 4 Sentences and their human score, old BLEU score, new BLEU score, respectively, and the plot is presented in Fig. 8

Sentence no.	1	2	3	4	5
Human score	1.000	1.000	1.000	1.000	1.000
Old BLEU	0.582	0.669	0.751	0.799	0.599
New BLEU	0.607	1.000	0.970	1.000	0.782

11 Conclusion

In this paper, the power of GRU has been shown to generate grammatically correct text and to translate English text to Hindi with a high BLEU score. Generating text makes it easy task with character-level RNN rather than word level. Our text generation module and grammar correction module have generated sentences that are more meaningful and easier to understand for any human. Our text generation had 1200 hidden units with much bigger RNNs with millions and billions of units and more fined-tuned grammar correction high standards of performance can be achieved. Translations to Hindi, our algorithm has achieved better BLEU score as seen in the results of the M-BLEU score. M-BLEU scores better scores because in Hindi language word has many synonyms. For the BLEU score, the exact word match has to occur; otherwise, it gets a low BLEU score. Sometimes after synonym replacement, sentence may be incorrect. So, shallow parsing is done to correct the meaning of each sentence. Through synonym replacement and shallow parsing, a better BLEU score has been achieved for text.

References

1. Bengio Y, Simard P, Frasconi P (1994) Learning long-term dependencies with gradient descent is difficult. IEEE Trans Neural Netw 5:157–166
2. Pollastri G, Przybylski D, Rost B, Baldi P (2002) Improving the prediction of protein secondary structure in three and eight classes using recurrent neural networks and profiles. Proteins Struct Funct Genet 47:228–235
3. Chung J, Gulcehre C, Cho K, Bengio Y (2014) Empirical evaluation of gated recurrent neural networks on sequence modeling. arXiv preprint arXiv:1412.3555
4. Souri A, El Maazouzi Z, Al Achhab M, El Mohajir B (2018) Arabic text generation using recurrent neural networks. In: Communications in computer and information science, pp 523–533
5. Gasthaus J, Wood F, Teh Y (2010) Lossless compression based on the sequence memoizer. In: 2010 data compression conference
6. Bahdanau D, Cho K, Bengio Y (2014) Neural machine translation by jointly learning to align and translate. arXiv preprint arXiv:1409.0473
7. Luong M-T, Manning CD (2015) Stanford neural machine translation systems for spoken language domain. In: International workshop on spoken language translation
8. Papineni K, Roukos S, Ward T, Zhu W (2001) BLEU. In: Proceedings of the 40th annual meeting on association for computational linguistics - ACL 2002

9. Dwivedi SK, Sukhadeve PP (2010) Machine translation system in Indian perspectives. J Comp Sci 6(10):1111–1116. https://doi.org/10.3844/jcssp.2010.1111.1116
10. Verwimp L, Renkens V, Wambacq P State gradients for RNN memory analysis. http://dx.doi.org/10.21437/Interspeech.2018-1153
11. Hochreiter S, Schmidhuber J (1997) Long short-term memory. Neural Comput 9:1735–1780
12. Rozovskaya A, Roth D Grammatical error correction: machine translation and classifiers. https://www.aclweb.org/anthology/P16-1208
13. Hermanto A, Adji T, Setiawan N (2015) Recurrent neural network language model for English-Indonesian machine translation: experimental study. In: 2015 International Conference on Science in Information Technology (ICSITech)
14. Castilho S, Doherty S, Gaspari F, Moorkens J (2018) Approaches to human and machine translation quality assessment. In: Machine translation: technologies and applications, pp 9–38
15. Bojanowski P, Joulin A, Mikolov T (2015) Alternative structures for character-level RNNs. arXiv preprint arXiv:1511.06303

Chapter 2
Destress It—Detection and Analysis of Stress Levels

Neha Udeshi, Nemil Shah, Urvi Shah, and Stevina Correia

1 Problem Statement

Elevated levels of stress are normal among college students, especially engineering students. Students report an absence of sleep, grades, competition, change in way of life, and other critical stressors all through their undergraduate training. Stress and uneasiness have appeared to adversely affect student understanding, scholarly execution, and maintenance. When a student thinks something is beyond his/ her capability and still consistent to accomplish the task unsuccessfully, it results in behavioral and psychological pressures to be stated as stress [1]. Continuous high mental workloads and nonstop technological development, which prompts consistent change and requirement for adjustment, make the issue progressively genuine for students [2]. Therefore, stress needs to be dealt with carefully and effectively. The current stress detection and management systems which include ECG and EEG signals are either excessively costly or tedious, making them ineffective for an eager, distressing student, on the verge of hurting himself. There is a requirement for an approach that acts as a solution to this problem. The other proposed approaches utilize various parameters to recognize stress; however, no approach integrates all these parameters to shape an exact and economical system, accessible to a student consistently.

N. Udeshi (✉) · N. Shah · U. Shah · S. Correia
Dwarkadas J. Sanghvi College of Engineering, Mumbai, India
e-mail: udeshineha9@gmail.com

N. Shah
e-mail: nemilshah1999@gmail.com

U. Shah
e-mail: upshah271999@gmail.com

S. Correia
e-mail: stevina.dias@djsce.ac.in

I. Jeena Jacob et al. (eds.), *Data Intelligence and Cognitive Informatics*,
Algorithms for Intelligent Systems,
https://doi.org/10.1007/978-981-15-8530-2_2

Stress changes pupil dilation, blink rate, and facial gestures [3] which can be used to compute stress economically. Our system incorporates different parameters, such as measurement of pupil dilation, emotion detection through face and voice and a unique test to assist them with finding their way out.

2 Literature Survey

2.1 Paper [4]: Mental Stress Detection Based on Soft Computing Techniques

Input: The input is in the form of a video. The genetic algorithm was used for the detection of the pupil.

Algorithms: Fuzzy filter was used for noise reduction, and edge detection was performed based on the fuzzy reasoning. Ellipse detection algorithm was utilized to extract the pupil dilation measure and check the stress concerning this measure.

2.2 Paper [5]: Detection of Stress Using Image Processing and Machine Learning Techniques

Input: A stress detection system is developed based on the analysis of facial expression. The acquired video frames are processed sequentially by the software modules.

Algorithms: The face acquisition module processes the video sequences captured by the camera. Pixel transformation is used. The deep learning module is the final module that consists of the sub-modules of the training dataset, learning linear regression algorithm, model and testing the dataset for prediction

2.3 Paper [6]: Stress Detection Through Speech Analysis

Input: Automatic video analysis is adopted for automatically detecting stress.

Algorithms: To identify different emotions in human speech, features like pitch, articulation rate, energy, or mel-frequency cepstral coefficients (MFCCs) are used. SVM and ANN were implemented in this model. Both algorithms have shown excellent results, with ANNs having slightly better scores than SVMs.

2.4 Paper [7]: Machine Learning to Differentiate Between Positive and Negative Emotions Using Pupil Diameter

Information: The first dataset contained information of student enlargement which was recorded utilizing the Tobii TX300 eye-following framework that deliberate eye developments, for example, eye stare. The second dataset contained the understudy widening of the correct eye.

Calculations: All debased student size information related with eye flicker areas or brought about by the subject's head or understudy development were expelled and preliminaries with over half of missing information were disposed of. To order positive and negative feelings, the k-closest neighbor (kNN) was applied.

2.5 Paper [8]: iCare-Stress: An Integrated Mental Health Software

Info: The student expansion was recorded at a 50 Hz inspecting rate utilizing a SMI RED 4 remote video eye tracker.

Calculations: Eye-squint relics were recognized and supplanted by direct introduction. Pupillary reactions following sound introductions (in t2 and t4) were assessed to affirm the presence of pupillary responses to upsetting improvements. The examination of difference signal guess extraction was performed, and the characterization was finished with neural systems.

2.6 Paper [9]: Machine Learning to Differentiate Between Positive and Negative Emotions Using Pupil Diameter

Info: The student expansion was recorded at a 50 Hz inspecting rate utilizing a SMI RED 4 remote video eye tracker.

Calculations: Eye-squint relics were recognized and supplanted by direct introduction. Pupillary reactions following sound introductions (in t2 and t4) were assessed to affirm the presence of pupillary responses to upsetting improvements. The examination of difference signal guess extraction was performed, and the characterization was finished with neural systems.

3 Dataset Description

3.1 *Dataset for Stress Recognition Through Voice*

The dataset utilized is the Ryerson Audio-Visual Database of Emotional Speech and Song (RAVDESS). This dataset joins 24 on-screen characters, and every entertainer has given 60 preliminaries. The sound documents are 16 pieces and 48 kHz records in .wav design. The discourse feeling incorporates quiet, glad, miserable, irate, dreadful, shock, and sicken articulations, and the tune's feeling contains quiet, upbeat, pitiful, furious, and frightful feelings. Every articulation is created at two degrees of passionate power (ordinary, solid), with an extra nonpartisan articulation.

3.2 *Dataset for Emotion Recognition Through Facial Recognition*

The dataset utilized is FER2013. The preparation set has 28,709 pictures and the approval set comprises of 3589 pictures. Each picture is of size 48×48. The dataset classifies each face dependent on feeling appeared in the classifications, irate, disturb, dread, cheerful, miserable, shock, and unbiased. It comprises of two sections, "feeling" and "pixels." The "feeling" segment contains a numeric code from 0 to 6. The "pixels" section contains a string encompassed in cites for each picture.

4 System Architecture and Implementation

Figure 1 shows the system architecture of our system.

The model consists of 5 major modules which are:

- Facial Emotion Recognition
- Measurement of Pupil Dilation
- Emotion Recognition through Speech
- Questionnaire
- Recommendation System.

Input from the user is taken in the form of video and audio recordings. The user is provided with a sample line on the mobile screen where the video and audio recording starts in the background. The video and audio recordings are sent to the server for further processing.

The video is divided into frames, and these frames are given as an input to the facial detection module. The facial detection module acts as an intermediate step. For facial detection, the frame is initially converted into a black and white image. This black and white image is processed by applying a histogram of oriented gradients

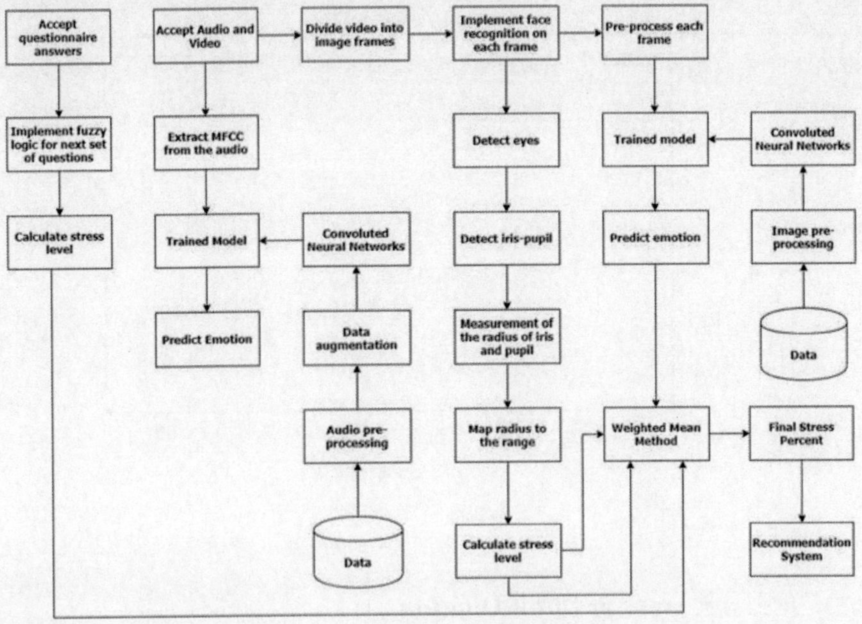

Fig. 1 System architecture

(HOG) wherein the gradients flow from a dark pixel to a neighboring light pixel. The accuracy of the facial detection module is 99.38%. The facial detection module is applied to all the frames.

4.1 *Facial Emotion Recognition*

The output from the facial detection module is given as an input to the facial emotion recognition module. Here, the CNN model is used for the prediction of facial emotion recognition. The CNN model implements 11 layers. Categorical cross-entropy acts as a loss function in this model. The model uses Adam as its optimizer. For training, the model is iterated for 200 epochs. The model attains an accuracy of 69%. This model is applied to 150 image frames, which are obtained by dividing the video file of the sample user. The outputs from 150 image frames are further used to compute the final negative and the positive facial emotion percentage of the sample user. This method is used for compensating the accuracy issues. The graph below (Fig. 2) represents the loss curves for the training and testing datasets of this module.

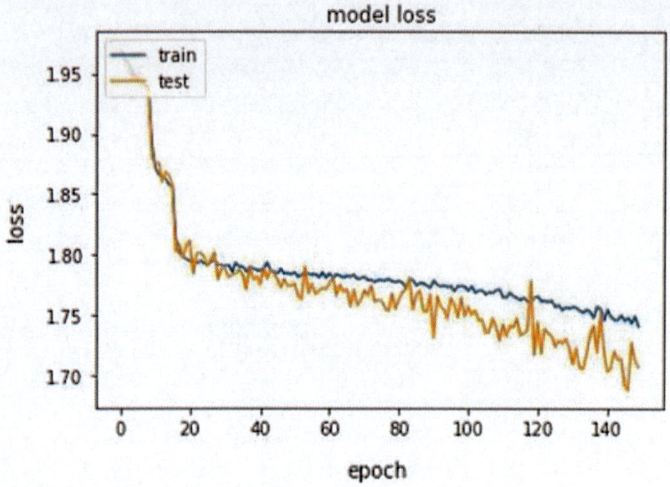

Fig. 2 Model loss

4.2 *Measurement of Pupil Dilation*

The output from the facial detection module is given as an input for this module. For iris and pupil detection, eyes are initially detected. Haar Cascade classifier is used for the detection of eyes. The iris and pupil detection are shown in the figures below (Figs. 3 and 4). The detected image of the eye is converted to a grayscale image. Median blur is applied on the grayscaled image where the central pixel of the mask is replaced by the median of the other present pixels. For iris detection, the Hough transform is used as the feature extraction technique. For forming Hough circles, 48 px is supplied as the max radius. With the help of the Hough circles, iris can be detected. The accurate radius of the iris can be calculated based on the Hough circles. The aforementioned procedure is followed for the detection of the pupil, but here max radius is set as the radius of the iris minus (−) 4 px for forming the Hough circles. After implementing the above procedure, the final accurate radius of the pupil is got. The defined range of radiuses of dilated (stressed) pupils is 4.0–8.0 mm. Thus, our results can now be compared with the defined range.

4.3 *Emotion Recognition Through Speech*

The recorded audio is given as the input to this module. For training the model, the RAVDESS dataset is being used which contains 1440 (audios of male and female) samples. For analyzing the audios, the spectrogram is constructed which helps in representing the signal strength. Through spectrogram, the presence of noise can be identified. Audio preprocessing includes trimming the silences present in the

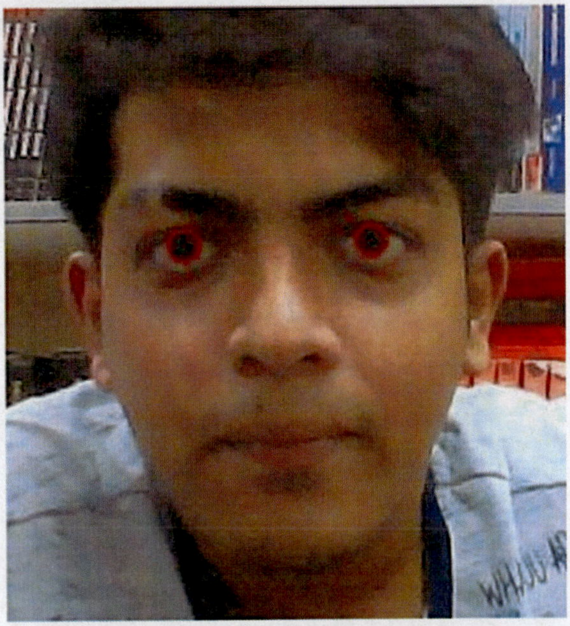

Fig. 3 Iris detection

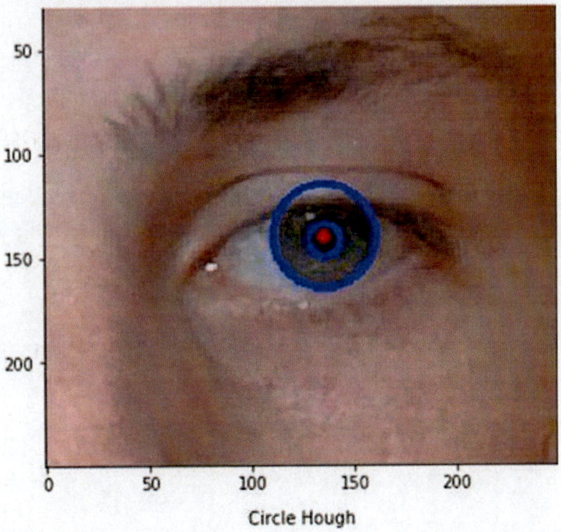

Fig. 4 Pupil and iris detection

audio. Figures 5 and 6) represent the spectrogram of the audio wave, before and after trimming the silences.

The most representative voice features for stress identification are mel-frequency cepstral coefficients (MFCC), the delta coefficients based on the MFCC [10, 11], voice pitch, and zero-crossing rates [10]. The MFCC is a signal with a set of 10 to 20

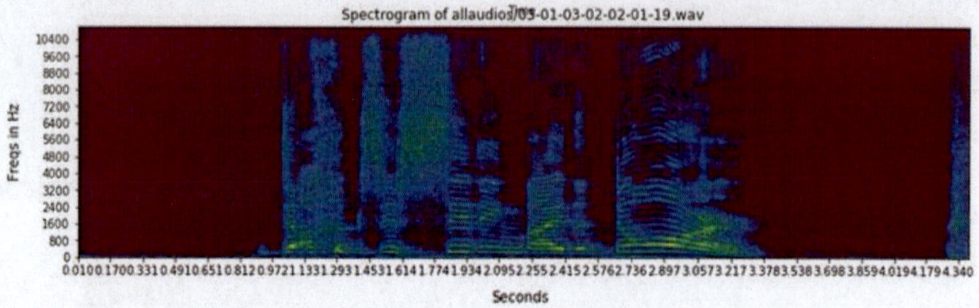

Fig. 5 Before trimming the silences

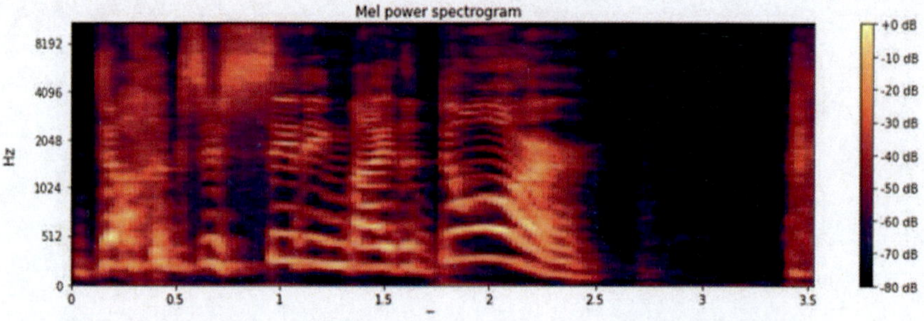

Fig. 6 After trimming the silences

features that accurately describes the overall shape of a spectral envelope. MFCCs are used for feature extraction of audio. Figure 7 represents the MFCC which signifies the signal strength of the audio.

Pitch-related, intensity-related (energy-related features and PowerDb-related features) are identified as the most important among all the features [12]. Thus, for precise prediction, the dataset needs to be split for the gender, i.e., a separate dataset

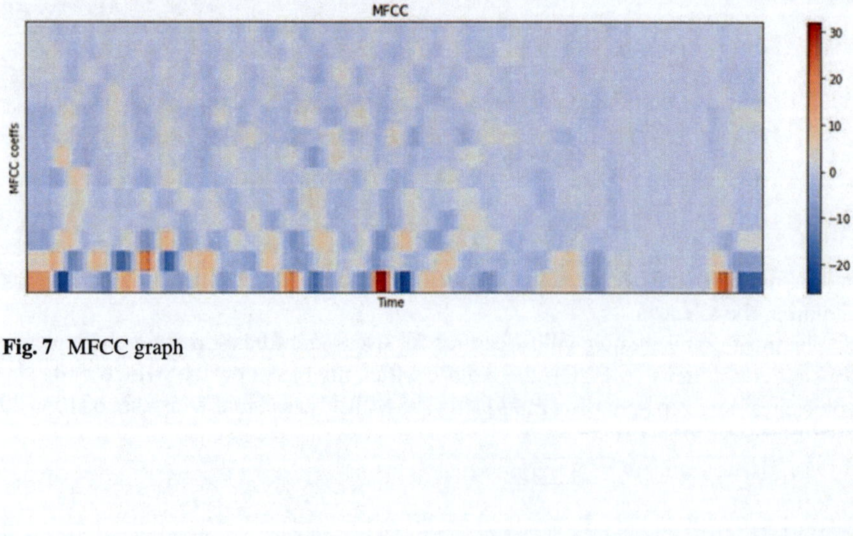

Fig. 7 MFCC graph

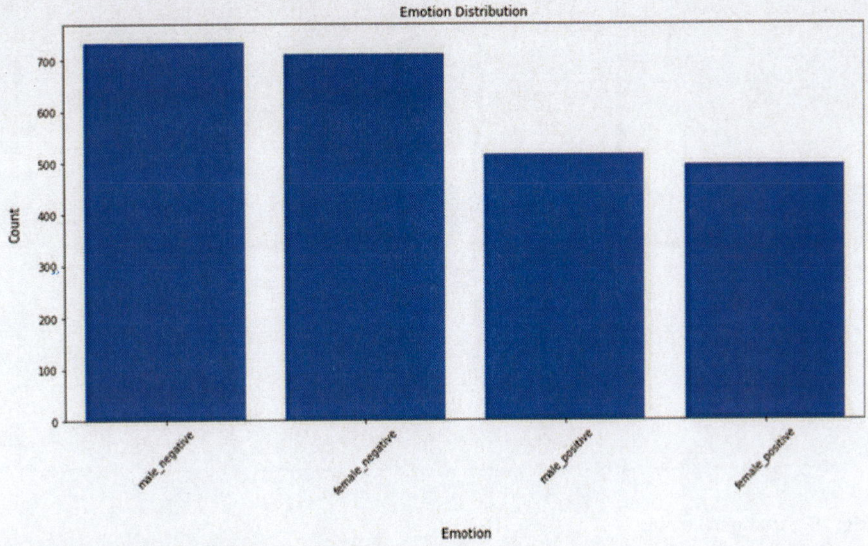

Fig. 8 Emotion distribution

for both males and females. The graph (Fig. 8) represents the emotion distribution of the samples present in the dataset.

Data augmentation is implemented for increasing the number of samples. Following are the methods that can be implemented for data augmentation (Figs. 9 and 10):

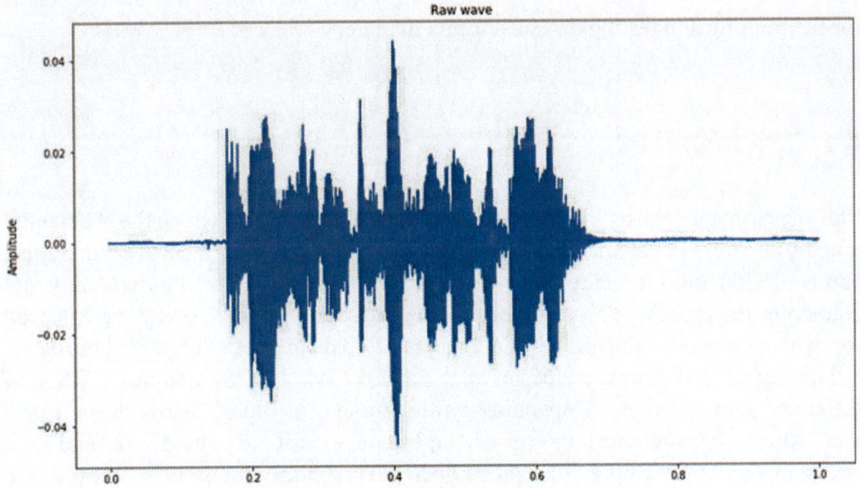

Fig. 9 Before pitch tuning

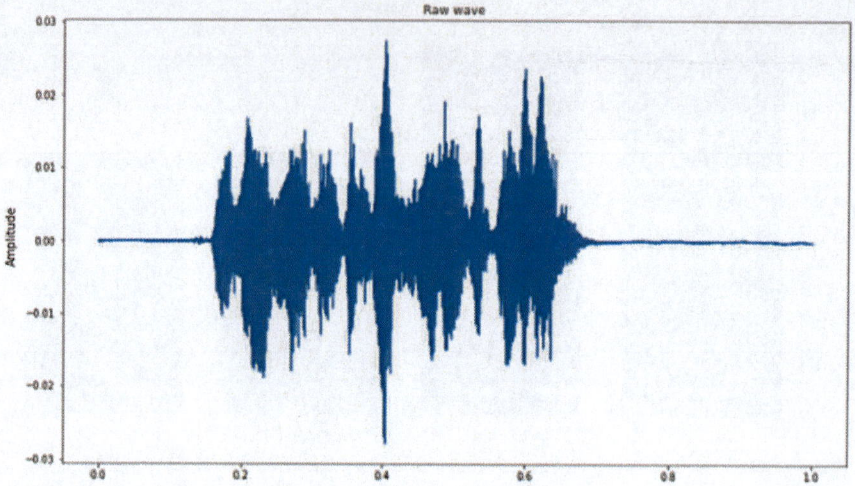

Fig. 10 After pitch tuning

- Pitch tuning
- Adding white noise
- Random shifting
- Stretching the sound.

After augmenting the data, the samples for each gender go up to 2400 samples. For training the model, CNN is applied on the MFCCs acquired. The model has been iterated for 700 epochs. The obtained training accuracy of the model is 99.84%, and validation accuracy is 92.29%. The graph below (Fig. 11) represents the loss curves for the training and testing datasets of this module.

4.4 Questionnaire

The questionnaire consists of the list of verified questions from which the substantial amount of stress is calculated. These questions are verified by a certified therapist, and hence, the questionnaire reduces the process of going to them individually and automates the process. There are four sets of questions in which fuzzy logic is used for dynamically changing the set of questions based on previous answers of the users. All questions have four types of answers, namely "Rarely," "Sometimes," "Most of the time," and "Always." Depending on the number of times, "Most of the time" and "Always" are selected, the set of questions for basic, low, moderate, and high stress appear. For example, if a student does not select the two options "Most of the time" and "Always" in the majority of questions, the next set of questions that appear might be low or moderate stress questions. If he selects the aforementioned options for all or majority of the low-stress questions, the next set which appears might be

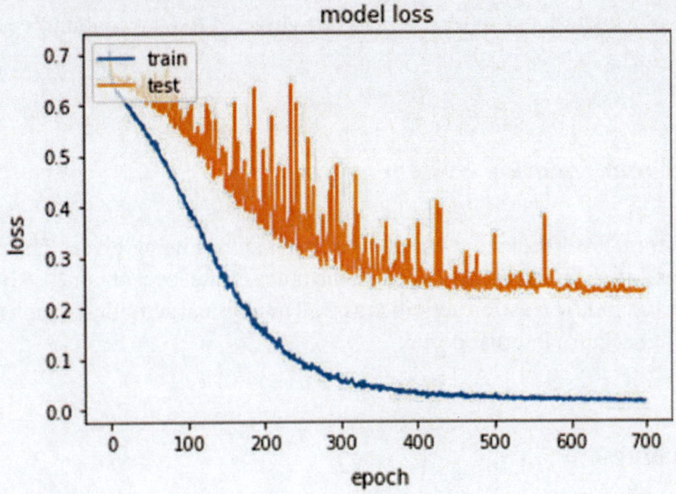

Fig. 11 Model loss

the medium level. The final stress level is calculated based on the answer selection for these questions.

According to Table 1, the type of current set and the answers obtained from the users are taken into consideration. The table represents the type of next step that will occur if certain conditions are met. For example, if a user is currently on a basic level and selects "Always" more than 2 times in the current set, he will be redirected to the set containing high-level questions. If at any point there does not exist any answer with more than 2 counts, the counts of "Rarely" and "Sometimes" are merged as well as the counts of "Always" and "Most of the time" are merged. If the merged counts of "Rarely" and "Sometimes" are more than 1, the user will be redirected to a

Table 1 Rules table

Rules		Type of current level			
		Basic	Low	Moderate	High
Type of answer selected more than 2 times in each set	Rarely	Low set	Moderate set	"Low level of stress detected"	Moderate set
	Sometimes	Low set	Moderate set	"Low level of stress detected"	Moderate set
	Most of the times	High set	High set	"Moderate level of stress detected"	"High level of stress detected"
	Always	High set	High set	"Moderate level of stress detected"	"High level of stress detected"

comparatively lower level, or else he will be redirected to a set containing high-level questions.

4.5 Recommendation System

The final stress level leads to a list of recommendations being given. These recommendations appear dynamically based on the range of the level obtained. The recommendations include relaxation activities as well as physical activities, which may help an individual acquire internal peace.

5 Outcome

The video of a female was taken as input. The following figures show the outcomes obtained for each module.

In Fig. 12, it can be observed that the outcome of facial emotion recognition was 20% negative and 80% positive. The video file of the sample user was given as an input to this module. Further, the video file was divided into 150 image frames and all the image frames were given as an input to the facial emotion recognition model.

In Fig. 13, the emotional state was classified to be positive based on the audio file that was given as an input by the sample user. Since the audio results are mapped to provide a Boolean output, i.e., 0 for a positive outcome and 1 for a negative outcome, the obtained result was further mapped to 0 for further calculations.

In Fig. 14, the pupil radius measurement is 2.08 mm. If the measurement is between 2.0 mm to 4.0 mm the Boolean output is 0, whereas if the measurement is between 4.0 mm to 8.0 mm the measurement is mapped to the Boolean outcome as 1. Hence for the obtained result, the measurement is mapped to 0 for further calculations.

In Fig. 15, the outcome of the questionnaire module is 18.75% negative.

Fig. 12 Facial emotion outcome

```
Negative percentage: 20.0%
Positive percentage: 80.0%
```

Fig. 13 Audio emotion outcome

```
   predictedvalues
0  female_positive
```

Fig. 14 Pupil and iris ratio outcome

```
Ratio of dilation in pixels is 8px : 25px
Ratio of dilation in mm is 2.08mm : 6.5mm
```

Stress percent on the basis of questionnaire: 18.75

Fig. 15 Questionnaire outcome

$$W = \frac{\sum_{i=1}^{n} w_i x_i}{\sum_{i=1}^{n} w_i}$$

w_i – the corresponding weight for each item

Fig. 16 Weighted mean formula

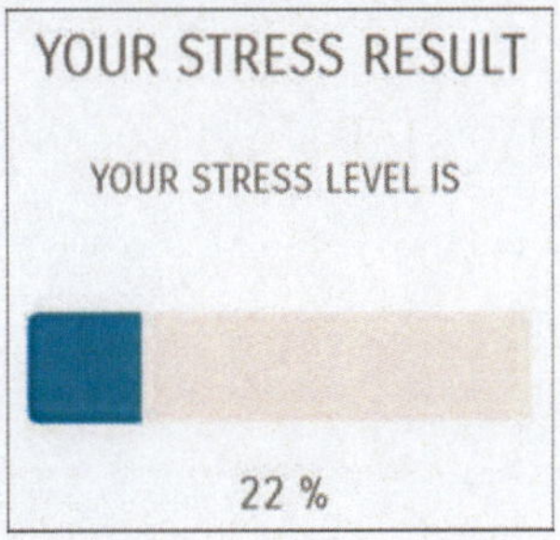

Fig. 17 Final stress outcomes

The above outcomes provide the final result shown below through calculation by the weighted mean method. The iris and pupil radiuses are mapped with a defined range to provide an intermediate stress level. While facial emotion and questionnaire modules provide a numerical value of stress level, the audio emotion module and pupil dilation measurement provide a Boolean result stating 0 for positive and 1 for negative. These values are then multiplied to their appropriate weights (determined through the help of professional therapists and experimentation), and their sum is divided by the sum of weights, as per the weighted mean formula below (Fig. 16).

The final stress level obtained is 22% as shown in the figure below (Fig. 17). Also, the set of recommendations for this stress level are presented in Fig. 18.

6 Conclusion

In this mobile application, Destress It is proposed as mobile and remote stress monitoring, alleviation, and management system. In this framework, the client's feeling of anxiety is quantitatively evaluated in a consistent territory and guided the clients to lighten the pressure utilizing different remedial methodologies. The framework gives a foundation of incorporated capacities to support the screening, diagnosis, and treatment. There is an algorithm to classify positive and negative emotions using pupil dilation measurement. The method is utilized to separate among positive and negative encounters in subjects. Facial emotion recognition and pupil dilation are captured

Fig. 18 Recommendation system

through video and emotion recognition through audio which is used. The video file is properly divided into frames at the server. Emotion recognition and pupil dilation detection are implemented at the individual frames. The facial and voice recognition using the CNN algorithm is incorporated which gives a palatable outcome. A questionnaire consisting of different weighted questions is created, and the new questions appear depending on the blend of every question answered before. The recommendation system is aptly developed to provide a dynamic recommendation based on the stress output percentage to build a good stress relief mechanism.

References

1. Prakash A, Bajpai P (2015) Stress among engineering students: Who are responsible. Adv Psychol Res
2. Alberdi A, Aztiria A, Basarab A (2016) Towards an automatic early stress recognition system for office environments based on multimodal measurements: a review. J Biomed Inform 59:49–75
3. Subhani AR, Mumtaz W, Saad MNBM, Kamel N, Malik AS (2017) Machine learning framework for the detection of mental stress at multiple levels. IEEE Access 5:13545–13556

4. Mokhayeri F, Akbarzadeh-T M-R, Toosizadeh S (2011) Mental stress detection using physiological signals based on soft computing techniques. In: 18th Iranian conference on biomedical engineering
5. Zhang J, Tang H, Chen D, Zhang Q (2012) deStress: mobile and remote stress monitoring, alleviation, and management platform. Hong Kong university of science and technology
6. Tomba K, Dumoulin J, Mugellini E, Abou Khaled O, Hawila S (2018) Stress detection through speech analysis. J. Inf Process Soc Jpn 50(3):1181–1191
7. Pedrotti M, Mirzaei MA, Tedesco A, Chardonnet J-R, Mérienne F et al (2014) Automatic stress classification with pupil diameter analysis. Int J Hum-Comput Interact 30:220–236
8. Mitrpanont J, Phandhu-Fung J, Klubdee N, Ratanalaor S, Mitrpanont T (2017) iCare-stress: an integrated mental health software. In: 2017 2nd International Conference on Information Technology (INCIT)
9. Raichur N, Lonakadi N, Mural P (2017) Detection of stress using image processing and machine learning techniques. Int J Eng Technol 9:1–8
10. Sandulescu V, Andrews S, Ellis D, Dobrescu R, Martinez-Mozos O (2015) Mobile app for stress monitoring using voice features. In: 2015 E-health and Bioengineering Conference (EHB), Iasi 2015, pp 1–4
11. Patil A, Rucha M, Nikita K, Viraj C, Sanket P, Ajinkya Y. Stress detection in IT professionals by image processing and machine learning
12. Rong J, Li G, Chen Y-P (2009) Acoustic feature selection for automatic emotion recognition from speech. Inf Process Manag 45(3):315–328

Chapter 3
On the Design and Applications of an Integrated Smart Home Automation

Thanaz Shafeer, M. Senthil Arumugam, and A. Sasikala

1 Introduction

1.1 Smart Homes

In a world, where almost everything is digitalized and updated every now and then it is important to note that along with devices, the next topic of concern is one's own home. Smart homes have proved to lead a more flexible life where almost every functionality is only a click away. Moreover, due to the advancement of technology and availability of modules at a lower cost smart homes have now become a trending topic. *IoT* also is known as the Internet of things paves as a pathway for communication within the system. *IoT* mainly focuses on using the Internet as a medium of communication between devices, processors and so on [1]. *IoT* is the foundation upon which systems such as smart homes are built. Basically, a smart home is a system that is designed to have Web-connected smart devices which operate on data collected from the environment along with the combination of processors

T. Shafeer (✉) · M. Senthil Arumugam
School of Engineering and Physical Sciences, Heriot Watt University, Dubai, United Arab Emirates
e-mail: ts46@hw.ac.uk

M. Senthil Arumugam
e-mail: m.senthilarumugam@hw.ac.uk

A. Sasikala
Department of Electrical and Electronic Engineering, Srisairam Institute of Technology, Chennai, India
e-mail: sasikala.eee@sairamit.edu.in

I. Jeena Jacob et al. (eds.), *Data Intelligence and Cognitive Informatics*, Algorithms for Intelligent Systems,
https://doi.org/10.1007/978-981-15-8530-2_3

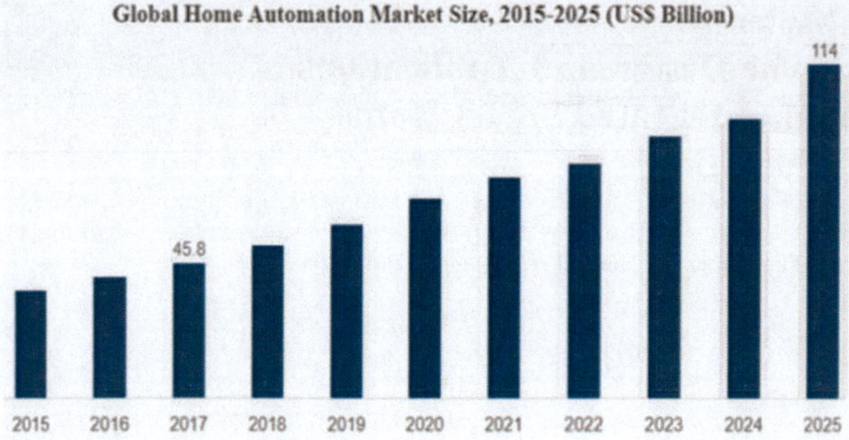

Fig. 1 Growing revenue on smart homes

to generate appropriate outcomes. The scope of smart homes has attracted a large number of investment prospects. With an increased revenue being spent on smart homes to 114 USD by 2025 as shown in Fig. 1, there is no doubt that smart homes are a topic of great potential [2].

1.2 Sustainable Development

Growing environmental concerns has led to the need for systems to be more sustainable and eco-friendlier. The demand for energy is expected to only increase over the coming years [3]. This expected increase in demand for energy will result in rapid depletion of fossil fuels and other non-renewable resources. Therefore, it must be kept in mind that sustainability concerns are of the highest priority while designing any type of system.

1.3 Target Audience

Along with the general public, more attention has been given to people with special needs or the elderly. Smart homes, in general, have been described as a pathway to a more secure and convenient lifestyle as every functionality is simply a 'click away'. The convenience of remote access allows the elderly and the disabled people in the house to access multiple sections of the house without physical movement. This ensures that the house caters to almost anyone and everyone thus providing socio-economic benefits. These applications could prove to be a life-changing improvement for those struggling with a long-term health condition.

2 Literature Review

2.1 Internet of Things [IoT]

Internet of things is a concept that brings in the revolutionary idea of smart devices. Traditional devices are designed to solely produce a user-defined output which is not ideal as it does not provide information on the performance of the system. Hence, what distinguishes a normal device from a 'smart' device is the fact that the latter provides information or else known as feedback every time the device is used. To access, record and manipulate such information it is important to note that data should travel from one point to another or from system to system. This requires a medium of communication. *IoT* mainly focuses on implementing Internet as a smarter mode of communication as opposed to traditional transmission cables [4]. By extending Internet connectivity, *IoT* opens up a market with endless possibilities.

2.2 Integrated Smart Systems

When smart devices combine along with sensors, processors and user-friendly graphical user interface (GUI) it constitutes a smart system [4]. By definition, a smart home is a combination of appliances, sensors and appropriate modules that work simultaneously to produce an output at the user end. Another point to note is the fact smart homes are incredibly flexible and customizable. Therefore, developers can bring about changes or updates in the system without disrupting the working of the system. This is mainly because each of the independently working units is interconnected via the Internet [5]. The main database is connected to the user's mobile device. Therefore, all information/data can be accessed and manipulated by the user. This always allows the user to control and access the system, and hence, reducing human effort. This means that smart homes are designed to cater to not only the general public but also occupants of the house that have certain limitations.

2.3 Security Concerns

One of the major vulnerabilities of smart homes is security [6]. This does not only involve the security of the house on its own but also security of the data that is used, as any unauthorized access can result in malicious practices. So, to combat the same, the security systems are preferred to be layered. Basically, the term layering suggests a sequential order of security-based tests rather than a normal pin code lock system [6]. Another accurate approach toward a security system is using biometric systems. Fingerprint scanners are used to sense wet or dry fingerprints which can then be assigned a unique ID by programming this module with the Arduino. These fingerprints are stored initially in the registration phase with a unique ID for each type of fingerprint. This enhances the security factor as the fingerprint of every individual is unique and cannot simply be guessed like pin code numbers.

2.4 *Smart Metering*

With growing ecological concerns, systems must be created with the inclusion of energy management. A smart system is also one that is designed to be efficient as well. Home energy management systems are known for either being energy conscious or energy efficient both of which contribute to a more ecologically balanced system. To foster an energy conscious environment, rather than using traditional meters it is advised to use smart meters. Smart meters are essentially power meters that measure the power consumption for each appliance with real-time display [7]. Smart meters can also be used to validate the device status such that the user need not switch off any appliance physically.

3 Proposed System

The proposed system, as shown in Fig. 2, has been designed to implement a smart home automation system to lead to an overall convenient and energy conscious.

The entire system is *IoT*-based with a biometric-based security system along with typical controls of a home automation system. The smart home designed also uses smart meters that are used to read, store and display the power drawn for each appliance. The whole system is integrated using *Wi-Fi* where all the data enters the main source which acts as a connecting medium for all the independently working sections of the house.

4 Methodology

The smart home automation system can be split into multiple phases which includes:

- Smart security system,
- Temperature, humidity, LPG and smoke controls,
- Device management,
- *Wi-Fi* integration.

The whole project focuses on using two modules namely the Arduino Nano and the *NodeMcu* as the controller unit. Basically, such modules are responsible for

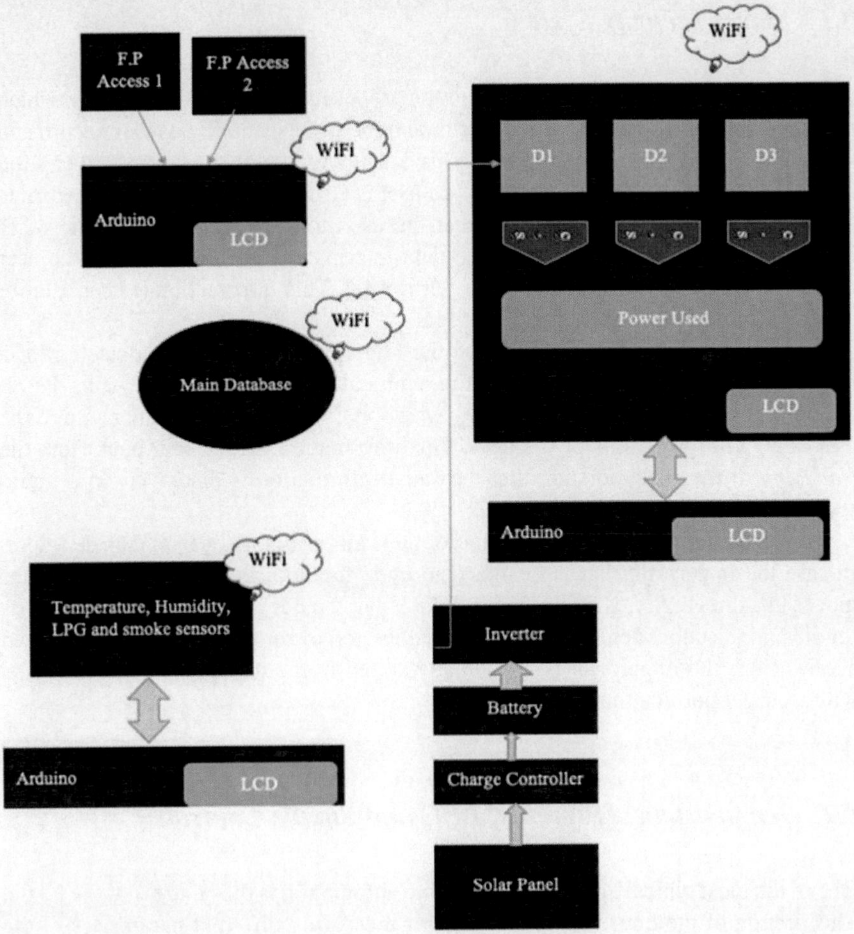

Fig. 2 Overall system in pictorial form

controlling, calibrating and monitoring all the data that is transmitted throughout the system. Arduino Nano is a microcontroller that operates at a voltage of 5 V and it is a user-friendly software development board. *NodeMcu* also known as ESP8266 *Wi-Fi* module is an essential part of the project. It is a board in itself which can be coupled with sensors for multiple *IoT*-based projects and has an operating voltage of 3.3 V. It is integrated with TCP/IP making it compatible with other microcontrollers as well [4].

4.1 Smart Security System

This part of the project is responsible for the proposed layered security system which involves a biometric system. A conventional fingerprint scanner is used to record and store data specific for each individual. This data is stored with the help of an Arduino and registered using a unique user ID. The user is also asked to store two fingerprints under the same name such that one can be used to open the door and the other to close. The *NodeMcu* uses the same data to register information about the user that corresponds to their respective fingerprint ID. This information is used later to display information about authorized users.

First, the user places their finger on the fingerprint sensor. Upon detection of a valid user the code redirects itself to the main database that is availed using *Wi-Fi*, the information gathered about the user which mainly constitutes—the name of the user along with the status of the door. This information is then sent back from the *NodeMcu* to the Arduino such that the same information is displayed on a liquid crystal display (LCD) screen.

The system also involves a servo motor that can be used to open and close a door. In case the fingerprint that is designed to open the door is used, the servo motor moves in a direction, i.e., calibrated to 1 such that the door is opened and same goes for closing the door such that the servo is calibrated to zero. In case of unauthorized access, since the fingerprint is not already stored in the database the user notified with a buzzer notification.

4.2 Temperature, Humidity, LPG and Smoke Controls

One of the most typical applications of home automation is the controls of the home. This section of the house is responsible for measuring different parameters of the house-like temperature and displaying the same with real-time display and notifying the user immediately in the case of any harmful circumstances. Three sensors are used to calibrate which include:

- DHT11 sensor—a low-cost sensor temperature and humidity sensor.
- LPG sensor (MQ6)—the sensor is highly sensitive to gases like LPG, iso-butane and propane. It also has a slight sensitivity to alcohol and smoke. Therefore, it can be used as a smoke sensor as well.
- Smoke sensor (MQ2).

These sensors together work to measure the varying temperature, humidity, gas and smoke conditions in the environment by converting the same into varying measures of voltage from 0 to 5 V. These varying voltages are then used by the Arduino to read these varying voltages and converting them into respective units.

These units are then again converted using respective formulas to obtain the temperature in degrees and humidity along with LPG and smoke values in percentages. These percentages are then also sent to the *NodeMcu* such that these values are stored in the server that is Web-based with a real-time display such that the user can access the parameters of the house at any time from any mobile device that is connected to the Internet. Values for gas and smoke percentages are set to certain limits such that if exceeded the user is notified with a buzzer until the system returns to normalcy.

4.3 Device Management

This section involves the smart meters that lead to an energy conscious environment. There are three appliances in general with the third one as a part of solar power metering. The first two devices showcase the different ways in which the device can be controlled along with the display of power consumption. The first appliance can be controlled using any laptop or mobile device while the second appliance can be controlled using a switch. A relay is a device which allows the user to control high voltage and current from a low signal controller like an *Arduino* board. The four channels solid-state relay module is an *Arduino* compatible relay module whose working is similar to that of a mechanical relay. The control of the status of the different appliances is implemented with the help of a solid-state relay module that is responsible for mimicking an electrical switch for the different appliances that are connected using respective line sockets.

Both the devices have the status entered into the main database where the status can be altered using the connection between Arduino and *NodeMcu.* Along with this the power consumed by both devices is measured simultaneously using a current sensor.

Since the voltage is constant, the varying values of current are measured to calculate the power consumed in each case. This power that is calculated via the current sensor on the Arduino is sent to the *NodeMcu* to access the database and then store the same for future reference. This power that is calculated is extracted from the database and displayed on the LCD at the user end which constitutes the smart metering and the control of the different appliances.

4.4 Wi-Fi *Integration*

The project is *IoT*-based due to the extension of the system to the Internet using *NodeMcu.* The *NodeMcu* is responsible for opening up the main source unit also known as the smart home which is responsible for keeping track of all the relevant data that is sent through the system. Since the system is accessed using *Wi-Fi* that is password protected hence adding a new layer of security, it easily accessible on any laptop, mobile device or personal computer (PC) that is connected to the Internet.

The main source unit provides an entire database with respective attributes for different sections of the house. This data is what is then send back to the Arduino Nano and then displayed on the LCD screen. Since all the data is stored with real-time display authorized users can go back to the main database and check on power consumption or information on any authorized or unauthorized access to the house at specific times.

5 Results and Observations

This section of the paper includes the results or images of the independently working sections of the home automation system that was implemented on a PCB and connected by the main source to work simultaneously to create optimum results.

5.1 Layered Security Model Prototype

The prototype of the proposed layered security system that is connected with a printed circuit board (PCB) is shown in Fig. 3. An adapter used to provide as the power supply

Fig. 3 Security model prototype

to the Arduino Nano which then finally powers the *NodeMcu* as well as the input section involving the fingerprint sensor and the output that is provided by the servo motor, buzzer and the 16 * 2 LCD screen. The fingerprint sensor as shown in Fig. 4 is used to validate whether access to the house is authorized.

The display of information of the valid user once the *Wi-Fi* connection is established is shown in Fig. 5. It enters the main source unit of the house that involves all the data recorded and chooses the corresponding user ID and displays the name as well as the status of the door lock system that corresponds to the fingerprint that is specifically designed to close and open the lock, respectively, and changes the directional move of the motor corresponding to the same.

Figures 6 and 7 represent the *IoT*-based section of the project by creating the main source unit where all information is stored which can be accessed by authorized users.

Fig. 4 Fingerprint validation

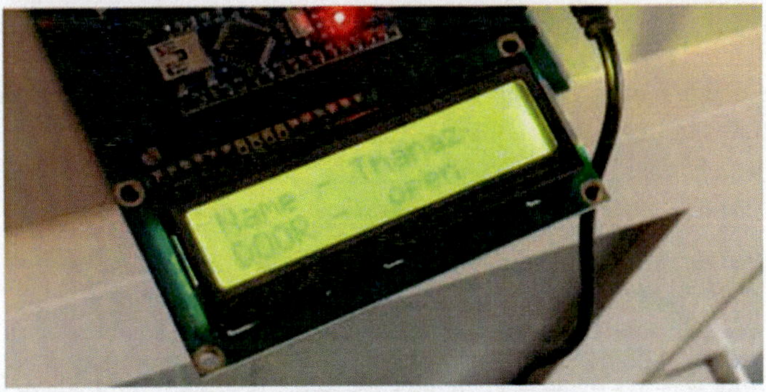

Fig. 5 Authorized user information

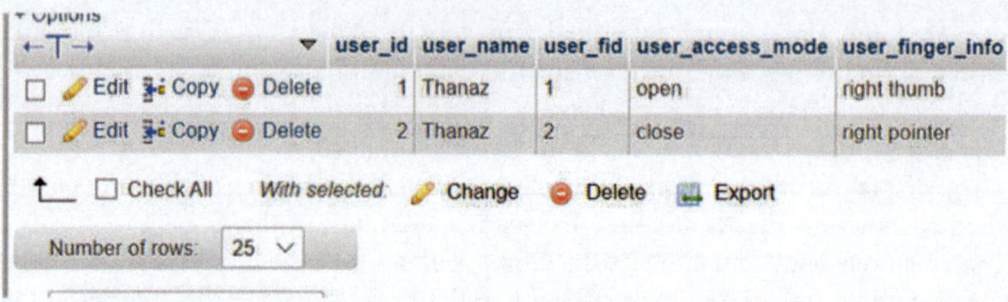

Fig. 6 Author registration

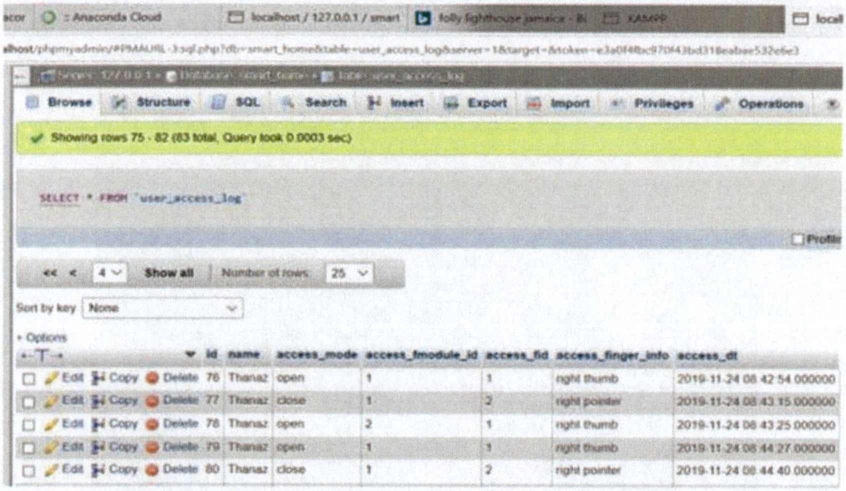

Fig. 7 Door access information

Figure 6 represents the registration of respective users with the fingerprint as well as a unique user ID for the same to validate access to the door lock system.

By creating a main database that is connected and created using a *Wi-Fi* module that is password protected there is another layer of security that is added to the system. As shown in Fig. 7, the user can able to check-in for information regarding the door lock system with real-time display as it registers which occupant at what time along with the status of the lock system.

5.2 *Temperature, Humidity, LPG and Smoke Alarms*

The overall PCB that contains all the respective sensors that are calibrated with the help of Arduino which makes a soldered connection to the *Wi-Fi* module (*NodeMcu*) and displays the same output on a 16 * 2 LCD screen is represented in Fig. 8 illustrates

Fig. 8 Display of parameters

the display of temperature in degrees as well as the humidity percentages with real-time display on the database that is available at authorized access on any mobile device.

Figure 9 describes the validation of the sensor functioning, for example, the LPG sensor is checked with using a lighter. If the calibrated sensor exceeds the value of optimum, the user is notified immediately with a buzzer notification and turns the buzzer off when the system returns to normalcy as shown in Fig. 10.

Fig. 9 LPG and smoke sensor testing

Fig. 10 Harmful detection levels

parameter_id	temperature	humidity	lpg	smoke	person_count	updated_dt
176	27.80	59.00	56.93	54.88	NULL	2019-08-04 16:45:04.000000
177	27.80	60.00	56.74	54.59	NULL	2019-08-04 16:45:12.000000
178	27.80	59.00	56.84	54.88	NULL	2019-08-04 16:45:20.000000
179	27.70	59.00	57.13	54.79	NULL	2019-08-04 16:45:28.000000
180	27.70	60.00	56.84	54.69	NULL	2019-08-04 16:45:36.000000
181	27.70	58.00	56.93	54.59	NULL	2019-08-04 16:45:45.000000

Fig. 11 Parameters available at database

As shown in Fig. 11, all the data that is evaluated using the respective sensors are recorded at the main source/database for future references. If the user wants to check the conditions of the house at any time the user simply can access the very same from any PC or mobile device that is connected to the Internet.

5.3 *SMART Metering Prototype*

The power metering section of the household is shown in Fig. 12. The first four connections made to the current sensor are connected to the board that is used to calculate the power while the five-point connection is made due to the relay module

Fig. 12 Power metering board

that is used. The line sockets that are connected to the board that is used as a connection medium between the appliance and the board have a respective line and neutral connections that then is connected to the relay module and respective current sensors (Fig. 13).

The power metering for the same is shown in Fig. 14. The control for the appliance can be controlled either by the switch case that is used or through any laptop that

Fig. 13 Device powered using *Wi-Fi* or switch

Fig. 14 Power metering of both devices

parameter_id	device_one_power	device_two_power	device_three_power	solar_output_voltage	solar_output_current	solar_output_power	updated_dt
202	0.00	0.00	0.00	0.00	0.00	0.00	2020-01-21 10:19:49.000000
203	0.00	79.66	0.00	0.00	0.00	0.00	2020-01-21 10:19:56.000000
204	0.00	52.98	0.00	0.00	0.00	0.00	2020-01-21 10:20:04.000000

Fig. 15 Power recorded in database

can change the device status from the main database that can be accessed only by authorized users as shown in Fig. 15.

5.4 Main Source

An *IoT*-based database is created on a server using the XAMPP control app. This app is opened in an administration mode which is password protected for solely authorized user purposes. Since the database and most of the controls are connected through *Wi-Fi*, it ensures that all the controls can be viewed or controlled at any point of time on any device that is connected to the Internet.

To extend Internet connectivity to the system, it is desired to use a *Wi-Fi* module. *NodeMcu* is a module that is used to interconnect all the units to the Internet. The database as shown in Fig. 16 contains multiple tables that each contribute to a separately working section of the house with its own unique attributes, wherein each row of information is characterized by a unique ID.

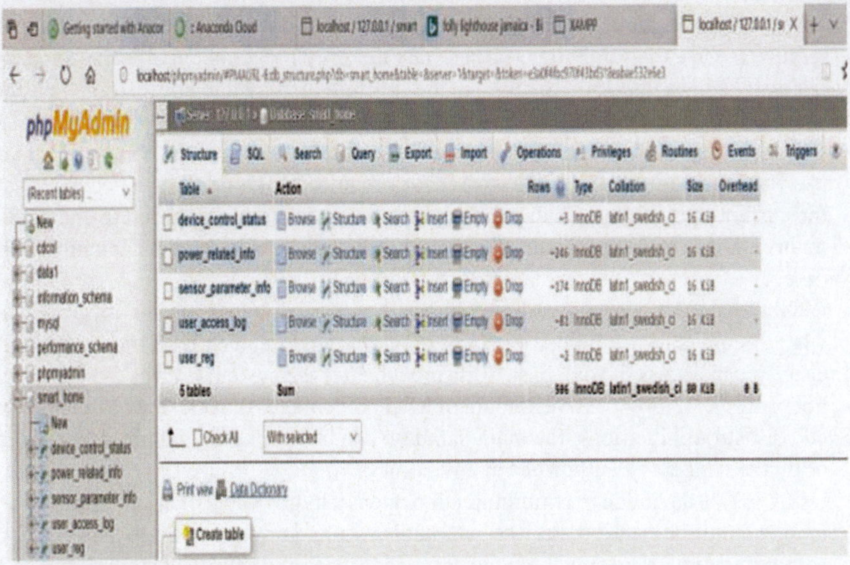

Fig. 16 Main database

6 Discussion

The independently working sections which include the PCB's can be further developed and used on multiple occasions some of which include:

- The fingerprint-based security model developed in this paper can be used in various areas of work outside home automation. Traditional hotel systems involve a security system based on locks that are opened using RFID card readers.
  Hotel systems are often demanding on the security part as it also requires the user access and entry timings to validate the check-out process. With an RFID card reader or any sort of pin number identification not only can the system be hacked by either cracking the code or find a replacement for the card, but it also means that keeping track of the customers access and exit is hard to keep track of for the hotel management.
- By replacing this same system with the smart security model, it offers the system with more security as no two fingerprints are alike. Unlike cards, fingerprints are next to impossible to steal.
  By extending the concept of *IoT*, the hotel management can track the user as well as the entry to the room which makes it so much easier to validate the check-in and check-out place. Along with which any sort of staff access to the same room all the information is readily accessible at all times at the help desk hence providing a secure hotel room stay.
- With the same fingerprint validation system, it is possible to use the same in offices as well as in academic institutions as an attendance record.

Any staff or employee or student that is saved in the main database is simply required to use the same fingerprint which not only prevents unauthorized access but also keeps a check on the user with real-time display and so it is much easier to track whether the individual is present on a certain day or not. Such systems not only ensure a higher level of security using biometrics but can be of great use for future reference. In any case, it is required to validate the occupants at the current or on any given day the administration is simply required to check the main database for all information regarding which employee/student at a supposed time.

- LPG and smoke alarms that are connected to an *IoT*-based can be of great use when it comes to fire industries. In case of any emergencies with the authorized users, it can access the database to ensure whether the system returns to normalcy. For example, if an excessive amount of LPG percentages that can lead to a gas leak and disastrous situations, the main database can be availed by the fire department or by the concerned authorities at any point of time.
- Using *IoT* as the mode of communication opens a larger range of application as the mode of communication, i.e., the Internet is vast and paves as a huge platform for communication. By doing the same, large industries and offices that are connected to the Internet can share and access data which is not limited to information from sensors alone.
  Usual home automation systems include a Bluetooth module as a mode of communication at the user end. However, Bluetooth has limitations when it comes to distance and has a range of a few meters.
- Smart meters can be of great use on an industrial level as power metering is something if great use when it comes to data analysis of energy consumption in general. By producing a real-time display of power consumed of each of the devices or sections in general, it is easier to evaluate the data and make necessary changes.
  For example, the energy that is used during peak hours can be analyzed and the respective measures can be taken to lead to a more economically viable system.

- Smart meters are essentially a fundamental building block that leads to smart buildings which apply to smart cities that are finally integrated using smart grids which are the very future of power transmission in the upcoming years.

7 Conclusion

An insight into a working user-friendly smart home automation has been designed and presented. Home automation processes always move toward improvement for better living. However, with the current advancement of modern technology, smart homes have proven to play a vital role in this era.

It was observed that due to platforms like *IoT*, smart homes are incredibly easy to customize, one can add multiple sections to the house by not disturbing the integrity of the system. It is also to be noted such systems need to be energy conscious such that it caters to the environmental conditions at the present to lead a convenient and sustainable lifestyle. By providing a friendly GUI that gives remote access to all functionalities of the house, the smart house ensures that it is catered to people with special needs hence leading to overall social benefits. The prototype built can be used for various other functionalities outside home automation alone. Multiple industries, hotels and school system can benefit from this built prototype.

By providing door lock systems that do not need RFID card readers, the system is made foolproof. By providing smart meters in a normal household, users can reduce their energy intake during peak time. Being aware of the energy consumed at all times, behavioral changes can be adopted. Individuals can be more aware of the tariffs and respectively reduce the energy intake. Thus, it can be said that the results on the build of the prototype have led to economical as well as sustainable benefits. It can finally be concluded that this paper not only gave insight to home automation prototyped but also opened numerous possibilities possible based on the prototypes built. Although the prototypes were built to fulfill all specifications, the built boards can be further developed to benefit society. The smart meters that were built about this project can be programmed to optimize the overall energy use which is the basis to the future of smart cities. Since the household consists of the main source unit in the form of a database, the information that can be attained from the database can be programmed with a voice controlling module to foster a voice-controlled smart home. It can finally be concluded that this paper not only gave insight to home automation prototyped but also opened numerous possibilities possible based on the prototypes built.

References

1. Vishwakarma SK, Upadhyaya P, Kumari B (2014) Smart energy efficient home automation system using IoT. IEEE
2. Fortunebusinessinsights.com (2020) Home automation market size, share, trends, analysis report, 2026
3. Balta-Ozkan N, Boteler B, Amerighi O (2014) European smart home market development: public views on technical and economic aspects across the United Kingdom, Germany and Italy
4. Mihalache A (2017) Wireless home automation system using IoT. Inform Econ 21(2/2017):17–32
5. Pinki Vishwakarma V (2017) A comparative analysis on smart home system to control, monitor and secure home, based on technologies like GSM, IOT, Bluetooth and PIC Microcontroller with ZigBee Modulation. IEEE Conference Publication
6. Meenakshi et al (2020) Arduino based smart fingerprint authentication system. IEEE Conference Publication

7. Louis J, Calo A, Leiviskä K, Pongrácz E (2015) Environmental impacts and benefits of smart home automation: life cycle assessment of home energy management system. IFAC-PapersOnLine 48(1):880–885
8. Griffiths S, Mills R (2016) Potential of rooftop solar photovoltaics in the energy system evolution of the United Arab Emirates. Energy Strat Rev 9:1–7

Chapter 4
Voice and Gesture Based App for Blind People

Tilak Satra, Manali Shah, Ajay Lad, and Stevina Correia

1 Introduction

Earlier people across the globe communicated with each other through time-consuming letters. However, with so many technological advents happening over the years, primarily the Internet, it is now possible to communicate with anyone in the world at a very low or no cost. The Internet provides instant communication with the help of e-mails and other messaging applications. They can be used via smartphones or over the web. Besides these e-mail and messaging apps, many other facilities are provided by our smartphones like calling, setting alarms and timers, listening to music, searching for nearby relevant places, and viewing locations that help in completing our daily tasks with ease. However, there is a special criterion with the help of which one can use these services-eyesight. Thus, it becomes challenging to provide these services to the visually impaired society, and hence, they rely on other individuals for assistance in completing their tasks. There are various systems developed trying to assist the blind or visually impaired people. However, these systems are either not blind-friendly as they require mediation from the third person

T. Satra (✉) · M. Shah · A. Lad · S. Correia
Dwarkadas J. Sanghvi College of Engineering, Mumbai, India
e-mail: ayushsatra@gmail.com

M. Shah
e-mail: manalishah69490@gmail.com

A. Lad
e-mail: ajaylad11098@gmail.com

S. Correia
e-mail: stevina.dias@djsce.ac.in

I. Jeena Jacob et al. (eds.), *Data Intelligence and Cognitive Informatics*,
Algorithms for Intelligent Systems,
https://doi.org/10.1007/978-981-15-8530-2_4

or completely based on voice interaction which makes the process monotonous and time-consuming. Google Assistant too is not blind-friendly and so requires mediation to use it certain functionalities. One of them is reading e-mails [1]. Moreover, it relies on speech-to-text detection for its proper functioning. If the input itself is not recognized accurately via the speech-to-text detector, then the entire process becomes sluggish [2]. Thus, to eliminate an intervention or mediation from a third party and the sluggishness caused due to speech-to-text detection, a gesture and interactive voice response (IVR)-based Android application is proposed, wherein, a gesture along with the voice-based input indicates which task is to be performed which eradicates the need to use voice-recognition repetitively and excessively to know which action is to be executed [3–5].

2 Literature Survey

2.1 Speaking E-mail-Mobile Application

Working: This is a mobile application which provides different e-mail functionaries. This app provides various features for composing an e-mail via voice input. It also reads out the received e-mails to the user.

Limitation: the system is not working properly as well as it is not useful for blinds. The system requires the assistance of a third person.

2.2 Be My Eyes: Helping the Blind Application

Working: The system enables blind to receive live assistance from sighted volunteers through video calls [6]. As soon as the blind requests for help, a volunteer is assigned to help that blind and a live audio-video connection is set up between the two.

Limitation: The system requires the help of the third person, which does not make the blind person self-reliable and cannot be used to perform day-to-day activities like setting alarms, performing basic calculations, making a call, sending a message, etc.

2.3 Eye-D for Visually Impaired

Working: The system can be used to help the blind to know the current location, evaluate the surrounding using images. Users can also explore the nearby places of interest. It can also assist the user by reading the text from the image.

Limitation: The major problem with the system is that the system does not guide the user to navigate through the system. The user requires the help of the third person to know where a particular button is located to perform a particular action and which button to be clicked to perform a specific activity.

2.4 Visually Impaired

Working: The system allows the user to access various different features of mobile such as turning the flashlight on/off, call a contact, turn Wi-Fi on/off, and turning Bluetooth on/off. This system is also used to read the text from the image.

Limitation: The system is not useful because any of the features do not function successfully as well as the blind person needs the assistance of the third person to help the user to navigate across the system.

2.5 Envision AI

Working: The system does a great job of describing scenes. It can describe what it sees around in an easy to understand way. It can read texts from any surface accurately and quickly.

Limitation: The system crashes down as soon as it is opened and does not work at all. It also requires the help of the third person to navigate from the first screen of the system.

2.6 Google Assistant

Limitation: This system is not completely blind-friendly, as, for certain functionalities, it does require the user to ask for mediation. For instance, for reading e-mails, it lists down the e-mails received, however, does not read them out, and, thus, making it incompatible for a blind to use the e-mail functionality. Also, it heavily relies on its speech-to-detection algorithm for its proper functioning. If the input is not recognized accurately, then the entire process becomes repetitive and sluggish.

3 System Architecture with Modular Description

The architecture consists of a two-stage approach which comprises gesture recognition and taking the voice input. Before performing any gesture, the system allows the blind to control the application [7]. It does so by allowing the blind to set the gestures for the functionalities provided. On completion of the gesture setting for all modules, the blind will then be permitted to use all the functionalities [8]. For using any functionality, the blind will have to perform the gesture set by themselves. On performing the gesture, the next step is gesture recognition. The gesture recognition maps the performed gesture to corresponding activity or task. On successful mapping, the corresponding activity gets executed and the second-stage is ready to be performed. The second step asks for the input, if necessary, which is mainly voice-based input taken through the speech-to-text detector (Fig. 1).

The architecture consists of following major modules which are:

- Performing gesture and its detection
- Text-to-speech engine
- Speech-to-text engine.

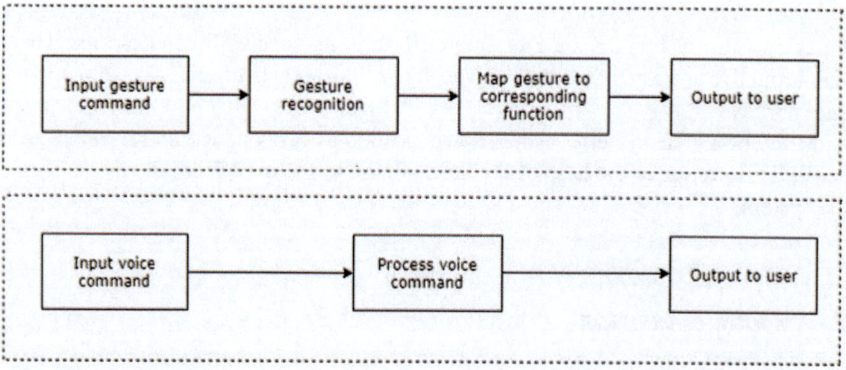

Fig. 1 System architecture

3.1 Performing Gesture and Its Detection

The entire application is primarily focused on performing and detecting gestures. The reason for it is centered around the aim to make the visually impaired society more self-reliant. The gestures are basic and easy to learn. Hence, performing them does not require any assistance. Thus, the simplicity involved in using gestures helps in automating the daily chores. The blind themselves assign gestures to various activities or functionalities which can be changed later as per the convenience. Thus, when a particular gesture is executed, the application identifies the functionality which is to be performed.

3.2 Text-to-Speech Engine

The text-to-speech engine enables the application to alert, instruct as well as guide the blind throughout the application. It simply does not read out text. It helps in instructing the user of whether an input is needed, for instance, while calling or sending SMS, a person's name or number is required. Moreover, what all users can perform while on a particular activity is also instructed. It also alerts by informing that a particular task is complete and now another task can be performed. Additionally, whenever confirmation is required for completing an action, it also asks for the final consent.

3.3 Speech-to-Text Engine

The speech-to-text engine helps in converting the voice into a text. Hence, it is used for taking necessary input wherever required. It helps in avoiding a third party's assistance. Thus, the blind themselves can provide or rather speak out the required information which will be converted to simple text and processed by the application and appropriate action will be taken depending on it. For instance, when a blind wants to call a person, the speech-to-text engine will help to interpret the contact's name or number, which will be provided to the application and a call will be made by the app if the contact exists.

4 Implementation Details

4.1 Read Aloud

A text-to-speech engine that reads out content and instructs the blind in a very concise and clear manner. This task is achieved by using the in-built text-to-speech library, also known as TTS library, within Android.

4.2 Setting Gestures Dynamically

This feature allows the blind to decide and set gestures for all the functionalities being provided, which is different from all other systems which have a pre-defined format or follow a fixed pattern. These gestures need not be remembered and can be changed as well, as per the requirement. Thus, the application turns out to be dynamic. Using the gesture listener interface allows the user to perform the various gestures which are then detected using the Gesture Detector Compat Library. Gesture Detector Compat enables the application to detect the gesture, mainly single tap, double tap, long press, and fling using, which detected swipe up, swipe down, swipe left, and swipe right.

4.3 Adjust Reading Speed

This feature allows setting the speed of the text-to-speech reader according to one's own pace.

4.4 Composing E-mails

This module allows the blind to send e-mails. The user can send e-mails by specifying the name of the person whose e-mail-id is extracted from contacts, subject, and the body. Before sending the e-mail, Google OAuth is used, only once, to ensure that the application using the account information is non-malicious and thus secured. After providing the consent, necessary input is taken via text-to-speech, which is later structured using java mail jars and finally sent using the Gmail API [9, 10].

4.5 Music

This module serves to be a source of entertainment for the blind. It provides automatic Bluetooth connectivity and, hence, helps to listen to the downloaded songs. The gesture, which is assigned to the module, is used to turn on and off the Bluetooth connectivity. Moreover, only those devices which have been paired previously can be connected again. Content URI helps to access the downloaded audio. Additionally, play, pause, next, and previous song call-back is achieved using the media player which is controlled via gestures only.

4.6 Clock

This module allows the blind to set alarms and timers. Alarm manager is used to access the alarm service. The time as input is taken via text-to-speech engine. The time is then structured using calendar, and alarm/timer is set using Intent via broadcast receivers. When the alarm or timer rings, the app if it is closed or terminated, opens again reminding the blind of the timer/alarm and allows it to pause or close the alert through the gestures. Moreover, they can even be canceled, which is done Intents only.

4.7 Emergency Service

This module helps in alerting the guardian about any exigency or trouble which the blind may be in via an SMS using the Intents in Android. Upon receiving the SMS, the guardian can login to the app and track the blind's movement. The location of the blind is periodically updated and sent to a database which is fetched and updated continuously on Google Maps [11].

4.8 Contacts Module

This module allows the blind to call as well as message any person from within or outside the phone contacts. Actual execution happens after performing the gesture which was set and providing the name or number of the person. The name is searched using content provider and resolver in Android with the contacts. All the matching contacts are instructed to the blind. Among these, the final contact name is again provided to which a call is made by Intents or an SMS is sent via SMS manager. Moreover, a confirmation is also asked, depending upon which the final action is executed [12].

4.9 Nearby Places

This module helps the blind browse and view places like hospitals, police stations, and other selected places and also provides instructions to navigate through them. All the details such as location and contact numbers are fetched using Google Places API. The functionality assigned to each gesture helps to either directly call the place or know the directions.

5 Outcome

5.1 E-mail Module

Figure 2 shows that the guardian needs to select an e-mail id through which the mail will be sent while entering the personal information of the blind. In this module, the user will have to give voice inputs for the recipient e-mail id, subject, and body of the mail as shown in Fig. 3. And then, the confirmation to send the e-mail is asked, and accordingly, the mail is sent as shown in Fig. 4.

5.2 Contacts Module

Figure 5 shows the various gestures to be performed for enabling call and message functionality. For calling functionality, voice input is taken from the person to whom they wish to call as depicted in Fig. 6. And then, the confirmation for the call is asked, and accordingly, call is made to the requested person as shown in Figs. 7 and 8. The message module works similarly to the calling module by asking for the message that the user wants to send to the contact, and finally, the message is sent as shown in Figs. 9 and 10.

Fig. 2 Choosing account

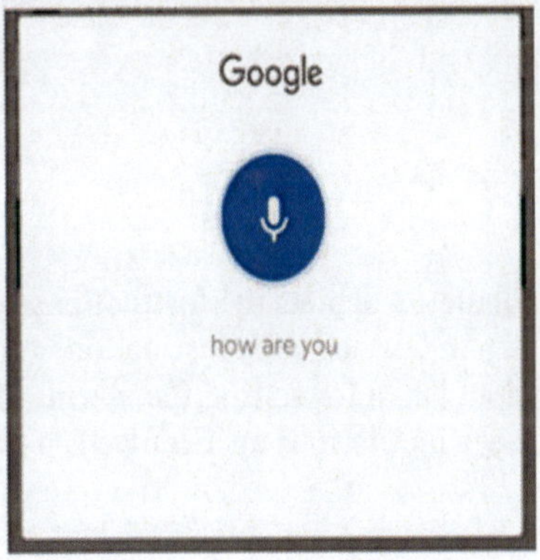

Fig. 3 Voice input from user for body field

5.3 *Emergency Services Module*

Figure 11 shows that the message is sent to the guardian whenever the user performs the defined gesture in case of emergency. After the guardian has received the message, the guardian can open the application and login to view the location of the guardian as shown in Fig. 12. The blind's location can be viewed in two formats as shown in Figs. 13 and 14.

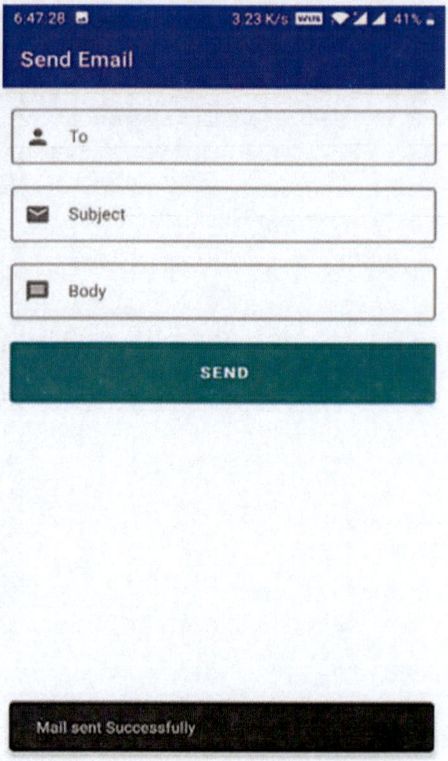

Fig. 4 E-mail sent successfully

5.4 Clock Module

Figure 15 depicts the instructions for the gestures to be performed for this module. Figure 16 shows the instructions for the alarm module, and Fig. 17 depicts the voice input taken for setting the alarm. Figure 18 shows the instruction to stop the alarm when the alarm is up. Similarly, a timer is set which is shown in Figs. 19, 20, 21, and 22.

5.5 Music Module

Figure 23 shows the instructions for this module. Figure 24 shows that the Bluetooth is turned on when the defined gesture is performed for automatic Bluetooth connectivity. Figure 25 shows the music being played from the music library of the device of the user. Accordingly, various gestures can be performed for pausing, playing next or previous song, or stopping the music being played.

Fig. 5 Instructions for contacts module

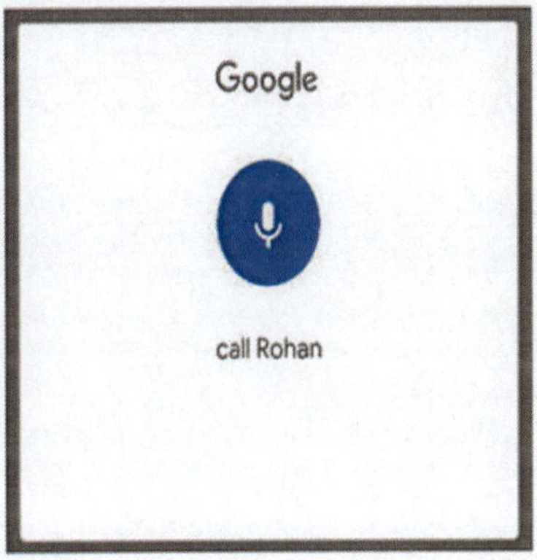

Fig. 6 Voice input for contacting a person

5.6 Nearby Places Module

Figure 26 shows the current location of the user. Performing various gestures, the blind can search for nearby hospitals, banks, police stations, and restaurants as shown in Fig. 27. Additionally, double-tap allows calling the respective place as shown in Fig. 28.

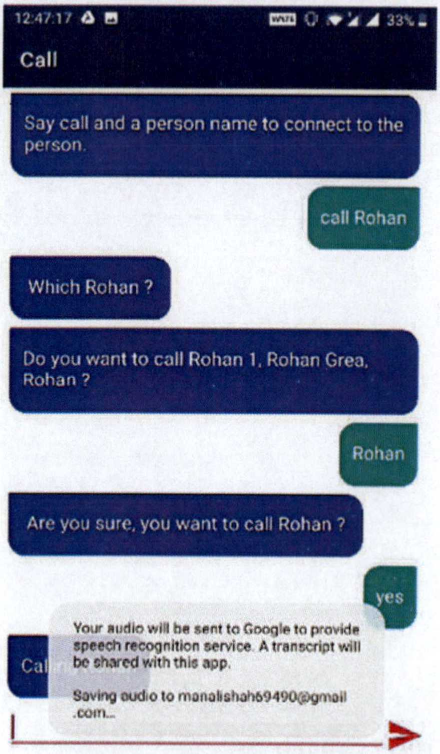

Fig. 7 Confirmation taken

Fig. 8 Final call made

Fig. 9 Message input from user

Fig. 10 Message sent

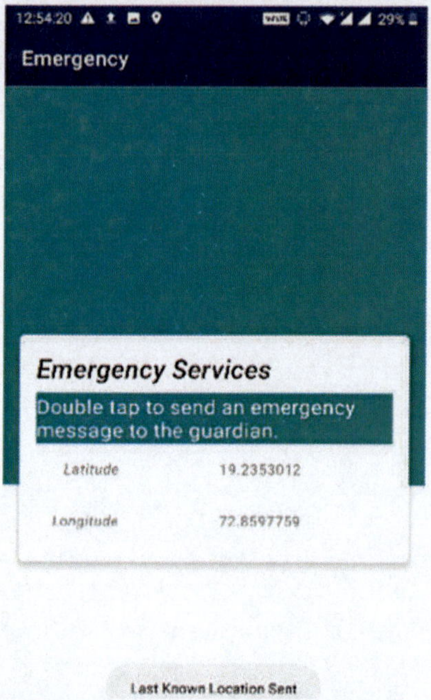

Fig. 11 Blindside emergency module

Fig. 12 Guardian side emergency module

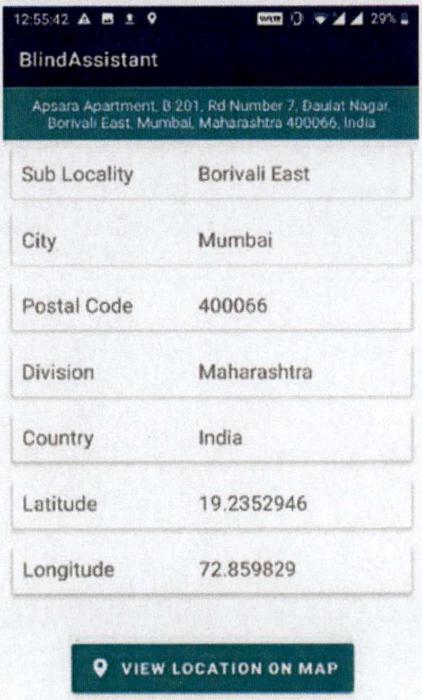

Fig. 13 Blind's location in address format

Fig. 14 Blind's location in map view

Fig. 15 Module instructions

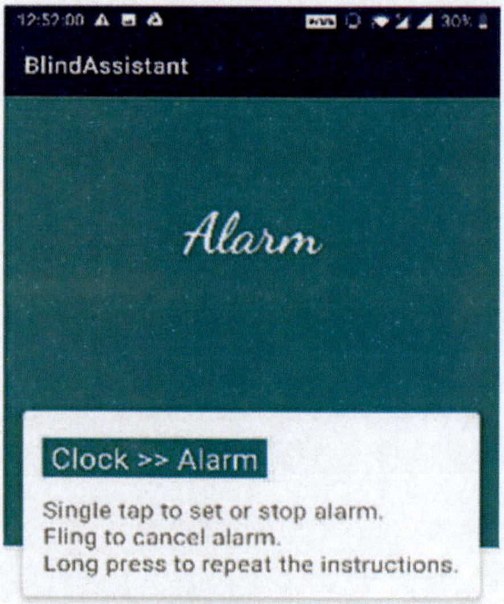

Fig. 16 Alarm instructions

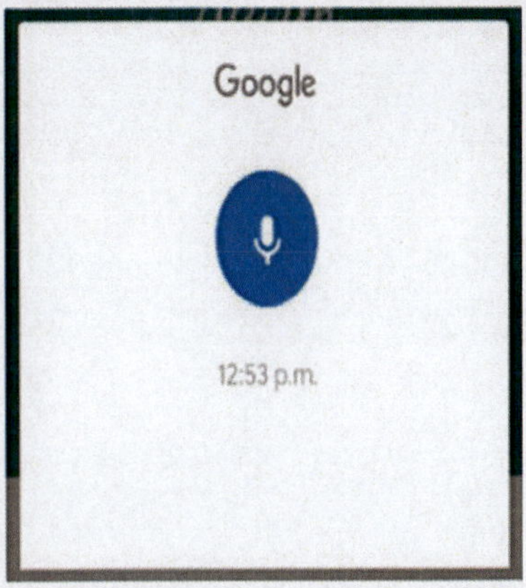

Fig. 17 Voice input to set an alarm

Fig. 18 Alarm ringing

6 Conclusion

In this mobile application, blind-assistant serves to help the visually impaired society in becoming more self-reliant by using their smartphones effectively to complete their

Fig. 19 Timer instructions

Fig. 20 Voice input to set timer

daily activities. The application widely makes use of gestures which can be easily

Fig. 21 Timer running

Fig. 22 Timer up

Fig. 23 Module instructions

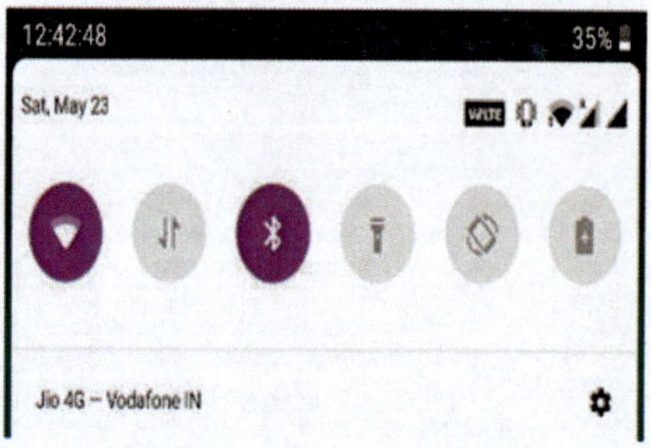

Fig. 24 Bluetooth turned on for connecting to a wireless device

performed by the visually impaired and a speech-to-text engine to take the input and confirmation for final action to be performed wherever required. Hence, it minimizes or eliminates the intervention required from a third party. Moreover, the mapping of a gesture to a particular activity is totally under the control of the blind, making the application dynamic. Additionally, the application with the help of a text-to-speech engine alerts visually impaired regarding activity completion, instructs what input is required, and also repeats instructions when required. In short, application converses in a smarter and an effective manner.

Fig. 25 Song being played either on speaker or Bluetooth

Fig. 26 Current Location

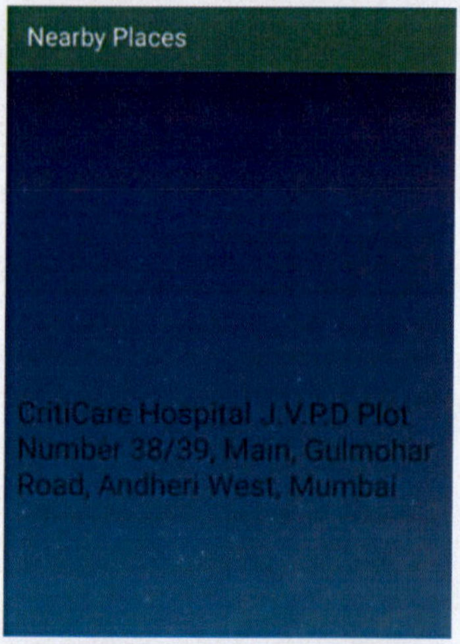

Fig. 27 Nearby place-hospital for instance

Fig. 28 Gesture performed to make a call

References

1. Ummuhanysifa U, Nizar Banu PK (2013) Voice based search engine and web page reader. Int J Comput Eng Res (IJCER) 1–5
2. Kasthuri R, Nivetha B, Shabana S, Veluchamy M, Sivakumar S (2017) Smart device for visually impaired people In: 2017 Third International Conference on Science Technology Engineering & Management (ICONSTEM), pp 54–59. IEEE
3. Sultan N, Siddiq K, Rashid T, Farooque M (2015) Evaluation of smart phone applications accessibility for blind users. Int J Comput Appl 127(3):9–16
4. Nilesh J, Alai P, Swapnil C, Bendre MR (2014) Voice based system in desktop and mobile devices for blind people. Int J Emerg Technol Adv Eng (IJETAE) 4:404–407
5. Paul IJL, Sasirekha S, Mohanavalli S, Jayashree C, Priya PM, Monika K (2019) Smart eye for visually impaired-an aid to help the blind people. In: 2019 International Conference on Computational Intelligence in Data Science (ICCIDS), pp 1–5. IEEE
6. https://www.bemyeyes.com
7. Csapó Á, Wersényi G, Nagy H, Stockman T (2015) A survey of assistive technologies and applications for blind users on mobile platforms: a review and foundation for research. J Multimodal User Interfaces 9(4):275–286
8. Sierra JS, Togores JS (2012) Designing mobile apps for visually impaired and blind users. In: The fifth international conference on advances in computer-human interactions, pp 47–52
9. Tripathi S, Kushwaha N, Shukla P (2019) Voice based email system for visually impaired and differently abled. Int J Eng Res Technol (IJERT) 08(07)
10. Dasgupta T, Anuj A, Sinha M, Ghose R, Basu A (2012) VoiceMail architecture in desktop and mobile devices for the Blind people. In: 2012 4th International Conference on Intelligent Human Computer Interaction (IHCI), Kharagpur, pp 1–6. https://doi.org/10.1109/ihci.2012.6481791
11. Tanveer MSR, Hashem MMA, Hossain MK (2015) Android assistant EyeMate for blind and blind tracker. In: 2015 18th International Conference on Computer and Information Technology (ICCIT), Dhaka, pp 266–271. https://doi.org/10.1109/iccitechn.2015.7488080
12. Kardyś P, Dąbrowski A, Iwanowski M, Huderek D (2016) A new Android application for blind and visually impaired people. In: 2016 Signal Processing: Algorithms, Architectures, Arrangements, and Applications (SPA), Poznan, 2016, pp 152–155. https://doi.org/10.1109/SPA.2016.7763604

Chapter 5
Digital Image Forgery Detection Using Ternary Pattern and ELM

D. Vaishnavi, D. Mahalakshmi, and M. S. Shawon Rahman

1 Introduction

Due to an evolution of mobile digital technology and the internet, a human can simply exercise multimedia devices, digital cameras, and mobile phones to capture the incidents as images. They also can edit such images with the assistance of image editing tools and spread them out over the internet. In recent times, many image altering occurrences have taken place [14], which undermines social equity and national security, consequently how to viably make out the legitimacy of images. Accordingly, it is imperative to bring out the effective method to discern such kind of image tampers from the spurious. The authentication of digital images can be branded into active and passive practices [5, 9, 10, 16]. The first practice called, active one requires some prior information, but the passive technique does not. The passive techniques are having more attention in recent days, and it works based on the concept of hypothesis; that is trying to fetch the underlying image statistical or textural changes (inconsistencies) that occurred in a tampered region of the image. There are two unremarkable tampering procedures. The first one is called copy-move and the second one is called splicing. An image undergone for copy-move forgery procedure can be detected by finding the akin regions exist in an image and

D. Vaishnavi (✉)
Department of CSE, SRC, SASTRA Deemed to be University, Kumbakonam, Tamilnadu, India
e-mail: vaishume11@gmail.com

D. Mahalakshmi
Department of IT, AVC College of Engineering, Mayiladuthurai, Tamilnadu, India
e-mail: dmahaa1985@gmail.com

M. S. Shawon Rahman
Department of CSE, University of Hawaii, Hilo, USA
e-mail: srahman@hawaii.edu

I. Jeena Jacob et al. (eds.), *Data Intelligence and Cognitive Informatics*,
Algorithms for Intelligent Systems,
https://doi.org/10.1007/978-981-15-8530-2_5

its endeavors to uncover them if any. On the other hand, splicing detection aims at discovering the amalgamated regions in the images [8].

2 Related Work

In general, image splicing forgery can be done by concealing the image elements or by composing at least two images together. As a result, an actual image is perceptibly changed and formed as a forged one.

There are a huge number of incidents in history to demonstrate. Among them, one incident is demonstrated in Fig. 1. India's official press information bureau published a photograph of Prime Minister Narendra Modi looking over the surge damage in Chennai by air [7] is displayed in Fig. 1b. The public, however, quickly spotted the sloppy photoshop work and discovered that the image of the flooded city had been superimposed in the window of Modi's plane as seen in Fig. 1a.

In recent years, there are numerous classes of passive image splicing falsification finding strategies that have been projected. The work in [9] used quadratic phase coupling to distinguish the spliced images, and it was achieved 72% of success rate. The authors extracted wavelet moment characteristics and Markov features of vector length 266 and secured 91.8% of accuracy [12]. The method presented in [17], fetched the gray level co-occurrence matrix (GLCM) and an edge image of the Markov features chain. Then, the low dimensional feature vectors are taken out to identify the forgery. The work presented in [4], the Markov features are taken out from DCT and DWT domains. The method proposed in [8], obtained the QDCT coeffecients of quaternion blocked images to get the expanded Markov features from the transition probability matrices based on the inter-block correlation.

The work presented in [10], specially designed the convolution neural network (CNN) to learn the forgeries, and network is initialized with basic high-pass filter set.

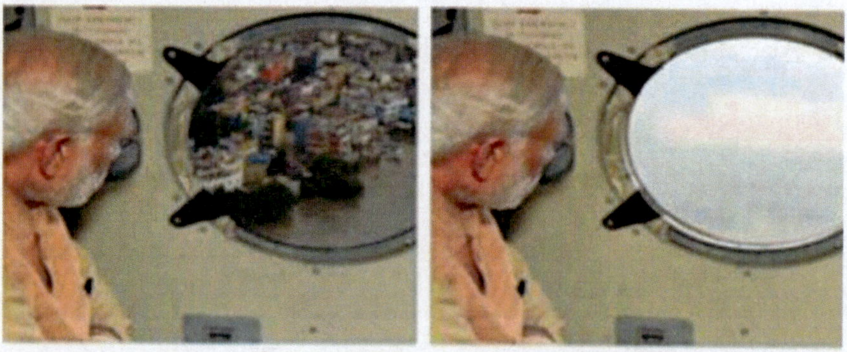

a) b)

Fig. 1 Illustration of image splicing forgery

Most of the proposed methods extracted the Markov features to disclose forgeries [4, 5, 8, 19].

Alternatively, nowadays, local patterns were obtained. In [11], the authors extracted the multi-scale Weber Law Descriptors (WLB) to distinguish the image from the forged one. From three divergent neighborhoods, the histogram and gradient orientation are computed and merged as a feature vector. The proposed method presented in [1] has used LPQ operator, and the phase information-based image internal statistics are taken out. The work in [3] extracted the DCT, LBP, Curvelet, and Gobar features and combined altogether to distinguish the spliced images. In the article [2], the authors used the Curvelet transform on the image's chrominance channel to disintegrate it as many scale wedges and orientation. Then, computed the LBP histogram from each wedge and combined altogether to obtain the ultimate feature vector. The authors in [15] computed the histogram of the local binary pattern (LBP) and employed support vector machine (SVM), backpropagation neural network (BPNN), and combined k-nearest neighbor classifiers to discover the forged images. The maximum of the above-mentioned state-of-the-art methods used the SVM algorithm to identify the authentic and forged images. But, this work utilized extreme learning machine (ELM) algorithm since it produces superior results than the SVM, and it can also get very less errors during the training phase. This paper also aims to extract the histogram of local ternary pattern to detect splicing forged images.

3 Proposed Work

The proposed method is formulated with three phases and they are pre-processing, feature construction, and classification. The meticulous explanation of workflow is illustrated in successive sessions.

3.1 Pre-processing

Pre-processing aids to advance the chromatic quality of the image and to interpret the images. The tampered images look realistic, and the traces of tampering are not easily detectable. These traces can be distinctly acquired over the chrominance channel than the RGB channel [18] and [19]. Therefore, this work also transforms an input image from RGB to YCbCr color space, which consists of luminance and chrominance components which represent the color information.

3.2 Feature Extraction

Generally, all digital images would have regularities with their neighborhoods such as statistical or textural domain. Splicing of images disturbed the statistical characteristics of the image and distorts the texture of the image. By extracting and analyzing these traces, the image can be distinguished as forged or an authentic one. Therefore, the histogram features from local ternary patterns are extracted. These features are input vector to the machine learning algorithm to unveil the genuineness of the images.

3.3 LTP

LTP is an extension of LBP, which is introduced by Tan and Triggs [13]. LTP stretches LBP to three values code. Instead of fixing a threshold based on the chief pixel value of the neighborhood, the user defines a threshold e and determines the pixel difference value. It assigns value 0 if the pixel difference lies in the interval of $-e$ and $+e$. It dispenses the value 1 if the value is greater than $+e$ and if it is below than $-e$, then it assigns the value -1. The following Eq. gives the same.

$$\mathrm{LTP}_{R,N} = \sum_{i=0}^{N-1} s(p_i - p_c)3^i,\ S'(y) = \begin{cases} 1 & y \geq e \\ 0 & y < e \\ -1 & y \leq -e \end{cases}. \tag{1}$$

where R and N refer to the radius and neighborhood pixels. Then, each ternary pattern is fragmented as positive parts and negative parts and these are managed as two separate channels of descriptors to generate the LTP code. It is depicted in Fig. 2. Afterward, the histograms of both channels are calculated separately and concatenated to get the final feature vector.

3.4 Classification Using ELM

Classification is a significant step in pattern recognition. It accepts the feature vectors as input, and it classifies the patterns based on inputs, which have been given. To identify the splicing forged images, the proposed method employs ELM algorithm.

ELM is a single-layer feed-forward neural network (SLFN) which randomly selects the output weights and the hidden nodes too. This random selection can be done if the hidden layer activation functions are enormously differentiable. The SLFNs output weight is a contrary procedure of the hidden layer output matrices. The pictorial representation of generalized SLFN is presented in Fig. 3. The learning speed of ELM is a thousand times more rapid than the classic feed-forward network

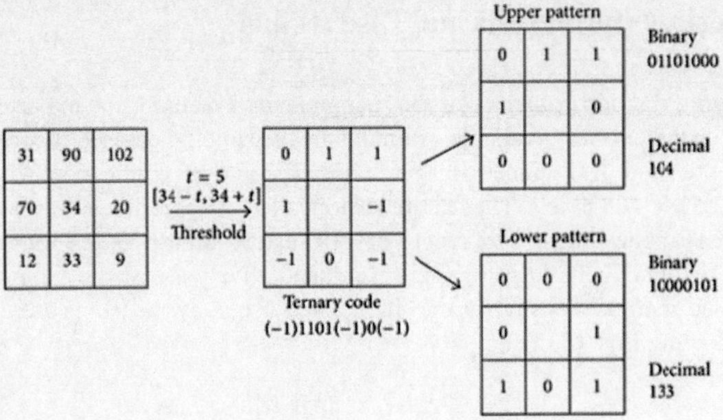

Fig. 2 LTP operator

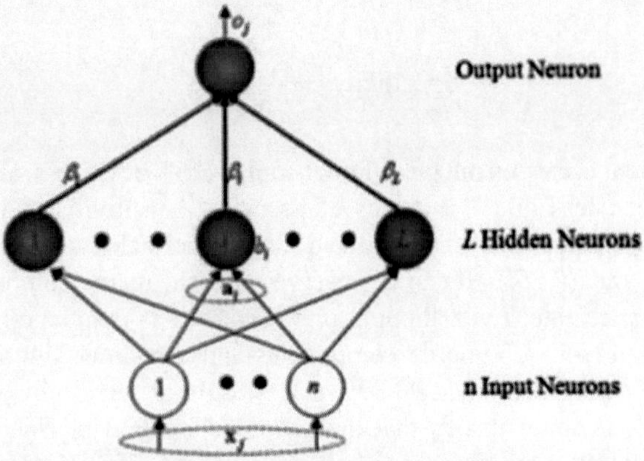

Fig. 3 Generalized SLFN

algorithm. This learning algorithm influences to get a very less training error with good classification accuracy. It is an extremely fast-learning algorithm and can be simply applied to ELM [6]. For N arbitrary distinct samples $(x_i, t_i) \in R^n \times R^m$, LFNs with L hidden nodes each with output task, $G(a_i, b_i, x)$. e statistically derived and shown as Eq. (2).

$$\sum_{i=1}^{L} \beta_i G(a_i, b_i, x_j) = t_j, \quad j = 1, \ldots, N \tag{2}$$

where (a_i, b_i) are hidden nodes parameters and β_i is the weight vector connecting the *ith* hidden node and the output node.

4 Experimental Results and Discussions

For justifying the performances of the recommended method, the experiments are exercised on the publicly available Columbia image splicing dataset. It encompasses 183 authentic and 180 spliced images whose resolution varies from 757 × 568 pixels to 1152 × 768 pixels. The feature vector called histogram features is extracted from the LTP code image. The length of 119 feature vectors is fed into the ELM algorithm to identify the forged images. To quantitatively assess the achievement of the proposed method, sensitivity, specificity, and accuracy measures are computed by the following Eqs. (3–5).

$$\text{Accuracy} = \frac{\text{tp} + \text{tn}}{\text{tp} + \text{fp} + \text{tn} + \text{fn}} \tag{3}$$

$$\text{Sensitivity} = \frac{\text{tp}}{\text{tp} + \text{fn}} \tag{4}$$

$$\text{Specificity} = \frac{\text{tn}}{\text{tn} + \text{fp}} \tag{5}$$

The experiment is carried out on different components of YCbCr, and their performances are given in Table 1 in terms of accuracy, sensitivity, and specificity. It shows that the performance of the Cb and Cr channel achieved a detection accuracy of 92.01% and 89.35%. It can be observed that chromatic components achieve a high-performance rate. Even though, one more test is conducted by fusing the features extracted from Cb and Cr components and it attained the detection accuracy of 94.29%. Also, it attained 93.89% and 94.69% of sensitivity and specificity, respectively. So, it could justify that the proposed method produces high results when combining both chrominance components and the same has been graphically exhibited in Fig. 4.

To observe the influence of LTP, the proposed method is effectuated to experiment with diverging parameters (*N, R*) (Here, *N* defines the number of neighboring pixels, and *R* refers its radius) and their outcomes are pictured in Figs. 5 and 6. It reveals that the best performances are attained with $N = 8$ and $R = 1$.

Furthermore, the proposed method is endorsed by constructing a new dataset, which contains 80 images which are taken from the internet. In which, 40 images

Table 1 Performances on Columbia dataset

Channel	Accuracy	Sensitivity	Specificity
Y	77.90	76.08	80.00
Cb	92.01	90.97	93.07
Cr	89.35	89.92	89.47
Cb and Cr	94.29	93.89	94.69

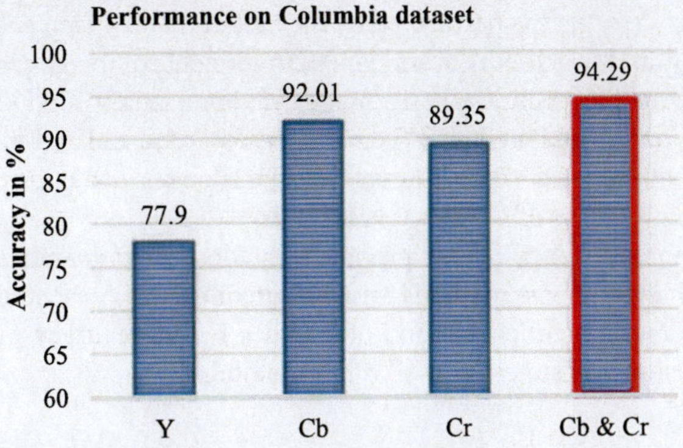

Fig. 4 Performance of the proposed method on Columbia dataset

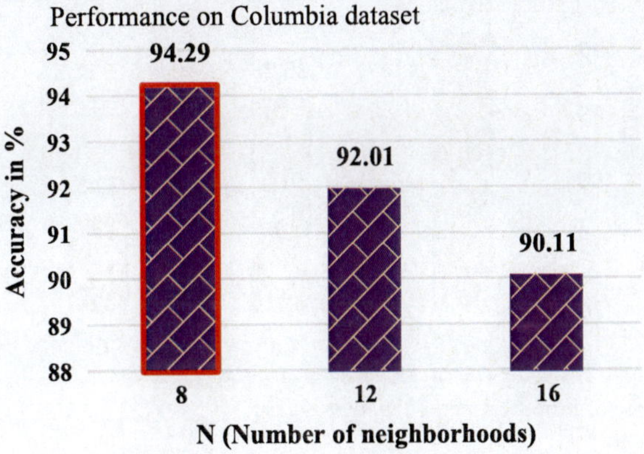

Fig. 5 Performance of the proposed method with varying neighborhoods

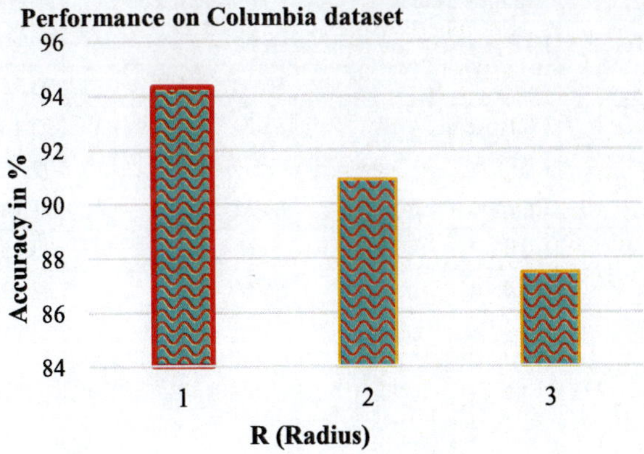

Fig. 6 Performance of the proposed method with a varying radius value

are forged images (forgery is performed by the authors) and 40 are authentic images. Features are acquired from both chrominance components of images and are provided for ELM classifier. The accuracy performance is appraised by conducting investigations on chrominance channels and their outcomes are visualized in Fig. 7. It also demonstrates that the proposed method secured a superior detection rate of 91.25% when combining features of Cb and Cr channels.

To certify the excellence of our proposed method, a comparative study is also carried out with state-of-art methods and its outcomes are presented in Table 2. It proves that the recommended method obtained a better accuracy of 94.29% with ELM classifier, and it is superior to all other methods.

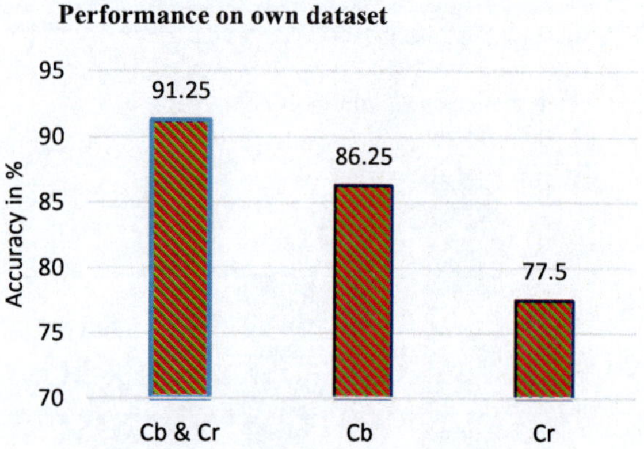

Fig. 7 Performance of proposed method on own dataset

Table 2 Performance comparison on Columbia dataset

Methods	Features used	Classifier used	Accuracy in %
Proposed method	LTP	ELM	94.29
[15]	LBP	SVM	93.75
[11]	Multi-scale WLD	SVM	93.14
[15]	LBP	Comb-KNN	92.50
[1]	Entropy + LPQ	SVM	91.14
[15]	LBP	BPNN	88.75
[4]	Markov	SVM	87.52

5 Conclusion

A method for detecting splicing forged images is proposed and implemented successfully. The histogram of a local ternary pattern is extracted, and extreme learning machine is exercised to perceive the forged images. The publicly accessible dataset is also utilized to validate the performance of the method which is proposed, and accuracy, sensitivity, and specificity measures are also figured. Various experiments are also conducted to find an optimal result and are shown that the proposed method obtained superior results when combining features of Cb and Cr channels. A comparative study is also done with state-of-the-art methods, and it is shown that the proposed method succeeded with a superior accuracy of 94.29% with other existing methods.

References

1. Agarwal S, Chand S (2015) Image forgery detection using multi scale entropy filter and local phase quantization. Int J Image Graph Signal Process 7(10):78
2. Al-Hammadi MH, Muhammad G, Hussain M, Bebis G (2013) Curvelet transform and local texture based image forgery detection. In: International symposium on visual computing, pp 503–512
3. Hashmi MF, Keskar AG (2015) Image forgery authentication and classification using hybridization of HMM and SVM classifier. Int J Secur Appl 9(2015):125–140
4. He Z, Lu W, Sun W, Huang J (2012) Digital image splicing detection based on Markov features in DCT and DWT domain. Pattern Recogn 45(12):4292–4299
5. Holliman M, Memon N (2000) Counterfeiting attacks on oblivious block-wise independent invisible watermarking schemes. IEEE Trans İmage Process 9(3):432–441
6. Huang G-B, Zhu Q-Y, Siew C-K (2006) Extreme learning machine: theory and applications. Neurocomputing 70(1):489–501
7. Indian Express (2015) Chennaiflood fake proven. http://indianexpress.com/article/india/india-news-india/pib-removes-pm-modis-chennai-photo-from-website-after-proven-fake/
8. Li C, Ma Q, Xiao L, Li M, Zhang A (2017) Image splicing detection based on Markov features in QDCT domain. Neurocomputing 228(2017):29–36
9. Ng T-T, Chang S-F, Sun Q (2004) Blind detection of photomontage using higher order statistics. In: Proceedings of the 2004 International Symposium on Circuits and Systems, ISCAS 2004, p V
10. Rao Y, Ni J (2016) A deep learning approach to detection of splicing and copy-move forgeries in images. In: 2016 IEEE International Workshop on Information Forensics and Security (WIFS), pp 1–6
11. Saleh SQ, Hussain M, Muhammad G, Bebis G (2013) Evaluation of image forgery detection using multi-scale weber local descriptors. In: International symposium on visual computing, pp 416–424
12. Shi YQ, Chen C, Chen W (2007) A natural image model approach to splicing detection. In: Proceedings of the 9th workshop on multimedia & security, pp 51–62
13. Tan X, Triggs B (2010) Enhanced local texture feature sets for face recognition under difficult lighting conditions. IEEE Trans İmage Process 19(6):1635–1650
14. Vaishnavi D, Subashini T (2015) A passive technique for image forgery detection using contrast context histogram features. Int J Electron Secur Digit Forensics 7(3):278–289
15. Vaishnavi D, Subashini T (2016) Recognizing image splicing forgeries using histogram features. In: 2016 3rd MEC international conference on big data and smart city (ICBDSC), pp 1–4

16. Vaishnavi D, Subashini T (2019) Application of local invariant symmetry features to detect and localize image copy move forgeries. J Inf Secur Appl 44(2019):23–31
17. Wang W, Dong J, Tan T (2009) Effective image splicing detection based on image chroma. In: 2009 16th IEEE international conference on image processing (ICIP), pp 1257–1260
18. Zhao X, Li J, Li S, Wang S (2010) Detecting digital image splicing in chroma spaces. In: International workshop on digital watermarking, pp 12–22
19. Zhao X, Wang S, Li S, Li J (2015) Passive image-splicing detection by a 2-D noncausal Markov model. IEEE Trans Circuits Syst Video Technol 25(2):185–199

Chapter 6
A Study on Identification and Cleaning of Struck-Out Words in Handwritten Documents

R. Dheemanth Urs and H. K. Chethan

1 Introduction

Document analysis and recognition is a process of extracting information present in the document (printed or handwritten). Nowadays, there is an increased interest in processing handwritten documents. Over centuries, paper documents are considered as an important source of information and communication [1]. Processing these documents helps in various domains like a study of old historical documents, archeological documents, personal letters, address on envelopes, bank documents, forensic document analysis, etc. These documents are converted to computer readable form using optical character recognition (OCR) applications. OCR is a software that recognizes characters from scanned documents.

Handwritten character recognition (HCR) is one of the domains in the area of digital document analysis where OCR is used frequently. When it is handwritten document, OCR faces various dynamic problems on recognition of texts. This is because if the individuality of each writer behaviors, i.e., writing styles (cursive, slant), skill, language, etc. Processing these types of documents is really challenging work. Currently, there exist various methods/techniques for handwritten documents, which are discussed in [2, 3]. These methods assume a fair copy of the document

R. Dheemanth Urs (✉)
Information Science and Engineering, DSCE, Banglore, India
e-mail: dheemu.urs@gmail.com

H. K. Chethan
Computer Science and Engineering, MIT, Mysore, Belavadi, India
e-mail: hkchethan@gmail.com

I. Jeena Jacob et al. (eds.), *Data Intelligence and Cognitive Informatics*,
Algorithms for Intelligent Systems,
https://doi.org/10.1007/978-981-15-8530-2_6

for processing, i.e., document containing no errors. But in reality, a free form handwritten document contains various types of noise (annotation, strike-out text, underlines, ink bleed, etc.). This survey paper discusses various methods that exist for the identification and recognition of such noise present in handwritten documents.

The common type of editing operation a writer does is striking out the text, super scripting on the stroked text, etc. Given this noise document for OCR, results in producing the meaningless converted document (Junk output string). Figure 1 depicts a sample document which contains various real-time noises like strike-out, scratched words, superscript texts, etc. Increase in noise makes OCR difficult to recognize texts in the document.

Among all these types of noise, striking operation is the most commonly used operation to avoid or hide irrelevant text written [4]. Figure 2 depicts the various types of strike-out strokes used in [4].

Identification and recognition of all these strokes in a document are challenging research work since each stroke differs. Identifying this stokes requires the development of unique procedures by observing its pattern.

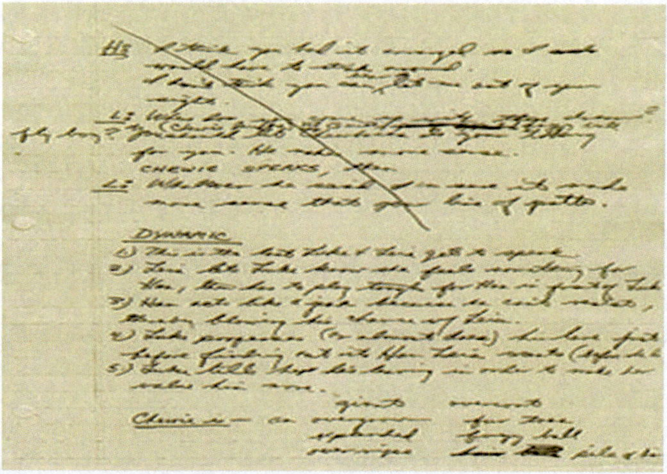

Fig. 1 Sample document with noise

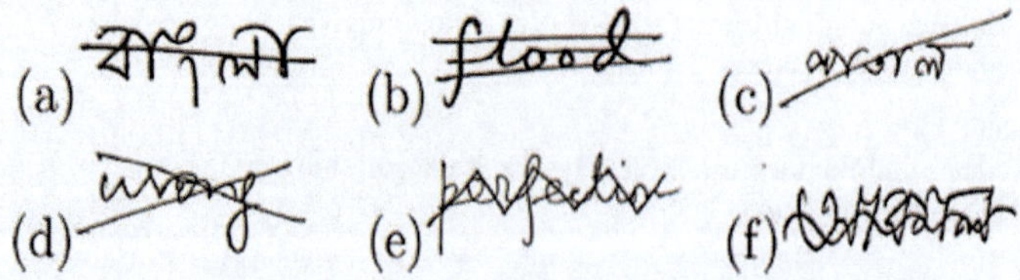

Fig. 2 Types of strike-out: (**a**) single, (**b**) multiple, (**c**) slanted, (**d**) crossed, (**e**) zigzag and (**f**) wavy [4]

Strike-out text processing helps various application domains like:

(a) For a better understanding of the behavior pattern of writers.
(b) For handwritten postal address recognition (superscript texts).
(c) For historical documents which include strokes, the annotation.
(d) Identifying strokes may give a clue for forensic crime detection.

Next sections of this paper are as follows; second section briefs the literature work done in this area. The third section discussing various methods/techniques proposed by other authors. Fourth sections contain details related to the performance of the proposed technique. The fifth section finally concludes giving room for future work and research opportunity.

2 Literature Survey

For identification and recognition of strike-out texts in a document, authors have proposed various methods and techniques. Very few research works are available related to this problem statement [4–8].

Chaudhuri et al. [4] proposed an approach for detection and cleaning of various strokes in an offline handwritten document written in English and Bengali. The approach starts with the identification of text line and connected components (CC), followed by removing small CC and converting big CC to medium size CC. This CC's is fed into support vector machine (SVM) classifier for the separation of normal and stroke text. The strokes are identified using graph-based approach (8-neighbor and shortest path). Both deletion and cleaning is done using image inpainting.

Adak et al. [9] in this paper experimented to find out the impact of struck-out text in the identification of the writer considering offline handwritten document images (English and Bengali). The approach first tries to detect all the struck-out texts in the entire document, later writer identification. For detection of struck-out text, features are extracted using CNN LeNet-5 (convolution neural network) architecture [10]. The extracted feature vector is fed into a classifier for the separation of normal and struck-out texts. Here, the authors use "contour-hinge" features [5, 11, 12] which track curvature and orientation strokes of individual writers. These feature vectors are further fed into SVM-RBF classifier for distinguishing normal texts and struck-out texts in the document. Experimentation was conducted on two different types of database namely NewISIdb: SoT [4] and their own generated database.

Alessandro et al. [6] here used hidden Markov model (HMM) and statistical language models (SLMs) for the recognition of unconstrained offline handwritten texts. Using SLM as a preprocessor detects, the regularities in natural language which increases the performance of natural language processing. The authors here used N-gram language models. The advantage of using the N-gram model is that it eases the incorporation of language model and recognition of the decoding phase. HMM is used for the recognition process. The models are trained with the Baum–Welch algorithm [7] and recognition is done using the Viterbi algorithm [13]. Authors here proposed a decoding technique that integrates recognition and language model. The advantage of going with the decoding technique is to avoid the segmentation process and path optimal matching is search space.

Adak et al. [14] here proposed an approach for the detection of strike-through text from a handwritten document (English, Bengali, Devanagari). The aim was to feed a fair copy of the document for optical character recognition (OCR) to perform text recognition. The above problem is solved by solving subproblems: (i) locating strike-out text, (ii) finding the strokes and (iii) removing the strokes. The approach first binarizes the image and labels the connected components in the binarized text. The line used for striking the whole word is marked as a connected component. The start/endpoints are considered using 8-connected neighbor, followed by detection of edges. Considering the structural features of stroke, the textual component is identified. For the conformation of strike-through Dijkstra's shortest path algorithm is used. To remove strike-through text, thinned text component is used.

Likforman et al. [8] used hidden Markov models (HMM) for the identification of wave-trajectory strokes and line-trajectory strokes in an offline handwritten document. The horizontal strokes were simulated based on delta lognormal model of rapid line movement and wave strokes were simulated based on spline curves. The models are trained using the Baum–Welch algorithm [7]. The recognition approach first binarizes the image to words and later removes slope and slant by applying normalization algorithm. These normalized words are converted into a sequence of feature vector using a sliding window approach.

Nilanjana et al. [15] considered various types of noise that can be present in a text document, i.e., overwriting, repetition, crossing-out texts in Bangla handwritten document. Even though the proposed approach is for online handwritten text documents, it can also be applied on offline handwritten text documents. The approach uses density-based features for extracting noisy parts in the document. For image binarization, the author here used grayscale images for the representation of stroke density. To extract the features, it is necessary to compare the densities of two neighboring regions. So, for this purpose, modified local binary patters (MLBP) are used [16]. These features are fed to HMM sequential classifier. The Viterbi algorithm is used for recognizing positions which point exact starting and ending positions of clean text and noisy text in the document.

Sanjoy et al. [17] proposed a technique which removes handwritten annotation lines from document image using digital geometric analysis and image inpainting technique. The detection of the annotation area was on the observation of sharp changes that may occur at the junction points and touching area of lines. The selection of the annotation object is by comparing its width. At first, the boundary edges are extracted using a binarized image. Fast marching method (FMM) [18] algorithm is used to propagate image information. Finally, the source image is subtracted from the mask which is detected from the image input. The proposed apply boundary tracking mechanism works without restoring any color information and it successfully identifies the area of annotation line irrespectively not bothering on whether it is touched or untouched lines by text characters.

Nicolas et al. [19] proposed a technique to extract the layout of a complex handwritten document using Markov random field (MRF). The MRF models are stochastic models of contextual interaction in bi-dimensional data. A Gaussian mixture is used to model the conditional probability density and MRF-MAP labeling framework is used for image segmentation. The experimentation is conducted on French Flaubert's manuscript. The decoding of images is implemented on several methods and compared global labeling rate (GLR) versus normalized labeling rate (NLR). The segmentation was obtained on page fragmentation using 4-state, 5-state and 6-state models.

Arlandis et al. [20] proposed rejection strategies and confidence measure using k-nearest neighbor (K-NN) classifier. In statistical classifier, probability density functions are computed for the training set. At the pre-processing stage, noise present in the document is removed and sent to K-NN classifier to differentiate between character recognition with low confidence and high confidence. Ambiguity threshold *Ta* is applied to reject those texts which are not clear. Distance threshold *Td* is for the different training set. Three systems are used for experimentation purpose, *SYS-A* with ambiguity rejection but no distance rejection option, *SYS-B* using both rejection text and *SYS-C* for a special class. Settings up different error rate against rejection rate results are obtained.

Banerjee et al. [21] proposed an approach to restore degraded document image which is independent of script, font and style. The method considers noise parameters like cuts, blobs, merges and vandalized document. At first, the entire document is segmented into words and characters. These segments are clustered to identify the shape of the document. Markov model is used for the restoration of characters. The Markov random filed-based framework is used in localizing the patches, and to improve the document image resolution, Viterbi algorithm has been used. The experimentation is conducted on degraded English books and achieved improving results. The proposed approach is script independent and can handle multiple scripts and fonts.

Brink et al. [22] proposed an approach for removing crossed-out text in a handwritten document hoping for the improvement is verification and identification of writer. The proposed approach uses the 8-connected component for the extraction of features (black components), which further discriminates into two components that is a small component and big component considering the text's width and height. These two features are skeletonized as branch features and size features.

Tuganbaev et al. [23] proposed an approach to recognize stricken-out characters in an English handwritten document and used feature-based classifiers for the recognition. Since it is a patented paper could not find detailed information on the method used and its experimental results.

3 Methodology

Few authors proposed various techniques for identifying, recognizing and cleaning the strike-out text in the handwritten text document. Author [24] has discussed a smart image processing algorithm for text recognition and [25] have discussed capsule neural network in various applications. Figure 3 depicts a sample convolution neural network, it is a deep learning algorithm which takes in an input image and assigns weights and biases to various objects in the image. The pre-processing required in a convolution neural network is minimum when compared to classification algorithms.

Convolution Layer

The kernel is the element involved in carrying out the convolution operation in the first part of a convolutional layer. Here, kernel is used as a matrix. The kernel shifts n times depending on the stride length, each time computing a matrix multiplication operation between the kernel and the portion of the image over which the kernel is flying. The filter moves to the right with a certain stride value till it parses the complete width. Moving on, it hops down to the left of the image with the same stride value and the same process continues for the entire image.

Pooling Layer

Pooling layer reduces the spatial size of the convolved feature. This been done to decrease the computational power required to process the data through dimensionality reduction. Also, it extracts dominant features which are rotational and positional invariant, also training of the model effectively. There are two types of pooling. One, max pooling returns the maximum value from the portion of the image covered by the kernel. On the other hand, average pooling returns the average. Max pooling is also used for noise suppressant. It discards the noisy activations and performs de-noising along with dimensionality reduction, whereas, average pooling performs

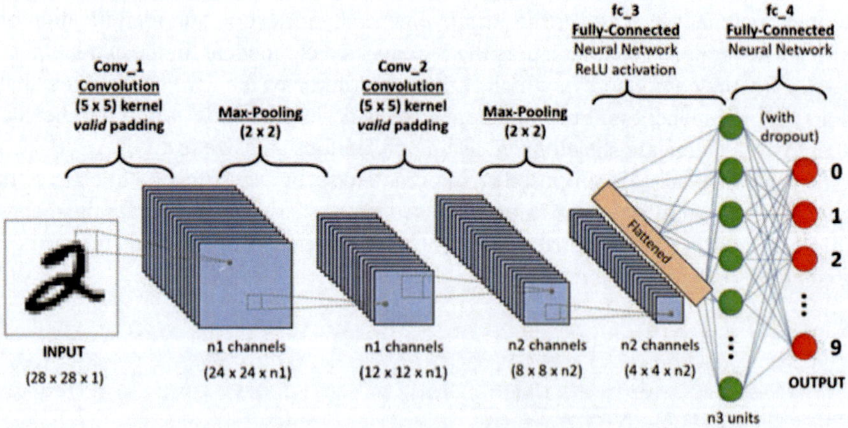

Fig. 3 CNN sequence that classifies handwritten digit

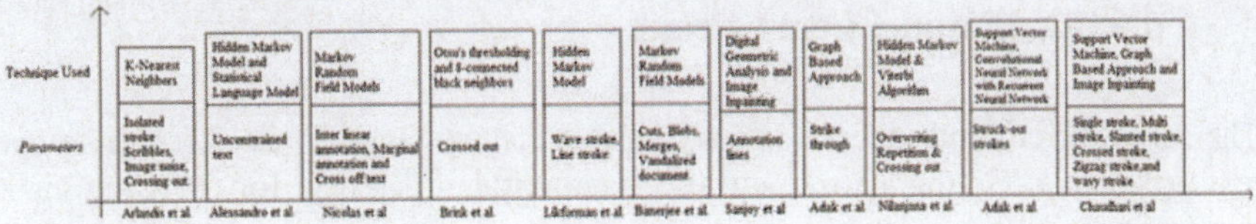

Fig. 4 Parameters and techniques used for identification

dimensionality reduction as a noise suppressing mechanism. So, max pooling does better than average pooling.

Figure 4 shows the related research works that were conducted in this area and also lists the type of parameter considered and technique of various works done so far in the identification, recognition and cleaning of strike-out words in a handwritten document.

The public available databases are NFI [26], IAM [27], CMATERdb [28], BH2M [29], ICDAR segmentation contest [30], MLS [31] and BFL [32] contain few strike-out texts which is not enough to carry out research work. Authors in [4] generated a new database called *NewISIdb* [4] which includes more strike-out texts which suit the topic of research work.

Figure 5 depicts the results of various methods that were used to identify and recognize strike-out strokes.

Since all the proposed method aims are different it cannot be compared, i.e., one method ends at identifying strike-out, whereas others identify and recognize and some other method identifies, recognizes and removes strike-out strokes.

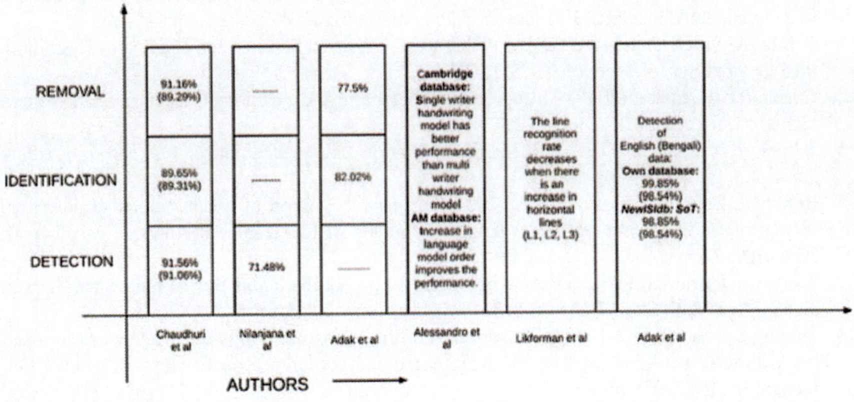

Fig. 5 Detection, identification and removal accuracy

4 Conclusion and Future Scope

The survey briefly introduces various methods that are currently used for strike-out text processing. Giving a strike-out text handwritten document for OCR returns a meaningless output, so it is very important to preprocess and restore the text to the document before giving it to an OCR. Working on this, interest makes OCR better to handle any sort of handwritten documents. Like striking a text, a writer has many editing operations like super scripting on top of strike-out text, inserting characters in between words (using arrow mark symbols), rewriting on the same character, striking out one-third of the written text, etc. This paper will be a good literature survey for researchers starting their work on strike-out text processing in a handwritten document.

All the proposed approach majorly contributes to English handwritten documents, so it can be further designed on other forms of language which makes OCR multi-lingual. Need for the development of a generic method which can handle all types of strokes plays a vital role in dealing with noise handwritten documents. Better algorithms to be designed for the improvement in recognition rate and accuracy in real-time scenarios.

References

1. Marinai S (2008) Introduction to document analysis and recognition. In: Studies in computational intelligence (SCI), vol 90, pp 1–20
2. Sonkusare M, Sahu N (2016) A survey on handwritten character recognition (HCR) techniques for English alphabets. In: Advances in vision computing (AVC), vol 3, pp 1–12
3. Purohit A, Singh Chauhan S (2016) A literature survey on handwritten character recognition. Int J Comput Sci Inf Technol (IJCSIT) 7:1–5
4. Chaudari BB, Adak C (2017) An approach for detecting and cleaning of struck-out handwritten text. Patten Recogn (PR) 61:282–294
5. Jain R, Doermann D (2013) Writer identification using an alphabet of contour gradient descriptor. In: ICDAR, pp 550–554
6. Bunke H, Bengio S, Vinciarelli A (2004) Offline recognition of unconstrained handwritten texts using HMM's and statistical language models. IEEE Trans Pattern Anal Mach Intell 26(6):709–720
7. Baum LE, Petrie T (1996) Statistical inference for probabilistic function of finite state Markov chains. Ann Math Stat 37:1554–1563
8. Sulem LL, Vinciarelli A (2008) HMM-based offline recognition of handwritten words crossed out with different kinds of strokes. In: International conference on frontiers in handwriting recognition (ICFHR), pp 70–75
9. Adak C, Chaudhuri BB, Blumenstein M (2017) Impact of struck-out text on writer identification. IEEE, pp 1465–1471
10. LeCun Y, Bottou L, Bengio Y, Haffner P (1998) Gradient-based learning applied to document recognition. Proc IEEE 86:2278–2324
11. Schomaker L, Bulacu M (2004) Automatic writer ıdentification using connected component contours and edge based features of uppercase western script. IEEE Trans Pattern Anal Mach Intell 26(6):787–798

12. Bulacu M, Schomaker L (2007) Text-independent writer ıdentification and verification using textural and allographic features. IEEE Trans Pattern Anal Mach Intell 29(4):701–717
13. Viterbi AJ (1967) Error bounds for convolution codes and an asymptotically optimal decoding algorithm. IEEE Trans Inf Theory 13:260–269
14. Adak C, Chaudhuri BB (2014) An approach of strike-through text identification from handwritten documents. In: Conference on frontiers in handwriting recognition (ICFH), pp 643–648
15. Bhattacharya N, Frinken V, Pal U, Roy PP (2015) Over writing repetition and crossing-out detection in online handwritten text. In: Asian conference on pattern recognition (IAPR), pp 680–684
16. Ferrer MA, Morales A, Pal U (2013) LBP based line-wise script ıdentification. In: ICDAR, pp 369–373
17. Pratihar S, Bhowmick P, Sural S, Mukhopadhyay J (2013) Removal of hand-drawn annotation lines from document image by digital-geometric analysis and ınpainting. IEEE
18. Sethian JA (1996) A fast marching level set method for monotonically advancing fronts. Proc Nat Acad Sci 93(4):1591–1595
19. Nicolas S, Paquet T, Heutte L (2006) Markov random field models to extract the layout of complex handwritten documents. In: Workshop on frontiers in handwriting recognition
20. Arlandis J, Pérez-Cortes JC, Cano J (2002) Rejection strategies and confidence measure for a K-NN classifier in an OCR. In: International conference on pattern recognition (ICPR), vol 1, pp 576–579
21. Banerjee J, Namboodiri AM, Jawahar CV (2009) Contextual restoration of severely degraded document images. IEEE, pp 517–524
22. Brink A, van der Klauw H, Schomaker L (2008) Automatic removal of crossed-out handwritten text and the effect on writer verification and identification
23. Tuganbaev D, Deriaguine D (2013) Method of stricken-out character recognition in handwritten text. Patent US8, 472, 719. http://www.google.co.in/patents/US8472719
24. Manoharan S (2019) A smart image processing algorithm for text recognition, information extraction and vocalization for the visually challenged. J İnnov İmage process (JIIP) 1(01):31–38
25. Vijayakumar T (2019) Comparative study of capsule neural network in various applications. J Artif İntell 1(01):19–27
26. Otsu N (1979) A threshold selection method from gray-level histogram. IEEE Trans Syst Man Cybern 9:62–66
27. Marti U-V, Bunke H (2002) The IAM-database: an English sentence database for offline handwriting recognition. Int J Doc Anal Recogn (IJDAR) 5(1):39–46
28. Sarkar R, Das N, Kundu M, Nasipuri M, Basu DK (2012) CMATERdb1: database of unconstrained handwritten Bangla and Bangla-English mixed script document image. Int J Doc Anal Recogn (IJDAR) 15(1):71–83
29. Fernández-Mota D, Almazán J, Cirera N, Fornés A, Lladós J (2014) BH2M: the Barcelona historical, handwritten marriages database. In: International conference on pattern recognition (ICPR), pp 256–261
30. Stamatopoulos N, Gatos B, Louloudis G, Pal U, Alaei A (2013) ICDAR 2013 handwriting segmentation contest. In: International conference on document analysis and recognition (ICDAR), pp 1402–1406
31. Surinta O, Holtkamp M, Karabaa F, Van Oosten JP, Schomaker L, Wiering M (2014) A path planning for line segmentation of handwritten document. In: International conference on frontier in handwriting recognition (ICFHR), pp 175–180
32. Freitas COA, Oliveira LS, Sabourin R, Bortolozzi F (2008) Brazilian forensic letter database. In: International workshop on frontiers in handwriting recognition (IWFHR)

Chapter 7
RFID-Based Railway Crowd Prediction and Revenue Analysis

Shraddha S. More, Gulshan Pathak, Simran Panchal, and Monika Patil

1 Introduction

Although ever-changing, the railways are not a foolproof system even after numerous advances have been introduced. The queues around ticket windows compete with themselves and, thus, a short queue seldom appears. These can be attributed to inefficiencies in paper processing for tickets, which can be frustrating technical problems such as a printer malfunction. Waiting and lack of patience leads to passengers boarding the train without tickets, which is a crucial revenue-generating problem. While ticket-checkers are present in broad daylight, at peak hours, i.e., at dusk and dawn, it is almost impossible to compete with the crowd in the train. Concentrating on these problems with the existing system, a system using the new technology has been suggested.

IoT is a robust system consisting of sensors, other elements, and the internet. The RFID [1] scanner, NFC [2] cards as paperless tickets, and Arduino board are some sensors used in this study. The Arduino board serves as a way of receiving commands and transferring instructions to RFID and NFC cards via the program. At ticket windows, an Arduino program must be mounted in the units. NFC cards are

S. S. More (✉) · G. Pathak · S. Panchal · M. Patil
Department of Information Technology, St. John College of Engineering and Management, Palghar, India
e-mail: shraddham@sjcem.edu.in

G. Pathak
e-mail: gulshan.pathak89@gmail.com

S. Panchal
e-mail: simran.panchal1998@gmail.com

M. Patil
e-mail: monika.patil2911@gmail.com

I. Jeena Jacob et al. (eds.), *Data Intelligence and Cognitive Informatics*, Algorithms for Intelligent Systems,
https://doi.org/10.1007/978-981-15-8530-2_7

checked by RFID scanner for time and destination. A dashboard designed using style sheets, JavaScript, and HTML is available for the analysis of railway commuters' real-time data. Raw data in a database is processed using machine learning libraries such as NumPy and pandas. Predominantly used for scientific computing are pandas and NumPy written in python language. Pandas is used for data manipulation and analysis, while NumPy supports multidimensional data and provides mathematical functions at a high level. Flask is a python language micro-framework that serves as a bridge between the dashboard and the database.

In this paper, Sect. 2 provides a literature survey which constitutes as the research base for the project. Further, Sect. 3 gives an overview of the system architecture that includes the hardware and software components involved in the implementation. In Sect. 4, the components have been set up and mentioned the results of the implemented project. Section 5 concludes the title and semantics of the purpose of the proposed system

2 Literature Review

In 2016, Khoeblal et al. [3] presented a system that focuses on problems arising in public city trams and developed an automated system based on RFID and NFC-based tickets that allow the passenger to enter into the valid station. Thus, they have improved the problem of NFC ticket [4] redundancy. In 2017, Grout et al. [5] proposed a system that discusses about the design and development of a contactless sensor interface using an RFID tag. Both RFID and NFC were considered in it. Not only is the working, but sensor stimulation in the overall implementation is also discussed in the proposed system. Thus, it overall discusses about working and implementation of RFID and NFC using sensor system control and sensor system data which completely eliminates the manual checking of individual tickets by authorized person.

In 2018, Xu et al. [6] presented a new passenger flow forecasting system for better organization of passenger equipment and services provided to them. In 2016, Ray et al. [7] proposed an IoT-based approach for thermal comfort measurement using various sensors in soil. Though the paper is based on IoT-based thermal comfort measurements, it shows the plotting of the graph using Plotly and also implements graph using ThingsSpeak cloud. Thus, a detailed explanation of the visualization and tools that can be used for the very same purpose is suggested in this system. The representation is shown with the help of an open-source cloud which can be replaced according to the developer's ease of use. Missenard ordering has been executed for estimating warm solace by IoT cloud-helped design. Constant graphical representations of warm solace are acquired from Plotly and ThingSpeak [8] cloud stages. Sequential information yield is likewise created while playing out the trial in the given framework. In 2018, Abhinay et al. [9] proposed a seat prediction system for metro networks using machine learning. The problem of seat availability and controlling the flow of passenger in the local trains is a challenging task. The passengers are not aware of time and seat availability which leads them to the waiting list. Also, most of the passengers try to book at the same time and it makes the server to slow down its process. In this paper, the prediction system for ticket availability and calculation of travel time is accomplished. The proposed model is tested with the collection of the several years of data set from the southern railways. The data set possesses details like train number, train name, total passengers, the total number of seats, date, and time. The proposed prediction system reduces customer's effort for easy booking of the train. The existing system focuses only on the train prediction system. The paper proposes an efficient train prediction system which allows passengers to easily search available train with the available tickets. It helps the passenger to get alert for ticket booking in future, so it will be useful for every passenger. Privacy and security is not a concern in the proposed work. In future, the work will be implemented with privacy and security using a various algorithms like AES, RSA, etc.

3 Proposed System

The railways, being the most opted and common means of transportation, is an essential part of generating high revenue for the government. But there are certain negative perks associated with the system concerning the railway system that might be acquainted with. Let me highlight some issues for you, as these points are the base of exactly why chosen to work this system for the better. Majority of the people, particularly the bread earners, travel to and from, and for business purpose. Now the timings of business may collide, thus resulting in rush hours. Rush hours are the peak time when most of the people have to travel at the same time, opting for the same trains, in the same direction. Due to this rush, many citizens without a ticket are not caught by the ticket checker. This proves the fall in revenue collection. As well creates health issues like claustrophobia and suffocation for travelers. To minimize these problems, the railway data analysis system is brought to you. It is basically NFC- and RFID-based system that generates paperless tickets and ensures a rigid collection of revenue. It monitors and analyzes the data and uses it efficiently. Since the proposed system allows only those passengers who have valid NFC tickets to enter the station, no bread earners will be allowed to enter the station. The ticket checker will also be having a handheld RFID scanner to check passenger's NFC ticket. That will automatically increase the generation of revenues.

The existing railway system has successfully proved to be the fastest and most convenient route for traveling across cities and states as well. However, as effective as this system used to be, it is not anymore. The sole purpose is an exponential growth of population that makes it a necessity to earn as much as possible, which increases travel to and from, for business purposes. Thus, the existing system faces the issue of over-populated trains. Due to this, many citizens get away without buying tickets and paying for it as well. Also, the safety measure of people traveling in such over-populated trains is at high risk. Risks associated with it are claustrophobia,

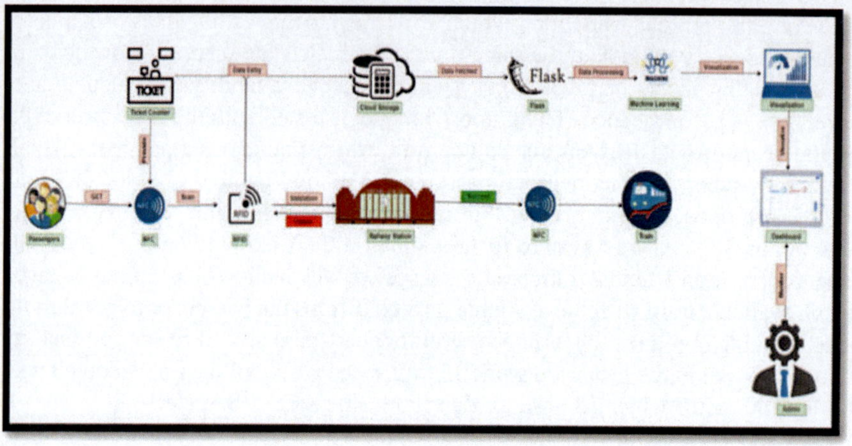

Fig. 1 Proposed system architecture

suffocation, respiratory problems, etc. The new and revised system will prove to overcome all the current issues.

With the use of various libraries, the crowd forecast is carried out. The predicted crowd is monitored by an administrator and along with it, admin can also keep a track of revenue generated. The crowd that is to be there will be displayed on a dashboard so that even naive users can check the results with some levels of abstraction. Various blocks of proposed system architecture are described in below Fig. 1.

- **Passenger:**
  The passengers who want to travel via the railway buys NFC ticket from the ticket counter for the respective destination and pays for the respective path covered only.
- **Ticket Counter:**
  Here, the ticket counter distributes NFC ticket to the passengers and stores the data about each NFC ticket distributed for further analysis purpose.
- **NFC Ticket:**
  NFC ticket is a programmable paperless ticket. The destination is programmed into the ticket and can be reused as many times as possible. It is obligatory that passengers collect an NFC ticket before entering the train. The NFC tag is a passive device with no proprietary control. The users then touch an NFC compatible device onto the tag when one is used. The NFC tag takes a small amount of power from the reader/writer to power the electronic tags. The tag then allows a small amount of information to be conveyed to the reader/writer. The data stored in the memory tag is transferred to the device that is enabled by NFC. Although usually only a small amount of data, this can be used to guide the user to a website URL; it can be a small amount of text or other information.
- **RFID Scanner:**
  RFID scanner scans the NFC ticket. [10] It is used as an authentication tool. The passenger is allowed to pass through once the scanner accepts the NFC ticket [11].

- **Cloud Storage:**
  The data stored in each railway database are stored on the cloud. This data can be accessed by the admin for analysis purpose to make strategic plans and take actions based on facts. Only admin can access the database.
- **Flask:**
  Flask is a micro web framework written in python. It does not require particular tools or libraries, thus it is classified as a micro-framework. However, a flask can add application features and supports various extensions.
- **Visualization:**
  The data is visualized using some visualization techniques and mathematical formulae and display it on the dashboard. This helps to track the progress of the proposed system. Visualization is transparent to the public so they get an idea of how the system runs in the backend.
- **Admin:**
  The admin is the administrator. The administrator manages the dashboard and handles all the backend activities related to the data.

3.1 Working of the Proposed Architecture

Passengers buy NFC ticket at the ticket counter to validate their entry at the station and board the train. NFC ticket is scanned by an RFID scanner at the gate. If a ticket is valid, i.e., if date and time of issue and expiry correspond within current date and time, a passenger is good to go. In case of an error in validation of NFC ticket, a passenger is obliged to buy a ticket from the ticket counter. RFID scanner sends same data appended with destination to admin database. Onboarding the train and reaching its destination, passenger dumps NFC ticket on the way out. Data stored in the database (cloud) is used for the purpose of revenue analysis and crowd prediction. This data is visible to admin and other officials via web dashboard created with the help of HTML5 and cascading style sheet and other style sheets. Dashboard fetches data from the cloud using flask, a micro web framework technology which is simple to use and easy to learn for the developers. Dashboard predominantly presents the statistics and data visualized in the form of flexible graphs. Visualization is performed using python libraries such as NumPy and pandas.

Hardware components with its specification which are used in the proposed system to achieve hardware connection to scan NFC card are shown in following Table 1.

Table 1 Hardware components with it's specification

Components/devices	Specifications
Arduino uno microcontroller	ATMega328P
PN532 NFC module	80C51 microcontroller core with 40 Kbyte ROM and 1 Kbyte RAM
RC522 RFID reader	13.56 MHz RF reader

Fig. 2 Circuit connections to scan NFC card

4 Results and Discussion

The proposed work is divided into 3 main phases. Phase 1 focuses on collecting data for any NFC ticket swipe. It involves circuit connections using node to scan NFC tickets and Google sheets to interpret the scanned data. Phase 2 is performed by analyzing information obtained by NFC swipes by passengers. Phase 3 consists of the prediction of crowds based on the phase 2 analysis.

4.1 Hardware Connections to Scan NFC Card

Node MCU is the microcontroller in the proposed framework, which is connected to RFID. NFC ticket, which contains traveling data of the customer, is read using RFID. These RFID's are installed at the entrance of the entire station, so the user has to swipe the card to confirm his/her entry. RFID readers are also spread through all the stations to each ticket checker to review any passenger tickets if necessary. Figure 2 shows circuit connections to scan NFC ticket.

4.2 Depositing Data

The data found in the NFC ticket is placed into Google sheets after scanning the NFC ticket. A Google script [12] is used to deposit data in Google sheets. NFC tickets have data such as travel fares and destination. While collecting data into Google sheets, more attributes including date, time, and number of the day are also added using Google script (Fig. 3).

1	No.	Date	Day	Time	Destination	Cost
2	1	11/03/2019	1	07:00:01	palghar	15
3	2	11/03/2019	1	07:01:55	andheri	20
4	3	11/03/2019	1	07:02:44	borivali	15
5	4	11/03/2019	1	07:04:21	vasai	5
6	5	11/03/2019	1	07:05:01	palghar	15
7	6	11/03/2019	1	07:06:11	churchgate	30
8	7	11/03/2019	1	07:09:19	churchgate	30

Fig. 3 Inserting data in google sheets

4.3 Analysis of a Passenger's Journey

Relevant data is obtained depending on the type of analysis to be done for the study of passengers traveling to and from, using visualization techniques. For analysis, data is loaded once and then displayed. Each time-specific columns are selected, such as cost or date. Since IoT devices are designed to receive data in a fixed format only, no refining is required before the data is loaded for analysis. For example, 'cost' has been used as a parameter for revenue collection graph, 'date' to find number of passengers traveled per day.

In Fig. 4 bar graph shows the total number of passengers arrived at a particular station till date. A scatter graph shows the total number of passengers departed from a particular station to date.

Following Fig. 5 shows an analysis of a passenger's journey per day along with the generated revenue. A bar graph shows the total number of passengers traveled using the railway to date. A pie graph shows the total revenue generated in percentage from a particular station to date.

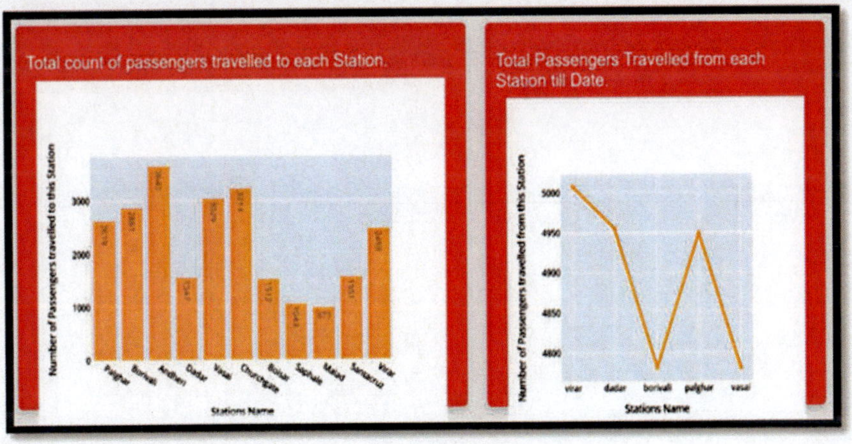

Fig. 4 Analysis of a passenger's journey to each station till date

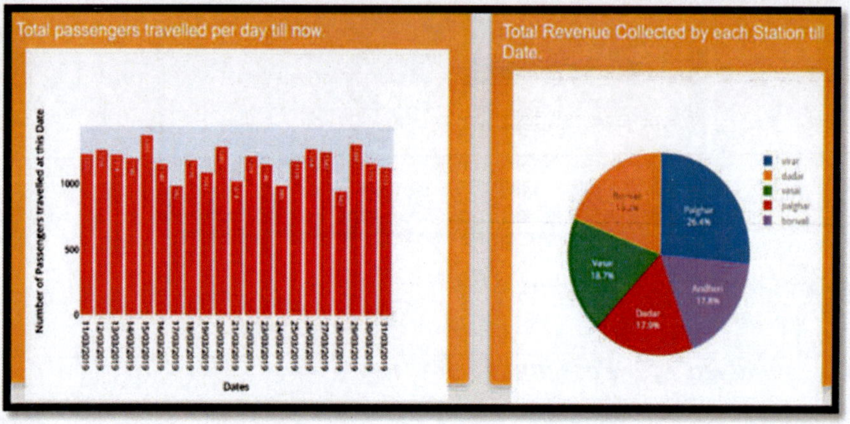

Fig. 5 Analysis of a passenger's journey per day and revenue collection

4.4 Crowd Prediction

Python libraries such as NumPy [13] and pandas [14] are used for predicting the volume of the crowd. Pandas is used to read and display data in a specific structure, while NumPy is used to perform arithmetical operations. Data is taken from Google sheets during this process, separated, and then arithmetic operations are performed on it. Thereafter, a list is generated that contains an average value and passenger count. Afterward, the list is passed to plot a graph. For example, to predict the crowd, mean of total number of passenger over a period is taken. Thus, crowd prediction is done on the basis of an average number of passengers traveled to and from a particular station previously. In Fig. 6 a bar graph shows the estimated number of passengers

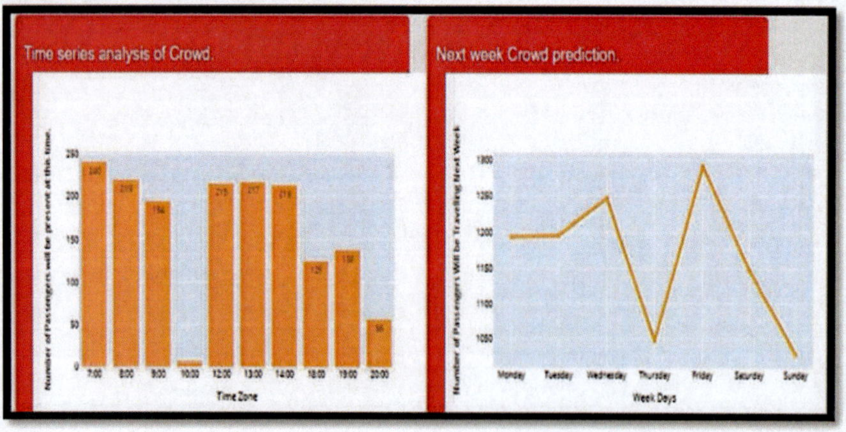

Fig. 6 Crowd analysis and prediction

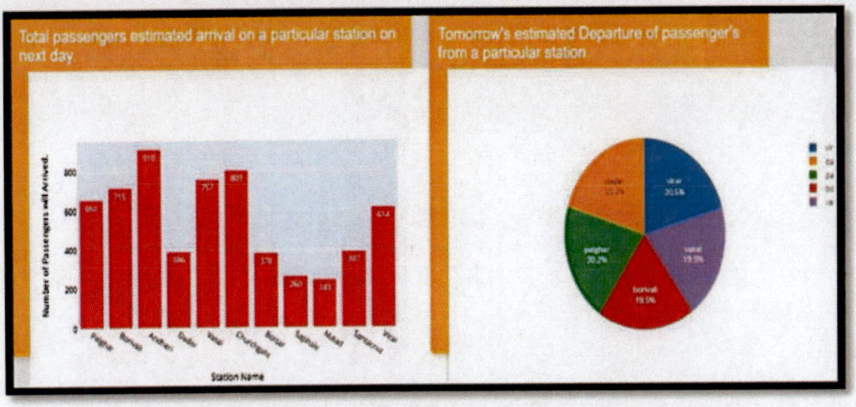

Fig. 7 Estimated arrival and departure of passengers

present tomorrow in a particular time frame. A scatter graph shows the estimated number of passengers traveling in next week at respective weekdays.

In the following Fig. 7, a bar graph shows the estimated number of passengers arriving at a particular station next week. A pie graph shows the estimated percentage of passengers traveling from a particular station on the next day.

4.5 Analysis for a Particular Station

Consider analyzing for station Virar. Parameters such as date and number of day and expense will be used for three different analyzes in this study. The count is recorded with each passenger using the NumPy library, and the graph is plotted for the same. Based on count, the most crowded day will be treated as the day that records the maximum count on a given day. In Fig. 8 a pie graph shows the total number of passengers departed from Virar station till date to respective destination stations in a percentage format. A scatter graph shows the total number of passengers traveled in a week at respective weekdays from Virar station.

In the following Fig. 9, a bar graph shows the total amount of revenue generated from Virar station with respect to other destination station.

4.6 Prediction for a Particular Station

For Virar station, graphs are plotted based on parameters like the number of day, date, and count calculated. On this basis, following logic is applied and gives expected results.

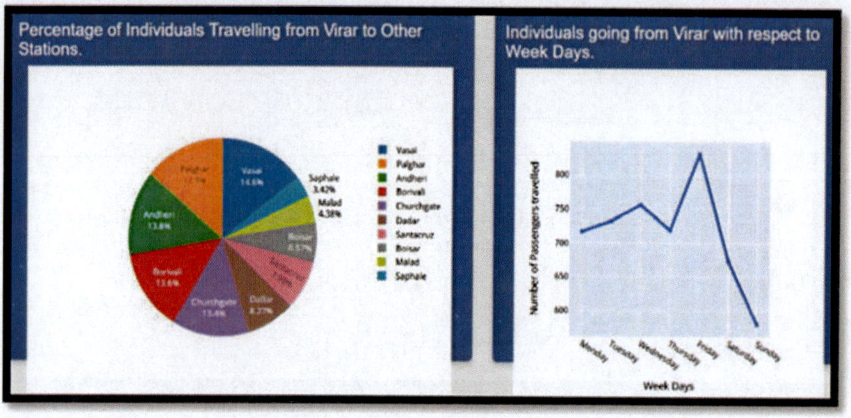

Fig. 8 Passenger's journey and count analysis for Virar station

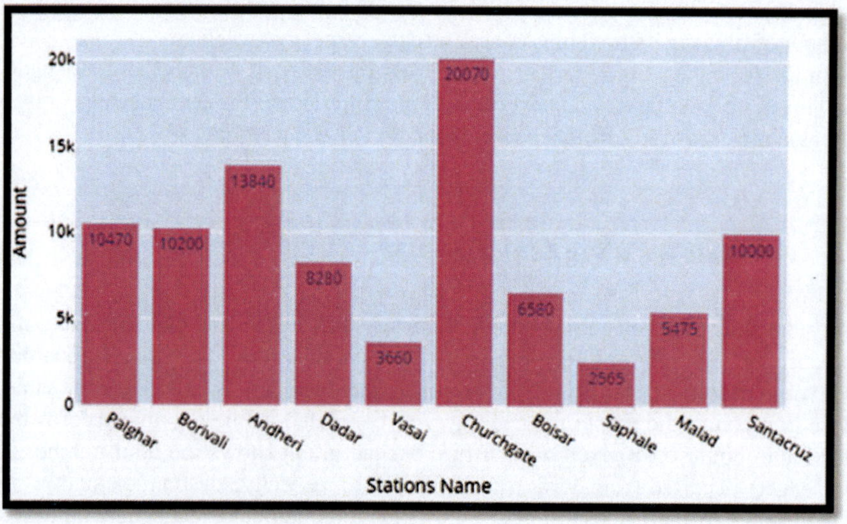

Fig. 9 Revenue generated from Virar station

```
r = requests.get('virar.xlsx')
data = r.content
virar = pd.read_csv(io.BytesIO(data),
index_col = 0)
r1 = requests.get('dadar.xlsx')
data1 = r1.content
dadar = pd.read_csv(io.BytesIO(data1),
index_col = 0)
r2 = requests.get('borivali.xlsx')
data2 = r2.content
borivali =
pd.read_csv(io.BytesIO(data2),
index_col = 0)
r3 = requests.get('palghar.xlsx')
data3 = r3.content
palghar =
pd.read_csv(io.BytesIO(data3),
index_col = 0)
r4 = requests.get('vasai.xlsx')
data4 = r4.content
vasai = pd.read_csv(io.BytesIO(data4),
index_col = 0)
```

Pseudo code for counting the number of passengers travelling

In the above pseudo-code, the count of passenger's traveled from Virar as a base station to respective multiple destinations is counted using pandas syntaxes as well as total revenue collected from the passengers is also calculated and displayed using graphs. In following Fig. 10, a scatter graph shows the estimated number of passengers traveling from Virar to other destination stations on the next day. A pie

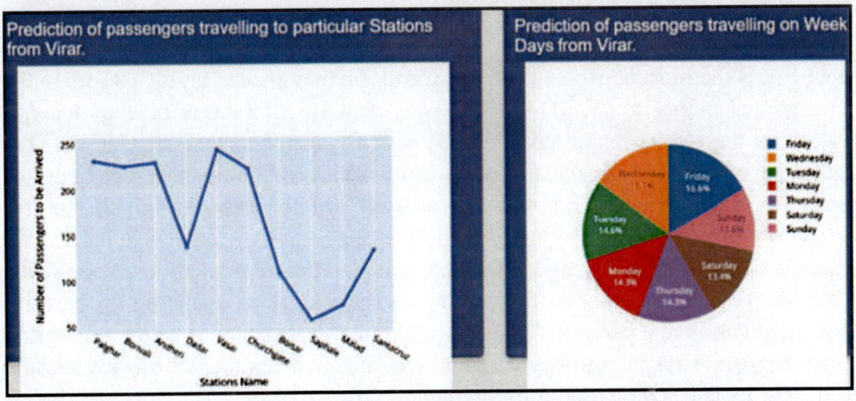

Fig. 10 Prediction of days and count of passengers traveling from Virar to other stations

graph shows the estimated number of passengers traveling in next week at respective weekdays from Virar station in percentage format.

```
r = requests.get('virar.xlsx')
        data = r.content
        virar =
        pd.read_csv(io.BytesIO(data), in-
        dex_col = 0)
        a=list(virar['Date'].unique())
        b=len(a)//7
        lvi=[]
        for i in range(1,8):
                c=virar[virar["Day"]==i].Day.
        count()
                c=c//b
        lvi.append(c)
```

Pseudo code for crowd prediction

The above pseudo-code is used to calculate how many passengers are going to travel on the next day at a particular time or next week or a particular destination. To calculate the values, average methodology is being used here. By calculating how many numbers of weeks data is present, the final number is divided and the average is taken out. In following Fig. 11, a bar graph shows the estimated number of passengers present tomorrow in a particular time frame at Virar station for traveling.

5 Conclusion and Future Scope

Since rail is the most convenient way to travel, there are some disadvantages to the conventional method too. RFID-based railway crowd prediction and revenue analysis system are therefore been proposed which not only overcomes the drawbacks of the traditional railway system, but also provides additional functionalities such as crowd prediction. Using this method, real-time analysis can be performed to help the user correct or adjust his/her traveling timings. The suggested approach would certainly demonstrate substantial changes in the current situation if applied on a wide scale. Thus, the proposed method will (a) carry a paperless ticket evolution (b) perform crowd prediction at a particular station (c) eliminates revenue losses by monitoring daily progress. Thus, in future when more secure and faster transactions are made to be done, big data can be used for the same purpose.

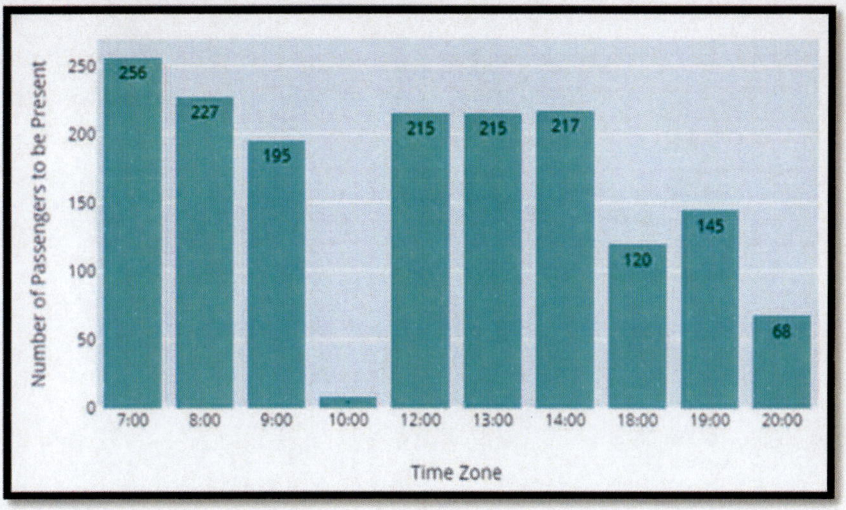

Fig. 11 Crowd prediction for Virar

References

1. Radio frequency identification (RFID). https://en.wikipedia.org/wiki/Radio-frequency_identification
2. Near field communication (NFC). https://nfc-forum.org/
3. Khoeblal R, Kamyod C, Laohapensaeng T, Chaisricharoen R (2016) Improved passenger ticket control in easily accessible public city trams. In: International symposium on intelligent signal processing and communication systems. ISBN 978-1-5090-0629-8
4. NFC ticket. https://www.ccv.eu/public-transport-payment-systems/contactless-nfcticketing/
5. Grout I, da Silva ACR (2017) RFID enabled sensor system design. In: International electrical engineering congress. ISBN 978-1-5090-4666-9
6. Xu X, Dou Y, Zhou Z, Liao T, Lu Y, Tan Y (2018) Railway passenger flow forecasting based on time series analysis with big data. In: Chinese control and decision conference. ISSN 1948-9447
7. Ray PP (2016) An internet of things based approach to thermal comfort measurement and monitoring. In: 3rd international conference on advanced computing and communication systems
8. Thingspeak. https://www.mathworks.com/help/thingspeak/
9. Abhinay R, Yella T, Devakunchari R (2018) Passenger flow analysis in metro network using machine learning technique. Int J Appl Eng Res 13:10604-10606. ISSN 0973-4562
10. Krishna KV, Selvarathinam S, Roopsai V, Kumar RR (2013) Modified ticketing system using radio frequency identification (RFID). Int J Adv Comput Res 3. ISSN 2277-7970
11. Foisal Mahedi Hasan M, Tangim G, Kafiul Islam M, Rezwanul Haque Khandokar M, Ul Alam A (2010) RFID-based ticketing for public transport system: perspective megacity Dhaka. In: 3rd international conference on computer science and information technology (2010)
12. Google Scripts. https://developers.google.com/apps-script/guides/sheets
13. Python library – Numpy. https://numpy.org/
14. Python library- Pandas. https://pandas.pydata.org/

Chapter 8
A Review on Efficient Scheduling Techniques for Cloud Computing

Venkateswara Rao Kumpati and Manish Pandey

1 Introduction

Cloud computing is a well-known innovation with benefits for industry as well as the academic community. It provides an adaptable and effective method to reach and store documents. This is a web-based technique where programming, assets, and data are imparted when requested to a PC or other gadgets. Using cloud gives an office to clients to get to data whenever and from anyplace. It isn't required for the client to be in a similar area as the equipment that stores information. At the point when the client has a web association, they can get to the administrations of the cloud. It conveys all the administrations progressively through the web as per the client fundamentals. At its core, cloud computing consists of two related parts—the front and back ends—which are connected through the web. Clients can use the front end which consists of interface systems and applications to reach the cloud computing stage. The back end is connected to a specialist organization through other interfaces and applications and enables access to the distributed computing stages. Thus, different users can access different cloud services. Different clients can access a given site or web application at a given time. This makes it difficult for the web application to deal with all these demands one after another. Now and again it might bring about framework breakdowns. Here, the load balancer assumes a significant job. Burden adjusting is used to separate the necessary work of a PC to be handled by at least two PCs to ensure that the task is correctly performed.

V. R. Kumpati (✉) · M. Pandey
Maulana Azad National Institute of Technology, Bhopal, India
e-mail: venkatesh.kumpati@gmail.com

M. Pandey
e-mail: manishpandey@yahoo.co.in

I. Jeena Jacob et al. (eds.), *Data Intelligence and Cognitive Informatics*,
Algorithms for Intelligent Systems,
https://doi.org/10.1007/978-981-15-8530-2_8

This method is faster and uses material goods more productively. As per the requirements of clients, cloud computing is Internet-based computing in which services are provided on demand. Three main services provided by cloud computing are (IAAS) infrastructure as a service is everything customer provides for application maintenance servers are customer side hosted, storage, network servers, etc. Amazon web services. Platform as a service (PAAS) is used for software development environment services java, dot net, Linux platforms example. Azure service, force.com, Google. Software as a service (SAAS) cloud provides ERP applications provide URL less cost example: salesforce.com, which are shown in Fig. 1. Cloud computing is based on pay per usage model. Customers will have to pay only for what they consume. In service, provisioning cloud computing provides scalability, mobility, and flexibility. The major objectives of the cloud are to reduce cost, upgrade the reaction time, and improve the execution. Thus, the cloud is comparable to a pool of administrations. A cloud refers to a unique IT system that intends to remotely provide estimated and adaptable IT assets. The word cloud in distributed computing is utilized as an illustration for the web so distributed computing can be characterized as the web-based registering in which various administrations like servers and applications are given to associations PCs and gadgets utilizing the web. So when contrasted with conventional "claim and use" strategy if distributed computing is used, the buying and support cost of the foundation is wiped out. It permits the clients to utilize assets as indicated by the appearance of their needs progressively. In this way, it can be distributed computing empowers the client to have helpful on-request access to a common pool of figuring assets, for example, system, applications, and administrations, and so on. National Institute of Standards and Technology (NIST) model of cloud computing is discussed in Fig. 1.

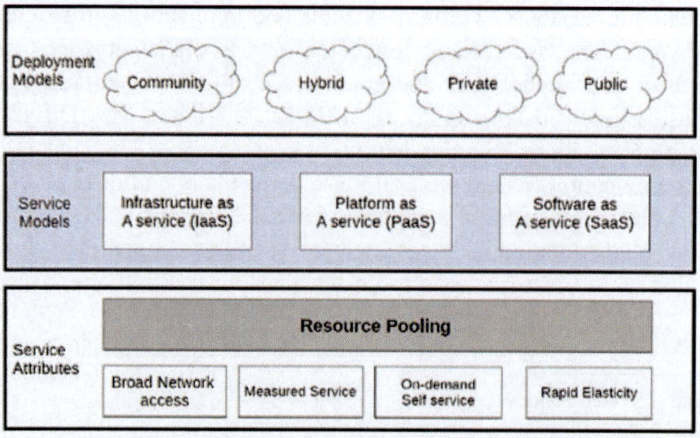

Fig. 1 NIST model of cloud computing Ref. [1]

2 Related Work

While research in this area is still in the development stages, quite a few areas of research in the cloud environment need more attention. These include, for example, load balancing, application scheduling, resource provisioning, consumption of energy, etc. Resource scheduling is an area where research is still in its infancy. Here, to discuss seek review papers related to our area of research; that is, papers on the topic of scheduling techniques and resource provisioning. Amiri and Khali et al. [2] have done an exhaustive survey to predict the future demand for different aspects of the applications. Reference [3] Some pertinent observations include that the results of one phase are useful for another one; there is a focus on parameters like virtual machine migration and availability. However, these two surveys do not touch in detail upon the difficulty of over-provisioning and under-provisioning. Reference [4] comparison of several QoS-based heuristics and metaheuristic algorithms in the cloud environment. The metrics met through the current algorithms are shown in a table, but it is found that most of these algorithms have not taken into account the parameter of energy efficiency which is vitally used for green computing. Besides, this survey is incomplete in scope in terms of the algorithms considered. Reference [5] five metaheuristic scheduling algorithms for the cloud environment and also for the grid. Their work discusses Pareto optimal theory as a means to solve the multi-objective issue. However, this survey is limited by the fact that it discusses only metaheuristic techniques. Reference [6] discusses the problem of workflow scheduling (NP-Complete) and proposed a solution called (particle swarm optimization) PSO which is based on a meta-heuristic algorithm. They have also studied several variants of this technique such as "multi-objective PSO" (MOPSO), jumping PSO, learning PSO, and bi-objective PSO. The aim was to enhance QoS parameters like time, reliability, and cost of cloud computing. Reference [7], this work discusses several important issues such as allocation based on QoS parameters, and meta-heuristic algorithm-based static and dynamic allocation strategy. Within a given field, each specific approach has its value but not in other instances. Therefore, the proposed model blends various methods and interchanges between suitable methods for load balancing according to the state of the device. Here, the idle status uses a multi-objective of ant colony optimization algorithm (MUL-ACO) which optimizes the multi-objective load balance on various scale factors such as good load efficiency, reduces power consumption, and enables the optimal solution to be found and at a reasonable time cost. The goal of the multi-objective ant colony optimization algorithm (MUL-ACO) is to simultaneously optimize objective functions. In this paper, the ant colony optimization algorithm (MUL-ACO) multi-objective algorithm is defined as a load balancing solution. Adapting this algorithm results in high throughput, response time, and process migration with the multi-objective parameters. Review and systematic comparison of different meta-heuristic algorithms will help guide future research in this field.

3 Categorization of Load Balancing Algorithms

Classification of task allocation algorithms in the cloud is done based on the present status of VM. Dynamic strategy refers to the situation when one has information about the current load of VMs before allocation.

3.1 Heuristic Scheduling Algorithm

These algorithms are issue ward and also provided great execution used for a particular area of issues, however, low execution for other people. Reference [8], typically, these algorithms give a careful answer for a particular space of issue boundless measure of time however cannot take care of hard reorganization issues. There be loads of heuristic algorithms, when future in a cloud situation that takes care of the work process booking issue just as autonomous assignments or various tasks. Here, different heuristic algorithms have been talked about by separating them into various classes dependent on the key maxim during the article, for example, Heterogeneous Earliest Finish Time (HEFT). These algorithms conveyed the undertakings on the virtual memory utilizing diverse scheduling approaches and attempted to upgrade the different QoS standards.

3.1.1 FCFS and RR Algorithm

FCFS algorithm and the following interest will be in line until, and the main undertaking is finished; however, the algorithm neglected to adjust the outstanding task at hand among the virtual machines (VMs). Reference [9]. An algorithm that chooses the briefest assignment served first at quality to lessen the time taken different QoS standards, yet the algorithm neglected to develop the QoS standards because of symmetrical burden dissemination at hubs. Round-robin also weighted-round-robin methods can be used to resolve the increased load balancing as well as effective response time and makespan time problems in the cloud Technology. Right now cut, a component is utilized to process information. Here time is partitioned into cuts and every hub is given a period interim, where the hub needs to play out the assignment. If the preparation isn't finished inside the time quantum, it needs to hang tight for the following opening.

3.1.2 Min-Min and Max–Min Algorithm

To conquer its constraints of the min-min algorithm, Ref. [10] an improved burden adjusting algorithm was suggested that diminishes the time and builds the asset usage proportion. The need of the client is likewise considered right now guarantees assurance of the administration. Min–max booking algorithm was suggested that appointed the bigger responsibilities to best resources at the beginning for effective makespan service, reaction time just as quality use proportion of client demand yet instruction follow have the issue of overusing the less use of resources and neglected to improve expected parameter to face the constraint of the max–min algorithm, an altered max–min algorithm have been recommended that depended on the idea of required execution time rather than the complete preparing time and span time yet incapable to improve different parameters. There are a few improved adaptations of the maximum min algorithm that have been proposed to advance the QOS standards in cloud computing however expected algorithm neglected to get the effective key execution marker parameters.

3.2 *Metaheuristic*

Metaheuristic computations are issue independent and fitting to deal with issues of various spaces with commendable execution. Metaheuristic strategies are individual of ordinary frameworks for handling NP-hard improvement issues. Metaheuristic þ randomization and diverse metaheuristic algorithms are existing in the cloud condition to find the evaluated (defective) game plan of the NP-Complete issue in a brief timeframe. Booking of assignments is one such NP-complete issue because of the huge arrangement space and sets aside a long effort to locate the ideal arrangement. The shifted decision of metaheuristic algorithms is similar to particle swarm optimization (PSO).

3.2.1 Particle Swarm Optimization

Particle swarm optimization (PSO) is a concept that Ref. [11] originated from observing the way shoal of fish and flock of birds navigate their path. In such algorithms, the load is balanced by transferring only extra tasks from a virtual machine that is overloaded rather than the entire overloaded virtual machine. Here, migration time is reduced since only extra tasks are migrated. Each particle's performance is measured through a fitness function that keeps changing based on the problem to be optimized. Performance of the schedule is observed through the fitness function; greater the value of the latter, lesser is the energy of the scheduling. The status of a particle on the search space is characterized by a further two factors: the position of the particle on the search space and its velocity. Reference [12] These factors may be manipulated to accurately match the dynamics of the social behavior of fish and birds based on the following equations to achieve superior performance of optimization issues. This paper proposes allocating resources using a model that processes applications anciently and a PSO-based scheduling algorithm called PSO-COGENT that reduces the energy consumed by cloud data centers and also optimizes cost and time required for execution. The PSO-COGENT algorithm developed has been simulated at CloudSim. It has been found to decrease the cost and time of execution, the task rejection ratio, and the amount of energy consumed. It has also enhanced the throughput as compared to the honeybee, PSO, and min-min algorithm.

3.2.2 Artificial Bee Colony (ABC) Algorithm

ABC may be a metaheuristic algorithm that is propelled by the brilliant grass of honeybee swarm proposed by Reference [13]. In ABC, unnatural honeybees fly around during a multidimensional inquiry space to locate the inexact arrangement of the advancement issue. There are three sorts of honeybees: utilized honeybees, scouts, and onlookers. Scouts control the examination and pick new nutriment sources unplanned not including user experience and used and observed artificial bees complete the abuse method reliant on the capability of themselves and their home mates. Used the artificial bee lights up assed by an artificial bee about the nutriment source then viewer artificial bee algorithms the well-being regard and picks the simplest nutrient source that's accumulated by Utilized artificial bee. A stack modifying count has been proposed reliant on the direction of the ABC algorithm, during which artificial bee function as undertakings and nourishment source fill during as virtual machines in a cloud situation. The future estimation modifies the extraordinary weight active and improves distinctive key execution pointer parameters, as an example, reaction time, makespan time, amount of moving jobs, etc., since necessities of assignments as QoS parameter.

3.2.3 (ACO) Ant Colony Optimization Technique

ACO is a metaheuristic methodology to utilize in the direction of locating the ideal arrangement of computational issues in the field of cloud computing just like framework processing. Reference [14] It is motivated by the searching conduct of genuine underground insects finding the briefest way between their region and a wellspring of nourishment dependent on the centralization of unpredictable preparation import pheromone. To increase make length time and parity, the outstanding burden in cloud bunch, a clever advancement algorithm named period ACO (PACO). PACO finds the ideal arrangement of the NP-hard issue (task planning for distributed computing) utilizing the pheromone force update technique yet the algorithm neglected to improve different QoS parameters because of the static modifying system. A two-level cloud scheduler has been proposed to execute the logical trials and improve the various QoS parameters (reaction time, throughput, and makespan time) in cloud conditions utilizing the ACO algorithm. To settle the difficulty of fleeting apex stacks in circulated figuring, improved ACO-based computation has been later proposed named pre-"Ant Policy" course of action that involves the desired model that chooses the simplest mapping of a job with virtual machines and improves diverse QoS parameters. Reference [14] "ACO" algorithm accessible its varieties is often won't to handle the difficulty of arranging, graph shade, vehicle directing.

3.2.4 Genetic Algorithm (GA)

GA is based on factors like the individual chromosome and the populace that may possess a variety of wellness esteems for improvements such as makespan, vitality utilization, and throughput. Reference [15] The GA algorithm is involved during every cycle, with Selection (for example Tournament choice, or Roulette wheel determination), Crossover (for example Single-point hybrid, Framework Crossover, etc.), Mutation (such as limit change, bit-wise transformation, etc.). Reference [16] Several scientists thought about the chromosome (in vector structure) size in terms of the all-out number of assignments which land at the framework. GA-based burden adjusting approach limits the makespan. They encoded the populace with parallel strings and the chromosomes experience an arbitrary single-point hybrid and considered 0.05 due to change likelihood (Table 1).

4 Conclusion

This paper presents a detailed review of recent research in scheduling algorithms for resource provisioning in cloud computing. Various algorithms are reviewed, used in scheduling in recent research, and categorized them. Our objective is to study the strategy used in the scheduling algorithms in cloud computing in the resource allocation considering the performance parameters such as cost, makespan time, response time, availability, reliability are impacted. The existing scheduling algorithms are not taking into consideration; many of these performance parameters are observed. So an efficient scheduling algorithm is concluded; considering the above-mentioned major performance parameter is much needed in present times.

Table 1 Review of heuristic–metaheuristic algorithms among their advantages and limitations

Year	Algorithm	Techniques	Advantages	Limitations/challenges
"2018 [12]"	PSO-COGENT algorithm	Proposed modified PSO algorithm	Non-dominance set of optimal solutions and increase the quality of parameters like time, cost, throughput, task accepting ratio	These challenges cannot use the real cloud environment to verify that the performance of the algorithm should be used with CloudSim simulation toolkit
"2018 (Srichandan et al. 2018)"	Genetic Algorithm	Genetic algorithm through bacterial Foraging	Recover scalability, makespan time	Future algorithm take extra time for mutation and crossover
"2017 [17]"	Heuristic–metaheuristic algorithm	Static and dynamic load balancing algorithm	Efficient utilization of resources in the cloud environment	Bigger tasks take a longer time to complete
"2019 [18]"	ACO, GA, PSO	Nature-inspired metaheuristic algorithms	Good at to manage the heterogeneity of the cloud environment, minimize the overhead and latency and improve the performance	Many factors such as energy consumption, service cost, and quality of service have not structuralized
"2018 (Sambit kumar, Bibhudatta, Pirti, 2018)"	Heuristic-based algorithm	Load balancing techniques	Can achieve the fault-tolerant property	Need to evaluate the proposed algorithms in real-world cloud deployment and implementation
"2019 (Arunarani, Manjula, Vijayan, [19])"	Scheduling algorithms	Different scheduling algorithms	Report solves three different perspectives: methods, applications and parameter-based measures utilized	To improve popular and classic scheduling new methods need to be developed

(continued)

Table 1 (continued)

Year	Algorithm	Techniques	Advantages	Limitations/challenges
"2017 [2]"	ACO	Prediction methods in different aspects	Presents a comprehensive report on the application prediction, the main characteristics and challenges	According to the dynamic nature of cloud and deficiencies of methods presented until now, need to improve the accuracy and the efficiency of the prediction methods
"2019 [11]"	PSO	Metaheuristic Algorithm	This algorithm uses the mechanism of the average absolute value of the velocity	The proposed algorithm is not using any machine learning approach
"2019 (Kaur, A., Kaur, B.,)"	PEFT-ACO, HEFT-ACO	HDD-PLB	Two important metrics-makes pans and cost	Hybrid PEFT-ACO approach for proposed HDD-PLBFW

Reference

1. Bohn RB, Messina J, Liu F, Tong J, Mao J (2011) NIST cloud computing reference architecture. In: IEEE world congress on services, pp 2378–3818
2. Amiri M, Khanli L (2017) Survey on prediction models of applications for resource provisioning in cloud. J Netw Comput Appl 93–113
3. Smanchat S, Viriyapant K (2015) Taxonimies of workflow scheduling problem and Technique in the cloud. Future Gener Comput Syst 1–12
4. Nandhakumar C, Ranjithprabhu K (2015) Heuristic and meta-heuristic workflow scheduling algorithms in multi-cloud environments—a survey. In: International conference on advanced computing and communication systems pp 1–5
5. Kalra M, Singh S (2015) A review of metaheuristic scheduling techniques in cloud computing. Egypt Inf J 275–295
6. Masdari M (2017) A survey of PSO-based scheduling algorithms in cloud computing. J Netw Syst Manag 122–158
7. Bittencourt L (2018) Scheduling in distributed systems: a cloud computing perspective. Comput Sci Rev 31–54
8. Topcuoglu H (2002) Performance-effective and low-complexity task scheduling for heterogeneous computing. IEEE Trans Parallel Distrib Syst 260–274
9. Samal P, Mishra P (2013) Analysis of variants in round Robin algorithms for load balancing in cloud computing. Int J Comput Sci Inf Technol 416–419
10. Chen H, Wang F, Helian N, Akanmu G (2013) User-priority guided min- min scheduling algorithm for load balancing in cloud computing. In: National conference on parallel computing technologies (PARCOMPTECH), pp 1–8.
11. Kumar, M, Sharma SC (2019) PSO-based novel resource scheduling technique to improve QOS parameters in cloud computing. Neural Comput Appl
12. Kumar M, Sharma SC (2018) PSO-cogent: cost and energy efficient scheduling in cloud environment with deadline constraint, sustainable computing. Inf Syst 147–164
13. Dokeroglu T, Sevinc E, Cosar A (2019) Artificial bee colony optimization for the quadratic assignment problem. Appl Soft Comput 595–606
14. Mahato D (2017) On scheduling transactions in a grid processing system considering load through ant colony optimization. Appl Soft Comput 875–891
15. Shishido H (2018) Genetic-based algorithms applied to a workflow scheduling algorithm with security and deadline constraints in clouds. Comput Electr Eng 378–394
16. Casas I (2018) GA-ETI: an enhanced genetic algorithm for the scheduling of scientific workflows in cloud environments. J Comput Sci 318–331.
17. Deepa T, Cheelu D (2017) A comparative study of static and dynamic load balancing algorithms in cloud computing. In: International conference on energy, communication, data analytics and soft computing (ICECDS), pp 3375–3378.
18. Milan ST, Rajabion L, Ranjbar H, Navimipour NJ (2019) Nature inspired meta-heuristic algorithms for solving the load-balancing problem in cloud environments. Comput Oper Res 0305-0548
19. Arunarani AR (2019) Task scheduling techniques in cloud computing: a literature survey. Future Gener Comput Syst 407–415

Chapter 9
A Synonym-Based Bi-LSTM Model for Machine Reading Comprehension

S. Anbukkarasi, N. Abinaya, and S. Varadhaganapathy

1 Introduction

Natural language processing is one of the domains of artificial intelligence (AI) which makes the system to interact with the human-understandable language in terms of texts. NLP has two broad categories, namely natural language understanding (NLU) and natural language generation (NLG). NLU provides the feature to a system which understands the human language and extracts some information from them. On the other hand, NLG makes the system to generate some text or information from the given model. Question answering (QA) is one such system which belongs to NLU. Earlier works of QA focus on information retrieval (IR) or knowledge-based QA [1]. Recently, factoid system which focuses on the facts brings promising results in QA system. Machine reading comprehension (MRC) is the extended version of the QA system which is the process of answering to a particular question from the given documents, thus, in turn, making the machine to understand the human languages. The understanding capability of a machine reading comprehension model is measured from the score which is calculated to predict the correct answer for the given context. These kinds of answering the questions help to develop various smart devices or techniques like Alexa and chatbot. Based on the complexity of MRC, it could be treated as,

S. Anbukkarasi (✉) · N. Abinaya
Nandha Engineering College, Erode, Tamil Nadu, India
e-mail: anbu.1318@gmail.com

N. Abinaya
e-mail: abi9106@gmail.com

S. Varadhaganapathy
Kongu Engineering College, Erode, Tamil Nadu, India
e-mail: varadhaganapathy@gmail.com

I. Jeena Jacob et al. (eds.), *Data Intelligence and Cognitive Informatics*,
Algorithms for Intelligent Systems,
https://doi.org/10.1007/978-981-15-8530-2_9

Passage: The ABC technology has arranged the one day workshop for student's beneficiary on the topic, 'Personality Development'. This session will be more informative to the students and could be an eye opener. It is handled by Ms. Madhu. The venue and timing of the workshop will be intimated later.

Question: Who take the session?

Answer: Ms.Madhu

Fig. 1 A sample passage with QA

1. Simple reasoning MRC task—In this task, the context could be of a single passage, a story or a news article.
2. Complex reasoning MRC task—In this task, the context consists of multiple passages like book chapters and web contents.

Figure 1 shows the simple example of a paragraph with question answering data. India has many regional languages like Tamil, Telugu and Bengali. Each language has a unique scripting. For example, if the Tamil language is considered, it has 12 vowels, 18 consonants, 216 compound characters and one special character known as aydha ezhuthu (.·.) totally 247 letters in Tamil. All these languages have an enormous amount of articles, documents to be processed with. It is the need of the hour to build a machine reading comprehension model for those language scripts too.

In recent days, deep learning techniques incorporate into the field of machine reading comprehension which deals with huge dataset. Many models are developed so far with sequential networks along with attention mechanism. The main crucial task in developing a network model is building an efficient encoding layer which contributes the better performance at the end [2–4]. More efficient methods developed using word-level embedding as well as character level embedding with pre-train vectors such as GloVe [5].

In this paper, the synonym-based bidirectional model with attention mechanism is proposed for the MCR task. Here character embedding is implemented, word embedding along with bidirectional model to obtain query-aware context information. Query-aware context information deals with the keywords available in the query term to be matched with the information occurred in the context or story. This would enhance the understanding ability of the model towards the accurate answer for the given query. To empathize this, a synonym dictionary is induced at the input layer along with the query data. This, in turn, expands the keywords of the questions to map the keywords in the context. The more matching keywords in the questions to the context help to increase the attention are assumed.

The kernel method is selected to give the juxtaposition of the words in context. As the model is bidirectional, the attention mechanism is used in both directions of query and context. Figure 2 represents the various levels of hierarchy for modeling the given paragraph along with the question.

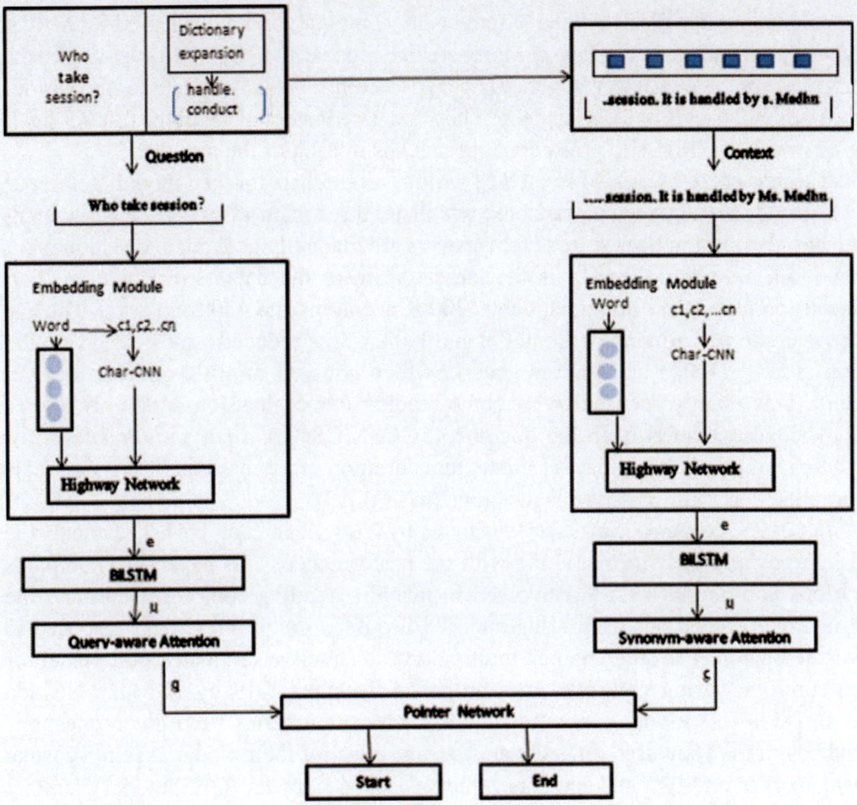

Fig. 2 A synonym-based bidirectional MRC model

2 Related Works

Machine reading comprehension was initiated well before the years, and the scope of working with MRC is still more. In 2017, LAMBADA dataset was used for reading comprehension which was observed that 83% of instances worked well and 17% of instances failed on this model. This model performed well on when selecting the name as an answer from the context when compared to other kinds of answers [6].

A novel approach named multilayer embedding with memory network (MEMEN) was proposed in 2017 [7]. They produced competitive results using Stanford Question Answering Dataset (SQuAD) in precision and efficiency. They also implemented their approach to TriviaQA dataset. The machine reading comprehension was enhanced by taking the position information [8]. Reference [8] deals with position information directly by using position encoder and position mapping embedding. They concluded that position-aware attention-based comprehension yields a better result than the other two position information.

A lot of studies are also done based on the techniques and models used for MRCs. A study on machine reading comprehension models (MRC) and tasks used until now were made by Chao Wang in 2020 [9]. Reference [10] also discusses the various methods and trends in MRC in deep. They discussed in detail about the four modules used in neural MRC and prominent approaches in each of the module.

The new corpus was also created by various researchers for their model. Reference [11] introduces a large-scale database which has three main advantages. Data sources are based on Baidu Search in which answers are manually generated, question types have rich annotation, and it is the largest scale of the dataset in Chineses. This dataset contains 1 million documents, 200 K questions and 420 K answers. The gap between the performance of a model and human was reduced atmost in [11]. Yibo Sun et al. [12] also build a new dataset which consists of 40 K question–answer pairs. This mainly focused on machine reading comprehension done only after a knowledge extracted from the document. COSMOS QA a large-scale dataset for reading comprehension that has contextual common sense reasoning. This dataset is formulated as multiple-choice-questions (MCQ) [13].

In 2019, a corpus comparison was made by Xing Zhan et al. [14]. They analyzed the corpus and their format along with the performance. This paper also compares various non-neural methods involved in machine reading comprehension and the recent neural methods used. This majorly discussed the progress that happened in MRC research. The study named multi-source abstractive summarization model for question answering (Masque) was proposed in paper 2019 by Kyosuke Nishida et al. [15]. This model generates the abstractive summaries from given questions and passages. They used different answering styles for their model. Again Kyosuke et al. worked on MRC and developed query focused extractor (QFE) model to extract the evidence [16]. RNN was used by having attention on a question and answer to extract the evidence. In our previous work, machine reading comprehension with long short-term memory deep neural network is implemented [17].

3 Dataset

In this work, the dataset is used from Facebook research. The dataset is known as 'The 20 QA bAbI task' dataset. It depicts 20 different tasks in which various types of stories, and queries are derived. This dataset is free of any noise. So could get 100% accuracy. The dataset has training and testing data separately [18]. Here paragraphs are mentioned as stories. It has 10,000 stories for training and 1000 for testing. This dataset is designed as follows: For a few lines, it has stories on various topics along with the number line. After a few lines, it consists question along with the answer and the number in which line the answer is presented. The stories cover 20 various tasks such as basic induction, size reasoning and time reasoning. The overall details of the task are given in Table 1.

Table 1 Tasks in dataset

Tasks [1–10]	Tasks [11–20]
Single supporting fact	Basic coreference
Two supporting facts	Conjunction
Three supporting facts	Compound coreference
Two argument relations	Time reasoning
Three argument relations	Basic deduction
Yes/No questions	Basic induction
Counting	Positional reasoning
Lists/sets	Size reasoning
Simple negation	Path finding
Indefinite knowledge	Agent's motivations

Example of the dataset is given below.
Yes/No questions
John moved to the playground.
Daniel went to the bathroom.
John went back to the hallway.
Is John in the playground? A: no
Is Daniel in the bathroom? A: yes

4 Preprocessing

The given dataset is preprocessed in terms of tokenization. The sentences with the words are tokenized including special characters such as punctuation. The dictionary with the input words along with the synonym based dictionary is passed to the model, to create the word indexes. Hence, each word is assigned with the given unique number. The maximum word length for stories and query is fixed with the help of padding. For our work, 10,000 training data is used, 1000 for testing purposes.

5 Model

Our model consists of hierarchical layers of various stages as shown in Fig. 1.

5.1 Embedding Layer

It contains character embedding as well as word embedding. Character embedding is achieved by character level CNN. It maps each word to the vector space. Word embedding maps the words to a high-dimensional vector space. Pre-trained vector known as GloVe is used which gives the fixed vector size for each word. This vector implements the unsupervised learning algorithm to give the vector representation for the given words. It handles the co-occurrence of the relationship among the words. The dictionary with various synonyms for the given query words is also encoded and append in the embedding layer. This enhances in mapping the story (context) words with the query words. The combined embedding from both the layers is passed to the high way network to get the final representation for query ($e_i{}^Q$) and story $e_i{}^S$. Highway network calculates the feature at different levels. Highway networks use transform gate $T(W_{\mathrm{T}}, X)$ as well as carry gate $C(W_{\mathrm{C}}, X)$. These gates use nonlinear transfer function, namely sigmoid function. This network is established to suppress the vanishing gradient problem hence optimizing the network. It is represented by the following equation.

$$Y = H(x, W_H).T(x, W_T) + x.C(x, W_C) = H(x, W_H).T(x, W_T) + x.(1 - T(x, W_T)) \tag{1}$$

5.2 Modeling Layer

Bi-LSTM of size d is implemented for each direction as modeling layer which takes e as input, helps to encode the context words based on the query words. This layer outputs the interaction between the context words with the query keywords. As the model works in two directions, it computes the representation u^{S} for story and u^{Q} for question with obtained question and context from previous layers. It is given as:

$$u_k^{\mathrm{Q}} = \mathrm{BiLSTM_Q}\left(u_{k-1}^{\mathrm{Q}}, e_i^{\mathrm{Q}}\right) \tag{2}$$

$$u_k^{\mathrm{S}} = \mathrm{BiLSTM}_C\left(u_{k-1}^{\mathrm{S}}, \mathrm{e}_i^{\mathrm{S}}\right) \tag{3}$$

This layer outputs the matrix M to the output layer to predict the answer for the given question. It contains the contextual information on the stories and query given. Bi-LSTM helps to combine two different RNNs work together. At every step, it allows the input to process in both the forward and backward directions. It means from past to future and vice versa. In terms of long short-term model (LSTM) which is unidirectional that helps to preserve some information from the future and with the help of hidden states, it preserves both past data as well as future data.

5.3 Output Layer

It accommodates Bi-LSTM with pointer network to predict the answer for the given question with start and end scores. To obtain the start and end probability, softmax function is applied. The start and end score with softmax function are given below.

$$S = \text{softmax}\left(W^T[e, M]\right) \tag{4}$$

$$E = \text{softmax}\left(W^T\left[e, M'\right]\right) \tag{5}$$

6 Experiments and Results

The bidirectional RNN model which being implemented here gives promising results as shown in Table 2. For the comparison purpose, the same synonym-based model is implemented with long short-term memory (LSTM) deep neural network which has the gated units to remember the previous inputs to the model. For machine reading comprehension task, exact match (EM) or the accuracy is used to evaluate the designed model. In our implementation, the accuracy is used as the matric to evaluate the model. It is defined as the closeness between the measured value and the known value. The results show that bidirectional RNN works far better than the LSTM network.

The result shows that Bi-LSTM shows a better result than one-way LSTM network, and the comparison is also performed between simple attention and synonym-based attention. The synonym-based dictionary eventually increases the performance of the system. In our approach, the base model (Bi-LSTM) with attention mechanism is implemented and evaluated. Further, the same model is extended with synonym-based dictionary and the results are compared with the previous results. The graphical representation of the performance of the model is shown in Fig. 3.

Table 2 Accuracy of Bi-LSTM and LSTM

Model	Bi-LSTM	LSTM
Base model with attention	79.5	72.1
Synonym based attention	84.4	76.4

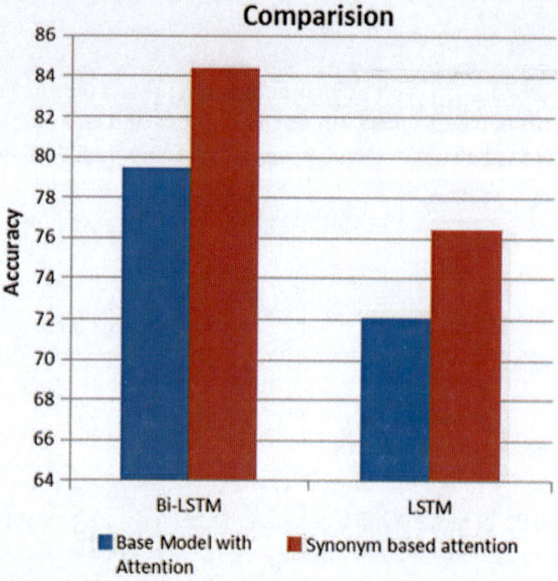

Fig. 3 Graphical analysis of Bi-LSTM with LSTM

7 Conclusions

In this paper, synonym-based Bi-LSTM model architecture is proposed with various stages for machine reading comprehension (MRC). The dictionary with various synonyms for the given context word in the input layer gives promising results. Our model is implemented in 20 QA bAbI task dataset. This dataset consists of twenty classes to analyze the ability of the model at various stages of answer retrieval. Our model can be further elaborated with the jig of attention layers. As Indian regional language-based NLP has wide scope for the researchers to explore, this model could be incorporated for achieving more state-of-the art results in pure Indian language scripts as being discussed earlier. There are various Indian languages such as Tamil, Telugu, Hindi which has huge scripts of their own style need a good comprehension model to understand the context and answer the given queries. So the future work may involve developing a model in regional Indian languages which require huge dataset.

References

1. Sujith V, Anand Kumar M, Soman (2019) A sequence-based machine comprehension modeling using LSTM and GRU. Emer Res Electroni, Comput Sci Technol
2. Minghao H, Yuxing P, Xipeng Q (2017) Mnemonic reader for machine comprehension. arXiv: 1705.02798

3. Wenhui W, Nan Y, Furu W, Baobao C, Ming Z (2017) Gated self-matching networks for reading comprehension and question answering. In: Proceedings of the 55th annual meeting of the association for computational linguistics (ACL)
4. Minjoon S, Aniruddha K, Ali F, Hannaneh H (2016) Bidirectional attention flow for machine comprehension. arXiv:1611.01603
5. Jeffrey P, Richard S, Christopher DM (2014) Glove: global vectors for word representation. In: Empirical methods in natural language processing (EMNLP), pp 1532–1543
6. Zewei C, Wang H, Kevin G, David M (2017) Broad context language modeling as reading comprehension, vol 3. arXiv:1610.08431
7. Booyuan P, Li H, Zhou Z, Bin C, Deng C, Xiaofei H (2017) MEMEN—multi-layer embedding with memory networks for machine comprehension, vol 1. arXiv:1707.09098
8. Yajing X, Weijie L, Guang C, Boya R, Siman Z, Sheng G, Jun G (2019) Enhancing machine reading comprehension with position ınformation. In: IEEE, special section on aı-driven big data processing: theory, methodology, and applications
9. Wang C (2020) A study of the tasks and models in machine reading comprehension, vol 1. arXiv:2001.08635
10. Shanshan L, Xin Z, Sheng Z, Hui Z, Weiming Z (2019) Neural machine reading comprehension—methods and trends. Appl Sci 9:3698. 10:3390. vol 2
11. Wei H, Kai L, Jing L, Yajuan L, Shiqi Z, Xinyan X, Yuan L, Yizhong W, Hua W, Qiaoqiao S, Xuan L, Tian W, Haifeng W (2018) DuReader: a Chinese machine reading comprehension dataset from real-world applications, vol 4. arXiv:1711.050
12. Yibo S, Daya G, Duyu T, Nan D, Zhao Y, Xiaocheng F, Bing Q (2018) Knowledge based machine reading comprehension. Work done during internship at Microsoft Research Asia, vol 1. arXiv:1809.04267
13. Lifu H, Ronan LB, Chandra B, Yejin C (2019) COSMOS QA: machine reading comprehension with contextual commonsense reasoning. In: Proceedings of EMNLP 9th International Conference on NLP, pp 2391–2401. ACL
14. Xin Z, An Y, Sujian L, Yizhong W (2019) Machine reading comprehension—a literature review, vol 1. arXiv:1907.01686
15. Kyosuke N, Itsumi S, Kosuke N, Kazutoshi S, Atsushi O, Hisako A, Junji T (2019) Multi-style generative reading comprehension. Research Gate (2019)
16. Kosuke N, Kyosuke N, Masaaki N, Atsushi O, Itsumi S, Hisako A, Junji T (2019) Answering while summarizing—multi-task learning for multi-hop QA with evidence extraction, vol 2. arXiv:1905.08511
17. Anbukkarasi S, Suganya G, Sakunthala J, Suguna A (2020) Machine reading comprehension using bi-directional LSTM sequence model. J Adv Res Dyn Control Syst 12:(3)
18. Jason W, Antoine B, Sumit C, Alexander MR, Bart van M, Armand J, Tomas M (2016) Towards AI-complete question answering—a set of prerequisite toy tasks. ICLR
19. Amrita S, Rahul, A., Mitesh MK, Karthik S (2018) Duorc: towards complex language understanding with paraphrased reading comprehension. In Association for compu-tational linguistics (ACL), pp 1683–1693
20. Jacob IJ (2020) Performance evaluation of caps-net based multitask learning architecture for text classification. J Artif Intell 2(01):1–10

Chapter 10
Designing of Arbiter PUF for Securing IP and IoT Devices

Swati Kulkarni, R. M. Vani, and P. V. Hunagund

1 Introduction

The embedded system on FPGAs is normally designed and built using intellectual property (IP). IP is already predefined, pretested, and reusable macro or register transfer level (RTL) code. It has two types, hardcore IP and softcore IP. Hardcore IP is technology independent and it takes less to market. All Xilinx ZYNQ 7 series consists of two Arm® Cortex™-A9 MP Core™ application processors are the example of hardcore IP [1]. Whereas softcore IPs are technology-dependent and flexible to use as they are customizable. Xilinx provides an IP integrator where all the IPs are present. Customized IP can be designed and add it to the user repository for future reuse. Many times, IP can be taken from a third party. As mentioned above, hardcore IPs are normally used to integrate into chips. Similarly, when the embedded systems are being developed it is recommended that the designing should be done using IPs. Now, the problem is that if IP is purchased from the third party, cannot guarantee that it does not consist of any hardware Trojan. Hardware Trojan is a malicious function and deliberately stealthy modification can be inserted into hardware. It is similar to software Trojan, which is not recommended in the design. Hardware Trojan is a mainly passive component that alters the functionality of the device. Normally, it is a small addition in hardware module, which is incorporated during design time. It is neither identified as a tiny component nor detected using any anti-virus. Hardware Trojan is waiting for a specific event and when an event gets trigger, it could leak some secrete information to the outside world through any network interface. This way it

S. Kulkarni (✉) · P. V. Hunagund
Department of Applied Electronics, Gulbarga University, Gulbarga 585106, India
e-mail: swatikulkarni494@gmail.com

R. M. Vani
University Science Instrumentation Centre, Gulbarga University, Gulbarga 585106, India

I. Jeena Jacob et al. (eds.), *Data Intelligence and Cognitive Informatics*, Algorithms for Intelligent Systems,
https://doi.org/10.1007/978-981-15-8530-2_10

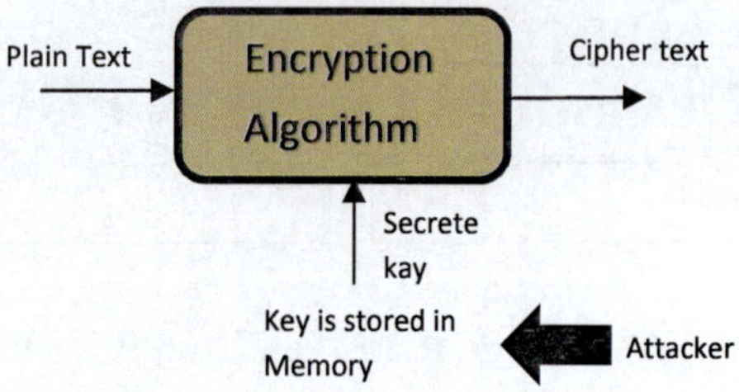

Fig. 1 Traditional way of cryptography

could spoil the trusted environment [2]. Security of IP is a serious issue in hardware designing. There are various levels in the chip fabrication process. Hardware Trojan may get inserted in any level of IC designing [3, 4].

Cryptography is known as an art of protecting information by using a secret key. In the cryptography technique, the secret keys are stored in volatile or non-volatile memory [5]. Figure 1 shows an encrypted algorithm that uses a secrete key and encrypts plain text and produces ciphertext. Later on, the key remains in memory and can be retrieved even when power is turned off. Such attacks are called side-channel attacks. Even volatile memory is susceptible to different types of attacks if one has physical access to it. Thus, the traditional approaches to key storage are not favoured, especially in high-security applications like defence or banking applications. Data security, authentication, data integrity, and confidentiality are the basic goals of cryptography techniques [6, 7].

PUF is only the promising innovative primitive solution for hardware security. Physical unclonable functions (PUFs) are new, hardware-based security primitive, which has been introduced just about a decade ago. The working principle of PUF is based on manufacturing process variations [8]. When any manufacturer is fabricating the same IC in the same fabrication unit using the same material and process still two identical ICs is having some difference in threshold voltage or gate delays [9]. PUF design uses these process variations by taking challenges as an input and produces responses as an output. Every challenges PUF generates a unique and random response. Further responses can be used as a secrete key to secure hardware designs where power, security, and device size are the important requirements. That's why, such design is called as unclonable. The designer will already prepare a challenge-response table. Once the chip will get fabricate designer can verify the functionality by applying challenges and verifying responses.

The computer is a sequential machine and cannot generate random numbers so any computer cannot be used to generate a unique unpredictable random key. Figure 2 shows PUF-based cryptography, and PUF is an inbuilt component of integrated chip (IC). According to applied challenges, it produces a unique response. These responses applied to the key extraction algorithm unit. Fuzzy extraction algorithm or helper data algorithms [10] are used to extract the unpredictable random key. That secrete key is used by the encryption algorithm to modify input into ciphertext. The difference between traditional cryptography and PUF-based cryptography is that in PUF-based cryptography no need to store secrete key.

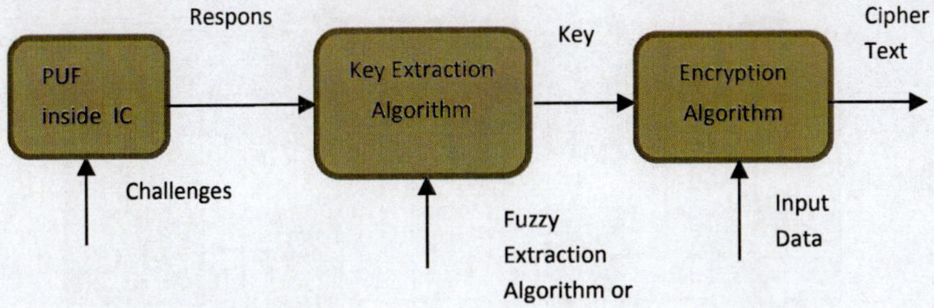

Fig. 2 PUF-based cryptography

PUF will generate a unique key for every new transition. PUF-based cryptography is not only used to secure information but also authenticate the device. As discussed earlier, every device has its unique physical characteristics, so the device will give a unique response. PUF itself is generating device signature is as unique as human fingerprints. PUF is not only used for securing IPs but also it is using in IoT applications where security is the main concern [11–13].

Weak PUFs and strong PUFs are two basic types of PUF. As the name suggests weak PUFs are less secure than strong PUFs. Also, weak PUF has few numbers of challenge-response pairs (CRP), whereas it is strong PUF has a greater number of CRPs. In this paper, Arbiter PUF and its types are being discussed. Arbiter PUF is strong PUF [14, 15]. PUF also has to suffer from different types of attacks like machine learning attacks, side-channel attacks, etc. The Arbiter PUF and Zedboard FPGA using Vivado 2018.3 have been implanted. Analysis of PUF is not the scope of this paper.

2 Typical Arbiter PUF

Every IC has its own propagation delay and that will be unique for each IC. As mentioned earlier, PUF will take advantage of this propagation delays and produce a unique and random response. Arbiter PUF is delay-based PUF which will utilize these propagation delays to produce random responses. Basic Arbiter PUF consists of multiplexers as a switching component and trail of multiplexers connected to the arbiter that is D-latch [16]. The data input of arbiter is the upper trail of switching component and clock input is a lower trail of switching component. The propagation delay of both trails will be different due to process variation. When clock input and data inputs are '1' then only PUF output will be '1' otherwise '0'. The select lines of multiplexers are the challenge inputs, and the output bit of arbiter is a response bit [16]. Encryption algorithms like advance encryption algorithm (AES) will require a minimum of 128 bits of the key. Multiple instances of the above design need to implement to get multiple responses.

Figure 3 shows the basic layout of Arbiter PUF. Here, two paths of switching components are formed, shown in black and blue colours. Enable is common to

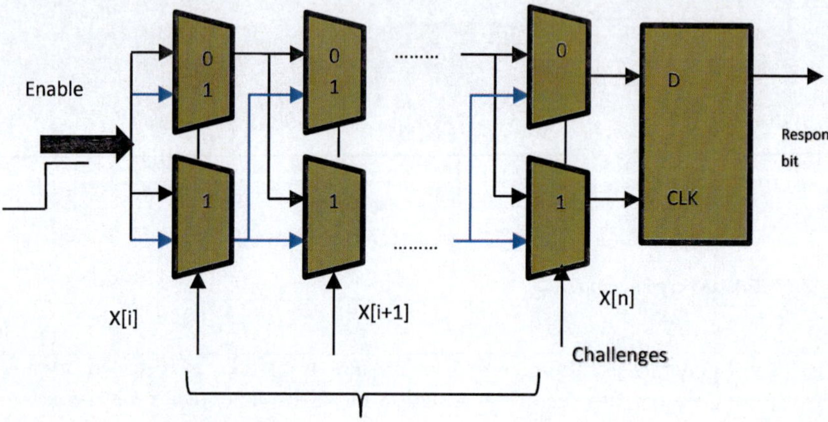

Fig. 3 Basic layout of Arbiter PUF

input to both paths. X[i] to x[n] is the challenges that select input to the multiplexers. Depending upon the propagation delays of switching element input will reach to arbiter and response bit will be generated [15, 17]. Arbiter PUF is prone to modelling attacks or machine learning attacks.

3 XOR Arbiter PUF

Typical Arbiter PUF is susceptible to modelling attacks. An attacker can easily clone to normal Arbiter PUF design. XORed PUF can avoid modelling. Figure 4 shows the basic layout of XOR Arbiter PUF. Here, responses generated from Arbiter PUF are XORed [18, 19]. Enable is common to input to all Arbiter PUF. X[i] to x[n] is the challenge that selects the input to the multiplexers. XOR Arbiter PUF is more secure than normal Arbiter PUF [20]. Depending upon the propagation delays of switching element input will reach to XOR gate and the response bit will be generated. As both

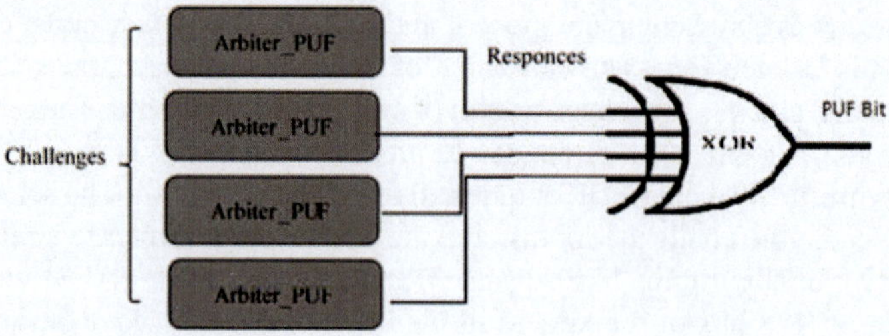

Fig. 4 Basic layout of XOR Arbiter PUF

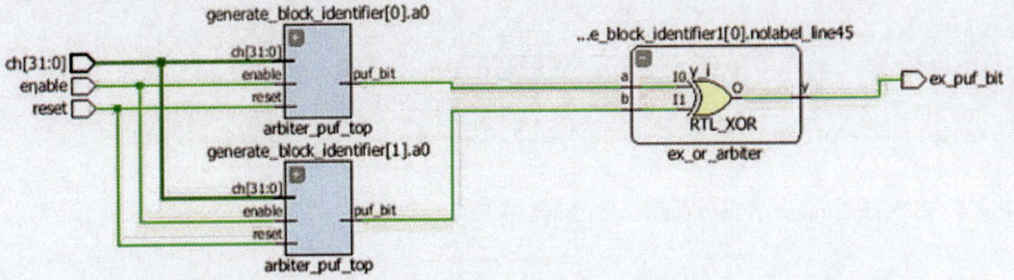

Fig. 5 RTL schematic of XOR Arbiter PUF

inputs will be logic '1' response will be '0' bit and when either of input is logic '1' then response bit will also be logic '1'.

Figure 5 shows the RTL schematic of 1-bit XOR Arbiter PUF. This is a prototype design of Ex-OR Arbiter PUF. In this design, gate, Arbiter is connected to an input to XOR gate and the final output is XORed output. The XOR PUFs evaluate the responses of individual PUFs and XOR the final output. This architecture is difficult and tedious. This is the reason XOR PUFs are accepted to be more secure than standard Arbiter PUFs.

4 Feed Forward Arbiter PUF

The above two types of Arbiter PUF are susceptible to modelling attacks. Even though Arbiter PUF can said to be strong, PUF still modelling such PUF is easy for attackers [21]. Figure 6 shows feed forward Arbiter PUF is a nonlinear design of

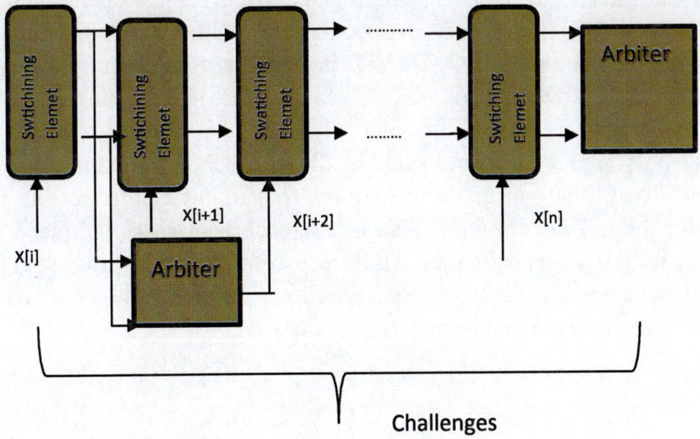

Fig. 6 Feed forward Arbiter PUF

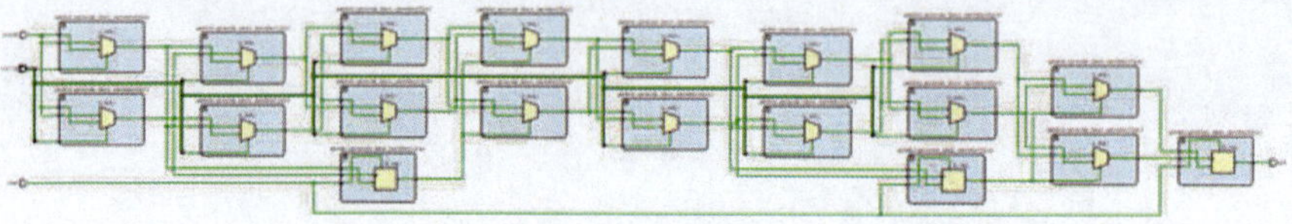

Fig. 7 RTL schematic of feed forward Arbiter PUF

Arbiter PUF. In this case, few challenges are feed forwarded to switching elements instead of applying externally to them [20].

Figure 7 shows the RTL schematic of feed forward Arbiter PUF. Two extra arbiters are used to apply the challenges of Arbiter PUFs. Input to these two arbiters is again taken from some switching elements and they are unpredictable. Positing of the arbiter is nonlinear so it is difficult to model the design. The architecture of feed forward Arbiter PUF increases the unpredictability and help to reduce machine learning attacks.

5 Results

All three types of Arbiter PUF design on Xilinx Vivado 2018.3 have been implemented and tested on two different Zedboard. Xilinx Vivado tool tried to optimize the design, the desired netlist was not able to be generated. To avoid this problem, Xilinx Vivado synthesis user guide was referred and used attribute "* dont_touch = "yes" *" in our design. This helps us to generate required netlist. One trail is of 32 switching elements that mean 32 different challenges can be applied. Digilat's Zedboard is having only 8 onboard LED's and switches which are not enough for testing our design. As having 32 challenges, enable and reset as an input port, and 64 output port in total 98 input-output ports are required. The design was interfaced with Xilinx virtual input-output (VIO) IP. VIO provides virtual input out ports to fulfil our requirements. 64 bits of responses are generated by instantiating 64 instances of design (Fig. 8).

Finally, the same bit has been loaded in all two Zedboards and applied the same challenges. Every time got unique responses for similar challenges but the same response for the individual board. That means each board was produced the same random key for the same challenges. All boards were produced different keys for the same challenges because delays of each board are different. The entire experiment is carried out at room temperature.

Sr. No	Types of PUF	FPGA Resource Utilization		
		LUT	FF	IO
1	Arbiter PUF	4160	64	98
2	XOR Arbiter PUF	8383	128	98
3	Feed Forward Arbiter PUF	4222	329	92

Fig. 8 Report utilization result

6 Conclusion

It is important to note that several key issues of IoT are mostly related to data security and privacy. PUF is used to authenticate device signature as well as to generate secrete random key to encrypt the plain text. Arbiter PUF is more sensitive to modelling attacks. Feed forward Arbiter PUF is the most promising type of Arbiter PUF compare to other arbiters. In the future, hardware optimization of PUF will be focused on because area and power consumption are overheads associated with existing IoT systems.

References

1. ZC702 Evaluation Board for the Zynq-7000 XC7Z020 SoC User Guide UG850 (v1.7), 27 March 2019
2. Chakraborty RS, Saha I, Palchaudhuri A, Naik GK (2013) Hardware Trojan insertion by direct modification of FPGA configuration bitstream. IEEE Des Test 30(2):45–54
3. Rührmair U, Busch H, Katzenbeisser S (2010) Strong PUFs: models, constructions, and security proofs. In: Sadeghi AR, Naccache D (eds) Towards hardware-intrinsic security. Information security and cryptography. Springer, Heidelberg
4. Guajardo J, Kumar SS, Schrijen G-J, Tuyls P (2007) FPGA intrinsic PUFs and their use for IP protection. In: Paillier P, Verbauwhede I (eds) CHES 2007, vol 4727. LNCS. Springer, Heidelberg, pp 63–80
5. Rührmair U, Devadas S, Koushanfar F (2012) Security based on physical unclonability and disorder. In: Tehranipoor M, Wang C (eds)
6. Sklavos N (2013) Securing communication devices via physical unclonable functions (PUFs). https://doi.org/10.1007/978-3-658-03371-2_22
7. Handschuh H, Schrijen GJ, Tuyls P (2011) Hardware intrinsic security from physically unclonable functions. In: Sadeghi A-R, Naccache D (eds) Towards hardware-intrinsic security, information security and cryptography. Springer Verlag, Heidelberg https://doi.org/10.1007/978-3-642-14452-3_2
8. Kulkarni S, Vani RM, Hunagund PV (2019) A study on physical unclonable functions based security for internet of things applications. In: Hemanth D, Shakya S, Baig Z (eds) Intelligent data communication technologies and internet of things. ICICI 2019. Lecture Notes on Data Engineering and Communications Technologies, vol 38. Springer, Cham, pp 607–614

9. Joshi S, Mohanty SP, Kougianos E (2017) Everything you wanted to know about PUFs. IEEE Potentials 36(6):38–46
10. Delvaux J, Gu D, Schellekens D, Verbauwhede I (2015) Helper data algorithms for PUF-based key generation: overview and analysis in IEEE Transactions on Computer-Aided Design of Integrated Circuits and Systems 34(6):889–902, June 2015, https://doi.org/10.1109/TCAD.2014.2370531
11. Kulkarni S, Vani RM, Hunagund PV (2018) Review on IoT based case study: applications and challenges. In: International conference on intelligent data communication technologies and internet of things. Springer, Cham
12. Johnson AP, Chakraborty RS, Mukhopadhyay D (2015) A PUF-enabled secure architecture for FPGA- based IoT applications. IEEE Trans Multi-scale Comput Syst 1(2)
13. Mugunthan SR (2019) Security and privacy preserving of sensor data localization based on internet of things. J ISMAC 1(02):81–91
14. Rührmair U, Schlichtmann U, Burleson W (2014) Special session: how secure are PUFs really? On the reach and limits of recent PUF attacks. In: Design, automation & test in Europe conference & exhibition (DATE), pp 1–4
15. Tajik S, Dietz E, Frohmann S, Seifert J-P, Nedospasov D, Helfmeier C, Boit C, Dittrich H (2014) Physical characterization of Arbiter PUFs. Springer, Heidelberg
16. Sahoo DP, Chakraborty RS, Mukhopadhyay D (2015) Towards ideal Arbiter PUF design on Xilinx FPGA: a practitioner's perspective. In: 2015 euromicro conference on digital system design, Funchal, pp 559–562
17. Anandakumar NN, Hashmi MS, Sanadhya SK (2017) Compact implementations of FPGA based PUFs with enhanced performance. In: 30th international conference on VLSI design and 2017 16th international conference on embedded systems (VLSID), pp 161–166
18. Dubrova E (2018) A reconfigurable arbiter PUF with 4 × 4 switch blocks. In: 2018 IEEE 48th international symposium on multiple-valued logic (ISMVL), Linz, pp 31–37
19. Alkatheiri MS, Zhuang Y, Korobkov M, Sangi AR (2017) An experimental study of the state-of-the-art PUFs implemented on FPGAs. In: IEEE conference on dependable and secure computing, Taipei, Taiwan
20. Kumar MA, Bhakthavatchalu R (2017) FPGA based delay PUF implementation for security applications. In: IEEE international conference on technological advancements in power and energy (TAP Energy), Kollam, India
21. Idriss T, Idriss H, Bayoumi M (2016) A PUF-based paradigm for IoT security. In: 2016 IEEE 3rd world forum on internet of things (WF-IoT), Reston, VA, pp 700–705

Chapter 11
Enhancements on the Efficacy of Firefighting Robots Through Team Allocation and Path-Planning

Sreesruthi Ramasubramanian, Senthil Arumugam Muthukumaraswamy, and A. Sasikala

1 Introduction

Firefighting robots are widely used to extinguish fires in indoor environments [1–3]. Most firefighting robots in the past used flame sensors for detection. However, fires beyond a certain threshold distance from the robot cannot be detected using sensors. Thus, other methods of fire detection such as *Haar cascade classifier* [4] and *deep learning models* [5–7] were investigated. These detection models can be used in firefighting robot applications. *Deep learning models* are preferred since they have better performance [14]. Using the deep learning model trained to identify fires, the fires can be detected along with their position. Thus, the robots do not have to keep patrolling the area to detect fires, and their battery power can be saved. The position of the fire on the floor can be obtained by using the coordinates of the bounding box surrounding the detected fire in real time.

To find the target position to which the robot should travel to, the *minimum distance* the robot should maintain from each fire was computed. The *minimum distance* can be calculated by considering the size and intensity of the fire. An algorithm that can be used to identify the *minimum distance* the robot should maintain from the fire based on the parameters discussed is presented in this paper. Although firefighting robots have been designed to have necessary protection such as mist sprays to cool

S. Ramasubramanian (✉) · S. A. Muthukumaraswamy
Heriot-Watt University, Dubai, United Arab Emirates
e-mail: snr1@hw.ac.uk

S. A. Muthukumaraswamy
e-mail: m.senthilarumugam@hw.ac.uk

A. Sasikala
Sri Sairam Institute of Technology, Chennai, India
e-mail: sasikala.eee@sairamit.edu.in

I. Jeena Jacob et al. (eds.), *Data Intelligence and Cognitive Informatics*,
Algorithms for Intelligent Systems,
https://doi.org/10.1007/978-981-15-8530-2_11

themselves down [1], it would be better to maintain an appropriate distance from the fire to reduce the usage of extinguishment resources such as water for self-protection. Since the distance that has to be maintained would be calculated, the robots need not position themselves further than needed from the fire. Hence, in this manner too, use of extinguishment resources would be reduced.

After the target positions of the fires detected by the *deep learning model* and the camera placed in a particular room is computed, the shortest path between each of the robot's current position and the target position should be computed in the fastest time possible. After the *local target position* to the room is converted to the *global target position* to the floor, path-planning algorithms such as grid-based or sampling-based algorithms can be used to compute the shortest path. In this paper, the probabilistically complete sampling-based path-planning algorithm *PRM* is investigated. In *PRM*, random sampling points or nodes are generated throughout the free space or region without obstacles during the *learning phase* [8]. Each node is connected to other nodes within the connection distance value specified. The *PRM* consisting of the system of connected nodes can then be used to identify the paths required depending upon the start and end positions in the *query phase*.

The path identified using *PRM* constructed using a different number of nodes is compared based on computational time and path length to choose a suitable number of nodes to construct the *PRM* in this paper. The identification of paths using *PRM* for a single-robot and multiple-robot situations is also compared to demonstrate the effectiveness of using multiple robots in fire extinguishment. Additionally, an algorithm is proposed to allocate robots into teams of different sizes to extinguish the different fires. In this manner, a larger number of robots could be used to extinguish the larger-sized and/or higher intensity fires compared to the smaller-sized and/or lower intensity fires. Thus, by following the methodologies discussed, the efficacy of the fire-extinguishing process can be greatly improved.

2 Literature Review

Various experiments have been performed to determine the path-planning algorithm that would give the least cost or distance path. The literature review section demonstrates the efficiency of path-planning algorithms through comparisons made in existing research.

The performance of grid-based algorithms *A** and *Dijkstra's* was compared for path-planning a firefighting hexapod while using the KRPAI map 2016 [9]. The objective was to find which algorithm had a better performance in such an environment. The performance of the algorithms was investigated based on two factors—*time complexity* and *space complexity*. *Time complexity* was the amount of time needed by the respective algorithm to find the path, and *space complexity* was the amount of memory used by the algorithm while processing. It was found that the *A** algorithm had an average execution time of 4 270.72 ms and used an average memory of 16 768 KB. *A** had a much better performance than the *Dijkstra's* algorithm which had an average execution time of 6 757 ms and used an average memory of 40 992 KB. Hence, $A*$ was chosen for path-planning the firefighting hexapod.

Kumar et al. simulated the *PRM* path-planning algorithm using *MATLAB* [10]. The binary occupancy map of the region where the *PRM* was to be implemented was created using laser readings obtained by the robot moving around the corridor world created in *Gazebo* simulator. The TurtleBot followed the waypoints specified to obtain the laser readings. The binary occupancy map was then used to implement *PRM*. From the results obtained, it was concluded that the most efficient path was created by using a large number of nodes and a small connection distance to connect the nodes. For the environment considered, the maximum number of nodes was set to 50 and the maximum connection distance used set to 10.

Simulation experiments to reduce radiation exposure of staff in nuclear plants using probabilistic roadmap method for path-planning, map model and dose calculations were carried out [11]. The results obtained from the $A*$ algorithm and *PRM* were compared to determine the most effective algorithm. In the first simulation, the *probabilistic roadmap* was used to find the path between the start and end nodes. Initially, the shortest path was computed without considering dose calculations. In the second case, the most feasible path was computed by taking into consideration the amount of radiation or cost. The computational time for both these methods did not have a significant difference. However, the shortest path had a radiation level almost three times greater than the most feasible path. Hence in this particular situation, the feasible path was preferred to the shortest path. In the second simulation, the most feasible path was computed using $A*$ and *PRM*. $A*$ produced slightly better paths compared to *PRM*. However, the time taken for computation using $A*$ was 11.35 times greater than the time taken for *PRM*. The computational time of $A*$ was likely to increase with the number of nodes. By following the path identified using *PRM*, staff could be evacuated much earlier, thus reducing radiation exposure. Hence, *PRM* was used for finding the optimized path in the nuclear plant environment since it could find the required path within a short period.

3 Methodology

3.1 Algorithm to Find the Minimum Distance the Robot Should Maintain from the Fire

A camera was placed in each room which was focused to capture most of the floor. Fires present on the floor were focused upon. The camera was placed at ground level at a horizontal angle of 90°. Figure 1 shows a sample scenario to demonstrate the working of the algorithm. In this scenario, each frame captured by the camera was represented by a pixel coordinate system spanning 600 pixels horizontally and 400 pixels vertically. The image represents a room in which multiple fires of different

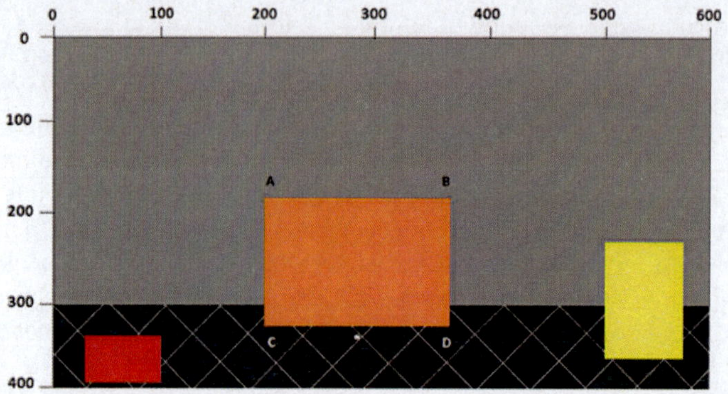

Fig. 1 Depiction of an image with bounding boxes surrounding the different fires detected

sizes and intensities were present. The red (fire 1), orange (fire 2) and yellow (fire 3) rectangles represent the different fires detected surrounded by the bounding boxes.

To find the intensity of each fire, the mean pixel value of all the pixels in each region enclosed by the bounding box was found from the image. The relative sizes of the fires were found by finding the area covered by each bounding box. By using the position of the fire and *safe distance* calculated, the *target position* of the fires can then be found. The pseudocode of the algorithm used to find the safe distance is presented in Fig. 2.

The maximum possible area that a bounding box could cover would be 600×400 or 240,000 pixels. In the same manner, the highest possible intensity was 255 which corresponded to the color white. The area enclosed by the bounding box and the mean pixel intensity in the enclosed region were scaled according to the largest possible area and the highest possible pixel value. The *ScaledParameter* obtained by taking the product of the *areaScaled* and *meanScaled* was multiplied by the appropriate parameter, 10 in this scenario, to obtain the *safe distance* in meters.

Fig. 2 Pseudocode for minimum safe distance algorithm

Pseudocode: Algorithm to find Safe Distance

```
CurrentFire ← The fire number currently under consideration
x1 ← x coordinate of the top left point of the bounding box
x2 ← x coordinate of the bottom right point of the bounding box
y1 ← y coordinate of the top left point of the bounding box
y2 ← y coordinate of the bottom right point of the bounding box
length ← x2-x1
height ← y2-y1
area ← length * height
areaScaled ← area/(600*400)
meanPixelIntensity ← mean pixel value of the fire region
meanScaled ← meanPixelIntensity/255
ScaledParameter ← areaScaled * meanScaled
SafeDistance ← ScaledParameter * 10
```

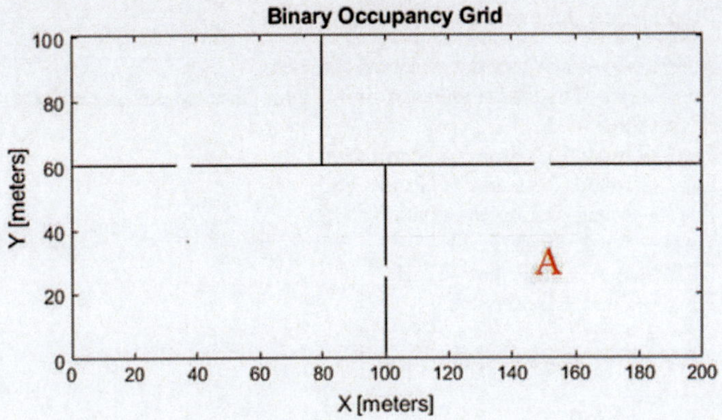

Fig. 3 Binary occupancy map of the floor

3.2 *Probabilistic Roadmaps*

The *probabilistic roadmap* was simulated on a computer system having Intel i7-5500U processor with 2.4 GHz and 8 GB RAM. The *robotic system toolbox* from *MATLAB* was used to implement path-planning using *PRM*. The robot was set to be a mobile differential drive robot which uses *PRM* for path-planning. The steps below outline the procedure followed to simulate the robot's path using *PRM* and *MATLAB R2019b* [12].

- The *binary occupancy map* shown in Fig 3. was used to create the *PRM* [13].
- To create a *PRM*, the *number of nodes* and the *maximum connection distance* were specified. The *maximum connection distance* was specified as 50 m.
- Nodes were generated randomly through the free space, and the nodes were connected based on the connection distance.
- From the roadmap created, the start and end locations were used to identify the path required from the *PRM*.
- The graphs obtained with the different number of nodes for the selected connection distance were analyzed to choose an appropriate number of nodes.
- The computational time taken for constructing the *PRM* and for finding the path between various start and end positions for the chosen number of nodes and connection distance was tabulated along with path length for single- and multiple-robot situations for comparison.

3.3 *Algorithm for Team Allocation*

The total number of robots available was allocated into teams based on the number of fires (three, four and five fires were considered). The pseudocode of the algorithm is shown in Fig. 4 for three fires. The working of the algorithm is outlined below.

```
Pseudocode: Algorithm to find Safe Distance

NumRobot ← The number of robots available
NumFires ← The total number of fires // three fires are considered here
if NumFires == 3
Fire1 ← round [0.5 times of NumRobot]
Fire2 ← round [0.3 times of NumRobot]
Fire3 ← round [0.2 times of NumRobot]
Total ← Fire1+Fire2+Fire3
Difference ← NumRobot - Total
Fire1 ← Fire1 + Difference
if Fire2 > Fire1
Swap the robot teams of Fire1 and Fire 2
end if
if Fire3 > Fire2
Swap the robot teams of Fire2 and Fire 3
end if
display the number of robots allocated for each fire
end if
```

Fig. 4 Pseudocode for team allocation algorithm when three fires are present

- The number of robots available (*NumRobot*) was split into teams using the percentages shown in Table 1.
- As in the previous algorithm, the product of the relative size and mean pixel intensity of each fire from the image can be used to judge which fires require more robots for fire extinguishment. For example, *Fire 1* would be the fire whose relative size and mean pixel intensity product give the largest value, and the second-largest product would correspond to *Fire 2*.
- Since finding percentages often introduced decimals, each number of robots obtained was rounded off to remove decimals.
- The total number of robots (*Total*) was found by adding the number of robots in each team responsible for extinguishing Fires 1 to *NumFires*. This number was compared with the number of robots available to obtain *difference*.
- If there was any difference between the number of robots, robots were added to Fire 1 if the *total* was less than *NumRobot* or subtracted from Fire 1 if the

Table 1 Percentage division of robots based on the number of fires

Fire number	Total number of fires		
	NumFire = 3	NumFire = 4	NumFire = 5
Fire 1	50%	40%	35%
Fire 2	30%	30%	30%
Fire 3	20%	20%	20%
Fire 4	–	10%	10%
Fire 5	–	–	5%

Table 2 Safe distance obtained for each fire using the safe distance algorithm

	Fire 1	Fire 2	Fire 3
AreaScaled	0.0146	0.0992	0.0400
MeanScaled	0.3935	0.5503	0.6497
SafeDistance (m)	0.0574	0.5457	0.2599

total was greater than *NumRobot* according to the difference, respectively. Since a large number of robots were already allocated to the robot, adding or subtracting a small number of robots did not have a profound effect in most cases.

- In case that during the above process, the number of robots allocated to Fire 1 became less than the number of robots allocated to Fire 2, the robots of Fire 2 and Fire 1 were swapped after comparison.
- The same procedure was repeated for the other robot teams until the team responsible for extinguishing the last fire was checked. This arrangement ensured that a greater number of robots always extinguished the bigger, more intense fires.

With these arrangements, the robots were allocated into teams of various sizes for fire extinguishment.

4 Results

4.1 *Algorithm to Find the Minimum Distance the Robot Should Maintain from Fire*

The results obtained by applying the algorithm in the sample environment are shown in Table 2, in particular the *areaScaled* parameter, the *meanScaled* parameter as well as the SafeDistance, obtained for each fire in the sample environment in Fig. 1.

Thus, the algorithm can effectively calculate the safe distance the robot should maintain from each fire while extinguishing it based on its size and intensity.

4.2 *Probabilistic Roadmaps*

The total computational time taken to identify the path was the sum of the computational time taken for generating the probabilistic roadmap and the time taken to identify the path between the start and end positions. The average total computational time taken to create a *PRM* using 50 to 200 nodes to find the path between (40, 20) and (180, 40) along with average path length is shown in Table 3. The average values were calculated by taking the mean of ten readings for every number of nodes.

Table 3 Computational time taken to identify path with *PRM* constructed using a different number of nodes

No. of nodes	Computational time (s)	Path length (m)
50	0.6506	176.0539
75	0.6510	159.8162
100	0.8657	158.4208
125	0.9474	155.3137
150	1.1878	154.9023
175	1.4273	153.4678
200	1.8528	150.0453

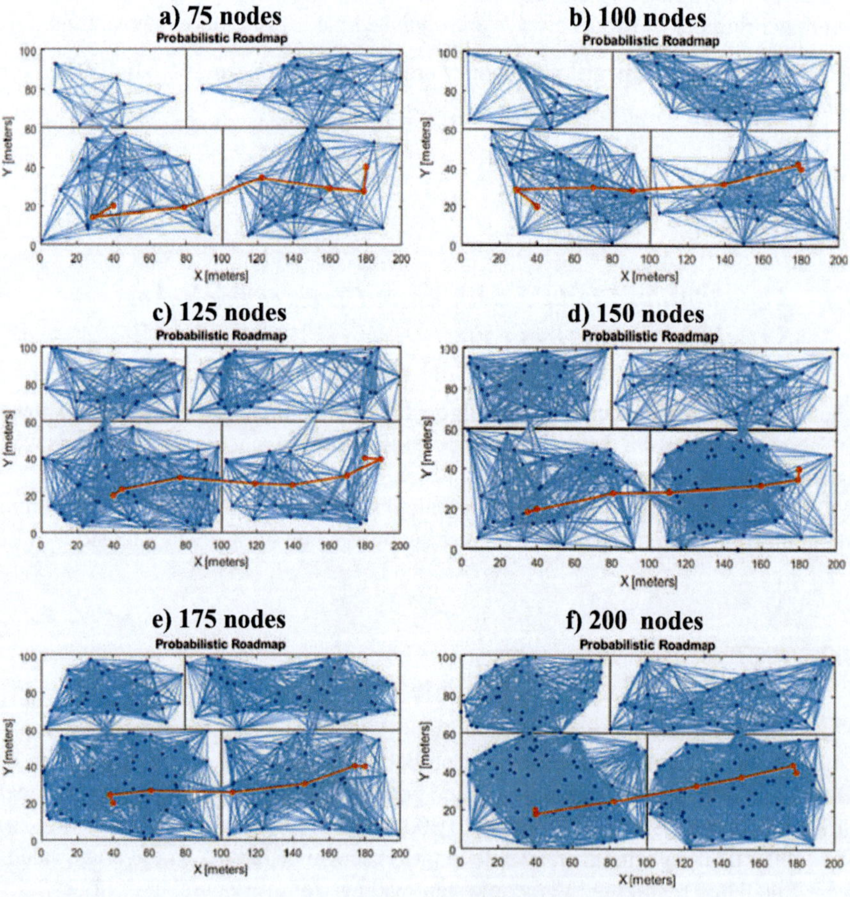

Fig. 5 Path produced by *PRM* constructed using a different number of nodes

Table 4 Computational time taken to identify the required path using *PRM*

Start	End	Average computational time taken (s)	Standard deviation
(120,80)	(120,20)	0.1940	0.0456
(120,20)	(160,10)	0.1843	0.0277
(160,10)	(180,40)	0.2281	0.0463

Figure 5 shows the path obtained for the different number of nodes. It can be observed from Fig. 5 and Table 3 that an increase in the number of nodes improves the quality of the path. The total computational time taken for path identification also increases with the increase in the number of nodes as shown in Table 3. However, when the number of nodes is increased from 175 to 200, even though the identified path using a *PRM* of 200 nodes shows improvement, the length of the path only changes by a small degree. Increasing the number of nodes beyond 175 will merely increase the computational time without giving a path of much greater quality. The difference in paths can be easily covered in the extra computational time taken to create the *PRM* using 200 nodes. Hence, 175 is chosen as the number of nodes to create the probabilistic roadmap. The average computational time required to create a *PRM* map alone with 175 nodes is 1.4174 s.

Suppose a robot is in position (120,80) and the robot has to move to the next room (Room A in Fig. 3) to extinguish the fire by targeting the global positions (120,20), (160,10) and (180,40). Table 4 shows the average computational time taken to identify the required paths from the *PRM* obtained using 175 nodes for the start and end global target positions mentioned. Table 5 shows the average length of the paths obtained. The average value was calculated by taking the mean of ten readings for every path computed.

The average time taken for computing the required path would be 1.4174 s + 0.1940 s + 0.1843 s + 0.2281 s = 2.0238 s. After the path is computed, the robot would have to cover 198.4797 m to reach the target positions and extinguish the fire. When multiple robots are available, the fire can be extinguished faster. Suppose three robots are present at the locations (120,80), (160,90) and (180,70). When the fires are detected at positions (120,20), (160,10) and (180,40), a robot will each go to the fire location directly from their respective locations. The average computational time taken for computing paths along with the average path length for the additional two robots in this scenario is shown in Tables 6 and 7, respectively.

Table 5 Path lengths obtained using *PRM*

Start	End	Average path length (m)	Standard deviation
(120,80)	(120,20)	105.0981	5.6649
(120,20)	(160,10)	48.3851	6.5201
(160,10)	(180,40)	44.9965	3.4195

Table 6 Computational time taken to identify the required path using *PRM* for multiple robots

Start	End	Average computational time taken (s)	Standard deviation
(160,90)	(160,10)	0.2010	0.0478
(180,70)	(180,40)	0.1982	0.0373

Table 7 Path lengths obtained using *PRM* for multiple-robot scenarios

Start	End	Average path length (m)	Standard deviation
(160,90)	(120,10)	93.6014	6.7970
(180,70)	(180,40)	90.7227	9.6513

Although the computational time taken shown in Table 6 is comparable to the computational time taken in Table 4, the path length shown in Table 7 is much greater than the path length shown in Table 5. However, these distances will be covered simultaneously; thus, each robot will cover 105.0981 m, 93.6014 m and 90.7227 m, respectively. Thus, the distance covered by each robot would be much less than the total distance that would have to be covered by one robot, thus greatly reducing the extinguishing time and increasing the efficacy of the fire-extinguishing process.

4.3 *Algorithm to Allocate Robots into Teams*

The results obtained by using the algorithm to allocate the available robots into teams are shown in Tables 8, 9 and 10 for a different number of available robots and a number of fires.

Table 8 Allocation of robots when NumFire = 3

	NumRobots = 3	NumRobots = 23	NumRobots = 45
Fire 1	1	11	22
Fire 2	1	7	14
Fire 3	1	5	9

Table 9 Allocation of robots when NumFire = 4

	NumRobots = 16	NumRobots = 34	NumRobots = 57
Fire 1	6	14	23
Fire 2	5	10	17
Fire 3	3	7	11
Fire 4	2	3	6

Table 10 Allocation of robots when NumFire = 5

	NumRobots = 9	NumRobots = 19	NumRobots = 30
Fire 1	3	6	10
Fire 2	3	6	9
Fire 3	2	4	6
Fire 4	1	2	3
Fire 5	0	1	2

When the number of robots was less than 10, some fires had no robots allocated to it initially. If some fires had no robots allocated initially, after a particular robot team accomplished the task of extinguishing the respective fire, the robots would then extinguish the unextinguished fires. The results demonstrate the ability of the algorithm to effectively allocate robots into teams of different sizes to extinguish the different size and intensity fires.

4.4 Performance Analysis of Proposed Model Against Existing Approaches

The advantages of implementing the proposed model for firefighting are emphasized in Table 11.

Table 11 Performance analysis of proposed model against existing approaches

Methodology followed in existing approach	Advantage of implementing proposed model
The firefighting robot does not calculate the required safe distance to be maintained from the robot. Thus, extinguishing resources such as water is often wasted for self-protection.	The safe distance finding algorithm proposed was able to find the distance that the robot should maintain from the fire based on its size and intensity. This ensures that water is not wasted.
Robots are not autonomous. They are often guided by a person who navigates the robot according to the position of obstacles in the environment observed from the camera.	By implementing path-planning using probabilistic roadmaps, the robots can operate autonomously and find the shortest path. This ensures that the fire will be extinguished as soon as possible even when a person is not available to control the robot.
The firefighting robots are not allocated into teams of different sizes to extinguish fires present at different locations. Thus, teams responsible for extinguishing the larger, more intense fires may have less robots than required.	The team allocation algorithm effectively allocates robots into teams of different sizes to tackle the different fires. This ensures efficient fire extinguishment.

5 Conclusion

Several methods to enhance the efficiency of firefighting robots were implemented, namely the algorithm to find the *safe distance* that the robot must maintain from the fire, team allocation and path-planning using *probabilistic roadmaps*. The algorithm developed to obtain the *safe distance* found the appropriate distance using the size and intensity of the fire. The team allocation used pre-set percentage values and comparison to determine the number of robots in each team. With these arrangements, the algorithms worked effectively.

The *PRM* generated nodes throughout the entire space so that paths could be identified as required. After analyzing the paths produced by *PRM* with a different number of nodes based on computational time and path length, 175 nodes were chosen to produce the *PRM* since increasing nodes beyond 175 increased the computational time without producing a much better path. When a multiple-robot system was used, the fires could be extinguished faster by following the paths identified from the *PRM* as robots could simultaneously target fires in all locations. Thus, the ability of multiple robots to improve the efficiency of fire extinguishing was highlighted.

6 Future Scope

When multiple robots are present, it is necessary to plan the passage of each robot in such a way that collision does not occur. Thus, when the paths of two robots intersect, a certain distance away from the intersection point, one of the robots must stop. Only after the first robot has passed, the second robot must move. The robots closest to the room where there is fire can be instructed to extinguish the larger fires since they would be able to reach the destination faster. In this manner, the efficiency of working can be even further improved.

References

1. Mittal S et al (2018). CeaseFire: the fire fighting robot. In: International conference on advances in computing, communication control and networking (ICACCCN), pp 1143–1146
2. Jia Y-Z et al (2018) Design and research of small crawler fire fighting robot. In: Chinese automation congress (CAC), pp 4120–4123
3. Suresh J (2017) Fire-fighting robot. In: International conference on computational intelligence in data science (ICCIDS), pp 1–4
4. Pranamurti H et al (2019) Fire detection use CCTV with image processing based Raspberry PI. In: International conference on electronics representation and algorithm (ICERA)
5. Shen D et al (2018) Flame detection using deep learning. In: 4th international conference on control, automation and robotics (ICCAR), pp 416–420
6. Jiao Z et al (2019) A deep learning based forest fire detection approach using UAV and YOLOV3. In: 1st international conference on industrial artificial intelligence (IAI), pp 1–5

7. Hao Wu et al (2019) An intelligent fire detection approach through cameras based on computer vision methods. Process Safety and Environmental Protection 127: 245–256
8. Kavraki LE, Svestka P, Latombe J-C, Overmars MH (1996) Probabilistic roadmaps for path planning in high-dimensional configuration spaces. IEEE Transactions on Robotics and Automation 12(4):566–580
9. Hidayatullah AA, Handayani AN, Fuady MJ (2017) Performance analysis of A∗ algorithm to determine shortest path of fire fighting robot. In: International conference on sustainable information engineering and technology (SIET), pp 53–56
10. Kumar N et al (2016) Robot path pursuit using probabilistic roadmap. In: 17th international symposium on computational intelligence and informatics (CINTI), pp 139–144
11. Wang Z, Cai J (2018) Probabilistic roadmap method for path-planning in radioactive environment of nuclear facilities. Progress in Nuclear Energy 109:113–120
12. Mathworks.com. Probabilistic Roadmaps (PRM)- Matlab & Simulink. https://www.mathworks.com/help/robotics/ug/probabilistic-roadmaps-PRM.html
13. Mathworks.com. Create occupancy grid with binary values. https://www.mathworks.com/help/nav/ref/binaryoccupancymap.html
14. Ramasubramanian S, Muthukumaraswamy SA, and Sasikala A (2020) Fire Detection using artificial intelligence for fire-Fighting robots. In: 4th International Conference on Intelligent Computing and Control Systems (ICICCS), pp 180–185

Chapter 12
Smart Meter Using Raspberry Pi for Efficient Energy Utilization

Chidrupa Manogna Mamunooru, Suseendra Attada, Sai Vipanchi Kodali, Mohan Reddy Tumma, Santanu Kumar Dash, and Y. Padma Sai

1 Introduction

The Internet of Things is being used to make everything smart today, smart cities, smart grids, smartphones, smartwatches, smart homes, etc. Collecting and monitoring various data is helping us to improve privacy, provide better access to things around us. Using single controllers, all smart appliances at home can be controlled which makes our lives a lot easier [1, 2]. The Internet of Things is now looked upon to solve large-scale problems running in the society like depletion of non-renewable sources. Collecting information regarding energy usage, clean water usage, gas usage will help us to evaluate the requirement and use these limited resources efficiently [3–6].

The Indian power sector is primarily facing three issues. The first one being power stealing that results in a huge shortage of power. Power stealing, which suggests ineligible consumption of power, is done by changing the state of the meters or sounding the electrical lines. The Indian Government is losing crores of rupees every year due to power theft. The second downside is in multiplied electrical bills, as a result of abnormal operation of appliances consuming a high quantity of power than its rated limit. The last downside is poor power management. The domestic power usage has increased drastically. This leads to an inevitable power shortage in the long run for the generations to come because the resources are decreasing exponentially by the day.

C. M. Mamunooru (✉) · S. Attada · S. V. Kodali · M. R. Tumma · S. K. Dash · Y. Padma Sai
Electronics and Communication Engineering, VNR Vignana Jyothi Institute of Engineering and Technology, Hyderabad, India
e-mail: mchidrupa01@gmail.com

S. Attada
e-mail: asuseendra26@gmail.com

I. Jeena Jacob et al. (eds.), *Data Intelligence and Cognitive Informatics*, Algorithms for Intelligent Systems,
https://doi.org/10.1007/978-981-15-8530-2_12

The electricity meters can be converted to smart meters to supply the user with information about their energy consumption to employ energy with efficiency and ultimately save energy. Smart meters will solve most of the preceding issues by the endless observance of the consumption. Various papers have dealt with these smart meters in various ways, with different components. While some have used wireless fidelity [7]; others have used GSM [8, 9], Zigbee [10, 11], Raspberry Pi, Node MCU, Arduino, etc. Various controllers have been used for different requirements. GSM, Raspberry pi has been decided to use for our application. Further, few researchers have also used big data to analyze the smart meter readings [12]. Research has been on building smart meters using embedded systems [10, 11, 13, 14]. There also has been integral research on smart grid technology in improving the power quality in the smart grid technology [15–17] (Fig. 1).

The present project intends to unravel these three problems associated with power larceny in domestic lines, high electricity consumption, and poor power management. A smart meter using Raspberry pi based on IoT is proposed which is connected to the electricity meter. A threshold value which is the total of the average power of all the appliances that is stored in the smart meter. The meter constantly monitors the power used, and if it is greater than the threshold value, then further action is taken. This will detect if there is any appliance which uses more power than rated value, and the house owner will be notified of this malfunction. If every appliance is consuming power within the required range, then this implies that there is power theft taking place. The server at the power station is notified of this abnormal power consumption.

The third problem deals with power management. The power rate is high during peak hours; the users will be notified during this time so that they decrease their consumption which will save both power and money. This network of electricity meter Raspberry pi and server will help us ensure transparency, where both the

Fig. 1 Power theft example

government and the user can be updated simultaneously. This project will help to save power and curb illegal practices to a large extent. Using this smart meter concept, the usage of power for domestic purposes can be optimized.

2 Conventional System

Nowadays, the device being employed includes a daily electricity meter that measures the voltage in volts and current in amperes. The measured power is then calculated into units with admire to joules or kilowatt-hours. An LCD or an LED show is installed in the meters to show the ate up strength reading.

Conventional energy meters do not provide any further information that can be understood by a common man; they do not update the user about their usage. There is no detection of electricity theft and unnecessary delay in the reception of power bills. This leads to losses in power as well as money from each of the suppliers and the consumers.

This conventional system will become obsolete within no time, as automation taking over the world currently. An energy meter that keeps up with the automation will be beneficial since smart grids are being adopted everywhere.

3 Proposed System

Our project system comprises of Raspberry pi, a relay module, and a GSM-based module (Fig. 2).

The system is used for constantly monitoring the power intake of the household. Any surprising and large difference in the intake of strength or deviation from the normal consumption traits suggests that there is some kind of overconsumption taking place. The user is immediately notified via a message followed by a power cut.

To implement this system, a Raspberry pi is used. In this proposed device, Raspberry pi 3 has been used. It is a Broadcom with 1.2 GHz and 64-bit CPU with 1 GB capacity RAM, and the UART module is used to enable an interface a GSM modem and energy meter. A smart energy meter works on the verbal exchange immediately with wireless records protocol, so there can be precise studying, and there may be no vital for a meter reader to manually take electricity meter studying in client premises. Smart energy meters can function in divergent methods with GSM module.

After the Raspberry pi and GSM modem is switched ON, the activated relay connects the meter to load. It reads and displays the present-day data. Raspberry pi tests the readings provided by the voltage and current sensors. If any distinction is observed between the linked sensors, the relay is turned OFF, thereby sending an SMS to the carrier provider. Also, this Raspberry pi facilitates the utility for electricity disconnection while the invoice isn't always cleared with the aid of the purchaser.

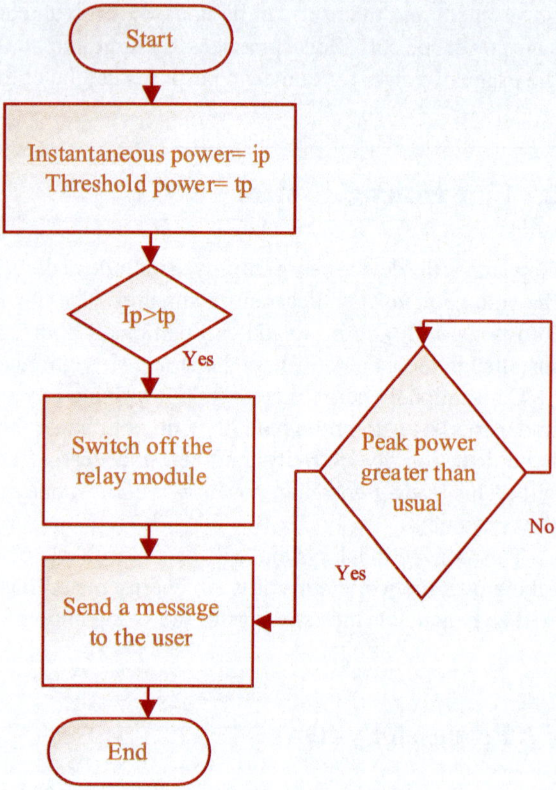

Fig. 2 Workflow of the proposed system

From the purchaser point that is again to monitoring their daily/month-to-month consumption, voltage fluctuation, and disconnection from the supply.

Every appliance will hold a power-rated value. If it is consuming beyond the rated value, the power consumption is higher, the user receives a message, and the power supply to those appliances is ceased until it is repaired. Overconsumption of electricity will gradually decrease the efficiency of the appliance until it completely stops working. Hence, restricting the power usage to a threshold value is a great way of sorting out which appliance is depicting abnormal behavior.

Notifying the user of the price per unit power variation will enable him/her to use power efficiently and restrict their usage.

4 System Architecture

This project comprises of a system which incorporates Raspberry pi 3 used as a development board. SIM 900A, a GSM module that is employed for communication with the user and a relay module as a switch to regulate the current flow to the appliances.

The entire procedure is defined by an if-else condition of whether the current usage is greater or lesser than the predefined threshold value. This threshold value is defined by the sum of the average power consumption of all the appliances in the household, which defines the maximum possible consumption within a household. This value can be updated with new additions in the list of household appliances. The threshold value is defined in joules or kilowatts.

The usual energy meter calculates the instantaneous voltage and current consumption.

Calculation of power using the formula:

$$P = V * I$$

Here,

P = Power consumed (joules or kilowatts).

V = Instantaneous (voltage volts or kilovolts).

I = Instantaneous current (amperes) (Fig. 3).

The calculated power indicates the instantaneous power. This value is constantly compared with the threshold value. It checks for the instant when the instantaneous power crosses the threshold power.

The instant at which the consumed power is greater than the threshold power; the relay unit is switched off, and the power supply is cut down. The controller is

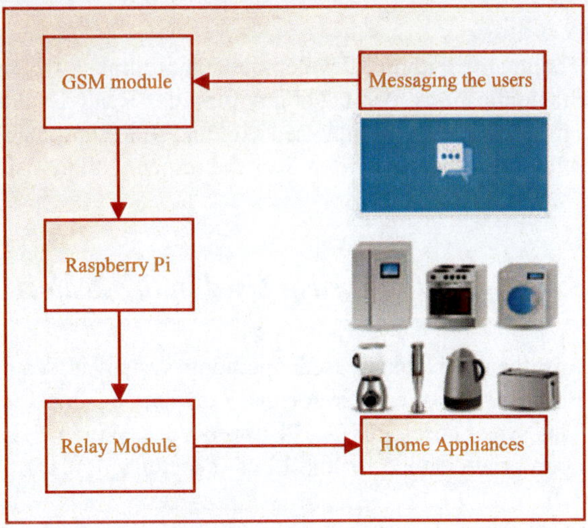

Fig. 3 System architecture of the project

programmed such that whenever the cross over happens, the GSM module sends a message to the user whose number is stored in the controller.

5 Software Implementation

5.1 Programming Languages in Raspberry

Raspberry Pi contains a distinctive feature of adapting to a variety of programming languages. Python is supported by the helpful resource of the approach of the Raspberry Pi basis, above all for the beginners. There are varied languages together with C, Java, Scratch, C++, and Ruby preset up in Raspberry Pi.

5.2 Programming Using Python

Python, a powerful, efficient programming tool, which is user-friendly. Interfacing Raspberry Pi with the Python connects a project with the real world. Python has a well distinguished and abbreviated syntax, because of the usage of standard English words increases readability. Python is designed to be highly readable. Unlike other programming languages using punctuation, Python uses English keywords. Some of the key features of Python include interpretation and interaction ability, portability, extensibility, broad standard library, etc.

5.3 Thonny

Thonny is a Python 3 development environment, which allows access to a desktop or the applications menu. Thonny provides Read–Evaluate–Print Loop (REPL). This novel feature is a prompt used to enter Python commands. Through REPL, the output of commands gets written over the screen without using print.

5.4 Operating System Used in the Project

Raspbian, a Debian-based operating system is designed for operating Raspberry Pi. Raspbian is suitable for the Raspberry Pi line's low-performance ARM CPUs. This system uses PIXEL, Pi Improved X Window Environment. Debian makes an appropriate choice for the Pi as it includes tools for browsing, GUI desktop, and programming. This environment is known as the "Lightweight X11 Desktop Environment" (LXDE) because of an interactive user interface that is built using the X Window System software.

Raspian OS is available for free to download and use. It is based on Debian Linux and enables to work efficiently with the Raspberry Pi computer. Pi is software with basic programs and utilities equipped to run on the hardware specified.

5.5 *Interfacing Raspberry Pi with Gsm Modem*

Dial-up modem, a wired network, is a common device in the landline telephone network. GSM modem is added to rectify the primary issue of the dial-up modem mainly based on its reputation for the usage of a SIM card. GSM requires a subscription to a mobile operator to operate; GSM is almost similar to a mobile communication system. As a high-end embedded device, Raspberry pi could be interfaced with a GSM modem, which in turn will result in the evolution of an intelligent, smart system.

Firstly, a SIM card is inserted in a slot provided by the GSM module. Connections should be made as shown inside the figure. Note that, the SIM used in the module should have a balance sufficient to send SMS and an appropriate range of a network. The transmission pins and reception pins are supposed to be connected in reverse order, and the ground pins should be shorted. For SIM initialization, wait for a few seconds after activating the GSM with the necessary power range.

Now, the working condition of the GSM can be verified by sending and receiving messages through a Python script. A script is basically a program uniquely designed by a user (Figs. 4 and 5).

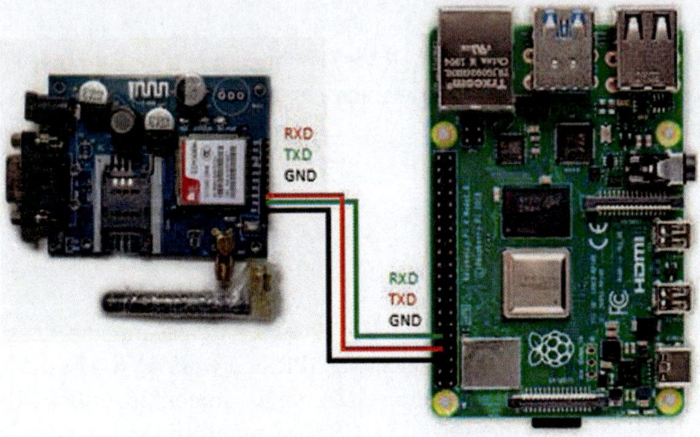

Fig. 4 Interfacing GSM module with Raspberry Pi

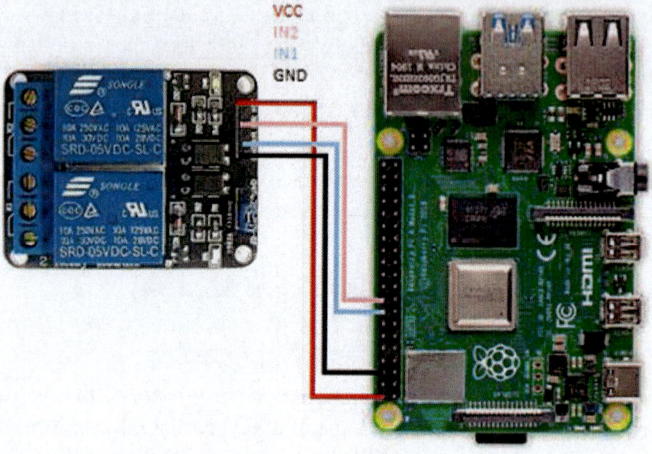

Fig. 5 Interfacing relay module with Raspberry Pi

6 Hardware Implementation

The basic hardware components used in the project are listed below:

A. Raspberry pi 3
B. GSM Modem
C. Relay control unit.

6.1 *Processor Used in the project — Raspberry Pi*

The Raspberry Pi is a computer based on a single controlled by a more recent version of Debian Linux optimized for the ARMv7 architecture. The Ethernet is a 10/100 Base T Ethernet socket. The CPU of the microcontroller allows a 700 MHz low-power ARMv7 applications processor.

6.2 *GSM Modem*

A GSM, also known as GPRS modem, is a class of wireless modem. It is primarily designed for communication over the GSM, GPRS network. A SIM card is required for communication. It has an International Mobile Equipment Identity (IMEI) unique number for the purpose of identification. GSM communication has been effectively used in several smart grid applications [8, 9].

The SIM900A modem has a SIM900A GSM chip. Which also has an RS232 interface; this enables easy to connect with computer through USB to serial connector.

SIM900A — GSM/GPRS module is designed purposefully for a global market. This dual-band GSM/GPRS module works on frequencies of 900 MHz for GSM network and 1800 MHz frequency for GPRS network in providing efficient performances of SMS, voice, and data.

6.3 RS-232 Communication (UART)

Communication with RS-232 allows point-to-point transmission of statistics. It is usually used to transfer data between the microcontroller and a PC in information retrieval applications.

Step 1. Insert a SIM card in the slot provided by the module.

Step 2. To connect an antenna, screw the RF antenna on the provided RF cable output.

Step 3. Connect the RS232 cable to the host device.

Step 4. Ensure no hardware handshaking and provide a power supply of 9 V AC/DC.

Step 5. In the case of DC, polarity should be center + ve and outer –ve on DC jack.

Step 6. A network LED indicates the status of GSM, which includes power activation, network registration, and GPRS connectivity.

Step 7. As soon as the modem registers the network, LED will blink in steps of 3 s. Now, as the modem has established the network connection, you can start using it for your application.

6.4 Relay Module

Relay, an electromagnetic switch is used to control a high voltage supply. The relay module used here has two channels. There are models with one, four, and eight channels. The module ought to be supercharged with 5 V, which is acceptable to use with a Raspberry pi.

7 Results

See Fig. 6.

This system has been tested with a series of appliances like bulbs and have given the threshold value as the total sum of the rated power values of all the bulbs. These series of bulbs were switched on; no difference was observed. To check the overconsumption case, a fan has been added to the series and switched it on, to see that the

Fig. 6 Hardware implementation of the project

power consumption was cut off, and a text message was sent to the registered user mobile number explaining that the consumed power was greater than the registered threshold value.

8 Conclusion and Future Scope

The novelty of our work lies in enhancing consumer convenience and rationalizing electricity consumption. The adoption of smart meters can cause huge total annual savings to the government and therefore also the buyer savings. The projected project will solve three issues associated with power felony in domestic lines, high electricity consumption, and poor power management.

The usage of smart meters will increase awareness relating to the importance of efficient and economical domestic power usage. The house owners can keep a track of their power consumption and thus take necessary measures to decrease, if necessary. This increased transparency can facilitate the government to take action, if there is any power felony going down. This smart meter will help us in following the malfunctions at the bottom level. Further, including machine learning models in these smart meters to forecast the demand will be greatly helpful in regulating power usage.

References

1. Babu VS et al (2016) An intelligent controller for smart home. In: International conference on advances in computing, communications and informatics (ICACCI), IEEE, September 2016
2. Langhammer N, Kays R (2012) Performance evaluation of wireless home automation networks in indoor scenarios. IEEE Trans Smart Grid 3:2252–3226
3. An IOT-Based Smart Utility Meter. In: 2018 2nd International conference on smart grid and smart cities. August 2018. https://doi.org/10.1109/ICSGSC.2018.854131
4. Suresh M, Muthukumar U, Chandapillai J (2017) A novel smart water-meter based on IoT and smartphone app for city distribution management. In: IEEE international symposium on technologies for smart cities. IEEE Region 10, India, pp 1–5
5. Mihajlović Ž, Milankov A, ivković L, Tolić M (2016) Implementation of wireless M-bus concentrator/gateway for remote reading of smart gas meters. In: 24th telecommunications forum (IEEE), pp 1–4
6. Dong S, Duan S, Yang Q, Zhang J, Li G. Tao R (2017) MEMS-based smart gas metering for Internet of things. IEEE Internet Things, J 4(5):1296–1303
7. Hlaing W, Thepphaeng S, Nontaboot V, Tangsunantham N, Sangsuwan T, Pira C (2017) Implementation of WiFi-based single phase smart meter for Internet of things (IoT). 2017 International electrical engineering congress (iEECON), Mar 8. IEEE, pp 1–4
8. Das H, Saikia LC (2015) GSM enabled smart energy meter and automation of home appliances. In: International conference on energy, power and environment: towards sustainable growth (ICEPE), IEEE
9. Mohammad N, Barua A, Arafat MA A smart prepaid energy metering system to control electricity theft. Department of Electrical and Electronic Engineering Bangladesh University of Engineering and Technology Dhaka, Bangladesh
10. Pereira R, Figueiredo J, Melicio R, Mendes VMF, Martins J, Quadrado JC (2015) Consumer energy management system with integration of smart meters. Energy Rep 1:22–29
11. Ranjith T, Sivraj P., Futuristic smart energy meter—design based on embedded perspective. Dept. of Electrical and Electronics Engineering Amrita School of Engineering, Coimbatore Amrita Vishwa Vidyapeetham, India. In: Proceedings of the second international conference on intelligent computing and control systems (ICICCS 2018) IEEE Xplore Compliant Part Number: CFP18K74-ART; ISBN:978-1-5386-2842-3.
12. Yao H-W, Wang X-W, Wu L, Jiang D, Luo T, Liang D (2018) Prediction method for smart meter life Based on Big Data. Procedia Eng 211:1111–1114
13. Maitra S (2008) Embedded energy meter—a new concept to measure the energy consumed by a consumer and to pay the bill. In: Joint international conference on power system technology and IEEE power India conference. pp 1–8
14. Dash SK, Panda G, Ray PK, Pujari SS (2016) Realization of active power filter based on indirect current control algorithm using Xilinx system generator for harmonic elimination. ELSEVIER Int J Electr Power Energy Syst 74:420–428
15. Dash SK, Ray PK (2017) Platform specific FPGA based Hybrid active power filter for power quality improvement. Int J Emerg Electr Power Syst 18(1):301–310. ISSN (Online) 1553-779X, ISSN: 2194-5756. https://doi.org/10.1515/ijeeps-2016-0143
16. Dash SK, Ray PK Novel PV tied UPQC topology based on a new Model Reference control scheme and Integral plus Sliding Mode dc-link controller. Wiley Int Trans Electr Energy Syst 28(7). https://doi.org/10.1002/etep.2564.
17. Dash SK, Ray PK Photovoltaic tied unified power quality conditioner topology based on a novel notch filter utilized control algorithm for power quality improvement. SAGE Trans Inst Measurement Control. https://doi.org/10.1177/0142331218790314

Chapter 13
An Unsupervised Content-Based Article Recommendation System Using Natural Language Processing

S. Renuka, G. S. S. Raj Kiran, and Palakodeti Rohit

1 Introduction

In today's world, where data is getting generated in amounts that are almost impossible to quantify, the text is one of the major sources of that data. Previously text was only available in the form of scripts, books, and newspapers, etc., but now with Internet revolution and its ease of access increased the range of sources for text generation immensely. Starting from social media apps, Web pages to search engine queries, digitalized e-books, and research paper platforms are major sources of text. Using this data, intelligent systems can be built with the help of statistical and learning algorithms. One such application where these intelligent systems can be employed is recommendation systems. The recommendation system is those which understand user's preferences and recommends relevant items which are of interest to the user. Text recommendation systems can be modeled in two different approaches. One being collaborative filtering and other being content-based filtering. In collaborative filtering, the model tries to identify users with similar tastes and preferences. It will recommend items to the current user which have been preferred by other users who have similar taste as the current user. In content-based filtering, the model tries to find similarity patterns between various text documents (articles) in the corpus and recommends the documents which are similar to the one which current user finds interesting.

Content-based filtering approach has both its advantages and disadvantages. The advantages are that it does not require user-specific data related to other users using

S. Renuka · G. S. S. Raj Kiran (✉) · P. Rohit
Department of Information Technology, GRIET, Hyderabad, Telangana, India
e-mail: gssrajkiran@gmail.com

S. Renuka
e-mail: renuka.csit@gmail.com

I. Jeena Jacob et al. (eds.), *Data Intelligence and Cognitive Informatics*,
Algorithms for Intelligent Systems,
https://doi.org/10.1007/978-981-15-8530-2_13

the system. Also, user-specific suggestions can be made as the data will be worked on which is very much specific to the user. The disadvantage of content-based filtering approach is that as data specific to a single user will be worked on, one needs domain-specific knowledge to related to the content present in the dataset for making meaningful suggestions to the user.

This paper discusses the second method that is content-based filtering approach to model an article recommendation system. The dataset consists of two hundred and thirty Web articles based on machine learning and its concepts. This was achieved with the implementation of unsupervised methods such as K-means clustering, agglomerative clustering, and cosine similarity (similarity measuring metric). The methodology of this paper discusses how these clustering algorithms and cosine similarity metric are put to work to make recommendations to the current user. The results and discussions area discusses the performances of various algorithms and their pros and cons.

2 Background Study

Kamath et al. [1] developed the e-news recommendation system using natural language processing. By an advanced search engine to find out the similarity between news articles by opting semantics-based bag-of-words approach. Observed results showed an improvement over the keywords-based- bag-of-words model.

Haruna et al. [2] proposed a collaborative filtering-based model to recommend research papers by implementing a collaborative filtering approach to give personalized recommendations without the data of user's expertise and area of research.

Gipp et al. [3] developed a hybrid research paper recommendation system and a robust replacement of existing academic-based search engines. This was achieved by using distance similarity index, in-text impact factor, and various other methods.

Aalharitha [4] proposed a content-based recommendation system that predicts a user's future interests by extracting elements from book texts based on learnings of the author's style of writing.

3 Methodology

The paper has gone through the following methodology (Fig. 1):

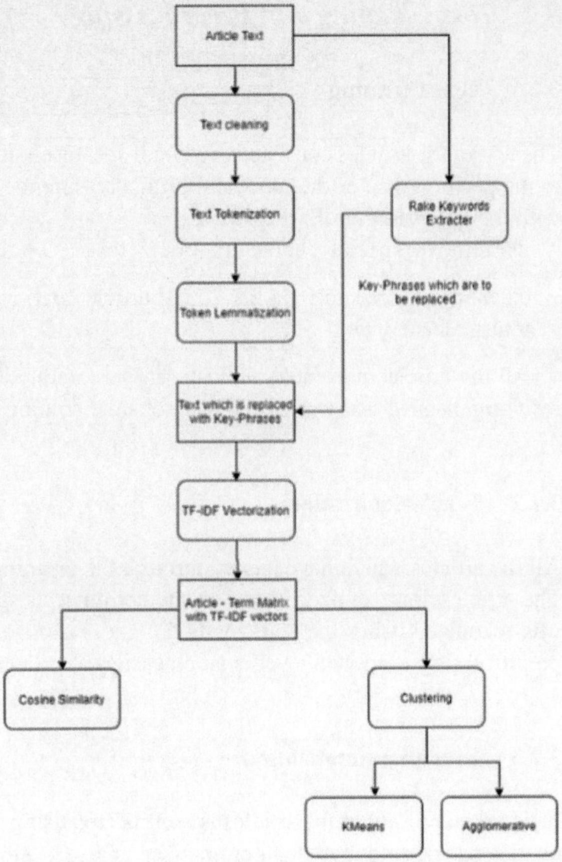

Fig. 1 Functional block diagram of model

3.1 Data Collection

The dataset consists of features such as the name of the article's author, claps (number of claps awarded by readers), reading time, hyperlink to the article, the title of the article, and text of the article. This dataset was acquired from the Kaggle Web site posted by Hsanaesara [5]. The Web articles present in the dataset are related to different subject areas of the fields machine learning and data science.

3.2 Text Cleaning and Preprocessing

3.2.1 Text Cleaning

The text of the articles consisted of several special characters which do not contribute to the performance of the models. For this reason, they were removed from dataset with the aid of RegularExpressions.

```
def remove_special_characters(article_text):

cleaned_text=re.sub(r'[^a-zA-Z]',r'',article_text)
return cleaned_text
```

All the articles present in the dataset were iterated through this method and the resulting cleaned text was stored as a separate column namely 'cleaned_text'.

3.2.2 Text Tokenization

All the articles which have passed through the cleaning process are now tokenized. The articles are tokenized based on the delimiter as space. The cleaned article text was tokenized using the split method. The reason for tokenization is performing preprocessing at a token level is much easier.

3.2.3 Token Lemmatization

Each token present in the articles is lemmatized using WordNetLemmatizer present in nltk package in Python. Lemmatizer analyzes all words morphologically and removes only the inflectional portion of the words [6].

3.3 Keyword Generation

Rapid automatic keyword extraction (RAKE) is employed for the purpose of extracting keywords from each article.

RAKE extracts the keywords by constructing a word co-occurrence matrix (Co-Mat). A co-occurrence matrix is a matrix which consists of rows and columns as words which are present in the text corpus.

Let us say there are N words in the corpus and the co-occurrence frequency of *i*th word and *j*th word in the corpus where $i < N$ and $j <$ N can be found in the Co-Mat$_{ij}$ cell of the matrix.

It later assigns a score to each word present in the corpus using the following formula,

$$\text{Word}_{\text{score}} = \frac{\text{Word co-occurrence frequency in corpus}}{\text{Word frequency in corpus}}$$

The RAKE also identifies key-phrases in the corpus. A phrase is a combination of words also called *n*-gram. When two or more individual keywords or key-phrases which are co-occurring together two or more than two times in the corpus it identifies them as a key-phrase and computes the score of that key-phrase using the similar principle mentioned in the above formula.

A word or a phrase is selected as a keyword or key-phrase if its corresponding score is present in the top T scores for words in the corpus.

The input to RAKE was the cleaned_text column from a dataset. Later the formed bigrams and trigrams would replace the words forming them in the original article text.

```
def extract_rake_keywords(df):

r = Rake()
r.extract_keywords_from_text(df)
return list(r.get_ranked_phrases())

df['rake_keywords'] = df['cleaned_text'].apply(extract_rake_keywords)
```

3.3.1 Modeling Bigrams and Trigrams

The intension behind the extraction of keywords from articles is to be able to formulate bigrams and trigrams on the keywords generated per article. The key-phrases returned by the RAKE are the ones which have high co-occurrence frequency in the article corpus. A key-phrase is a combination of words which make more sense together as *n*-grams rather than individually. *n*-grams add more context and meaning to text rather than unigrams. Therefore, the obtained *n*-grams formed on top of the key-phrases generated by RAKE are replaced as their *n*-gram form in the actual article text in place of separate words or phrases.

3.3.2 Stopwords Removal

A custom stopwords list was prepared which consisted of all the vanilla stopwords provided by nltk.words.stopwords('english') along with other common English words. All the bigrams and trigrams which consisted of stopwords were not considered to replace the words present in the original article text.

3.4 TF-IDF Word Vectorization

Term frequency–inverse document frequency (TF-IDF) vectorizer is used to generate word vectors. The article text which is cleaned and replaced with bigrams and trigrams, whenever frequently co-occurring keywords appear together in the input given to the vectorizer's fit_transform() method. The fit_transform() method generates word vectors and returns the TF-IDF weighted document-term matrix, where each row is article and column a word belonging to the corpus. The corresponding value present in the *i*th row and *j*th column is indicative of the TF-IDF value of the jth word of the corpus in the *i*th article. This document-term matrix is the main data source for the recommending algorithms as now articles are represented in their word vector forms [7].

$$W_{i,j} = \mathrm{tf}_{i,j} \times \log\left(\frac{N}{\mathrm{df}_i}\right)$$

where $W_{i,j}$ is the weighted TF-IDF value which is computed. $tf_{i,j}$ is the frequency of term i article j. N is the total number of articles and df_i is the number of documents containing term i.

3.5 Dimensionality Reduction Using TruncatedSVD

TruncatedSVD is a dimensionality reduction method which performs linear dimensionality reduction with a main fundamental principle of truncated singular value decomposition. It is also known as latent semantic analysis. The difference between principle component analysis and latent semantic analysis is that in latent semantic analysis the data is not centered before dimensionality reduction is applied. This feature helps it work more efficiently with sparse data [8]. For selecting the ideal number of components to which the actual data is to be reduced, the main factor to keep in mind is how much amount of variance is explained by the newly reduced data in comparison with the actual data as the objective is to reduce the features and make each feature more independent from each other. For this purpose, a cumulative scree plot between the number of components and explained variance was plotted [9] (Fig. 2).

As it is visible by reducing the actual data to 200 components 90% of the variance of the actual data is retained. As a result, the number of components is now reduced to 200 from which were previously 55,335. This reduced data is employed in training the algorithms used for recommendation.

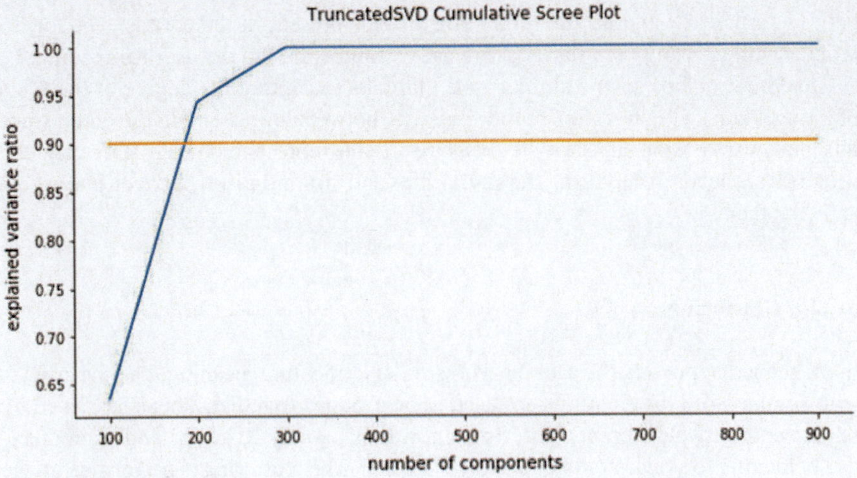

Fig. 2 Truncated SVD cumulative scree plot

3.6 Approaches

Two approaches were modeled in this paper which is based on the principles of content-based recommendation system. The two approaches employed here are:

3.6.1 Cosine Similarity

Similarity measuring metrics are the statistical methods by means of which the degree of closeness or similarity can be measured among various objects present in the *n*-dimensional space. Cosine similarity is one of such similarity measuring metrics used widely to measure similarity among various objects [10].

$$\text{Similarity}(A, B) = \frac{A \cdot B}{\|A\| \times \|B\|} = \frac{\sum_{i=1}^{n} A_i \times B_i}{\sqrt{\sum_{i=1}^{n} A_i^2} \times \sqrt{\sum_{i=1}^{n} B_i^2}}$$

Here A and B are the word vectors present in the TF-IDF matrix where vector A is the vectorized representation of article A and B is the vectorized representation of article B. A dot product is computed in between them, and the resulting value is normalized by dividing the value with the sum of squares of the individual vectors.

As now the articles are represented by the matrix whose dimensions are reduced to 200×230 consisting of TF-IDF vectors of words present in that document one article can be compared with another. To recommend the articles which might be of interest to the reading user, one should first know what kind of articles a user reads. Therefore, say if a user reads an article X_1 pairwise cosine similarity is computed between article X_1 and all the remaining articles present in the corpus. The top 15 articles with very high similarity score are recommended to the user for reading.

One drawback of cosine similarity is that fails to capture the degree of the scale of rating when trying to estimate the similarity between one user and the other, since this paper uses cosine similarity to measure articles represented in TF-IDF weighted vectors it succeeds in capturing the similarities and dissimilarities between one article and the other.

3.6.2 Clustering

In the second approach, the articles with similar contextual meaning, i.e., the articles with similar word distributions are tried to be grouped together. This is achieved by implementing two different clustering algorithms namely K-means and agglomerative belonging to sklearn.cluster package [11, 12]. The clustering is performed on the reduced version of the TF-IDF document-term matrix. Articles discussing similar topics or aspects consist of similar words with identical distribution patterns, as a result, they will be located closely in the *n*-dimensional space. This creates a cluster like structures in the space.

One of the major drawbacks of clustering when performed on very high-dimensional data. Reason for this the values of the features present in the dataset can range from very low values such as 10^{-3} to 10^{3}, this situation is very commonly found in text datasets. This can have an effect while applying to a cluster. For this reason, being TF-IDF vectorization was performed which in the end normalizes the values of the samples.

K-means Clustering

K-means algorithm is a clustering algorithm which computes K centroids from the given data points. All the data points are assigned cluster based on their distance from cluster centroids. Cluster centroids are updated with mean of all points present in respective clusters. This is repeated iteratively. Elbow method is widely used to compute the ideal value K for a given sample of data into which the clusters are to be differentiated. Elbow plot is plotted between inertia which is the sum squared sum of intra-cluster distances and a range of K values over which the K-means is trained iteratively [13]. In the plot, inertia (intra-cluster distances) is on the y-axis and corresponding range of k values on the x-axis. The value K is to be chosen up which the variation in the inertia is linear or almost linear in nature. The threshold for the variation of slope of the line was chosen to be 0.001 that is 0.1% of the variation of inertia from the previous iteration. In the scenario of this paper on performing elbow method 23 value was chosen as the optimal K value.

KMeans(algorithm='auto', copy_x=True, init='k-means++', max_iter=300, n_clusters=23, n_init=10, n_jobs=None, precompute_distances='auto', random_state=1, tol=0.0001, verbose=0)

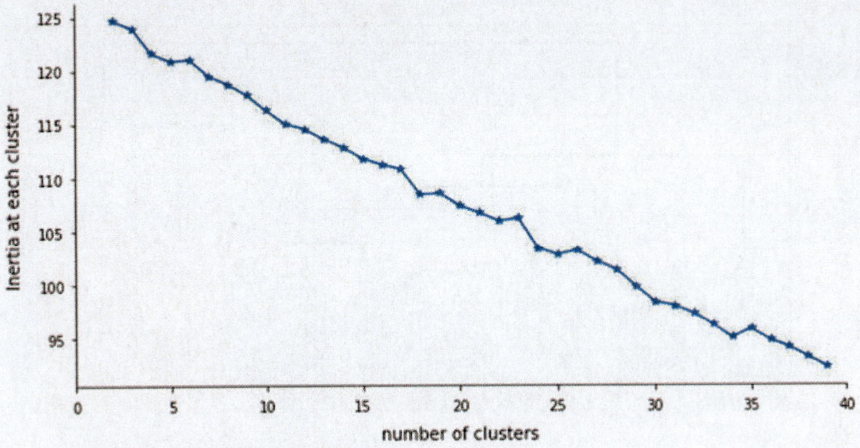

Fig. 3 Elbow plot for intra-cluster distances

Instance of the trained KMeans algorithm (Fig. 3).

K-means clustering trained with the training set. Post-fitting the algorithm, whenever user reads a new article the model computes the Euclidean distance from the cluster centroids $X_c = \{X_0, X_1, \ldots.X_\mathrm{k}\}$ of the trained K-means algorithm. Say the article is closest to the cluster centroid Xn, the model next computes pairwise Euclidean distance between all the articles belonging to that particular cluster and the article read by the user. Therefore, all the articles are sorted in ascending order of distance of separation. The fifteen closest articles belonging to cluster n are recommended to the user.

K-means clustering algorithm has a couple of cons. One of them is that K-means struggles during clustering high-dimensional data as the high-dimensional data is separated by large distances. As text data are being worked with which is very high in dimensions and is also sparse in nature this would affect the performance of the K-means clustering algorithm. To overcome this, the dimensions of the dataset are reduced using TruncatedSVD.

Agglomerative Clustering

Agglomerative clustering is a bottom-up-based clustering algorithm in which initially every single data point is assumed to be a cluster and later iteratively similar clusters are merged. It is a hierarchical-based clustering algorithm [12]. It constructs a hierarchical tree structure which represents the hierarchy of topics present in the text corpus. In the tree, every node is given an ID and all the nodes with ID's less than the number of training samples are the actual articles and are leaf nodes, whereas all the nodes with an ID greater than a number of training samples are non-leaf nodes and have ID—number of training samples number of child nodes. Dendrograms

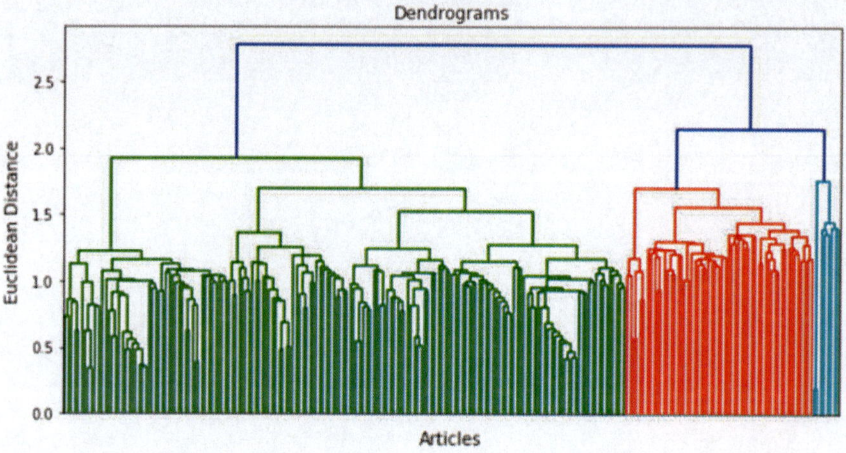

Fig. 4 Dendrogram agglomerative clustering

are used for computing n_clusters for agglomerative clustering. A horizontal line is drawn through the dendrogram in such a manner that it intersects the largest vertical portion of the line in the plot where that section does not intersect with any cluster splitting. N_clusteres is set to the number of vertical lines that this horizontal line intersects [14] (Fig. 4).

AgglomerativeClustering(affinity='euclidean', compute_full_tree='auto', connectivity=None, distance_threshold=None, linkage='ward',memory=None,n_clusters=2, pooling_func='deprecated')

As it is visible from the above plot that when a line is drawn from the largest section of the non-splitting region it intersects with two long lines. So as a result, n_clusters is set to two.

Once the training of the model is complete, the following algorithm is implemented.

Algorithm ()

Step 1: Parse the hierarchical tree and locate the article read by the user in the tree
Step 2: If the node ID of the article's sibling is less than the number of training examples recommend the sibling else goto step 3
Step 3: Backtrack to one level higher of the article and check if any actual articles are if any nodes have ID's less than number of training samples. If there are any then recommend those articles else repeat step 2 and step 3.

4 Results and Discussion

The performance of the models was evaluated based on their recommending abilities in two use case scenarios where Scenario I is when a user reads an article on a generic topic and Scenario II is a more specific topic.

Scenario I

In this case, the article read by the user is 'Machine Learning: how to go from Zero to Hero' from freeCodeCamp. This article discusses about how artificial intelligence (AI) has taken leaps and now being applied in various fields. It also discusses some interesting facts about how AI trained models are replacing humans and defeating humans in various domains such as games like GO. As the title hints, it discusses various open-source resources available on the internet to learn AI.

Scenario II

In this case, the article read by the user is titled 'Reinforcement Learning from scratch' from Insight Data. This article discusses the applications and areas where reinforcement learning is applied such as building games and creating AI apps. It discusses how neural networks help in constructing a network of reinforcement learning model.

4.1 Results of Cosine Similarity-Based Model

4.1.1 Results for Scenario I

The above articles mentioned in the Table 1 are the top fifteen recommendations suggested by the cosine similarity-based model. All these articles discuss very much similar aspects discussed in our base generic article. They also cover aspects on how to get started with machine learning for beginners who are willing to explore the field of data science. The articles such as Machine Learning is Fun!, its series and how to code neural nets are the evidence.

4.1.2 Results for Scenario II

The results of the Scenario II speak for themselves. It is quite clear that the cosine similarity did a great job by extracting articles which are very coherent to the base article 'Reinforcement Learning from scratch' of Scenario II. The recommendations consist of articles such 'Write an AI to win at Pong from scratch with Reinforcement Learning', 'Explained Simply: How an AI program mastered the ancient game of Go' and 'Using Deep Q-Learning in FIFA 18 to perfect the art of free-kicks' describe how reinforcement is applied in designing AI models which are capable of playing

Table 1 Top fifteen recommendations of cosine similarity Scenario I

S. no.	Title	Similarity
1	What I learned from interviewing at multiple AI companies and start-ups	0.728
2	Explained Simply: How an AI program mastered the ancient game of Go	0.727
3	Machine Learning is Fun!—Adam Geitgey—Medium	0.690
4	Deep Learning Achievements Over the Past Year—Stats and Bots	0.650
5	What worries me about AI—François Chollet—Medium	0.646
6	In defense of skepticism about deep learning—Gary Marcus—Medium	0.644
7	Machine Learning is Fun Part 5: Language Translation with Deep Learning and the Magic of Sequences	0.641
8	Deep Learning the Stock Market—Tal Perry—Medium	0.630
9	Machine Learning for Humans, Part 2.1: Supervised Learning	0.625
10	Learning How To Code Neural Networks—Learning New Stuff—Medium	0.624
11	Write an AI to win at Pong from scratch with Reinforcement Learning	0.621
12	Machine Learning is Fun! Part 3: Deep Learning and Convolutional Neural Networks	0.617
13	The future of work—Oxford University—Medium	0.617
14	Ultimate Guide to Leveraging NLP & Machine Learning for your Chatbot	0.616
15	Rohan & Lenny #1: Neural Networks & The Backpropagation Algorithm, Explained	0.614

games played by humans. Articles such as 'Simple Reinforcement Learning with Tensorflow' and its series describe how to code reinforcement learning models in Tensorflow environment (Table 2).

4.2 *Results of KMeans*

4.2.1 Results for Scenario I

The K-means-based approach did in the Scenario I did a good job as the articles which are recommended to the user are similar to the base generic article that is 'Machine Learning: how to go from Zero to Hero'. Articles like 'Learning How To Code Neural Networks—Learning New Stuff—Medium', 'Machine Learning in a Week' and 'Machine Learning in a Year' discuss how to begin to learn the fundamentals of Data Science and Machine Learning. K-means was also successful in extracting the other parts of the base generic article like 'Machine Learning is Fun Part 6' (Table 3).

Table 2 Top fifteen recommendations of cosine similarity Scenario II

S. no.	Title	Similarity
1	Write an AI to win at Pong from scratch with Reinforcement Learning	0.835
2	Formalizing indirect normativity—AI Alignment	0.795
3	Deep Learning the Stock Market—Tal Perry—Medium	0.775
4	Rohan & Lenny #1: Neural Networks & The Backpropagation Algorithm, Explained	0.771
5	Simple Reinforcement Learning with Tensorflow Part 0: Q-Learning with Tables and Neural Networks	0.740
6	Deep Learning Achievements Over the Past Year—Stats and Bots	0.736
7	Netflix Recommendations: Beyond the 5 stars (Part 1)	0.721
8	Understanding Capsule Networks—AI's Alluring New Architecture	0.720
9	Machine Learning is Fun! Part 3: Deep Learning and Convolutional Neural Networks	0.714
10	Using Deep Q-Learning in FIFA 18 to perfect the art of free-kicks	0.714
11	Explained Simply: How an AI program mastered the ancient game of Go	0.713
12	Ultimate Guide to Leveraging NLP & Machine Learning for your Chatbot	0.711
13	Traffic Sign Recognition with TensorFlow—Waleed Abdulla—Medium	0.702
14	How to solve 90% of NLP problems: a step-by-step guide	0.700
15	Simple Reinforcement Learning with Tensorflow Part 8: Asynchronous Actor-Critic Agents (A3C)	0.700

4.2.2 Results for Scenario II

The recommendation made by K-means in a Scenario II which is a scenario where recommendations are to be made for a given specific topic is somewhat similar to the ones made by the cosine similarity. It suggests articles which like 'Write an AI to win at Pong from scratch with Reinforcement Learning', 'Simple Reinforcement Learning with Tensorflow part 0 and part 4' and 'Understanding Capsule Networks' are very much related to reinforcement learning (Table 4).

4.3 Results of Agglomerative

4.3.1 Result for Scenario I

In the case of a base article with a general topic, the recommendation made by the agglomerative clustering is 'Modeling Madly—Data Engineering' in which author gives tips to succeed in hackathons such as what kind of tools to be employed and which is somewhat similar to what our base article speaks about that how to get started with machine learning and what tools to be used.

Table 3 Top fifteen recommendations of K-means Scenario I

S. no.	Title	Distance
1	Explained Simply: How an AI program mastered the ancient game of Go	0.714
2	Learning How To Code Neural Networks—Learning New Stuff—Medium	0.853
3	What Could Happen If We Did Things Right: An Interview With Kim Stanley Robinson, Author of Aurora	0.907
4	Machine Learning in a Week—Learning New Stuff—Medium	0.923
5	You can build a neural network in JavaScript even if you don't really understand neural networks	0.931
6	I Let IBM's Robot Chef Tell Me What to Cook for a Week	0.946
7	Machine Learning is Fun Part 6: How to do Speech Recognition with Deep Learning	0.948
8	Thoughts after taking the Deeplearning.ai courses—Towards Data Science	0.949
9	Coursera vs Udacity for Machine Learning—Hacker Noon	0.976
10	Blockchain is not only crappy technology but a bad vision for the future	0.986
11	Machine Learning in a Year—Learning New Stuff—Medium	0.996
12	Why you just can't black box an a.i.—[weird things]	1.023
13	The mind-blowing AI announcement from Google that you probably missed.	1.025
14	Facebook M — The Anti-Turing Test—Arik's Blog	1.026
15	The 3 Tricks That Made AlphaGo Zero Work—Hacker Noon	1.035

4.3.2 Result for Scenario II

In Scenario II the input to the model was 'Reinforcement Learning from scratch—Insight Data' and the article recommended by the model was 'Using Deep Q-Learning in FIFA 18 to perfect the art of free-kicks' which is an article where the author gives a demo on how to build an AI model using reinforcement learning and deep learning algorithms which is capable of scoring free-kicks in FIFA 18. This is an indication that agglomerative clustering's hierarchical approach to clustering is successful in capturing specific features of a type of article.

4.4 Performance Evaluation of Approaches

The cosine similarity-based approach was successful in recommending in both scenarios. Its simple approach to compute the similarity index by applying dot product between various articles represented in the form of TF-IDF vectors proved to be effective.

The K-means-based approach also did a decent job of suggesting relevant articles in the Scenario I and Scenario II. The performance of K-means is very similar to that

Table 4 Top fifteen recommendations of K-means Scenario II

S. no.	Title	Distance
1	Write an AI to win at Pong from scratch with Reinforcement Learning	0.542
2	Formalizing indirect normativity—AI Alignment	0.638
3	Deep Learning the Stock Market—Tal Perry—Medium	0.647
4	Rohan & Lenny #1: Neural Networks & The Backpropagation Algorithm, Explained	0.652
5	Machine Learning is Fun! Part 3: Deep Learning and Convolutional Neural Networks	0.713
6	Netflix Recommendations: Beyond the 5 stars (Part 1)	0.718
7	Simple Reinforcement Learning with Tensorflow Part 0: Q-Learning with Tables and Neural Networks	0.726
8	Understanding Capsule Networks—AI's Alluring New Architecture	0.726
9	How you can train an AI to convert your design mockups into HTML and CSS	0.727
10	Simple Reinforcement Learning with Tensorflow Part 4: Deep Q-Networks and Beyond	0.750
11	How Artificial Intelligence Will Make Technology Disappear	0.752
12	How to solve 90% of NLP problems: a step-by-step guide	0.757
13	Machine Learning for Humans, Part 2.1: Supervised Learning	0.762
14	Ultimate Guide to Leveraging NLP & Machine Learning for your Chatbot	0.767
15	Machine Learning is Fun! Part 4: Modern Face Recognition with Deep Learning	0.775

of cosine similarity. It is very clear as the recommended articles by both approaches for the same pair of articles are have several articles in common. This can be attributed to the fact that similarity measuring metrics used in cosine similarity and K-means are very similar in nature. K-means uses Euclidean distance to calculate the distance between articles and cosine similarity measures the angle between articles. Here, distance is basically the distance covered by the angle between articles. Therefore, their results are comparable and coherent [15].

Agglomerative clustering was not very successful in recommending for the generic topic in Scenario I. But this cannot be said when it comes to Scenario II in which it did a very good job of recommending an article which basically is a practical and implementation version of the base article related to reinforcement learning. This can be attributed to the fact that it uses a hierarchical-based approach to cluster data. It is more successful in capturing deeper sublevels of a particular topic cluster.

5 Conclusion

The cosine similarity-based model was the overall allrounder capturing the patterns of similarity at both generic level and topic-specific level and recommend articles to the reader. The K-means approach did an appreciable job at suggesting generic articles and also did a great job at recommending articles which are specific to a topic related to machine learning. The results of K-means and cosine similarity were almost similar. Agglomerative clustering was not so successful in suggesting article related to the generic topic, whereas it was on point while suggesting topic-specific article as its fundamental principle of clustering involves modeling in a hierarchical fashion of clustering. This helps it understand better what the different hierarchies of topics present in the text corpus.

References

1. Kamath SS, Kanakaraj M (2015) Natural language processing-based e-news recommender system using information extraction and domain clustering. Researchgate. https://www.google.com/url?q=https://www.researchgate.net/publication/281337114_Natural_language_processing-based_enews_recommender_system_using_information_extraction_and_domain_clustering&sa=D&source=hangouts&ust=1602675437394000&usg=AFQjCNETOYZsnHsgC6KmpNyjG9BgSMxMxA
2. Haruna K, Akmar Ismail M, Damiasih D, Sutopo J, Herawan T (2017) A collaborative approach for research paper recommender system. Journals.plos. https://www.google.com/url?q=https://journals.plos.org/plosone/article?id%3D10.1371/journal.pone.0184516&sa=D&source=hangouts&ust=1602675454947000&usg=AFQjCNH8IymrhCMN8SovXLSAzpo7yjMvkg
3. Gipp B, Beel J, Hentschel C (2009) Scienstein: a research paper recommender system. Researchgate
4. Alharthi H (2019) Natural language processing for book recommender systems. Uottawa
5. Hsankesara (2019) A collection of articles on ML, AI and data science. Medium Articles
6. Stanford NLP: Stemming and lemmatizarion
7. Mägi I (2016) Applying TF-IDF algorithm in practice. Plumbr
8. Sklearn (2009) TruncatedSVD. https://scikit-learn.org/stable/modules/generated/sklearn.decomposition.TruncatedSVD.html
9. Brems M (2017) A one-stop shop for principal component analysis, medium
10. neo4j. The cosine similarity algorithm. Github
11. Sklearn. KMeans. https://scikit-learn.org/stable/modules/generated/sklearn.cluster.KMeans.html
12. Sklearn. agglomerative clustering. https://scikit-learn.org/stable/modules/generated/sklearn.cluster.AgglomerativeClustering.html
13. Wikipedia: Elbow method (clustering)
14. Wijaya CY (2019) Breaking down the agglomerative clustering process. Medium
15. Emmery C (2017) Euclidean vs. cosine distance. Github

Chapter 14
A Survey of the Exemplary Practices in Network Operations and Management

K. M. Majidha Fathima

1 Introduction

A network is defined as a collection of nodes, viz., client, server and middleware. These nodes communicate with each other and exchange information. The components (repeaters, hubs, bridges, switches and routers) occupy their position in various layers of the ISO-OSI (Open Systems Interconnection) networking model. The seven layers are physical, data link, network, transport, session, presentation and application. The application layer helps in serving applications like File Transfer Protocol (FTP), HyperText Transfer Protocol (HTTP) and Simple Mail Transfer Protocol (SMTP). The presentation layer takes care of presentation and compatibility of various images like jpeg, png and bmp. The session layer helps in maintaining sessions between the users. The transport layer provides end-to-end connectivity between hosts. The network layer helps in Internet connectivity and routing. The data link layer provides error control and checksum. The physical layer transmits information in the form of bits.

As shown in Table 1, repeaters and hubs work in layer 1. Repeaters amplify the signal for greater distances. Hubs just provide connectivity. Switches and bridges function in layer 2. Bridges provide connectivity between two networking devices. Switches are intelligent devices compared to bridges. Routers and brouters offer network connectivity between devices and they operate in layer 3.

As shown in Fig. 1, the data when it travels from the source, encapsulation takes place, i.e., wrapping of data form (placing one inside another). The Layer 4 PDU, TCP segment consists of TCP Header and upper layer PDU (Data). The next Layer 3 PDU, IP Datagram contains IP Header, TCP Header and upper layer data. The data

K. M. Majidha Fathima (✉)
Sri Krishna College of Engineering and Technology, Coimbatore, India
e-mail: majidhafathimakm@skcet.ac.in

I. Jeena Jacob et al. (eds.), *Data Intelligence and Cognitive Informatics*,
Algorithms for Intelligent Systems,
https://doi.org/10.1007/978-981-15-8530-2_14

Table 1 Devices at the seven layers

OSI Model	DoD Model	protocols		devices/apps
layer 5, 6, 7	application	dns, dhcp, ntp, snmp, https, ftp, ssh, telnet, http, pop3... others		web server, mail server, browser, mail client...
layer 4	host-to-host	tcp	udp	gateway
layer 3	internet	ip, icmp, igmp		router, firewall layer 3 switch
layer 2	network access	arp (mac), rarp		bridge layer 2 switch
layer 1		ethernet, token ring		hub

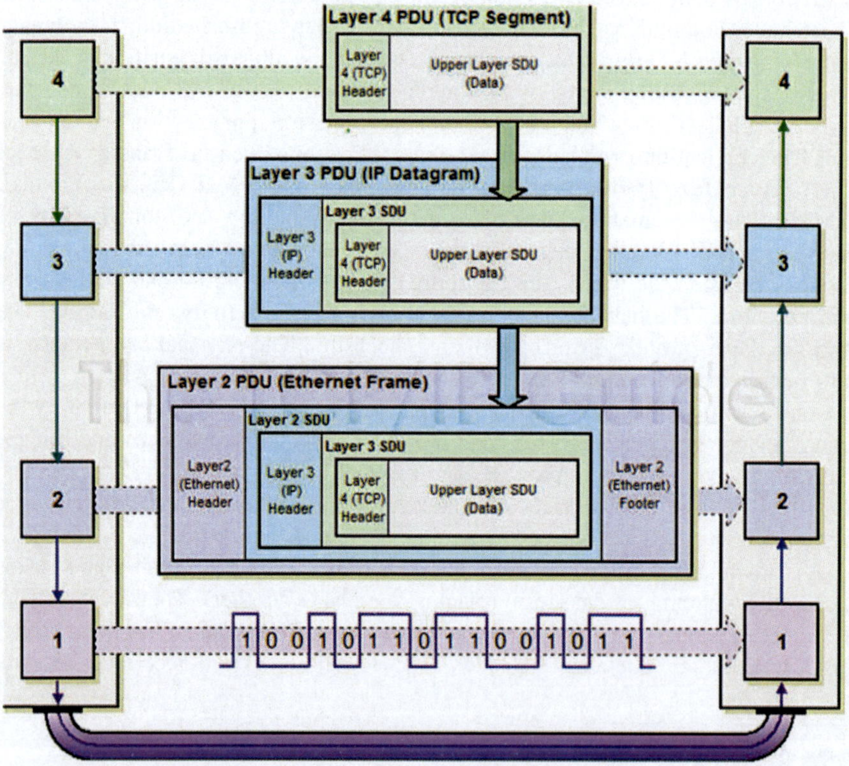

Fig. 1 Encapsulation and decapsulation of data

field at this layer will be the upper layer header and data. The next layer 2 PDU, Ethernet frame consists of Layer 2 Ethernet Header, Layer 3 IP Header, Layer 4 TCP Header and the upper layer PDU. Layer 2 also includes Ethernet Footer consists of checksum for error detection. The physical layer finally transmits information in the form of bits. The reverse process of data transmitting from physical layer to application layer is called as decapsulation. The data is unveiled from Layer 1 to Layer 7 in the destination end. Besides, communication happens at each layer as peer-to-peer communication. The network management occurs at each of the layers portrayed above.

2 Literature Survey

The related works on the network operations, monitoring and management are discussed in Table 2. The various network monitoring tools and protocols are NetFlow, SNMP, CDP, Wireshark, NetBIOS, etc., They assist in packet capturing, analysis, IP traffic, source address, destination address, network protocol, TCP or UDP port numbers, session IDs, application layer port numbers, etc. The additional features provided by them are identifying spurious traffic generated by malicious nodes, bursty traffic in cases of peak times, link utilization both at last mile and at the aggregation, core and backbone routers. They also separately monitor the Virtual Private Network (VPN) and Internet VPN (IVPN) traffic based on the IP provisioning and Committed Access Rate (CAR). The link utilization is being monitored for the

Table 2 Comparison of different types of network performance monitoring tools

TYPE	Passive or active	Network layer of operation	Basic operation
Integrated SNMP platform [1]	Active	Data link and network	A management station polls agents on network devices to gather information about the network
Packet capture/Sniffing [2]	Passive	Data link	Listens for and captures packets on a network device for analysis
Application/service monitoring [3]	Active	Application	Actively polls applications and application servers to provide performance information
Flow monitoring [4]	Active	Transport and higher	Monitor connections and flows on the network for higher layer trends

bandwidth calculation, i.e., billing purpose. The peer ISP traffic is also being monitored by the protocols. If the customer is dual-homed, i.e., he has connections to two different ISPs for auto-fallback, both the forward and reverse traffic are captured.

As shown in Table 2, the tools and the example implementations are

- SolarWinds' Orion Network Performance Monitor—Integrated SNMP
- Ethereal—Passive Capture Tool
- WebMetrics GlobalWatch—Application and Service Monitoring
- Cisco IOS NetFlow—Network Flow Monitoring

3 Network Operations Center

The network operations center (NOC) and security operations center (SOC) are responsible for identifying, investigating, prioritizing, escalating and resolving network issues. Based on the priority of the customers such as A*, the issue is escalated to the next level as technical support manager, regional manager and the general manager. This escalation can be automatically done based on a timely basis.

As shown in Fig. 2, the network management team resides inside the restricted access area. The display in front shows the monitoring responses of various devices like routers, firewalls, switches in the network diagram of the entire Internet Service Provider (ISP).

In case of any abnormal alarms or notifications for any particular networking device (local or remote) in the display, the corrective action is taken immediately by the network team [1].

Fig. 2 Network operations center (NOC)

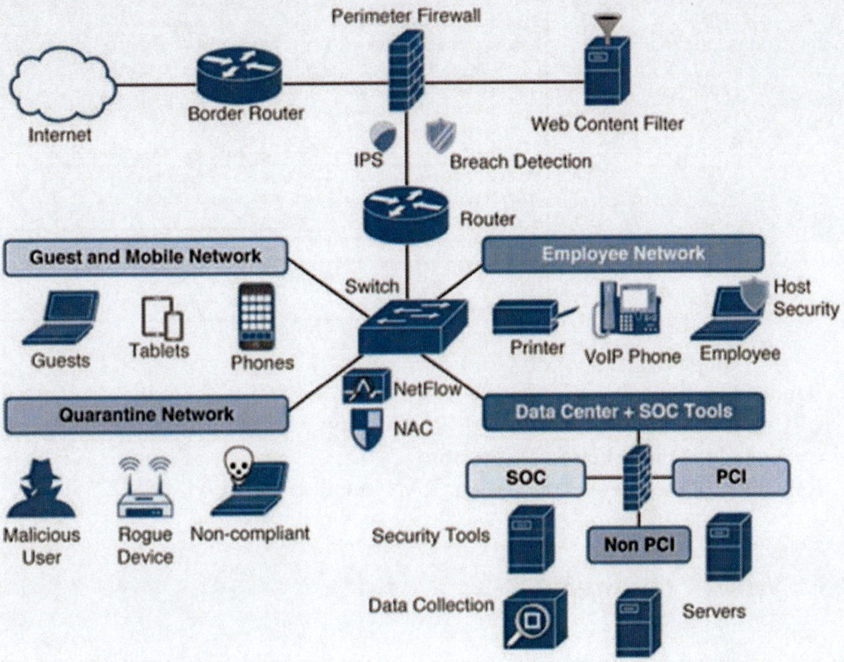

Fig. 3 Security operations center

3.1 Security Operations Center

As shown in Fig. 3, the customer lands on the last-mile connectivity of the ISP. The ISP then carries forward the network traffic across the entire world. It also ensures that data is transmitted securely throughout the communication and guarantees prevention from malicious nodes and security breaches [2]. The security tools, NetFlow, intrusion prevention system (IPS) and breach detection are done at the firewalls.

3.2 Technical Support—Tier 1

A separate portal as info desk is designed. It looks into issues like

- Attending calls from the customer
- Logging in the complaint and issuing Trouble Ticket number for tracking purpose
- Identifies the type of connectivity as leased line, ISDN or broadband from the info desk knowledge base.
- The info desk knowledge base is populated by the corporate provisioning team (CPE) [3].
- Troubleshoots online and updates the customer in case of
- Physical link going down

Fig. 4 Access list configuration at the router

```
Router#show access-lists SecureBackbone
Standard IP access list SecureBackbone
    deny 200.0.0.0 0.0.0.127 (8 match(es))
    permit host 200.0.0.130
    permit host 200.0.0.131
    deny 200.0.0.128 0.0.0.63
Router#clear access-list counters SecureBackbone
Router#show access-lists SecureBackbone
Standard IP access list SecureBackbone
    deny 200.0.0.0 0.0.0.127
    permit host 200.0.0.130
    permit host 200.0.0.131
    deny 200.0.0.128 0.0.0.63
Router#
```

- Bandwidth utilization is high
- CPU utilization is high
- Cyclic redundancy check (CRC) errors
- Reliability at the connected WAN or LAN interface (serial, viz., 251/255)

3.3 Network Operations

After the first-hand troubleshooting is done by the front end Tier 1 team, the issue is isolated and forwarded to the next level NetOps team. It takes care of tasks such as

- Change Management
  If any core device has gone down and due to that a number of customers need to be shifted to other switch or router and network operating system up-gradation to support the new router or switch series, viz., 2 900 [4].
- Configuration Management
  Any new router, switch or firewall is introduced in the network and new configuration, viz., assigning IP addresses, routing protocols has to be done or old configuration needs to be changed.
- Access List Modifications
  The access list at a switch or router decides the traversal of traffic with restrictions between a particular source and destination as shown in Fig. 4.

3.4 Network Engineering

It is the core team which takes care of the entire backbone network as per Fig. 5.

- Network engineering (NEG) team implements the end-to-end connectivity of the entire backbone.
- The configuration from the last mile to the aggregation router.
- The connectivity from aggregation to the core router.

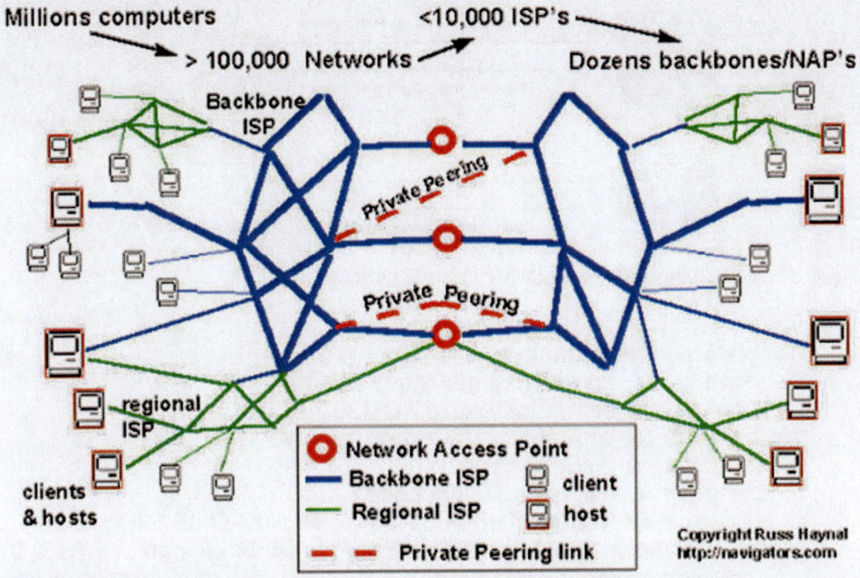

Fig. 5 ISP backbone network

- Peering configurations between the different ISP routers identified by their autonomous system (AS) numbers.
- Internal firewalls placed at the ISP network.
- Packet shapers, if any inside the protected network [5].
- SNMP server configurations at the routers [6].

4 Network Management

4.1 Simple Network Management Protocol (SNMP)

SNMP is used for monitoring and managing a network. The SNMP Agent may be a router or a switch [7]. A server is configured as SNMP Manager. The SNMP Agent generates values in the form of a Management Information Base (MIB). They get and set MIB values in the server. The server sends back responses and traps as shown in Fig. 6.

As shown in Fig. 7, every group of hosts is identified by a community string, viz., 14all4$$ and gds4chv1. The IP addresses allowed in the Access Control List (ACL) 10 and 30 are only configured for capturing logs. ACL 10 has been given with SNMP READ WRITE access, whereas ACL 30 has been configured with SNMP READ only [8].

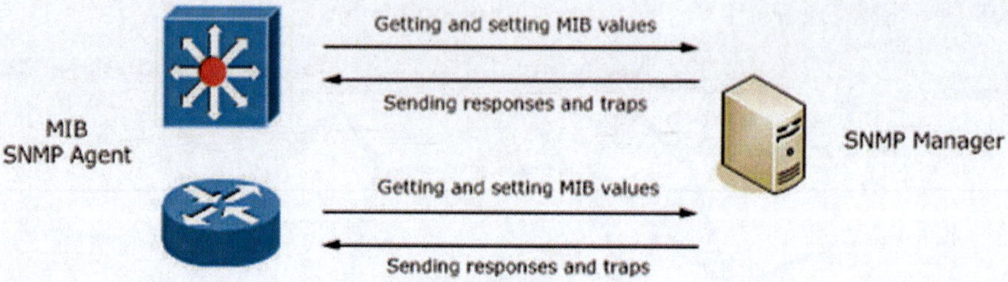

Fig. 6 Communication between an SNMP agent and manage

```
snmp-server community 14all4$$ RO 30
snmp-server community gds4chv1 RW 10
snmp-server community mrtg RO 1300
snmp-server community VBCCrep0rting RO 1333
snmp-server ifindex persist
snmp-server trap-source Loopback0
snmp-server location THHQCE1-3845: Facility-Code THHQ
snmp-server contact network operations 66-6428 xxxx
snmp-server enable traps tty
snmp-server enable traps config
snmp-server host 172.22.71.201 Voyence  config
snmp-server host 172.22.71.201 config
snmp-server host 172.22.9.201 config
!<----- SNMP SETTINGS
!<----- ACL 10 for Read Write, ACL 30 for Read Only
!
! ACL 10 - SNMP READ WRITE
!
no access-list 10
access-list 10 permit 172.27.124.18 log
access-list 10 deny any log
!
!
! ACL 30 - SNMP READ ONLY
!
no access-list 30
access-list 30 permit 172.40.46.89 log
access-list 30 permit 172.40.46.114 log
access-list 30 deny any log
!
```

Fig. 7 SNMP configuration in a cisco router

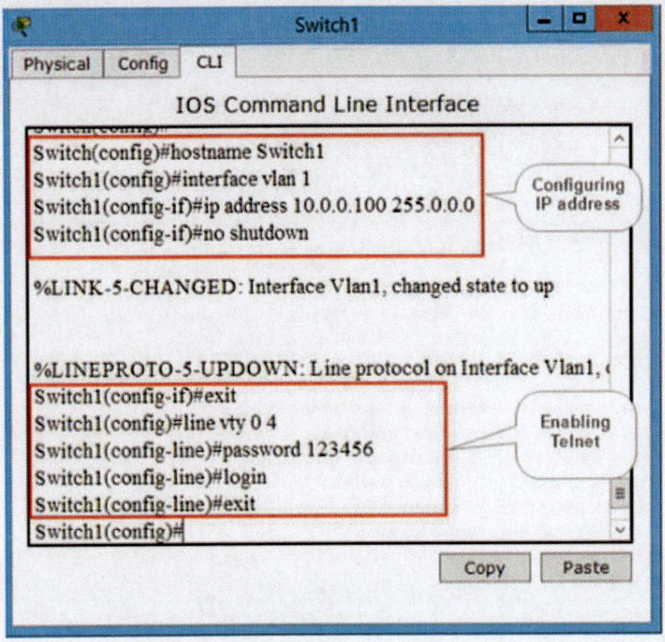

Fig. 8 Telnet configuration on a switch

4.2 *TELecommunication NETwork (Telnet)*

Telnet is an application protocol to remotely log into a network host using a virtual terminal connection. It works on Transmission Control Protocol (TCP) port number 23. So port number 23 has to be allowed in the access list to login to a host [9].

As shown in Fig. 8, interface vlan1 (virtual LAN) is configured with IP address 10.0.0.100/24. The line protocol is changed to 'up' state by giving no shutdown. 'Line vty 0 4' symbolizes that the number of Telnet sessions is allowed to 4 maximum. The password 123 456 is also for the remote login.

4.3 *Secure Shell (SSH)*

As shown in Fig. 9, the putty client is installed on a system and putty access is allowed to a host on port number 22. It carries transmission securely through a crypto key [10].

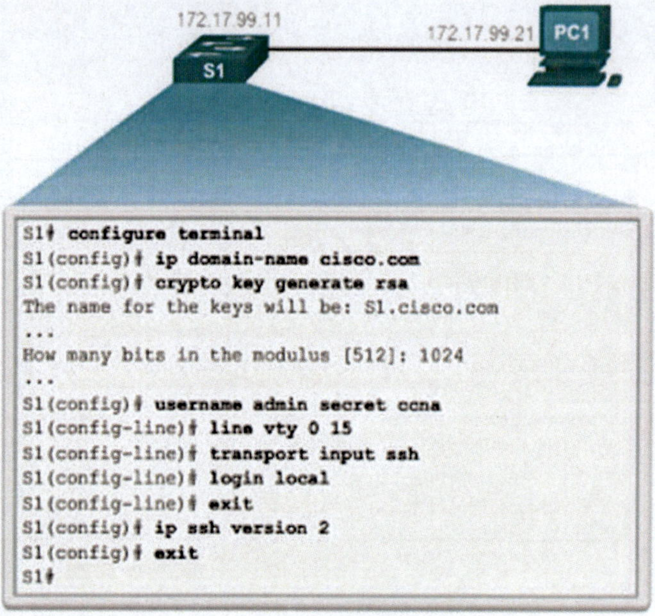

Fig. 9 SSH configuration on a router (command line interface)

4.4 Terminal Access Controller Access Control System (TACACS)

Authentication, authorization and accounting services (AAA) are provided separately under TCP port number 49 [11]. The key is also provided as 'mys3cr3t!' as shown in Fig. 10.

5 Algorithm

1. If two terminals are connected, then they have the same signal.
   $\forall t1, t2\ \text{Terminal}(t1) \wedge \text{Terminal}(t2) \wedge \text{Connected}(t1, t2) \Rightarrow \text{Signal}(t1) = \text{Signal}(t2)$
2. The signal at every terminal is either 1 or 0:
   $\forall t\ \text{Terminal}(t) \Rightarrow \text{Signal}(t) = 1 \vee \text{Signal}(t) = 0$
3. Connected is commutative:
   $\forall t1, t2\ \text{Connected}(t1, t2) \Leftrightarrow \text{Connected}(t2, t1)$.
4. Three types of devices:
   $\forall g\ \text{Device}(g) \wedge k = \text{Type}(g) \Rightarrow k = \text{Router} \vee k = \text{Switch} \vee k = \text{Firewall}$

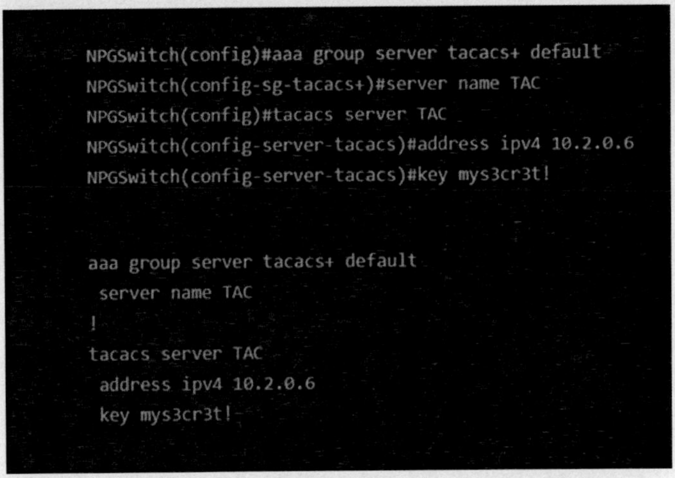

Fig. 10 TACACS configuration

5. A router can understand only IP address
   $\forall$g Device(g) $\wedge$ Type(g) = Router $\Rightarrow$
   Address(g) = IP Address—32 bits
6. A switch can understand only MAC address
   $\forall$g Device(g) $\wedge$ Type(g) = Switch $\Rightarrow$
   Address(g) = MAC Address—48 bits
7. A firewall permits or deny traffic to and from a network host
   $\forall$g Device(g) $\wedge$ Type(g) = Firewall $\Rightarrow$
   Action(g) = allow/deny
8. A packet shaper restricts the upper limit of a traffic to fall below the bursty level
   $\forall$ g Device(g) $\wedge$ Type(g) = Packet Shaper $\Rightarrow$
   Action(g) = cutoff limit

6 Result and Discussion

Multi Router Traffic Grapher (MRTG) is used for the visual interpretation of network traffic management.

The inbound traffic is the forward traffic from the ISP network to the customer network [12]. The outbound traffic is the reverse traffic from the customer network to the ISP network. As shown in Fig. 11, there are three graphs generated by MRTG for the number of packets transmitted named as traffic load, the CPU utilization of the router and the nodes with new TCP connections.

The free memory space available, the average network data transmission in case of upload and download, the established TCP connections are shown.

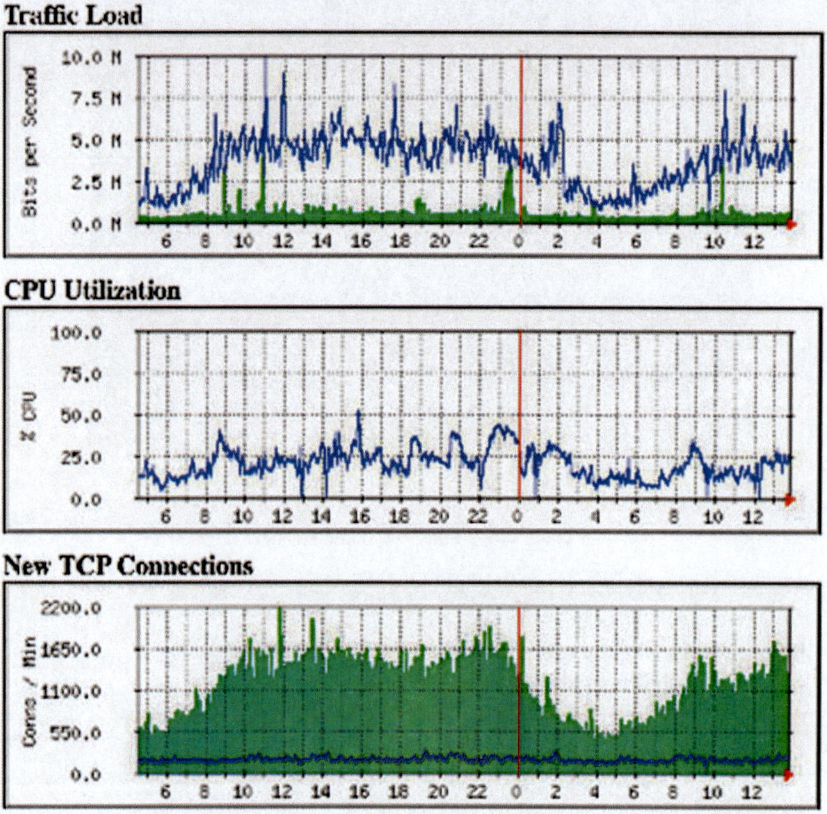

Fig. 11 SNMP—MRTG graph—traffic, CPU, TCP

The colors used for the alert symbols in the network architecture diagram are red, yellow and green as per Fig. 12 [13].

- Green indicates the router has normal expected traffic flow through it without any abnormalities
- Red symbolizes that the device has gone down due to
  - Hardware link failure in case of wired last mile
  - Access point (AP) or subscriber unit (SU) failure in case of broadband networks
  - Security breach by any unauthorized cybercrime
- Yellow indicates that care has to be given for the device to check for
  - Bandwidth utilization
  - The redundant link between two devices.

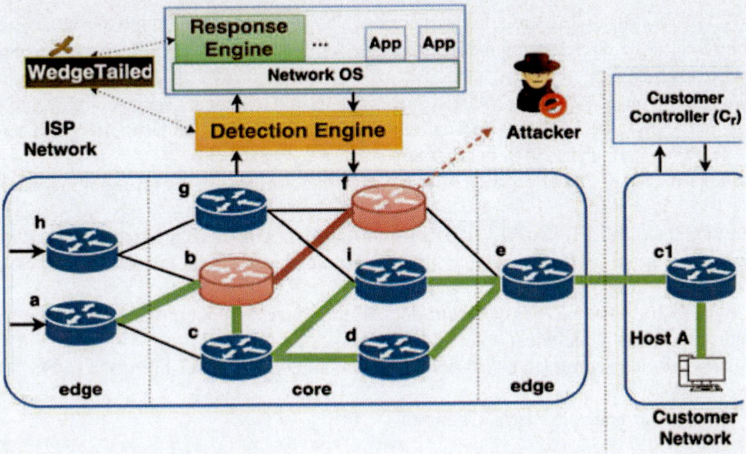

Fig. 12 Alerts in an ISP network

7 Conclusion

Thus, the knowledge engineering in first-order logic can be applied to a network domain. It is integrated with SNMP for monitoring and managing the network. The various teams in a network domain, their operations, and access rights are discussed. The results are shown with the focus of SNMP, Telnet, SSH, TACACS as monitoring and management tools. MRTG graph generated with SNMP is also displayed for the interpretation of results.

References

1. Savic M (2016) Bridging the SNMP gap: simple network monitoring the ınternet of things. Facta Univ-Ser Electr Energ 29(3):475–487
2. Xiao Y (2010) Physical path tracing for IP traffic using SNMP—a comprehensive solution for end-to-end and multicast streams monitoring. In: Proceedings of 2nd international conference on education technology and computer, pp V5-310-V5-317
3. Pandey S, Choi M-J, Won YJ, Hong JW-K (2010) SNMP-based enterprise IP network topology discovery. Int J Netw Manag 21(3):169–184
4. Jun L, Xuefeng Z, Weihong S, Qilin Z (2013) Network topology discovery based on SNMP. In: Proceedings of 9th international conference on computational ıntelligence and security, pp 194–199
5. Han YQ, Zhang Y, Li T, Cao L (2013) Research of network monitoring based on SNMP. In: 2013 third international conference on ınstrumentation, measurement, computer, communication and control, Shenyang, pp 411–414
6. Liu J, Liu G (2011) Research and implementation of SNMP-based network management system. In: 2011 4th international conference on ıntelligent networks and ıntelligent systems, Kunming, pp 129–132

7. Lu Y, Qian S (2010) Research on the theory of SNMP and the technology of SNMP programming. In: 2010 3rd international conference on computer science and information technology, Chengdu, pp 23–25
8. ZWang ZQ, Wang Y, Shao G (2009) Research and design of network servers monitoring system based on SNMP. In: 2009 first international workshop on education technology and computer science, Wuhan, Hubei, 2009, pp 857–860
9. Stallings W (1998) SNMPv3: a security enhancement for SNMP. IEEE Commun Surv 1(1):2–17
10. Han YQ, Zhang Y, Li T, Cao L (2013) Research of network monitoring based on SNMP. In: 2013 third ınternational conference on ınstrumentation, measurement, computer, communication and control, Shenyang, pp 411–414
11. https://www.cisco.com/c/dam/en/us/td/docs/telepresence/endpoint/mxp-series/white_papers/white_paper_snmp_technical_manual.pdf
12. https://www.cisco.com/c/en/us/support/docs/availability/high-availability/15114-NMS-bestpractice.html
13. https://www.bitpipe.com/tlist/SNMP.html

Chapter 15
Utilization of Fish Excrete for Plant Growth Using SVM

C. Gnana Kousalya, T. U. Gowdhami, and R. Mayavinothini

1 Introduction

Agriculture is an important source of livelihood in most parts of the world. It plays a vital role in increasing the income of developing countries like India. Jasmine which is considered the queen of flowers is commercially cultivated in various districts of Tamil Nadu, Kerala, Andhra Pradesh, Assam and West Bengal. Tamil Nadu is the leading producer in India, cultivating nearly 10,000 ha of Jasmine, and the production is 45% more than any other states. India exports large amounts of Jasmine to Malaysia, Singapore, Sri Lanka and middle-east countries. The demand for Jasmine is increasing nowadays. Due to diseases, production rate and product quality decrease gradually.

Diseases have the potential to destroy a large number of plants, if not detected and controlled in time. Diseases affect the plant's leaves, stem and roots, but leaves are more liable to detect the disease. Farmers concentrate on the cause of the disease in the plant during its growth, but it is not easy to recognize the disease on the spot. The Jasmine plants with the disease are shown in Fig. 1. The diseases which affect plants are Alternaria Alternata, Bacterial Blight, Anthracnose and Cercospora leaf spot. Before the evolution, plant diseases were observed by expert farmers, and all the detection and irrigation processes were manually facing a lot of problems associated

C. Gnana Kousalya (✉) · T. U. Gowdhami · R. Mayavinothini
Department of Electronics & Communications, St. Joseph's Institute of Technology, Chennai, India
e-mail: hodecestaffaffairs@gmail.com

T. U. Gowdhami
e-mail: gowdhamichand26@gmail.com

R. Mayavinothini
e-mail: vino23maya@gmail.com

I. Jeena Jacob et al. (eds.), *Data Intelligence and Cognitive Informatics*,
Algorithms for Intelligent Systems,
https://doi.org/10.1007/978-981-15-8530-2_15

Fig. 1 Jasmine plant with disease

with these methods. Nowadays, automatic detection is used to detect the affected plants in a more precise manner.

By using this method, diseases can be identified and controlled more quickly. As per the Central Water Commission, 85.3% of the total water consumed was for agriculture in 2010. The objective of the paper is to decrease the water scarcity, and providing an efficient way of farming is done in agricultural fields. Disease identification of cultivated Jasmine plants is done, and depending upon the plants, the irrigation system is carried out. As the fishponds water is used, it is both efficient and eco-friendly. By using this method, diseases can be identified and controlled more quickly.

By smearing the right amount of water on the farm, wrong time of irrigation can be avoided. Reducing the usage of more water will reduce the saturation of soils and increase crop performance. Advanced tools like IoT technologies are a method for accurate soil moisture control in specialized production. The quality of the crops is based on the data collected from the field such as soil moisture, ambient temperature and humidity. This technique enhances the productivity and excellence of the crops. The system identifies the diseases of the Jasmine plants with the help of machine learning technique. The algorithm used in this system is the support vector machine.

2 Literature Survey

Diseases of plants destroy a large number of crops in the field which is insignificant to farmers. Moreover, diseases affect plants leaves, stem and root, but leaves are very effective. For miles of fields, the farmer is unable to conduct the survey to observe the plants. In addition to the remote areas, expert farmers are not present. Many developing countries organize the awareness program for farmers about various pests and diseases.

In this approach, the RGB images of leaves of a plant are captured, and by color transformation structure, the green pixels of the image are masked. Using Sobel and Canny with homogenous operator techniques, edge detection is found out to range function to configure diseases in the image of the plant [1]. The plant of maize diseases is identified by the neural network of backpropagation algorithms, and the RGB images of leaves are converted into Hue Saturation Intensity (HSI) color space representation. The masking green pixels are done to generate matrices for disease detection [2].

Content-based image classification is used to extract features of image, and further the features are classified by Naïve Bayesian classifier. By applying Baye's theorem, the data images are discretized and normalized to provide the color mean features [3]. By this approach, the image acquisition of plants leaves is done, for the input RGB image, a color transformation structure is created, and then by preprocessing techniques, noise is removed. The genetic algorithm segments image by selecting a pair of chromosomes from a randomly generated population [4]. Grape leaf images are acquired by a digital camera, and preprocessing is done to improve the quality of image leaves. K-means clustering algorithm is used to segment the image into clusters of smaller sets of similar, and then, SVM classifier divides the sets of data to identify the disease of plants [5].

Convolutional neural network (CNN) for deep learning is done for analyzing the visual image of plants leaves and extracts appropriate features from multilayer hierarchy. Data is aggregated for training which converts an image into binary for image preprocessing. The softmax activation function is computed, and diseases are determined [6]. IoT comprises deployment of a sensor at the agricultural application and storage of their data cloud for processing [7].

In this approach, research based on machine learning for various techniques is done. A survey based on the techniques like decision tree learning, association rule learning, support vector machine and k-nearest neighbor for the various disease of plants is applied [8]. Researchers process image acquisition, preprocessing, segmentation and extraction of the plants, and then, gray level co-occurrence matrix (GLCM) is calculated to represent pixels image of plants into matrices. Texture features such as energy, contrast, dissimilarity, correlation, mean and standard deviation are calculated using GLCM. SVM classifier plots the dimensional space in a coordinate to identify tomato plants leaf [9].

The overall survey on plant disease detection using various of types of machine learning is done, the methods like region growing segmentation, hierarchical clustering, genetic algorithm-based segmentation, k-means clustering, Gaussian filter, a histogram of oriented gradients(HOG), Laplacian filter, hybrid algorithm and kernel features for future perspective to make an efficient way of identifying diseases of plants[10]. On strawberry cultivation, cloud-based automatic drip irrigation is applied based on cumulative free surface water evaporation that is estimated by a various sensor like humidity, temperature and soil moisture are done [11].

The diseases affected leaves are estimated by preprocessing, feature extraction, training of classifier and classification. In this paper, the histogram conversion is done by HOG extraction to plot healthy and diseased leaves [12].

Automatic irrigation process observes moisture level, temperature level in the crop area. The paper aims a high precision monitoring for the data and control agriculture automation system with IoT technologies. Raspberry Pi and cloud-based IoT system to monitoring the real-time data come from the crop field. The system mainly focuses on moisture variations correlate with temperature changes data by smart sensors and controls the irrigation system [13]. In this paper, SVM classification on the tested images of tomato leaves with Extended Region of Interest (EROI), Typical Region of Interest (TROI) and Regional Region of Interest (RROI) segmentation processed to find the dissimilarity features of leaves images [14].

In plant disease detection, a review on machine learning classification techniques for various classifier like SVM classifier, ANN classifier, KNN classifier and Fuzzy classifier is done for providing better efficiency and estimating the accuracy [15].

3 Proposed System

The design of the system is to detect the disease by image preprocessing, segmentation and classification. The dissimilarity features improved classification performance using support vector machine (SVM).

3.1 Image Acquisition

Acquiring an image is the first stage of vision proposed system. The image of the given plant is digitized and stored in the system. Acquiring images is done using a camera from the field. Then, the stored images are applied for various method of processing. The images of Jasmine leaves are shown in Fig. 2.

The Jasmine leaf images are taken from various sources for analysis. After the images have been obtained, various methods of processing are applied to the image to perform the many different vision methods.

3.2 Image Preprocessing

The image preprocessing is to improve the quality of the images. Then, the collected images are enhanced to clarify the collected images. Digital image processing is used for a various operation like image re-sizing and filtering.

Calculating the median of each window leads to more computational effort and time because the filter must process entry by entry in the signal because it preserves edges while removing noise. The image after filtration is shown in Fig. 3. In this proposed system, preprocessing is done in the frequency domain, to reduce the noise to an average level.

Fig. 2 Raw images of Jasmine leaves

Fig. 3 Image after filtration

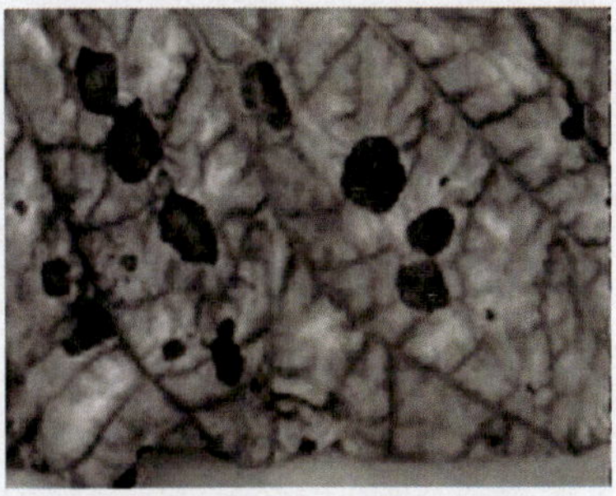

3.3 Image Segmentation

In image segmentation, K-means clustering is used. During this process, the affected parts of the leaf can be detected. The images of collected data are converted from RGB to GRAYSCALE. Color thresholding conversion of the image from the RGB to GRAYSCALE leads to the thresholding of the image. Foreground and background separation are done to disease part of the leaves, and the image is segmented into various parts according to the region of interest (ROI).

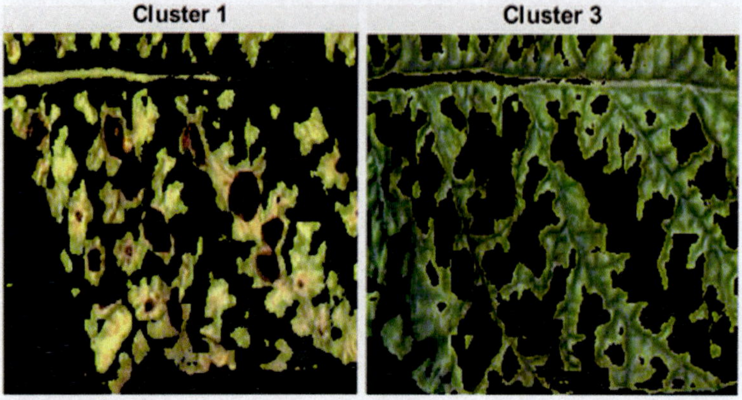

Fig. 4 Clusters images

3.3.1 K-means Clustering

K-means is one of the simplest algorithms and classifies a given dataset through the clusters of datasets. The clusters images are shown in Fig. 4. The input data is of a huge collection of data and leads to complex processing. The inputs given are bound together and represented as a set of features.

K-means clustering is to partition n-dimensional vectors into k clusters with the nearest mean. It minimizes within clusters variance that is least squared Euclidean distance. The procedure for clustering of dataset assuming three clusters follows and is to define k center, one for each cluster. Then, the next step is to take each point dataset to its nearest center. Specifying the number of cluster K, Initial means of centroids by shuffling the dataset and then randomly partition places of selecting K data points close to the center of the data, then iterating the K value until there is convergence is reached, then calculating the sum of the squared distance between data points and all centroids and finally taking an average of it.

$$S_i^{(t)} = \left\{ x_p : \left\| x_P - m_i^{[t]} \right\|^2 < \left\| x_p - m_j^{(t)} \right\|^2 \forall j, 1 \leq j \leq k \right\} \tag{1}$$

3.4 Feature Extraction

In this system, color features are obtained for diseases identification. Gray levels in an image are calculated through the gray level co-occurrence matrix (GLCM). The gray level can be determined by the number of columns and rows in GLCM. GLCM matrices are in the four special orientation (0°, 45°, 90° and 135°). Another matrix is built as the average of earlier matrices. Suppose the co-occurrence matrix

$$
\text{GLCM} = \begin{bmatrix} 871 & 556 & 112 & 52 & 21 & 2 & 0 & 0 \\ 547 & 2403 & 1407 & 516 & 274 & 19 & 1 & 0 \\ 118 & 1567 & 4269 & 3064 & 577 & 81 & 0 & 0 \\ 48 & 480 & 3087 & 7715 & 3562 & 327 & 9 & 0 \\ 24 & 101 & 661 & 3371 & 6344 & 1920 & 47 & 0 \\ 12 & 27 & 58 & 364 & 1566 & 2113 & 305 & 0 \\ 0 & 0 & 0 & 6 & 53 & 251 & 853 & 7 \\ 0 & 0 & 0 & 0 & 0 & 0 & 7 & 0 \end{bmatrix}
$$

Fig. 5 GLCM matrix

Table 1 Calculation of GLCM

S. No.	GLCM feature	Formula
1	Contrast	$\sum_{i,j} \lvert i-j \rvert^2 p(i,j)$
2	Correlation	$\sum_{i,j} \frac{(i-\mu i)(j-\mu j)P(i,j)}{\sigma i \sigma_j}$
3	Energy	$\sum_{i,j} p(\mathrm{i}, j)^2$
4	Entropy	$\sum_{i_1 j}^{N-1} -\ln(P_{ij}) p_{ij}$
5	Homogeneity	$\sum_{i,j} \frac{P(i,j)}{1+\lvert i-j \rvert}$
6	Mean	$\sum_{i_1 j}^{N-1} i p_{ij}$
7	Variance	$\sum_{i,j}^{N-1} p_{i_1 j}(i-\mu_i)^2$

$$
\text{Normalised GLCM} = \begin{bmatrix} 0.017 & 0.011 & 0.0025 & 0.001 & 0.0004 & 0.004 & 0 & 0 \\ 0.0109 & 0.0482 & 0.0282 & 0.0103 & 0.0055 & 0.0003 & 0 & 0 \\ 0.0023 & 0.031 & 0.085 & 0.0615 & 0.0115 & 0.0016 & 0 & 0 \\ 0.0009 & 0.0096 & 0.062 & 0.1550 & 0.07115 & 0.0065 & 0.0001 & 0 \\ 0.0004 & 0.0020 & 0.132 & 0.0677 & 0.1274 & 0.0385 & 0.0009 & 0 \\ 0.0002 & 0.0005 & 0.0011 & 0.0073 & 0.0031 & 0.0424 & 0.0610 & 0 \\ 0 & 0 & 0 & 0.0012 & 0.00110 & 0.005 & 0.0171 & 0.0001 \\ 0 & 0 & 0 & 0 & 0 & 0 & 0.0001 & 0 \end{bmatrix}
$$

Fig. 6 Normalized GLCM matrix

is $L(i, j)$ and the matrix size is $(N \times N)$. Each element (i, j) denotes the frequency of pixel at a gray level where (i) is spatially related to the pixel with gray level (j). GLCM construction of a grayscale image is shown in Fig. 5 [9], and the features are calculated using Table 1 formula (Fig. 6).

3.5 SVM Classifier

Many systems have used different techniques to get high accuracy. In this proposed system, SVM classifier technique is used for accurate classification and regression

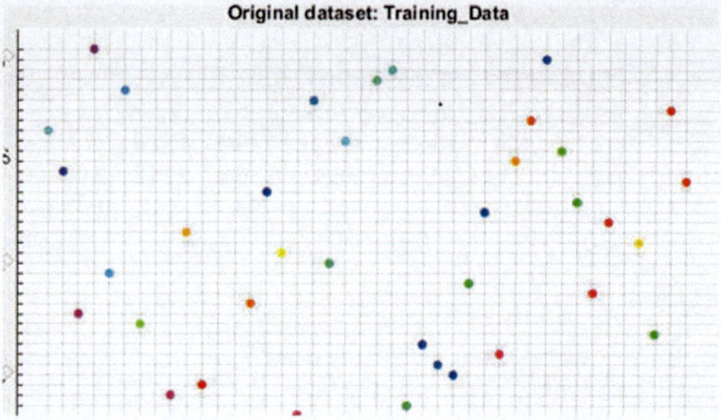

Fig. 7 Scatter plot of datasets

analysis of the disease. SVM is a boundary classifier that uses hyperplane called the decision boundary between two classes. The sets of data are plotted in *n*-dimensional space in a coordinate according to its class.

SVM vectors mark a plot to either one or other category and representation of points in space by a separate clear gap. Maximum margin hyperplane in an *n*-dimensional space is taken at which it divides into two categorized features. The above process repeated for both training dataset and testing dataset, and the datasets taken are the theoretical values of GLCM features analyzed from various leaves of the plants. In SVM predicting the testing, sets are done by maximizing the loss function. The scatter plot of the trained dataset is shown in Fig. 7 SVM graph.

The output of the linear function is taken within the range [0, 1] in sigmoid function. If the value is greater than half of maximum value, then it assigns it to label 1, else it assigned to label 0. In SVM, if the output is greater than 1 that it belongs to one and if the output is -1, then it belongs to other class. The range of values ([−1, 1)] acts as a margin in the SVM algorithm.

3.6 Irrigation Process

Finally, depending upon the intensity of the disease, water from fishpond will be provided, which contains nitrogen, phosphorus and the substances that are essential for plants. This will reduce the usage of artificial fertilizers. The proposed system helps to improve the quality of the agricultural field by detecting temperature, soil moisture value and water level of the tank without human intervention. The automated system helps to optimize water more effectively. The sensors are used to get the data from the humidity and the temperature of the soil, the humidity, the temperature and the water level of the fish tank. These data are processed and estimates the amount of water required to irrigate into the field.

After the disease is detected in Jasmine plants, water from fishpond is poured in agriculture land. Jasmine plants must be placed in soil depth of 15 cm × 30 cm, moisture depth of 40–60%, the quantity of irrigation water in 15–20 L for 4 days. The pH range of a soil is 7.8 and has chemical properties of nitrogen 220 kg ha^{-}, phosphorous of 20.1 kg ha^{-} and potassium of 229 $kgha^{-}$. The fishpond is placed of height 11 feet and square feet 1500 × 2000 ft^2. The farm field area will be 20 × 16 ft^2. The fishes in the pond are Tilapia, pomfrets fish and ana fish. Fish are typically raised in indoor tanks, troughs or outdoor ponds, where they produce excrement.

4 Flowchart

The flowchart of the proposed system is shown in Fig. 8. After identification whether the plant is affected by diseases or not, the fishpond farm or normal water flows into the fields. This will help to improve the growth of the plants and reducing the diseases in the early stage itself. Preprocess the images, segmentation of testing datasets happens and then the classification of diseases with GLCM features. Then, the irrigation process continues until it reaches the water level threshold.

5 Results and Discussions

From the existing approach, SVM gives better accuracy with less training data. The various diseases of the plant have been identified using correlation, contrast, energy and homogeneity. The analysis of disease identification is shown in Fig. 9. The correlation, contrast, energy and homogeneity vary with each leaf. The features give the distance and angular spatial relationship over an image subregion scale value.

6 Conclusion

The goal of this research is to develop an advanced system to identify disease that affects the leaf of the Jasmine plant using machine learning. This is done using MATLAB, which input data images of the plant and the disease are sampled. Then, the samples are converted into segmented images, by which diseases are classified. The comparison of machine learning classifier used for plant disease detection is shown in Fig. 10. The SVM classifier gives 96% accuracy better than other classifiers. For higher accuracy, SVM technique is used, and it performs better with a low amount of training data and large features. Proper scaling can be provided for fair treatment among features. The irrigation system provides high rich nutrients to crop fields. The system reduces the pesticides as well as to increase the crop yields.

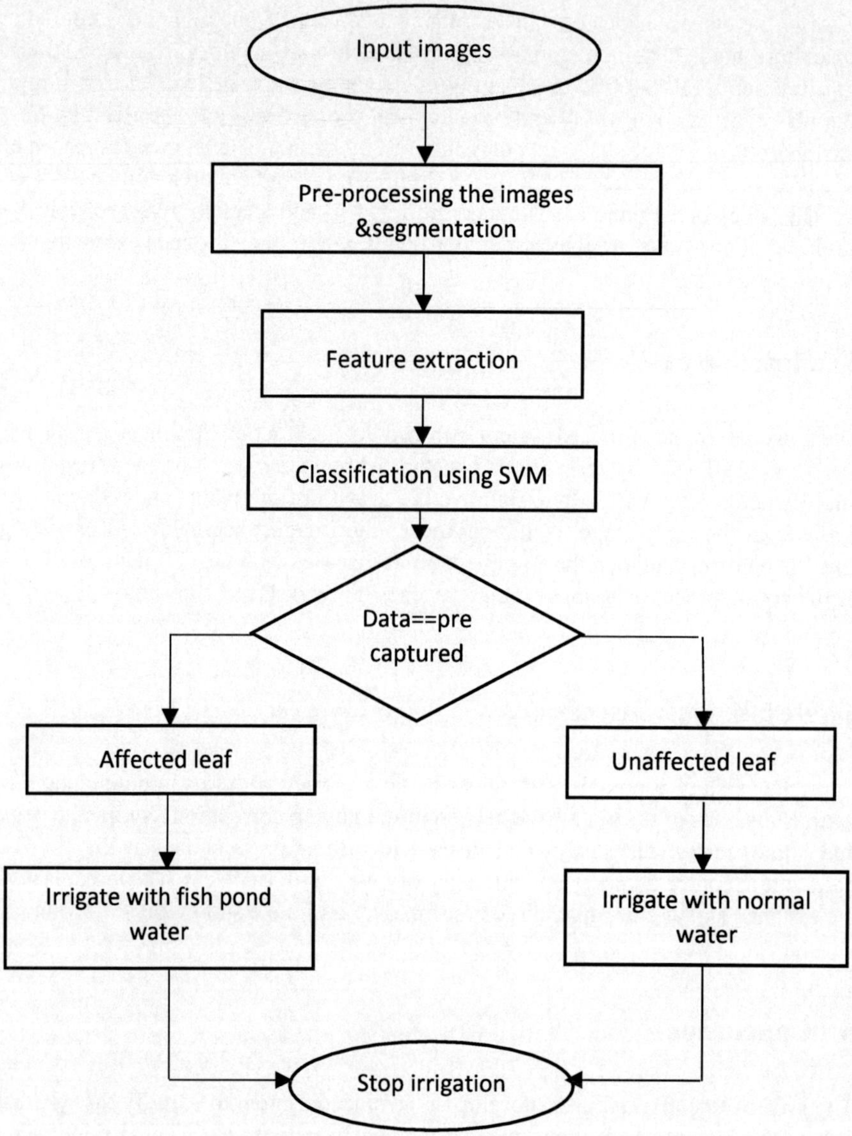

Fig. 8 Flowchart

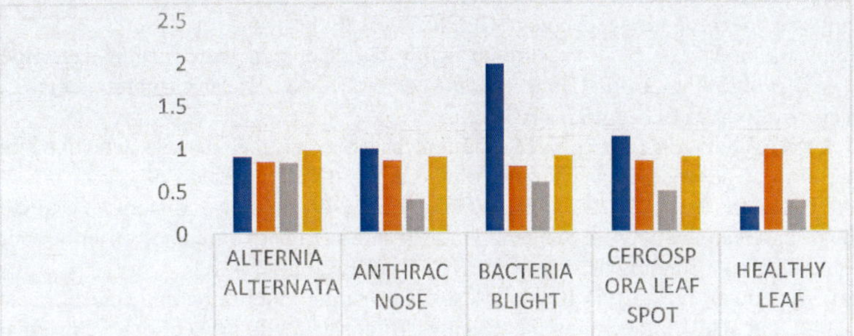

	ALTERNIA ALTERNATA	ANTHRACNOSE	BACTERIA BLIGHT	CERCOSPORA LEAF SPOT	HEALTHY LEAF
CORRELATION	0.889	0.9785	1.9723	1.1178	0.2878
CONTRAST	0.826	0.839	0.77	0.8306	0.957
ENERGY	0.818	0.387	0.5818	0.482	0.3632
HOMOGENETY	0.956	0.881	0.896	0.876	0.9572

Fig. 9 Analysis of GLCM features with various diseases

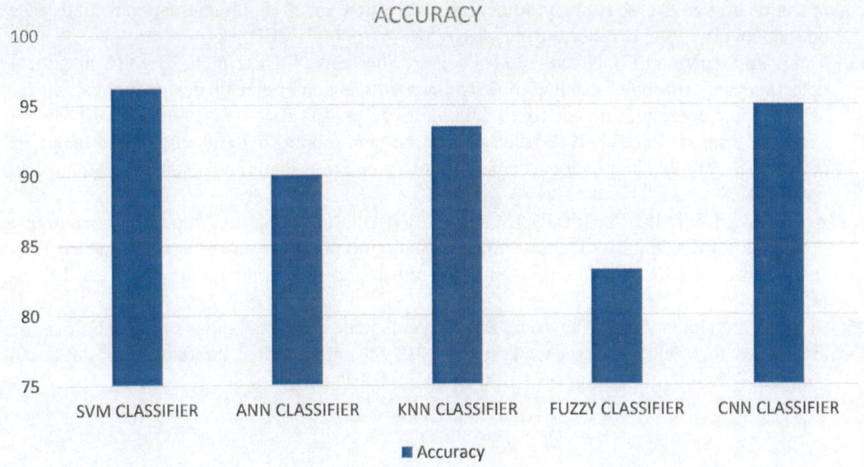

Fig. 10 Comparative results with existing approaches

References

1. Revathi P, Hemalatha M (2012) Classification of cotton leaf spot diseases using image processing edge detection techniques. In: 2012 international conference on emerging trends in science, engineering and technology (INCOSET). IEEE, pp 169–173
2. Landge PS, Patil SA, Khot DS, Otari OD, Malavkar UG (2013) Automatic detection and classification of plant disease through image processing. Int J Adv Res Comput Sci Softw Eng 3(7):798–801
3. Shinde SR, Sabale S, Kulkarni S, Bhatia D (2015) Experiments on content based image classification using color feature extraction. In: 2015 International conference on communication,

information & computing technology (ICCICT). IEEE, pp 1–6
4. Singh V, Misra AK (2015) Detection of unhealthy region of plant leaves using image processing and genetic algorithm. In: 2015 international conference on advances in computer engineering and applications. IEEE, pp 1028–1032
5. Padol PB, Yadav AA (2016) SVM classifier based grape leaf disease detection. In: 2016 conference on advances in signal processing (CASP). IEEE, pp 175–179
6. Park H, Eun JS, Kim SH (2017) Image-based disease diagnosing and predicting of the crops through the deep learning mechanism. In: 2017 international conference on information and communication technology convergence (ICTC). IEEE, pp 129–131
7. Saraf SB, Gawali DH (2017) IOT based smart irrigation monitoring and controlling system. In: 2017 2nd IEEE international conference on recent trends in electronics, information & communication technology (RTEICT). IEEE, pp 815–819
8. Shirahatti J, Patil R, Akulwar P (2018) A survey paper on plant disease identification using machine learning approach. In: 2018 3rd international conference on communication and electronics systems (ICCES). IEEE, pp 1171–1174
9. Ud Din Z, Adnan SM, Ahmad RW, Aziz S, Ismail W, Iqbal J (2018) Classification of tomato plants leaf diseases using image segmentation and SVM. Tech J 23(02):81–88
10. Poojary H, Shabari SB (2018) A survey on plant disease detection using support vector machine. In: 2018 international conference on control, power, communication and computing technologies (ICCPCCT). IEEE, pp 292–295
11. Avşar E, Buluş K, Saridaş MA, Kapur B (2018) Development of a cloud-based automatic irrigation system: a case study on strawberry cultivation. In: 2018 7th international conference on modern circuits and systems technologies (MOCAST). IEEE, pp 1–4
12. Maniyath SR, Vinod PV, Niveditha M, Pooja R, Shashank N, Hebbar R (2018) Plant disease detection using machine learning. In: 2018 international conference on design innovations for 3Cs compute communicate control (ICDI3C). IEEE, pp 41–45
13. Rao RN, Sridhar B (2018) IOT based smart crop-field monitoring and automation irrigation system. In: 2018 2nd international conference on inventive systems and control (ICISC). IEEE, pp 478–483
14. Abdu AM, Mokji MM, Sheikh UU, Khalil K (2019) Automatic disease symptoms segmentation optimized for dissimilarity feature extraction in digital photographs of plant leaves. In: 2019 IEEE 15th international colloquium on signal processing & its applications (CSPA). IEEE, pp 60–64
15. Shruthi U, Nagaveni V, Raghavendra BK (2019) A review on machine learning classification techniques for plant disease detection. In: 2019 5th international conference on advanced computing & communication systems (ICACCS). IEEE, pp 281–284

Chapter 16
A Novel Approach of Resource Scheduling Algorithm to Improve QoS in Green Cloud Computing

P. Geetha and Dr C. R. Rene Robin

1 Introduction

The ever-growing solicitation is managed through gigantic degree datacenters, which consolidate hundreds and an enormous number of servers with other establishments, for instance, cooling, and amassing and framework structures. Various web associations, for instance, Google, Amazon, eBay, and Yahoo are working such huge data centers around the globe.

Normally, the scheduling methods in Green Cloud Computing Environments are Resource scheduling, Task scheduling and Workflow scheduling. The designation of framework assets to different assignments is known as occupation/asset planning. The framework keeps up the organizes lines of occupations hanging tight for CPU time and must choose which occupation to take from the line and how much time to allot to it so all occupations are finished in a reasonable and opportune way. Recently, vivacity productivity has raised as one of the most significant plan necessities for current registering frameworks, for example, server farms and Clouds, as they proceed to devour huge actions of electrical force. Isolated from high working costs realized by handling resources, this prompts tremendous releases of carbon dioxide into the condition.

In Load based resource scheduling algorithms likely, First Come First Serve (FCFS) scheduling, Round Robin (RR) scheduling, Shortest Job First (SJF) scheduling, Priority based resource scheduling algorithms [1]. In Temperature based

P. Geetha (✉)
Dhanalakshmi College of Engineering, Chennai, Tamilnadu, India
e-mail: geetha_mails@yahoo.com

Dr C. R. Rene Robin
Jerusalem College of Engineering, Pallikaranai, Chennai, Tamilnadu, India
e-mail: crrenerobin@gmail.com

I. Jeena Jacob et al. (eds.), *Data Intelligence and Cognitive Informatics*, Algorithms for Intelligent Systems,
https://doi.org/10.1007/978-981-15-8530-2_16

resource scheduling algorithms are core level and socket level. As today the primary issue is realized that the produces are getting ready such gadgets which are anyway progressively proficient and precise however they utilize more vitality and advance very toxics, hazardous gases and synthetic concoctions. Numerous gadgets organizations particularly in PC use lead, mercury, cadmium and different toxics synthetic compounds. As indicated by another examination it is evaluated that PCs and different gadgets make up two-fifth of all lead in land-fills on the earth. Because of this explanation contamination is expanding quickly [2]. Server farm servers utilize 50 times more the vitality per square foot as in an office. Server farms are the fundamental vitality utilization sources.

The vitality utilization isn't just dictated by the proficiency of the physical assets; however, it is likewise subject to the asset the executive's framework sent in the foundation and productivity of uses running in the framework. Vitality productivity impacts end clients regarding asset utilization costs [3], which are regularly dictated by the Cost of Ownership (CO) brought about by an asset supplier. Lessen the hazard existing in the PCs, for example, a concoction known to cause malignant growth, nerve harm and resistant responses in people [4]. Vitality proficient asset the executives have been first present with regards to battery feed cell phones, where vitality utilization must be decreased so as to improve the battery lifetime. The vitality utilization is characterized by the normal force utilization over some stretch of time. Subsequently, the genuine vitality utilization by a server farm doesn't influence the expense of the foundation. Then again it is reflected in the power cost devoured by the framework during the time of activity which is the fundamental segment of a server farm's working expenses.

In this proposed system mainly focused on, to diminish booking job cost by limiting vitality utilization, effectively improve the resource utilization of the MPM and reduce energy consumption [5], to utilize on same VM requests on a single-core machine as well as the multi-core physical machines were running on below the dangerous temperature. MPMs are somewhat equalized or less than the temperature of execution among nodes and uniformly increased the resource cost obtained by varying the arrival rate of the jobs.

The remaining paper is structured as follows. Section 2 describes the review study of resource scheduling algorithms in green cloud computing. Section 3 describes the Proposed work of the given problem statement, GRUD-S algorithmic steps with its architecture. Section 4 follows simulation environments and their results. Section 5 shows the comparative results of resource scheduling algorithms into green cloud computing. Section 6 gives the conclusion and future work of a given proposed algorithm.

2 Related Reviews

Cao et al. [6], Vitality productive logical work process booking calculation to confine vitality use and CO_2 outpouring while at the same time satisfying certain Quality of Service (QoS). The distinctive development resource plan and assignment calculation apply Dynamic Voltage and Frequency Scaling (DVFS) advancement to reduce vitality use inside good execution cutoff points, and breaking point the Virtual Machine (VM) overhead for extra decreased vitality use and higher resource use rate. The offer of various server ranches from the vitality and execution adequacy perspectives is in like manner evaluated [7]. The reenactment results show that our calculation can achieve an ordinary up to 30% of vitality save assets and add the advantage use rate for about 25% provoking higher advantage and less CO_2 outpourings. It is a multi-step heuristic workflow scheduling algorithm, in particular, EARES-D to address the different destinations including ensured QoS, energy proficiency, CO_2 outflow by a gathering of accessible server farms. Careful examinations are performed to exhibit the effectiveness of EARES-D. Energy utilization, energy cost, CO_2 emanations are diminished, and asset usage rate is improved.

Singh, et al. [8], Vitality based productive asset booking calculation says, Taking imperativeness as a QoS parameter, a benefit is used for the arranging of Cloud exceptional job that needs to be done execution on the thought of essentialness use of the advantages. Essentialness is a basic segment to be contemplated at the hour of advantage arranging [9]. Power use is recognized by the unit of advantages that are eaten up by the Cloud purchaser for the execution of their Cloud remaining jobs needing to be done. After the minimization of the imperativeness use of the benefits, the resource would be reserved. Hence, essentialness based QoS can be given to resource customer and it is discovered to calculate the power usage of the server ranch by the Heterogeneous Cloud Workload Consolidation. At the point when essentialness usage is restricted, the cost of advantages will normally be diminished.

Shu et al. [10], This paper proposes an Improved Clonal determination calculation subject to time cost and imperativeness usage models in dispersed registering condition. Clonal selection algorithm (CSA) is a remarkable class of safe calculations which are persuaded by the clonal choice theory to make incredible methods for search and improvement. Clonal selection hypothesis yet, in addition, speaks to an astute abuse of a heuristic hunt in a tremendous achievable arrangement space. When another solicitation for asset shows up, the framework will run the ICSA to alter the general assignment of the assets. ICSA for asset distribution, the clonal administrator is a counteracting agent arbitrary guide prompted by the partiality. In the natural immune system, cloning infers that a social affair of unclear cells is made from a lone essential antecedent and only antibodies with high proclivity will be cloned to attack the pathogens. The antibodies are assessed over a partiality work and arranged in diminishing request of fondness [11]. Right off the bat, the partiality of every immune response is assessed, and the ones with higher proclivity are chosen for the people to come. By then the picked antibodies increase into explicit copies, and the copied and extraordinary ones are reproduced in the current people. A short time later, the antibodies in the populace will actualize change activity.

Yong et al. [12], In view of the ideas and standards of burden adjusting, a novel vitality effective burden adjusting worldwide advancement algorithm, called resource-mindful burden adjusting clonal algorithm for task scheduling, is proposed to manage the issue of vitality utilization in green distributed computing. Right off the bat, the issue is planned as a combinatorial advancement issue that expects to advance both vitality utilization and burden adjusting. At that point, the resource-mindful scheduling algorithm is proposed dependent on load adjusting technique and clonal selection standard. At last, recreation contemplates showing that the proposed algorithm can viably lessen vitality utilization in green distributed computing, and its investigation and abuse capacities can be improved and even. Clonal Selection Resource Scheduling Algorithm (CSRSA) in view of clonal selection algorithm (CSA) in green distributed computing has the most ideal capacity of the decreasing vitality utilization of server farms.

Kaur et al. [13], Temperature aware resource scheduling can give green improvements inside distributed computing condition. It is suggested that current and basic temperature of physical machines must be considered before circulating remaining task at hand among hubs. Part of research work has just been improved cooling framework. However, these frameworks devour a great deal of vitality in cooling the hubs. This paper proposes the strategy for temperature aware resource scheduling and contrasts it and other scheduling strategies, for example, FCFS and Round-Robin scheduling. Temperature aware resource scheduling calculation is better than existing techniques as it thinks about the basic temperature of a physical hub before making Virtual Machine (VM) allotment choice, in this manner lessening the requirement for cooling frameworks in a cloud domain. Investigation of the consequences of the recreation try appears that this calculation can successfully improve the normal resource use of the server farm and decrease vitality utilization. Right now is on the temperature of the physical machines in the cloud condition. At the point when the outstanding task at hand of the machine is expanded then the temperature of the Physical machine might surpass its basic temperature.

Kapur et al. [9], the paper presents a new Load Balanced Resource Scheduling Algorithm (LBRS) which performs the resource scheduling in a balanced banner. LBRS categorize the resources based on-demand as well as reservation. The LBRS calculation considers load adjusting as a significant QoS parameter, plays out a check for its need and whenever required (just when demands are very enormous or burden gets expanded with a section of time), does the heap adjusting and upgrades the exhibition just as the general asset cost. The uniform increment in the asset cost is likewise in light of the fact that the assets are constantly dispensed as for a steadily fixed proportion if there should arise an occurrence of the LBRS calculation. Further analysis of LBRS algorithm against other QoS parameters like Response Time, Throughput and CPU Utilization etc. is possible.

Geetha et al. [14], This paper focuses on VM planning and provisioning calculations, those are utilized for enhancement of various elements like Time, Cost, Energy and Security. There are fewer calculations are accessible that arrangement the VMs with the point of view of the security of neighboring VMs or hubs. The investigation uncovers that there are numerous vitality proficient systems for distributed computing and server farms that make cloud figuring a Green distributed computing. Green booking plans have appeared to lessen the transmit power (for example by 90%) and improve the Energy Efficient of the framework (for example by 94.2%) contrasted and conventional booking.

Tang et al. [15], DEWTS can use the valuable leeway time repetitively after cuts off are blended. DEWTS right off the bat figures the underlying planning request everything being equal and acquires the entire makespan and cutoff time dependent on Heterogeneous-Earliest-Finish-Time (HEFT) calculation. Through turning the processors with their running undertaking number and vitality use, the underutilized processors can be converged by shutting the last hub and redistributing the relegated assignments on it. At long last, in the assignment loosen stage; the undertakings can be circulated in the inert openings under a lower voltage and recurrence utilizing DVFS system, without disregarding the reliance limitations and expanding the loosened makespan. DEWTS has three significant stages: introductory undertaking mapping stage, processors consolidating stage, and undertaking drawing stage. To lessen the quantity of processors being utilized, the fitting schedule vacancies are right off the bat chose to put the undertakings which originate from other low-used processors.

3 Proposed Work

In Green Cloud Computing, Resource scheduling objectives are limited makespan, higher throughput, expanded asset use, load adjusted mapping. Especially, the energy consumption of any computing resource based on CPU, Memory, Storage and Network. QoS parameters are asset cost, reaction time, CPU utilization, throughput, holding up time, pivot time. And it is a three-step process of distribution, conglomeration and planning.

Resource Distribution includes distinguishing what assets are expected to finish the work. On account of consumable assets, it is basically the amount required. On account of re-usable assets, it is the all-out exertion required and the quantity of individual assets.

Once resource distribution and planning are complete, the resources can be conglomerated on day by day, week by week or month to month premise as suitable. The amassed information is typically introduced in a histogram that represents the fluctuating utilization of assets against time. On account of consumable assets, a combined bend is additionally used to show the aggregate sum devoured anytime. Barely any re-usable assets are boundless, so the time plan must be changed in accordance with considering the restricted accessibility of assets after some time. There are two ways to deal with accommodating asset cutoff points and time imperatives; asset smoothing (or time-restricted asset booking) and asset leveling (or asset constrained planning).

Proficient asset the board will improve conveyed registering execution by decreasing essentialness use, e-waste, and costs. In Green Cloud Computing, asset the board implies utilizing heterogeneous and geologically conveyed assets to meet customers' solicitations with the base negative impact on the environment. Green IT is named to be a significant advance. It incorporates a colossal number of center zones for example to give appropriate administration of intensity, virtualization of servers, structure of server farms, reusing strategies, eco-naming, condition supportability plan and vitality effective assets and so forth.

3.1 Green Resource Usage Distributor Scheduling (GRUD-S) Algorithm

3.1.1 Problem Statement

The main problem statement ensures that it will not provide sufficient resources for the execution of cloud remaining task at hand of jobs. The utilization of the same VM requests on the single-core machine was not running on below the dangerous temperature compared with the multi-core machines. Un-consistently expanded the asset cost got by changing the appearance pace of the occupations.

Proficient asset the board will improve conveyed registering execution by decreasing essentialness use, e-waste, and costs. In green cloud computing, asset the board implies utilizing heterogeneous and geologically conveyed assets to meet customers' solicitations with the base negative impact on the environment. It incorporates a colossal number of center zones for example to give appropriate administration of intensity, virtualization of servers, structure of server farms, reusing strategies, eco-naming, condition supportability plan and vitality effective assets and so forth.

3.1.2 Objectives of GRUD-S

The objectives of Green Resource Usage Distributor Scheduling (GRUD-S) are to make sure that this calculation will condition property for implementation of the cloud remaining task at hand of jobs. It facilitates to:

1. Measure at a reasonable level help execution and diminish booking job cost by limiting vitality utilization.
2. Successfully progress the supply consumption of the MPM and reduce force use.
3. Utilized on same VM requests on a single-core machine as well as the multi-core physical machines were running on below the dangerous temperature. MPMs are somewhat equalized or less than the temperature of execution among nodes.
4. Uniformly increased the reserve cost obtained by unreliable the coming rate of the jobs (Fig. 1).

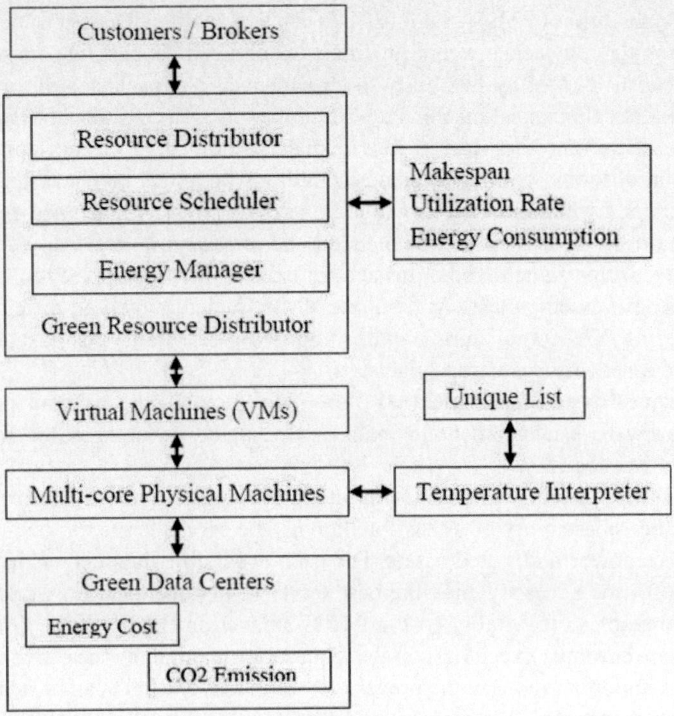

Fig. 1 Green resource usage distributor architecture

Customers/Brokers—Cloud clients or their middle people submit organization requests from wherever on the planet to the Cloud. It is basic to observe that there can be a differentiation between Cloud purchasers and customers of passed on organizations.

Temperature interpreter—The basic temperature is the perilous temperature which is set by one of a kind rigging creator. Right when a machine is running on or past fundamental temperature, by then the risk of hardware likewise, programming frustration will be more and machine requires a vital proportion of cooling, to perform computation in a perfect world. At the present time, save the machine from going to its essential temperature and extra power use by diminishing the enthusiasm of the cooling system. The essential temperature is assorted for different focuses.

Green Resource Distributor—Goes about as the interface between the Cloud framework and Customers. Green Resource Distributor contains Resource Distributor, Resource Scheduler, Energy Consumption. By supporting the advancement of VMs between physical center points, it enables dynamic migration of VMs according to QoS essentials. When VMs don't use all gave resources, they can be reasonably resized and hardened on an irrelevant number of physical center points, while inactive center points can be killed.

- **Virtual Machines (VMs)**—Various VMs can be dynamically started and stopped on a singular physical machine to meet recognized requesting, consequently giving most noteworthy flexibility to structure various packages of advantages on the identical physical machine to different express necessities of organization requests. Different VMs can in like manner at the same time run applications reliant on different working structure conditions on a lone physical machine. In like manner, by continuously moving VMs across physical machines, remaining jobs that need to be done can be blended and unused resources can be put on a low-power state, slaughtered or masterminded to work at low-execution levels in order to save essentialness. At first, one VM is fueled on and all different VMs are sleeping. VM combination requires improving the vitality cost just as the use of cloud assets of occupations viably.
- **Multi-core Physical Machines (MPMs)**—The essential physical figuring servers give hardware establishment to making virtualized advantages for satisfying organization needs.
- **Green Data Centers**—Choose server farm area calculating the server farm reason just as the accessibility of assets for running the server farm, for example, less expensive power and gifted assets. For each accessible data center that fulfills the cutoff time necessity, pick the best server farm dependent on work process consummation time, vitality cost and CO_2 emanation (ECC). If the Green data centers are currently executing service requests or maintaining the active sessions, it is first important to make the process server inactive, which can be achieved in two unique manners. The primary methodology is to just hold up until existing service requests complete and all active sessions terminate. The subsequent methodology is to relocate the execution of services and active sessions onto another data centre.

Resource consolidation gives adaptability of sharing the accessible assets and dispensing then relying upon the business need. Asset solidification by virtualization assists with lessening the quantum of vitality devoured. Different ways by which asset advancement can be accomplished are closing down assets when not required, reuse and reusing.

4 Methodology

In this framework, guaranteed the Quality of Service (QoS) parameters of Green Cloud Environments are initialized as well as computed the energy cost, average utilization rate, energy consumption, CO_2 emission rate, makespan, temperature interpreter. Resource scheduling is a three-step process of distribution, conglomeration and planning. The announcement is that there is a lot of ordinary jobs(tasks) and each activity has numerous sub-jobs with priority limitations. Each sub-job is permitted to be prepared on some random accessible resources.

4.1 Resource Scheduling

Resource scheduling demonstrates the way toward sorting out the assets among the diverse cloud clients as indicated by specific principles and guideline of resources use under a predefined cloud condition. Resource planning for an asset the board is the fundamental innovation of distributed computing. The hazardous interest of distributed computing has prompted the need for cautiously dealing with the assets that offer types of assistance to the clients. Resource Scheduling manages this thought. It alludes to the procedure of suitable age of the calendar that chooses which undertakings will be mapped on to which assets.

A cloud resource has given the purpose of ability (e.g., CPU, Memory, Network, Storage). A sub-job is taken care of on every benefit thusly, and the given resources are available continually. In Green Resource Distributor, Resource distributor will maintain a list of resource jobs, how to allocate the jobs to Customers/Brokers based on their needs, configuration, prior allocation of jobs to the customers.

Let Resource of jobs (RJ) = $\{RJ_1, RJ_2, RJ_3, \ldots\ldots. RJ_n\}$ be the set of m available assets which should process j independent jobs denoted by the set $R = \{R_1, R_2, R_3 \ldots\ldots R_j\}$ where m = 1,2,3,…..n, j = 1, 2, 3, ….i.

Resource distributor will categorize the resources as a condition scheme and uncondition scheme. The condition scheme of resources are already jobs are allocated to the Customers/Brokers, the uncondition scheme of resources of jobs are allocated based on the Customers/Broker's need.

Resource scheduler, categorize the resources of jobs are scheduled into VMs. Then calculate the cost per job of each condition scheme then it will arrange it in order. If all condition scheme of resources is completed, then find the total number of jobs left to be served. Calculate the required number of uncondition scheme of resources from the respective available jobs.

(i) Makespan: Makespan indicates the finishing time of the last job. The planned makespan is the time that contains as a rule task zenith on resources including tolerating, taking care of, and holding uptime. The completing time of employment T_j on asset RJ is meant as F_{ij}. The key point is to reduce makespan that can be as follows:

makespan = $\max\{F_{ij} \mid T_j\}$, Where $i = 1, 2, 3 \ldots.. m$, $j = 1, 2, 3, \ldots\ldots n$.

MPM System Resource Utilization is the percentage used job resources compared with total jobs. Resource usage proportion is the level of utilized assets contrasted and all-out assets. This measurement is the fundamental element that right now have been considered paper as it can coordinate mirror the asset use effectiveness of a calculation.

(ii) Average resource utilization $= \sum \frac{\text{Time taken by resource } i \text{ to finish all jobs}}{\text{makespan} * n}$, Where n is the number of resources, $i = 1$ to n.

(iii) Resource Cost($) is computed based on the request arrival of jobs. The resource cost also depends on the allocated jobs to the nodes.

Temperature interpreter is utilized to anticipate the temperature of the Multi-center Physical Machine(MPMs) before it comes to the (maximum)dangerous temperature while assigning the VMs solicitation to it. Perilous temperature is the temperature which is fixed by exceptional. At the point when an MPMs is progressively on or past perilous temperature, then the danger of equipment and programming disappointment will be more or MPMs requires a significant measure of cooling, to perform turning out ideally. The dangerous temperature(T_{dt}) and the execution temperature(T_{et}) in the virtualized cloud system are assumed here.

(iv) Temperature interpreter (T_{ti}) $= T_{dt} - T_{et}$

Temperature interpreter is to calculate the temperature of each MPMs and create a unique total list. Initially, it estimates the temperature T_n of each MPMs and stored it in a unique total list formed as a queue. Then it finds $T_{dt} - T_n$ if MSMs reaches the maximum value then VM is allocated.

Energy Manager will maintain the consumed energy based on static as well as runtime-independent. The runtime-independent energy consumption of a task/job is the amount of energy used by the hardware while executing a particular job. For a given job, it is well known about the relation between the execution time and energy consumption equals the time multiplied by the power. It can be computed as,

(v) Energy Consumption (EC) = Power (CP) $\times T$

Where power = W/T, CP is the Computing power of an MPM, T is the execution time of a job.

In Green Datacenter, takes the outstanding jobs on processors which are used to reduce the clock frequency and also increase the task execution time then automatically reduce the energy costs.

Green Cloud Computing is the blend of essentialness capable figuring and circulated registering. Disseminated figuring is the rising advancement that uses server farms to keep up the data. The cutting edge and academic establishment are being on a cloud structure. In this way, the greater part of the association and organizations heading towards distributed computing to make sure about their significant information, this raise to the making of enormous scope Data Centers. Yet, data focuses devour huge measure of electrical force. As the registering framework devours overabundance measure of intensity which brings about emission rate of carbon dioxide (CO_2) and this prompts influences our condition.

The process of calculating the energy cost, CO_2 emission rate based on these parameters the data center has been chosen. To calculate the energy cost of a Green data center,

(vi) Energy Cost (ECost) = E_P * C_P where C_P is the Electricity cost ($/kWh) and E_P is the Executing price ($/hour)

(vii) Carbon Emission of a Green Data Center is calculated by, CO $E_P = E_P * R_K$, Where R_K is the CO_2 emission rate (kg/kWh)

4.2 GRUD-S ALGORITHM

1. ***Input:*** *Total Number of Job Requests;* ***Output:*** *Find Makespan, Resource Utilization, Resource cost*
2. *begin*
3. *Get the number of requested jobs from the Customers / Brokers*
4. *For each request of job*
5. *Divide the jobs based on the Customer's condition.*
6. *Categorize the resources of jobs are scheduled into VMs.*
7. *On completion of jobs, to compute makespan, average resource utilization, resource cost.*
8. *Call - Algorithm for finding Temperature interpreter for each MPM. Go to step 5*
9. *Relocating VMs across physical machines, remaining tasks at hand can be merged and unused assets can be put on an off state. Otherwise on the state.*
10. *Choose a Green Datacenter*
11. *Find Energy Consumption, Energy Cost, CO_2 emission rate.*
12. *Residual idle slot of GDC among on processors to reduce the energy costs.*
13. *End for*
14. *end*

Algorithm for finding Temperature interpreter for each MPM

1. *Allocation of VM to the MPMs according to its temperature interpreter*
2. *For each MPM*
3. *Calculate Temperature interpreter value*
4. *Find initial temperature of MPMs*
5. *Find the execution temperature of MPMs (T_{et})*
6. *Calculate T_n for all MPMs*
7. *Store it in unique list of all MPMs*
8. *Fix the dangerous temperature (T_{dt}) by the User*
9. *Then if (max< T_{dt} - T_{n})*
   a. *Find Temperature interpreter (T_{ti}) = T_{dt} - T_{et}*
   b. *Allocate job to the first MPM in the list*
   c. *Else remove the MPM in the list*
10. *Endif*
11. *End for*

5 Experimental Setup and Evaluation

The experimental setup of GRUDS algorithm of Green Cloud Computing platform is that 6 Multi-core Physical Machines of 8 GB RAM of 6 TB storage that has a capacity power of 10,000 MIPS. Simulation Environment of Cloudsim toolkit to modal the Green Cloud Computing Infrastructure with time-dependent cloud resources. It is also used to speed up the process, to avoid data center overloads, to reduce the cost of energy consumable resources are taken. This proposed algorithm based is evaluated based on the number of completed jobs and allocation of jobs to the VMs.

Case 1: Comparison of results with GRUD-S, DEWTS, LBRS, and CSRS of Average Resource Utilization from 100 to 500 requested jobs (Fig. 2).

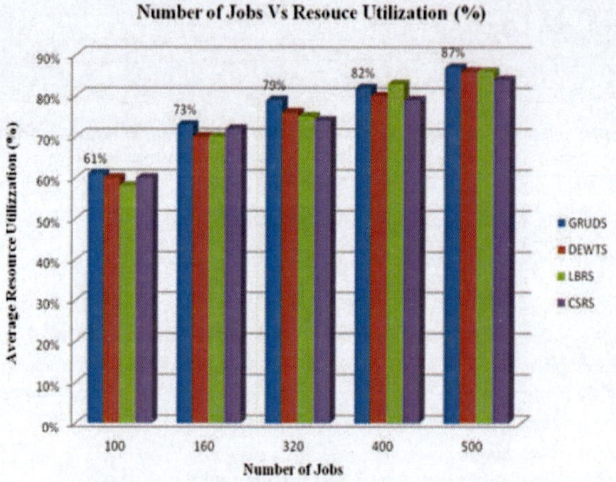

Fig. 2 Comparison results of resource utilization (%)

Case 2: Comparison of results with GRUD-S, DEWTS, LBRS, and CSRS of Resource Cost from 100 to 500 requested jobs (Fig. 3).

Case 3: Comparison of results with GRUD-S, DEWTS and CSRS of Energy consumption ratio of up to 500 requested jobs (Fig. 4).

Case 4: Comparison of results with GRUD-S, DEWTS and CSRS of Energy consumption ratio of up to 500 requested jobs (Fig. 5).

Case 5: Comparison of results with Temperature average of using 4 Single-core machines using TARS to 4 Dual Core machines (MPMs) ratio of GRUD-S up to 500 requested jobs (Fig. 6).

Case 6: The process of calculating the energy cost, CO_2 emission rate based on these parameters the data center has been chosen. Usually, data centers consume

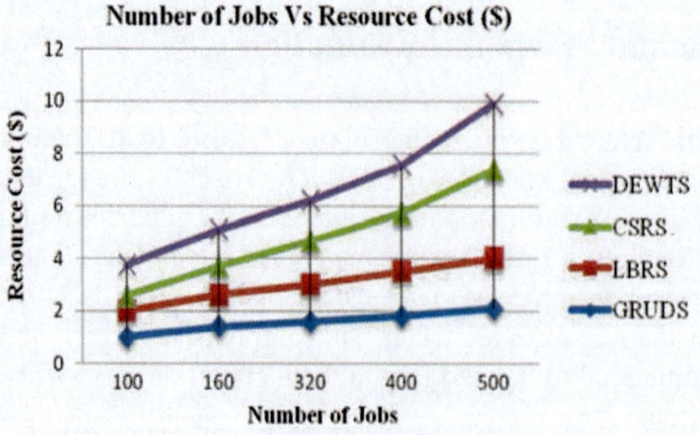

Fig. 3 Comparison results of resource cost ($)

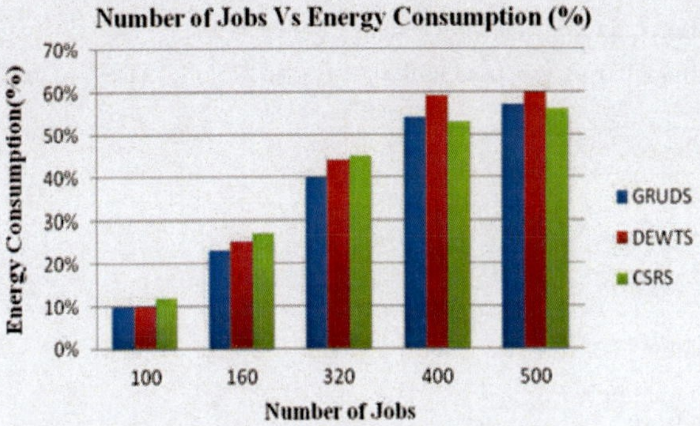

Fig. 4 Comparison results of energy consumption (%)

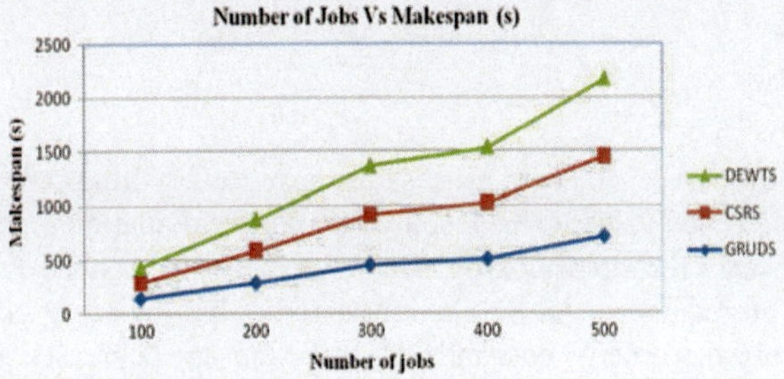

Fig. 5 Comparison results of Makespan(s)

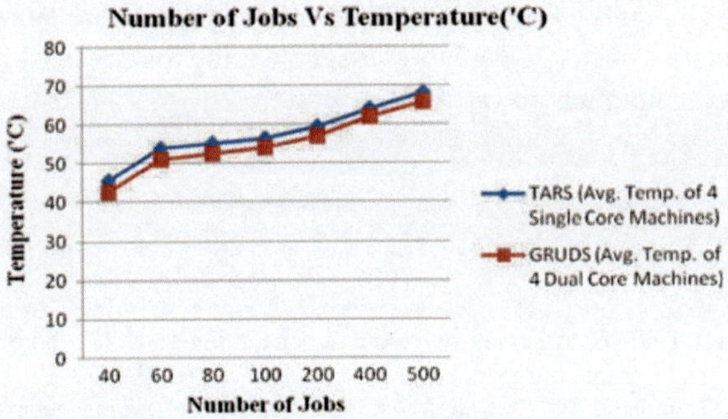

Fig. 6 Comparison results of Temperature (°C)

more electrical power. As our green data center consumes an extra amount of power will results in the emission of CO_2 as well as reducing energy consumption.

Case 7: Comparison of various resource scheduling algorithms with GRUD-S (Table 1).

Table 1 Comparison of green resource scheduling parameters

S.no.	Green computing parameters	Proposed (GRUD-S)	LBRS	TARS	DEWTS	CSRS
1	Makespan(s)	√			√	√
2	Resource utilization (%)	√	√		√	√
3	Cost-effective($)	√	√		√	√
4	Energy consumption (%)	√			√	√
5	Carbon emission rate(kg/kWh)	√		√	√	
6	Temperature interpreter(°C)	√		√		

6 Conclusion

In this paper, as the pervasiveness of processing despite everything keeps on rising, the need of difference and evaluation of resource booking calculation in the green cloud can help to choose the resource booking calculation dependent on the outstanding task at hand's QoS prerequisites. As the demonstrated output of energy consumptions, temperature interpreter, energy cost and CO_2 emissions are decreased and effectively improved resource utilization rate based on the allocation of resources. Remaining jobs at hand should be executed productively in order to be versatile, adaptable and to keep away from over stacking and under stacking of resources.

Future research in this paper, to test this scheduling algorithm on real-time applications. It is stretched out to QoS-mindful asset planning for Green Cloud Computing and improves a commitment towards an in future.

References

1. Singh S, Chana I (2016) A survey on resource scheduling in cloud computing: issues and challenges. J Grid Comput https://doi.org/10.1007/s10723-015-9359-2
2. Dabbagh M, Hamdaoui B, Guizani M, Rayes A (2015) Towards energy-efficient cloud computing: prediction, consolidation, and overcommitment. IEEE Netw 29:56–61
3. Kamble RS, Nikam DA (2013) Green cloud computing- new approach of energy consumption. Int J Latest Trends Eng Technol (IJLTET) **3**(2). ISSN 2278-621X
4. Luo L, Wu W, Di D, Zhang F, Yan Y, Mao Y (2012) A resource scheduling algorithm of cloud computing based on energy efficient optimization methods
5. Younge AJ, Von Laszewski G, Wang L, Lopez-Alarcon S, Carithers W (2010) Efficient resource management for cloud computing environments. In: IEEE international conference on green computing. https://doi.org/10.1109/greencomp.2010.5598294
6. Cao F, Zhu MM (2013) Energy efficient workflow job scheduling for green cloud. In: 2013 IEEE 27th international symposium on parallel & distributed processing workshops

7. Hussain A, Aleem M, Iqbal MA, Islam MA (2019) Investigation of cloud scheduling algorithms for resource utilization using cloudsim. Comput Inform 38:525–554 (2019). https://doi.org/10.31577/cai20193525
8. Singh S, Chana I (2014) Energy based efficient resource scheduling: a step towards green computing. Int J Energy Inf Commun 5(2):35–52. http://dx.doi.org/10.14257/ijeic.2014.5.2.03
9. Kapur R (2015) A workload balanced approach for resource scheduling in cloud computing. IEEE
10. Shu W, Wang W, Wang Y (2014) A novel energy-efficient resource allocation algorithm based on immune clonal optimization for green cloud computing. EURASIP J Wirel Commun Netw 2014:64. http://jwcn.eurasipjournals.com/content/2014/1/64
11. Ma T, Chu Y, Zhao L, Ankhbayar O Resource allocation and scheduling in cloud computing: policy and algorithm. IETE Tech Rev. http://dx.doi.org/10.1080/02564602.2014.890837
12. Lu Y, Sun N (2019) An effective task scheduling algorithm based on dynamic energy management and efficient resource utilization in green cloud computing environment. J Clust Comput https://doi.org/10.1007/s10586-017-1272-y
13. Kaur A, Kinger S (2013) Temperature aware resource scheduling in green clouds. In: IEEE transactions on 2013 international conference on advances in computing, communications and informatics (ICACCI)
14. Geetha P, Rene Robin CR (2019) A performance - review of cloud scheduling algorithms into green environments (JARDCS). http://www.jardcs.org/archivesview.php?volume=1&issue=3&page=5
15. Tang Z, Qi L, Cheng Z, Li K, Khan SU, Li K (2016) An energy-efficient task scheduling algorithm in DVFS-enabled cloud environment. J Grid Comput. https://doi.org/10.1007/s10723-015-9334-y

Chapter 17
Machine Learning Approach for Analysis of the Company Performance Based on Fundamental Data

Pratik Rathi, Madhur Kabra, Rahil Sheth, and Jignesh Sisodia

1 Introduction

One of the major factors that play an important role in determining the company's growth and performance is the influence of fundamental ratios. With a large number of financial ratios available, it is important to determine the set of ratios that are significant and have a major impact on the company's performance in the future. There is a need to understand the company's performance [1] and finds out the attractiveness of the company based on its fundamentals. This paper sheds a light on the aforementioned subjects and provides a method to find the company's performance through the financial ratios. The performance of the model is analyzed using machine learning algorithms [1, 2] to get the most optimized results. The company's performance can be gaged based on a variety of different ratios that belong to various categories.

This paper uses the following categories of ratios for the analysis of the companies: valuation ratios, profitability ratios, liquidity ratios, and per share ratios.

These categories of ratios [2–4] and their significance on a company's performance are discussed below:

P. Rathi · M. Kabra · R. Sheth (✉) · J. Sisodia
Sardar Patel Institute of Technology, Bhavans Campus, Old D N Nagar, Munshi Nagar, Mumbai 400058, India
e-mail: rsheth972@gmail.com

P. Rathi
e-mail: pratikrathi14@gmail.com

M. Kabra
e-mail: madhur.kabra@spit.ac.in

J. Sisodia
e-mail: jsisodia@spit.ac.in

I. Jeena Jacob et al. (eds.), *Data Intelligence and Cognitive Informatics*, Algorithms for Intelligent Systems,
https://doi.org/10.1007/978-981-15-8530-2_17

- Per Share Ratio: Per share ratio is used to find out the net earnings of a company that is available for payment to the common shareholders of the company. Per share, the ratio includes several ratios. The dividend per share ratio is one such ratio under this category which indicates the dividends issued by a company for every outstanding share. Another important ratio is "Basic earnings per share," which is used to indicate how much of a firm's net income was assigned to each share from the total number of issued shares.
- Profitability Ratios: Profitability ratios are used to find out the earnings of the company about other costs incurred by the company such as assets, operating costs, shareholder's equity. This set of ratios are used to evaluate the profitability of a company with other attributes.
  It includes several ratios, one such ratio is net profit margin which indicates the profit margin of the company which is the percentage of revenue when expenses are removed from the sales.
  Return on assets is a profitability ratio that indicates the amount of profit the company generates from its asset.
- Liquidity Ratios: Liquidity ratios determine the ability to pay the debtors off their debts without raising external capital. It includes some prominent ratios like quick ratio, current ratio, inventory turnover ratio. The current ratio is the ability of a company to pay off its liabilities with the help of its current assets such as cash, accounts, and other such assets. A quick ratio determines a company's ability to meet its short-term commitments with the liquid assets.
- Valuation Ratios: It determines the worth or valuation of the company. It serves as a useful tool for finding investment potential.
  It includes price to book value which is used to estimate the value of a company. Consider, for example, a company is eventually going to be shut down. So all the assets will be cleared, and debts will be paid off. Whatever assets valuation remains is called the book value of that company.
  PBV ratio is used to indicate the value of stocks. More is the ratio value, more high priced will be the value of a stock.

The algorithms discussed by this paper identifies the most significant ratios [5–7] from the above categories and uses it to compare the different companies and also to analyze the performance of any particular company based on its past years.

2 Methodology

The method derived in this paper analyzes the performance of the company and also determines the grade of the company among its peers using the following algorithms. The assumptions for the algorithms are stated as follows:

- This paper considers a total of 35 financial ratios and finds their significance score.
- The dataset for the algorithm consists of data for the ten companies over the last ten years in each sector.

2.1 Selection K Best Algorithm

The above-discussed ratios cover a very wide array of topics, and thus, there is a need to find the most significant ratios with utmost importance. This paper uses selection K best algorithm to find the best attributes justifying the growth of the company.

The 35 different ratios are given as input to the system and the ratios which correlate the most toward the company's growth are identified using this algorithm. The top ten attributes which affect the company's growth which are obtained from this algorithm. Selection K best is an attribute selection method that determines the attributes which impact most on our output or predictor variable.

2.2 Random Forest Classification

This is a type of classification algorithm [8]. It is based on the decision trees, and it combines the output of all the decision trees. The more the number of decision trees the more is the accurate outcome. It is based on the rule-based system. Given the training dataset with targets and features. So, getting the prediction depends on the test dataset can be derived using the same rules.

This algorithm takes the input of the ten ratios which are generated from the selection K best algorithm and uses that to predict the future growth of the company. This algorithm obtained an accuracy of 78% for the dataset that was tested with. With this algorithm, the future growth of the company can be predicted based on current fundamental analysis of the data with high accuracy.

2.3 Linear Regression for Cumulative Grading

Linear regression is used to plot the trend line of the company's previous performances plotted against the years. This provides the trend in which the company is moving. This helps to find the cumulative grade of the company since it will be calculated by considering the trend of the company.

This algorithm considers the company's previous performance and then generates the best fit line of the company's periodic growth [9]. This data helps us to identify whether the company is moving in an increasing trend or a decreasing one.

The slope of the line helps in identifying this trend and plays an important factor in deciding the grade of the company. The more positive the scope, the better the company's performance and vice versa. Thus, this algorithm helps to find the growth trajectory of the company.

2.4 Multiple Linear Regression for Attractiveness Score

This algorithm is used to determine the relationship between the dependent and independent variables. It can also be used to predict the outcome of variables that rely on others. The coefficient for determination (R-square) is used to measure the variations in an outcome which can be attributed to the variation in independent variables. R-square always lies between 0 and 1; the higher the value better is the predicted outcome.

The formula for MLR is: $y = a + b_1x_1 + b_2x_2 + b_3x_3.....b_nx_n$.
where

a y-intercept (constant).
b slope coefficient for independent variables.
y dependent variable.

In our case, independent variables are taken as the following financial ratios:

- Dividend/Share
- PBDIT/Share
- Asset turnover ratio (%)

And the predictor variable is PE Ratio since the company's growth and performance depend significantly on the value of this ratio.

PE ratio is formulated as (Price of share/Earning per share). If the predicted PE ratio with the help of the MLR algorithm is lower than the actual PE ratio, it indicates that the current price of the company is overvalued, which indicates a sell signal on that specific stock.

If predicted PE ratio is higher than the actual PE, it indicates that the current price of the company is less than the price which the algorithm has generated, i.e., the company is undervalued creating a buy signal on that particular stock.

Since the PE ratio is sector specific, the data have been considered of only that particular sector, and the algorithm is trained only on the stocks of that sector. This provides greater accuracy of the algorithm.

Then, this paper determines the ATTRACTIVENESS MEASURE OF THE COMPANY which is the difference between the actual and predicted value of PE ratio; thus, it can be derived that if the company is overvalued or undervalued.

The attractiveness index of the company is defined as the ratio of attractiveness score and the actual PE ratio.

- Dividend Per Share: The DPS ratio refers to the total dividend given out by the company for one year divided by the outstanding shares. The DPS value can indicate the profitability of the company. It can tell about financial stability and its current health.
- PBDIT/Share: PBDIT per share stands for profits before depreciation, interests, and taxes which can be used to describe as a substitute for the total income. Simply, it is a measure of profitability.

- Asset Turnover Ratio(%): This ratio indicates the ratio of company sales to the underlying assets. It can indicate the efficiency of the company. The higher the value, the more efficient the company is.

3 Implementation

The user first uploads the balance sheet data of the company in the excel format. After uploading, the model computes various financial ratios like the current ratio, the quick ratio [5, 6] to mention a few. The system shows the significance of the ratios and the ideal values for the same.

The user next uploads the data of financial ratios [5–7] over the past ten years to get a year-on-year grade of the company and overall grade determined by the linear regression model. The different grades of the company denote the class of the company. The overall grade of the company is calculated using four different attributes. The mean of the previous performance of the company and the grade generated by the model is taken into consideration.

The trend line is generated, and the slope of it is also taken to find the overall trend of the company. The initial values are also taken by considering the y-intercept of the trend line. The number of years to calculate the overall grade also influences it. The various factors give a grade which is a reflection of the company's performance and the consistency of the company to deliver it year-on-year. A better overall grade shows the company's dedication toward achieving excellent performance.

For comparison between companies in a sector using multiple linear regression model, the predictor variable, i.e., PE ratio is calculated by considering the three ratios mentioned above for the ten companies in the FMCG sector, and the result is displayed in the form of graphs to get an overview about the user's company to see where does it lie as compared to other companies and what does it's PE ratio value indicate about its financial health.

Figure 1 Indicates an analysis of current ratio and what is the significance for the ratio.

The attractiveness measure of the company is calculated based on the company's past performance [8] and the trend in the company's growth derived from the regression model. The below image displays a bar graph of the company's quarter-wise performance, and based on this performance and also the MLR output which is achieved the overall attractiveness measure of the company is calculated. This grade is having a value between 0 and 1. This value is converted to percentage form and displayed as a result to the user. The higher the percentage, the better is the company's quality.

Figure 2 shows the year-wise grade assigned to each company based on its previous performance derived using a regression model.

Figure 3 shows the grade assigned to each company based on its year-wise percentage growth. Higher grade signifies better growth.

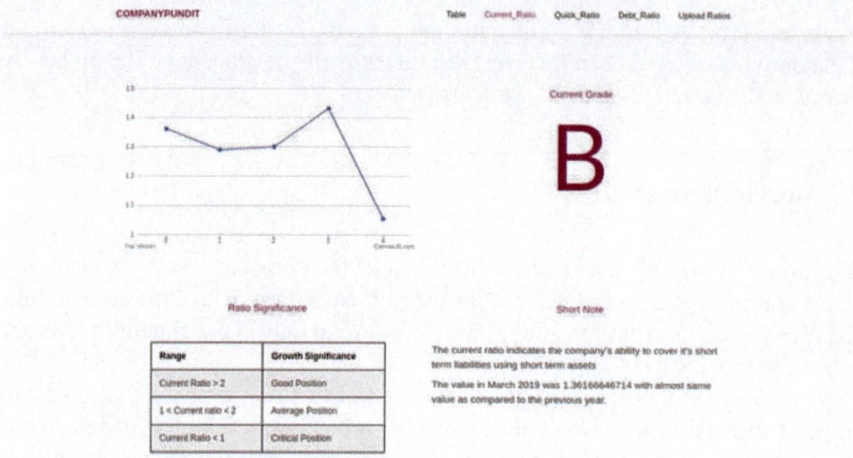

Fig. 1 Ratio analysis

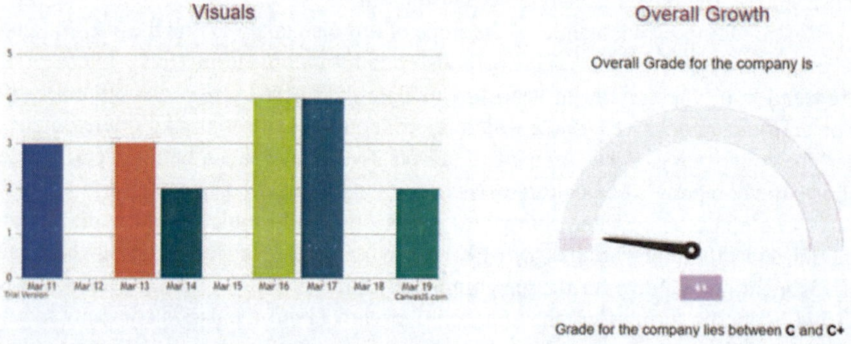

Fig. 2 Cumulative grade and year-wise grade

Figure 4 depicts the comparison of a user's company with its top peers in the FMCG industry. The ratio value for the company is well above the average sector value of the ratio which signifies the company's ability to generate more sales from its underlying current assets. The comparison can be made similarly for the other ratios as well which can indicate if the company is strong or weak as compared to its competitors in the same market sector.

Figure 5 indicates the attractiveness index of the company and other metrics such as predicted and actual PE.

By looking at the comparison, it can be said that the company is slightly overpriced which is evident from the negative value of the attractiveness score. The more positive the difference between predicted and actual PE, the more attractive the company is for the investors as the company can be claimed as underpriced.

Grade for each year

Mar 11	B
Mar 12	A+
Mar 13	B
Mar 14	B+
Mar 15	A+
Mar 16	C+
Mar 17	C+
Mar 18	A+
Mar 19	B+

Grade	Significance
A+	Exemplary Position
A	Great Position
B+	Good Position
B	Comfort Position
C+	Average Position
C	Poor Position
D+	Bad Position
D	Critical Position

Fig. 3 Class of the company

ASSET TURNOVER RATIO

Comparisons

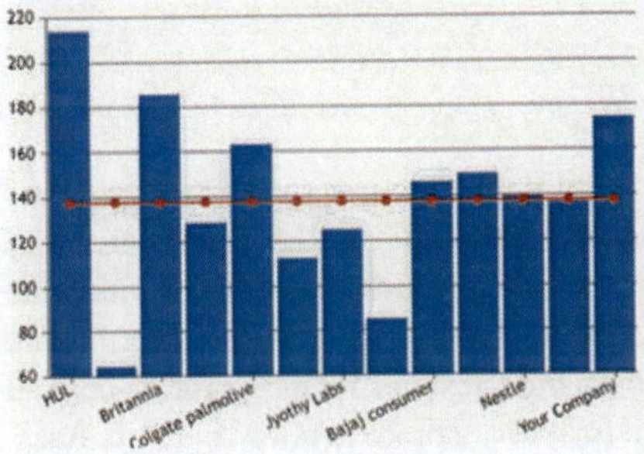

Ratio Significance

The asset turnover ratio is an efficiency ratio that measures a **company's ability to generate sales from its assets** by comparing net sales with average total assets.

The company ratio is 26.41858253991311 percent above the optimal ratio of the sector.

The higher the Ratio, the more likely it is that a business uses its assets efficiently to generate sales.

Fig. 4 Asset turnover ratio analysis

Parameters	Value
Attractiveness Index	-0.0020285184189917497
Attractiveness	-0.09858599516299904
Predicted PE	48.501414004837
Actual PE	48.6

Fig. 5 Attractiveness metrics

Fig. 6 Accuracy results

ALGORITHM NAME	ACCURACY
KNN algorithm	33%
Naïve Bayes Classifier	38%
Support Vector Machine(SVM)	55%
Random Forest Classification	78%

4 Results

Figure 6 indicates different algorithms tried along with their accuracy. From the table, it is evident that the random forest classification algorithm [8] has the highest accuracy reason being it considers random samples from data and combines the output of all the decision trees. For all the algorithms tried, a set of ten financial ratios are taken [1–5] and their data for the last ten years. This result is used for the prediction of PE ratio, and it predicts whether the company is overvalued or undervalued based on it.

5 Conclusion

Thus, the model provides an in-depth analysis of the company performance and growth in a particular market sector which can additionally be used as a comparability measure with companies in the same sector. The attractiveness measure can indicate the current financial status of the company based on the price-to-earning ratio. The important aspect of the model is to consider only the subset of financial ratios instead of considering all the ratios which will lead to further accurate results of our analysis. Altogether, one will be able to obtain its company performance and its comparison with the top peers in the same industry.

6 Future Work

The model considers only the financial and fundamental aspects [10–12] of the company which sometimes cannot be regarded as a true measure of the company's performance. The attractiveness measure is based on the multiple linear regression algorithm wherein only the important ratios are considered. The model incorporates year-on-year data and not quarterly results for computing the metrics. The system doesn't collaborate with the scores of the company based on its past performance and in comparison to its peers. A proper collaboration of both scores is difficult to achieve. Technical and sentiment [3, 4] aspects of the company can be effectively included in the model to derive accurate predictions. A consolidated excel sheet can be uploaded by the user to generate the performance and growth results. The model does not compute the ratios but takes them directly. An effort can be made to derive the ratios from the balance sheet, cash flow statements, and other financial documents.

References

1. Ratto AP, Merello S, Oneto L, Ma Y, Malandri L, Cambria E (2018) Ensemble of technical analysis and machine learning for market trend forecasting. In: 2018 IEEE Symposium Series on Computational Intelligence (SSCI)
2. Gupta A, Chaudhary DK, Choudhury T (2017) Stock market prediction using Functional Link Artificial Neural Network (FLANN). In: 2017 3rd international conference on Computational Intelligence and Networks (CINE)
3. Ren R, Wu DD, Liu T (2018) Short term market prediction using deep learning. In: IEEE Systems Journal, vol 111. CMP Books
4. Khare K, Darekar O, Gupta P, Attar VZ (2017) Predicting stock market movement direction using sentiment analysis. In: 2017 2nd IEEE international conference on Recent Trends in Electronics, Information Communication Technology (RTEICT)
5. Yunus NM, Malik SA (2018) Financial performance evaluation of construction industries. In: 2018 international conference on innovation management and technology
6. Loke KS (2017) Impact of financial ratios and technical analysis on stock price prediction using random forests. In: 2017 International Conference on Computer and Drone Applications (IConDA)
7. Yanhong L, Peng L, Zheng Q (2006) An analysis model of financial statements based on data mining. In: 2006 3rd international IEEE conference intelligent systems
8. Lin W, Wu Z, Lin L, Wen A, Li J (2017) An Ensemble random forest algorithm for insurance big data analysis. IEEE Access 5:16568–16575. https://doi.org/10.1109/ACCESS.2017.2738069
9. Loke KS (2017) Impact of financial ratios and technical analysis on stock price prediction using random forests. In: 2017 International Conference on Computer and Drone Applications (IConDA), Kuching, 2017, pp. 38–42. https://doi.org/10.1109/ICONDA.2017.8270396
10. Tan CS, Yong CK, Tay YH (2012) Modeling financial ratios of Malaysian plantation stocks using Bayesian Networks. In: 2012 IEEE conference on Sustainable Utilization and Development in Engineering and Technology (STUDENT)
11. Raj JS, Ananthi JV (2019) Recurrent neural networks and nonlinear prediction in support vector machines. J Soft Comput Paradigm (JSCP) 1(01):33–40
12. Suma V (2019) Towards sustainable industrialization using big data and internet of things. J ISMAC 1(01):24–37

Chapter 18
Evaluation of Deep Learning Models in the Prediction of Lung Disease(Pneumonia)

Adusumilli Rohit, B. Padmaja, K. Vinay Kumar, T. Chandana, and M. Madhu Bala

1 Introduction

The computer and tech world have been evolving with some amazing improvements and transformations. One such stream which changed the course of many applications in various fields of research and problems is machine learning. Machine learning algorithms work well with processed data sets, i.e., pre-processed data sets which are structured in a good way as they can be used to train a model efficiently. But when it comes to unstructured data like images and videos, they are not much efficient. Therefore, deep learning algorithms like neural networks have evolved which foster good performance. Deep learning algorithms include some networks which are trained continuously and which require a lot of computational power and time [1].

Deep learning algorithms can as well be used in the medical field. One such application is detecting lung diseases like pneumonia using chest x-rays. According to Pranav et al. [2], these kind of applications are quite effective when deep learning methods are deployed. Pneumonia is an infection which is generally caused in the lungs. Bacteria, fungi, viruses cause it. This infection leads to inflammation in the

A. Rohit (✉) · B. Padmaja · K. Vinay Kumar · T. Chandana · M. Madhu Bala
Department of CSE, Institute of Aeronautical Engineering, Hyderabad, India
e-mail: arohit5636@gmail.com

B. Padmaja
e-mail: b.padmaja@gmail.com

K. Vinay Kumar
e-mail: vinaykumar10026@gmail.com

T. Chandana
e-mail: chandanachowdaryy@gmail.com

M. Madhu Bala
e-mail: baladandamudi@gmail.com

I. Jeena Jacob et al. (eds.), *Data Intelligence and Cognitive Informatics*, Algorithms for Intelligent Systems,
https://doi.org/10.1007/978-981-15-8530-2_18

air sacs of the lungs. Many deep learning algorithms can be used to perform this task like multilayer extreme machine or online sequential extreme machine or other networks [3]. However, building convolutional neural networks from scratch or using custom hyperparameters is not useful since it involves many parameters to choose from, such as layers, filters, weights, etc. So, using a pre-trained network of transfer learning [4, 5] which provides state-of-the-art algorithms can be considered as the best approach for classifying different kinds of images. There are several architectures in the transfer learning developed over the years which can produce better results in the predictions. In this paper, comparison of various architectures of transfer learning is done by building a model and training it with multiple sets of images classified into pneumonia and normal ones and efficiency and performance of various models in transfer learning is compared and studied.

2 Review of Literature

There are many applications in medical science which are being considered to be implemented using deep learning frameworks. Deep learning has a large scope in the field of medical science and radiology [6]. The analysis of chest radiography has a vital role in medical diagnosis and application [7]. Latest improvements in deep learning frameworks and models are facilitating algorithms to outperform medical experts in various medical imaging tasks such as skin cancer classification, diabetes detection, heart disease detection [8], lung disease detection. The diagnoses utilizing the chest radiographs and x-rays are gaining good popularity and preference. The analysis of chest radiography has a vital role in medical diagnosis and application [7]. These kinds of algorithms are being used to predict various lung diseases involving classification [9]. The performance of various convolutional models on varied factors relying on data sets which are publicly available found that similar deep convolutional network architecture differs in performance across abnormalities. Ensemble models improved various factors like classification, accuracy and finally, deep learning method improved accuracy when compared to normal methods like rule-based methods [10].

In various categories of learning algorithms and techniques, CNN proved to be one of the best algorithms for image classification, analysis, etc. [6]. Transfer learning is one such technique which mingles well with the applications of deep learning when large, multi-class data sets are available. Using standard architectures in transfer learning, various pooling and flattening operations can be explored. When pre-trained networks have utilized, the accuracy of results was found to be more promising [11]. However, these techniques do not work on small data sets and fail abruptly. The main aim here is to provide a novel ensemble approach based on various transfer learning techniques in modelling our application and prediction of lung disease [10, 12, 13].

3 Working Methodology

3.1 Introduction to Transfer Learning

Convolutional neural networks usually are used in the case of image classification. This neural network is a combination of convolutional and max-pooling layers. These layers are finally combined or attached to dense layers. The main operational characteristic of these convolutional networks is to extract several features from the images by using various kinds of filters of different sizes and combinations. Transfer learning architectures contain many layers which include convolution layers and max pool layers followed by flattening layers and dense layer. The dense layer is connected to the output prediction. Pre-trained models are the core of many deep learning applications. In other words, transfer learning is a model used for optimization for improved performance while modelling another task. Here, the knowledge gained from the previous task is transferred into the current one which makes the process easier and efficient. Transfer learning is applied to multi-task learning. It is popular in deep learning when there is a need to train humongous and challenging data sets. Transfer learning is only useful in deep learning if the features of the model learned from the first task are general. The stack of convolution and max-pooling layers is the core of the transfer learning architectures. These pre-trained networks consist of different layers like max pool and convolution layers followed by a dense layer which connects to output. Therefore, transfer learning is an optimizer to save some unnecessary time on training.

The transfer learning approach is very much appreciated in deep learning when there are image data sets. One such competition is ImageNets in various networks that are built, which are capable of categorizing more than 1000 different objects. These are considered as the state-of-the-art algorithms as they are balanced with the right amount of weights and optimizations. Every year, this competition is held known as ImageNet classification. There will be different problem statements and new CNNs are designed to cope up with the solution which can classify 1000 different images and the one with higher accuracy will be awarded prize money. Some of the algorithms that are designed for image classification with weights trained on ImageNet are:

- Xception
- VGG 16/VGG 19
- ResNet/ResNetV2/ResNet
- InceptionV3/Inception Resnet V2
- MobileNet/MobileNetV2
- DenseNet, etc.

In this paper, experiments are conducted with various architectures of transfer learning which include VGG 16, VGG 19, Inception V3, and ResNet.

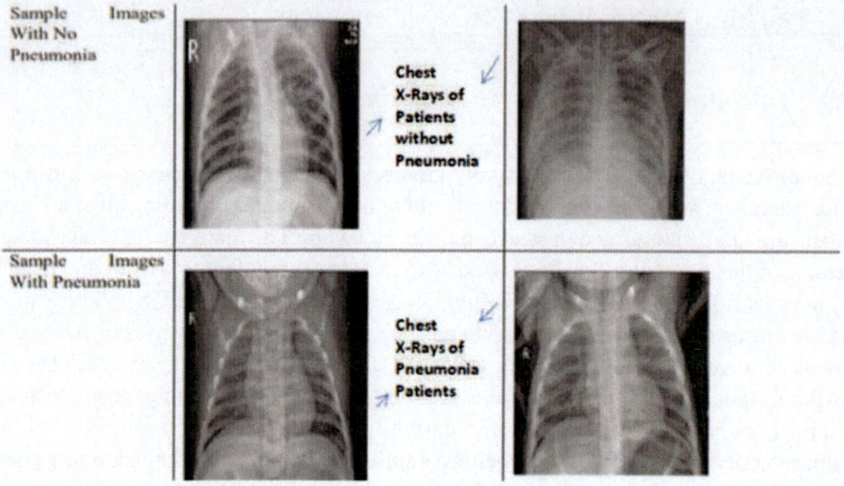

Fig. 1 Sample images of chest X-rays from the data set

3.2 Datasets

Initially, various data sets are collected and are fine-tuned to make them suitable to train the network. The original data set mainly consists of three folders. They are—training, testing, and validation. In addition to these, it contains two subfolders that contain pneumonia (P) and normal (N) chest X-ray images, respectively. Approximately, 5000 X-ray images of anterior–posterior are used to experiment which are publicly available on the data collection platform known as Kaggle [14]. The entire data is rearranged into only two sets, which are training set and validation set. About 3722 images were assigned to the training set and 2134 images were kept in the validation set. The validation set is different from the test set and is a part of a training set. Validation sets are chosen randomly from training sets. The trained model's performance is evaluated on validation sets. Then, the model with the best validation set performance is chosen. Later, this chosen model is evaluated on the test set. This addition of images to the validation set improves the validation accuracy. The sample images of both x-rays with and without pneumonia can be seen in Fig. 1.

3.3 Proposed System Methodology

The model is based on the classification problem of classifying two classes which include pneumonia and normal labels. In this classifying problem reduction of the loss, the function is done using several sets of training data sets in pairs using a specific loss function. Various architectures of transfer learning framework are utilized to study the model in terms of efficiency and performance. The entire model is focused

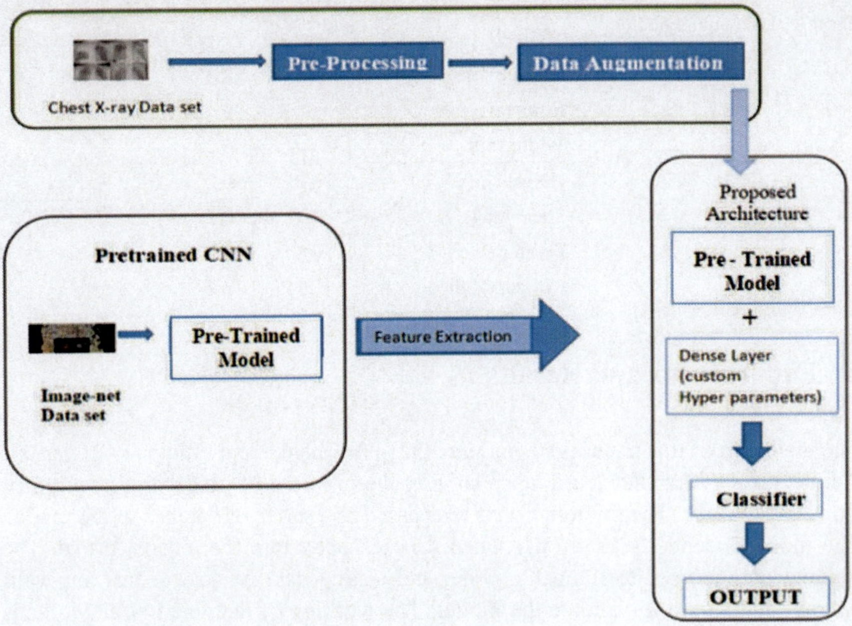

Fig. 2 Block diagram of the proposed system

on architectures of VGG 16, VGG 19, ResNet, and Inception V3. In order to adapt the neural network for predicting pneumonia, one should change/update the last dense layer of the original architecture with a new dense layer matching the number of labels and add an appropriate activation function for our binary-label classification problem. The block diagram of the proposed system can be seen in Fig. 2.

The pre-processing stage includes a collection of various chest x-rays and classifying them into two classes, namely pneumonia and normal ones. In the validation phase, the trained model's performance is evaluated on validation sets. Then, the model with the best validation set performance is chosen. Data augmentation is performed and custom hyperparameters are utilized from the pre-trained architecture. The settings in the data augmentation can be seen in Table 1.

The resulting neural network with custom hyperparameters performs well as it is equipped with good parameters. It describes the utilization of the entire architecture of the VGG 16 model with modified outputs that suit the predictions of our proposed system. The number of outputs depends upon the number of distinct classes to identify or classify using our model. The model mainly consists of feature extractors and classifiers. The final layer is a dense layer from which the output layer arises. The dense layer is fully connected. In this model, the architecture of a pre-trained model is being utilized. There is no necessity of modifying weights but the last layer shall be modified to suit the requirement of binary classification.

Table 1 Data augmentation

Method	Setting
Rescale	1/255
Rotation range	40
Width shift	0.2
Height shift	0.2
Shear range	0.2
Zoom range	0.2
Horizontal flip	True

4 Experiments and Results

The main aim of this paper is to compare the performance and efficiency of various architectures of transfer learning in solving the problem of predicting pneumonia from chest x-rays. The input data sets are passed in a series of batches to the model. The model accepts the data only when it is reshaped into the tensor format. The training data is used to train the model and testing data is used to test and validation data is used to validate the model. The working of deep neural networks is mostly dependent on the hyperparameters. In this case, all the architectures which are being used are pre-trained models with certain hyperparameters. The performance is compared by running various neural networks, by training them with a large number of chest x-ray data sets and ultimately training the model to accurately predict the disease.

4.1 Implementation

The experiment is conducted by considering the architectures of VGG 16, VGG 19, ResNet, and Inception V3. Various batch sizes of images are considered to perform the experiment in each case. However, the data set size was not varied due to the fact that large numbers of data sets are required to get better results. So, all the available data sets are utilized instead of varying data set sizes while conducting the experiment.

The first neural network is constructed by utilizing the VGG 16 architecture. In this model, images of size 224×224 with three color channels, namely RGB(Red, Blue, Green) are being considered. Here, 2D images are considered with various network parameters. The format of the data, which is tensor format modifies and changes into different forms during the course of the process. Different batches of images are considered to study the model in more depth. The softmax activation function is used in the dense layer. The default input size for this model is 224×224. The model if fit by running 5 epochs and considering batch sizes of 16,32,64,128 each consecutive time. The total params were found to be 14,764,866 out of which 50,178 are trainable params and 14,714,688 are non-trainable params.

Next, the neural network is constructed by utilizing the VGG 19 architecture. In this model, images of size 224 × 224 with three color channels, namely RGB(Red, Blue, Green) are being considered. Here 2D images are considered with various network parameters. Different batches of images are considered to study the model in more depth. A softmax activation function is used in the dense layer. The default input size for this model is 224 × 224. The model if fit by running 5 epochs and considering batch sizes of 16,32,64,128 each consecutive time. The total params were found to be 20,074,562 out of which 50,178 are trainable params and 20,024,384 are non-trainable params.

Next, the neural network is constructed by utilizing the ResNet architecture. The default input size in this model is 224 × 224 with three channels. Different batches of images are considered to study the model in more depth. A softmax activation function is used in the dense layer. The total params were found to be 23,788,418 out of which 200,706 are trainable params and 23,587,712 are non-trainable params. Finally, the neural network is constructed by utilizing the Inception of V3 architecture. The default input size in this model is 299 × 299 with three channels. Different batches of images are considered to study the model in more depth. The softmax activation function is used in the dense layer. The softmax activation function is used in predicting great probability limiting the interpretation to 1 or 0 so that it can be more helpful in binary classifying problems like a prediction of Pneumonia. The total params were found to be 22,064,930 out of which 262,146 are trainable params and 21,802,784 are non-trainable params.

4.2 Results

Table 2 summarizes the performance and evaluation of various models which were utilized to conduct the experiment. In total, 16 different experimental setups were evaluated with varying batch sizes and other hyperparameters.

The neural networks should not be over trained which could lead us to the problem of overfitting as well as the neural networks should not be trained very less which would lead to the problem of underfitting. Considering the fact that loss should be reducing and accuracy should be increasing many models in this experiment resulted in different patterns. One more important thing to be considered is to observe the values of validation loss and validation accuracy. If the validation loss increases and validation accuracy decreases, the model is not ready to learn, i.e., the model is not learning. If the validation loss increases and validation accuracy increases, it may lead to the problem of overfitting. In this experiment, ResNet performance was very bad and is not suitable in this application. If more computation power and training epochs are provided then VGG 16 and VGG 19 performance is quite suitable in this type of medical application. Even Inception V3 initially started to perform well but its validation accuracy states that it is not suitable in this type of application. So,

Table 2 Model performance on chest x-ray data set

Model	Batch size	Training accuracy (%)	Validation accuracy (%)	Training loss	Validation loss
VGG 16	32	97.07	86.20	0.0748	0.8101
	64	96.78	90.85	0.0817	0.3077
	128	96.23	91.33	0.1010	0.2324
VGG 19	32	95.73	87	0.1179	0.8723
	64	96.11	89.09	0.0994	0.3560
	128	95.94	92.13	0.1056	0.1705
ResNet 50	32	97.39	62.44	0.6700	16.6825
	64	97.72	62.44	0.3697	14.8602
	128	98.01	62.44	0.2521	7.3505
Inception V3	32	93.75	76.08	1.2244	3.7245
	64	94.85	64.69	0.6634	6.4815
	128	95.05	69.34	0.3838	3.0134

in this particular application of predicting Pneumonia from chest x-rays, the VGG models have shown good results.

5 Discussion

This paper presents a varied and systematic evaluation of the performance of various approaches to transfer learning for CNN-based X-Ray classification to predict Pneumonia from chest X-rays. The results produced are satisfactory and proposes a varied scope in this field of application. Some best results are obtained as the networks are optimized on ImageNet data set. However, the results produced signify distinctive differences when various architectures are utilized to experiment. Considering all the factors like training time, validation accuracy, training accuracy, and loss function the VGG 16 and VGG 19 models are found to be effective in this particular application of medicine. The models can effectively and accurately predict Pneumonia in the chest X-rays. Deep learning frameworks are playing a vital role in the field of medicine. The medical field is advancing at a faster pace and there are developments due to the utilization of technology particularly those based on classification and data science.

6 Conclusion

The proposed system has shown better accuracy and validation than existing systems as the novel concept of transfer learning framework in modelling our data sets is used. The proposed model is highly effective in the diagnosis of Pneumonia through chest x-rays and can be implemented easily with some good computing requirements showing a good amount of accuracy. The results obtained are very much valuable and focuses on the opportunity that the training of deep neural networks in the medical field is a viable and good option as more number of public data sets are becoming available. In the future, the model can be upgraded to even predict the type of virus and kind of various infections with the greater availability of open-source data sets of x-rays. Thus, in the future, the number of categorical classes can be increased to identify various kinds of infections or diseases from one x-ray itself after the network is trained with the required parameters. However, some powerful computing support is required to facilitate future applications based on this problem definition.

References

1. Vijaykumar T (2019) Comparitive study of capsule neural network in various applications. J Artif Intell 1(01):19–27
2. Rajpurkar P et al (2017) ChexNet: radiologist-level pneumonia detection on chest x-rays with deep learning. arXiv preprint arXiv:1711.05225
3. Vijendran S, Dubey R (2019) Deep online sequential extreme learning machines and its application in pneumonia detection. In: 8th international conference on industrial technology and management (ICITM), Cambridge, United Kingdom. IEEE
4. Pan SJ, Yang Q (2009) A survey on transfer learning. IEEE Trans Knowl Data Eng 22:1345–1359
5. Manikonda SKG, Gaonkar DN (2019) A novel islanding detection method based on transfer learning technique using VGG16 network. In: 2019 IEEE international conference on sustainable energy technologies (ICSET). IEEE
6. Taylor AG, Mielke C, Mongan J (2018) Automated detection of moderate and large pneumothorax on frontal chest X-rays using deep convolutional neural networks: a retrospective study. PLoS Med J 15:e1002697
7. TaPasa F et al (2019) Efficient deep network architectures for fast chest x-ray tuberculosis screening and visualization. J Sci Rep 9(1):1–9
8. Poplin R et al (2018) Prediction of cardiovascular risk factors from retinal fundus photographs via deep learning, Nat Biomed Eng 2(3):158
9. Rawat W, Wang Z (2017) Deep convolutional neural networks for ımage classification: a comprehensive review. Neural Comput 29:2352–2449
10. Biswas M et al (2019) State-of-the-art review on deep learning in medical imaging. Front Biosci Landmark 24:392–426
11. Peng YL et al (2020) Fully hardware-implemented memristor convolutional neural network. Nature 577(7792):641–646
12. Sahiner B et al (2019) Deep learning in medical imaging and radiationtherapy. Int J Med Phys Res Pract 46(1):e1–e36.A
13. Baltruschat IM et al (2019) Comparison of deep learning approaches for multi- label chest X-ray classification. Sci Rep 9(1):1–10
14. Paul M (2018) Chest X-Ray Images. www.kaggle.com/paultimothymooney/chestx-ray-pneumonia

Chapter 19
An Automated Invoice Handling Method Using OCR

Pranay Kumar and Dr. S. Revathy

1 Introduction

The number of invoices issued in different sectors for various purposes is increasing day by day. These contain information related to the vendor, customer, invoice number, date, amount, and many others. These data could be used by us to analyze and manipulate what people are attracted toward nowadays and help improve the production and selling to gain more profit.

Within easy reach, various technologies and studies have been reviewed and taken into account, providing us with equal knowledge and information clarifying the methods used to produce the most favorable outcome. To recognize and retrieve complete alphanumeric handwritten or printed components from characters present in images containing text or documents obtained through scanning, an engine called OCR is very helpful. Famous open source and optical character recognition engine until now is Tesseract.

Tesseract's basic undergone process is a step-by-step pipeline approach, in which a connected component analysis is carried out. The connected component analysis is the first stage for the object outlines to be placed in. Outlines are collected into blobs, simply through nesting. Blobs are arranged into lines of text, and the lines and regions are evaluated for either fixed pitch or proportional text. The lines are divided into terms or words depending on the sort of distance between characters. Set pitch text is automatically chopped by the character cells. Proportional text is split into terms using specific and precise spaces or blurred spaces. After that digital texts are

P. Kumar (✉) · Dr. S. Revathy
Sathyabama Institute of Science and Technology, Chennai, India
e-mail: pranayk649@gmail.com

Dr. S. Revathy
e-mail: revathy.it@sathyabama.ac.in

I. Jeena Jacob et al. (eds.), *Data Intelligence and Cognitive Informatics*,
Algorithms for Intelligent Systems,
https://doi.org/10.1007/978-981-15-8530-2_19

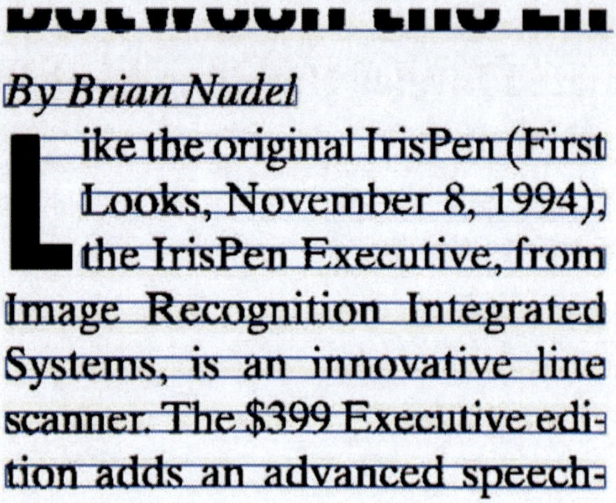

Fig. 1 Line segmentation

produced. The segmentation of text images is usually used for finding objects and boundaries in pictures. Text segmentation is a method via which digital images are partitioned into a couple of segments. Segmentation is typically implemented for text-based images, aimed at extracting different records from the entire image.

The technique for segmentation rates are as follows:

A. Line Segmentation—Segmentation of the line is the initial step to segment the images. Here, the engine scans the image horizontally. Figure 1 suggests the technique of identification and segmentation traces from an image.
B. Word Segmentation—The vertical test is the second one step after the horizontal scan, which is better known as segmentation of words. Figure 2 suggests a method for splitting lines into words.
C. Character Segmentation—The final standard for segmentation of text-based images is segmentation of character. Vertical scanning of the photograph is achieved on this step. Figure 3 suggests the single segmentation process, and the words are segmented into characters.

This paper has been considered and adopted into the mainstream because the concept of processing and extraction of textual content from a pdf is sufficiently tempting. Additionally, this paper clears the hidden possibilities of further studies with enough information, which is the next large thing in this rapidly transforming technological age. The paper goals to extract information or records from the payments contained in a pdf document, right here concentrating only on bills and invoices, which may later allow us to locate what human beings opt for these days within the marketplace and the way their alternatives and likings range over geography, time, etc. The method begins with evolution by using a pdf and attempting to extract the textual content from it by different strategies and ended up using Tesseract OCR which could be an impressive tool in its area. In essence, regular expressions

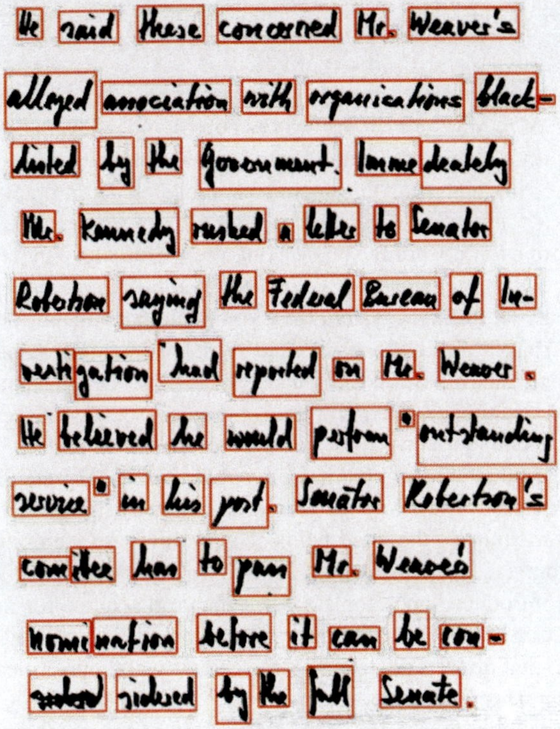

Fig. 2 Word segmentation

Fig. 3 Character segmentation

are a small, highly specialized programming language embedded within Python and made accessible via the re-module. Use this small language to define the rules for the collection of possible strings you want to match; this set may contain english sentences, e-mail addresses, TeX commands, or anything you like. Regular expression patterns are compiled into a collection of bytecodes, which a corresponding engine is written in C executes. It may be important to pay close attention to how the engine would execute a given RE for advanced use, and write the RE in a certain way to generate bytecode that runs faster. The language of a regular expression is relatively limited and constrained so that not all possible string processing tasks can be done with regular expressions. Some functions can be done with regular expressions, but the expressions are quite complicated. You may be better off writing Python code to do the processing in such cases; although Python code may be slower than a complicated regular expression, it is also likely to be more understandable.

2 Literature Survey

As the economy keeps on rising strong, growing numbers of invoices are being issued [1]. The method of reimbursement of invoices comes under severe constraints when observed under time complexity and imposes a strain on staff from the accounting department, requiring both reimbursement and signatures from financial personnel, invoice analysis, documentation and backup, as well as approval from a manager. The current process for a refund through invoice often constitutes an unnecessary waste of material resources. The sustainable creation and reliability of invoice reimbursement require an easy and convenient system for invoice reimbursement.

Delie et al. [2] have submitted a response to boost the operating output of economic personnel in which the Chinese and Arabic numerals have been mechanically entered from the bill into a computer. Their proposed approach used smooth scanning, image angle checking, and image slope correction to preprocess the invoices. The processed image knowledge obtained from the invoice could then be expressed and incorporated into the device, and the key features of the invoice image are extracted and specifying the functions of the fixed fields. The preliminary step in the process of enhancing reimbursement performance is classifying the various forms of invoices. Alippi et al. [3] introduced an automated classification scheme for invoice records for the classification of invoices. The program evaluated graphical details such as the document's logo and trademarks to assess the category and performed well in open and closed world classification tasks.

A regular expression is a series of characters describing a pattern for the quest which was used by Wang et al. [4]. It is a methodology built-in theoretical computational science and theory of formal languages. The regular expression on steroids can be thought of as wildcards like*.txt in order to find all text files in a file manager.

The wand is a simple ImageMagick binding to Python based on ctypes. It actually does not cover any of the MagickWand API functionalities. This is based on Python 2.6, 2.7, 3.2, 3.3, and PyPy applied by Peng et al. [5]. Computer algorithms are used in digital image processing to create, process, communicate, and view digital images. Convert the signals to digital images from an image sensor. Improve visibility and minimize noise and other things. Extract the scale, size, or number of objects inside a scene. Prepare pictures for printing or display. Compress pictures into a network for communication [6]. Sun et al. [7] presented a technique for classification of the invoice using deep learning to improve reimbursement execution proficiency. The system can recognize different invoice images directly within a short time. A new approach for checking the continuity and originality of the Quranic text is proposed by Alotaibi et al. [8]. A mechanism is based on an integration of techniques for character recognition and Quranic diacritics where a region-based algorithm is used to detect diacritics. To enhance the ratio of recognition, an optimization technique is applied. The system proposed by Zhu et al. [9] is composed of deep neural and conditional adversarial generative networks where they proposed a new network architecture with bi-skip connections for the discriminator network to fix hard training and information that lose problems. Multi-objective optimization is extracted within the generative network to overcome the question of non-identical common conditions. Bartoli et al. [10] suggested a technique whose goal was to reduce the effort of user annotation. The user notifies only one extraction required and then only answers machine created extraction queries. Smart algorithm for processing of image through text recognition, knowledge retrieval, and visually impaired vocalization was presented in the work by Manoharan [11]. Onboard, the system uses the LattePanda Alpha system which processes the photos which were scanned. After preprocessing, segmentation, extraction of features, and post-processing of the scanned or image-based information, the image is classified into its corresponding alphanumeric character. In addition, a text to speech synthesizer is utilized for material interpreted by vocalization.

Furthermore, due to its multiple achievement applications in many fields, deep learning is a likely response to find out smart invoice reimbursement technology. However, the generation of deep learning-based invoice reimbursement may also cause excessive costs due to high computational power requirements. These issues are dealt with in the preprocessing phase in this article. The invoice pdf is first preprocessed to do away with the noise and transform into a high-resolution image, then the picture is given to optical character recognition (OCR) engine [12] to become aware of the characters and bring text document containing all the data, and finally, the extracted outcomes of invoice records are exported to the excel form inside the identical format.

In this day and age, the bill remunerate or compensate procedure of most groups is performed manually, which includes signatures of each reimburses and economic body of workers, a summary, recording, and backup of the invoices, which inevitably wastes in numerous human and material resources. Therefore, by replacing the staff with computers, the suggested novel intelligent solution in this paper would successfully resolve this difficulty with high precision and rapid speed.

3 Proposed System

In this section, a way is recommended to recognize invoice records. In Fig. 4, the architecture of the proposed system is shown representing the complete approach technique together with enter scanned pdf, high-resolution image, optical character recognition, extracting statistics, and exporting data into the database.

A Preprocessing

The original pdf of the invoice taken into consideration in this paper is shown in Fig. 1. Document of this type cannot be used directly to recognize the required information due to its asymmetric nature, hence preprocessing is highly essential.

The preprocessing step chiefly incorporates conversion into a high-resolution image. This technique uses a Python library module which is widely used to transform a simple pdf to produce a fine, clear, and crisp image containing all the information better known as Wand. The preprocessed invoice image is provided in Fig. 5.

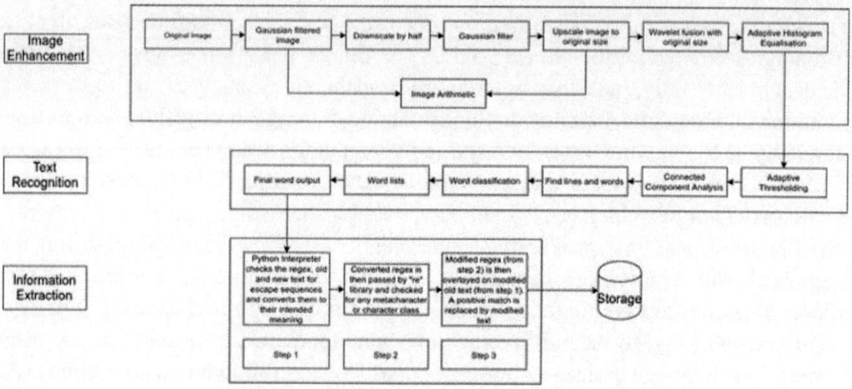

Fig. 4 Architecture of proposed system

The above functioning will produce high-quality images; however, the text is required instead of images to carry forward this process. Therefore, a method, which can convert images containing information into text, is required.

Optical character recognition allows converting images into the text at fast speed and with high accuracy when there is a (very) clean segmentation of foreground text from a background as shown in Fig. 6. It distinguishes the characters in an image as maintained by the brightness, and transforms the characters into the text. The image processed is passed to the Tesseract OCR engine [13], and the algorithm assists in recognizing text or drawing out textual content from the image. Segmentation here is essentially a method in which a piece of text is broken down into a series of words and then into characters and further into individual parts. Recognizing the final characters and allowing concrete interpretation is the need to do so.

Each word considered reasonable is then made through as training data to the adaptive classifier. OCR engines benefit from the application of an adaptive classification because in generalizing to any form of the font, the static classifier must be good. The ability to differentiate between various characters or between non-characters is diminished. Tesseract has a font-sensitive adaptive classifier that is trained using static classifier outputs that are widely used to achieve greater diversity in every document where the number of fonts is reduced. Important deviation from the training data between the two classifiers is that the isotropic baseline/x-height normalization is applied in adaptive classifier while the centroid character for a position and anisotropic size normalization for second moments are normalized by a static classifier. The upper and lowercase characters could be differentiated with the assistance of baseline normalization and enhance the tolerance of specks of noise. The key advantage of normalizing character moment is eliminating font aspect ratio and a certain degree of font stroke width. This also makes it easier to identify subscripts and superscripts but includes an extra feature to differentiate between the upper and lower case characters. The adaptive classifier's basic functionality is to identify the page present at the lower side more accurately, in order to produce an improved or

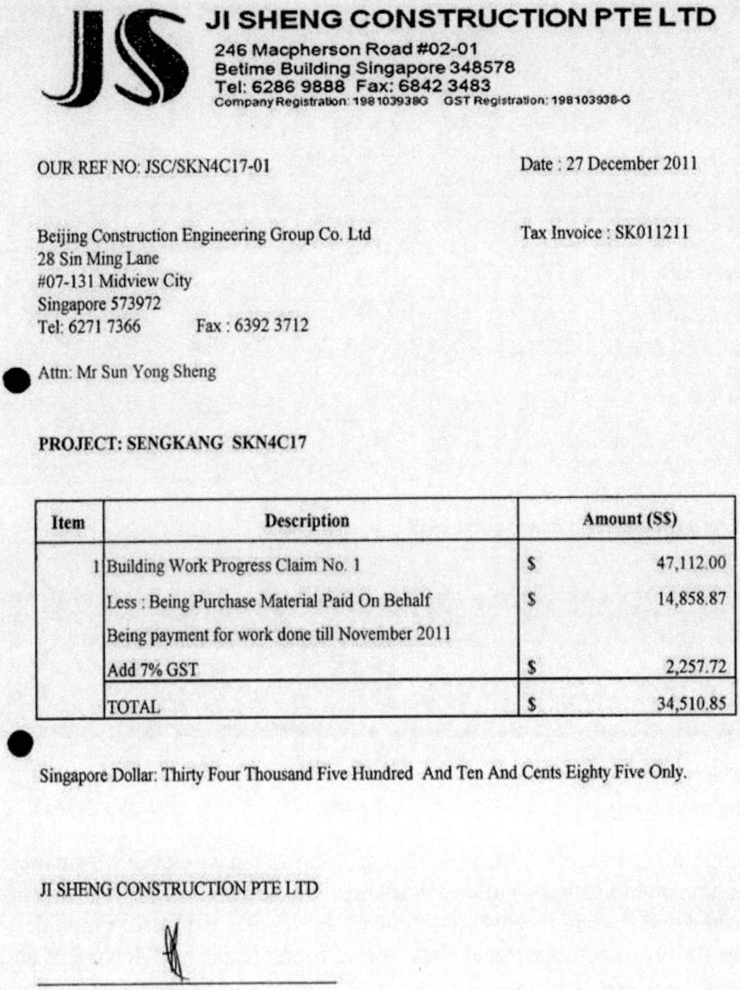

JI SHENG CONSTRUCTION PTE LTD
246 Macpherson Road #02-01
Betime Building Singapore 348578
Tel: 6286 9888 Fax: 6842 3483
Company Registration: 198103938G GST Registration: 198103938-G

OUR REF NO: JSC/SKN4C17-01

Date : 27 December 2011

Beijing Construction Engineering Group Co. Ltd
28 Sin Ming Lane
#07-131 Midview City
Singapore 573972
Tel: 6271 7366 Fax : 6392 3712

Tax Invoice : SK011211

Attn: Mr Sun Yong Sheng

PROJECT: SENGKANG SKN4C17

Item	Description	Amount (S$)	
1	Building Work Progress Claim No. 1	$	47,112.00
	Less : Being Purchase Material Paid On Behalf	$	14,858.87
	Being payment for work done till November 2011		
	Add 7% GST	$	2,257.72
	TOTAL	$	34,510.85

Singapore Dollar: Thirty Four Thousand Five Hundred And Ten And Cents Eighty Five Only.

JI SHENG CONSTRUCTION PTE LTD

Fig. 5 Processed invoice image

extraordinary performance. It could be important for the adaptive classifier to figure out anything special that was not found previously at the page present at the top, so a second search for similar segmentation is given to the full page. When this is completed, it performs a final search to search ambiguous spaces and small caps.

B Extraction of Information

The extraction process involves the text file being read and finding similar matches or patterns that follow the given criteria. The criteria are set for each information type differently using pre-defined regular expressions. For example, "\bA[B-CX-Y]\b" is the regular expression used for finding invoice numbers. This expression looks for strings that begin and end with blank space (meaning a word within the text),

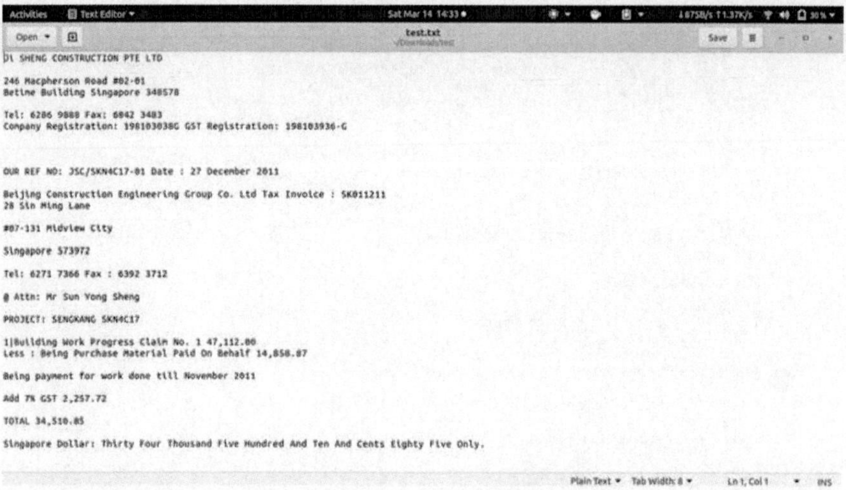

Fig. 6 Extracted text from image using OCR

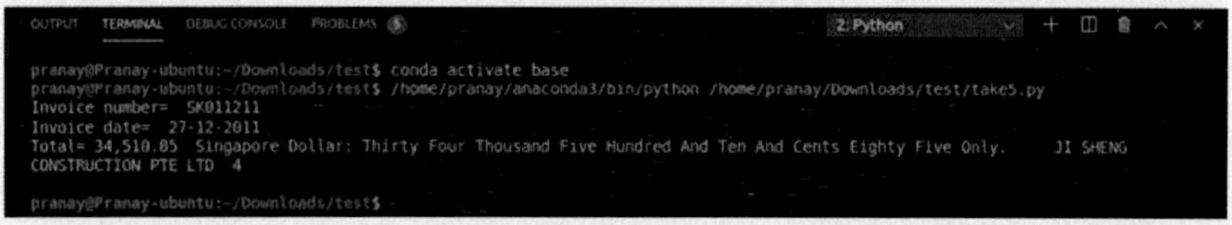

Fig. 7 Data extracted using Regex

alphanumeric, starting with letter A. Figure 7 shows the extracted information from the text document using a regular expression. The output obtained in the text format out of the OCR engine is given as input to the regular expression module, and it searches for all possible matches that can be found in the text document and then write into a separate text document.

C Storage

The storage is the final step in the process as the data exported into the database can be used for various purposes like data analysis, manipulation, and finding the hidden patterns. This process mainly concerns the output being received into the database. The database can act as a dataset for many manipulation models. It is fantastic because it ensures that data can be analyzed in more depth, and eventually gets better insights.

D Proposed Algorithm

Step 1: Collect the input portable document format invoices which are similar in pattern and printed

Step 2: Preprocessing input pdf to convert it into a high-resolution image with sharp and clear textual content
Step 3: Recognition of each and every character present in the image and extract the textual content and generate a text document with all the information which are present in the image
Step 4: Taking out all the necessary information like invoice number amount, date, etc. with a specific matching pattern criteria using a regular expression from the text document
Step 5: All the extracted information from the invoice is then exported into a database which could be used for various purposes.

4 Results and Discussion

This work has contributed to some impressive results but has also demonstrated the directions and complexity of progress.

4.1 Dataset

The sample invoice dataset has been collected from a company. The dataset consists of 100 invoices. This subsists of various kinds of information like vendor and customer details, addresses of vendor and customer, fax number, invoice number, date of purchase, a quantity of items, description of the product, the final amount. Figure 8 shows a sample invoice image which is the input for the system.

4.2 Performance Analysis

No OCR software is accurate to 100 percent. The number of errors, including the font used, depends on the content and form of a document. Errors occurring during OCR include misreading characters, missing unreadable characters or combining text from adjacent columns, or captions of images. When high precision is needed, such as converting digital books to electronic format, then cleanup of the electronic text would be needed. The time complexity depends on the performance of the CPU and its ability to perform the same tasks repeatedly. For the proposed system, a GPU is much suited because of its ability to render similar simultaneous mathematical equations with precision. Figure 9 shows a graph of the surge in CPU while using a 3.1 GHz processor, the peak achieved during the text recognition. The surge is observed to be of 95% for the duration.

The time complexity depends on the performance of the CPU and its ability to perform the same tasks repeatedly. For the proposed system, a GPU is much suited

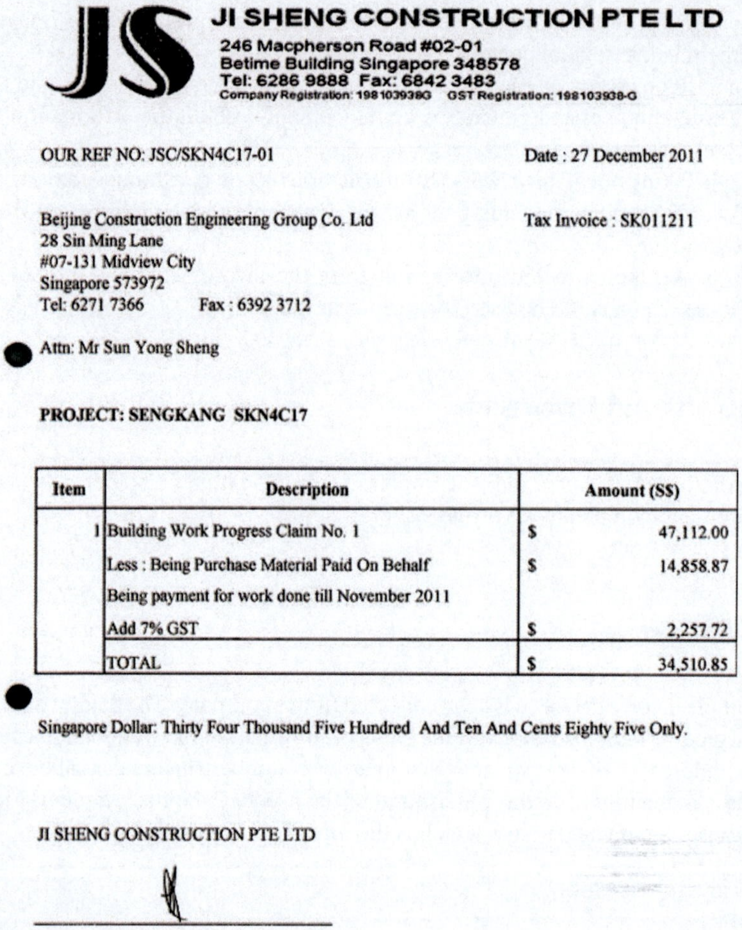

JI SHENG CONSTRUCTION PTE LTD
246 Macpherson Road #02-01
Betime Building Singapore 348578
Tel: 6286 9888 Fax: 6842 3483
Company Registration: 198103938G GST Registration: 198103938-G

OUR REF NO: JSC/SKN4C17-01 Date : 27 December 2011

Beijing Construction Engineering Group Co. Ltd
28 Sin Ming Lane
#07-131 Midview City
Singapore 573972
Tel: 6271 7366 Fax : 6392 3712

Tax Invoice : SK011211

Attn: Mr Sun Yong Sheng

PROJECT: SENGKANG SKN4C17

Item	Description		Amount (S$)
1	Building Work Progress Claim No. 1	$	47,112.00
	Less : Being Purchase Material Paid On Behalf	$	14,858.87
	Being payment for work done till November 2011		
	Add 7% GST	$	2,257.72
	TOTAL	$	34,510.85

Singapore Dollar: Thirty Four Thousand Five Hundred And Ten And Cents Eighty Five Only.

JI SHENG CONSTRUCTION PTE LTD

Fig. 8 Sample invoice image

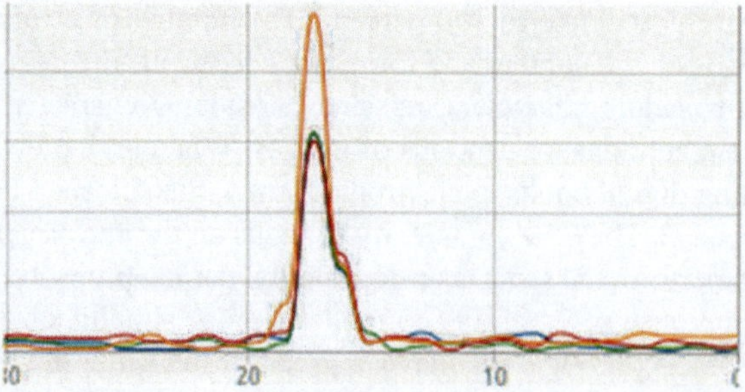

Fig. 9 CPU surge—Tesseract

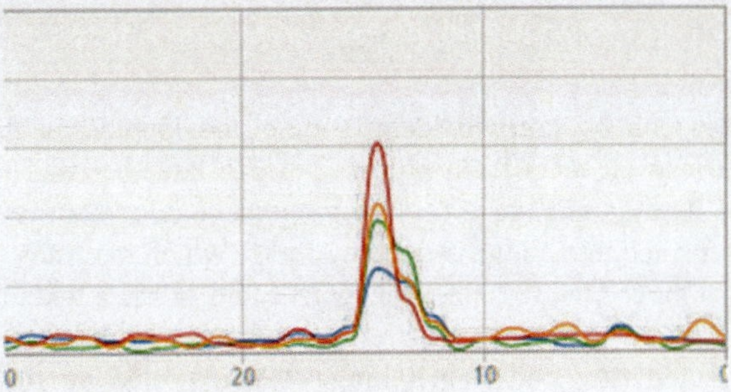

Fig. 10 CPU surge—model

because of its ability to render similar simultaneous mathematical equations with precision. Figure 10 shows a graph of the surge in CPU while using a 3.1 GHz processor, the peak achieved during the data extraction. The surge is observed to be of 60% for the duration.

Figure 11 displays the database output which is the final result of the entire system. The applied methodology has some drawbacks to it. Scanner-generated images use most memory space. Photos lose clarity during the process of scanning and digitization. The accuracy of the final image depends on the initial picture quality. The above work allowed the text to be extracted from the bills which are printed but goes wrong for bills and invoices which are handwritten because the Tesseract OCR engine uses the technique of decomposition of text that first recognizes characters appearing in the row or line and then it attempts to interpret the entire line according to the location of the first character, so with regards to handwritten or transcribed text the movement is not in a way that the characters could be understood correctly and extracted. Since the extraction of information is based on regular expression and just a few characters can make such a big difference in performance. The longer regex is the better it performs. They use particular character/character classes and have greater and better structure. That causes faster running of good regular expressions as they more accurately predict their input.

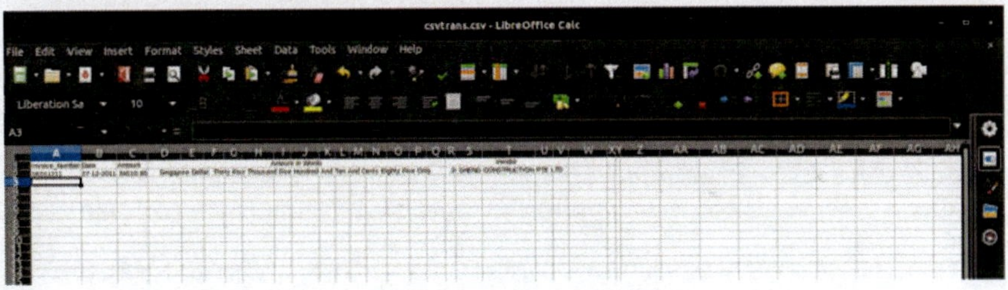

Fig. 11 Database output

5 Conclusion

The methodology can be used to identify bills and invoices which are further processed to retrieve all texts from bills which can later be used to analyze and manipulate data. At the initial step, the input invoice in document format is preprocessed and transformed into a high-resolution image which in turn is the input to the Tesseract OCR engine. The Tesseract recognizes and extracts texts from an image and generates a text document which acts as input for the regular expression, and the information extracted using pattern matching is then exported into the database. The technique adopted in the paper has helped increase the accuracy of the test results and the reliability of the study.

References

1. Bayar S (2016) Invoice processing performance analysis on e-archive on various embedded platforms. In: October 2016 in proceedings of IEEE 10th international conference on application of information communication technology (AICT), Baku, Azerbaijan, pp 1–4
2. Delie M, Jian L, Jinwen T (2002) Designing and implementing a Chinese framework for the identification of financial invoices. In June 2002 in proceedings of international symposium on VIPromCom video/image processing multimedia communication, Zadar, Croatia, pp 79–82
3. Alippi C, Pessina F, Roveri M (2005) Adaptive program for the classification of automated invoices –documents. In: September 2005 in proceedings of international conference on image processing, Genova, Italy, pp II-526
4. Wang X, Hong Y, Chang H, Park K, Langdale G, Hu J, Zhu H (2019) Hyperscan: for modern CPUs, a fast regex matcher with multi patterns. In: 2019 in 16th {USENIX} symposium on the development and implementation of networked systems ({DINS} 2019), pp 631–648
5. Peng KC, Singhose W (2009) Machine vision and wand following crane control. In: 2009 IEEE international conference on mechatronics, Malaga, pp 1–6
6. Chui C, Parmar M (2019) Outward Inc. Image processing techniques based on machine learning. U.S. Patent Application 16/056,110
7. Sun Y, Zhang J, Meng Y, Yang J, Gui G (2019) Intelligent, smartphone-based invoice classification system using deep learning. IEEE Access 7:118046–118054
8. Alotaibi F, Abdullah MT, Abdullah RB, Rahmat RW, Hashem IA, Sangaiah AK (2017) Optical character recognition for matching similarity of quranic image. IEEE Access 6:554–562
9. Zhu C, Chen Y, Zhang Y, Liu S, Li G (2019) ResGAN: a low-level image processing network to restore compressed images to the original quality of JPEG. In: 2019 Data Compression Conference (DCC). IEEE, pp 616
10. Bartoli A, De Lorenzo A, Medvet E, Tarlao F (2017) Active learning of regular expressions for entity extraction. IEEE Trans Cybern 48(3):1067–1080
11. Manoharan S (2019) A smart image processing algorithm for text recognition, information extraction and vocalization for the visually challenged. J Innov Image Process (JIIP) 1(01):31–38
12. Tesseract For Open Source http://code.google.com/p/tesseract-ocr. Accessed November 2017
13. A Description of the Tesseract OCR Engine—Analysis at Google. https://research.google.com/pubs/archive/33418.pdf

Chapter 20
Performance Evaluation of Clustering-Based Classification Algorithms for Detection of Online Spam Reviews

N. Krishnaveni and V. Radha

1 Introduction

The social Web has created a drastic change in the life of everyone nowadays through expressing their views online [1, 2]. The size of the user-generated content is growing rapidly. E-Commerce provides a sophisticated one for online users. The Web sites allow their customers to provide their feedback as reviews in their sites. More than 90% of the consumers before purchasing any product or using any service has become a habit of reading online reviews for decision making purpose [3, 4]. The new customers, as well as the existing customers, consider the reviews as their important source of information. All the reviews written are not true. Some of the reviews are spam as they are written for some benefits such as advertisement for their product, promotion of their product or service, simply spreading information or sometimes out of these fake reviews may even get financial profits [2].

Detection of these fake reviews is really a tedious task as it involves huge content and a lot of people involved. These reviews should be carefully handled and categorized based on many features. Dimensionality reduction should be done in order to increase the performance of the result [5]. Classification of spam and ham review is a text classification process. Some limitations in the supervised approaches can be solved by introducing a semi-supervised approach through automatic labeling and dimensionality reduction. Selection of the best solution should be based on facts. And the text content can be classified by nominating tags depending upon the given passage [6]. Natural language processing (NLP) helps in doing so. Classification

N. Krishnaveni (✉) · V. Radha
Avinashilingam Institute for Home Science and Higher Education for Women, Coimbatore, India
e-mail: krishnaveni.adu@gmail.com

V. Radha
e-mail: radharesearch@yahoo.com

I. Jeena Jacob et al. (eds.), *Data Intelligence and Cognitive Informatics*,
Algorithms for Intelligent Systems,
https://doi.org/10.1007/978-981-15-8530-2_20

process in data mining is a powerful tool in analyzing the dataset and classifying into data classes. Enormous algorithms have been developed and used worldwide in machine learning for different applications [7].

Dimensionality is the curse of any classification process leads to overfitting, and it has to be reduced to improve the efficiency of a classifier model [5]. When text classification is carried, the features are sparser and sparser which obviously entertain to increase the cost and time. Only the significant features should be added for building a model. If the clustering process before classification is applied, then it gives a noteworthy result. This research work aims to find out the combination of clustering and classification algorithms, which produces considerable improvement in the result.

The remainder of the paper is structured as follows. Section 2 reviews the state of the art on fake review detection on other domains. Section 3 discusses the clustering-based classification. Afterwards, Sect. 4 presents the followed methodology. Experimentation is detailed in Sect. 5. Finally, Sect. 6 highlights and discusses the mainly obtained results and future directions.

2 Background Study

Hadeer Ahmad proposed a classification approach by using linear support vector machine (SVM) model. N-Gram analysis has been used for text preprocessing and TF-IDF. Fake news dataset from Kaggle.com and political news articles were used for analysis purpose, and 92% of accuracy was achieved. Other SVM models can also be utilized by using larger size datasets [8]. Cagatay Catal & Suat Guldan proposed an ensemble classifier using libLinear, libSVM, sequential minimal optimization, random forest, J48 classifiers based on majority voting system. T-test was done to check the efficiency, and 88.125% of accuracy was achieved. Hotel reviews dataset from TripAdvisor was used for evaluating the ensemble classifier. Different classifiers with different parameters can be tried as there are many more classifiers are available in the literature. User messages from social media also to be considered [9]. Rout proposed a semi-supervised algorithm for handling labeled and unlabeled features. Labeled features were handled by classification algorithms, and unlabelled features were handled by clustering algorithms. The proposed algorithms were compared with the existing methods, and F-score was 0.837 which is relatively higher than the existing methods. Gold standard dataset was used for the experiment. Data collection from real-world and by integrating minimal meta-data can be considered in future. Multimedia content can also to be considered in future [10]. Chengai sun proposed an ensemble algorithm using the ensemble supervised algorithms such as SVM, CNN applying through bagging strategy. Gold standard dataset was used for the experiment and the found that the error rate was reduced considerably. Only limited features were considered, and it can be recognized with many more features in future [11]. Wu schemed a semi-supervised hybrid PU-learning-based spammer detection (hPSD) algorithm by undertaking Amazon dataset for experimentation. F-measure was found as upgraded convincingly. Only homogeneous data was involved for experiment, and further heterogeneous data can be evolved [12]. Shao proposed an approach for detecting spams by using unstructured dataset through a hybrid algorithm. Dataset used to contain both images as well as text data. Clustering, as well as classification algorithms, was used [13]. Osman propounded new techniques with feature weight optimization K-means clustering, logistic regression for e-mail spam detection. The accuracy was attained 98.41%, and the error rate was found reduced up to 1.59%. This technique can also be used for online review dataset, and the other features can be extracted [14]. Elssied came up with a hybrid K-means clustering SVM classifier for e-mail spam dataset and brought about 98.1% of accuracy. This algorithm can be effectively used for online review dataset. The other classifiers also can be used with clustering [15]. Lu proposed an ensemble classifier based on clonal selection, Naive Bayes, AdaBoost and ensemble decision tree-based classifier(C4.5) and obtained 0.9261 of F-measure and 0.89 of accuracy. Web spam datasets such as WEBSPAM-UK2006, WEBSPAM-UK2007 were used for the experiment. Further unbalanced datasets can also be used, and links and user information may be considered [16]. Pandey anticipated an algorithm using particle swarm optimization with improved cuckoo search method, differential evolution, genetic algorithm with cuckoo search method and K-means clustering and achieved higher accuracy. Yelp Hotel, Yelp Restaurant and Twitter Reviews were used for carrying out the trial. It can be expanded utilizing more optimization and feature selection techniques [17]. Nagwani estimated the latent dirichlet allocation (LDA) and nonnegative matrix factorization (NMF)-based clustering, Naive Bayes classification, SVM classification algorithms and accomplished improved F-measure. SMS spam dataset was used for experimentation. Other classifiers and clustering approaches can also be tried [18]. Chakraborty has given a survey on the developments in social spam detection and combating techniques for different datasets and many types of features used in the literature. The significant features and methods for classification of spam and ham are analyzed [19]. Julien carried out review-based and reviewer-based feature analysis through supervised algorithms for Yelp dataset and achieved 80.6% of accuracy. Detecting singleton reviews is the limitation of this work [20]. Abu Hammed demonstrated the spam detection method by using supervised algorithm with review-based features. The classification algorithms such as ID3, SVM, Naive Bayes and KNN were used to model the classifier, and the result obtained as 99.20% accuracy and 99.59% of F-measure. The TripAdvisor dataset was used and found as limited one and a synthetic dataset [21]. Nair experimented reviewer-based supervised SVM algorithm and Naïve Bayes algorithm using Amazon online reviews dataset and attained 92.5% of accuracy. Only limited numbers of features were used, and more features can be utilized [22].

3 Clustering-Based Classification

Clustering process groups the data into clusters, the data in the corresponding clusters are homogeneous among themselves and heterogeneous to other clusters. Each cluster is assigned with a label. Clustering is a treasure trove procedure of defining subgroups through careful investigations appliances extensively like science, technology, biology, social science economics, medicine, smart farming, geospatial, stock market and many more [6]. Clustering is widely exploited in various fields such as biological analysis, dimensionality reduction, pattern recognition and statistical data analysis. Even though clustering is an unsupervised machine learning technique, the clusters can be used as features in a supervised machine learning model [4, 23].

There exist clustering-based classification in the literature. Clustering-based classification is mainly divided into two categories such as supervised and unsupervised. Classification algorithm can be applied after clustering the data into multiple clusters. There is a good chance to get better improvements in the accuracy or whatever is focused on when the classification model on each cluster separately is built separately. Clustering is done on unlabelled data and provides a label for each cluster. Classification requires labels. Therefore, clustering is done on the data first and save the resulting cluster labels. Then, train a classifier using these labels as a target variable. By saving the labels, the steps of clustering and classification can be effectively separated.

This enables to use of any classification algorithm such as random forest, SVM and Naive Bayes. The problematic part of this pipeline is to choose the best suitable clustering algorithm. Therefore, the result after the clustering is also to be evaluated. Alternatively, other methods such as using other clustering algorithms and comparing their results to find out the right choice can be adopted. Improving supervised learning algorithms with clustering and enhancing a model performance can be challenging at times. So, hit upon the machine learning algorithms that are precise and ideal to accomplish higher accuracy is the goal of this research work. There is a large number of classification algorithms. But, only a few algorithms outperform in incomparable situations for diverse datasets. This research work focuses on catching the immune classifier as well as the right choice of the clustering algorithm when used as a preprocessing tool. The following classification algorithms and clustering algorithms have been selected based on the existing literature for the evaluation purpose.

The following text classification algorithms are used for evaluation purpose. (1) **Naïve Bayes Algorithm**: A Naïve Bayes classifier is a probabilistic machine learning algorithm used for employing the classification procedure [6]. The inference of the Bayes classifier postulates strong and independence hypothesis amidst the features. Bayes Theorem is as follows: $P(A/B) = P(B/A)P(A) / P(B)$. Here, the probability of A is to be originated when B is taken place [6]. In this, A is the lemma, and B is the manifestation. The theorization is formed as that the features are self-determined. There is no potential prevailing of one feature on the other. Hence, it is called naive. It is so simple and fast. Highly scalable too. But, the prior probability should be calculated and turns out relatively high error rate sensible on the input data form. (2)

SVM Classifier: Support vector machine (SVM) algorithm is an important classifier used to find a hyperplane in an N-dimensional space [23]. Here, N is the number of features that distinctly classify the data points. There are populous hyperplanes can be conceivable in order to set aside the data points just for two classes. The intention is to catch sight of preeminence margin for a plane. SVM classifier can handle nonlinear data efficiently, but choosing the right kernel function and interpreting the final model of the classifier are the tedious process. (3) **Decision Tree Classifier**: A decision tree is a simple representation for classification but can be extended to a large structure. The data is prolonged for separation endowed based on a certain state of affairs. Decision tree classifier does not involve normalization, and data preprocessing requires less time and analysis. Model training time and computational complexity are high for large-sized data.

These text classification algorithms are applied to the dataset to evaluate its performance. Each algorithm has its own merits and demerits. These algorithms are compared with the following clustering algorithms combined with the above classifiers to ascertain the suitable algorithm for analyzing online reviews. Clustering algorithms used are as follows. (1) **K-means Clustering**: It works by reiterating the process of partitioning the dataset into k pre-defined distinct and discongruence clusters. K-means clustering is easy to implement but harder to decide the value of K. When scaling is required, one instance can produce a lot of changes in the clustering process. (2) **One-Pass clustering**: It works by reading the input at a single time in a clear-cut way on the outside any disengaged buffering. The time taken for this process is too less, but it is critical when a number of clusters have to be changed [24] (3) **DBScan Clustering**: Density-based spatial clustering of applications with noise (DBSCAN) is a type of clustering algorithm that is used in machine learning to uncouple the clusters of immense and inferior density [6, 24, 25]. These classification algorithms and clustering algorithms are analyzed with its performance for the spam review detections.

4 Methodology

4.1 Dataset Preparation

This is the first step for any text classification process. Dataset is loaded and basic preprocessing operations such as Sentence Tokenization, Word Tokenization, Stemming and Noise Removal are performed. Sentence tokenization is a step which splits longer strings of text into smaller pieces as sentences. Sentences can be tokenized into words [26]. Stemming and bag of words are used as normalization process. Text Preprocessing: Converting the text documents into a commodious portrayal for the clustering task. The context of text archives has some prodigality in its content such as special characters, punctuations, stop words and other irrelevant bearings. To carry out the word stemming process, the stop words are taken away with the given content. Word stemming is the process of suffix removal to generate the common word stems. There are plenty of stemming algorithms exist such as Porter stemmer, Snowball stemmer and Lancaster stemmer. Stemming algorithms are rule-based algorithms. Here, Porter stemmer algorithm is used for stemming purpose. It eliminates the prevailing morphological endings of words. It is accelerated, and there is no complexity in using this algorithm compared to other stemmers.

4.2 Feature Engineering

The text data is the high dimensional dataset, which should be converted to features that could be used with the machine learning process. The text data, which is actually unprepared, could be turned over new leaf as flat features. Term frequency-Inverse document frequency (TF-IDF) is used in tokenization process, the tokens are further converted as vectors based on its occurrence frequency. Here, one hot vectorization is done for creating a feature vector. These vectors are extremely sparse when large corpora are used. These vectors can also be compared to create new features.

4.3 Model Training

The classifier model building is the ultimate stage, in which the machine learning algorithms can be applied to the prepared dataset. The dataset is then split into training and validation sets. This experiment considers 80% of data for training and 20% for testing purpose for both the datasets. **Text Classifier**: The performance of text classifier is improved in many ways. It has been proved that the clustering process is included for dimensionality reduction, which can further improve the performance of the classifier. Figure 1 gives an overview of the clustering-based classification model of the proposed work.

Online Reviews are taken as inputs. This work focuses on improving the performance of the classifier based on clustering. In machine learning and artificial intelligence, the text classification and clustering are considered as high demanding research fields. In this research work, text classification algorithms used are Naïve Bayes algorithm, SVM classifier, decision tree classifier are considered. These text classification algorithms are applied to the dataset to evaluate its performance. These algorithms are compared with the following clustering algorithms combined with the above classifiers to ascertain the suitable algorithm for analyzing online reviews. Clustering algorithms used are K-means clustering, One-Pass clustering, DBScan clustering for combining with the clustering process.

These three clustering algorithms are used with the above-mentioned classification algorithms, and comparisons are made based on the evaluation criteria. Applying the classification precisely to the dataset is not advantageous. So, first, the data has to be preprocessed. Preprocessing may also involve dimensionality reduction. The

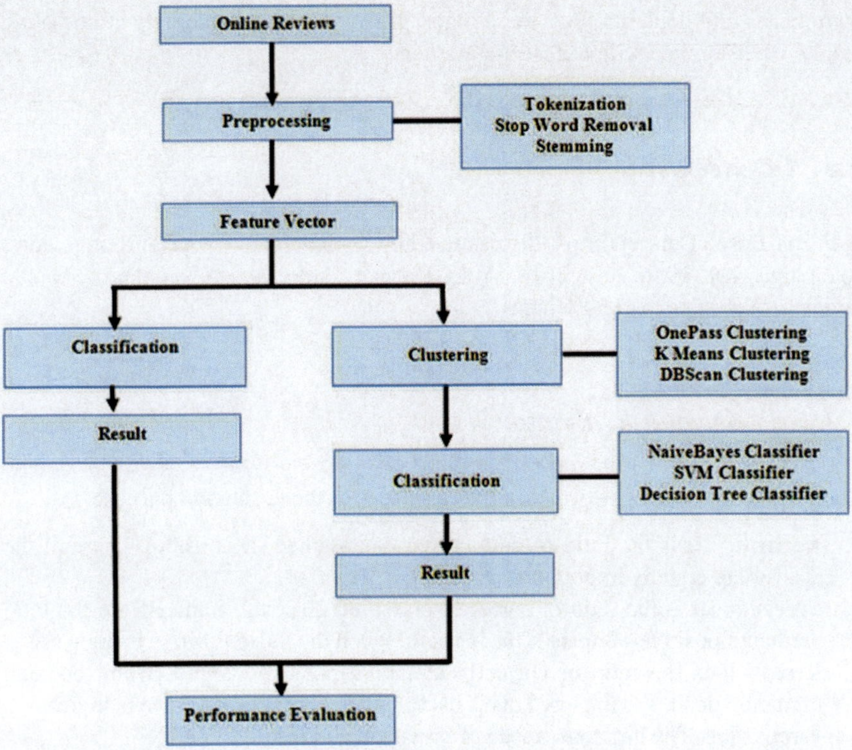

Fig. 1 Clustering-based classification model

proposed method is to deal with the clustering process, and then, carrying through the classification process thereby the accuracy of the classifier is improved.

Existing literature proves that there are so many classifiers and clustering algorithms, which can be used for text classification. But identifying the finest classifiers up to snuff clustering for upgrading the efficacy of the classifier is demanding specifically. So, this research work spotlights on contemplating different and opportune classifiers and clustering algorithms with all possible combinations, which provides a more advanced result.

The classification process is carried out by a single classification algorithm to investigate its expertise. The Naïve Bayes classifier, SVM classifier and decision tree classifier are the famous and verified methods for reputable text classifications. These three algorithms are separately applied to the given dataset for classification process, and the results are obtained. The performance of classifiers can be improved in many ways. But for reducing the dimensionality of the dataset, clustering can be a part of the classification process. So that, the first rated clustering algorithms such as One-Pass clustering, K-means clustering and DBScan clustering are combined with the above-mentioned classification process. For each classifier, all these three clustering algorithms are used for dimensionality reduction purpose. Each algorithm has its own merits and demerits. But, these algorithms produce convincingly culminating results when applied with different classifiers.

4.4 Dataset Used

Amazon-Based Dataset (https://liu.cs.uic.edu/download/data/): 42.8 million reviews spanning May 1996–July 2014. Yelp Dataset (https://www.yelp.com): 16,282 reviews (50% spam and 50% ham).

4.5 Metrics and Evaluation

The following are the performance metrics used for the evaluation purpose.

1. **Accuracy**: It is the ratio of data correctly classified. It is useful when all the classes are equally important.
2. **Precision**: It is the ratio of correctly classified positive samples from the total predicted positive samples. This is useful when the False Positive is high.
3. **Recall**: It is the ratio of correctly classified positive samples from the total predicted positive samples. This is useful when the False Negative is high.
4. **F-measure**: The harmonic mean of precision and recall.

5 Results and Discussions

The following results are discussed. (1) Impact of applying a simple text classifier directly. (2) Impact of applying to cluster prior to classification. (3) Comparison of all possible combination of selected clustering-based classification. The performance evaluation is done based on the parameters such as accuracy, precision, recall, F-measure. Table 1 shows the accuracy, precision, recall, F-measure of Yelp dataset, which contains considerably smaller size, when compared to the Amazon dataset.

Tables 1 and 2 represent the momentous diversification in the results. Table 1 provides the result comparison of Yelp dataset. The outcome of the clustering-based classifier is relatively upgraded in contrast to the classifier on the outside clustering. The SVM classifier, when improved with K-means clustering, produces 90.02% of accuracy, which is more significant than the other results in Table 1.

The same kind of result is obtained, when mobilized with the Amazon dataset, which is comparably populous, can be seen from Table 2. The SVM classifier outperforms than the other classifiers for one and the other datasets. The same SVM classifier put through superior result when reckoning up with all clustering algorithms. Even though all three clustering algorithms when applied before classification to get

Table 1 Performance analysis for Yelp dataset

Model	Algorithm	Accuracy	Precision	Recall	F-measure
Classifier only	Naive Bayes	82.15	83.22	81.88	0.81
	SVM	89.02	86.88	87.45	0.87
	Decision tree	87.58	84.11	85.39	0.85
K-means clustering + classifiers	K-means + Naive Bayes	84.79	84.04	85.12	0.83
	K-means + SVM	90.02	88.25	89.21	0.89
	K-means +decision tree	89.01	86.67	86.22	0.87
One-pass clustering + classifiers	One-pass + Naive Bayes	84.25	83.97	83.83	0.83
	One-pass + SVM	89.32	87.01	88.55	0.87
	One-pass + Decision Tree	88.94	86.29	85.98	0.86
DBScan clustering + classifiers	DBScan + Naive Bayes	83.50	83.58	82.34	0.82
	DBScan + SVM	89.92	87.23	89.03	0.88
	DBScan + decision tree	88.13	84.13	85.85	0.85

Table 2 Performance analysis for Amazon dataset

Model	Algorithm	Accuracy	Precision	Recall	F-measure
Classifier only	Naive Bayes	85.21	84.32	81.71	0.85
	SVM	86.03	85.18	86.45	0.87
	Decision tree	81.80	82.15	79.31	0.79
K-means clustering + classifiers	K-means + Naive Bayes	86.00	85.08	85.29	0.87
	K-means + SVM	88.25	87.28	89.16	0.89
	K-means + decision tree	84.22	84.08	81.45	0.83
One-pass clustering + classifiers	One-pass + Naive Bayes	85.66	84.71	82.11	0.85
	One-pass + SVM	87.29	86.76	88.00	0.88
	One-Pass + Decision Tree	82.11	82.56	79.45	0.8
DBScan clustering + classifiers	DBScan + Naive Bayes	85.87	84.78	84.25	0.86
	DBScan + SVM	87.19	86.55	87.26	0.87
	DBScan +decision tree	83.19	83.11	80.33	0.82

Table 3 Comparison of clustering algorithms

K-means clustering	One-Pass clustering	DBScan clustering
Gives the highest accuracy than the other clustering algorithms	Provides highest accuracy than the DBScan clustering and lesser than the K-means clustering	When examining in contrast, gives minor inferior result than the other classifiers

All these three clustering algorithms provide significant improvement in the performance of the classifier when combined with all the classifiers for both the datasets. Whereas SVM classifier outperforms than the other classifiers

features to bring forth a reputable result, the K-means clustering with SVM produces the higher result in all selected parameters as 88.25%.

Table 2 gives a comparison of different classifiers used for this research work using the Amazon dataset.

Table 3 discusses about the performance of different clustering algorithms when combined with classifiers. One-Pass clustering with SVM takes the second position, and DBScan with SVM takes next place. Naive Bayes classification is in the next level with smaller differences for Amazon dataset, and decision tree classifier is in the last position when dataset size is increased. Whereas, decision tree algorithm provides better result when used for the Yelp dataset which is smaller in size.

6 Conclusion and Future Work

The clustering process is unsupervised machine learning technique; by the way, it can also be adopted for setting up clusters as features. This helps to improve classification models through dimensionality reduction. The classification accuracy can be improved when clustering is used as features. The accuracy of the classifier is improved significantly when clustering is introduced before the classification process. K-means clustering for SVM classifier provides improved outcome. The clustering-based classification algorithm can further be improved by incorporating ensemble-based or improved classifiers. Other feature selection algorithms such as filter-based, wrapper-based or hybrid algorithms can be used with the classifier model for optimum results. Feature engineering can be done further to select the features for reviewers also. Datasets can be collected through crawlers and apply the algorithm to check its efficiency.

References

1. Lin HC, Bruning PF, Swarna H (2018) Using online opinion leaders to promote the hedonic and utilitarian value of products and services. Bus Horiz 61(3):431–442
2. Bajaj S, Garg N, Singh SK (2017) A novel user-based spam review detection. Procedia Comput Sci 122:1009–1015
3. Dewang RK, Singh AK (2018) State-of-art approaches for review spammer detection: a survey. J Intell Inf Syst 50(2):231–264
4. Rout JK, Singh S, Jena SK, Bakshi S (2019) Deceptive review detection using labeled and unlabeled data. Multimedia Tools Appl 76(3):3187–3211
5. Krishnaveni N, Radha V (2019) Feature selection algorithms for data mining classification: a survey. Indian J Sci Technol 12(6). https://doi.org/10.17485/ijst/2019/v12i6/139581
6. Towards Data Science, https://towardsdatascience.com
7. Martinez-Torres MR, Toral SL (2019) A machine learning approach for the identification of the deceptive reviews in the hospitality sector using unique attributes and sentiment orientation. Tour Manage 75:393–403
8. Ahmed H, Traore I, Saad S (2017) Detection of online fake news using N-gram analysis and machine learning techniques. In: International conference on intelligent, secure, and dependable systems in distributed and cloud environments. Springer, Cham, pp 127–138
9. Catal C, Guldan S (2017) Product review management software based on multiple classifiers. IET Software 11(3):89–92
10. Rout JK, Dalmia A, Choo KKR, Bakshi S, Jena SK (2017) Revisiting semi-supervised learning for online deceptive review detection. IEEE Access 5:1319–1327
11. Sun C, Du Q, Tian G (2016) Exploiting product related review features for fake review detection. Math Probl Eng
12. Wu Z, Cao J, Wang Y, Wang Y, Zhang L, Wu J (2018) hPSD: a hybrid PU-learning-based spammer detection model for product reviews. IEEE Trans Cybern
13. Shao Y, Trovati M, Shi Q, Angelopoulou O, Asimakopoulou E, Bessis N (2017) A hybrid spam detection method based on unstructured datasets. Soft Comput 21(1):233–243
14. Osman AH, Aljahdali HM (2017) Feature weight optimization mechanism for email spam detection based on two-step clustering algorithm and logistic regression method. Int J Adv Comput Sci Appl (IJACSA) 8(10):420–429
15. Elssied NOF, Ibrahim O, Osman AH (2015) Enhancement of spam detection mechanism based on hybrid ***k***-mean clustering and support vector machine. Soft Comput 19(11):3237–3248
16. Lu XY, Chen MS, Wu JL, Chang PC, Chen MH (2018) A novel ensemble decision tree based on under-sampling and clonal selection for web spam detection. Pattern Anal Appl 21(3):741–754
17. Pandey AC, Rajpoot DS (2019) Spam review detection using spiral cuckoo search clustering method. Evol Intel 12(2):147–164
18. Nagwani NK, Sharaff A (2019) SMS spam filtering and thread identification using bi-level text classification and clustering techniques. J Inf Sci 43(1):75–87
19. Chakraborty M, Pal S, Pramanik R, Chowdary CR (2016) Recent developments in social spam detection and combating techniques: a survey. Inf Process Manage 52(6):1053–1073
20. Fontanarava J, Pasi G, Viviani M (2017) Feature analysis for fake review detection through supervised classification. In: 2017 IEEE international conference on data science and advanced analytics (DSAA). IEEE, pp 658–666
21. Abu Hammad AS (2014) An approach for detecting spam in arabic opinion reviews. An approach for detecting spam in arabic opinion reviews
22. Nair A, Phapale A, Yagnik V, Bathe K (2016) Opinion spam mining. Int Res J Eng Technol (IRJET) 3(4):1855–1859
23. Kiwi Data Science, https://kiwidatascience.com
24. Kokate U, Deshpande A, Mahalle P, Patil P (2018) Data stream clustering techniques, applications, and models: comparative analysis and discussion. Big Data Cogn Comput 2(4):32

25. Brown D, Japa A, Shi Y (2019) A fast density-grid based clustering method. In: 2019 IEEE 9th annual computing and communication workshop and conference (CCWC). IEEE, pp 0048–0054
26. Kafi A, Alam MSA, Hossain SB, Awal SB, Arif H (2019) Feature-based mobile phone rating using sentiment analysis and machine learning approaches. In: 2019 1st international conference on advances in science, engineering and robotics technology (ICASERT). IEEE, pp 1–6

Chapter 21
Product Recommendation Systems Based on Customer Reviews Using Machine Learning Techniques

J. S. Shyam Mohan, Hanumath Sreeman Vedantham, Venkata Chakradhar Vanam, and Nagendra Panini Challa

1 Introduction

Online e-commerce platforms like Amazon, Flipkart, Snap deal, Paytm, etc., often provide costumers to share their opinions and experience quality on purchasing products online. Reviews describe the product information, service quality, and the voice of customers experience quality. For example, the reviews on a women's clothing might refer to the features such as name, price, color, size, and style. Reviews are the important part of any e-commerce platform because the sales person can have a detailed report of the particular product from the customer perspective. Since the number of products increases, the reviews of products also increases. It is difficult to parse the product reviews manually. Based on the customer reviews, the best products will be recommended to customer. The techniques and the tools used vary according to the organization or individual. Based on the findings of analyzing data, the importance of every feature in the data set will be found. Data is growing at an exponential pace daily. Retrieving information according to the user's requirement

J. S. Shyam Mohan · H. S. Vedantham (✉) · V. C. Vanam
Department of CSE, SCSVMV, Enathur, Kanchipuram 631561, India
e-mail: hanumathsreeman@gmail.com

J. S. Shyam Mohan
e-mail: jsshyammohan@kanchiuniv.ac.in

V. C. Vanam
e-mail: vanamvenkatachakradhar@gmail.com

N. P. Challa
Department of IT, SVECW, Bhimavaram, Andhra Pradesh 534202, India
e-mail: paninichalla123@gmail.com

I. Jeena Jacob et al. (eds.), *Data Intelligence and Cognitive Informatics*,
Algorithms for Intelligent Systems,
https://doi.org/10.1007/978-981-15-8530-2_21

is a difficult task. The retrieved information should be relevant to the requirement of user. For example, if the customer needs the information regarding a gadget in the e-commerce, it should retrieve according to the user requirement. So, in this case, the information retrieval is very essential to bring out the information related to the search instead of displaying the information which is irrelevant to the search. The retrieved information includes the review text which was given by the costumers. To meet and overcome the difficulties involved in the text processing, natural language processing is added to the machine learning model which helps to classify the product reviews are good or bad [1].

2 Existing Work

Existing researches mainly focus on product quality assessment, sentimental analysis and text analysis based on online reviews. The reviews are the common platform between the customers and the company. Based on the reviews, the company can enhance the quality of a product and the sales of a product in an online market [2]. Text analysis of the user reviews based on the NLP techniques is implemented. Web scraping for extracting data in an efficient way that helps for data retrieval in the required formats. Thereby, these data are used for further findings like data analysis and text analysis [3]. Based on the text mining techniques, the information can be extracted and categorize the extracted knowledge from the text for best results. Similarly, the task of text clustering is to group the similar words which give the same meaning to a common group called cluster and there must be some difference between two clusters that analyzes the word net and sentiment analysis [4]. Sentiment analysis is a process of extracting opinions that have different polarities, i.e., positive, negative, or neutral. It is also known as opinion mining and polarity detection. Sentiment analysis is a type of classification where the data is classified into different classes. These classes can be binary in nature (positive or negative). Natural lnguage processing is used to remove the punctuations, html tags, and stop words from the text data and performs techniques like stemming and lemmatization for the text data. Stemming is a process of reducing words to their root form. Lemmatization is used for grouping the review text into similar forms which can be suitable for further usage [5]. In supervised learning algorithms, input features are independent variables and output features are dependent variables. The term supervised learning comes from the idea that an algorithm is learning from a training dataset. In supervised learning, the algorithm teaches the machine how to achieve a specific goal by showing the machine what the goal is. Classification algorithms used to map input features to a particular class either a binary class classification or multi-class classification. Classification problems basically deals with target value as discrete in nature it belongs to some class. Binary class classification deals with two possible predictions either true or false, male or female, yes or no. Multi-class classification deal with more than one than two classes but the predicted value or outcome will assign in single classes from these classes [6].

3 Description of the Proposed Work

Fetching data from the corporate websites is a Text Mining operation probably extracted in excel format. After scraping the data from websites, the data will be analyzed. The data preprocessing operations have to be performed with the data which, scraped from the source to analyze the patterns present in the data. Now, the natural language processing (NLP) techniques on the dataset will be performed. Text mining is a process that uses NLP to form the vectors from the unstructured data. This text mining technique focuses on identifying the extraction of entities, attributes, and their relationships from semi-structured or unstructured texts. After completing text, pre-processing performs the vectorization process for the text data. There are two methods in vectorization process: one is CountVectorizer and the other one TFIDF Vectorizer. CountVectorizer only counts the number of times a word appears in the review text which results the count of frequent words by ignoring rare words. Tfidf Vectorizer considers the full review text to find the frequency of a particular word over all words in the text. Thereby, the frequency count of useless words can be reduced like stop words. Split data for training and testing after the vectorization process. Training data is the part of the dataset which is used to train the model and make the hypothesizes more accurate. For testing process, the data is split in the form of 8:2 that is 80% of data is sent for training and 20% of data will be sent for testing. After splitting the data, scikit-learn models are used for predictions. This paper is organized into the following sections: Sects. 4–6 explains the process of the proposed work. Sections 7 and 8 discuss about the tested environment and result analysis. The later section is about conclusion.

4 Data Pre-processing and Analysis

Data preprocessing is a process to redefine the data which were obtained from the source data set. This raw data has to be changed with respective to our analysis. Since any operations can be performed with raw data for our need, as far as business concern, the data plays everything regarding the enhancement of sales. So data analysis should be done in the pre-processed data to understand every detail in the data. This leads to making a decision for the right investment in the business. The e-commerce online market data is scraped from the website and arranged the data on the excel as required. After creating the excel file, the dataset is assigned in our workspace according to our requirements and dropped the unwanted columns from the table. Then, convert the rating column into two classes. Another column is created after the rating named as rating class which tells the rating either a good one or bad one. Here, rating less than or equal to 3 is a bad one and the remaining are good one. Now, some of the columns have been analyzed for better understanding of correlation this resulting impacts the creation of machine learning models. First, good and bad ratings are analyzed for products which tells the number of count of good rating and bad ratings that results the following Fig. 1.

The recommended indicators are counted according to the age interval. Figure 2 tells which age interval has the most recommended indicators.

Counting division name and the rating class give the analyzation about the division names and their good count and bad count. For this observation, Fig. 3 shows the co-relation between both fields.

Analyzing the rating class with respect to age intervals. It counts number of goods and bads for each age interval. The results are shown in Fig. 4.

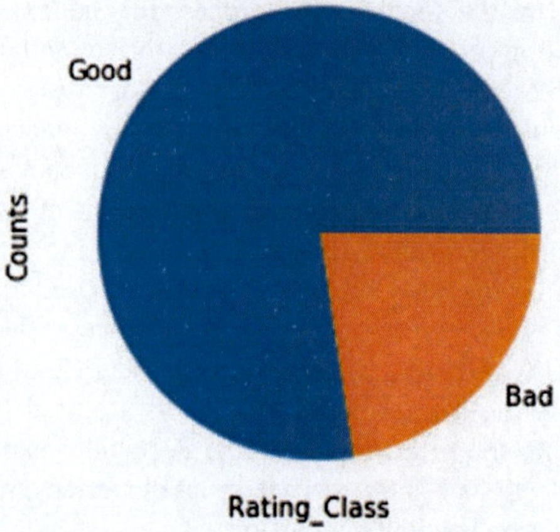

Fig. 1 Rating_Classs versus count

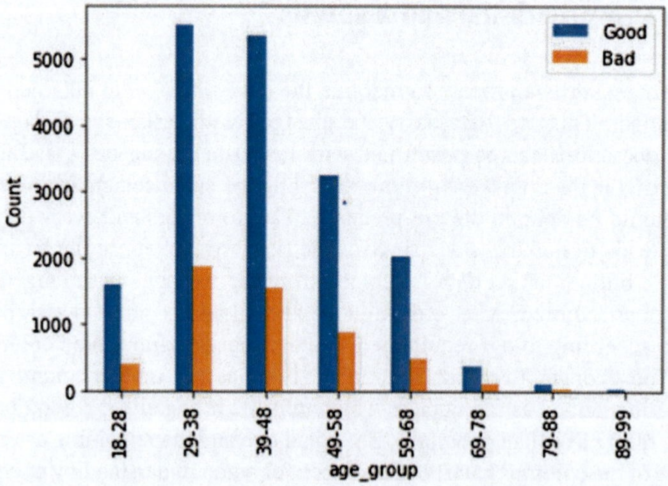

Fig. 2 Age_Group versus recommended indicators

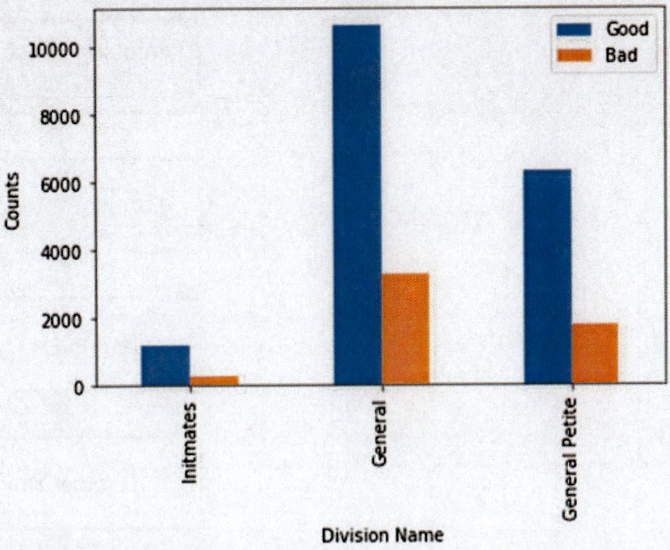

Fig. 3 Division_Name versus Rating_class

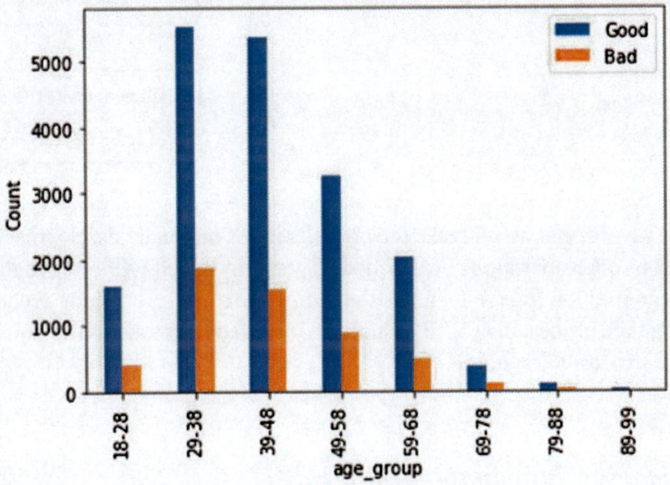

Fig. 4 Rating_Class Respective to Age_Group

5 Text Pre-processing

Text data are an unstructured data may contain some special characters like punctuations, some html, tags, etc., and also it may contain some words which have less importance in analysis since these words do not have the capacity to change over the text analysis, these are said to be stop words. These types of data present in the text

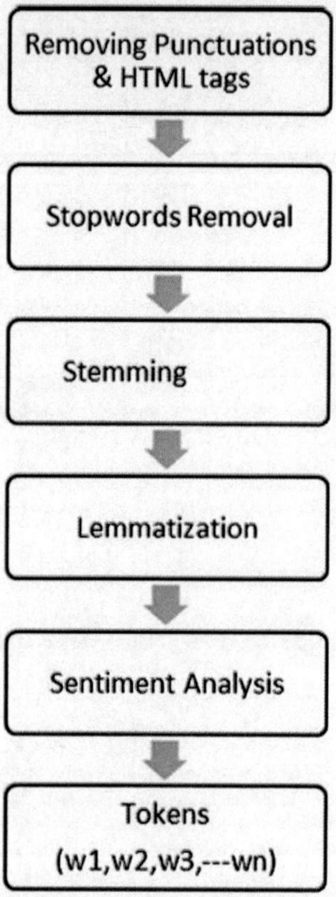

Fig. 5 Steps in text preprocessing

which is in raw format ay increase the complexity of our analysis. So to achieve the text analysis, some operations are needed to perform which will remove the unnecessary things in data. Figure 5 shows the various text pre-processing techniques of natural language processing (NLP) which are involved to extract the meaningful data from the unstructured data.

5.1 Removing Punctuations and HTML Tags

The raw data obtained from a review text column is an unstructured data which contains the punctuations, html tags, digits, etc. A model cannot be created using this type of unstructured data. So, here, the punctuations will be removed and html tags with a certain condition for doing the further text pre-processing techniques. Define methods for both html and punctuations removal, in that method using the re.compile and re.sub functions will remove the html and punctuation marks. Here, "re" represents the regular expression. After the creating, the methods that cleaned text will be returned.

5.2 Stopwords Removal

Stop words are generally the most common words in a language. Stop words are simply words that have very little meaning and are mostly used as part of the grammatical structure of a sentence. Words like "the", "a", "an", "in", etc., are considered as stop words. Removing the stop words in the review text is must because stop words present in the text may result in the complexity. To perform this removal of stop words operations, the sentences have to be splitted into the words using word_tokenize or str.Split () method. After splitting the sentences, the removal of stop words can be performed.

5.3 Stemming

Stemming is a process that can use to obtain the base form or root form of a particular word by chopping the word endings. In natural language processing, there several predefined libraries for stemming such as PorterStemmer, SnowballStemmer, and LancasterStemmer. Among all these stemmers, potter stemmer provides the best results to obtain the root form of the word. Consider also the words that are closely related, such as democracy, democratic, and democratization. For example, consider connected, connecting, and connected, by using the stemmers the base word can be found of these words as connect. The output of a stemmer is called the stem, which is the root word [7].

5.4 Lemmatization

Lemmatization is a process which is very similar to the stemming. It is used to group the words which are different forms of a word into the root form so they can be analyzed as a single word. In the lemmatization, the term lemma means it groups together different infected forms of a word, called lemma. Behind the extraction of the suitable lemma, Morphological analysis plays the main role. Morphological analysis means the exploring of all the possible outcomes of a particular thing. With the help of lemmatization process, the root words can be able to found which relate to verbal speech since lemmatization have a capacity to work on parts of speech. For example, a lemmatizer should map gone, going and went into go. The main methodology behind the lemmatization is the study of a particular language. It is very important to have a dictionary of a particular language which some operations are going to be perfomed using lemmatization.

5.5 Sentimental Analysis

Sentiment analysis is used to study the customer sentiment through reviewing text; the main use of sentimental analysis is, it is big enough for finding the objective, emotion, and attitude of the customer with the help of the review which he posted for the product.

There are some logical formulas which are mentioned below to calculate absolute proportional difference, relative proportional difference, logit scale.

The total number of words = $P + N + O$.

- P is Positive count-coded sentiment words.
- N is Negative count-coded sentiment words.
- O is the count of all other words.

1. Absolute proportional difference. Bounds: [0, 1]

$$\text{Sentiment} = (P - N)/(P + N + O)$$

2. Relative proportional difference. Bounds: [−1, 1]

$$\text{Sentiment} = (P - N)/(P + N)$$

3. Logic scale. Bounds: [−infinity, +infinity]

$$\text{Sentiment} = \log(P + 0.5) - \log(N + 0.5)$$

This tends to have the smoothest properties and is symmetric around zero. The 0.5 is a smoother to prevent log(0). It assigns scores to the emotions expressed in the document in the form of a positive or negative emotion. Finally, it assigns labels to the document on the basis of the aggregate score. This sentiment is quantified with a positive or negative value, called polarity. The polarity score is dependent upon the polarity dictionary used. Extract sentiments from the text data using a powerful Python library called "TextBlob"[8, 9] (Fig. 6).

Product reviews are collected from the dataset, after performing the above text preprocessing techniques like punctuations and html tags removal, stop words removal, stemming, lemmatization and parts of speech tagging. Now, these product reviews are sent to the next step called sentiment identification. After identified sentiments, the further step will be gone as feature selection such as "Fine-grained Sentiment Analysis", "Emotion detection", "Aspect-based sentiment analysis", etc. Now, sentiment classification classifies whether the sentiment is a positive or negative by using sentiment dictionary. A sentiment dictionary specifies sentiment indicators and rules to be used in the analysis. This process is called as sentiment polarity.

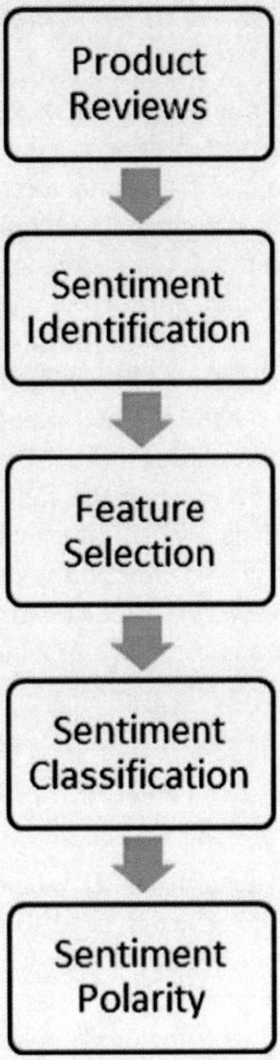

Fig. 6 Steps in sentimental analysis

6 Vectorization Process

Vectorization is a process in natural language processing (NLP) used to totally the words or phrases from vocabulary to a corresponding vector of real numbers. These real numbers are used in the machine learning model for predictions. In the vectorization process mainly two types of vectors have, one is "Count Vectorizer" and the other one is "TF-IDF Vectorizer." This process of converting NLP text into numbers is called vectorization in ML. There are multiple ways to convert text into vectors.

- **Counts**. Count the number of times each word appears in a document.
- **Frequencies**. Calculate the frequency that each word appears in a document out of all the words in the document.

Based on how this scoring is created, there are following methods.

6.1 Count Vectorizer

CountVectorizer is a process used to tokenize a text. The results of stemming and lemmatization are the input for CountVectorizer. The results of the count Vectorizer are in the vector format, i.e., $(n \times 1)$ the elements in this vector are real numbers. In the previous sections, the earlier text is an unstructured data and, therefore, it is difficult to give the text feature as the independent variable for the machine learning model. Vectorizer method is used to transform the unstructured data into vectors. These vectors are used to obeying the difficulty in creating a model. There is a predefined library in the Python for CountVectorizer.

After the text preprocessing, split the data into 8:2 ratio, i.e., 80% of the data is for training and 20% of the data is for testing. Training data is sent to the CountVectorizer using the function called fit (), test data is transformed using the function transform (). Fit function is used to learn vocabulary from the review text, whereas transom function is used to transform the respective review text into vectors. The vectors returned from the transform function are sparse vectors and these vectors can transform back to numpy array by using toarray() function.

6.2 TF-IDF Vectorizer

The term TF-IDF stands for Term Frequency-Inverse Document Frequency. TF-IDF vectorizer is used to transform a text to feature vectors which are used as input to estimators. The weight of a word is a statistical measure used to calculate the importance of a particular word in the review text of the collection of words for classifying the good or bad reviews. The term "Term Frequency" is used to measure the frequency of a particular term that occurs in the review text. Since a chance may have been to repeat the particular word two or many times in one review. So term frequency will estimate the frequency of occurrence of a specific word in the product review text. The term "Inverse Document Frequency" will measures the importance of a specific term. However, TF will show equal importance to all the words in the review texts which have less usage so IDF will weigh down for some of the repeated words and scales up the rare once. The first term is the term frequency means the number of times a word appears in a text, divided by the total number of words in that text. The second term is the inverse document frequency (IDF), computed as the log of the number of the text in the corpus divided by the number of texts where the particular term appears [10]. The mathematical formula for calculating the TF-IDF Vectorizer is mentioned below:

$$tf - idf(t, D) = tf(t, d).idf(t, D)$$

$$tf(t, d) = f_{t/d}$$

$$idf(t, D) = \log\left(\frac{N}{|\{d \in D : t \in d\}|}\right)$$

The importance of a particular word can be computed by using this TF-IDF formula. Since the importance of a word is directly proportional to the count of a word which appears in the review text. The later steps are similiar to the process discussed in Sect. 6.1.

7 Implementation and Tested Environment

After the vectorization, the next step is to implement a model based on dependent and independent variables. Independent variables are the review text and dependent variables are the rating class. If the continuous results are goint to be predicted, then will be going for regression algorithms. Classification algorithms maps the input features into a specific category. From here, whether the review of a product is good or bad have to be classified. To obtain the results in the categorical format for text data some of the classification algorithms have with:

- Naive Bayes algorithm.
- Logistic algorithm.
- Support vector machine (SVM).

7.1 Naive Bayes Algorithm

The Naive Bayes classification algorithm is a supervised learning algorithm that uses Bayes theorem to classify input features. It is mostly used for text classification. Both binary and categorical data can be used as the input to this type of classification naive is an assumption in a naive Bayes algorithm which uses Bayes' theorem for its predictions it exempts that there should be no dependency between the attributes of the dataset, i.e., he attributes of the dataset are independent. The Naive Bayes algorithm is called naïve because it makes the assumption that the occurrence of a certain feature is independent of the occurrence of the features [11]. The main use of Bayes' theorem is when new data are used by the specific event, and then, the probability of that particular event can be adjusted. Bayes' theorem is stated as probability of the event *B* given *A* is equal to the probability of the event a given *B* multiplied by the probability of *A* upon the probability of *B*.

$$P\left(\frac{A}{B}\right) = \frac{P\left(\frac{B}{A}\right)P(A)}{P(A)}$$

P(*A*/*B*): Conditional probability of occurrence of event *A* given that event *B* is true.

P(*A*) and *P*(*B*): Probabilities of the occurrence of event *A* and event *B*, respectively.

P(*B*/*A*): Probability of occurrence event *B* given that event *A* is true.

Sklearn has inbuilt multinomial naïve Bayes classifier package. Using this package, our model can be trained directly with the matrix obtained from TF-IDF transformer. By creating the instance for the MultinomialNB (), the predefined methods of that particular class can be used. Multinomial Naive Bayes classifier is given by the formula as shown below:

$$P(j) \propto \pi_j \prod_{i=1}^{|V|} P(i|j)^{f_i}$$

where

j is a class.

i is a specific word.

f_i is a word frequency of i.

Naive bayes classifiers are imported from Sklearn libraries. After importing the classifier, now, fit the vectorized x_train data and y_train data into the classifier by calling fit() method with the help of instance that have created for MultinomialNB (). Now, use the predict() method for recording the predictions. In the predict() method pass the vectorized x_test data. These predictions are used for classification report.

7.2 *Logistic Regression*

Logistic regression is used to find the discrete output, i.e., either 0 or 1. In this case, logistic regression model is used to classify whether the review text is good or bad since the output for our requirement is in discrete type. In logistic regression, the main struggle is to find the hypothesis. Since the output is despite, our hypothesis should lie in between 0 and 1. Logistic regression uses an hypothesis function representation similar to linear regression. Since the hypothesis of linear regression is a continuous function which may use the predictions less than 0 or greater than 1. The hypothesis function of logistic regression must be discrete. For this reason, the hypothesis function of linear regression fails. To avoid this type of problem, sigmoid function can be used. Sigmoid function is a mathematical function used for getting the hypothesis function between 0 and 1. Sigmoid function is useful for transforming an arbitrary value function into a better suited for classification [12].

$$g(z) = \frac{1}{1 + \mathrm{e}^{-z}}$$

The function $g(z)$ maps any real numbers to the (0, 1) interval. Now, add the sigmoid function $g(z)$ to the hypothesis function. Therefore, our hypothesis function is stated as given below:

$$g(\theta^t x) = \frac{1}{1 + e^{-\theta^t x}}$$

$$h_\theta(x) = g(\theta^t x)$$

The accuracy of hypothesis function can be measured with the help of cost functions. If the cost function is used which is similar to the linear regression, then, it may give the non-convex wave. Since non-convex wave has many local minima. It is difficult to find global minima for gradient descent. To avoid this, convex wave logarithmic terms have to be added for our hypothesis function [13].

$$j(\theta) = \frac{-1}{m} \sum_{i=1}^{m} \left[y \log(h_\theta(x)) + (1 - y) \log(1 - h_\theta(x)) \right]$$

where

m is total number of training examples.

x is the input feature of training example.

y is respective value of input feature.

$h_\theta(x)$ is hypothesis function with sigmoid curve.

$j(\theta)$ is cost function.

Sometimes, overfit problem may be faced, i.e., the hypothesis may be fit for the given data with some complex function, but fails for classifying new data other than existing data this is said to be over fitting. This overfit problem can be avoided with the help of regularization. Regularization is used to decrease the magnitude of parameter vector θ. So add a regularization factor to the existing cost function.

$$j(\theta) = \frac{-1}{m} \sum_{i=1}^{m} \left[y \log(h_\theta(x)) + (1 - y) \log(1 - h_\theta(x)) \right] + \lambda \sum_{j=1}^{n} \theta_j^2$$

Here, λ is the regularization parameter and n is the number of features. The Naive Bayes classifier is imported from "from sklearn.linear_model import LogisticRegression". After importing the classifier, now fit the vectorized x_train data and y_train data into the classifier by calling fit() method with the help of instance which LogisticRegression are created for. Now, use the predict() method for recording the predictions. In the predict() method, pass the vectorized x_test data. These predictions are used for classification report.

7.3 Support Vector Machine (SVM)

Similar to the above models, support vector machine model is used for the classification type of problems. The functionality and the accuracy of this model are mainly dependent on the kernel function. The kernel mainly requires the parameter vector theta which was updated by the user. Simply can be said this as hyper parameter. Similar to the learning rate alpha in gradient descent this hyper parameter has to control the learning process. So the learning rate of a model is completely dependent on this type of hyper parameters mathematically the hyper parameter process uses trail an error method to find the best fit parameter vector [14]. Similar to the cost function of logistic regression the cost function can be computed for support vector machine. If in logistic regression hypothesis function is $y = 1$ then almost $h_\theta(x) = 1$. Then, $\theta^t x$ is much greater than 0. Similarly, If in logistic regression hypothesis function is $y = 0$, then almost $h_\theta(x) = 0$. Then, $\theta^t x$ is much less than 0. So, $y = 1$ the cost function of logistic regression may compute into $ct_1(z)$. Similarly, $y = 0$, then $ct_0(z)$.

$$j(\theta) = \frac{-1}{m}\sum_{i=1}^{m}\left[y\log(h_\theta(x)) + (1-y)\log(1-h_\theta(x))\right]$$

$$+\lambda\sum_{j=1}^{n}\theta_j^2$$

$$ct_1(z) = -\log\frac{1}{1+\mathrm{e}^{-z}}$$

$$ct_0(z) = -\log\left(1-\frac{1}{1+\mathrm{e}^{-z}}\right)$$

where

$ct_1(z)$ is a cost function where $y = 1$.

$ct_0(z)$ is a cost function where $y = 0$.

With the help of the newly computed $ct_1(z)$ and $ct_0(z)$ cost function of support vector machine can be computed.

$$j(\theta) = \min_\theta c\sum_{i=1}^{m}\left[y.ct_1(\theta^t x) + (1-y)ct_0(\theta^t x)\right] + \frac{1}{2}\sum_{i=1}^{n}\theta_j^2$$

where, c is a hyper parameter constant.

After importing the classifier, now fit the vectorized x_train data and y_train data into the classifier by calling fit() method with the help of instance which SVM is created for. Now, use the predict() method for recording the predictions. In the predict() method, pass the vectorized x_test data. These predictions are used for classification report.

8 Experimental Setup

For implementing model preparation, data analysis, and text pre-processing techniques, the working environment have been setup in Jupyter Notebook. Jupyter notebook is open source and easy to access because it has all the required libraries pre-installed and the results can get very quickly and easily. Jupyter notebook works on the basis of Anaconda platform.

9 Results and Performance Validation

9.1 *Classification Report*

Classification report is a report which can tell us the quality of the model which was trained. It is used to check how good our algorithm for our data like how many predictions are correct and how many went wrong there by some detailed information can get about accuracy, precision, recall, and f1-score.

9.2 *Accuracy*

Accuracy is said to be how much accurately our model is getting predictions when testing data is passed. So the model results can be evaluated using accuracy.

$$\text{accuracy} = \frac{\text{TP} + \text{TN}}{\text{TP} + \text{TN} + \text{FP} + \text{FN}}$$

9.3 *Precision*

Precision is used to calculate the percentage of relevancy of the result which was predicted by model by obtaining the ratio of correctly true predictions and the total production

$$\text{precision} = \frac{\text{TP}}{\text{TP} + \text{FP}}$$

9.4 Recall

Recall gives the result which the percentage of model correctly predicted. Proportion of correct positive classifications (true positives) from cases that are actually positive.

$$\text{recall} = \frac{\text{TP}}{\text{TP} + \text{FN}}$$

where,

TP is True Positive.

TN is True Negative.

FP is False Positive.

FN is False Negative.

9.5 F*1-Score*

$f1$ score is the balanced score of both precision and recall. To understand the better situations in the model, the $f1$ score can be used because it can compute the uneven class distribution (Table 1).

$$f1\_\text{score} = \frac{2(\text{precision} * \text{recall})}{(\text{precision} + \text{recall})}$$

The total reviews considered for classification are approximately 23,000. The Naive Bayes algorithm has the highest accuracy rate comapred with other algorithms. While comparing the $F1$-score with the other classifiers, logistic regression had 93% of good score and 72% of bad score. Therefore, Naive Bayes classifier is recommended as the best one compared with other models.

Table 1 Summary of all machine learning models

Model name	Precision		Recall		$F1$-score		Accuracy (%)
	Good (%)	Bad (%)	Good (%)	Bad (%)	Good (%)	Bad (%)	
Naive Bayes	90	78	95	66	92	71	89
Logistic regression	90	79	95	66	93	72	88
Support vector machine	90	77	94	64	92	70	87

9.6 Confusion Matrix

Confusion matrix is simply a matrix format of the classification report, which can tell us the count of True Positives, True Negatives, False Positives, and False Negatives, and also helps us to calculate the accuracy and $f1$ score. Figures 7, 8, and 9 show the results of the three machine learning models that show the count of True Positives, True Negatives, False Positives, and False Negatives, i.e., the actual goods, actual bads, predicted goods and predicted bads. These three confusion matrices contains the three different values from each other.

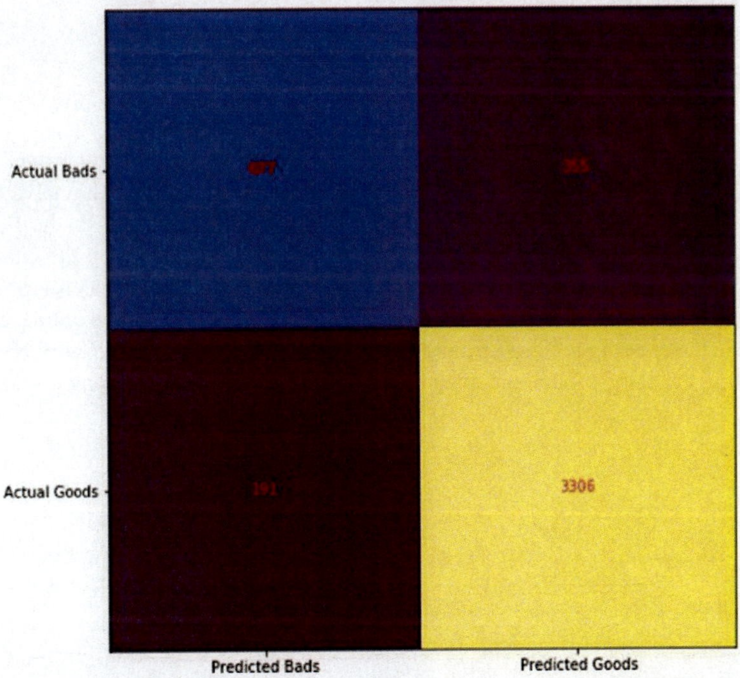

Fig. 7 Confusion matrix for Naive Bayes classifier

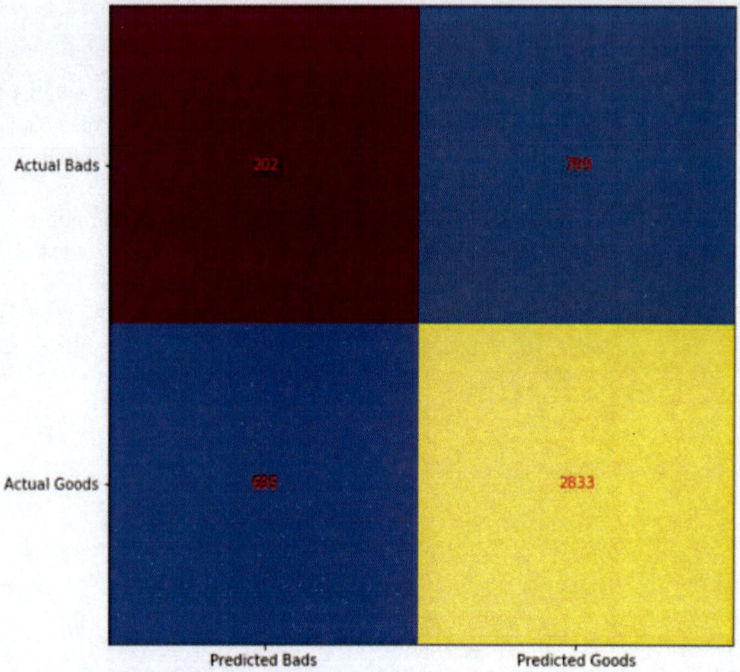

Fig. 8 Confusion matrix for logistic regression

10 Conclusion

The results obtained shows that Naive Bayes gives the best accuracy than other algorithms like a support vector machine and logistic regression. The naive bayes algorithm has been used for text classification for our dataset. With the help of rating_class, it classifies the review of a product which was given by the customer into positive class and negative class. For example, if a customer need a product, then our proposed model recommends the best product in e-commerce online market based on the existing reviews of a product given by customers. Future work involves extending the same model to deploy in cloud for many online shopping portals.

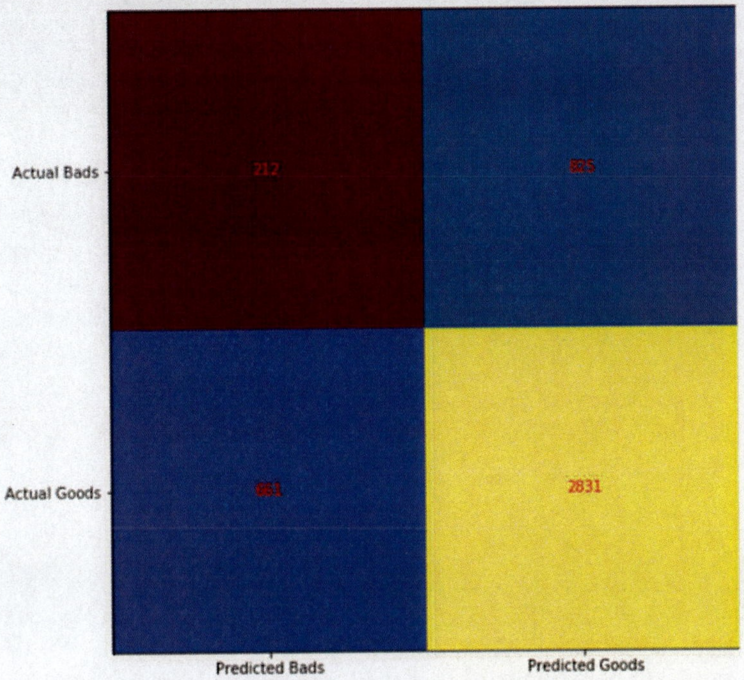

Fig. 9 Confusion matrix for support vector machine

References

1. Joby PP (2020) Expedient information retrieval system for web pages using the natural language modeling. J Artif Intell 2:100–110
2. Li C et al (2019) Product quality assessment based on online reviews. Int J Decis Support Syst Technol (IJDSST). IGI Global
3. Bianchi G et al (2018) Identifying E-commerce in enterprises by means of text mining and classification algorithms. Mathematical Problems in Engineering, Hindawai
4. Salloum SA et al (2018) Using text mining techniques for extracting information from research articles. In: Intelligent natural language processing. Springer, Berlin, pp 373–397
5. Mali D et al (2016) Sentiment analysis of product reviews for E-commerce recommendation. Int J Manage Appl Sci 127–130
6. Osisanwo FY et al (2017) Supervised machine learning algorithms: classification and comparison. Int J Comput Trends Technol (IJCTT) 128–138
7. Singh J et al 2016 Text stemming: approaches, applications and challenges. ACM Comput Surv
8. Haider S et al (2016) Lexicon-based sentiment analysis of teachers' evaluation. Appl Comput Intell Soft Comput
9. Furini M et al (2018) Sentiment analysis and Twitter: a game proposal. In: Personal and ubiquitous computing. Springer, Berlin, pp 771–785
10. Qaiser S et al (2018) Text mining: use of TF-IDF to examine the relavance of words to documents. Research Gate
11. Subramanian S (2014) Document classification using multinomial Naive Bayesian classifier. Research Gate, pp 1557–1563
12. Sperandei S (2014) Understanding logistic regression analysis. Research Gate

13. Iduseri A et al (2016) On maintaining high levels of ımmunization coverage in edo state using binary logistic regression model. Research Gate
14. Tian Y et al (2012) Recent advances on support vector machines research. Research Gate

Chapter 22
Mobile Application for Classification of Plant Leaf Diseases Using Image Processing and Neural Networks

K. S. Chethan, Sumanth Donepudi, H. V. Supreeth, and Vachan D. Maani

1 Introduction

Agriculture is the art and science of cultivating the plants, growing crops and raising livestock. The production of agriculture activity is about $375.61 billion; India stands in second to be the largest producer of agriculture products. India accounts for 7.4% of total global agricultural output, and agriculture is the backbone for our Indian economy. About 60% of people in the world depend on agriculture for their income according to the Food and Health Organization. Many crops are grown in India, with rice and wheat being the most significant ones. Many other consuming and non-consuming crops such as tea, coffee, coconut, pulses, oilseeds, sugarcane, rubber are also widely grown. About one-third of the agricultural product is getting wasted because of various reasons. There are many reasons for the cause of agricultural waste some of them are contaminated water, soil erosion and sedimentation, livestock, pests and weed and many more but one of the major reasons for wastage of agricultural produce is plant diseases.

Generally, farmers don't have much knowledge about different types of plant diseases, and they are not able to identify through their naked eyes, and sometimes it may result in incorrect decisions, they may apply the wrong pesticide which may

K. S. Chethan · S. Donepudi (✉) · H. V. Supreeth · V. D. Maani
Department of ECE, PES University, Bengaluru, India
e-mail: sumanth1341999@gmail.com

K. S. Chethan
e-mail: chethanks@pes.edu

H. V. Supreeth
e-mail: supreeth.haval@gmail.com

V. D. Maani
e-mail: vachanmaani@gmail.com

I. Jeena Jacob et al. (eds.), *Data Intelligence and Cognitive Informatics*,
Algorithms for Intelligent Systems,
https://doi.org/10.1007/978-981-15-8530-2_22

destroy crops and result in agriculture waste. Identifying and detecting plant disease at an early stage and guiding the farmers by giving them suitable treatment pieces of advice can help them to produce more agricultural products without any damage or wastage.

This project aims to identify the disease name of the plant and gives suitable pesticide for that particular disease plant suffering through a mobile application. To achieve this dataset are necessary for different plant diseases so in this project, let us aim for the main basic eight types of plant diseases those are Apple Scab, Rust, Early blight, Late blight, Bacterial spot, Mosaic virus, Powdery Mildew, Black rot. The first step is to do image processing techniques on the dataset to enhance contrast. Image segmentation is to partition the image into interesting regions. Image feature extraction to extract the disease feature based on texture and colour, at last, these features are fed to a neural network, where the dataset is used to train by CNN model. All these image processing techniques and deep neural networks are done using python. To make it easy for farmers', mobile application is developed so that they can capture the photo of a diseased plant leaf and immediately it will show the result which disease it is and the pesticide name which can be used to cure that specific disease. To develop a mobile application, the Kivy library is used. This library is an open-source Python library for developing mobile applications and other multi-touch application software with a natural user interface (NUI). The connection is established between the client (farmer) and the server in order to exchange information, for that transmission control protocol (TCP) connection is used.

2 Literature Survey

Prakash et al. [1] in their paper checked whether a citrus leaf is affected or not. This was performed using a dataset of 60 leaves containing both affected and diseased leaves. This paper used k-means clustering for segmentation of images and GLCM for feature extraction. The classification was done using SVM, and accuracy of 90% was achieved.

Dhaware et al. [2] also focused on the detection of plant leaf diseases using a similar approach of segmentation, feature extraction and classification. The dataset used had 120 images of healthy plants and 120 images of diseased plant leaves. The diseased leaf dataset consisted of various diseased leaves such as sunburn, yellow mosaic and grasshopper. Similar to the above paper, this paper also uses support vector machine (SVM) to classify diseases.

Shruthi et al. [3] mainly focused on different algorithms used for classification. The authors used various classifiers like SVM, ANN, Fuzzy classifiers, CNN, etc. to classify the diseases and did a comparative analysis between them. CNN was found out to be the best classifier with an overall accuracy of 99.5%.

Qiao et al. [4] focused on one of the three data mining technologies clustering. Several common clustering algorithms like two-step clustering, KNN, Kohonen, k-means clustering were discussed and were compared based on the quality and execution time. When the above four algorithms were compared, it was found out that k-means has a faster execution time, and the quality of k-means is better than any other clustering algorithms. The paper also discusses the simulation of k-means clustering for different k-values.

Singh et al. [5] emphasized one of the oldest techniques used for feature extraction, i.e. Grey-level co-occurrence matrix. Grey-level co-occurrence matrix represents relative frequencies of grey levels of an image present at a certain distance and a particular angle. It discusses how the grey-level co-occurrence matrix is obtained, and how the calculations are performed. The second part of the paper focuses on the features obtained from the GLCM like contrast, correlation, energy, homogeneity.

Sarangdhar et al. [6] used a support vector machine (SVM) to classify five different cotton leaf diseases. The authors also built a system to check soil quality. The system is controlled by a mobile application. Overall classification accuracy of 83.26 was achieved.

Gavhale et al. [7] also used SVM to classify different diseases of a citrus plant leaf. The author used polynomial kernel and RBF kernel for SVM classification.

Khirade et al. [8] classified different plant leaf diseases with the help of artificial neural networks.

Badnakhe et al. [9] focused on k-means clustering and artificial neural networks to identify crop diseases.

Chouhan et al. [10] presented a new method named bacterial foraging optimization-based radial basis function neural network to identify and classify different diseases. The author also compared the results with k-means and genetic algorithm.

Singh et al. [11] analyzed various methods that were used for detecting plant diseases.

Kumar et al. [12] discussed about various feature extraction techniques and give a brief idea about the best-suited extraction techniques for a different type of data.

Vincent et al. [13] discussed three different methods for colour texture analysis. All the methods discussed are an extension to GLCM.

3 Plant Diseases

The proposed work focuses on detecting and classifying eight different plant leaf diseases. Around some 8000 datasets are considered for eight plant leaf diseases so each of the plant leaf diseases has some average of 1000 datasets (Fig. 1).

- Apple Scab: It is the most common disease found in apple and crabapple trees in Minnesota. The scab layer is caused by a fungus that infects both leaves and fruit. Infected leaf spots are round and have olive green to brown spots. Scabby spots on fruits and leaves are present and may have smoothness in the centre.
- Rust: The aerial parts of the plant are affected by the group of fungal diseases called rust. Most commonly the leaves are affected, but rust can also be found

Fig. 1 Different plant diseases [14]

rarely on stems, flowers and fruit. Rust disease is caused by a fungal parasite that damages plant tissue and then uses tissue as food.

- Early Blight: The pathogen that causes early blight is the fungus *Alternariasolani*. Found on potatoes and tomatoes, the white fungal growth may be seen at the edges of the damaged part on the underside of the leaves during humid conditions. On the stems, brown lesions may develop
- Late Blight: Usually found on tomato plants and potato plants, *Phytophthorain-festans* causes late blight which causes rotting. Most common in wet weather, leaves have pale green spots, often beginning at leaf tips or edges.
- Bacterial Spot: The bacteria that causes bacterial spots are called xanthomonads. Bacterial spots develop on young and mature plants. On younger plants, infections cause severe defoliation. Infections occur mainly on older leaves and appear as water-soaked areas on older plants.
- Mosaic Virus: Tobacco mosaic virus (TMV) affects a large variety of plants. It has its name because it creates patterns on the leaves. They are often discoloured and mottled, and you can see a mosaic-like structure. The colouration is mostly yellow, white or light and dark green. It is very hard to find the early stage in plant leaves so if ignored the plant gets infected and likely to die.
- Powdery Mildew: It is a fungus type resembling white powder. This powdery growth on the surface of leaves, young shoots, buds, fruits and flowers occurs around the world. *Podosphaera*, *Sphaerotheca*, *Uncinula,Phyllactinia* and *Erysiphe Microsphaera* are some of the fungal species that cause powdery growth on the plant leaf.

- Black Rot: Generally, cabbage and related crops are affected. Xanthomonas Campestris, the kind of bacteria which causes *Black rot*. It infects most cauliflower, cabbage-related plants at any growth stage in its life cycle. During humid and warm conditions, any rainy weather causes early crop development this disease which results in yield loss.

4 Methodology

The process involves multiple steps right from capturing the image from a mobile application to processing it and performing several steps on the server-side to obtain the required result (Fig. 2).

4.1 Image Preprocessing

Image pre-processing: Image pre-processing refers to performing mathematical operations with images at the basic level, the input is the normal image, and output may be increased intensity images or resizing images depending on further analysis. The primary objective of pre-processing is to improve image data that compresses unwanted distortion or to develop other important image features for further processing. There are some image pre-processing techniques like enhancing the image, resizing the image, increasing the contrast, blurring the image, sharpening the image, isolating colours, binarizing the image and many more. Let us consider only useful pre-processing techniques that are related to this project:

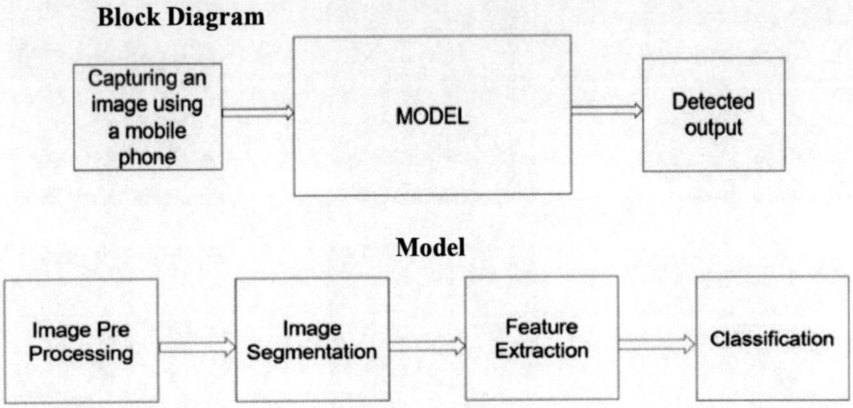

Fig. 2 Proposed methodology

- Image Enhancement: Image Enhancement: The short explanation is that it transforms the image so that it uses a wider range of pixel intensities. It is used for brightness purpose, contrast and to equalize an image. An image histogram is helpful in image enhancement. The histogram is a graph that shows frequency. Image histogram shows the frequency of pixels intensity values.

Histogram equalization is a method to enhance the contrast of an image so it aims to convert dark, bright and low contrast to the good quality image that is a high-contrast image. There are many different techniques for image enhancement in that advanced histogram equalization (AHE) and contrast limited AHE (CLAHE) are considered.

Advanced histogram equalization (AHE): Contrast can be improved using this technique. Over amplification happens to the contrast in near-constant regions of the image since the histogram in such regions is highly concentrated. So, it is not well suited for image enhancement.

Contrast Limited AHE (CLAHE): It is an improved type of AHE in which the contrast amplification is limited, so the noise problem can be reduced using this contrast limited AHE technique.

Results of image enhancement (Figs. 3 and 4):

- Resizing of an Image: Resizing images are a common task in image pre-processing for two reasons. First, images come in all shapes and sizes, and to be used as features, images must have the same dimensions. This standardization of image size does come with costs; however, images are matrices of data and to reduce the size of the image the size of that matrix is reduced and the information it contains. Second, machine learning can require thousands or hundreds of thousands of images so resizing reduces the data cost to store. When those images are very

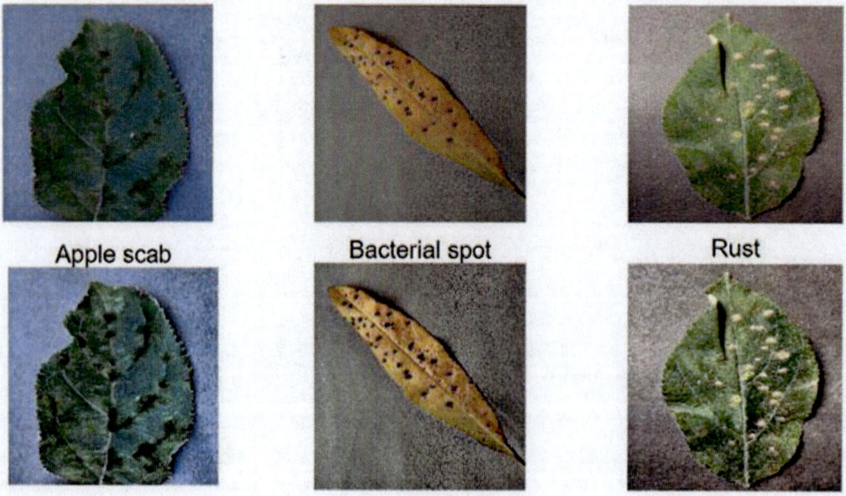

Fig. 3 Enhanced images [14]

Fig. 4 Enhanced images [14]

large, they can take up a lot of memory, and by resizing them memory usage can be reduced dramatically. Some common image sizes for machine learning are 32 × 32, 64 × 64, 96 × 96, and 256 × 256. However, in this project, input images are captured with the help of a mobile phone. Hence, the images captured might be of different sizes. So, the images have to be resized before being fed into the neural network.

4.2 *Image Segmentation*

This is an essential step in analyzing images and extracting data. Using this method, one can obtain the region of interest from the existing image. Then, further analysis like feature extraction can be performed on it to obtain the features. In this project, the subdivision aims to separate the image from the leaf area healthy and disease. The important parts can only be processed, i.e. made (disease) part of the image instead of processing the whole image.

K-means is a clustering or grouping algorithm. That means that based on their neighbourhood pixels can 'group' points. When a lot of data points or pixels are nearby, then it can be marked as one cluster. The whole data set is into k-different clusters. If k-value is chosen as 5, the dataset is divided into 5 clusters. Using this method, centre points can be found for these clusters to form. For image segmentation, clusters are different image colours. The disease part of the leaf will be in different colours and texture so this k-means cluster will help to identify the affected (disease) part of the plant leaf so that the affected part can be used for further process.

Fig. 5 Input image segmented into five different clusters

So, in a plant leaf data, the diseased part will have similar data, and the healthy part will have similar data so it will be easy to partition two different data using this k-means clustering. From this, the data of the diseased part can be obtained.

Steps for k-means clustering:

- Value K is predefined, which refers to the number of centroids you need in the image data points. A centroid is placed randomly at first representing the centre of the cluster.
- Every data point is allocated to each of the nearest centroid points. It forms a k number of clusters.
- Compute the mean for each cluster so that a new centroid data point is created.
- Reassign the data point to the newest centroid.
- Keep repeating the process until no longer a new centroid point is created.

Results of image segmentation (Figs. 5 and 6):

After segmentation, the diseased part of the plant is extracted. The colour and texture features are found out on the diseased part (Figs. 7 and 8).

4.3 Feature Extraction

Feature extraction has always been an integral part of computer vision. Usually, various image pre-processing techniques are applied before the feature extraction process. In this project, image enhancement and resizing of an image have been

Fig. 6 Input image segmented into five different clusters

Fig. 7 Shows the extracted part

Fig. 8 Shows the extracted part

applied. After that, image segmentation has been implemented to partition the plant leaf image into a healthy and diseased part. From the diseased part, disease features can be extracted. Then, classification is performed based on the extracted features. Many different kinds of features like colour, texture, shape can be obtained using different extraction methods. Let us concentrate on colour-based and texture-based feature extraction.

1. Colour-based features: One of the important features of an image is colour. Colour space represents the colour in the form of intensity value. Specify, visualize and create the colour can be done by using a colour space method. There are different colour feature extraction methods like mean and standard deviation.

Mean—The mean can be defined as the average of all pixel values of an image. It gives out the average intensity of the image. Mean is calculated for all three layers—red, green and blue. The average intensity of the above three colours is got.

$$\text{Mean} = \frac{1}{n} * \sum \text{pixel intensity}$$

n = total number of samples.

Standard Deviation—Standard deviation measures the dispersion of values from its mean value. It is measured as the square root of variance.

$$\text{Standard deviation} = \sqrt{\frac{\sum (x_i - \mu)^2}{N}}$$

X_i = value of each input from $i = 0$ to N.

μ = mean of all values.

N = total no of samples.

2. Texture Based: One of the main aspects of feature extraction is to gather texture features. Feature extraction is a way to capture graphic content for indexing and optimizing images. Properties can vary with positional variations that arise from the depth of the object's surface. In our work, a mathematical method is considered to analyze the texture. Information about the distribution of pixels can be obtained. The mathematical method reveals details about the texture features, whether they are smooth, rough, hard, etc. Feature extraction is implemented using the grey-level co-occurrence matrix (GLCM). A grey-level co-occurrence matrix is a histogram of co-occurring greyscale values of an image. Features like Correlation, Contrast, Energy, and Homogeneity were extracted from the GLCM. So for this grey-level co-occurrence matrix, the properties are computed as follows:

   - Contrast—Contrast is the difference in intensity that makes an object distinguishable and measures the spatial frequency of an image.

$$\sum_{i,j=0}^{\text{levels}-1} P_{i,j}(i-j)^2 \tag{3}$$

   - ASM—Angular second moment (ASM) is one of the most popular methods for facial feature extraction.

$$\sum_{i,j=0}^{\text{levels}-1} P_{i,j}^2 \tag{4}$$

   - Energy—Measures the texture uniformity that is pixel pair repetitions.

$$\sqrt{\text{ASM}} \tag{5}$$

ASM—Angular second moment.

- Entropy—Entropy measures the randomness of grey-level distribution and measures the disorder or complexity of an image.

$$H(X) = -\sum_{i=1}^{n} P(x_i)\log_b P(x_i) \tag{6}$$

- Homogeneity—Homogeneity measures the closeness of values in the diagonal of GLCM and measures the variations in the image intensity.

$$\sum_{i,j=0}^{\text{levels}-1} \frac{P_{i,j}}{1+(i-j)^2} \tag{7}$$

- Correlation—Measures the linear dependency of the grey level of neighbouring pixels.

$$\sum_{i,j=0}^{\text{levels}-1} P_{i,j}\left[\frac{(i-\mu_i)(j-\mu_j)}{\sqrt{\left(\sigma_i^2\right)\left(\sigma_j^2\right)}}\right] \tag{8}$$

In all the above equations,

$P_{i,j}$ is the (*i*, *j*)th entry of the normalized GLCM where *i* represents rows, and *j* represents columns.

$P(X_i)$ is the *i*th entry in the marginal probability matrix obtained by summing the rows of $P_{i,j}$.

μ_i = mean of row *i*.

μ_j = mean of column *j*.

σ_i = Standard deviation of row *i*.

σ_j = Standard deviation of column *j*.

Results of feature extraction:

In Fig. 9, each ID represents different diseases, and features for all the eight diseases are extracted.

4.4 Classification

Classification is a machine learning approach, where a machine learns from a set of input data and uses that learned information to classify a new observation. It is used to predict a discrete class out of many other classes. In the classification algorithm, input data is usually associated with a specific output class. So, from the information obtained from the previous stages of image processing and feature extraction, a model is trained to predict the correct disease when tested on new images. There are a lot of classification algorithms based on different types of applications. In this project, for the classification process, a comparative analysis is done between two algorithms—Support Vector Machine (SVM) and Convolutional Neural Networks (CNN).

- Support Vector Machine (SVM): Support Vector Machine (SVM) is one of the supervised learning algorithms usually used in classification problems. Compared

1	id	red_mean	green_me	blue_mear	red_var	blue_var	green_var	ContrastSt	Correlatio	Homogene	EnergyStat	ASMStats
2	0	127.3985	137.9915	118.0587	59.13906	49.19662	39.03474	1.43045	0.908125	0.783033	0.346605	0.120316
3	0	118.1312	112.8921	99.27719	48.91958	35.33208	30.66145	1.754813	0.819441	0.746073	0.322967	0.104407
4	0	121.2937	130.562	97.68236	40.62279	38.26699	35.09207	2.464881	0.771149	0.68724	0.326327	0.106595
5	0	111.1455	123.0923	127.2832	58.72405	59.58892	59.24385	8.352534	0.688869	0.515226	0.26509	0.07028
6	0	117.7513	118.3954	103.016	46.94026	40.28614	33.2571	1.249829	0.879828	0.776088	0.334817	0.112243
7	0	83.90877	109.8031	97.57623	64.47977	56.39767	52.4512	1.551697	0.934693	0.820965	0.365415	0.133613
8	0	117.0522	106.8743	100.5399	46.95649	42.9482	41.65224	5.041396	0.671773	0.561937	0.263721	0.069686
9	0	110.2466	138.2029	125.1646	45.82691	40.78373	36.81716	1.482308	0.8653	0.770331	0.348602	0.121673
10	0	123.2757	132.8417	124.1575	61.92545	59.65984	55.98236	12.38556	0.521272	0.419568	0.215908	0.046618
2924	1	117.3676	133.1857	128.8287	58.19517	52.76457	46.92843	7.588725	0.609037	0.472418	0.189478	0.03591
2925	1	103.6734	122.3076	118.0402	48.75873	48.05265	46.18708	6.088151	0.637168	0.539342	0.260041	0.067683
2926	1	136.8788	148.0073	149.848	63.08828	56.32161	47.38283	6.338556	0.69459	0.512414	0.232924	0.05427
2927	1	127.9818	138.5802	137.3325	48.46869	43.84696	34.30819	2.409188	0.784252	0.644597	0.252347	0.063719
2928	1	109.7334	122.1436	123.1141	46.58617	44.97752	38.75855	4.053366	0.703249	0.545432	0.241554	0.058405
2929	2	102.1506	120.7628	114.5571	60.43849	60.38995	54.1258	6.879382	0.714271	0.515325	0.234517	0.055012
2930	2	109.6937	124.9018	117.6559	53.20133	53.5885	46.85785	8.815086	0.521917	0.416119	0.206511	0.04265
2931	2	127.7296	133.1897	122.8101	55.03097	50.69942	46.2672	8.23048	0.540559	0.487736	0.198868	0.039555
2932	2	140.8152	147.6407	140.9333	64.38304	61.96821	52.50787	9.677594	0.616133	0.486004	0.209259	0.043803
4542	3	107.3555	139.3353	133.0287	58.83217	56.61491	55.57841	12.32586	0.50772	0.413131	0.209814	0.044024
4543	3	112.0481	136.9268	131.7549	61.77679	59.4362	58.93834	9.663054	0.645416	0.497023	0.231283	0.053523
4544	3	110.5357	143.6077	138.6351	60.8945	57.12635	56.6923	7.768205	0.700634	0.534095	0.22701	0.051571
4545	3	108.4292	137.3526	134.4662	61.15941	56.93693	57.06058	10.23703	0.600617	0.522174	0.252544	0.063803
4546	3	108.1645	145.6915	154.3642	56.44619	50.5711	51.47009	7.779191	0.63353	0.597156	0.277714	0.077172
4547	3	119.1388	138.0853	130.089	61.36135	57.20804	55.55066	9.264634	0.630813	0.511548	0.249127	0.062089
4548	3	119.8259	136.8787	128.2063	62.14521	59.19212	57.55428	9.0404	0.657664	0.527981	0.24818	0.061625
4549	3	119.8527	137.4612	134.2539	65.81013	61.59946	61.01097	9.935661	0.652613	0.491422	0.21537	0.046397
4550	4	120.0682	128.268	131.149	43.31586	41.84376	37.91526	3.702489	0.708663	0.640624	0.327002	0.107043
4551	4	120.6489	130.7872	123.7593	46.48785	45.46724	33.2579	4.425476	0.620291	0.544723	0.259397	0.06734
4552	4	122.3313	136.9504	130.4341	49.9173	46.57474	41.35039	5.044303	0.601661	0.524783	0.18927	0.035851
4553	4	123.958	132.1943	129.0179	50.21548	47.22499	43.81476	6.622455	0.604433	0.524119	0.237032	0.056214
4554	4	113.2436	129.5313	142.0448	55.15327	55.28339	52.67249	12.62376	0.440725	0.391619	0.210424	0.044281
4555	4	119.4702	127.82	141.9258	57.05562	56.68829	55.1786	8.753805	0.654457	0.481234	0.219441	0.048169
5915	5	123.9291	128.8966	132.2558	53.69694	52.20875	48.95398	14.16978	0.335895	0.368967	0.201669	0.040672
5916	5	129.8649	121.8061	125.848	49.15701	47.98696	45.91584	8.691136	0.438504	0.460552	0.249874	0.06245
5917	5	128.2009	123.201	130.9109	41.97212	41.11087	38.37388	7.714299	0.392523	0.540864	0.305391	0.093297
5918	5	131.4979	129.9721	136.4163	54.05087	52.8115	49.36066	11.79619	0.427458	0.417703	0.203788	0.041537
5919	5	131.9373	125.8714	122.7949	47.9524	46.35587	44.19483	8.727985	0.462079	0.469567	0.231488	0.053607
5920	5	122.1955	123.8311	126.6508	45.34792	44.91185	42.18059	7.046131	0.51085	0.517942	0.276965	0.076756
5921	5	122.4269	128.7635	130.1929	55.49156	54.05498	51.10473	15.97024	0.280019	0.338202	0.205128	0.04208
5922	5	136.3078	141.4846	142.8737	54.55965	53.47167	49.86681	12.20199	0.402052	0.384914	0.207206	0.042936
5923	6	89.03419	138.1908	110.2531	40.19055	35.37622	31.61701	3.744856	0.565114	0.598199	0.260666	0.067966
5924	6	103.0958	133.5171	109.9172	52.10413	50.65944	49.37766	9.577273	0.537938	0.446362	0.201213	0.040515

Fig. 9 Represents the various colour and texture features extracted from the diseased part

to neural networks and logistic regression, SVM sometimes gives a cleaner and more powerful way to work with non-linear functions. Few cases where SVM works well compared to other learning algorithms is when dealt with high-dimensional data such as images and when the number of samples is less (and dimensions are more). SVM is also memory efficient as the decision function utilizes only a part of the training set. The primary goal of an SVM is to find a hyperplane in an N-dimensional space that classifies the data points. It chooses the hyperplane such that the distance between the hyperplane chosen and the input data points is the largest.

To solve non-linear problems, SVM uses a function known as the kernel. The kernel transforms linearly non-separable data into linearly separable ones. The basic function of a kernel is that it maps linearly inseparable data to an N-dimension space, where it can be separable. In this project, an SVM model is built with three different kernels and checked which one works best for our application—Linear Kernel, Polynomial Kernel and Radial basis function (RBF) Kernel.

- Convolutional Neural Networks: Another machine learning approach used to classify is convolutional neural networks. With more data to work with and higher computational speed, CNN's gained traction over the last decade. In recent times, convolutional neural networks have been proven very effective in the areas of image recognition and classification. Unlike other algorithms, where input is in the form of a vector, CNN can directly take an image as an input. The main advantage of this is that hand engineering of features is no longer required, as CNN can learn features on its own. To get more intuition on how this happens, different layers will be looked upon and working of the CNN model.

  Different layers of CNN include:

Convolution layer: It is the first layer of a CNN. As discussed above, the CNN model takes an image as an input (usually a multi-channelled image—Red, green and blue). As the name suggests, the convolution operation is performed in this layer.

Pooling layer: The convolutional layer of a CNN is usually followed by a pooling layer. Decreasing the spatial size of the convolved feature is the primary objective of the pooling layer. When the spatial size is reduced, it takes computationally less power to process the data. This is helpful when high-dimensional data are dealt with or when large datasets are dealt with.

Fully Connected layer: Fully connected layer usually arrives at the end of CNN after multiple convolutional and pooling layers and functions in a similar way as that of an N input neural network. Here, all neurons are connected to all activations of the previous layer. To perform this, the values are flattened in output volume to a column vector. Fully connected layer takes this column vector as input.

In the final layer, a softmax activation function is used to distinguish different classes. The softmax function is generally used when more than two classes wanted to be classified. The output layer consists of n units, where n is the number of classes, and those n units give out the probability for n classes. The probability of each unit is calculated by taking the exponent of the value at that unit and then dividing it by the sum of exponents of all the units. Softmax activation can be represented as

$$f(X_i) = \frac{\mathrm{e}^{X_i}}{\sum \mathrm{e}^{X_j}} \tag{9}$$

where X = set of input values.

X_i = ith element in set of input values, and j = total number of input values.

Results of CNN (Figs. 10 and 11):

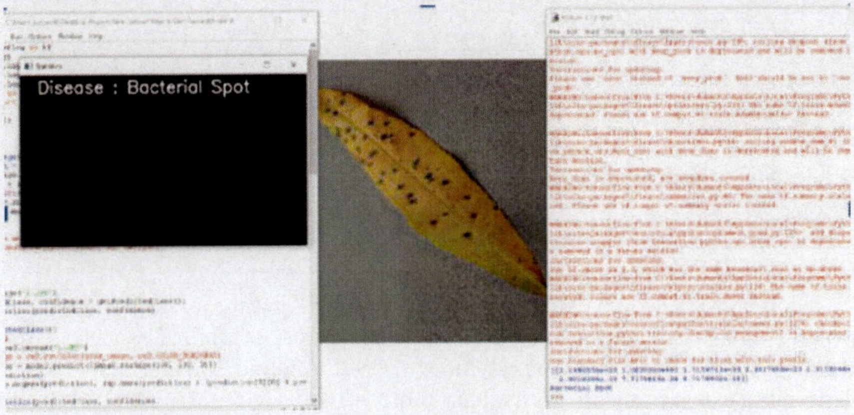

Fig. 10 Output of CNN detecting bacterial spot disease

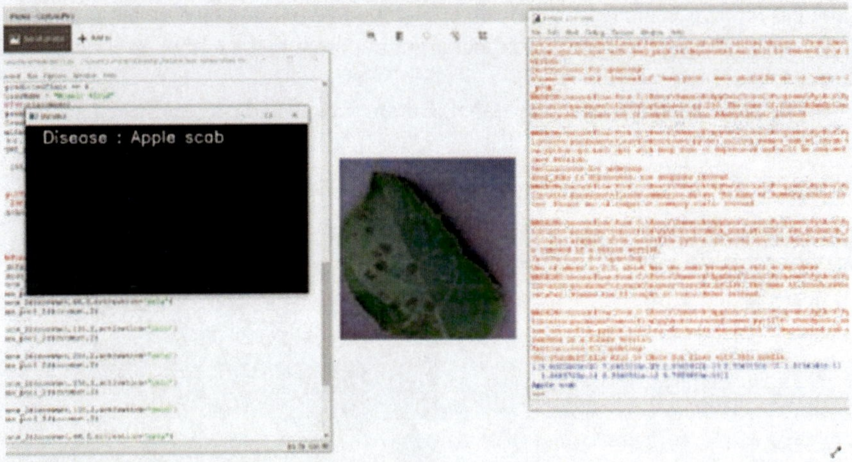

Fig. 11 Output of CNN detecting apple scab disease

5 Python App Development

Python is a high-level programming language, and it is more user friendly. Python uses the interpreter to run a programme which means it executes the programme line by line. Python is mainly in Data Science, Object-Oriented Programming (OOP), etc. Apart from that it is also used for building applications for Android, Windows, Mac, iOS, Web applications development.

5.1 Android Application

Client:

Python doesn't have built-in capability to build mobile applications; it uses third-party packages such as Kivy and Buildozer to compile from Python code to Android Package.

A client is a piece of computer software which can access the service provided by the server. It initiates the communication with the server.

Kivy is an open-source Python library under an MIT licence for the development of multi-touch applications. It is used to develop cross-platform Graphical User Interfaces (GUIs). It runs on Linux, Windows, OS X, Android, iOS and Raspberry Pi platforms. It can natively use most inputs, protocols and devices MacOS X Trackpad and Magic Mouse, Mtdev, Linux Kernel HID, TUIO. Its graphical engine is built over OpenGL ES 2 and uses fast graphical pipelines. The toolkit comes with more than 20 widgets, all highly extensible. Many parts are written in C using Cython. Kivy has its own language for writing GUI called Kivy Language which is used to easily design custom widgets, and then, it got imported to the Python programme to render GUI.

In our project, widgets such as Label, Button, Image and camera have been used (Figs. 12 and 13).

Server:

Fig. 12 Capture screen

Fig. 13 Analysis screen

The server is a computer programme that provides service to clients. The client will establish a connection with the server using the IP address and port number provided and then sends the image to the server. The server receives the image from the client and performs image pre-processing, contrast enhancement, image segmentation and classification, and it sends the result back to client (Fig. 14).

5.2 *Client–Server Model*

The basic setup in the client–server model is one device; the server that runs a service and patiently waits for clients to connect and make requests to the service.

A typical computing example is a web server. The server listens on a TCP port for clients that need its web pages. When a client (Web Browser) requires a web page that the server hosts, it connects to the server then requests that page. The server replies with the content, and then, the client disconnects the connection.

In this project, server programme will be waiting to give service to the client. If a client wants to identify plant leaf disease, then it will open a socket and then connect

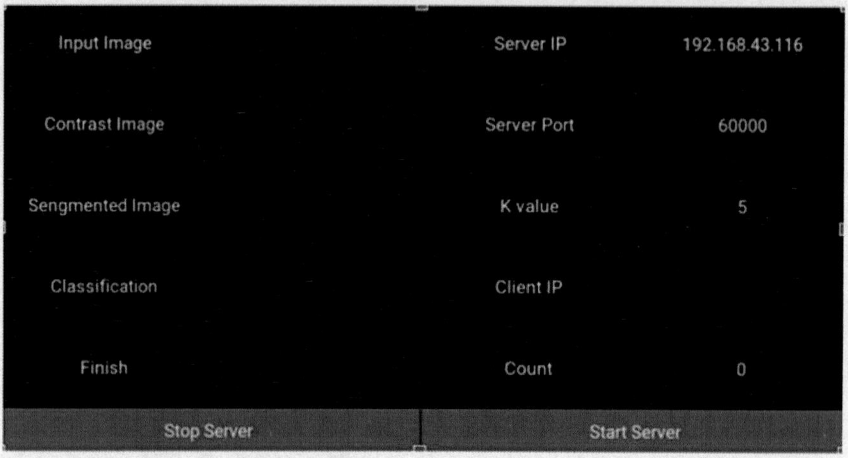

Fig. 14 Server window

Table 1 Accuracy of SVM classifier

K-value	Kernel	Accuracy
3	Linear	83.74
3	Polynomial	89.72
3	RBF	68.99
4	Linear	86.25
4	Polynomial	90.16
4	RBF	69.21
5	Linear	86.25
5	Polynomial	90.78
5	RBF	70.16

to the server via TCP connection. Then, the client will send the captured image to the server. Then, the server will convert the image from PNG format to JPG format then perform image pre-processing, contrast enhancement, image segmentation and classify according to the trained model. Then, the server sends back a contrast-enhanced image, segmented image and result of classification to the server, and the client closes the connection.

Table 2 Comparison of accuracy of SVM and CNN

Classification method	Accuracy
SVM	90.78

6 Results

Table 1 depicts the accuracy of the SVM classifier when tested with different values of and K and different Kernels. Best results were obtained when the image is clustered into five different parts, and when a polynomial kernel is used.

When tested with CNN, an accuracy of 99% was achieved (Table 2).

7 Conclusion and Future Work

A mobile application was successfully able to build to classify eight different plant leaf diseases. Over the course of this project, machine learning algorithms have come across such as k-means clustering, support vector machine and convolutional neural networks which have been a new learning experience for us. Many image processing techniques also come across used for resizing, image enhancement and extraction. Building and working with a mobile application has also been a good experience. All the programming parts were implemented using python, which had many libraries such as TensorFlow, NumPy, OpenCV which made the implementation easier.

In this project, the classification was done for eight different plant leaf diseases. Maximum accuracy of 90.78% could be achieved using the SVM algorithm with polynomial kernel and when k equals 5 and 99% using CNN algorithm.

In the future would like to work with much larger datasets and try identifying and detecting many more plant leaf diseases. Pesticide suggestions could also be provided to the farmers on the mobile application. Other things like precautions to be taken to avoid the disease or to prevent the disease can also be displayed on the application. The next focus would be not only detecting diseases on plant leaves but also on other parts like stems, branches, etc.

References

1. Prakash RM, Saraswathy GP, Ramalakshmi G, Mangaleswari KH, Kaviya T (2017) Detection of leaf diseases and classification using digital image processing. In: 2017 international conference on innovations in information, embedded and communication systems (ICIIECS). IEEE, pp 1–4
2. Dhaware CG, Wanjale KH (2017) A modern approach for plant leaf disease classification which depends on leaf image processing. In: 2017 international conference on computer communication and informatics (ICCCI). IEEE, pp 1–4

3. Shruthi U, Nagaveni V, Raghavendra BK (2019) A review on machine learning classification techniques for plant disease detection. In: 2019 5th international conference on advanced computing & communication systems (ICACCS). IEEE, pp 281–284
4. Qiao J, Zhang Y (2015) Study on K-means method based on data-mining. In: 2015 Chinese automation congress (CAC). IEEE, pp 51–54
5. Singh S, Srivastava D, Agarwal S (2017) GLCM and its application in pattern recognition. In: 2017 5th international symposium on computational and business intelligence (ISCBI). IEEE, pp 20–25
6. Sarangdhar AA, Pawar VR (2017) Machine learning regression technique for cotton leaf disease detection and controlling using IoT. In 2017 international conference of electronics, communication and aerospace technology (ICECA). IEEE, pp 449–454
7. Gavhale KR, Gawande U, Hajari KO (2014) Unhealthy region of citrus leaf detection using image processing techniques. In: International conference for convergence for technology-2014. IEEE, pp 1–6
8. Khirade SD, Patil AB (2015) Plant disease detection using image processing. In: 2015 international conference on computing communication control and automation. IEEE, pp 768–771
9. Badnakhe MR, Deshmukh PR (2011) An application of K-means clustering and artificial intelligence in pattern recognition for crop diseases. In: 2011 International conference on advancements in information technology, IPCSIT, vol 20
10. Chouhan SS, Kaul A, Singh UP, Jain S (2018) Bacterial foraging optimization based radial basis function neural network (BRBFNN) for identification and classification of plant leaf diseases: an automatic approach towards plant pathology. IEEE Access 6:8852–8863
11. Singh V, Varsha, Misra AK (2015) Detection of unhealthy region of plant leaves using image processing and genetic algorithm. In: 2015 International conference on advances in computer engineering and applications. IEEE, pp 1028–1032
12. Kumar G, Bhatia P (2014) A detailed review of feature extraction in image processing systems. In: 2014 fourth international conference on advanced computing & communication technologies. IEEE, pp 5–12
13. Vincent A, Debain C, Michel B (2004) In generalization of the co-occurrence matrix for colour images: application to colour texture classification. Image Anal Stereol J 23:63–72
14. https://github.com/spMohanty/PlantVillage-Dataset

Chapter 23
Energy-Efficient System-Based Algorithm for Maximal Resource Utilization in Cloud Computing

Arvind Dagur, Mishra Adityakumar Virendra, Mayank Pratap Singh, Deepak Gupta, Chetanya Sharma, and Rahul Chaturvedi

1 Introduction

In this modern era, demand for cloud computing has significantly increased because the cost of software and hardware which is too high to be afforded by many organizations for the proper use of resources. Cloud computing emerged as a solution to this problem, by providing resources such as (hardware, operating systems, software, etc.) at affordable prices to the users, so that they do not have to buy resources themselves. In cloud computing, data and services are stored in common space called data centers in a scalable amount, which is accessible through authentication [1]. A new age concept known as green computing enforces for making data centers environmentally sustainable and eco-friendly, i.e., make them able to run with minimal

A. Dagur (✉) · M. A. Virendra · M. P. Singh · D. Gupta · C. Sharma · R. Chaturvedi
Department of Computer Science and Engineering, Krishna Engineering College, Ghaziabad, Uttar Pradesh, India
e-mail: arvind.dagur@krishnacollege.ac.in

M. A. Virendra
e-mail: adityam525@gmail.com

M. P. Singh
e-mail: mayankpratap440@gmail.com

D. Gupta
e-mail: deepakgoel691999@gmail.com

C. Sharma
e-mail: chetanya1281998@gmail.com

R. Chaturvedi
e-mail: rahul.chaturvedi@krishnacollege.ac.in

I. Jeena Jacob et al. (eds.), *Data Intelligence and Cognitive Informatics*, Algorithms for Intelligent Systems,
https://doi.org/10.1007/978-981-15-8530-2_23

electrical power consumption. The implementation of this concept minimizes emission CO_2. Therefore, the primary goal is to make data centers as efficient with taking care of the environment. Public, private, and hybrid cloud computing can be used according to the limit of affording. In a public cloud, services are provided free over the Internet, whereas private cloud provides its services to only specific users. In a hybrid cloud, computing is done using the combination of public cloud and private cloud [2].

2 Related Work

Several virtual machine algorithms are based on schedules that are already designed during the current years. A scientist implemented a very popular algorithm of genetic which programs virtual machines of resource for achieving load balancing [3].

Quan-sheng guan et al. proposed two algorithms, shortest job buffering and min-min best scheduling algorithms, i.e., SJF-MMBF, was introduced specially for finding an optimal result. The two algorithms which are based on the schedule are the main aim to avert the process of starvation in these scheduling algorithms [4]. The authors have been introduced a voltage of dynamic and scaling of frequency (DVFS) which enabled energy-efficient workflow scheduling of tasks algorithm (DEWTS). One of the scheduling which is already proposed exploits the resources and consumption of energy by developing the virtual machine to allocate and using a policy of consolidation [5, 6].

In [7] Samah Alshathri et al. proposed a special type of algorithm that can apportion the resources in the environment of cloud computing which is dependent on the escalation of energy called distribution with the live transfer. In [8], Inderveer Chana and Sukh Pal Singh introduced very important analytical analysis literature management of resources in the cloud area and cloud resource scheduling. In [9], a method is improved to utilize the resources and decrease the consumption of power of servers which is better as compared to the technique of dynamic voltage. In these scheduling of algorithms, the main principle is for applications in data intensive that has introduced in for decreasing of consumption of energy by decreasing the rate of its parameter. The introduced technique was based on the management of resources only to save energy [10–12]. A policy of virtual machine in scheduling has a very important implement in the life cycle of computing of cloud and has various types of steps that are to be used for allocation and scheduling performances, which may affect the cloud surroundings process and working. In this article, the survey and start of the presenting scheduling methods are processed using Cloud Sim. Also, an approach for process the loaded cloud of light surrounding is done by arranging jobs and virtual machine based on heuristics function. This has shown process tuning based on parameters like implementation time, manufacturing, and throughput of scheduled tasks. In [13], author was the first person to propose the opposition-based learning. This method of optimization's beginning step is to initialize the population which can improve the solution of optimization. If the criterion is satisfied, the search terminates and haphazard guesses are implemented in the absence of prior information opposite to random numbers utilization which is the new method of this technique. Thus, the opposite random number might provide another chance for finding the solution.

Authors introduced the latest techniques which are hybrid called optimizer algorithm of a gray wolf which is based on opposition learning. The main contribution of this research paper is to reduce MAKESPAN in the environment of cloud computing. Dong et al. introduced an algorithm called QOS which is a similar algorithm to the scheduling of grouping of priority. This type of an algorithm which is to consider the last line and accepting various tasks and also to MAKESPAN includes important factors for scheduling of task in the whole system. When two algorithms are compared that is a min-min algorithm and min-min with QOS guided, the scheduling algorithm of QoS of grouping some priority achieves a better acceptance rate and completion time for the submitted tasks. Authors implemented an algorithm for scheduling when there is a refresh task which is given called suffrage of QOS only for environment mainly grid [14, 15].

3 Services of Cloud Computing

There are mainly three services provided by the cloud:

1. **Software as a service (SaaS)**: On-demand service, a complete application is provided to the customers in this model. Multiple end users are serviced through the running of a single instance on the cloud. Few companies which offer SaaS are Salesforce, Microsoft, Google, Zoho, etc.
2. **Platform as a service (PaaS)**: In this service, a layer of development environment or software is offered and encapsulated upon which other services can be built which are at high levels. Some popular PaaS examples are Force.com, Google's App Engine, etc.
3. **Infrastructure as a service (IaaS)**: Storage and computing capabilities are provided to multiple users in IaaS as standardized services. To manage storage systems, workloads, servers, networking equipment, data center space, etc., are pooled and made available. The customers would typically put their software on the infrastructure. Amazon, Go-Grid, 3Tera, etc., are some common examples.

4 Proposed Algorithm

Using the advantages of the two powerful and traditional min-min, max-min, an optimal algorithm has been developed to reduce the waiting time for jobs with maximum utilization of resources. The overall function of this algorithm is run via a queue of tasks that contain overall information of task to execute at given time and stack of recourses (processors) which contain overall information of given resources.

Algorithm:

1. Based on the previous performance data of each resource, estimate the threshold value of each resource such that

$$R_j T_h = f(R_j U, R_j T)$$

where

(a) $R_j T_h$ is estimated threshold value of resource R_j, $R_j U$ is a minimum load at which resource R_j goes underperformance, and $R_j T$ is the maximum time to complete last assign the task to resource R_j. A resource R_j is in the idle state if it is in stack otherwise in the active state.
(b) Put these resources in resource stack.
(c) Calculate the execution time of each task and put them in the task queue.
(d) With the help of midpoint formula, calculate the number of tasks having small execution time and a number of tasks having large execution time.

2. Based on execution time of tasks, i.e., if the number of tasks having small execution time exceeds the number of tasks that have larger execution time, then use the min-min algorithm, otherwise use the max-min algorithm.

(a) Pop resource R_j from resource stack.
(b) Copy their threshold value if temp variable.
(c) Assign tasks T_i from task queue to R_j and subtract its execution value from temp variable t.
(d) If temp t value reached nearby zero, then pop another resource and repeat step b and c.

3. Pop the resource R_j from resource stack and assign jobs such that,
Else if t value reached to a threshold value of R_j, then push R_j in resource stack.

Our proposed algorithm worked with some previous knowledge like the threshold value of each resource that is used to find maximum makespan on which a particular resource can work efficiently and effectively. First of all, the threshold value of each given resources have to be found for assigning tasks. Store all incoming tasks at a time in a task queue and chose the best-fit task-scheduling algorithm from min-min and max-min. Min-min and max-min are two very useful and traditional task-scheduling algorithms that help to schedule the tasks based on their execution time. The min-min algorithm allows tasks having minimum execution time first to resourcesm whereas the max-min algorithm allows tasks having maximum execution time first. Herem our scheduling algorithm is chosen based on the number of tasks having large execution time (execution time more than midpoint) and the number of tasks having small execution time (execution time more than midpoint) (Fig. 1).

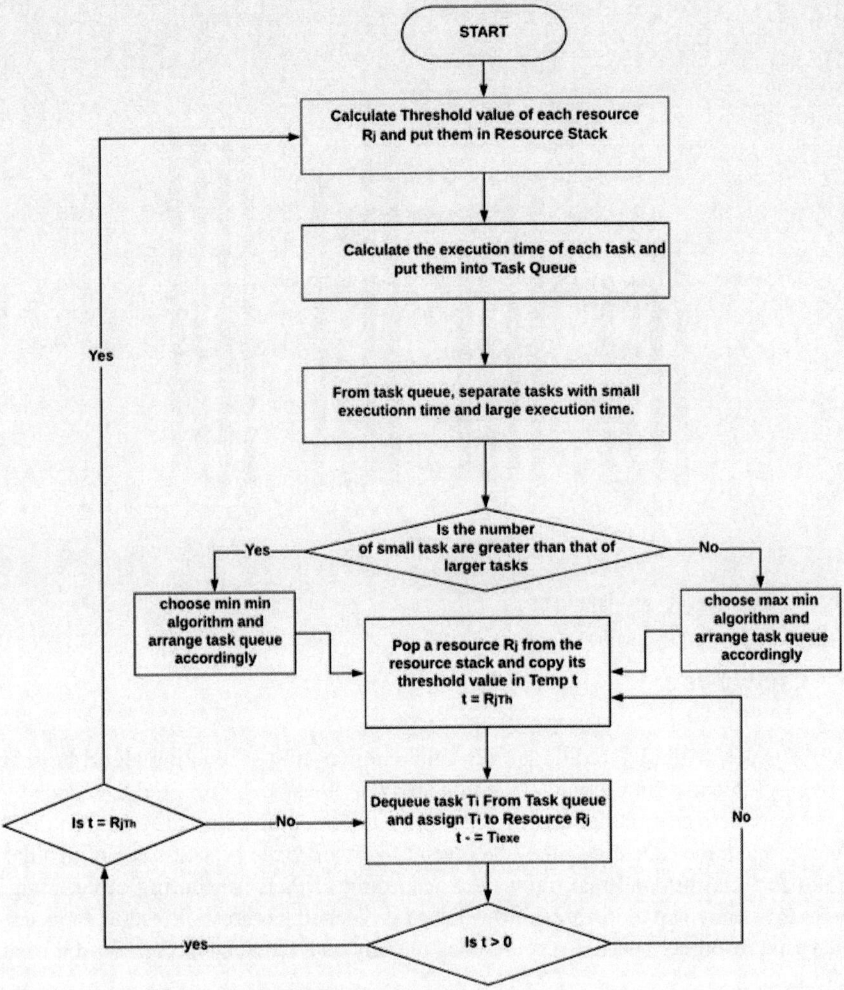

Fig. 1 Flowchart of proposed algorithm

5 Algorithm Performance Result

Comparison of the number of resources required with and without EEBA: In this section, the chart shows the comparison of the number of resources required to successfully execute a given task in a given table with and without EEBA using min-Min and max-min scheduling algorithm (Fig. 2).

According to the above discussion, it can be said that if EESBA is used, that is based on threshold values of each resource, it will reduce the number of resources required to execute given tasks. Reduction in the number of resources shows the least number of resources active at a time that leads less energy to complete execution.

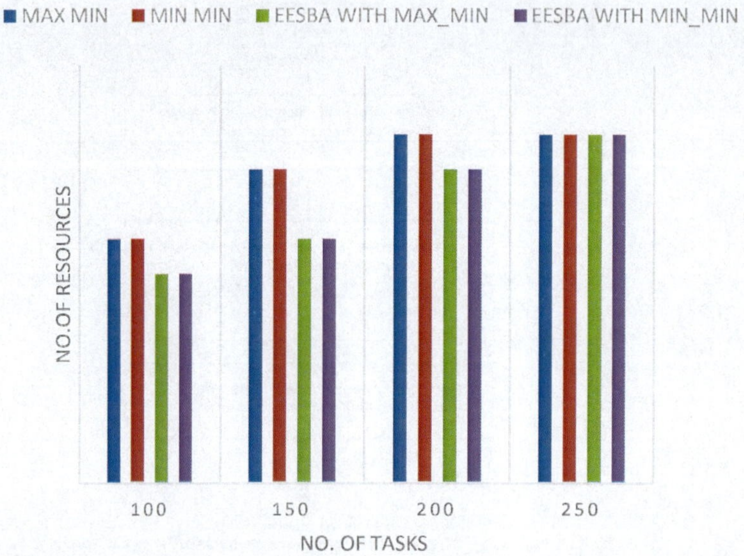

Fig. 2 Comparison of all algorithms

6 Conclusion

In this paper, with two traditional min-min and max-min scheduling algorithms, an energy-efficient system-based algorithm (EESBA) has been proposed to reduce the number of active resources at a time for tasks in the cloud computing environment. According to the execution time of a given task, choose the best-fit scheduling algorithm for min-min and max-min; then according to their scheduling order, assign tasks to resource up to its threshold value. If it reached its threshold value, then activate next resource; and if a resource has not any task to execute, then send it to an idle state.

The results of the experiments indicate that the min-min and max-min algorithm with EESBA are superior to min-min and max-min without EESBA. Also, given that, the makespan of the two traditional algorithms mentioned is low and that is increasing with the EESBA algorithm, but here, the number of active resources is less that requires less amount of energy to execute tasks and fully utilize the resources.

References

1. Mell P, Grance T (2011) The NIST definition of cloud computing
2. Kumar A, Alam B (2018) Task scheduling in real time systems with energy harvesting and energy minimization. J Comput Sci 14:1126–1133

3. Guo M, Guan Q, Ke W (2018) Optimal scheduling of VMs in queueing cloud computing systems with a heterogeneous workload. IEEE Access 6:15178–15191
4. Tang Z, Qi L, Cheng Z, Li K, Khan SU, Li K (2016) An energy-efficient task scheduling algorithm in DVFS-enabled cloud environment. J Grid Comput 14(1):55–74
5. Sharma Y, Dagur A, Chaturvedi R (2019) Automated bug reporting system with keyword-driven framework. In: Soft computing and signal processing. Springer, Singapore, pp 271–277
6. Allsmail SM, Kurdi HA (2016) Green algorithm to reduce the energy consumption in cloud computing data centres. In: 2016 SAI computing conference (SAI). IEEE, pp 557–561
7. Alshathri S, Ghita B, Clarke N (2018) Sharing with live migration energy optimization scheduler for cloud computing data centers. Futur Internet 10(9):86
8. Singh S, Chana I (2016) A survey on resource scheduling in cloud computing: issues and challenges. J Grid Comput 14(2):217–264
9. Kumar A, Alam B (2015) A dynamic approach for fault tolerance with voltage scaling. In: 2015 ınternational conference on green computing and Internet of Things (ICGCIoT). IEEE, pp 1522–1525
10. Ali SA, Alam M (2016) A relative study of task scheduling algorithms in cloud computing environment. In: 2016 2nd ınternational conference on contemporary computing and ınformatics (IC3I). IEEE, pp 105–111
11. Zhao Q, Xiong C, Yu C, Zhang C, Zhao X (2016) A new energy-aware task scheduling method for data-intensive applications in the cloud. J Network Comput Appl 59:14–27
12. Kumar A, Alam B (2014) Real time scheduling algorithm for fault tolerant and energy minimization. In: 2014 ınternational conference on ıssues and challenges in ıntelligent computing techniques (ICICT). IEEE, pp 356–360
13. Mehra PS, Doja MN, Alam B (2018) Stable period enhancement for zonal (SPEZ)-based clustering in heterogeneous WSN. In: Proceedings of first international conference on smart system, innovations and computing. Springer, Singapore, pp 887–896
14. Kumar A, Yadav RS, Ranvijay AJ (2011) Fault tolerance in real time distributed system. Int J Comput Sci Eng 3(2):933–939
15. Alam B, Kumar A (2014) A real time scheduling algorithm for tolerating single transient fault. In: 2014 ınternational conference on ınformation systems and computer networks (ISCON). IEEE, pp 11–14

Chapter 24
Energy Enhancement of WSN Using Fuzzy C-Means Clustering Algorithm

Arvind Dagur, Nidhi Malik, Priyanshi Tyagi, Rishu Verma, Riya Sharma, and Rahul Chaturvedi

1 Introduction

In the current scenario of the wireless sensor network (WSN), the size of the sensor node and energy is a big challenge. WSN has transformed essential research in the area of technology due to self-configured and infrastructure-less networks. These networks are formed or built with a collection of sensor nodes, and these sensor nodes broadcast the information to the sink or base station, but this is also true that every technology faces some pros and cons [8]. The energy consumption is one of the major challenges for the survival of a WSN node. Therefore, it needed to enhance energy usage and the failure of the sensor nodes because these sensor nodes hold very little storage capacity, less power bank, and communication capacity.

A. Dagur (✉) · N. Malik · P. Tyagi · R. Verma · R. Sharma · R. Chaturvedi
Department of Computer Science and Engineering, Krishna Engineering College, Ghaziabad, Uttar Pradesh, India
e-mail: arvinddagur@gmail.com

N. Malik
e-mail: maliknidhi98@gmail.com

P. Tyagi
e-mail: priyanshi.tyagi29@gmail.com

R. Verma
e-mail: rishuriyaverma@gmail.com

R. Sharma
e-mail: itsriya26@gmail.com

R. Chaturvedi
e-mail: rahul.chaturvedi@krishnacollege.ac.in

I. Jeena Jacob et al. (eds.), *Data Intelligence and Cognitive Informatics*, Algorithms for Intelligent Systems,
https://doi.org/10.1007/978-981-15-8530-2_24

Sensor nodes collect the data from the physical surroundings and connected to the controllers. These sensors can connect, communicate, and broadcast the sensed information (data) to the base station or sink through a wireless medium, but this process consumes lots of energy through sensor nodes. The introduced approach (fuzzy c-means) helped in reducing energy expenditure by the sensor nodes. The FCM approach gave better results than k-mean because it can deal with the overlaid data set. The proposed approach became more static using membership value than the other one.

In WSN, multiple clustering ways have been proposed to handle the problem of power usage, in which the huge area of the network was split to form various sets called clusters. All sensors broadcast the assembled data to the respective base station (CH), and then, each cluster head sends this data to cluster member, which was sent to BS for the final step. The method of clustering for energy enhancement was pounded on these two methods of cluster making, and the other was the selection of the head of a cluster. There is a list of authors who have proposed both these methods [9, 10].

Many researchers provide an enhanced method for clustering that is known as 'low-energy adaptive clustering hierarchy' (LEACH) [11] to select heads of a cluster that it used different levels of energy. But the problem was that these methods include irregular clusters and also the nodes containing fewer gaps from the sink result as the selected cluster head, which gave a huge gap from the cluster head and respective nodes of clusters. Hence, energy consumption is high during transmission due to this large gap.

In this proposed algorithm, the main concern was to reduce the gap between cluster head and cluster nodes, because they need to be closer to a cluster head or BS. Therefore, FCM clustering is used [2] considering both reasons, i.e., to reduce the gap between the cluster head and their nodes and gap between the cluster head and base station [3]. The FCM is the method that allocates the degree of membership to a cluster for all sensors, which turn us to solve the obstacle of uneven allocation. The FCM has been proved to increase network capability in terms of energy consumption [5]. The main concern was to get the maximum chance of collision upon a non-uniform medium with the minimum usage of energy of the nodes in sensors network with the creation of energy-efficient clusters and by opting for the best CH among all clusters.

2 System Model

1. **Network model**
   The nodes present in sensor networks collect data through their surroundings and forward it to the base station through one-to-one transmission or multipath transmission. The base station is an interface between the outside world and the network to analyze data collected by the base station.
   Data collection is based on the clustering technique to make an individual cluster and sends the information from the nodes to the sink. This technique used to avoid

network blockage and the distribution of excess data that makes the network more robust. This technique also helped in reducing the power usage of sensor nodes [7, 13, 14].

There are mainly two clustering techniques:

(a) **Static clustering**
Static clustering is the technique in which the formation of clusters is in static nature. Therefore, there is no change in the collection of nodes after the initial iteration. This can reduce the performance of the algorithm and sensor network.

(b) **Dynamic clustering**

The dynamic clustering technique changes the adjustment of the clusters after each iteration of communication. The cluster head selected in one iteration became one of the nodes sending signals to another cluster head in other iteration, so that there can be a steady energy usage in the whole network, that is more beneficial as the number of dead nodes might decrease.

2. **Energy model**:
   (a) **Communication in free space:**
   The free space propagation model assumes that communication is taking place in an empty environment. Thus, no spreading factor and absorption coefficient are to be considered.
   (b) **Multipath propagation model:**
   The multipath propagation in radio communication results in received signal reached via multiple paths within the network. Thus, the absorption coefficient and spreading factors are to be considered. The number of bits is 'n,' and 'd' is the distance of communication, and then, the energy of transmission is as follows:

$$E_{\text{tx}}(n, d) = E_0 * n + E_{\text{fps}} * d^2, \text{ when}(d < d_0) \tag{1}$$

or

$$E_{\text{tx}}(n, d) = E_0 * n + E_{\text{mps}} * T_{\text{p}} * d^{v+4} * 10^{(d/10)}, \text{ when}(d \geq d_0) \tag{2}$$

And the energy in receiving 'n' bits is given as:

$$E_{\text{rx}}(n, d) = E_0 * n \tag{3}$$

where E_0 is the energy per bit consumed.

E_{tx} is the energy spent in sending information.

E_{rx} is the energy used in receiving information.

E_{fps} is the energy spent in free space propagation per unit distance.

E_{mps} is the energy spent in multipath propagation per unit distance.

'T' is the transmission delay, and 'V' is the energy spreading factor of the channel.

3 Energy Computation and Utilization

If the number of nodes in the network is taken as 'N' and the number of clusters found by the algorithm as 'K,' now, (N/k) is the average number of nodes in every cluster, and for each cluster, there is one cluster head and there are {(N/k) – 1} normal nodes. Now, the energy of each cluster head is:

$$E_{\text{ch}} = E_{\text{CHrx}} + E_{\text{CHtx}} + E_{\text{DA}} \tag{4}$$

E_{CHrx} is the energy utilized by each cluster head in receiving a message from normal nodes and is given by:

$$E_{\text{CHrx}} = E_0 * \left(\frac{N}{k} - 1\right) \tag{5}$$

Cluster head uses E_0 energy in receiving a message from one source. Since it receives a message from {(N/k)-1} sources. E_{CHtx} is the energy utilized in transmitting the message to the base station [11, 12].

$$E_{\text{CHtx}} = E_0 + E_{\text{mps}} * T * d_{\text{BS}}^{V+4} * 10^{d_{\text{BS}}/10} \tag{6}$$

And E_{DA} is the energy for merging and choosing the message to be transmitted further. Therefore, the energy of each cluster head is formulated as:

$$E_{\text{ch}} = E_0 * \left(\frac{N}{k} - 1\right) + E_{\text{mps}} * T * d_{\text{BS}}^{v+4} * 10^{d_{\text{BS}}/10} \tag{7}$$

Now, energy of each normal node E_{nch} is given by:

$$E_{\text{nch}} = E_0 + E_{\text{NCHtx}} \tag{8}$$

where E_{NCHtx} is given as:

$$E_{\text{NCHtx}} = E_{\text{fps}} * d_{\text{CH}}^2 \tag{9}$$

Therefore, the energy of (N/k-1) of nodes is given by:

$$E_{\text{nchTOTAL}} = E_{\text{nch}} * \left(\frac{N}{k} - 1\right) = E_0 * \left(\frac{N}{k} - 1\right) + E_{\text{fps}} * d_{\text{CH}}^2 * \left(\frac{N}{k} - 1\right) \quad (10)$$

Therefore, energy used in each cluster in one round is given as:

$$E_{\text{cluster}} = E_{\text{ch}} + E_{\text{nchTOTAL}} \quad (11)$$

$$E_{\text{cluster}} = E_0 * \left(\frac{N}{k}\right) + E_{\text{mps}} * T * d_{\text{BS}}^{v+4} * 10^{d_{\text{BS}}/10} + E_{\text{fps}} * d_{\text{CH}}^2 * \left(\frac{N}{k} - 1\right) + E_0 * \left(\frac{N}{k} - 1\right) \quad (12)$$

As the network has k clusters, the total energy is given by:

$$k * E_{\text{cluster}} = \left(2Nk - k^2\right)E_0 + \left(kN - k^2\right)E_{\text{fps}} * d_{\text{CH}}^2 + k^2 E_{\text{mps}} * d_{\text{BS}}^{v+4} * T * 10^{(d_{\text{BS}}/10)} \quad (13)$$

To find an optimum number of clusters, the equation is derived concerning k, got (k_{optimum}), assumed the radius of the network is R_{net} is units:

$$k_{\text{optimum}} = \sqrt{\frac{2 * N * E_{\text{fps}} * R_{\text{net}}^2}{\pi * E_{\text{mps}} * T * d_{\text{BS}} * 10^{(d_{\text{BS}}/10)} * N - 4 * E_0}} \quad (14)$$

4 Proposed Algorithm

The technique for clustering which provides a sample of information to get a place with at least two clusters is known as fuzzy c-means [1]. This computation works by allotting membership value to each node that directly coincides with each cluster center based on the gap between the cluster head and the information point (nodes).

Accordingly, the near the information was to the center of the cluster the high was its membership toward the particular cluster. The membership addition of each node or information point would always be equal to one [6, 15].

1. **Data generation and setup**
   (a) Label the centers.
   (b) Define n cluster sigmas in x and y.
   (c) Create a test.
   (d) Imagine test data.

2. Cluster formation
   This method was often used in pattern recognition by allocating membership to all data points resembling each cluster center, where the addition of membership for all data points would always be equal to one [4]. After each repetition, membership and cluster centers are changed according to the formula:

$$\mu_{ij} = 1/\sum_{k=1}^{c}\left(\frac{d_{ij}}{d_{ik}}\right)^{\left(\frac{2}{m}-1\right)} \tag{15}$$

$$vj = \left(\sum_{i=1}^{n}(\mu_{ij})^{m}x_i\right)/\left(\sum_{i=1}^{n}(\mu_{ij})^{m}\right)$$
$$\forall j = 1, 2, 3, \ldots, c \tag{16}$$

where the number of data points is'n'.

'v_j' represents the cluster center.

'm' represents the fuzziness index (0–1).

'c' shows the no. of the cluster center.

'μ_{ij}' is membership, and 'd_{ij}' is Euclidean distance $\|X_i - V_j\|$.

$$J(U, V) = \sum_{i=1}^{n}\sum_{j=1}^{c}(\mu_{ij})^{m}\left\|x_i - v_j\right\|^2 \tag{17}$$

3. Prediction
   (a) Recreate homogeneously examined data in x and y.
   (b) Predict new cluster membership.
   (c) Plot the classified uniform data. Note the maximum for visualization.
   (d) Membership value is taken at each point.

5 Simulation and Result

The undivided network life span was based on energy-efficient utilization in a network. The method of clustering was one of the best techniques to efficiently utilize the energy of the wireless network.

The various other routing protocols that were applied for the other networks cannot be used in this case because the power of sensor nodes depends on the battery, that is why it was required to make these networks more energy efficient. In the current work, they used fuzzy logic and other methods for the cluster head formation to enhance the life span of sensor nodes in WSN. For the selection of the cluster head, the distance

Table 1 Parameters for simulation

Parameter	Value
Network area	200 * 200
Position of the base station	(100,100)
Nodes(N)	100
E_0	50nJ/bit
E_{mps}	0.0013 pJ/bit/m^4
E_{fs}	10 pJ/bit/m^2
Packet size	500 B
Node's initial energy	2 J
Transmission delay(T)	0.2 s
Energy spreading	2
Factor (v) (medium: water)	2

of the base station and residual energy concepts were used. The parameters used in simulation and results are given in Table 1.

Energy usage by each sensor node of the network was calculated as the addition of energy used in gathering, accumulation, and passing the sensed information to the remaining intermediary nodes or the base station. Throughput was defined as the addition of unique packets of data created in every round.

In the proposed algorithm, a fuzzy c-means method was used for the formation of clusters. Primarily, the deployment of cluster head in the middle of every cluster was done, and then, the selection of another cluster head was based on residual energy and distance with the base station. This enhances the network life span and improvement in the energy consumption of each node to communicate efficiently.

To generalize the results, all the four algorithms (LEACH, DEEC, K-MEANS, and FuzzK) were implemented on the same hundred deployments. Averaged results obtained from them have been used to present the corresponding account of the algorithms (Fig. 1).

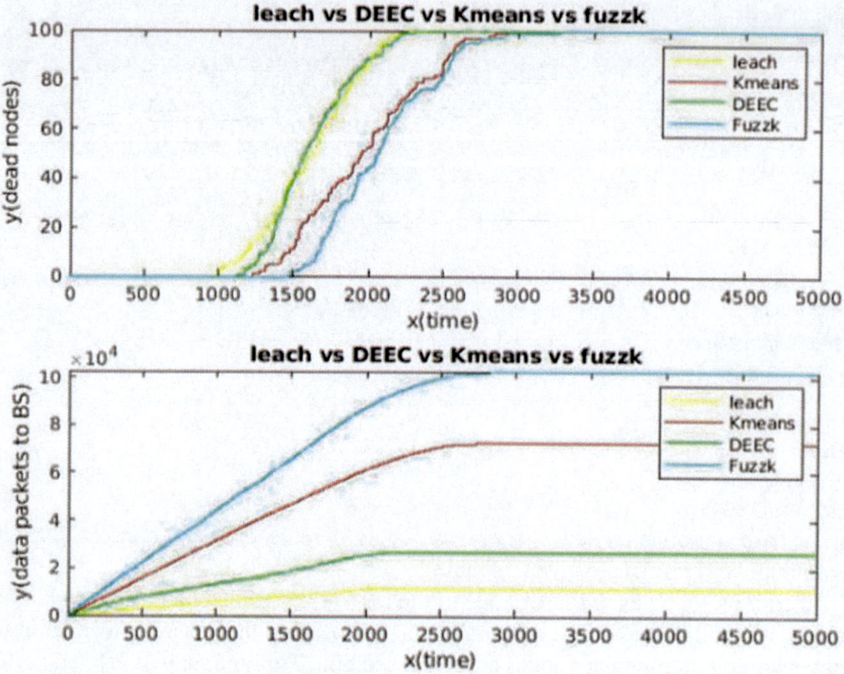

Fig. 1 Comparison between algorithms based on the formation of dead nodes and data sent per unit time

6 Conclusion

Wireless sensor network (WSN) made up of tiny devices collects appropriate and precise data (information) by the method of cooperating. In this paper, an energy-efficient algorithm (Fuzzk) has been designed and implemented to enhance the WSN lifespan using the fuzzy c-means clustering algorithm. Through experimental results, it has been analyzed that the consumption of residual energy is less than the same as DEEC, LEACH, and K-means by having successful and appropriate cluster head selection in less time.

References

1. Mehra P, Doja M, Alam B (2019a) Stability Enhancement in LEACH (SE-LEACH) for Homogeneous WSN. EAI Endorsed Trans. Scalable Inf. Syst. 6(20):156592
2. Mehra PS, Doja MN, Alam B (2019b) Enhanced Clustering Algorithm based on Fuzzy Logic (E-CAFL) for WSN. Scalable Comput. Pract. Exp. 20(1):41–54
3. B.Chen, H. Yao, M.Yang, B.J. LI, L.C. He, A Inter-Cluster Multi-Hop Routing Protocol Improved Based on LEACH Protocol, Chinese Journal of Sensors &Actuators 27(3) (2014).

4. Jain, Abhilasha, and Ashok Kumar Goel. "Energy Efficient Algorithm for Wireless Sensor Network using Fuzzy C-Means Clustering." International (IJACSA) International Journal of Advanced Computer Science and Applications (2018).
5. Yadav, Rama Shankar, Arvind Kumar, and Smriti Agrawal. "Energy management for energy harvesting real time system with dynamic voltage scaling." In Trends in Network and Communications, pp. 536–548. Springer, Berlin, Heidelberg, 2011.
6. Shreya Patel, Jayesh Munjani and Jemish Maisuria "A review of fuzzy related clustering protocol" International Journal of Computer Application (2250–1797) Volume 5– No. 3, April 2015.
7. Su S, Zhao S (2018) An optimal clustering mechanism based on Fuzzy-C means for wireless sensor networks. Sustainable Computing: Informatics and Systems 18:127–134
8. Kumar A, Alam B (2018) Task Scheduling in Real Time Systems with Energy Harvesting and Energy Minimization. J. Comput. Sci. 14:1126–1133
9. Sakkari, Deepak S., and T. G. Basavaraju. "Extensive Study on Coverage and Network Lifetime Issues in Wireless Sensor Network." International Journal of Computer Applications 52, no. 8 (2012).
10. Kumar, Arvind, and Bashir Alam. "Real time scheduling algorithm for fault tolerant and energy minimization." 2014 International Conference on Issues and Challenges in Intelligent Computing Techniques (ICICT). IEEE, 2014.
11. E.PRABASHINI, D.SIVAKUMAR "Energy Efficient Cluster-based Routing in WirelessSensor Networks" IJCSNS International Journal of Computer Science and Network Security, VOL.16 No.3, March 2016.
12. W. Heinzelman, A. Chadrakasan and H.Balakrishnan, "Energy-efficient communicationprotocol for wireless microsensor networks", in Proceedings of the 33rd Annual HawaiIInternational Conference on System Sciences, Jan 4–7, 2000.
13. Kumar, Arvind, and Bashir Alam. "Improved EDF algorithm for fault tolerance with energy minimization." 2015 IEEE International Conference on Computational Intelligence & Communication Technology. IEEE, 2015.
14. Sharma Y, Dagur A, Chaturvedi R (2019) Automated bug reporting system with keyword-driven framework. Singapore, Soft Computing and Signal Processing. Springer, pp 271–277
15. Mehra PS, Doja MN, Alam B (2018) Stable Period Enhancement for Zonal (SPEZ)-Based Clustering in Heterogeneous WSN. In: Innovation S (ed) Systems and Technologies, vol 79. Singapore, Springer, pp 887–896

Chapter 25
Optimization of Queries in Database of Cloud Computing

Arvind Dagur, Ankit Kaushik, Adamya Rastogi, Amritpal Singh, Ashish Kumar, and Rahul Chaturvedi

1 Introduction

Cloud computing is an IT service that provides storage and other facilities to various consumers that are using its services. User can access the data over server using computer notebook, tablet computer, smartphone, or with any device that supports cloud computing. In this computing framework, the local memory does not contain any data rather it is stored under a cloud server [1].

The uses of resources can be examined, controlled, and reported. This provides transparency over the usage of the resources. Users can access the server using the Internet, i.e., system can be accessed using a web browser. Cost reduction is the major feature of cloud computing, that is why it is more efficient than a locally hosted drive server. It is more secure than a locally hosted drive. Since the cloud computing applications need not be installed on each user's device, the maintenance

A. Dagur (✉) · A. Kaushik · A. Rastogi · A. Singh · A. Kumar · R. Chaturvedi
Department of Computer Science and Engineering, Krishna Engineering College, Ghaziabad, Uttar Pradesh, India
e-mail: arvind.dagur@krishnacollege.ac.in

A. Kaushik
e-mail: informtoankit@gmail.com

A. Rastogi
e-mail: kartikthegr845@gmail.com

A. Singh
e-mail: amritpalsinghwilkhu@gmail.com

A. Kumar
e-mail: ashishkumarsixplus@gmail.com

R. Chaturvedi
e-mail: rahul.chaturvedi@krishnacollege.ac.in

I. Jeena Jacob et al. (eds.), *Data Intelligence and Cognitive Informatics*, Algorithms for Intelligent Systems,
https://doi.org/10.1007/978-981-15-8530-2_25

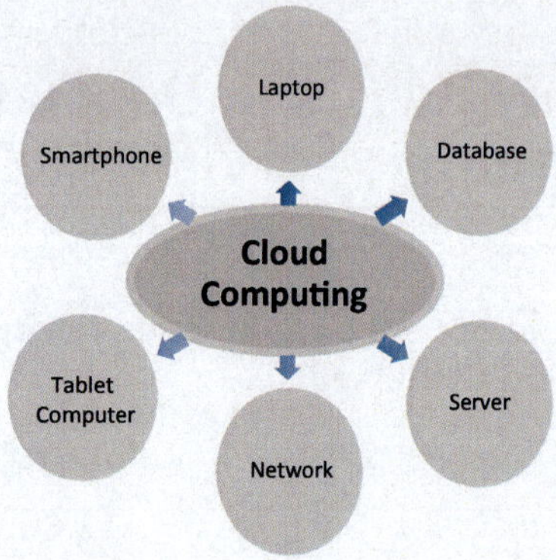

Fig. 1 Cloud computing

staff can retrieve admissible information wherever they are located. This results in easier maintenance. Costs and resources are shared among a large pool of users, which is generally referred to as multi-tenancy. Multi-tenancy of data and its efficient retrieval is the major concern to optimize. Cloud services become more complex in a heterogeneous computing environment [2, 3]. The problem is to optimize the queries in a database of cloud computing. A simple query efficient database architecture is proposed for the SaaS architecture model of cloud. A simple architecture of cloud computing is presented in Fig. 1.

Rest of the paper is organized as follows: Literature survey is presented in Sect. 2; Sect. 3 presents the proposed approach. In Sect. 4, simulation results have been presented, and the paper is concluded.

2 Literature Survey

The multi-tenant/user database uses our system model where the major data is controlled through the provider and the customers subscribe to the services that are made available to him by the provider. Few applications which are using use multi-tenant/user database management are BigCommerce and DocuSign [4, 5].

Mathematical expressions and Greek or other symbols must be written clearly with ample spacing. Operating models of cloud computing are presented in Fig. 2.

2.1 Operating Models

Software as a service (SaaS) is a phenomenon in which software is provided over the web. This software does not require to be installed in every device. Software

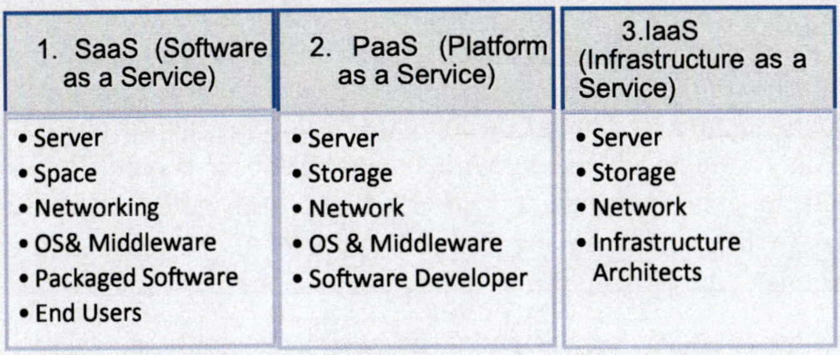

Fig. 2 Service operating models

as a service is an on-demand service that provides an independent platform. The providers of the software application are responsible for running it on cloud server architecture. The client can use a variety of services through a program interface and various interfaces such as the browser [6, 7].

Platform as a service (PaaS) can be explained as a phenomenon in which a platform is provided over the web to create a program or an application. The cloud service provider supports all the applications which are created using a coding language, present libraries and also many new services accessed by the provider. The consumer has no control over the architecture of the cloud which holds all networks OS and servers. The consumer has control over apps and apps configuration settings [8–10].

Infrastructure as a service (IaaS) can be explained as a model in which virtualized resources are provided over the Internet. This provided range of services over the Internet for organizations, and these services may include billing, account management, etc. The provider gives processing power, storage capacity, network connection, and other fundamental resources. The consumer is not having any control over the underlying architecture of the cloud, including all networks and server OS [1, 11].

2.2 Database Management System

It is a software technology which acts as a communication link between applications and the database, and this database is used to keep track of information in an organized and efficient manner. Database structures can be hierarchical, networked, relational, object-oriented, object-relational, and distributed [12, 13].

Few terms relating to DBMS are detailed as below—data independence, the data and program should be independent of one another so that the database could be accessed by another user without needing the same program. Data normalization, the process of organization the database into logical tables for minimum repetition and redundancy. Data modeling, to identifying and organizing a database's data, both logically and physically [14, 15].

3 Proposed Approach

Our proposed approach is based on the extensible table approach (ETA). In this approach, only a single table along with a single database is mutually being shared between all the consumers\tenants. The ETA approach holds the data for all the consumers. To differentiate among the information of all the consumers and to keep every consumer's data isolated from others in the servers, a mechanism is used.

1. The concept of XML has been used (Extensible Markup Language).
2. Introducing an Extensible Markup Language is a database with the help of which the size of the extended table can be reduced and also it eliminates the need for the primary table.
3. This approach helps in creating more than one table for a consumer which is stated in the system and helps in efficient optimization.
4. The protocol that is used in the approach for authenticating and authorizing is the Kerberos protocol.

Every transaction is associated with a t_id and r_id (t table, r record) that provides the transaction with a unique record in the database. In the table, the XML object holds the data of a whole row of a table. The name of the criteria in a table is referred by the tags that are present is an XML object. The t_id is used to represent the tenant id of a particular table in which there is a record that is entered by the customer. Thus, an Extensible Markup Language document is totally dynamic.

3.1 *Creating a Table*

It is up to the consumer or the tenant that customer can create any number of tables that customer wants, which is also supported by the database system on which customer is working. When the tenant creates a new table, it provides a name to the table and the new name is updated in the database. This table gives a special id of the consumer. This table gives a special id to a respective name for that unique tenant. t_id column in this table helps in the extraction of the names of the tables which are associative to that particular tenant. A metadata table is shown in Table 1.

Table 1 Metadata table

t_id	128	126	124	125
Table_name	Fund	Package	Cash	Customer
Table_id	8	7	6	2

3.2 Insertion of an Element onto a Table

If a customer wants to add some new value to the table, then it has to specify the identity in tables and then input the entries. Our system has a very systematic approach to add the values in the table. There are the following steps.

1. First, the t_id of the particular tenant is gathered from the metadata table.
2. Then the metadata is accessed so that what data type that is used is known.
3. For this particular transaction, a record id is gathered and is stored in the column that right Id.
4. After performing the above steps, an XML file is created.
5. After that XML creates tags which are associated with the column names of the database tables that hold the data stored by the user.

3.3 Updating the Database

To update an entry in the database, the user just has to specify the table name and the field where it wants to update the database.

It is done with the following process:

1. First of all, it checks metadata from which it then retrieves the particular id for that particular tenant.
2. Then finds the rows that belong to that t_id from the table.
3. After this, it develops a query that helps in parsing the XML document using the Python XML parser.
4. After this, it updates that document then it saves and retrieves the particular id for that particular tenant.

3.4 Developing New Table

A new table is given in Table 2, an individual t_id which is useful in the future, it keeps the track record of the information that is put into this table.

The following table has information like data types and external tables, etc.

Table 2 Developing new tables

Tenant_id	Ex_id	Ex_lib	Datatype	T_id
125	6645	expire	Bool	1
126	4574	update	Int	5
127	2234	reject	String	1

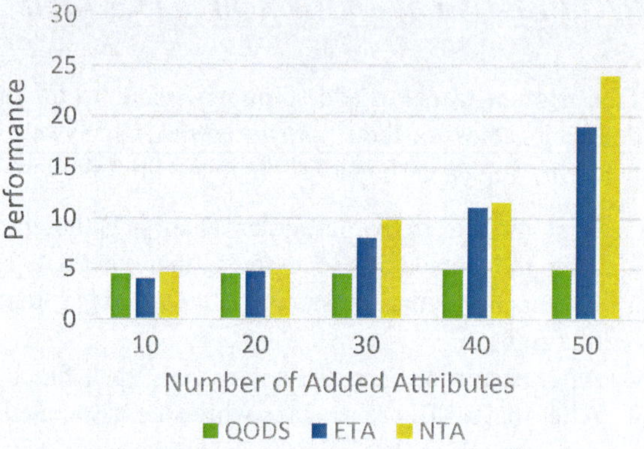

Fig. 3 Average performance for added attributes

4 Simulation Results

The factors used for simulation are RAM, CPU, network, and disk space usage, etc. The RAM usage is being less consumed in comparison with the extension table approach. The CPU usage is less due to less memory consumption, and the network traffic is also reduced with a major margin as compared to existing approaches.

The average response time for the execution of inserted queries is almost constant for our proposed approach but in the extension table approach, it rises gradually. The response time in the extension table approach reduced in five constant steps of ten attributes. This involves the execution of new tables that is a greater number of relations are needed to satisfy a query. This concludes in a previous method there is an increase in the response time with the addition of new attributes; whereas, our approach uses XML attributes which results in a smaller number of relations required for query execution. Due to the use of XML fields, our approach gives better performance and handles newly added attributes better as shown in Fig. 3.

The memory requirement in the extension table method is much greater than our proposed approach. The space requirements increase linearly in the proposed system because of using the XML fields.

The efficiency gained in space is massive. In our approach, the XML file size is bounded by which it can be embedded in the field like attributes by the native DBMS technology selection which is done taking the help of caching for better performance as shown in Fig. 4.

5 Conclusion

In this database, all the tenants are allowed to develop multiple tables, and also, they have an option to have various sets of attributes for its application. The proposed approach used XML attributes to store a large number of attributes of a database to

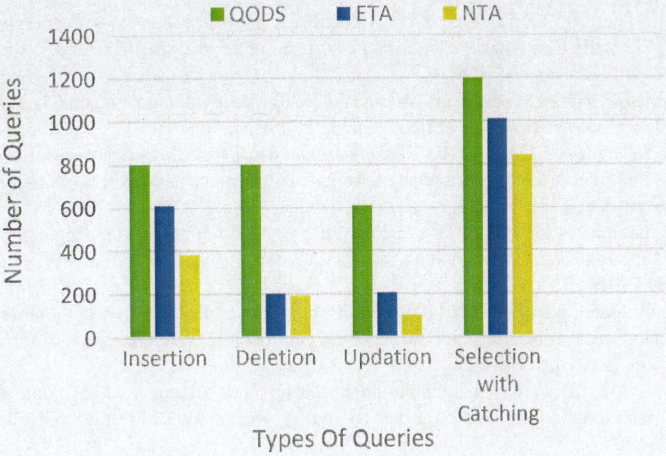

Fig. 4 Number of queries executed

the same field and saved in the form of XML. Extension table approach (ETA) has been used to optimize the queries to hold multiple values of all the consumers/tenants. Simulation results show that the proposed approach performs better as compared to existing approaches such as ETA and NTA where the number of attributes increased up to 50 queries. The null value problem is also solved in our approach as compared to ETA and NTA. The multi-tenant database architecture that has been developed in our research is highly efficient in terms of the execution of the query.

References

1. Mell P, Grance T (2011) The NIST definition of cloud computing
2. Aulbach S, Jacobs D, Kemper A, Seibold M (2009) A comparison of flexible schemas for software as a service. In: Proceedings of the 2009 ACM SIGMOD international conference on management of data, pp 881–888
3. Ren K, Wang C, Wang Q (2012) Security challenges for the public cloud. IEEE Internet Comput 16(1):69–73
4. Li J, Li YK, Chen X, Lee PPC, Lou W (2014) A hybrid cloud approach for secure authorized deduplication. IEEE Trans Parallel Distrib Syst 26(5):1206–1216
5. Kiefer T, Lehner W (2011) Private table database virtualization for dbaas. In: 2011 fourth IEEE international conference on utility and cloud computing. IEEE, pp 328–329
6. Kumar A, Alam B (2018) Task scheduling in real time systems with energy harvesting and energy minimization. J Comput Sci 14:1126–1133
7. Mehra PS, Doja MN, Alam B (2018) Stable period enhancement for zonal (SPEZ)-based clustering in heterogeneous WSN. In: Proceedings of first international conference on smart system, innovations and computing. Springer, Singapore, pp 887–896
8. Krebs R, Momm C, Kounev S (2012) Architectural concerns in multi-tenant SaaS applications. Closer 12:426–431
9. Kumar A, Yadav RS, Ranvijay AJ (2011) Fault tolerance in real time distributed system. Int J Comput Sci Eng 3(2):933–939

10. Kumar A, Alam B (2015) Improved EDF algorithm for fault tolerance with energy minimization. In: 2015 IEEE international conference on computational intelligence & communication technology. IEEE, pp 370–374
11. Gupta P, Raj P, Tiwari S, Kumari P, Mehra PS (2020) Energy efficient diagonal based clustering protocol in wireless sensor network. Available at SSRN 3565781
12. Sharma Y, Dagur A, Chaturvedi R (2018) Automated bug reporting system in web applications. In: 2018 2nd international conference on trends in electronics and informatics (ICOEI). IEEE, pp 1484–1488
13. Alam B, Kumar A (2014) A real time scheduling algorithm for tolerating single transient fault. In: 2014 international conference on information systems and computer networks (ISCON). IEEE, pp 11–14
14. Xue W, Qingzhong L, Lanju K (2011) Multiple sparse tables based on pivot table for multitenant data storage in SaaS. In: 2011 IEEE international conference on information and automation. IEEE, pp 634–637
15. Kumar A, Alam B (2015) A dynamic approach for fault tolerance with voltage scaling. In: 2015 international conference on green computing and internet of things (ICGCIoT). IEEE, pp 1522–1525

Chapter 26
Low-Cost IoT-Enabled Smart Parking System in Crowded Cities

Rahul Chaturvedi, Shubham Kumar, Utkarsh Kumar, Tanisha Sharma, Zeba Chaudhary, and Arvind Dagur

1 Introduction

Finding a parking space is really a tiresome job as urban areas are overpopulated due to the increased number of vehicles. This results in wastage of time and fuel which also leads to the pollution of air. There are vacant spaces in which vehicles can be parked but it is difficult for drivers to look for the same. As the areas are crowded and drivers do not find any space, they park the vehicles in an illegal manner which results in further consequences. N number of solutions have been proposed to solve this problem but they are not as reliable and secure. The existing approaches [1, 2] are less reliable to prevent accidents and maintain a certain amount of distance as it cannot detect the motion of the vehicle. Authors made a system which could not give real-time values and also it is not cost-efficient. This situation motivated us to design a system which will direct you to the nearest parking space easily. In this proposed approach, it has kept two factors which are most important for designing an efficient and reliable system that is time and space. The system created has the least

R. Chaturvedi · S. Kumar · U. Kumar · T. Sharma · Z. Chaudhary · A. Dagur (✉)
Department of Computer Science and Engineering, Krishna Engineering College, Ghaziabad, UP, India
e-mail: arvind.dagur@krishnacollege.ac.in; arvinddagur@gmail.com

R. Chaturvedi
e-mail: rahul.chaturvedi@krishnacollege.ac.in

U. Kumar
e-mail: utkarsh_fcb@hotmail.com

T. Sharma
e-mail: tanishasharma.ts24@gmail.com

Z. Chaudhary
e-mail: zeba1098@gmail.com

I. Jeena Jacob et al. (eds.), *Data Intelligence and Cognitive Informatics*, Algorithms for Intelligent Systems,
https://doi.org/10.1007/978-981-15-8530-2_26

human intervention thus giving the better result. Driver's need not to waste the time in finding the parking space. This paper also includes hardware devices which help us to easily allot the spot to the drivers, at this point, IoT comes into the picture. Internet of things can connect various devices in the system to the Internet. This connectivity plays an important role as it increases the capturing of data thus making the system more efficient and secure. IoT plays a major role in the transportation system, i.e. driver, vehicle or the user. It enables inter and intravehicle communication this way it helps users to track the real-time status and make the desired decision. This can only happen by using IoT as it gives seamless connectivity among the devices. IoT has modified the world in every aspect and every field. It converts a general device into a smart device by using the sensors. Without even human intervention, these devices are linked to each other. It has improved our life by making everything work like a human brain thus making it smart [3, 4]. Be it our smart home or eldercare, it has spread its wings in every field. IoT supports deep reinforcement learning, as teaching the agent in a traditional learning way such as supervised learning is not enough, so agents learn by observing the environment and receiving rewards in the longer term. This gives an agent an interactive environment. However, some risks are associated with this technology, like data privacy and security. Research and enhancements have been followed to overcome this issue. According to the report, the Internet of things (IoT) market is expected to reach $ 9 billion by 2020. By the end of 2020 more than 1.9 billion, devices are expected to be connected in India which will lead to growth in the market 31 times from the current market share of 5.6 this year [5].

Rest of the paper is organized as follows: Sect. 2 includes the related works. Section 3 includes the solution which has been proposed, Sect. 4 includes the result and discussion that have obtained and in the last section, and the paper is concluded.

2 Related Works

In today's time, it is difficult to keep on searching for an available parking spot as it consumes a lot of time and fuel. This also creates traffic congestion at peak time. Due to this driver illegally park the cars and therefore leading to towing of the vehicle. It creates chaos and the one who really is in need is not able to utilize the space [6, 7]. In today's scenario which is rapid population growth, the number of parking spaces are either constant or decreasing. Vehicles are outnumbering the available parking space thus clogging the roads. The reason behind the lesser number of parking spaces is the low tariff that is being charged for parking areas. Unregulated tariff leads to decreasing parking spaces. This low tariff encourages more vehicles to come on the road which results in problems such as environmental issues and congestion. So, the tariff needs to be in a regulated manner. Parking becomes more difficult when it comes to occasions. A tremendous number of vehicles come up in the street which creates clogging, quarrels and long queues that results in wastage of time and fuel adding to the pollution of the environment. This results in parking the vehicle in parks or an inappropriate manner. This increasing demand for parking is seeing no signs of stopping. So instead of increasing the number of parking spaces, on efficiently using the available parking space should be focused by turning them into smart online parking systems by using modern technology and sensors that can guide the driver to the available parking space. Several works have been proposed for the implementation of the parking system [8, 9]. The automation [6] of parking spots using the machine learning technique is a good approach towards the idea but does not meet the accuracy factor. The entry through [10] QR code scanning approach has been followed which saves plenty of time but it is not reliable as the scanning of vehicle number plates can lead to inaccurate results as plates can be deformed. This led to a need for a better and efficient system. The system is designed to look for an available parking space near the driver area. So that the time consumed is minimized and fuel is not burnt unnecessarily. The modern Internet of things technology is used to meet the demands that are real-time availability of data [11–13]. It also focuses on having an android application to have a hustle free access to the parking spot. Pre-booking of the parking spot is available so that the driver can book it before time and can also track the same. Sensors are used on the parking spot which helps in conveying messages directly to the Web servers to know more about the spot allocated and also help the driver in a meaningful way. The sensor helps the user in spotting the free space and adding to it an LCD to guide the driver from point A to Z. The system returns this value to the mobile application [14–16].

3 Proposed Solution

The existing IOT-based parking systems have been surveyed and proposed the better system which is more efficient than the existing ones. During our survey, numerous hardwares come across that were implemented which had certain loopholes. The existing system [1] has used infrared sensors which use infrared light which can only determine whether the object is present or not. Arduino [4] only provides a limited library and large structure Arduino is a pain for complex circuits. An IoT device is ought to be proposed for the parking system that not only will help return the real-time data but also is very cost-efficient. The Ultrasonic sensors that are used are very effective because not only is it cost-efficient but also their range can be altered which in turn is a plus point for us [6, 17]. As it uses sound waves and tells how far you are away from the object, which gives a more accurate result and prevents accidents. Node MCU has been used instead of Arduino as it is cost-efficient and a better replacement. The device is connected directly through the database to the android application. The designed system thus ensures there should not be any kind of loopholes further. Android application is the user application with which users can search, book and pays for the parking slot in the nearby parking available in the app. This application has provided different features for users. Users can book for new parking, it can check availability of the parking area, and it can also check nearby parking and the important feature, i.e. users can track their booking status. For a business user interaction, it is not the only thing which is important but management also plays a crucial role. The proposed system has kept the CRUD operation in mind, i.e. create, read, update and delete. This system has defined different panels for individual parking. Vendor panel has been created for a better management system. The flowchart explains what happens when the slot is available or not (Fig. 1) .

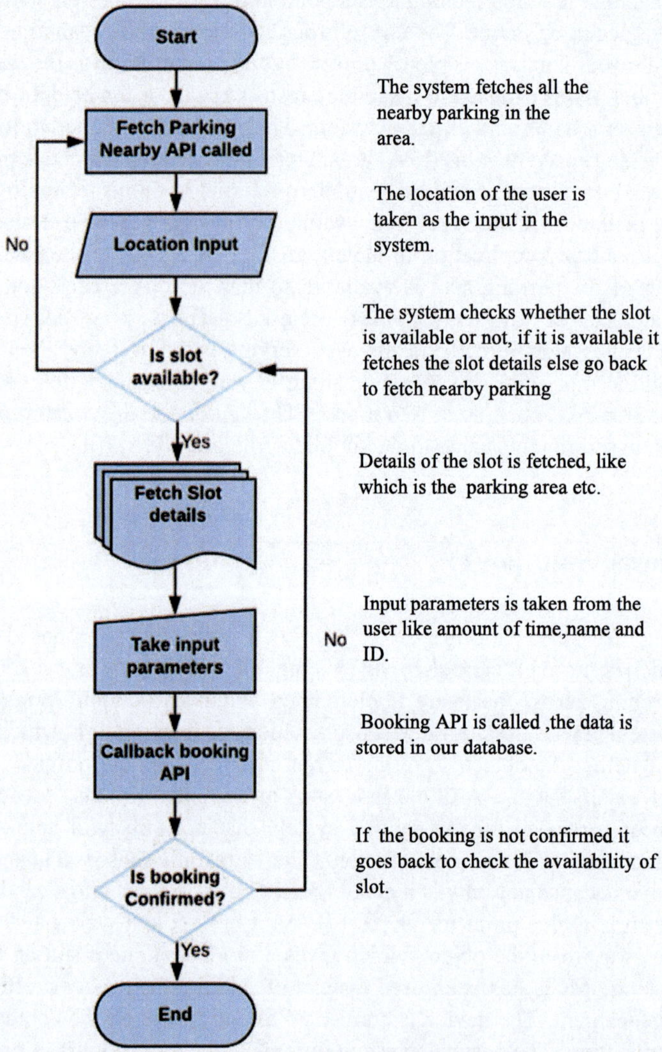

Fig. 1 Flowchart describing the user cycle to locate nearby parking and slot booking

4 Result and Discussion

The designed system is better than the existing system. As it has kept different factors for hardware and software in mind, a control management system has been made to make the system more efficient and give a better way to access the system. This flowchart explains what happens when a slot is available or not. Different panels are made for individual parking for a smooth management system. When the designed system is compared with the existing one, it comes out to be much better in a few factors. As ultrasonic sensors are used instead of infrared sensors to give more accurate results, Node MCU is used instead of Arduino as it is more cost-efficient.

In future, more features can be added in the designed system to enhance the performance. The result which has been made for the designed system gives a better performance as shown in the figure below. In Fig. 2, the existing system is defined as the systems which are already present and working, and the proposed system is the system which has been proposed by us and the comparison is shown between the ideal, proposed and existing. The number of workers required in a parking area has been reduced in the proposed system. In Fig. 3, the existing system is being compared to the proposed and ideal system and it clearly shows that the amount of time required for parking a vehicle has been reduced in the system which has been proposed.

So, the result clearly states that the manpower has been reduced as the system has been automated, and an error has been reduced because less human intervention is involved and time taken has been reduced, which was the most important factor in the system.

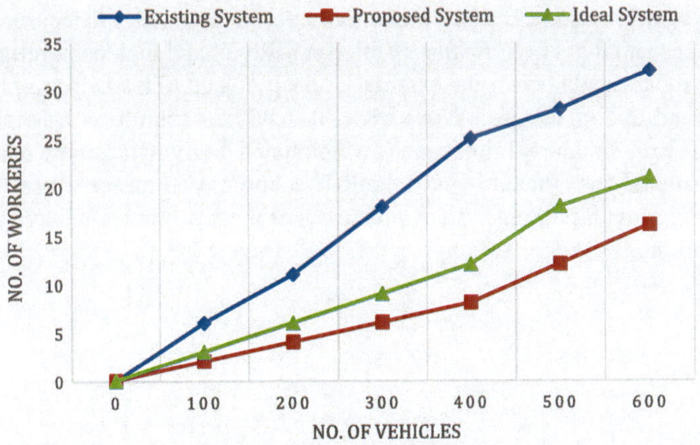

Fig. 2 Comparative graphical analysis of manpower needed with respect to parking

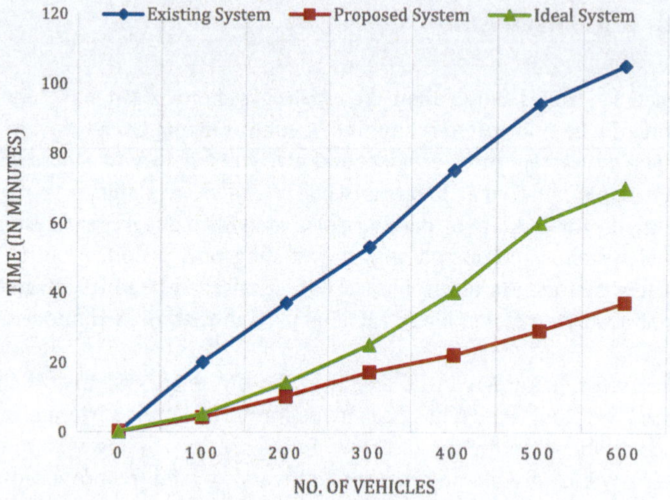

Fig. 3 Graph shows the comparison of time consumed with respect to no. of vehicles parked in parking system

5 Conclusion

There is a big problem faced by people to search for available parking near them. Looking for an appropriate parking space is not easy, which leads to wastage of time and fuel. Taking this problem into consideration, this system is designed in an efficient manner which helps in avoiding the congestion of traffic. As existing system have been compared, ultrasonic sensors are found to be better than infrared sensors, and this system has used ultrasonic sensors as it uses sound waves to measure how far the vehicle is and it has high frequency, high sensitivity and high penetrating power, therefore, it can easily detect the external or deep objects. This helps in preventing accidents and this system uses Node MCU instead of Arduino, as it is more cost-efficient. It works with real-time values which helps in giving accurate results. The system designed has efficiently used the IoT technology to meet the user demands and creating a smart parking system which makes it easy to find and navigate to the nearest parking.

References

1. Meenaloshini M, Ilakkiya J, Sharmila P, Sheffi Malar JC, Nithyasri S (2019) Smart car parking system in smart cities using IR. In: 2019 3rd international conference on computing and communications technologies (ICCCT). IEEE, pp 178–182
2. Karthi M, Harris P (2016) Smart parking with reservation in cloud based environment. In: 2016 IEEE international conference on cloud computing in emerging markets (CCEM). IEEE, pp 164–167

3. Al-Turjman F, Malekloo A (2019) Smart parking in IoT-enabled cities: a survey. Sustain Cities Soc, 101608
4. Subramani T (2012) Parking study on main corridors in major urban centre. Int J Mod Eng Res 2(3)
5. Cai BY, Alvarez R, Sit M, Duarte F, Ratti C (2019) Deep learning-based video system for accurate and real-time parking measurement. IEEE Internet Things J 6(5):7693–7701
6. Sharma Y, Dagur A, Chaturvedi R (2018) Automated bug reporting system in web applications. In: 2018 2nd international conference on trends in electronics and informatics (ICOEI). IEEE, pp 1484–1488
7. Mahmood Z, Haneef O, Muhammad N, Khattak S (2018) Towards a fully automated car parking system. IET Intel Transport Syst 13(2):293–302
8. Kamble P, Chandgude S, Deshpande K, Kumari C, Gaikwad KM (2018) Smart parking system. Int J Adv Res Dev 3(4):183–186
9. Malaiya AR, Kumar N, Kansagara A, Kadam A (2018) Smart parking system based on cloud computing using android and IoT. Int J Innov Sci Res Technol 3(3):233–235
10. Kumar A, Alam B (2015) An efficient checkpointing approach for fault tolerance in time critical systems with energy minimization. In: International conference on computing, communication & automation. IEEE, pp 704–707
11. Alam B, Kumar A (2014) A real time scheduling algorithm for tolerating single transient fault. In: 2014 International conference on information systems and computer networks (ISCON). IEEE, pp 11–14
12. Bessghaier N, Zargayouna M, Balbo F (2012) Online localized resource allocation application to urban parking management. In: 2012 IEEE/WIC/ACM international conferences on web intelligence and intelligent agent technology, vol 2. IEEE, pp 67–74
13. Mudaliar S, Agali S, Mudhol S, Jambotkar C (2019) IoT based smart car parking system. Int J Sci Adv Res Technol 5(1), 270–272
14. Thangam EC, Mohan M, Ganesh J, Sukesh CV (2018) Internet of Things (IoT) based smart parking reservation system using raspberry-pi. Int J Appl Eng Res 13(8):5759–5765
15. Pham TN, Tsai M-F, Nguyen BB, Dow C-R, Deng D-J (2015) A cloud-based smart-parking system based on Internet-of-Things technologies. IEEE Access 3:1581–1591
16. Kumar A, Alam A (2015) A dynamic approach for fault tolerance with voltage scaling. In: 2015 international conference on green computing and internet of things (ICGCIoT). IEEE, pp 1522–1525
17. Tripathi AM, Singh AK, Kumar A (2012) Information and communication technology for rural development. Int J Comput Sci Eng 4(5):824

Chapter 27
A Critical Review on Educational Data Mining Segment: A New Perspective

Randhir Singh and Saurabh Pal

1 Introduction

The disposition of adequate magnitude of data in approximately every field as well as the desire to excerpt helpful information along with knowledge from it substantiates as most important motivation with the intention of pull the eyes of researchers in new past toward data mining [1]. The information and knowledge extract can be significantly helpful for the application ranging from little business management to multifaceted engineering design to science exploration [2]. Data mining is the analysis in addition to scrutiny of mammoth datasets with an aspire to discover important pattern and rules that were until that time unidentified. Data mining tools carry out analysis of data and settle on the previously anonymous and very significant patterns. As each arena of human life has been converted into data-intensive, which stemmed in manufacture data mining as an essential constituent [3]. Applications of data mining methods are a way to evaluate the features applied to a large amount of data to find the hidden patterns and relations ready to lend a hand in decision making. The forecast of student presentation by way of high correctness is beneficial to identify the students with low academic achievements initially [4]. It is necessary that the recognized students can be assist additional by the teacher so that their presentation is enhanced in the future. Data mining techniques used to change the unprocessed data to a useful acquaintance by processing the historical storage of on or after which new acquaintance is capable of being made. Educational mining concerns with developing methods for discovering information from data that move

R. Singh (✉) · S. Pal
Dept of Computer Applications, VBS Purvanchal University, Jaunpur, India
e-mail: randhirs76@gamil.com

S. Pal
e-mail: drsaurabhpal@yahoo.co.in

I. Jeena Jacob et al. (eds.), *Data Intelligence and Cognitive Informatics*,
Algorithms for Intelligent Systems,
https://doi.org/10.1007/978-981-15-8530-2_27

toward on or after educational field in addition to it helps to take out of sight pattern along with to find out new knowledge as of great educational databases by way of the make use of data mining techniques along with tools.

2 Literature Review

Connected to this, it is the learning of Bhardwaj et al. [5] in which a data model was second hand to predict student's performance by means of importance on identifying the distinction of high learners in addition to slow learners using classification. The construction of data mining is motionless for the nearly everyone parts a matter of guidelines, best practices and undocumented expert knowledge. To be actually valuable, data mining must be integrated into the classification, KDM, and decision tree right from system inception [6].

Data mining means an involuntary user are used to trouble-free and improve the system beneath any state of affairs. Due to adding the rate of difficulty and its trouble-free operation, all the valuable data and dangerous increase rate of data issue in addition to its easy operation. All the expensive assets of a business and dangerous mission are stored in the data warehouse. Literature survey reveals that methods to take away trouble in data mining approaches are incredibly expensive in terms of time, money, as well as efforts [7].

The survey on educational data mining should be applied with the application of intelligent computing system discussed and presenting a basic conception of the data mining along with the prominent algorithms of the data mining and the classification of its techniques, further, the survey concludes with the challenges included in the overview of the survey done along with the future enhancement in the research that analyzes the data mining techniques in the intelligent computing applications[8].

Data mining may be ready to lend a hand to observe and accessing the excellence of data by means of classification approach. In this paper, a crucial point on the methodology which provides help to decrease the trouble in a large database system. Many experts and researchers in the area recommended that data mining is the important concept of a big data or component to continue normal operation despite the evaluation with performance and the similar is concise in Table 1.

3 Data Mining: A New Perspective

To get a better understanding of the learning process to a crucial point on identifying, extracting as well as evaluate variables connected to the learning process of students using data mining techniques are explained by Thakar [1].

Data mining methods applied in educational environment is called educational data mining is discussed by Mirza [6] describes data mining software to allow the users to analyze data from dissimilar dimensions, sort out it and sum up the relations

Table 1 Various previous studies

Experts	Year	Concept	Contributions	Gap analysis
Singh and Pal [15]	(2020)	Passive aggressive classifier (PAC), support vector machine (SVM), linear discriminant analysis (LDA), radius neighbor classifier (RNC), extra tree (ET)	Developed a model to predict the students' performance	Only classifiers performance measures feature selection and ensemble techniques are not applied
Sunday et al. [16]	(2020)	ID3 J48	Developed a model to measure the success and failure of students	Only 6 attribute and 239 instances are taken for prediction
Kunjumon et al. [17]	(2019)	Association rule Decision tree Naive Bayes Regression	Developed a model to predict the students' performance	No description is given how they developed the model. Only a comparison of results has done
Deng et al. [18]	(2019)	OEGP DEGP DEIA OEHA	Study the correlation between question items from assignments and exams and develop an interface for interactive exploration of student course performance data	PerformanceVis, a visual analytics tool is used and only single subject chemistry is chosen for analysis
Guerra et al. [19]	(2018)	Classification J48 Decision tree	Developed a model to identify the reasons of student dropouts	The accuracy achieved is 75% only, which is very less
Madni et al. [2]	(2017)	Decision tree, association rules, fuzzy logic	A critical review on data mining approach	Only review part
Ramírez-Gallego et al. [9]	(2017)	Dimensionality reduction, FS algorithms, weighted forgetting (LWF) algorithm	Survey on data mining concept	Implementation level is not presented
Mirza et al. [6]	(2016)	Clustering, classification, mining tools	Highlighted the data mining techniques	More work on sampling is needed

(continued)

Table 1 (continued)

Experts	Year	Concept	Contributions	Gap analysis
Sumitha et al. [11]	(2016)	Naive Bayes, multilayer perception, decision table, REP tree	Developed the approach to predict the outcome student performance	When the dataset is large enough, the improvement may be not achieved by only increasing the size of the training data
Thakar et. al.[1]	(2015)	All methodology comparisons	Evaluated the performance analysis	No specific methodology highlighted
Ragab et al. [10]	(2014)	Classification techniques used for 5260 instances with 8 attributes	Highlighted to a comparative analysis of data Ming techniques	Improve the efficiency in the information finding process
Sharma and Jain [3]	(2013)	Waikato Environment for Knowledge Analysis (WEKA), classification, decision tree	Developed the new approach for data mining concepts	Unstructured classification
Nagy et al. [12]	(2013)	Dataset collected from 2000–2012, k-means clustering, machine learning	An educational data mining system for advising higher education students	New automated mechanisms are required
Yadav and Pal [2]	(2012)	C4.5, ID3, and CART decision tree algorithms used for B. tech student data	Explained the performance improvement of engineering students using classification	Focus on specific structural patterns
Manis and Guo [13]	(2009)	Voluminous geographic dataset and associate to classification and prediction, spatial association rule mining, spatial cluster analysis	Provide the introductory part of the data mining concept	Traditional data mining techniques are used

which are recognized for the period of the mining process. Data mining introduce a new method to recognized as educational mining [14]. In educational data mining, the applications of different data mining classifiers are applied to data that is related to the field of education. Educational data mining is applied to the huge amount of data collected by an institute/university in previous years. Educational data mining helps to find out the hidden patterns of the students for analyzing for performance, drop-out analysis, online learning, etc. Data mining concept has presented the various view in Fig. 1.

Fig. 1 Data mining concepts

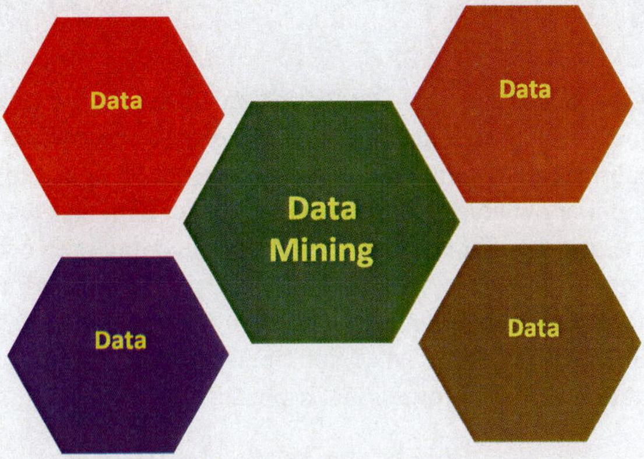

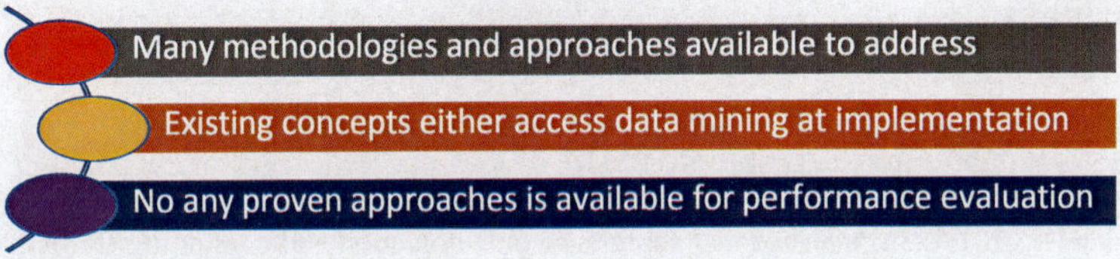

Fig. 2 Educational data mining issues

4 Research Issues

Several approaches are given and presented by various experts in term of data mining. In this article, various issues are found at the various level which is presented in the following manners as shown in Fig. 2. Data mining may be ready to lend a hand to observe and accessing the excellence of data by means of classification approach.

5 Research Enlightenment

Data mining is specifically designed to assist in a big data environment. Data mining giving a relationship between data element with values and their terminology. During the literature review, a number of various article attention to mining concept founded but not specified these questions presented in the following manner.

- Which is the information that directly influences data mining concepts?
- Is there any approach based on fuzzy logic concepts?
- Are any qualitative values may increase the accuracy rate of prediction?
- A model can be developed to check the efficiency of algorithms?
- Can a model be got, which can produce the graphical presentation of techniques and concept?

6 Conclusion

Data mining has until the end of time been a complex concept and its straight involvement is a complicated task. The paper accessible a revision of literature with data mining, a technique used to ascertain out of sight and practical patterns from a vast number of datasets. This research paper completes that how will give a view to show the effect of data mining concepts. The critical review will complete supporting the data mining properties for data systems at an assortment of level successfully. The work will be further extended out by designing the student model analyzing records of higher students on various perspectives by which data mining concept can be developed in a professional aspect.

References

1. Thakar P (2015) Performance analysis and prediction in educational data mining: a research travelogue. arXiv preprint arXiv:1509.05176.
2. Madni HA, Anwar Z, Shah MA (2017) Data mining techniques and applications—A decade review. In: 2017 23rd ınternational conference on automation and computing (ICAC). IEEE, pp 1–7
3. Sharma TC, Jain M (2013) WEKA approach for comparative study of classification algorithm. Int J Adv Res Comput Commun Eng 2(4):1925–1931
4. Yadav SK, Pal S (2012) Data mining: a prediction for performance improvement of engineering students using classification. arXiv preprint arXiv:1203.3832
5. Baradwaj BK, Pal S (2012) Mining educational data to analyze students' performance. arXiv preprint arXiv:1201.3417
6. Mirza S, Mittal S, Zaman M (2016) A review of data mining literature. Int J Comput Sci Inf Secur (IJCSIS) 14(11)
7. Cheng R, Kalashnikov DV, Prabhakar S (2003) Evaluating probabilistic queries over imprecise data. In: Proceedings of the 2003 ACM SIGMOD international conference on management of data, pp 551–562
8. Joseph SIT (2019) Survey of data mining algorithm's for intelligent computing system. J Trends Comput Sci Smart Technol (TCSST) 1(01):14–24
9. Ramírez-Gallego S, Krawczyk B, García S, Woźniak M, Herrera F (2017) A survey on data preprocessing for data stream mining: current status and future directions. Neurocomputing 239:39–57
10. Ragab AHM, Noaman AY, Al-Ghamdi, A. S., & Madbouly, A. (2014, June). A comparative analysis of classification algorithms for students college enrollment approval using data mining. In: Proceedings of the 2014 workshop on ınteraction design in educational environments, pp 106–113
11. Sumitha R, Vinothkumar ES, Scholar P (2016) Prediction of students outcome using data mining techniques. Int J Sci Eng Appl Sci (IJSEAS) 2(6)
12. Nagy HM, Aly WM, Hegazy OF (2013) An educational data mining system for advising higher education students. World Acad Sci Eng Technol Int J Inf Sci Eng 7(10):175–179
13. Mennis J, Guo D (2009) Spatial data mining and geographic knowledge discovery—an introduction. Comput Environ Urban Syst 33(6):403–408
14. Jackson D, Chapman E (2012) Non-technical skill gaps in Australian business graduates. Educ Train 54(2/3):95–113
15. Singh R, Pal S (2020) Application of machine learning algorithms to predict students performance. Int J Adv Sci Technol 29(05):7249-7261

16. Sunday K, Ocheja P, Hussain S, Oyelere SS, Samson BO, Agbo FJ (2020) Analyzing student performance in programming education using classification techniques. Int J Emerg Technol Learn (IJET) 15(02):127–144
17. Kunjumon LT, Shaji S, Saji ST, Naushad T, Joseph N (2019) An ıntelligent system to predict Students academic performance using data mining. Int J Inf 8(2)
18. Deng H, Wang X, Guo Z, Decker A, Duan X, Wang C, Ambrose A, Abbott K (2019) Performance Vis: visual analytics of student performance data from an introductory chemistry course. Vis Inf 3(4):166–176
19. Guerra MS, Neto HA, Oliveira SA (2018) A case study of applying the classification task for students' performance prediction. IEEE Latin Am Trans 16(1):172–177

Chapter 28
Exploration and Visualization of the Hidden Information from the Congestive Heart Failure Patients Data in MIMIC-III Database

S. Gayathri, M. Anitha, S. Nickolas, and S. Mary Saira Bhanu

1 Introduction

In any data analytic process, exploration of data is considered to be the primary step which helps in identifying the hidden patterns, characteristics of the data, summarization of the data and many more. The usage of computer systems to digitize and store all clinical examinations and medical records of patients as electronic health record (EHR) has grown enormously, and the data are freely available for healthcare data analysts and medical professionals [1]. Nowadays, there are needs for handling huge healthcare data in the perspective of big data [2]. In fact, some scholarly articles have projected that much attention is given only in designing frameworks for collecting and storing such data but not for exploring and visualizing it [3, 4]. One such big data repository is MIMIC-III data. This database consists of administrative, clinical and demographic details of 15,632 unique diagnoses with ~60,000 unique intensive care patients' admissions during the time period from 2001 to 2012 that includes both adult and neonatal data. The MIMIC-III database is available as 26 CSV files, and the same can be integrated through PostgreSQL. Irrespective of the availability of such huge repositories of medical data as EMR, the usage of it by various medical researchers who have lesser knowledge in programming is finding it to be difficult in

S. Gayathri (✉) · M. Anitha · S. Nickolas · S. Mary Saira Bhanu
National Institute of Technology Tiruchirappalli, Tiruchirappalli 620015, Tamil Nadu, India
e-mail: sgayathri@nitt.edu

M. Anitha
e-mail: anitham@nitt.edu

S. Nickolas
e-mail: nickolas@nitt.edu

S. Mary Saira Bhanu
e-mail: msb@nitt.edu

I. Jeena Jacob et al. (eds.), *Data Intelligence and Cognitive Informatics*,
Algorithms for Intelligent Systems,
https://doi.org/10.1007/978-981-15-8530-2_28

exploring the information which is hidden within it. Hence to help such community of people, exploration and visualization of various hidden information have been made by considering only the patients admitted under the diagnosis of congestive heart failure (CHF). Exploration and visualization are done using python packages like Matplotlib and Seaborn. As future work, developing an interactive tool is in the progress of in which exploration and visualization of any considered dataset along with the handling of missing values (imputation) and feature engineering will be possible.

The paper is structured as follows: Sect. 2 discusses some of the existing exploration and visualization tools, Sect. 3 discusses the method used for data exploration and visualization, and Sect. 4 discusses the result and discussions of the exploration, followed by a conclusion and future enhancement in Sects. 5 and 6, respectively.

2 Related Works

Many visualization tools are being proposed by researchers. In this section, some of them are discussed. Swapna Throve et al. have proposed EpiViewer, which is a web-based time series visualization tool, in which the user can load any epidemiological dataset in CSV format and visualize temporal data as line charts or bar charts [5]. EpiViewer also helps the user to perform any statistical analysis and visualize the results. Lee et al. proposed a web-based visualization tool that can be used to explore the MIMIC-II database which is designed using SQL. This tool has two primary functions, namely explore and compare. Various features of MIMIC-II dataset are first explored using the explore function, and the results are compared using the compare function [6]. Tushar et al. designed flusight, interactive open-source software for exploring, visualizing and forecasting model about the outbreak of influenza in the USA on weekly basis and indicates the geographic regions of the spread of the disease [7]. Orange is yet another open-source data exploration and visualization tool builds using a python programming language in Anaconda environment [8]. Tiziana Catarci et al. have proposed a framework for the exploration and visualization of haemodialysis data [9]. Despite the availability of many data exploration and visualization tools, our motivation is to design an effective exploration and visualization tool for epidemiological data along with the features of data imputation and feature engineering [10]. In this paper, our focus is on the exploration and visualization part of the proposal.

3 Method

3.1 Study Design

A retrospective study on patients registered in the emergency department with the chances of congestive heart failure in MIMIC-III v1.4 data [11] is performed. MIMIC-III consists of details of patients admitted in the critical care unit of Beth Israel Deaconess Medical Center (BIDMC) (Boston, MA, USA) during the period of 2001 and 2012. More than fifty thousand patients are admitted with 15,682 unique diagnoses and were treated in various critical care units. The details of the patients admitted are de-identified for hiding their personal identification. The permission to access this data is obtained after completing a basic course on 'Data or Specimens Only Research' conducted by Collaborative Institutional Training Program (CITI).

3.2 Study Population

Among 15,682 unique diagnoses, congestive heart failure which covers 1.6% of the admitted patients is considered for this retrospective study. There are 21 CSV files which record the patient's admission, demographics, prescriptions, microbiology events, ICU stay details during the period of stay on the hospital and five data dictionary files. Each file in MIMIC-III dataset is identified using identifier attribute which usually ends with suffix 'ID'. For this study, admission, patients, ICU stay, Diagnosis_ICD and prescription files are merged using 'HADM_ID' as the merging key in a python programming language. After merging the aforementioned files, the records of a patient diagnosed with CHF are filtered which is resulted in a total 1,674,447 tuples.

3.3 Study Variables

The demographic details like age, gender, religion, marital status and ethnicity of the patients based on 'Expiry_Flag' within the hospital who suffered from CHF are investigated. The length of stay in ICU, emergency department and in-hospital stay are analyzed. Time taken to get admitted to the hospital as an inpatient from the emergency department is identified. The drugs that are majorly used for the treatment are retrieved.

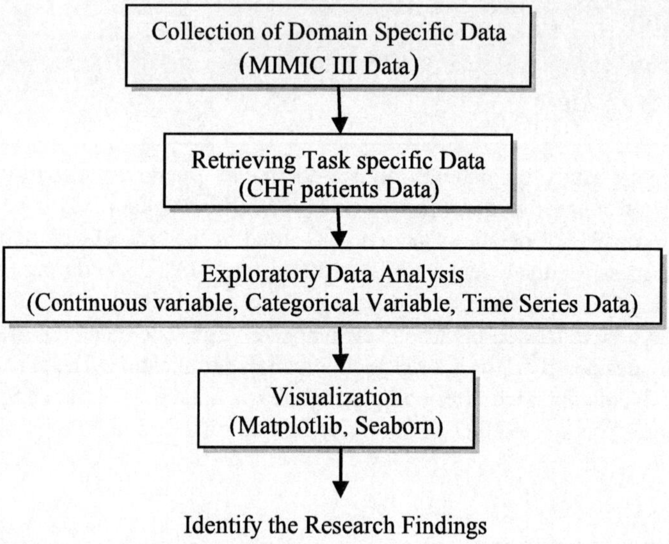

Fig. 1 Flow diagram of the research article

3.4 *Exploration and Visualization of Data*

Flow diagram of this research article is shown in Fig. 1. Initially, MIMIC-III dataset was downloaded using PostgreSQL and stored in local storage. The required CSV and data dictionary files are uploaded in the python environment. Then the exploration and visualization of the data were done using the python packages like Matplotlib and Seaborn. Matplotlib is a complete library for creating static, interactive and animated visualizations in python, whereas Seaborn constitutes a high-level interface for drawing informative and attractive statistical graphics [12, 13]. Statistical analysis was done using the NumPy, Pandas package of python. The results are represented in values of counts, mean, median, max and min for continuous variable and percentage for categorical variables.

4 Results and Discussions

4.1 *Patient's Demographic Data Analysis*

There were 810 unique patients admitted under the diagnosis of CHF. The demographic details of the patients were analyzed on the dataset in which CHF patient's admission, patient, ICU_Stay and Diagnosis_ICD files were merged. After merging, there were 15,332 rows and 41 columns in the dataset.

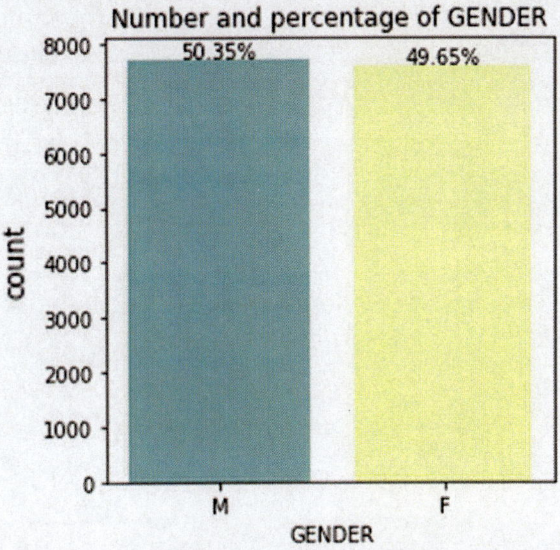

Fig. 2 Gender wise CHF affected patients

4.2 Gender

Among 15,332 CHF records, 50.35% was male patients and 49.65% was female patients with a count of 7719, 7613 respectively as shown in Fig. 2. The results obtained showed that 62.69% of the patients expired and 37.81% were not expired as shown in Fig. 3. The in-hospital mortality rate of male and female patients was obtained based on the attribute 'Expiry_Flag' which showed a count of 5108 (53.15%) and 4503 (46.85%) for male and female respectively as shown in Fig. 4.

4.3 Age

There is no separate age attribute in MIMIC-III dataset. The details of the patients are de-identified, and thereby, the 'DOB' attribute of patients above 89 age is added with 300. Hence for the purpose of analysis, the actual age is calculated and stored in 'age' attribute by finding the difference between the 'DOB' attribute and 'AdmitTime' attribute. Then the age is binned into various age bins according to the specification given by Medical Subject Headings (Mesh). The age bins created are young_adult (19–24), Adult (25–44), Middle_Aged (45–64), Aged (65–79) and Aged_Plus (>79). The range of age of patients admitted under CHF was 20–100 with a mean of 73. Accordingly, the count and percentage of patients diagnosed with CHF under various age groups are Aged_Plus (5886, 38.39%), Aged (5744, 37.46%), Middle_Aged (3338, 21.77%), Adult (327, 2.13%) and young_adult (37, 0.23%) as shown in Fig. 5. The count and percentage of patients died in each age category are Aged_Plus

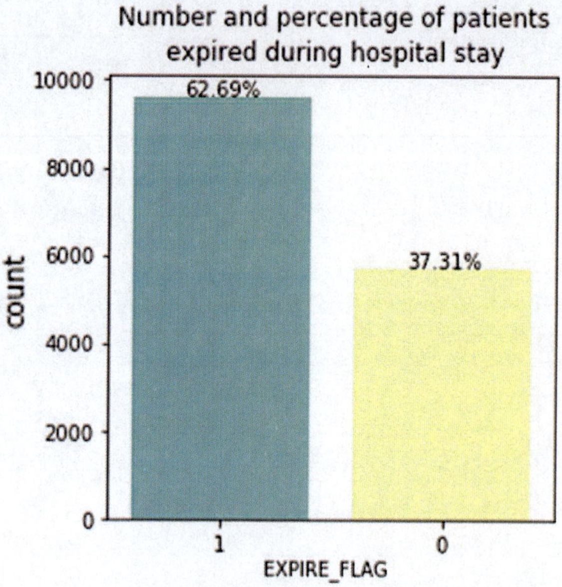

Fig. 3 Expiry flag of CHF patients

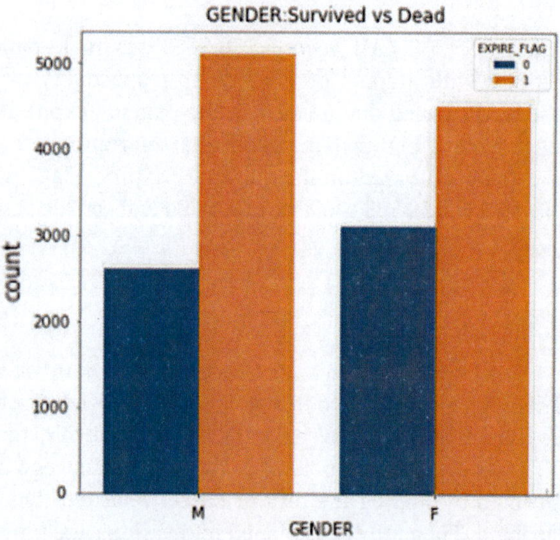

Fig. 4 Gender wise expiry flag of CHF affected patients

(4403, 45.81%), Aged (3711, 38.61%), Middle_Aged (1363, 14.18%) and Adult (134, 1.37%) as shown in Fig. 6.

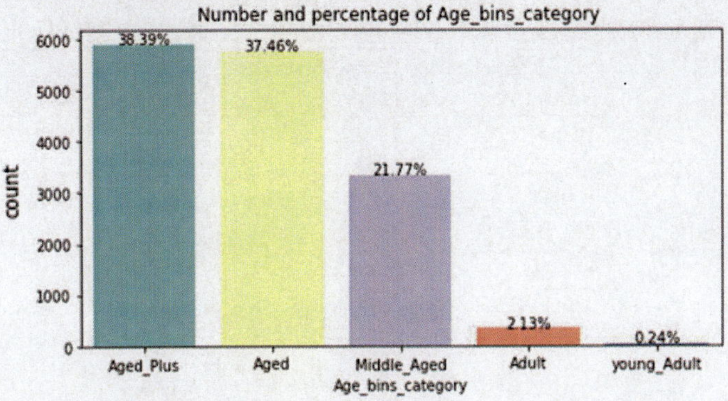

Fig. 5 Age category wise CHF patients

Fig. 6 Age category wise expiry flag

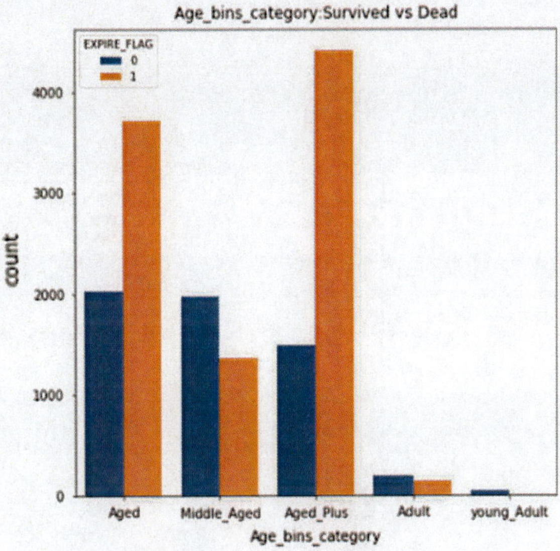

4.4 Marital Status

The marital status of the patients was registered under the attribute 'marital_status' in the various categories of married, widowed, single, divorced, separated, life partner and unknown. The percentage of patients admitted and expired due to CHF in various categories of marital_status are married (42.6%, 28.45%), widowed (26.3%, 19.07%), single (22.17%, 11.1%), divorced (5.2%, 2.5%), separated (1.7%, 0.8%), life partner (0.2%) and unknown (0.2, 0.2) as shown in Figs. 7 and 8.

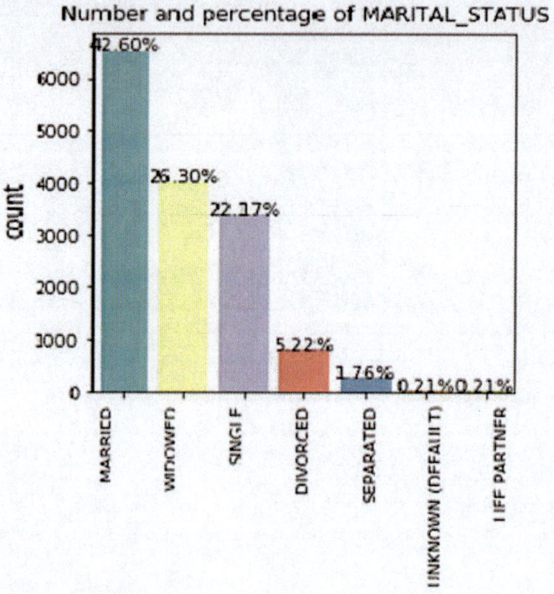

Fig. 7 Marital status wise % of CHF patients

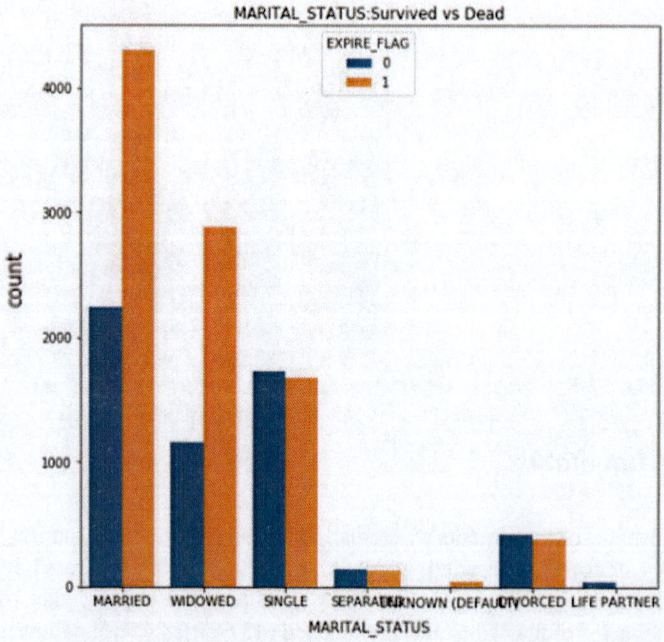

Fig. 8 Marital status wise expiry flag

4.5 Religion

The religion of the CHF patients is registered in the attribute 'religion' where they are categorized under 16 different religions and unknown are registered under others. The percentage of patients registered and expired under each category are Catholic (31.3, 19.4), Jewish (20.05, 14.8) and Protestant Quaker (19.07, 11.2), and rest are distributed under various other religions as shown in Figs. 9 and 10.

Fig. 9 Religion wise pie chart

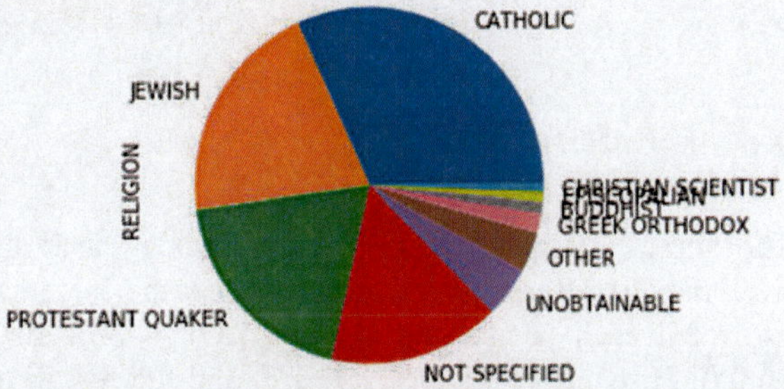

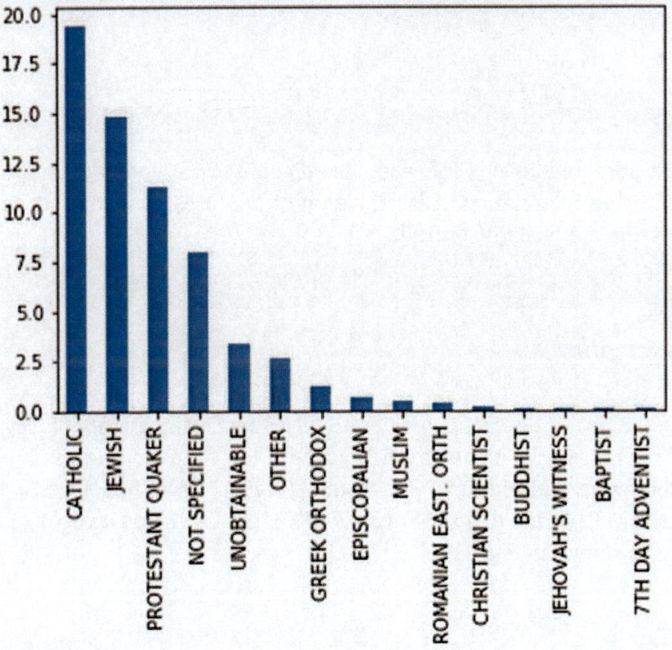

Fig. 10 Religion wise CHF patients

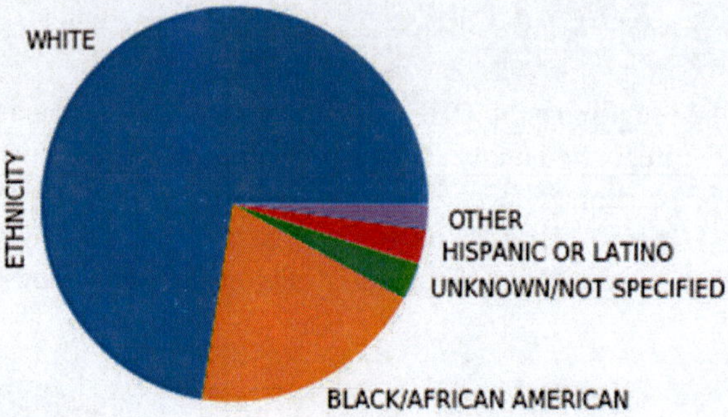

Fig. 11 Ethnicity wise pie chart

4.6 *Ethnicity*

The ethnicity of the patients is registered under the attribute 'ethnicity' and is categorized into 20 different types. Among those, the percentage of patients registered and expired in each category are White (67.06, 44.35), Black/African American (18.23, 10.05), Hispanic or Latino (2.7, 1.7) and the rest in other categories as shown in Figs. 11 and 12.

4.7 *Diagnosis_ICD*

The patients admitted under CHF were also diagnosed with other disorders and are represented using International Classification of Diseases, revision 9 codes, called ICD9 codes and are listed in Table 1.

4.8 *Prescriptions*

To analyze the drug prescriptions given to CHF patients, prescription.csv file was merged. Percentage of patients are treated with the following drugs: Furosemide (7.5%), Potassium Chloride (5.7%), Insulin (4.9%), D5W (4.6%), 5% Dextrose (2.3%), Sodium Chloride (0.9%), Flush (2.03%), Iso-Osmotic Dextrose (1.9%) and many more as shown in Fig. 13.

4.9 *Exploration of Time Series Data*

Exploring time series data plays a vital role in identifying some hidden information that is not directly provided in the dataset. In the MIMIC-III dataset, the age of the

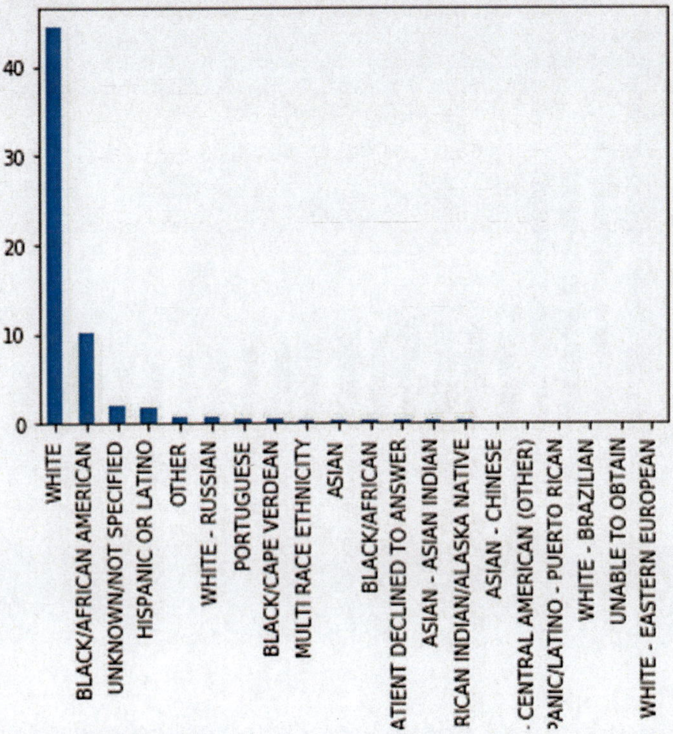

Fig. 12 Ethnicity wise CHF patients

Table 1 CHF patients affected with other IC9 code cohorts

ICD9_Code	Long titles of the disease	Percentage (%)
4280	Congestive heart failure, unspecified	5.9
42,731	Atrial fibrillation	3.2
5849	Acute kidney failure, unspecified	2.89
41,401	Coronary atherosclerosis of native coronary artery	2.5
4019	Unspecified essential hypertension	2.2
25,000	Diabetes mellitus without mention of complication, type II or unspecified type, not stated as uncontrolled	1.9
51,881	Acute respiratory failure	1.8
2724	Other and unspecified hyperlipidemia	1.5
42,833	Acute on chronic diastolic heart failure	1.48
5859	Chronic kidney disease, unspecified	1.34

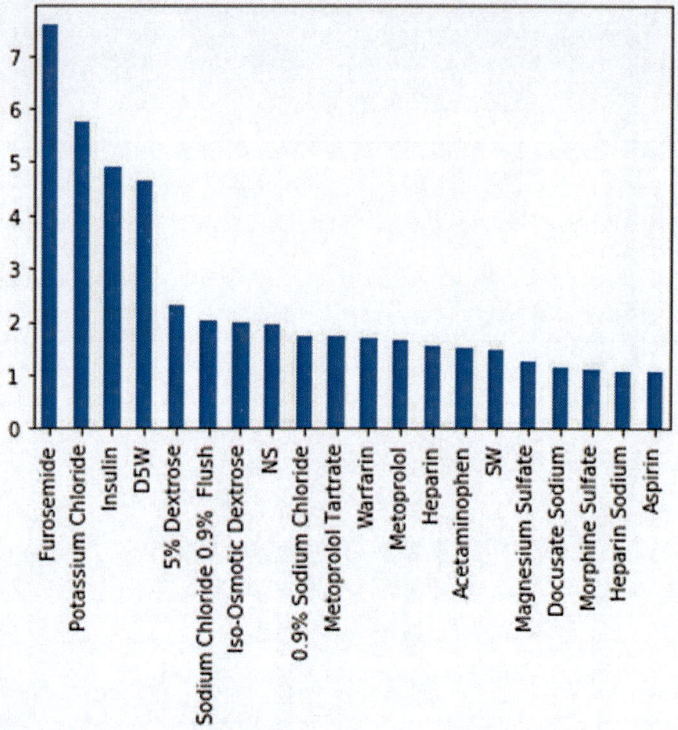

Fig. 13 Drug prescriptions for CHF patients

patient is derived by finding the difference between the Date of Birth attribute and the AdmitTime attribute. Similarly, predicting the length of stay (LoS) in different wards like emergency department, ICU, a total duration of stay in the hospital and the waiting time of the patient for hospital admission after arriving in the emergency department is very important in planning for the number of beds, drugs and various other resources needed during an emergency admission [14].

In this regard, Table 2 depicts the maximum, minimum and average length of stay in the ICU department of a congestive heart failure patients is 61, 0 and 7.3 days,

Table 2 LoS of CHF patients under various departments

Department	Length of stay (Los) in days		
	Maximum	Minimum	Median
ICU	61	0	7.3
Emergency	2	0	0.5
Total in-hospital stay	63	0	19
Waiting time between emergency dept and admission	1	0	0

respectively. Similarly in an emergency department, it is 2, 0 and less than 0.5 days, respectively. The total length of stay (LoS) in the hospital is 63, 0 and 19 days, respectively. The waiting time of the patient affected in congestive heart failure in between emergency department and admission is 1, 0 and 0 days, respectively.

5 Conclusion

There were 810 unique patient admissions with a total of 926 records including the readmissions. Through an effective exploration in the considered dataset, the new attributes like age, Age_category_bins, length of stay (LoS) in various ICUs from the existing attributes are identified. Among those, 50.35% were male and 49.65% were female. The in-hospital mortality percentage was 62.69% in which 33.33% was male and 29.36% was female. Mean age of the patient admitted with CHF was 73 and with minimum and maximum age of 20 and 100, respectively. Based on MeSH age group, there were 38.39% (Aged_Plus), 37.46% (Aged), 21.77% (Middle_Aged), 2.13% (Adult) and 0.24% (young_adult) admitted patients. 97.3% of patients were admitted in the emergency category in the admission_type. Among the various drugs, 7.56% Furosemide, 5.76% Potassium Chloride, 4.89% Insulin, 4.63% D5W, 2.3% 5%dextrose were supplied. Mean length of stay (LoS) was 19 days, 7 days and 0 days in hospital, ICU and emergency department (ED), respectively. Mean waiting time taken for admission from the ED was 1 h and 75 min. Such consolidated results can be obtained only through an efficient data exploration and visualization tool, and it shows the importance of developing such interactive tools.

6 Future Work

The process of developing the interactive exploration and visualization tool given any medical dataset is under progress using the python programming language and Dash web framework. As the additional work, the features for handling the missing values (imputation) in the dataset and generating the new needed features from the existing attributes in the dataset for effective analysis will also be added.

Acknowledgements This data exploration and visualization are made for our project 'Data Curation and Feature Engineering in Epidemiology Data analysis', supported by DST-ICPS, and would like to thank the authorities of a MIMIC repository of datasets.

References

1. Dash S, Shakyawar SK, Sharma M et al (2019) Big data in healthcare: management, analysis and future prospects. Big Data 6:54. https://doi.org/10.1186/s40537-019-0217-0
2. Smys S (2019) Big data business analytics as a strategic asset for health care industry. J ISMAC 1(02):92–100
3. Brinck T, York G (1998) User interfaces for computer-based patient records. In: Proceedings of the international ACM conference on human factors in computing systems (CHI'98), p 214
4. Plaisant C, Mushlin R, Snyder A, Heller D, Shneiderman B (1998) Lifelines: using visualization to enhance navigation and analysis of patient records. In: American Medical Informatic Association Annual Fall Symposium, Orlando, USA, pp 76–80
5. Thorve S, Wilson ML, Lewis BL, Swarup S, Vullikanti AKS, Marathe MV (2018) EpiViewer: an epidemiological application for exploring time series data. BMC Bioinformatics
6. Lee J, Ribey E, Wallace JR (2016) A web-based data visualization tool for the MIMIC-II database. BMC Bioinf Decis Making
7. Tushar A, Reich NG (2017) Flusight: interactive visualizations for infectious disease forecasts. J Open sources Softw
8. https://orange.biolab.si/
9. Catarci T, Santucci G, Silva SF (2003) An interactive visual exploration of medical data for evaluating health centres. J Res Pract Inf Technol
10. Mahadevan A, Arock M (2020) Review rating prediction using combined latent topics and associated sentiments: an empirical review. Serv Orient Comput Appl 14(1):19–34
11. Johnson A, Pollard T, Shen L (2016) MIMIC-III, a freely accessible critical care database. Sci Data 3:160035. https://doi.org/10.1038/sdata.2016.35
12. https://matplotlib.org/tutorials/introductory/pyplot.html
13. https://seaborn.pydata.org/
14. Baek H, Cho M, Kim S, Hwang H, Song M, Yoo S (2018) Analysis of length of hospital stay using electronic health records: a statistical and data mining approach. PLoS ONE 13(4):e0195901. https://doi.org/10.1371/journal.pone.0195901

Chapter 29
Neuromemorize—Mobile Application for Alzheimer's Patients Using AI-Based Face Recognition

Aziz Presswala, Srushti Pathak, Gunjan Jhanwar, and Neha Katre

1 Introduction

Alzheimer's disease is an irreversible disease of the brain that affects a person's memory, thinking and other abilities. It is progressive, meaning symptoms get worse, and more functions are lost the longer an individual has Alzheimer's. As the disease progresses, Alzheimer's symptoms become even more prominent and may include problems with processing new information, trouble concentrating, difficulty recalling the names of objects and important dates, recognizing their loved ones, losing things and being unable to locate them, poor judgment, detachment from social activities and change of mood/personality [1].

Worldwide, at least 50 million people are believed to be living with Alzheimer's disease or some form of dementia. In India, more than 4 million people have some form of dementia. By 2050, nearly 131.5 million people worldwide could be living with the disease, unless scientists develop new approaches to prevent or cure it [2]. As mentioned earlier, Alzheimer's is a medical condition, with symptoms gradually worsening over a number of years. In its early stages, memory loss is mild, but individuals lose the ability to carry on a conversation and adapt to their environment with late stage Alzheimer's. Over the years, research and advisory initiatives from

A. Presswala (✉) · S. Pathak · G. Jhanwar · N. Katre
Dwarkadas J. Sanghvi College of Engineering, Mumbai, India
e-mail: apresswala52@gmail.com

S. Pathak
e-mail: srushtipathak16@gmail.com

G. Jhanwar
e-mail: gunjan.jhawar21@gmail.com

N. Katre
e-mail: neha.mendjoge@djsce.ac.in

I. Jeena Jacob et al. (eds.), *Data Intelligence and Cognitive Informatics*,
Algorithms for Intelligent Systems,
https://doi.org/10.1007/978-981-15-8530-2_29

around the world have created a new market for software developers and professionals in the healthcare sector. Research has shown that technology instruments such as iPods help stimulate those with dementia. Innovative devices such as iPads and tablets, which are mainstream and easy to use, can not only help determine the stage of dementia but also provide stimulation to improve cognitive functioning [3]. More use of technology by older adults has also brought about a new age. Therefore, technology can be used for both people suffering from Alzheimer's and their caretakers who care for them [4].

Therefore, authors propose an android application that uses artificial intelligence-based face recognition that helps the patients to identify their loved ones, live location tracking that keeps the caretaker informed about the patient's location at any moment, timely reminders that remind the patient of their day-to-day activities and photo sharing which helps patients refresh their old memories, thereby relieving fear, developing rhythm and easily connect with their family and friends.

2 Literature Survey

This section briefly discusses the technologies that are available for people with Alzheimer's and the existing applications that are available, their advantages and drawbacks. The key objective of the research is to provide Alzheimer's patients with an application which can assist them in doing their daily tasks with a little help from the caretaker. Artificial intelligence can help analyze the patterns of how technology can enhance the improvement in the conditions of the patient using music therapy and sleeping patterns [5].

2.1 Technologies for People with Alzheimer's

Alzheimer's can be frightening and overwhelming—for both caretakers and people suffering from the disease. However, new technologies can help ease anxiety, establish routine and improve the quality of life of the patients [6]. Here is a list of some of the proven technologies for Alzheimer's patients and their caretakers:

Photo Albums. Photo albums hold pictures of loved ones, each with a short description of the photo along with the date on which the photo was captured. This can be a great way to remember loved ones or fun times in the past and help patients to ease anxiety and agitation [7].

Reminders. Reminders can help keep assets and loved ones safe when the caregiver cannot. Reminders can be recorded/set on the computer and played back loud at the appropriate time. For example, a caregiver may record a video message that advises a person to take a prescription at the right time [7].

Location Tracking. As the disease progresses, wandering of the patient becomes a problematic behavior, which can occur due to a variety of reasons. Location tracking is the answer to this problem. They include GPS devices, so the patient can easily be located and avoids getting lost. These devices can also be programmed to alert caretakers if the patient goes outside a specific area.

Picture-based Phonebook. They are reprogrammed with important phone numbers so that the patient need not remember them or have them written down. It features actual pictures of loved ones along with large, easy-to-use buttons corresponding to each person [7].

Music. Music helps to heal, so any system designed to play favorite songs for someone with Alzheimer's can be genuinely beneficial. After 20 min of listening to music, people with Alzheimer's in one study saw an immediate, measurable increase in happiness, eye contact and talkativeness and a decrease in fatigue [8].

Face Recognition. As discussed earlier, Alzheimer's patients tend to forget their family, friends and loved ones. Therefore, artificial intelligence-based face recognition can be used to help them to identify and recall their loved ones [9].

2.2 *Mobile Applications for Alzheimer's*

Currently, very few mobile applications are available for Alzheimer's patients and their caretakers. They have made it easier for patients, particularly in earlier stages, to communicate and pass the time. Some of the applications that are surveyed are listed below:

MindMate. The Mindmate application has daily activities and games designed to improve brain health and can be customized for your loved one depending on the level of thinking impairment. The app has mental exercises and even physical exercises and tracks progress. It includes nutrition advice and clips of classic movies, music and TV from decades past [10].

It's Done. It's Done is an application that provides a checklist for critical day-to-day tasks such as locking doors, feeding pets, taking medication and turning off the stove. The It's Done application is suitable for those with short-term memory loss due to brain disorder, Alzheimer's or other medical conditions [10].

CareZone. CareZone is a helpful application for managing all medical care, useful for someone in the early stages of dementia and caregivers during the later stages. The app easily organizes lists of medications, sets reminders of when to take them, orders refills and keeps track of doctor appointments [10].

Lumosity Mobile. The Lumosity application features a combination of more than 25 cognitive games aimed at "exercising" the brain. The games test memory and attention by getting the user interested in specific cognitive and neuropsychological activities [10].

Timeless. Timeless is a simple, easy-to-use application for Alzheimer's patients to recall things, stay connected and engaged with friends and family and recognize people using face recognition technology based on artificial intelligence [11].

2.3 Observations on Existing Systems

After an exhaustive survey of existing systems, it is observed that MindMate and Lumosity Mobile applications aim to reduce anxiety and agitation in Alzheimer's patients through interactive memory games. The application, It's Done, provides reminders for day-to-day tasks, whereas CareZone helps caretakers to manage medications and appointments of patients. Timeless is an application based on face recognition that helps the patient to recognize their loved ones [10]. A summary of all the applications is shown in Table 1.

Therefore, it is observed that there is an absence of a single application that provides all the required features and functionalities for Alzheimer's patients as well as their caretakers.

3 System Architecture and Implementation

The system architecture is depicted in Fig. 1. The application is based on a three-tier architecture. The first layer consists of the users of the application, and they are caretaker, patient and family/friends. The second layer consists of all the major modules of the application, and they are discussed in detail below. The third layer consists of the database where the user data is stored. The database used in the

Table 1 Summary of mobile applications

Application	Features	Drawbacks
Timeless	Face recognition, reminders and photo sharing	For iOS users only, no location tracking
MindMate	Brain games, classic movies and music	Lacks in other necessary features
CareZone	Lists of medicines, reminders and tracking of doctor appointments	Only useful in the early stages of the disease
It's Done	Checklist for day-to-day tasks	Only provides reminders
Lumosity Mobile	Cognitive and brain games	Paid, lacks in providing other essential features

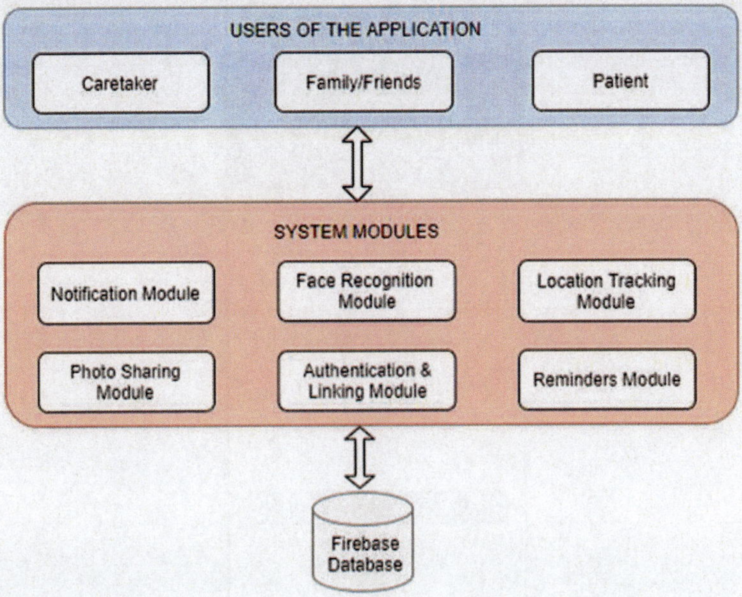

Fig. 1 System architecture

application is Firebase Realtime Database for storage of user data and Firebase Storage for storing media such as images and videos.

The system consists of seven major modules which are:

- Authentication and linking module,
- Photo sharing module,
- Notification module,
- Location tracking module,
- Reminder module,
- Picture-based phonebook,
- Face recognition module.

Let us discuss each of these modules in brief.

3.1 Authentication and Linking Module

It is responsible for managing the registration, signing-in and linking of the user accounts. Registration/sign-in in the application is managed via Firebase Authentication as shown in Figs. 2, 3 and 4. The caretaker and patient accounts are linked to each other so that the caretaker can manage and monitor the activities of the patient. Linking of accounts is done by scanning the QR code generated at the patient's account by his/her caretaker.

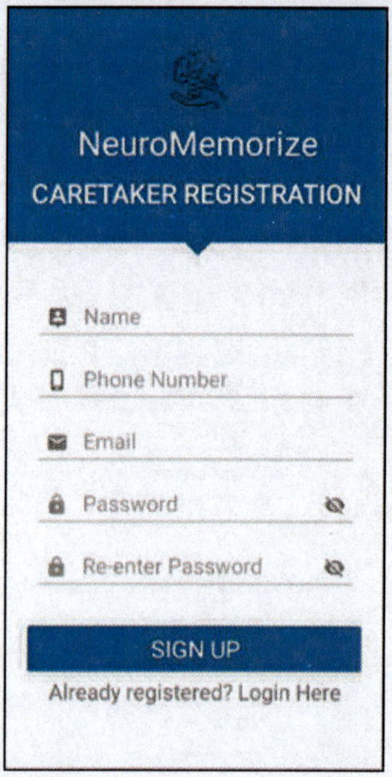

Fig. 2 Caretaker registration

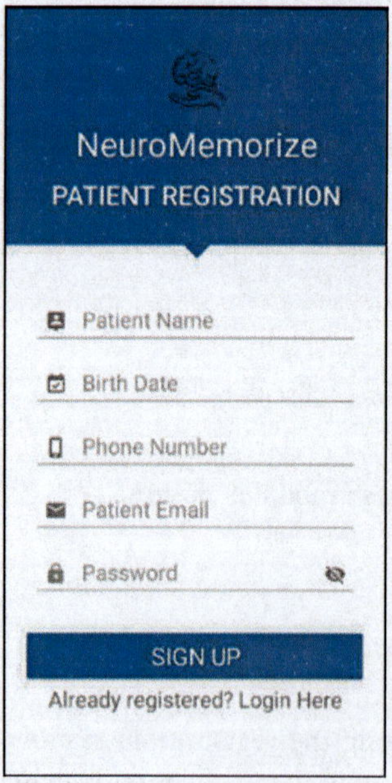

Fig. 3 Patient registration

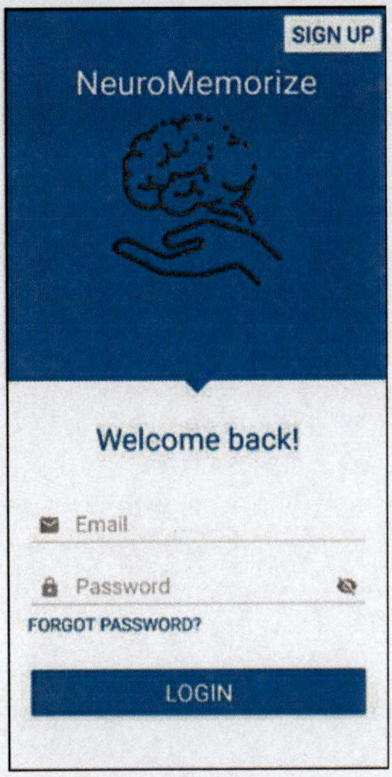

Fig. 4 Sign-in page

3.2 Photo Sharing Module

This module enables sharing of photos with the patient. The caretaker can share memories with the patient by uploading a photo along with a short description and the date on which the photo was taken, as shown in Fig. 6. All these photos are stored in Firebase Storage and will be displayed to the patient in the form of a scrollable list as shown in Fig. 5.

3.3 Location Tracking Module

This module tracks the live location of the patient and is displayed to the caretaker on a map as shown in Fig. 7. It also notifies the caretaker if the patient's location is outside of a safe zone. This feature is implemented by using Google Maps API that helps in displaying the map on the user screen and Google Play Services Location that tracks the live location of the patient.

Fig. 5 List of memories

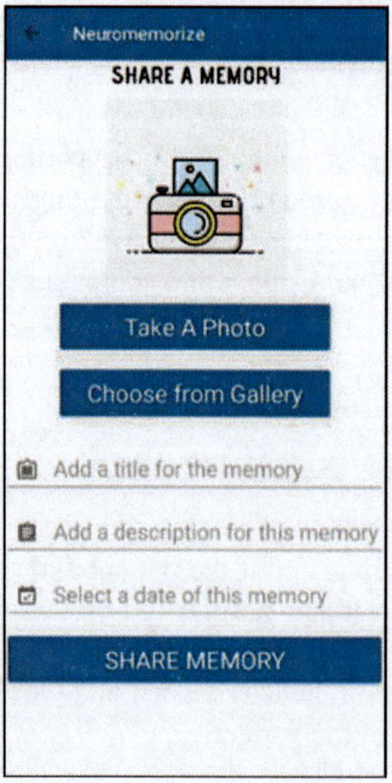

Fig. 6 Add memory page

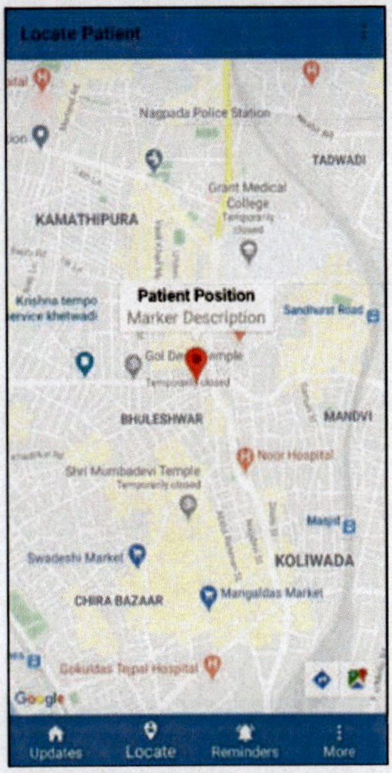

Fig. 7 Map view of patient's location

3.4 Reminders Module

This module is responsible for displaying and notifying the patient of the daily reminders set by the caretaker. Reminders may include medicines to be taken, appointments, meetings to be attended, important events, etc. The setting and receiving reminders are shown in Fig. 8.

3.5 Picture-Based Phonebook

It enlists the contact numbers of all the important people along with their pictures and large buttons to call or message them instantly.

The caretaker can directly call or text any of the contacts which are displayed to him in the form of pictures as shown in Fig. 9.

3.6 Face Recognition Module

As discussed before, Alzheimer's patients tend to forget their loved ones. In such a scenario, the patient can make use of the face recognition functionality, they just

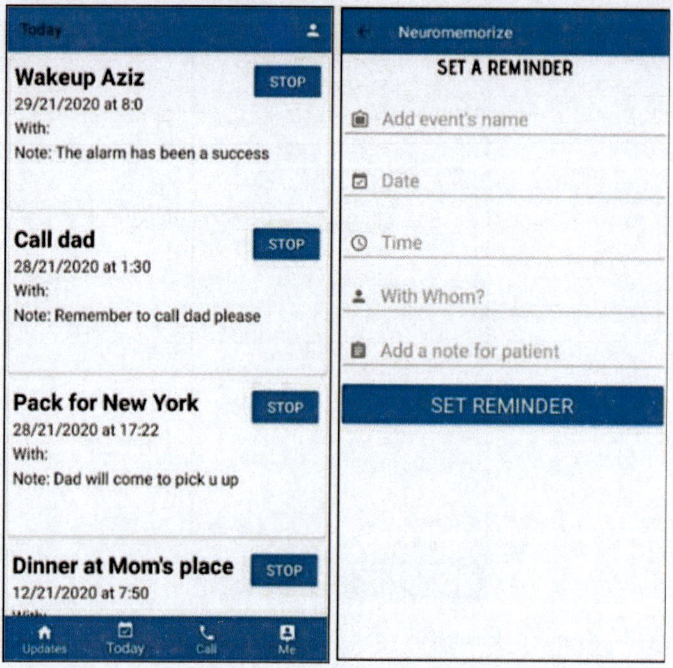

Fig. 8 Setting and receiving reminders

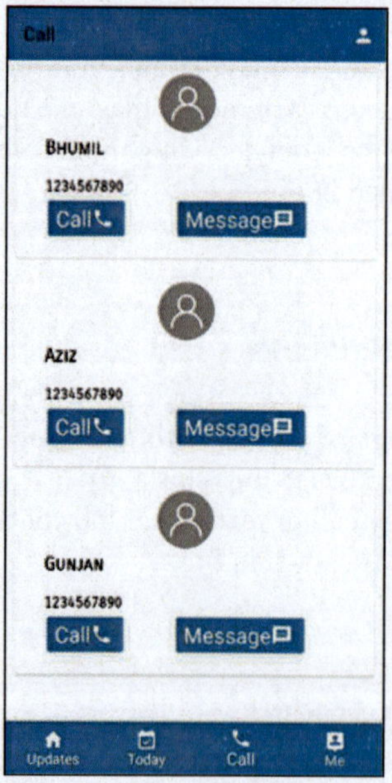

Fig. 9 Picture-based contacts page

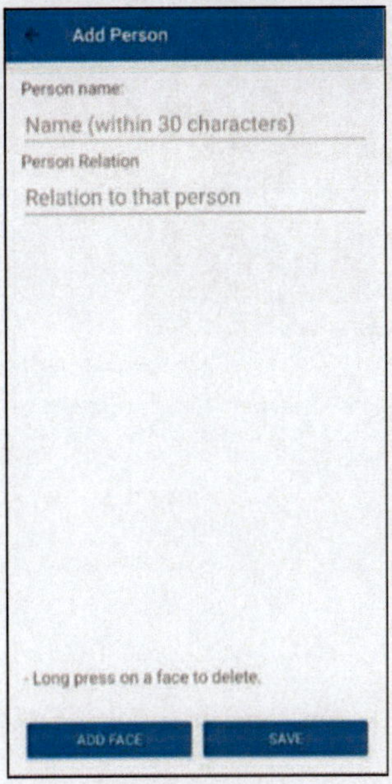

Fig. 10 Add person page

have to click the image of the person they want to identify, and the face recognition system will tell them who they are and how are they related to the patient. Microsoft Azure Face API has been integrated with the application which incorporates state-of-the-art cloud-based face algorithms to detect and recognize human faces in images [12]. Figures 7 and 8 show the working of the face recognition module in this system. This system can also recognize multiple faces in an image.

Now, let us understand briefly how the face recognition system works in the application. Initially, the caretaker has to add images of different people along with their names and their relation to the patient. These images will be used for training the face recognition system. Once the training is complete, the system is ready for identification. Now when an image is supplied to the face recognition system, it will compare the image with already trained images, and if a match occurs, the system will return the details of the identified person (Figs. 10, 11 and 12).

4 Results

To test the usability of the application and ensure its proper working, the application is shared with 10 Alzheimer's patients and their caretakers for a period of 1 month. After a month, feedback is taken from the caretakers of the corresponding patients which are depicted in the form of graphs below.

Fig. 11 List of persons

Fig. 12 Identified image

The graph shown in Fig. 13 can be observed that there was some problem as far as adaptability to the application is concerned. From Fig. 14, it is clear that around 80% of patients were satisfied by the face recognition feature. As in Fig. 15, location tracking feature was most satisfactory since it made tracking of patient easy. Overall, it can be seen that application implemented had satisfactory results.

Fig. 13 Overall adaptability of the application by the patient

Fig. 14 Usability and effectiveness of the face recognition feature

Fig. 15 Usability and effectiveness of the location tracking feature

5 Conclusion and Future Scope

Few applications are available in the market for Alzheimer's patients and their caretakers. In this paper, the authors briefly discussed the existing systems their features, advantages and drawbacks and proposed an android application that caters to the needs of Alzheimer's patients and their caretakers. The proposed applications make use of proven technologies such as artificial intelligence-based face recognition, GPS-based live location tracking, timely reminders, photo sharing and photo-based phonebook with a simplified user interface.

Hence, this system will help Alzheimer's patients to remember events, stay connected and engaged with friends and family and recognize people through artificial intelligence-based face recognition technology, thereby reducing their anxiety and agitation levels and ease the work of their caretaker.

The application is still in early beta stage and aims to be added more features in the near future. Constant efforts are being made to enhance the overall user experience and performance of the application so that it can be used effortlessly by the patients as well as the caretakers. Some of the features that aim to be added are voice-based reminders, wherein the patient will be notified of a particular reminder by a built-in assistant similar to that of Google Assistant. A preliminary quiz can be taken by the patient to know the severity of the disease. An Activity Tracker can be implemented which would track the daily activities of the patient and notify the caretaker on a timely basis.

References

1. National Institute on Aging. Available: https://www.nia.nih.gov/health/what-alzheimers-disease
2. Alzheimer's disease, Facts & Figures. Available: www.brightfocus.org/alzheimers/article/alzheimers-disease-facts-figures
3. Yamagata C, Kowtko M, Coppola JF, Joyce S (2013) Mobile app development and usability research to help Alzheimer's and Alzheimer patients. IEEE
4. Treatments. Available: www.alz.org/alzheimers-Alzheimer's/treatments
5. Pandian MD (2019) Sleep pattern analysis and improvement using artificial intelligence and music therapy. J Artif Intell 1(02):54–62
6. Technological innovations for those with dementia. Available: https://www.alzheimers.net/9-22-14-technology-for-dementia/
7. Best ways to stimulate memories through photos. Available: https://www.alzheimers.net/10-6-14-memories-photos/
8. Adoption and use of a mobile health application in older adults for cognitive stimulation. Research and Development Department, Paris (2016)
9. Flicker C, Ferris SH, Crook T, Bartus RT (1990) Impaired facial recognition memory in aging and dementia. Alzheimer Dis Assoc Disord. 4(1):43–54. https://doi.org/10.1097/00002093-199040100-00005

10. Great apps for people with Alzheimer's or Alzheimer's. Available: seniordirectory.com/articles/info/9-great-apps-for-people-with-Alzheimer's-or-alzheimers
11. Timeless. Available: https://www.timeless.care/
12. How Microsoft Azure simplifies face recognition. Available: https://community.nasscom.in/uncategorized/how-microsoft-azure-simplifies-face-recognition.html

Chapter 30
Brain Tumor Classification Using PCA and PNN of T1 and T2 Weighted MRI Images

Bhakti Gangan, Manjusha Deshmukh, and Dhananjay Borse

1 Introduction

Image processing has found various applications in real-world applications which pattern recognition, color identification, edge detection, character recognition, forensic analysis, material inspection, fault analysis and many more [1, 2]. Recently, image processing applications have evolved to assist—clinical experts, doctors and technicians in the field of medical research for critical decision making and analysis purposes as most of the images today are available in digital format [3, 4]. With the help of the information technology sector, which has witnessed massive exploration of artificial intelligence for data interpretation and decision making, image processing has penetrated deep into our lives. Healthcare workers are now using image processing to provide better and fast test facilities for various medical-related issues. One such critical problem of brain tumor classification which usually involves the opinion of a dedicated expert in the field is taken up to be assisted with the use of advanced image procession techniques in this research. As the number of patients increases with symptoms of brain tumor, it has been increasingly difficult to get quality diagnostic facilities available for everyone. Accurate image processing-based classification seems to be a better alternative for assistance to experts than to only rely on manual classification which may fail from time to time.

Abnormal and uncontrolled growth of cancerous tissue in the brain is deemed as brain tumor and is usually classified as benign (non-harmful) and malignant (harmful or infected). Brain tissue can be seen using MR images which is now a widely

B. Gangan · M. Deshmukh
Saraswati College of Engineering, Kharghar, Navi Mumbai, India

D. Borse (✉)
Radium Engineering, Pune, India
e-mail: borse25@gmail.com

I. Jeena Jacob et al. (eds.), *Data Intelligence and Cognitive Informatics*,
Algorithms for Intelligent Systems,
https://doi.org/10.1007/978-981-15-8530-2_30

accepted method for its visualization [5]. A benign type of brain tumor has uniformity in the image while a malignant type of brain tumor has non-uniformity in its tissue structure indicating abnormality and can be concentrated in one part of the brain or may have spread in its different regions. Brain tumor may appear in an individual at any age and is irrespective of gender; however, its impact on regular body functions of the individual may not be the same. Moreover, the brain is probably the most complex structure present in our body which certainly makes it difficult to accurately diagnose for any abnormality relying only on experts' vision. To avoid the eventuality, timely detection, information related to tumor location and classification at earlier stages is very important [6, 7].

In this research work, gray-level co-occurrence matrix (GLCM) feature extraction using discrete wavelet transform (DWT) along with probabilistic neural network (PNN) classifier is used to segment brain tumor between benign and malignant type. MATLAB-based GUI is developed for easy deployment and to provide quick classification. This research paper is organized as follows: Sect. 1 introduction is followed by a literature survey and related research in Sect. 2, Section 3 presents the proposed methodology used for this research work while Sect. 4 primarily focuses on the implementation of the proposed scheme and construction of GUI using MATLAB, and Section 5 carries out discussion on observations and results and performance of the proposed scheme for a case study carried out as a part of this research work to test the developed algorithm while Sect. 6 concludes the research work with points out guidelines for future work.

2 Literature Survey

Processing of MR images and analyzing them for various abnormalities and possible disease detection have been of interest for many researchers. Brain tumor classification using various image processing methods and machine learning and artificial intelligence methods has seen tremendous growth in recent years.

Popularly, the methods used to classify brain tumor range in categories and shown in Fig. 1. Each method has its unique feature of implementation and relies on complex mathematical algorithms for the computation of results and analysis.

K-means clustering algorithm [8] based segmentation of MRI brain images with morphological filtering and localized fuzzy clustering [9] for image segmentation using spatial information extraction has an accuracy of 83%.

Support vector machine is a part of a machine learning tool which is widely used for effective classification of segmented images by many researchers with non-linear data. Berkeley wavelet transform (BWT) [10], Fourier transform (FT) [11], contrast enhancement [12] and PCA based on radial basis function [13] have varying accuracies in classification. All these algorithms are able to achieve accuracies above 80%.

Sharma et al. [14] offered a reliable solution to classify brain tumor with 100% accuracy by making use of an artificial neural network tool by using textual primitive

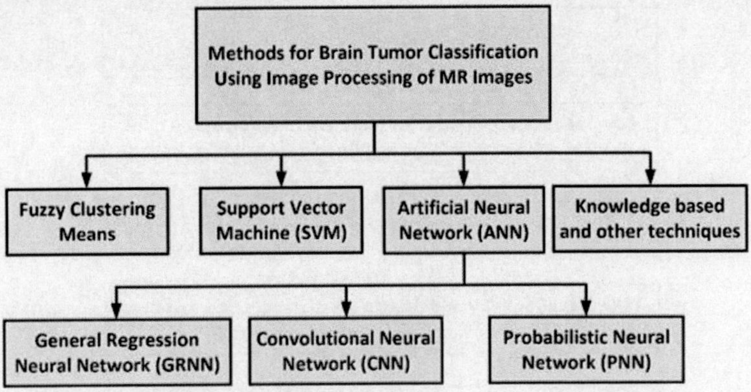

Fig. 1 Classification methods for brain tumor

features while [15] further extended this research. ANN is further classified to GRNN, CNN and PNN depending on the underlying training method. GRNN was developed by [16] for classification of brain tumor which like other neural networks (NNs) need very few training samples to work with. Additional knowledge required for GRNN is very small, but the accuracy depends on training dataset; hence, it is lesser as compared to other NNs. CNN utilizes feed-forward system unlike GRNN and has definite advantages but lacks adaptability [17, 18]. PNN, on the other hand, has advantages as its training speed is many times faster than a GRNN and CNN network and is easy and instantaneous. Bayes optimal result can easily be met, and it is robust of noise in the images. Moreover, speed of PNN is also very fast with high accuracy [19, 20].

A full survey of brain tumor detection techniques is carried out by [21]. Varuna and Kumar [22] propose a GLCM using DWT technique for image segmentation and feature extraction and [23] proposes PCA and PNN method for classification which is used in this research work.

3 Proposed Methodology

The proposed method involves segmentation of the test image to extract features from it and training the PNN to classify the extracted features into the type of brain tumor. The steps involved in doing so are shown in Fig. 2 in the form of a flowchart. The input image first goes through a pre-processing process in which it is converted to gray scale for its enhancement. After this step, the image goes through segmentation wherein its signal-to-noise ratio and appearance are improved and background noise and undesired regions are omitted from it. Segmentation keeps the overall aspect ratio of the image same and connected to its logical sense whereas taking only the

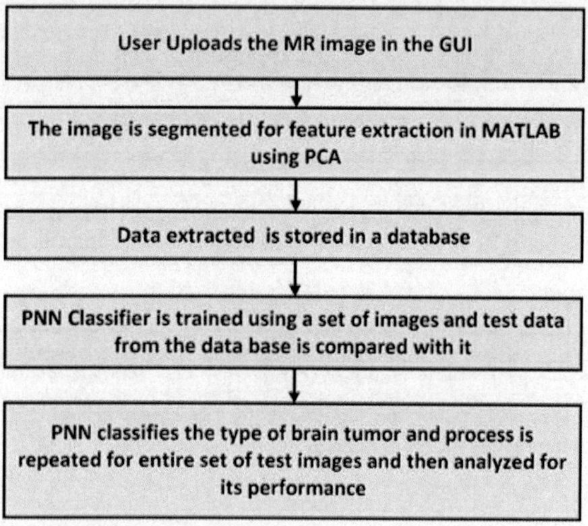

Fig. 2 Flowchart for the proposed scheme

irrelevant parts out of the context for further processes. Later operation is divided into two main categories: feature extraction and tumor classification.

3.1 Feature Extraction Using PCA with DWT and GLCM

The features extracted from an image include its texture, color, shape and contrast. In this process, DWT aids with the extraction of wavelet coefficients while GLCM helps with generating statistics of those features.

DWT is a powerful tool to analyze different frequencies at different scales within an image. It localizes frequency information to extracts coefficients of signal function. This information is important for classification. This approach involves the use of 2D wavelet transform to form four subbands that are LL (low–low), HL (high–low), LH (low–high) and HH (high–high), thus decomposing the image in area or region of interest (ROI), and approximates the image in its low- and high-frequency contents, respectively. The approximations used wavelets are represented by LL1 and LL2 which form the first and second level of low-level part of the images respectively while LH1, HL1, HH1, LH2, HL2 and HH2 formulate the high-level part of the image giving information about horizontal, vertical and diagonal axis of the image at first and second level, respectively. The decomposition process is repeated until a resolution is achieved up to the desired level. The expression for DWT is given as:

$$\text{DWT } p(s) = \begin{cases} di.j = \sum p(s)h * i(s - 2ij) \\ di.j = \sum p(s)g * i(s - 2ij) \end{cases} \tag{1}$$

where

$p(s)$	signal as a function,
coefficients $d_{i,j}$	component attribute in signal $p(s)$ corresponding to the wavelet function,
$h(s)$ and $g(s)$	high-pass and low-pass filters coefficients and
i and j	wavelet scale and translation factors, respectively.

It is a texture of the image which is used for differentiating between normal tissue and a defective or abnormal tissue according to our vision and for the machine learning process to understand it. Machine learning may understand finer details in an image which human eye can fail to notice, thereby increasing chances of accurate detection at earlier stages which prove effective in combating brain tumor.

3.2 *Classification Using PNN*

PNN is a feed-forward neural network and was derived from kernel Fisher discriminant analysis. It composes of four nodes which are also called as its layers such as an input layer, hidden layer, pattern layer and output layer and formulates neural network of weighted neighbors. The ‘P’ number of neurons in the input layer depends on features extracted from GLCM, and their weights are kept 1 to feed into the hidden layer. RBF is calculated in pattern layer after which summation layer adds the values of each class from the hidden layer to be fed to the output layer which chooses the class with the highest probability: 1 means positive for that class type and 0 means negative for that class type. The general depiction of PNN structure is shown in Fig. 3.

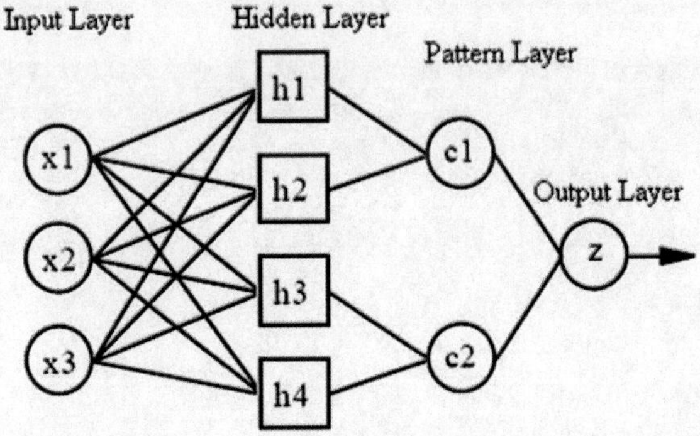

Fig. 3 General structure of PNN

4 MATLAB Implementation

The proposed scheme was implemented, i.e., coded using MATLAB which is a versatile software used for computer simulation study and mathematical analysis. It contains various predefined functions for image processing which are easy to use and help in reducing development time. Moreover, it gives smooth integration with machine learning and artificial intelligence algorithms which is convenient to use and develop the code rapidly.

The images used for training the PNN classifier are shown in Fig. 4. A dataset of 10 images of a benign and malignant type of brain tumor was created and loaded during the process of classification. The data extracted from the input images is stored in a workspace in MATLAB for use in the algorithm.

The developed GUI for the proposed method of brain tumor classification using PCA and PNN is shown in Fig. 5. The user is given an option to load image from desktop or computer and is displayed on the GUI screen. The image can be of various different formats, e.g., *.png, *.jpg/jpeg and *.bmp, to give the developed GUI a wider acceptance. Both T1 and T2 weighted images are accepted and can be accurately classified using the developed software.

The extracted features of the image plotted graphically in Fig. 5 are in the following sequence: (1) contrast (difference in luminance), (2) correlation (relation between pixels in an image), (3) energy (a measure of localized change of the image), (4) homogeneity (uniformity), (5) mean (gives the idea of pixel color of the image), (6) standard deviation (diversity of pixels), (7) entropy (intensity level of individual pixels), (8) RMS (used to give an idea for the similarity of the image), (9) variance (squares of standard deviation), (10) smoothness (smoothness of the surface of the image), (11) kurtosis (information about noise and distortion), (12) skewness (gives an idea about the brightness and type of surface of an image) and (13) inverse difference moment (texture of the image). These features are extracted for each test image

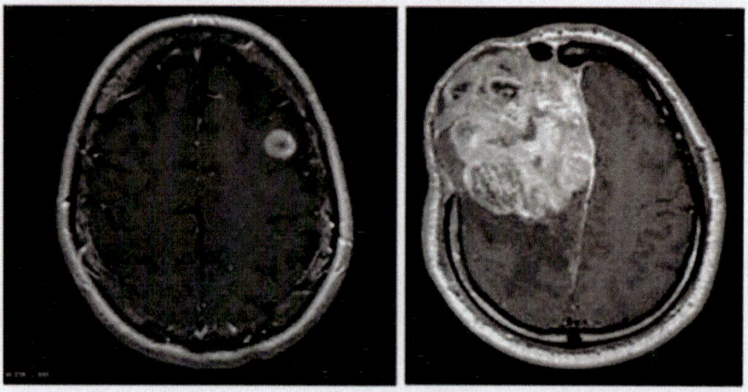

(a) (b)

Fig. 4 Brain MR images used for testing and analysis: **a** benign type of brain tumor, **b** malignant type of brain tumor

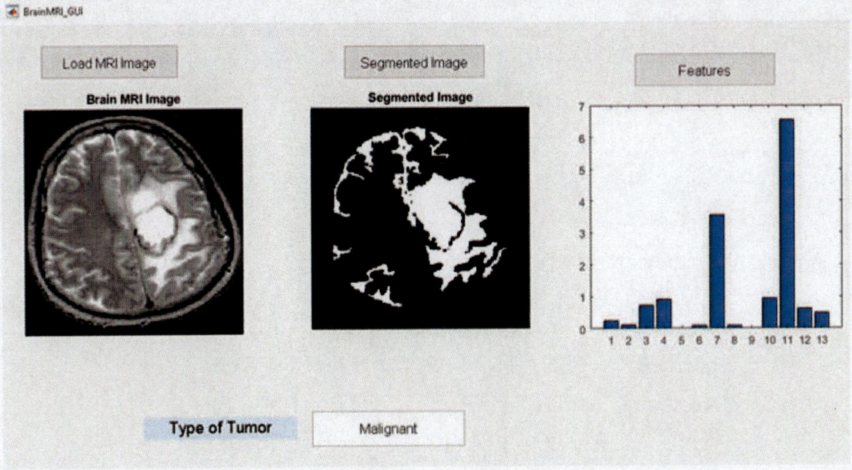

Fig. 5 Graphical user interface developed using MATLAB GUIDE for brain tumor classification using PCA and PNN showing: **a** selection of image (load image: extreme left), **b** segmented image (segmented image: middle) and **c** graph showing features (features: extreme right)

and compared with the dataset used to train the PNN classifier. The classification is displayed based on its probability after evaluation of each parameter.

5 Results and Discussion

The developed algorithm was tested using two types of datasets, viz. training dataset (consisting of 10 MR images of benign type and 10 images of malignant type brain tumors) and test dataset (consisting of 15 MR images of benign type and 15 images of malignant type brain tumors in a mix of T1 and T2 weighted images), to determine the accuracy of classification. Each section of the algorithm was separately validated to arrive at the final full algorithm along with GUI.

The statistical terms used to describe the performance of the developed algorithm are sensitivity, specificity and accuracy. These terms are defined in terms of true positive (TP), true negative (TN), false negative (FN) and false positive (FP) as follows:

$$\begin{aligned} \text{Sensitivity} &= \text{TP}/(\text{TP} + \text{FN}) \\ \text{Specificity} &= \text{TN}/(\text{TN} + \text{FP}) \\ \text{Accuracy} &= (\text{TN} + \text{TP})/(\text{TN} + \text{TP} + \text{FN} + \text{FP}) \end{aligned}$$

This case study consists of a total of 30 test images, and the characteristics found are tabulated in Table 1.

Table 1 Cases distributed among TP, TN, FN and FP

Type	Number of cases
TP	13
TN	12
FN	2
FP	3

Accordingly, the parameters resulted are as follows:

$$\begin{aligned}\text{Sensitivity} &= \text{TP}/(\text{TP} + \text{FN}) = 13/(13+2) = 0.8667.\\ \text{Specificity} &= \text{TN}/(\text{TN} + \text{FP}) = 12/(12+3) = 0.8.\\ \text{Accuracy} &= (\text{TN} + \text{TP})/(\text{TN}+\text{TP}+\text{FN}+\text{FP})\\ &= (12+13)/(12+13+2+3) = 0.833.\end{aligned}$$

This indicates that the accuracy of the proposed algorithm is around 83.33% which is noted to be excellent among similar comparisons. Moreover, the images were used as mixed type, i.e., T1 and T2 weighted. If either category is used to train the PNN and used to detect only that type of image, accuracy can be further enhanced.

6 Conclusion

In this paper, a MATLAB-based GUI was built to segment brain MR images into a benign (not harmful) and malignant (harmful) type of tumor using advanced image processing techniques and neural networks. The noise in the images is removed using preprocessing to smoothen the image and also to improve the signal-to-noise ratio. Then, a discrete wavelet transform is used to extract the features from the GLCM by decomposing the image into different bands. PNN is used to classify the type tumor from the images by training it using 10 images of each pattern. The observations prove that the classification of the tumor is very fast, satisfactory and accurate and can replace the need to be dependent on clinical experts for manual detection. The factors used to gauge the performance include sensitivity, specificity and accuracy all of which are found to be above 90% based on 20 test images of each type (excluding train images) used for analysis which was a mix of T1 weighted and T2 weighted. It was found that T2 weighted images had more accuracy than T1 weighted due to less noise and its high signal strength. The software developed is easy to use and install and can be recommended for regular clinical practice. As future work, this method can be extended for other issues as well.

References

1. Shih FY (2010) Image processing and pattern recognition: fundamentals and techniques, 1st edn. IEEE. Wiley
2. Parker JR (2010) Algorithms for image processing and computer vision, 2nd edn. Wiley
3. Breteler MM, Stöcker T, Pracht E, Brenner D, Stirnberg R (2014) MRI in the Thineland study: a novel protocol for population neuroimaging Alzheimer's Dementia. J Alzheimer's Assoc 10(4):92
4. Mohan A, Poobal S (2018) Crack detection using image processing: a critical review and analysis. Alexandria Eng J 57(2):787–798
5. Alfaro-Almagro F et al (2018) Image processing and Quality Control for the first 10,000 brain imaging datasets from UK Biobank. NeuroImage 166:400–424
6. Bindhu V (2019) Biomedical image analysis using semantic segmentation. J Innov Image Process (JIIP) 1(02):91–101
7. Vijayakumar T (2019) Neural network analysis for tumor investigation and cancer prediction. J Electron 1(02):89–98
8. Joseph RP, Singh CS, Manikandan M (2014) Brain tumor MRI image segmentation and detection in image processing. Int J Res Eng Technol 3(1):1–5
9. Cui W, Wang Y, Fan Y, Feng Y, Lei T (2013) Localized FCM clustering with spatial information for medical image segmentation and bias field estimation. Int J Biomed Imag
10. Bahadure NB, Ray AK, Thethi HP (2017) Image analysis for MRI based brain tumor detection and feature extraction using biologically inspired BWT and SVM. Int J Biomed Imag Hindawi
11. Alfonse M, Salem M (2016) An automatic classification of brain tumors through MRI using support vector machine. Egypt Comput Sci J 40:11–21
12. Yao J, Chen J, Chow C (2009) Breast tumor analysis in dynamic contrast enhanced MRI using texture features and wavelet transform. IEEE J Sel Top Signal Process 3(1):94–100
13. Kumar P, Vijayakumar B (2015) Brain tumor MR image segmentation and classification using by PCA and RBF kernel based support vector machine. Middle East J Sci Res 23(9):2106–2116
14. Sharma N, Ray A, Sharma S, Shukla K, Pradhan S, Aggarwal L (2008) Segmentation and classification of medical images using texture-primitive features: application of BAM-type artificial neural network. J Med Phys 33(3):119–126
15. Kharat KD, Kulkarni PP et al (2012) Brain tumor classification using neural network based methods. Int J Comput Sci Inf 1(4)
16. Thara KS, Jasmine K (2016) Brain tumour detection in MRI images using PNN and GRNN. In: IEEE ınternational conference on wireless communications, signal processing and networking (WiSPNET), Chennai, pp 1504–1510
17. Badisa H, Polireddy M, Mohammed A (2019) CNN Based brain tumor detection. Int J Eng Adv Technol 8(4):2249–8958
18. Malathi M, Sinthia P (2019) Brain tumour segmentation using convolutional neural network with tensor flow. Asian Pac J Cancer Prev 20:2095–2101
19. Sumathi SU, Geetha S (2015) Brain tumor classification using probabilistic neural network. Int J Innov Res Sci Eng Technol 4(9):7959–7964
20. Arun M et al (2014) Brain tumor classification using PCA& PNN. Int Conf Signal Process Embedded Syst Commun Technol Their Appl Sustain Renew Energy 3(3):326–331
21. Kapoor L et al (2017) A survey on brain tumor detection using ımage processing techniques. In: IEEE 7th ınternational conference on cloud computing, data science & engineering—confluence, pp 582–585
22. Varuna SN, Kumar TNR (2018) Identification and classification of brain tumor MRI images with feature extraction using DWT and probabilistic neural network. Brain Inf 5:23–30
23. Othman MF, Basri MAM (2011) Probabilistic neural network for brain tumor classification. In: IEEE second ınternational conference on ıntelligent systems, modelling and simulation, pp 136–138

Chapter 31
Workload-Driven Transactional Partitioning for Distributed Databases

R. D. Bharati and V. Z. Attar

1 Introduction

Data partitioning is one of the techniques where execution time can be reduced and optimize database access costs. It improves the workload performance of the database systems. In databases, systems partitioning is one among the useful access technique which enhances the availability, performance, and reliability of applications. Partitioning allows us to access data if and as per need, and therefore, the time needed to get a query time is often reduced. A distributed database is a collection of logically interrelated databases across multiple locations and connected by high-speed networks. As per the client demand, the database can be partitioned in different methods to apply on different query execution strategies whose aim is minimizing the amount of distributed accesses. Databases partitioning techniques [1] are widely utilized in the context of DBMS where the various approaches are to identify feasible design schemes and evaluate them using different methodologies.

Transactional partitioning analyses transaction workload for the availability of data towards application developer to increase the scalability of the database. To represent the workload model which will increase the scalability for the OLTP database [2], which monitors transactional logs using graph representation [3].

R. D. Bharati (✉) · V. Z. Attar
College of Engineering, Pune, India
e-mail: rdb15.comp@coep.ac.in

V. Z. Attar
e-mail: vahida.comp@coep.ac.in

I. Jeena Jacob et al. (eds.), *Data Intelligence and Cognitive Informatics*,
Algorithms for Intelligent Systems,
https://doi.org/10.1007/978-981-15-8530-2_31

The data partition may be a process data fragmentation of a logical and physical database of its design elements distributes in independent parts. For accessing, storing and maintaining the relevant items like rows and columns makes data partition, which may reduce the quantity of data access execution time by query processing. On the base of records and domains, partitioning is divided into access patterns. The first method is horizontal, vertical and hybrid partitioning.

1.1 Horizontal Partitioning

Horizontal data partitioning where tables and objects are partitioned into sets of tuples is physically stored and accessed. During this technique, data are partitioned horizontally and store on different partitions. Today, most relational databases follow horizontal data partitioning techniques. This system supports the static as well as dynamic data partitioning principles.

1.2 Vertical Partitioning

In vertical partitioning, data divide the table or object vertically into a number of columns. Vertical partitioning fragments table, index convert into sets of columns. Queries access only a small subset of columns during data processing. This system is additionally called column-based partitioning where partitioning is done on the domain wisely.

1.3 Hybrid Partitioning

Hybrid partitioning techniques were data partitioned based on the required query plan. This method is a combination of horizontal and vertical partitioning. Many Web applications store and retrieve data from servers, and they partitioned based on access patterns. This technique improves the availability of databases. In addition to these approaches the basis of availability of data to be partitioned both ways, dynamic as well as static. There are many challenges for data partition and replication in the big data era and that to develop a replacement strategy is required to tackle the issues which are caused by concurrency.

Transaction Management in Distributed Databases
In distributed systems, transaction is a sequence of reading and/or write operations performed on a single logical unit of work [1]. It is a collection of operations performed against the data object of the database. To preserve the integrity

of database, ACID properties of transactions should be ensured like atomicity, consistency, isolation and durability.

1. **Atomicity**: The atomicity property works like an all-or-nothing strategy, which means that all of the transactions work should be complete or not at all.
2. **Consistency**: Consistency property ensures the state of the database prior to initiation and after transactions. Data should be consistent when transactions were completed. Transactions are moving one state of the database to a consistent state.
3. **Isolation**: Property ensures independent transactions while performing operations without interference.
4. **Durability**: Ensures the transaction results permanently stored in the database.

Commit Protocols

The commit protocol is a transaction algorithm optimized in distributed databases to coordinate all processes in transaction processing. Two-phase commit protocol consists of two phases: prepare and commit phase. There are challenges of 2PC like blocking and state inconsistency. When the coordinator fails, the blocking may occur up to the participants get a message to release the resource. Inconsistency occurs when the global state vector contains commit as well as abort message.

In three-phase commit protocol in a distributed database, all nodes agree to commit. In addition with two phases in two-phase commit protocol, the pre-commit message gets from coordinator to participants. There is a problem with three-phase commit protocol when the situation occurs while multiple sites may fail. A number of messages occur while committing causes network congestion and overhead.

2 Related Work

As a research work, number of researchers carried out a study on data partitioning include horizontal, vertical and mixed (hybrid). The most related research work is cited here. As per literature, related work is classified in different approaches.

2.1 Random Approach

Sauer and Hao [4] represented the data partitioning approach with data mining techniques on a horizontal cloud database. In this design, clustering done on the base of database logs and data mining used to partition the database on a server.

2.2 Transaction Approach

Turcu et al. [5] presented automated data partitioning of scalable transactions. Granola model implemented the rules where main memory utilized for data storage. They perform static byte code analysis for partitioning. Due to which limits the synchronization by scalable transactions. Das et al. [6] proposed schema level partitioning. In schema level, data partitioning reduces the scalable transaction. The concerned rows resorted one partition. Which assessed by a set of nodes under EC2. TPC-C data model is utilized to assess execution on Amazon's cloud. This methodology focuses on the improvement of scalability only. To eliminate scalability bottlenecks, the key value data stores use for decreasing distributed transactions. ElasTras [6] concentrated on scalability compelling reason to focus on replication. ElasTras is suitable for static partitioning for Web applications. Kamal et al. [7] implement a workload aware data partitioning for online transactional processing databases which reduce distributed transactions.

2.3 Cloud-Based Approach

This approach implements cloud-based distributed databases, and for storage medium, cloud file is a system. But the system does not support to maintain ACID properties in case of machine failure. In the partitioning approach which support cloud based salable transaction system. It splits the transaction manager into the number on a node where each node considers as one partition. Cloud TPS [8] presented for an adaptable exchange for designs on top of two different families BigTable and SimpleDB for web applications. The read-write and read only transactions support Cloud TPS. Cloud TPS contains a static partitioning technique to increase scalability. For reliable applications, Wang et al. [9] presented an online partitioning method by MapReduce Environment on cloud-based partitioning. They have defined block placement techniques to reduce repeated I/O cost. In the schema level, the key is associated with data partitioning to avoid redundancy. In this partitioning, once the partitions are formed, it remains the same so-called static partitioning. Schema level partitioning limits transactional access to only one database partition. Schema level partitioning has used the technique that accesses a large number database schema, but transactions access only less number of associated data items which is spread over different tables.

In addition to above related work, Schism [10] presents graph-based transactional workload data partitioning. FiDoop-DP [11] model is based on frequent data item with data mining techniques. Megastore [12] produces database ACID semantics within the fine-grained data partitions.

3 System Model

Consider a database D and transactions on different warehouses $W = \{w_1, w_2, w_n\}$ and set of frequent queries $Q = \{q_1, q_2, \ldots, q_r\}$, so number of desired partitions k are $P = \{p_1, p_2, \ldots, p_n\}$.

Figure 1 shows the workload-driven architecture where client request through query which turns issues by a distributed transaction manager (DTM), which is responsible for all data items in warehouses to find distributed transactions. Workload-driven partitioning consists of distributed transaction manager and data partitioner at the server-side. A set of warehouses and query workload resides on the application side with a load balancer that balances data load at the server-side.

Distributed transaction manager reduces the distributed transactions. A two-phase commit protocol (2PC) is applied for transactions. In the initial phase, the coordinator requests all for correct operation executed or not, and the second phase commits transactions or abort. Distributed transactions to be partitioned by workload partitioner. While partitioning mechanism guarantees all ACID properties satisfied.

Load balancer distributes workload across concurrent nodes [14]. It enhances the availability of the application and manages transactions between warehouses and distributed transaction processes. Also, it handles varying load traffic in applications. Data partitioner reduces distributed transactions to improve the performance of the system through partitioning.

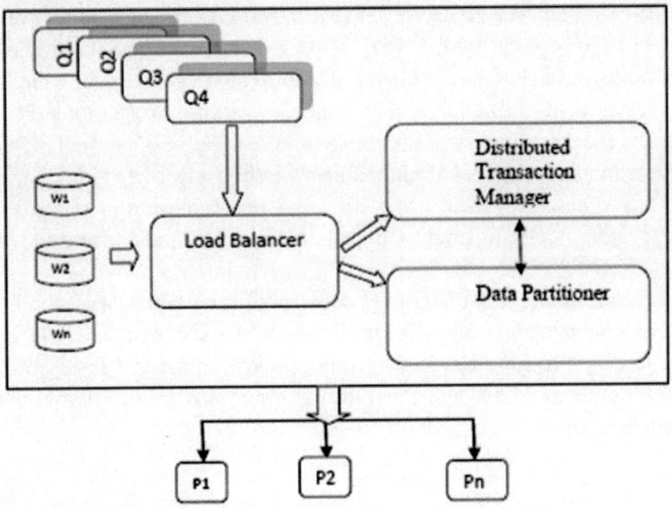

Fig. 1 Workload-driven architecture

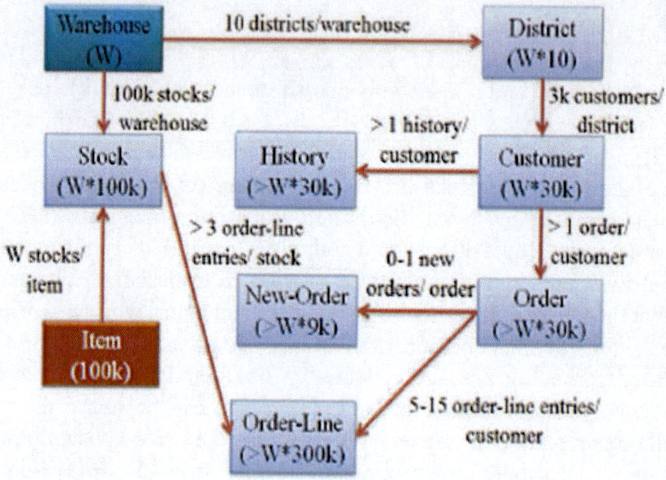

Fig. 2 TPC-C schema [13]

4 Implementation

The scalability of workload-driven data partitioning with two layers is performed and evaluated under workload derived from TPC-C with HBase NoSQL database as one layer and another Amazon EC2 general purpose t2.micro compute unit, 16 GB DDR and 350 GB storage with MySql cluster distributed database. The set-up of one server node and five nodes is clients. Hadoop HBase system in which one node works as a server with 8 GB DDR and five nodes works as workstations. The architecture decreases the distributed transactions while data processing. In TPC-C schema, design of relational database is of nine tables warehouse, stock, item, district history, customer, order, new-order and order-line, and five transactions execute an order-processing system within differently distributed districts and concerned warehouses (Fig. 2).

The relational model of TPC-C [13] consists of nine tables accessed by transactions that have been taken. Order-line and stock tables are big in records. For logical partition initially, it follows range partitioning. New entries are inserted into order-line and order table of 44.5% workload. Remaining transactions distributed with the round-robin method.

5 Result

The result is compared with no partitioning state with different data set the size of 12,000 data items. Figure 3 represents the total response time for random query workload for no partitioning takes more time than workload data partition (WDDP). For

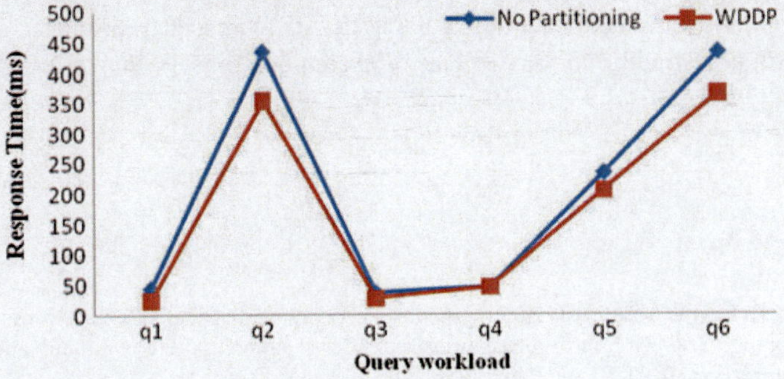

Fig. 3 Response time

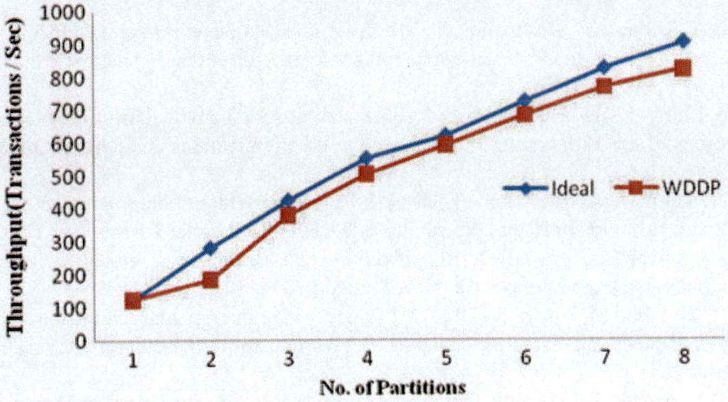

Fig. 4 System throughput

execution, six queries have been taken to execute the workload. Figure 4 represents the number of distributed transaction executed on partitions.

6 Conclusion and Future Work

To support scalability, such workload model is used to investigate distributed transaction management system which guarantees ACID properties. The purpose of this paper to increase the performance of a distributed database using workload driven data partitioning. The result shows better performance of the system using average response time on different partitioning techniques and transactions executed on a number of partitions using a distributed database. The presented model can be helpful for application developers for building distributed and cloud-based applications which need transactions consistency.

The implementation of NEWSQL databases with increased transactions and partitions to support traditional as well as Web applications is in process in future work.

References

1. Bharati RD, Attar VZ (2018) A comprehensive survey on distributed transactions based data partitioning. In: 2018 Fourth international conference on computing communication control and automation (ICCUBEA), Pune, India, pp 1–5
2. Chen Z, Yang S, Tan S, Zhang G, Yang H (2013) Hybrid range consistent hash partitioning strategy—a new data partition strategy for NoSQL database. In: Proceedings of the 12th IEEE international conference on trust, security and privacy in computing and communications (TrustCom)
3. Quamar A, Kumar KA, Deshpande A (2013) Sword: scalable workload- aware data placement for transactional workloads. In: International conference on extending database technology, pp 430–441
4. Sauer B, Hao W (2015) Horizontal cloud database partitioning with data mining techniques. In: Proceedings of the 12th annual IEEE consumer communications and networking conference (CCNC)
5. Turcu A, Palmieri R, Ravindran B, Hirve S (2016) Automated data partitioning for highly scalable and strongly consistent transactions. IEEE Trans Parallel Distrib Syst 27(1):1–14
6. Das S, Agrawal S, Agrawal D, El Abbadi A (2009) ElasTraS: an elastic, scalable and transactional data store in the cloud. HotCloud, pp 131–142
7. Kamal J, Murshed M, Buyya R (2014) Workload-aware incremental repartitioning of shared-nothing distributed databases for scalable OLTP applications. Elsevier, FGCS Special Issue: UCC2014
8. Zhou W, Pierre G (2012) CloudTPS: scalable transactions for web applications in the cloud. IEEE Trans Serv Comput 5(4):525–539
9. Wang YX, Luo JZ, Song AB, Dong F (2013) Partition-based online aggregation with shared sampling in the cloud. J Comput Sci Technol 28(6):989–1011
10. Curino C, Jones E, Zhang Y, Madden S (2010) Schism: a workload driven approach to database replication and partitioning. In: 36th International conference on very large data bases
11. Xun Y, Zhang J, Qin X, Zhao X (2017) FiDoop-DP: data partitioning in frequent itemset mining on Hadoop clusters. IEEE Trans Parallel Distrib Syst 28(1):101–114
12. Baker J, Bond C, Corbett JC, Furman J, Khorlin A, Larson J, Leon J-M, Li Y, Lloyd A, Yushprakh V (2011) Megastore: providing scalable, highly available storage for interactive services. CIDR 11:223–234
13. Transaction Processing Performance Council: TPC Benchmark. http://www.tpc.org/tpcc/
14. Shi Y, Qian K (2019) LBMM: a load balancing based task scheduling algorithm for cloud. In: Future of information and communication conference, pp 706–712. Springer, Cham

Chapter 32
Start-Up Leagility Assessment Using Multi-grade Fuzzy and Importance Performance Analysis

Edrion Chacko, M. Suresh, and S. Lakshmi Priyadarsini

1 Introduction

Start-ups have become a new trend in today's economy. A large number of start-ups are coming up in every segment of the business. Governments and NGOs all over the world are providing support for the development of start-ups in terms of technology and financial aids, as they play a vital role in economic development. But amidst this, the number of failures of start-ups is also highly correlated. According to the reports, about 90% of start-ups in India shuts down within 5 years of operation [1].

The reason for the failures is many and varies across start-ups. Taking a glimpse on the failure stories of these start-ups leads the focus of attention towards the term leagility. Naylor et al. [2] coined the term leagility which is the combination of lean and agile concepts. Leanness focuses on the development of value stream by the elimination of wastes and ensuring a level schedule, while agility focuses on adapting to the dynamic market conditions. Naim and Gosling [3] suggested that leagility can help organizations to enhance their performance to improve market share and better financial performance.

E. Chacko (✉) · M. Suresh
Amrita School of Business, Amrita Vishwa Vidyapeetham, Coimbatore, India
e-mail: edrionchacko@gmail.com

M. Suresh
e-mail: drsureshcontact@gmail.com

S. Lakshmi Priyadarsini
Govt Victoria College, Palakkad, Kerala, India
e-mail: lakshpriyadarsini@gmail.com

I. Jeena Jacob et al. (eds.), *Data Intelligence and Cognitive Informatics*,
Algorithms for Intelligent Systems,
https://doi.org/10.1007/978-981-15-8530-2_32

The motivation of this study was to

- Identify factors that affect the leagility of start-ups which was done by literature review and by collecting expert's opinion.
- Assess the leagility index of a case start-up applying the developed conceptual model through MGF approach.
- Classify the attributes based on its importance and the performance of case start-up through IPA analysis.

This is a pioneer study initiated to develop a framework for assessing leagility of start-ups using MGF approach and IPA analysis.

2 Literature Review

In [4], the significance of corporate culture in developing agile practices is highlighted. The study proposed that the culture of the organization must be supportive of working in a collaborative environment. Nguyen-Duc et al. [5] explained strategic management as one of the attributes that help start-ups to achieve agility in an early stage. Strategic management refers to strategic thinking as well as decisions by the directors of the company that direct the business. Patil and Suresh [6] explained collaboration as the ability of communication as well as sharing knowledge within groups in the same or different organization. Collaboration can help in bringing experts from different domains together avoiding duplication. Modi and Thakkar [7] demonstrated various principles in lean manufacturing. The philosophy of lean manufacturing emphasizes low cost and lesser time without compromising in quality. Merkaš et al. [8] define the importance of implementing a lean start-up methodology. The paper defines the five stages of a new product development identifying potential customer and understanding market.

Fuzzy logic resembles human reasoning, and it takes advantage of vast conceptual knowledge. It is the theory of fuzzy sets that calibrate vagueness. There are several examples of scientific studies which illustrate the efficient application of multigrade fuzzy in various fields of managerial applications in the manufacturing and service sectors. Vinodh and Aravindraj [9] used a fuzzy logic approach to develop a framework for measuring leagility index in the supply chain. The study combined the paradigms of lean and agile into the concept of leagility for the development of a conceptual model suitable for industries in managing their supply chain. Matawale et al. [10] used a fuzzy approach for assessing leanness, agility and leagility indexes in the supply chain for automotive industries. The study used a fuzzy set theory to rout the uncertain assessment information because of variances in human perception related to subjective evaluation. Susilawati et al. [11] used fuzzy logic to assess the leanness in manufacturing industries. Fuzzy logic can be applied in both quantitative as well as qualitative data where vagueness or imprecision exists.

Based on the literature review, it was identified that few studies have contributed to the leagility assessment using certain approaches. Many of those approaches are not validated with the industry. Also, it was found that agile, lean and leagile assessment was predominantly performed over manufacturing and service industries using MGF approach. In this context, this research aims to present a study where the conceptual model is developed using the literature review, and this model is practically validated with start-up industry to assess its leagility using MGF approach.

3 Conceptual Model for Assessing Start-Up Agility

The conceptual model developed for measuring leagility of start-ups is given in Table 1. This conceptual model is evolved based upon three aspects of start-up leagility which include management [12], technology [13], workforce [14] and Market [15].

Table 1 Conceptual model for assessing Start-up agility

Enablers	Criteria	Attributes
Management (Z_1)	Nature of management (Z_{11})	Flat organizational structure (Z_{111})
		Adaptability towards market dynamics (Z_{112})
		Employee autonomy (Z_{113})
	Organzational culture (Z_{12})	Employee involvement (Z_{121})
		Degree of stability (Z_{122})
	Strategic practices (Z_{13})	Anticipating changes in business (Z_{131})
		Degree of responsiveness (Z_{132})
Technology (Z_2)	Innovation (Z_{21})	Using latest technologies (Z_{211})
		Value added activities (Z_{212})
		Efficient methodologies (Z_{213})
	Product (Z_{22})	Developing new products (Z_{221})
		Upgradation of existing products (Z_{222})
	Waste elimination (Z_{23})	Continuous improvement (Z_{231})
		Cost management (Z_{232})
		Time utilization (Z_{233})
Workforce (Z_3)	Flexibility (Z_{31})	Employee adaptability (I_{311})
		Multi-skilled employees (Z_{312})
	Collaboration (Z_{32})	Team working (Z_{321})
		Cross-learning in organization (Z_{322})
Market (Z_4)	Market specification (Z_{41})	Quality and standard products (Z_{411})
		Meet market demand (Z_{412})
	Customer sensitivity (Z_{42})	Meeting real needs of customers (Z_{421})
		Achieving customer delight (Z_{422})

The model contains three levels. The first level indicates enablers, the second level includes criteria, and the third level is the respective attributes. For illustration, the enabler management has three criteria which include nature of management, organizational culture and strategic practices. The criteria nature of management consists of three attributes which include flat organizational structure, adaptability towards market dynamics and employee autonomy.

4 Research Methodology

Literature review on leagility assessment was carried out to identify suitable criteria based upon which conceptual model was formed. The weightage for assessing start-up leagility was carried out through expert opinion based on the conceptual model developed. A relevant start-up was identified for performing the case study. The attribute rating for leagility assessment was done by interviewing the proprietors and employees of the start-up. Leagility was calculated based upon the rating assessment and expert's weightage followed by IPA analysis which was done to identify attributes for improving leagility of a start-up.

5 Case Study

5.1 About Case Company

The study has been carried out in one case start-up from Cochin, India (hereafter referred to as ABC Start-up) which is into IT consulting and services. The firm specializes in technology and functional consultants to their clients and focuses on Web development, software development, online digital marketing, social media promotions and other marketing solutions. The company targets to provide unparalleled value to its clients.

5.2 Leagility Assessment Using Fuzzy Logic

The leagility of start-up is represented by Z which is the product of average weightage of the expert opinions denoted by W and the overall rating factor R. Therefore, the expression for leagility is equated as

$$Z = W * R$$

Five experts from similar industry contributed in the discussion conclave for weighting leagile enablers, criteria and attributes for IT and marketing-related start-ups. Five employees were evaluated from case company ABC to rating the attributes in order to assess leagility in their start-up. Table 2 shows the average normalized weightage of experts' opinion of enabler, criteria and attribute represented by W, W_i, W_{ij}, respectively, and rating for attributes collected from employee's case start-up represented from R_1 to R_5.

The leagility index indicates the leagility level of the start-up. The calculation starts from attributes level to the enabler level scaling down to fuzzy value thus enabling to find the precise leagile value. The primary assessment is applied over the leagile attributes where the employee rating for attributes and the average normalized weightage of attributes by the experts are assessed to give the fuzzy index value for the secondary assessment. The fuzzy indexed value obtained from primary assessment

Table 2 Single-factor rating and expert's weightage

Z_i	Z_{ij}	Z_{ijk}	R_1	R_2	R_3	R_4	R_5	W_{ij}	W_i	W
Z_1	Z_{11}	Z_{111}	7	8	8	8	9	0.36	0.35	0.25
		Z_{112}	8	7	8	9	8	0.33		
		Z_{113}	6	6	7	5	7	0.31		
	Z_{12}	Z_{121}	7	6	6	5	7	0.51	0.31	
		Z_{122}	8	7	7	6	7	0.49		
	Z_{13}	Z_{131}	7	7	8	7	8	0.49	0.34	
		Z_{132}	7	8	9	8	8	0.51		
Z_2	Z_{21}	Z_{211}	8	9	7	8	9	0.29	0.32	0.26
		Z_{212}	8	7	7	8	8	0.34		
		Z_{213}	8	8	6	6	7	0.37		
	Z_{22}	Z_{221}	6	6	7	7	6	0.51	0.33	
		Z_{222}	7	8	8	7	8	0.49		
	Z_{23}	Z_{231}	10	9	9	8	8	0.35	0.36	
		Z_{232}	8	8	8	7	8	0.32		
		Z_{233}	7	7	9	8	7	0.33		
Z_3	Z_{31}	Z_{311}	8	7	6	7	7	0.54	0.47	0.25
		Z_{312}	8	8	7	8	7	0.46		
	Z_{32}	Z_{321}	9	9	9	10	9	0.52	0.53	
		Z_{322}	8	9	9	8	9	0.48		
Z_4	Z_{41}	Z_{411}	9	8	8	9	8	0.53	0.47	0.24
		Z_{412}	8	7	7	8	7	0.47		
	Z_{42}	Z_{421}	9	8	8	9	8	0.49	0.53	
		Z_{422}	10	9	8	8	9	0.51		

has been evaluated with the average normalized weightage of criteria by the experts that gives the fuzzy indexed values for tertiary assessment. The obtained values when evaluated with the average normalized weightage of enabler have given the leagile index.

As the estimation of leagility factor involves fuzzy determination, the computation has been divided into five classes, $Z = (10, 8, 6, 4, 2)$ wherein the group 8 to 10 indicated that the start-up is 'extremely leagile', group 6 to 8 indicated 'leagile', group 4–6 indicates 'moderately leagile', group 2–4 indicated 'non leagile' and group 0–2 indicated 'extremely non leagile'.

5.2.1 Primary Assessment

The assessment used for criteria 'Nature of management' is described below.

The weightage for criteria 'Nature of management' $W_{11} = (0.36, 0.33, 0.31)$.

Rating vector for criteria 'Nature of management' was calculated as

$$R_{11} = \begin{bmatrix} 7\,8\,8\,8\,9 \\ 8\,7\,8\,9\,8 \\ 6\,6\,7\,5\,7 \end{bmatrix}$$

Index was calculated for the criteria 'Nature of management' as

$$Z_{11} = W_{11} * R_{11} \ \ [16]$$
$$Z_{11} = (7.02, 7.05, 7.69, 7.4, 8.05)$$

The same concept was applied for calculating the index of other leagility criteria.

$$Z_{12} = (7.49, 6.49, 6.49, 5.49, 7)$$
$$Z_{13} = (7, 7.51, 8.51, 7.51, 8)$$
$$Z_{21} = (7.92, 7.88, 6.56, 7.18, 7.84)$$
$$Z_{22} = (6.49, 6.98, 7.49, 7, 6.98)$$
$$Z_{23} = (8.37, 8.02, 8.68, 7.68, 7.67)$$
$$Z_{31} = (8, 7.46, 6.46, 7.46, 7)$$
$$Z_{32} = (8.52, 9, 9, 9.04, 9)$$
$$Z_{41} = (8.53, 7.53, 7.53, 8.53, 7.53)$$
$$Z_{42} = (9.51, 8.51, 8, 8.49, 8.51)$$

5.2.2 Secondary Assessment

The assessment used for enabler 'Management' is described below.

The weightage for enabler 'Management' $W_1 = (0.35, 0.31, 0.34)$.

Rating vector for enabler 'Management' is calculated as

$$R_1 = \begin{bmatrix} 7.02 & 7.05 & 7.69 & 7.4 & 8.05 \\ 7.49 & 6.49 & 6.49 & 5.49 & 7 \\ 7 & 7.51 & 8.51 & 7.51 & 8 \end{bmatrix}$$

Index is calculated for the enabler 'Management' as

$$Z_1 = W_1 * R_1 \quad [17]$$
$$Z_1 = (7.16, 7.03, 7.6, 6.84, 7.7)$$

The same concept is applied for calculating the index of other leagility criteria.

$$Z_2 = (7.69, 7.71, 7.7, 7.37, 7.57)$$
$$Z_3 = (8.28, 8.28, 7.8, 8.3, 8)$$
$$Z_4 = (9.05, 8.05, 7.78, 8.51, 8.05)$$

5.2.3 Tertiary Assessment

The assessment used for finding the leagility index of ABC Start-up is described below. Overall weightage$W = (0.25, 0.26, 0.25, 0.24)$.

Overall rating vector is given by R

$$R = \begin{bmatrix} 7.16 & 7.03 & 7.6 & 6.84 & 7.7 \\ 7.69 & 7.71 & 7.7 & 7.37 & 7.57 \\ 8.28 & 8.28 & 7.8 & 8.3 & 8 \\ 9.05 & 8.05 & 7.78 & 8.51 & 8.05 \end{bmatrix}$$

Leagility index

$$Z = W * R$$
$$Z = (8.12, 7.84, 7.8, 7.83, 7.9)$$
$$Z = \frac{1}{5}(8.12 + 7.84 + 7.8 + 7.83 + 7.9)$$
$$Z = 7.90$$

5.3 *Classification of Leagile Attributes Based on IPA*

The IPA model has been used to classify attributes based on the importance and performance of the company. IPA model is represented graphically in a framework categorized by four sectors [18]. The four sectors and their explanations are discussed in Table 3. The zones of IPA model are categorised based on the mean of average weightage of experts' opinion and mean of average assessment rating of case Start-up ABC [19].

6 Results and Discussions

Based on leagility assessment using the developed model, agility index for Start-up ABC has been evaluated to be 7.90. It falls in the category 6–8 which indicated the Start-up ABC is leagile. Group 8–10 is the category that represents extremely leagile. This shows that ABC has not yet attained a high level of leagility, and there are still scopes for advancement.

IPA was performed to figure out barriers in achieving a high level of leagility. The IPA model is shown in Fig. 1. The results of IPA analysis classified the attributes based on the importance of these attributes which was weighted by experts' and the performance of case start-up over these attributes which was obtained from the rating of employees of case start-up. The attributes developing new products, use of efficient methodologies, value added activities, anticipating changes in business and time utilization, were classified into zone 1. The attributes cost management, flat organizational structure, degree of responsiveness, meeting the real need of customers, cross-learning in an organization, continuous improvement, achieving customer delight and team working were classified into zone 2. The attributes employee involvement,

Table 3 Sectors and explanation of IPA model

Sector	Inference
Zone 1 Target sector	Attributes present in this sector have been assessed as highly important ones by experts. The performance seemed to be low for the company in this area indicating the need to put attention on attributes present in this sector
Zone 2 Preserve sector	Attributes present in this sector were assessed as highly important ones by the experts, and the start-up is performing on these attributes. Start-up needs to focus on maintaining its performance in this sector
Zone 3 Low-priority sector	This sector was assessed for lower importance. Even if start-up performs low in this sector, it may not reflect in their overall performance
Zone 4 Unmerited sector	Start-up performs well in this sector, but experts assessed these attributes as of lesser importance. The company should rethink on resource allocation in this sector

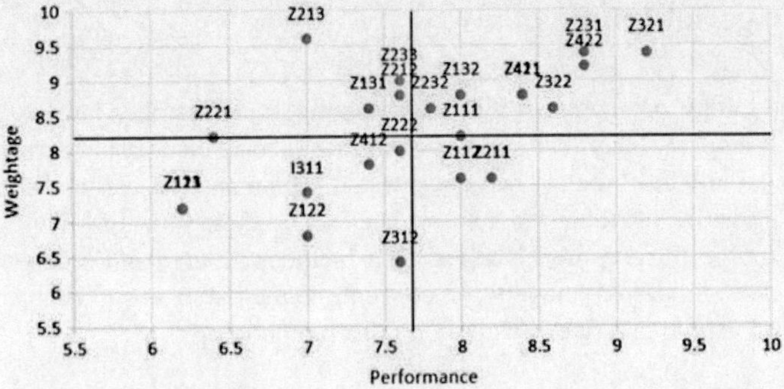

Fig. 1 IPA for leagility assessment

degree of stability, employee adaptability, meeting market demand and hiring multi-skilled employees were classified into zone 3. The attributes adaptability towards market dynamics and using latest technologies were classified into zone 4.

6.1 Improvement Proposal for the Weaker Attributes

The areas for advancement include developing new products, use of efficient methodologies, value added activities, anticipating changes in business and time utilization. The start-up should identify the needs of their targeted segment and develop products to cater to their needs. A manager should study the competitive situation in the market and estimate demand as well as all the associated costs before the launch of a new product. Using efficient methodologies can help start-ups to cut their cost and improve time utilization. Identification of the alternative ways of processes could be done and choose the best from available choices to make the best use of available resources. Proper allocation of resources and skillsets can help in reducing the non-value added activities in the company. Perform a value stream mapping to identify activities that do not provide value to customers. Managers should take necessary actions to reduce or eliminate those activities. Managers should be keen on the changes in the business environment. Today, with the advancement in technology, business is dynamic, and the needs of customer keep changing. So, the start-ups should keep track of changes in the market behavior to identify opportunities and challenges and should make an optimal plan to properly respond to these changes. Performing a time audit and planning can help in improving time utilization. Managers should ensure that the tasks are performed within a stipulated time.

7 Managerial Implications

The methodology presented in this study for assessing leagility helps start-ups to calculate its leagility level. The study enables managers to identify barriers in leagility improvement and attributes to be concentrated. The managers can make use of this model to prioritize attributes to achieve leagility in their start-ups. This study is a helpful resource for corporate persons as it includes inputs from industry experts. In this context, modern managers can concentrate their start-ups to be successful in making sure leagile practices by applying this methodology.

8 Conclusion

The prime motive of this study was to assess leagility level of start-ups. Leagility incorporates the concept of lean and agility into a single paradigm. Leagility focuses on utilizing lean philosophies by responding to changes in business by meeting the market demand and achieving customer delight. This study presents a model to measure the leagility index of start-ups. The model framed using MGF approach incorporated the opinions of experts from start-ups. The paper analyzes the case study of a start-up measuring its leagility index using the framed conceptual model. IPA is applied to find the barriers in leagility advancement of start-up. Improvement on weaker attributes can enable start-ups to attain higher leagility level and thereby achieve a competitive advantage. The proposed conceptual model can further be upgraded by incorporating factors which are not discussed in this study. The case study was performed in a single start-up which is into IT consulting and services. The study can be extended further to various start-ups from different business segments to make the conceptual model more generic and validate.

References

1. Business Line (2017) 90% start-ups in India fail within 5 years: IBM. https://www.thehindubusinessline.com/info-tech/90-startups-in-india-fail-within-5-years-ibm/article9704251.ece. Assessed on 1st Nov 2019
2. Naylor JB, Naim MM, Berry D (1999) Leagility: integrating the lean and agile manufacturing paradigms in the total supply chain. Int J Prod Econ 62(1–2):107–118
3. Naim MM, Gosling J (2011) On leanness, agility and leagile supply chains. Int J Prod Econ 131(1):342–354
4. Misra SC, Kumar V, Kumar U (2009) Identifying some important success factors in adopting agile software development practices. J Syst Softw 82(11):1869–1890
5. Nguyen-Duc A, Weng X, Abrahamsson P (2018, Oct) A preliminary study of agility in business and production: cases of early-stage hardware startups. In: Proceedings of the 12th ACM/IEEE international symposium on empirical software engineering and measurement, pp 1–4
6. Patil M, Suresh M (2019) Modelling the enablers of workforce agility in IoT projects: a TISM approach. Global J Flex Syst Manage 20(2):157–175

7. Modi DB, Thakkar H (2014) Lean thinking: reduction of waste, lead time, cost through lean manufacturing tools and technique. Int J Emerg Technol Adv Eng 4(3):339–344
8. Merkaš Z, Gregurek M, Viktor G (2016, Jan) Addressing market risks through lean startup methodology. In: 1st International business & entrepreneurial—BEE conference, 2016
9. Vinodh S, Aravindraj S (2013) Evaluation of leagility in supply chains using fuzzy logic approach. Int J Prod Res 51(4):1186–1195
10. Matawale CR, Datta S, Mahapatra SS (2015) Evaluation of leanness, agility and leagility for supply chain of automotive industries. Int J Agile Syst Manage 8(2):85–115
11. Susilawati A, Tan J, Bell D, Sarwar M (2015) Fuzzy logic based method to measure degree of lean activity in manufacturing industry. J Manuf Syst 34:1–11
12. Patri R, Suresh M (2019) Agility in healthcare services: a systematic literature exploration. Int J Serv Oper Manage 32(3):387–404
13. Suresh M, Patri R (2017) Agility assessment using fuzzy logic approach: a case of healthcare dispensary. BMC Health Serv Res 17(1):394
14. Rahimnia F, Moghadasian M (2010) Supply chain leagility in professional services: how to apply decoupling point concept in healthcare delivery system. Supply Chain Manage Int J
15. Chan FT, Kumar V (2009) Performance optimization of a leagility inspired supply chain model: a CFGTSA algorithm based approach. Int J Prod Res 47(3):777–799
16. Ganesh J, Suresh M (2016, Dec) Safety practice level assessment using multigrade fuzzy approach: a case of Indian manufacturing company. In: 2016 IEEE International conference on computational intelligence and computing research (ICCIC). IEEE, pp 1–5
17. Sridharan V, Suresh M (2016, Dec) Environmental sustainability assessment using multigrade fuzzy—a case of two Indian colleges. In: 2016 IEEE International conference on computational intelligence and computing research (ICCIC). IEEE, pp 1–4
18. Hemmington N, Kim PB, Wang C (2018) Benchmarking hotel service quality using two-dimensional importance-performance benchmark vectors (IPBV). J Serv Theory Pract 28(1):2–25
19. Feng M, Mangan J, Wong C, Xu M, Lalwani C (2014) Investigating the different approaches to importance–performance analysis. Serv Ind J 34(12):1021–1041

Chapter 33
Assessment of Leagility in Healthcare Organization Using Multi-grade Fuzzy Approach

V. Vaishnavi and M. Suresh

1 Introduction

The health sector is also now facing problems such as growing competition, technological advancement, a changing demand and consumer requirement [1]. Continuous and systemic changes are important for health care to remain economically effective, improve treatment and provide excellent quality of service [2]. The managing task and the achievement of performance levels will require continued attention and contribution to this goal. The two paradigms commonly used are leanness and agility [3]. Lean focuses mainly on eliminating waste during the entire process of organization and agile focuses mainly on rapid response to market changes with customer satisfaction [4, 5]. Agility means making use of business awareness and a virtual company in a competitive position to grab profitably. Leanness is the creation of a value stream to reduce and ensure that all waste is disposed of, including time [6]. At the very same time, the necessity to attain a lean and agile goal has expected to bring a hybrid strategy called leagility [7]. At the same time, the need to achieve a lean and agile objective has led to the emergence of a hybrid strategy known as leagility [7]. Leagile combines both lean and agile principles by responding quickly to the current demand for a variety of services by eliminating waste in the overall process. The adoption of a leagile strategy in health care is a new area of research, and much less exploration leads to the motivation of our study. The unique contribution of the current study is to utilize a multi-grade fuzzy framework to evaluate the leagility of

V. Vaishnavi (✉) · M. Suresh
Amrita School of Business, Amrita Vishwa Vidyapeetham, Coimbatore, India
e-mail: vaishnacsca@gmail.com

M. Suresh
e-mail: drsureshcontact@gmail.com

I. Jeena Jacob et al. (eds.), *Data Intelligence and Cognitive Informatics*,
Algorithms for Intelligent Systems,
https://doi.org/10.1007/978-981-15-8530-2_33

dispensaries that boost service quality and have a volatile market. The multi-grade fuzzy used to formulate the leagility assessment model in the following research question:

RQ1: How to measure the leagility of a healthcare organization?
RQ2: What are the attributes that influence leagility in a healthcare organization?
RQ3: How weaker attributes are improved to enhance leagility?

The study was performed to answer the questions mentioned above. Leagility will help the dispensary to respond quickly to all emergencies and eliminate all types of waste through the entire process. The assessment of the leagility level of the dispensary periodically helps to improve quality and service continuously.

The rest of the article is structured as follows. Section 2 contains the concise introduction about leagility and literature of healthcare. Section 3 includes the research methodology with multi-grade fuzzy approach and IPA. Section 4 is about the case study and the analysis of dispensary. Section 5 contains the results and discussion with recommendations for improvement. Section 6 includes practical implication and Sect. 7 includes a conclusion.

2 Literature Reviews

Naylor et al. [6] have attempted to describe "leagile" as a particular manner of lean and agile paradigms to be exploited. Leagile has been defined as "the combination of the lean and agile paradigms within a total supply chain strategy by positioning the decoupling point to better suit the need for responding to a volatile demand downstream, yet providing level scheduling upstream from the decoupling point" [6]. Mason-Jones et al. [8] have declared a leagile model in the supply chain of manufacturing where the lean and agile systems operate at various stages. A "decoupling point" is a central component in this model that separates the lean processes of agile supply chain procedures. Determination of the supply chain systems is determined by the decoupling point's position, and thus, one may determine whether and where to adopt lean or agile. Leagility can be accomplished by aggregating the three dimensions which include efficiency, effectiveness and responsiveness [9].

In health care, [7] have described a supply chain strategy for leagility application in medical care, in particular chronic treatment such as cardiovascular. To view the discrepancies of selected inventory categories, multi-dimensional scaling (MDS) methods are used to suggest that selections of the strategy of the supply chain based on demand and lead time variability alone are insufficient. Denavi et al. [10] have identified and ranked leagility variables which are utilized on work satisfaction in M.R.I Hospital of Shiraz, Iran. The study is applied research from the element. Two processes, such as the analytical hierarchy process (AHP), for measuring the incoherence rate and classifying the parameters of work satisfaction are used in this research and the fuzzy TOPSIS for classifying lean–agile requirements are used. Rahimnia and Moghadasian [11] have exposed in professional services, and leagility may be used, in particular in hospitals. The approach of case studies is used to consider the leagility concept in professional services. As a medical services provider, it also studies a speciality hospital in Iran.

3 Research Methodology

3.1 Multi-grade Fuzzy

Fuzzy logic modeling makes it possible to look beyond the binary meanings of the scale for wrong statements such as the usable verbally of expert information [12]. In the viewpoint of the professional, the fuzzy logic assists to constitute manufacturer's problems easily and practically, because this depends on less regulations and variables and does not sacrifice the objectivity of the problem [13]. The imprecise, vaguely, partly true and multi-possible responses can be overcome by the fuzzy technique. The method of fluctuation uses truth as a mathematical approach for vagueness and ignorance probability. Moreover, this technique can accommodate the unclear boundary in meanings and permits evaluators to utilize linguistic terminology to test meters in natural language terminology [14, 15].

The study utilized multi-grade fuzzy to assess the leagility in the dispensary, and Fig. 1 includes steps taken during the analysis. The current research begins with an analysis of the literature on the healthcare leagility evaluation. A new conceptual model is framed to assess the leagility index with 3 enablers, 11 criteria and 24 attributes which are shown in Table 1. A questionnaire is used with a 10-point Likert scale to collect data from experts and employees of case dispensary. Every interview is 30 min, and a brief introduction was given on the leagility enablers, criteria and attributes to respondents.

Fig. 1 Steps of the multi-grade fuzzy approach for assessment of leagility

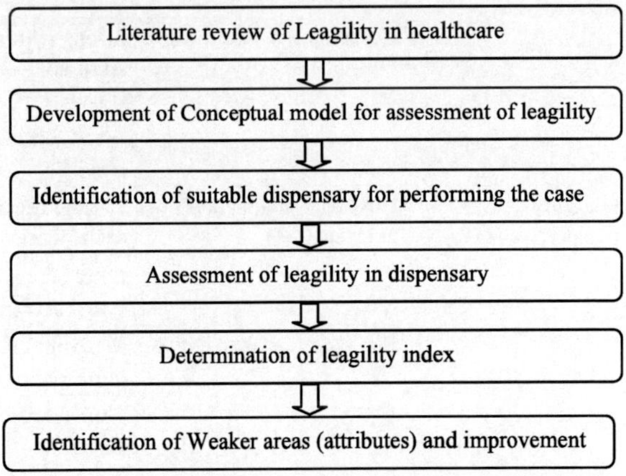

Table 1 Assessment model for evaluating leagility

Enabler	Criteria	Attributes	Reference/expert opinion
Management perspective (I_1)	Top management commitment (I_{11})	Frequent interval of meeting (I_{111})	[16]
		Sharing the management decisions (I_{112})	[17]
		Built trust among caregivers (I_{113})	[18]
	Multiple tasking and quick decisions (I_{12})	Multiple tasking (I_{121})	[19]
		Transparent and quick decision (I_{122})	[20]
	Transparent communication (I_{13})	Smooth information flow (I_{131})	[21]
		Transparent in decisions (I_{132})	[19]
		Create positive attitude (I_{133})	[22]
Organizational perspective (I_2)	IT integration (I_{21})	Electronic records for patient history (I_{211})	[19]
		Elimination of paper work (I_{212})	[23]
	Technology advancement (I_{22})	Technological upgradation of equipment's in dispensary (I_{221})	[19]
		Provide proper training for automation system (I_{222})	Expert opinion
	Collaborative relationship (I_{23})	Strong employee spirit and cooperation (I_{231})	[21]
		Work sharing culture (I_{232})	[19]
Leagility perspective (I_3)	Customer and market analysis (I_{31})	Current requirement of dispensary (I_{311})	[24]
		External environment analysis (I_{312})	Expert opinion
	Strategic management (I_{32})	Maximum utilization of resource (I_{321})	[19]
		Time management (I_{322})	[24]

(continued)

Table 1 (continued)

Enabler	Criteria	Attributes	Reference/expert opinion
	Flexible service (I_{33})	Mutual shifts and manage work responsibility (I_{331})	[19]
		Multi-skilled caregivers, nurses and lab technicians (I_{332})	[21]
	Innovativeness (I_{34})	Creative thinking (I_{341})	[19]
		New idea to service (I_{342})	[25]
	Continuous improvement (I_{35})	Minimization of non-value added activities (I_{351})	[19]
		Minimization of patient process time (I_{352})	[24]

3.2 *Importance-Performance Analysis (IPA)*

IPA is broadly applied for the appropriate method for practitioners to measure the current approach, identify urgent situation improvements for service elements and formulate a new strategy for the current market [26]. IPA analysis is the method that has two proportions, such as importance and performance pattern, where values are plotted crosswise attributes based on importance and performance and are normally divided into four quadrants. The diagrammatic representation of the quadrant [27, 28] is shown in Fig. 2 and explained in Table 2. In service firms, IPA was mainly

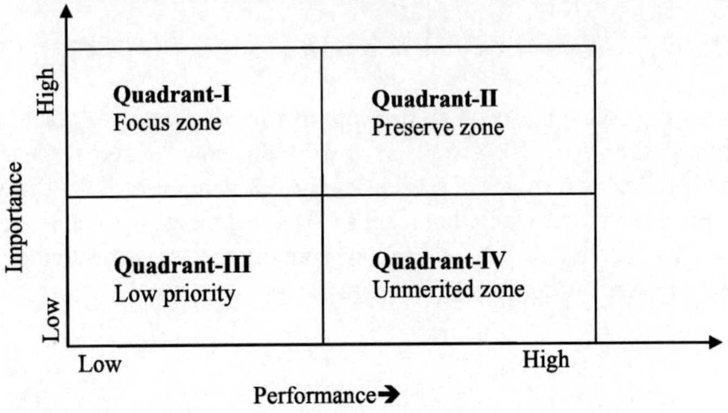

Fig. 2 Diagrammatic representation of IPA

Table 2 Details description of four quadrants of IPA

Quadrant	Description
Quadrant I: Concentrate here	The attributes are located in this quadrant which do not meet the respondent's expectation. There should be emphasis on development efforts
Quadrant II: Keep up the good work	The attributes in this quadrant meet the expectation and should be maintained
Quadrant III: Low priority	Attributes are recognized as poor in importance and poor performance
Quadrant IV: Possible overkill	This cell contains poor importance attributes, but with fairly high efficiency

used to reveal service attributes in four quadrants that help managers to categorize performance and also identify the weakest areas that need improvement [29]. The horizontal axis deals with the performance of attributes, and the vertical axis deals with the significance of attributes.

4 Case Dispensary

In India, the University dispensary is based. 5000 students and 1500 staffs are accessible. The university participates interactively in the exchange programs, for students and visiting professors of various universities and universities from different countries. The university dispensary should be enhanced to provide medical emergency services. The current study is therefore used to improve the leagility of the dispensary to improve service.

The leagility assessment index of a dispensary is denoted as I. This is the result of the overall assessment level dependent of ratings on each driver (R), and the experts give the overall weight (W). The equation for leagility index is

$$I = W \times R$$

The assessment scale has been graded into five levels since every factor involves fuzzy determination. $I = (10, 8, 6, 4, 2)$. 8–10 represents "extremely leagile," 6–8 represents "leagile," 4–6 represents "Fair leagile," 2–4 represents "not leagile" and less than 2 represents "extremely not leagile." The weightage has been given by five experts from the different dispensary. Table 3 contains the assessment model weights and the performance rating from the experts.

Table 3 Weights and performance rating from experts

I_i	I_{ij}	I_{ijk}	R_1	R_2	R_3	R_4	R_5	W_{ij}	W_i	W
I_1	I_{11}	I_{111}	8	6	2	5	5	0.281818	0.301887	0.364583
		I_{112}	9	3	5	5	6	0.354545		
		I_{113}	8	7	7	7	6	0.363636		
	I_{12}	I_{121}	9	8	8	8	5	0.523077	0.339623	
		I_{122}	7	9	8	5	7	0.476923		
	I_{13}	I_{131}	8	5	7	5	6	0.329787	0.358491	
		I_{132}	7	7	8	5	4	0.308511		
		I_{133}	7	10	9	8	7	0.361702		
I_2	I_{21}	I_{211}	5	3	4	4	3	0.521127	0.307692	0.364583
		I_{212}	5	7	8	8	7	0.478873		
	I_{22}	I_{221}	8	5	7	5	6	0.5	0.336538	
		I_{222}	5	5	5	8	9	0.5		
	I_{23}	I_{231}	8	10	8	8	8	0.525641	0.355769	
		I_{232}	8	9	8	7	8	0.474359		
I_3	I_{31}	I_{311}	7	8	5	5	4	0.442623	0.189542	0.270833
		I_{312}	7	5	5	5	4	0.557377		
	I_{32}	I_{321}	7	8	8	8	8	0.486111	0.215686	
		I_{322}	8	9	9	7	7	0.513889		
	I_{33}	I_{331}	9	9	8	7	8	0.486842	0.169935	
		I_{332}	9	9	8	8	7	0.513158		
	I_{34}	I_{341}	8	8	7	5	6	0.532258	0.176471	
		I_{342}	7	6	5	6	6	0.467742		
	I_{35}	I_{351}	8	3	5	5	5	0.460526	0.248366	
		I_{352}	8	8	5	8	8	0.539474		

4.1 Primary Assessment Calculation

The primary calculation for the "Top Management Commitment" given below is completed. Weightage relating to the criterion "Top Management commitment" is

$$W_{11} = [0.28, 0.35, 0.36]$$

The performance assessment of the criterion "Top Management Commitment" is given below as

$$R_{11} = \begin{bmatrix} 8\ 6\ 2\ 5\ 5 \\ 9\ 3\ 5\ 5\ 6 \\ 8\ 7\ 7\ 7\ 6 \end{bmatrix}$$

Index concerning of "Top management Commitment" criterion is provided as

$$I_{11} = W_{11} \times R_{11}$$
$$I_{11} = [8.35, 5.3, 4.88, 5.73, 5.72]$$

Utilizing the above principle, the index concerning the following criteria in leagility assessment is obtained and given below.

$$I_{12} = [8.05, 8.48, 8, 6.60, 5.95]$$
$$I_{13} = [7.33, 7.43, 8.03, 6.09, 5.75]$$
$$I_{21} = [5, 4.92, 5.92, 5.92, 4.92]$$
$$I_{22} = [6.5, 5, 6, 6.5, 7.5]$$
$$I_{23} = [8, 9.53, 8, 7.53, 8]$$
$$I_{31} = [7, 6.33, 5, 5, 4]$$
$$I_{32} = [7.51, 8.51, 8.51, 7.49, 7.49]$$
$$I_{33} = [9, 9, 8, 7.51, 7.49]$$
$$I_{34} = [7.53, 7.07, 6.06, 5.47, 6]$$
$$I_{35} = [8, 5.70, 5, 6.62, 6.62]$$

4.2 Secondary Assessment Calculation

The calculation for "Management Perspective" enabler is given below.

Weightage is given as $W_1 = [0.30, 0.34, 0.36]$ concerning the "Management perspective" enabler.

The "Management Perspective" enabler performance assessment is given as below

$$R_1 = \begin{bmatrix} 8.35 & 5.3 & 4.88 & 5.73 & 5.72 \\ 8.05 & 8.48 & 8 & 6.60 & 5.95 \\ 7.33 & 7.43 & 8.03 & 6.09 & 5.75 \end{bmatrix}$$

Index concerning of "Management perspective" enabler is provided as

$$I_1 = W_1 \times R_1$$
$$I_1 = [7.88, 7.14, 7.07, 6.14, 5.81]$$

Utilizing the above principle, the index concerning the following enabler in leagility assessment is obtained and given below.

$$I_2 = [6.57, 6.58, 6.69, 6.69, 6.88]$$

$$I_3 = [7.79, 7.23, 6.46, 6.45, 6.35]$$

4.3 Tertiary Assessment Calculation

The assessment value of the case university dispensary has been calculated as follows

Complete weight $W = [0.36, 0.36, 0.27]$

Complete assessment vector $R = \begin{bmatrix} 7.88\ 7.14\ 7.07\ 6.14\ 5.81 \\ 6.57\ 6.58\ 6.69\ 6.69\ 6.88 \\ 7.79\ 7.23\ 6.46\ 6.45\ 6.35 \end{bmatrix}$

Leagility index $I = W \times R$

$$I = [7.3804, 6.9611, 6.7634, 6.4226, 6.3458]$$

The final leagility index is the average of $I = 6.7747$.

4.4 IPA

Quadrant I (concentrate here): The attributes in the quadrants must be taken into account by the dispensary managers to improve their performance. The attributes are sharing the management decisions (I_{112}), electronic records for patient history (I_{211}), technological up-gradation of the equipment's in the dispensary (I_{221}) and provide proper training for automation system (I_{222}).

Quadrant II (Keep a good job): The characteristics of quadrants need to be maintained in the same way, and the attributes are built trust among caregivers (I_{113}), minimization of patient process time (I_{352}), strong employee spirit and cooperation (I_{331}), multi-skilled caregivers, nurses and laboratory technicians (I_{332}), work sharing culture (I_{232}), time management (I_{322}) and mutual shifts and manage work responsibility (I_{231}).

Quadrant III (low priority): The attributes in these quadrants are of low importance and low performance. Attributes include minimization of non-value-added activities (I_{351}), external environment analysis (I_{312}), a frequent interval of the meeting (I_{111}), smooth information flow (I_{131}), the current requirement of the dispensary (I_{311}), transparent in decisions (I_{132}) and new idea to service (I_{342}).

Quadrant IV (Possible overkill): The attributes in these quadrants are low in importance but high in performance. The attributes are the maximum utilization of resource (I_{321}), elimination of paperwork (I_{212}), creative thinking (I_{341}), transparent and quick decision (I_{122}), multiple tasking (I_{121}) and built trust among caregivers (I_{133}).

5 Results and Discussions

In the case of a university dispensary, the leagility index is calculated at 6.7747. This belongs to the range (6–8) which is specified as "leagile." The current level of case university dispensary is leagile, which required improvement to achieve "extremely leagile" by concentrating the weaker attributes and taking the necessary action by management to improve the dispensary. The assessment model would help to periodically investigate the improvement of leagile levels in the university hospital. Each assessment will require a different action plan to improve the leagile level. Maintenance of the "extremely leagile" by monitoring the performance of the periodic assessment model. The dispensary has the strength of teamwork between caregivers, multiple tasks, efficient use of resources and time management.

The current case study used IPA to identify the weaker attributes by analyzing performance with relevance. Improving technology upgrades, sharing management decisions, sound automation training and electronic records are needed. Management needs to focus its strength and special attention on the critical characteristics of improvement. Suggestions for improving the critical attributes are given in Table 4.

Table 4 Suggestions for weaker attributes

Weaker attributes	Suggestions for improvement
Sharing management decisions	• Discuss the important decision with all caregivers • Create an open meeting among caregivers for their opinions • The weekly meeting would help to deliver the management decisions effectively
Electronic records for patient history	• Create electronic records of the patients • Transfer all paperwork into electronic • Give special training for all caregivers on electronic change
Technological upgradation of the equipment's in a dispensary	• Update all the electronic equipment and computer for better service • Purchase new machine by replacing old machines • Installation and training would require caregivers to update themselves on technology updates
Provide proper training for automation system	• Special training session weekly organized in shifts for caregivers • Continuous training during their work time • Monthly review on the performance and change the training program accordingly

6 Practical Implications

The current IPA has identified weaker attributes including sharing the management decisions, electronic records for patient history, technological upgradation of the equipment's in the dispensary and provide proper training for automation system. The management team formed a team of experts who worked on the improvement attributes by developing a new feasibility strategy daily. Suggestions for improvement from experts will help them reach "extremely leagile." The team's experts will periodically evaluate the performance and continue to take the necessary action to improve the dispensary. All the employees are trained to improve their efficiency in automation and electronic records. The team would report to the management directly about the process of improvement and the requirements for leagility.

7 Conclusion

Leagility is primarily used in health care to improve customer satisfaction through better use of resources, better use of resources, increasing the productivity of workers through reducing all the existing costs and not adding value to their work. The measurement of leagility performance would indicate the current organizational position and help improve performance with service quality. In this aspect, an assessment of leagility is done utilizing a multi-grade fuzzy approach. To evaluate the current level of leagility and analysis, the case of the university dispensary will be selected as "leagile." The new approach aims at enhancing their efficiency and attaining their goals. Management needs to electronically move the whole program and upgrade the following equipment at the university. The dispensary that regularly checks the model of assessment will help improve its Leagility level in its unit. The separate team responsible for taking the appropriate steps to boost leagility should deliver better results. In the future, the evaluation will be performed using fuzzy logic and a DSS and artificial intelligence with neural network and fuzzy neurons included.

References

1. Tseng YH, Lin CT (2011) Enhancing enterprise agility by deploying agile drivers, capabilities and providers. Inf Sci 181(17):3693–3708
2. Singh S, Remya T, Shijo TM, Nair D, Nair P (2014) Lean six sigma application in reducing nonproductive time in operation theaters. J Natl Accred Board Hosp Healthc Providers 1(1):1–6
3. Huang YY, Li SJ (2010) How to achieve leagility: a case study of a personal computer original equipment manufacturer in Taiwan. J Manuf Syst 29(2–3):63–70
4. Matawale CR, Datta S, Mahapatra SS (2013) Development of a decision support system for leagility assessment in fuzzy environment. Asia-Pacific international congress on engineering and natural sciences (APICENS 2013), Bangkok

5. Smys S, Joe CV (2019) Big data business analytics as a strategic asset for health care industry. J ISMAC 1(02):92–100
6. Naylor JB, Naim MM, Berry D (1999) Leagility: integrating the lean and agile manufacturing paradigms in the total supply chain. Int J Prod Econ 62(1–2):107–118
7. Mishra V, Samuel C, Sharma SK (2019) Lean, agile and leagile healthcare management—a case of chronic care. Int J Healthc Manage 12(4):314–321
8. Mason-Jones R, Naylor B, Towill DR (2000) Engineering the leagile supply chain. Int J Agile Manage Syst 2(1):54–61
9. Okongwu U, Lauras M, Dupont L, Humez V (2012) A decision support system for optimising the order fulfilment process. Prod Plann Control 23(8):581–598
10. Denavi HD, Mirabi M, Rezaei A (2017) Ranking of leagility factors based on job satisfaction through a combinatory model of fuzzy TOPSIS and AHP (case study: MRI Hospital, Shiraz, Iran). Open J Bus Manage 6(01):21
11. Rahimnia F, Moghadasian M (2010) Supply chain leagility in professional services: how to apply decoupling point concept in healthcare delivery system. Supply Chain Manage Int J 15(1):80–91
12. Türkşen IB, Zarandi MF (1999) Production planning and scheduling: fuzzy and crisp approaches. In: Practical applications of fuzzy technologies. Springer, Boston, pp 479–529
13. Sreedharan VR, Raju R, Sunder MV, Antony J (2019) Assessment of lean six sigma readiness (LESIRE) for manufacturing industries using fuzzy logic. Int J Qual Reliab Manage 36(2):137–161
14. Lin CT, Chen CT (2004) A fuzzy-logic-based approach for new product Go/NoGo decision at the front end. IEEE Trans Syst Man Cybern Part A Syst Hum 34(1):132–142
15. Narayanamurthy G, Gurumurthy A, Subramanian N, Moser R (2018) Assessing the readiness to implement lean in healthcare institutions—a case study. Int J Prod Econ 197:123–142
16. Dubey R, Singh T, Gupta OK (2015) Impact of agility, adaptability and alignment on humanitarian logistics performance: mediating effect of leadership. Glob Bus Rev 16(5):812–831
17. Aravindraj S, Vinodh S (2014) Forty criteria based agility assessment using scoring approach in an Indian relays manufacturing organization. J Eng Des Technol 12(4):507–518
18. Shinwon S, Sunguk P, Mihyun G, Namgyu K, Sunguk L (2015) Key factors of change readiness for the success of management innovation in manufacturing industry. Int J u- e-Serv Sci Technol 8(10):179–192
19. Suresh M, Patri R (2017) Agility assessment using fuzzy logic approach: a case of healthcare dispensary. BMC Health Serv Res 17(1):394
20. Zain M, Rose RC, Abdullah I, Masrom M (2005) The relationship between information technology acceptance and organizational agility in Malaysia. Inf Manage 42(6):829–839
21. Vinodh S, Vimal KEK (2012) Thirty criteria based leanness assessment using fuzzy logic approach. Int J Adv Manuf Technol 60(9–12):1185–1195
22. Dubey R, Gunasekaran A (2016) The sustainable humanitarian supply chain design: agility, adaptability and alignment. Int J Logistics Res Appl 19(1):62–82
23. Vinodh S, Kumar VU, Girubha RJ (2012) Thirty-criteria-based agility assessment: a case study in an Indian pump manufacturing organisation. Int J Adv Manuf Technol 63(9):915–929
24. Matawale CR, Datta S, Mahapatra SS (2013) Interrelationship of capabilities/enablers for lean, agile and leagile manufacturing: an ISM approach. Int J Process Manage Benchmarking 3(3):290–313
25. Vinodh S, Aravindraj S (2015) Benchmarking agility assessment approaches: a case study. Benchmarking Int J 22(1):2–17
26. Hansen E, Bush RJ (1999) Understanding customer quality requirements: model and application. Ind Mark Manage 28(2):119–130
27. Martilla JA, James JC (1977) Importance-performance analysis. J Mark 41(1):77–79

28. Tzeng GH, Chang HF (2011) Applying importance-performance analysis as a service quality measure in food service industry. J Technol Manage Innov 6(3):106–115
29. Shieh JI, Wu HH (2009) Applying importance-performance analysis to compare the changes of a convenient store. Qual Quant 43(3):391

Chapter 34
A Unified Model for Face Detection Using Multiple Task Cascaded Neural Network Prepended with Non-local Mean Denoising Algorithm

Sapna Rathore and Luv Sharma

1 Introduction

The face detection technique is a modern computer technique which detects the position and magnitude of the human in a modern aspect. Generally, it refers to face only and eliminates everything, for example, houses and plants. It is generally taken as a very important case when dealing with face location. While locating face, the main job is to understand the magnitude and the position of the values of face, which are not recognized. During the face detection process, the face is made and sorted bitwise according to the following reflection image in the master database [1].

In the meantime, the face detection may also be a mental demand process, majorly identifying a position and giving proper attention to the target. Different factors like colour and orientation point out the likeliness to accepted faces and how they are interfering with. The face is a very crucial factor in identity and to give a clear feeling. The face generally performs this important role [1, 2]. It is too uncommon to lose memory on the face that a person had realized before, and this is very critical even then there are different changes in the optical stimulus objects to ease vision. It can be influenced by the changing of conditions, old age and debit of attention mostly in classes and in an account of the lifestyle of an individual.

MTCNN model comprises three networks viz., P-Net, R-Net and O-Net cascaded in its design, respectively. The initial network tries to identify human feature point, and the later two apply regression-based learning to filter out the correct bounding box enclosing the face features.

S. Rathore (✉) · L. Sharma
Department of Electronics and Communication, SS College of Engineering, Udaipur, India
e-mail: sapna.rathore22@gmail.com

L. Sharma
e-mail: engineerluv@gmail.com

I. Jeena Jacob et al. (eds.), *Data Intelligence and Cognitive Informatics*, Algorithms for Intelligent Systems,
https://doi.org/10.1007/978-981-15-8530-2_34

Some of the gripping computational replicas of face recognition are that it not only contributes theoretical information but practical application too. Regions of employment of the computers in which function is detection and realization of faces contain the film processing, criminal identification, image processing, security system, tagging purposes and identity verification, human–computer intervention [1, 3]. As faces of the human are complicated, and having multi-dimensions and also has a meaningful sight, develops the computational replica for detection of face. Few Websites using pictures like Picasa, Photobucket and Facebook normally use detection of face. The automatically tagging features in these Websites are used to tag the individual in the picture to have some idea about that person.

The image acquisition deals with the data from the face, and recognition of face is used in these areas to recognize face and edit information which is same as that with facial features. A picture is then reformed and gives correction which includes drawing to edit the background and to the cases which cannot be a rhyme to use ideally to have it for easier verification. Face recognition technique is normally made up of processing, classification of sub-module and extraction of characteristics of it. The input power of recognition of face is normally the picture of face, and it is obtained from the camera to the storage database of the user [1–4]. All pictures need to be pre-processed to generate the geometric and photometric standard of a picture of the face, and it shows vector feature results when there is the extraction of features present in the picture. Usually, the mixing of the number of clients and the detection of the individual in the storage database of the user are necessary to have a category of detection, and it includes

1. The pre-processing module is normally decrease and edit the category of factors when the picture of the face reflects as the illumination result.
2. Scanning the picture twice is very important and then store the output picture.

The modern task of deep learning is introduced in the system to provide the recognition of face and provides clarity method which is used as a benchmark. Different methods are used one by one over different datasets, while another method works at very random method relative to outcomes from experiment and acts. Face recognition with detection methods provides consecutive outcomes in system deploying Haar-like features which gives their best to the system; while it is connected with wrong recognition as compared to LBP which has to give priority in the diagram for parts of Gabor is given value because technically, its quality provides good database [5].

2 Literature Survey

Face detection task is handled via image processing approaches, machine learning and currently using deep learning strategies, and few pertinent approaches are discussed here.

The analysis of the rate of face recognition of the present several face recognition algorithms is extremely essential for the development of new and better algorithms which are more robust. In this report, the performance analysis of principal component analysis (PCA) and linear discriminate analysis (LDA) for face recognition is reported [2]. The mentioned analysis was performed on several current LDA- and PCA-based algorithms, and standard public databases were used. Several PCA algorithms were analysed, and among them, the best rate of face recognition of 100% was given by manual face localization, which was utilized on SHEFFIELD AND ORL database which consisted of 100 components. The second-best face recognition rate of 99.70% was achieved from immune networks (PCA-IN) [5] which is PCA based and was used on ORL database. A lot of LDA-based algorithms were analysed, and among them, the best rate of 98.9% of face recognition was given by illumination adaptive linear discriminate analysis (IALDA) on the CMU PIE database. The second-best rate of 98.125% was displayed by fuzzy fisher face on ORL database using genetic algorithm.

The article [6] contains a mixture of integrated BPNNs and optimized LDA and discrete cosine transform. The process depending on DCT can certainly reduce the size of data of true original signals. THE OPTIMIZED D-LDA pattern introduced by us can handle important data and can bear the effect and consequences of edge class caused due to restructuring the inside-class scatter matrix and in-between-class scatter matrix and improve the defining rate with lowering dimension. It also possesses the ability to solve smaller quantity issue keeping useful information safe. Past that, the pointing vectors are combined for the practice set and are kept to guide the important networks of the human race.

Considering the attractive qualities of the spatial region and introduction selectivity of the Gabor channel, channels are outlined for separating facial peculiarities from the nearby picture. The gimmick vector in light of Gabor channels is used as the classifier's data which is an feed-forward neural system (FFNN) [7] on a decreased peculiarity subspace which is adapted by a strategy that is much more straightforward and less demanding than PCA (foremost part examination). The proposed methodology is exceptionally successful, and its adequacy is depicted by the consequences of the investigations, on testing an enormous bunch of pictures and the examination with the most recent frameworks. The picture will join alongside Gabor channels by augmentation of the picture by Gabor channels in recurrence area [8]. They have been put away in recurrence area to spare time. Peculiarities are a cell exhibit which contains the consequence of the convolution of the picture with each of the forty Gabor channels. The information vector of the system will have substantial qualities, which implies a lot of reckoning. So, the matrix size is decreased to one-third of its original size by removing some columns and rows. To delete is not the best option, but it saves a lot of time as compared to other methods such as PCA. Face recognition and detections are applied in a plethora of different fields such as videoconferencing, security systems and identification. This research aims to implement a classifier which is based on neural networks (multi-layer perception) for face recognition/detection. The ANN helps in classifying non-face and face patterns.

Face discovery from a long database of face pictures with distinctive foundations is not a simple undertaking. In this work, the face location arrangement of shaded face pictures is exhibited which is invariant to the foundation and adequate enlightenment conditions. An edge level is situated to reject the non-human face pictures and the obscure human face pictures which are not shown in the data database of face pictures. In this paper, the worldwide gimmicks extraction is finished utilizing PCA-based eigenface processing technique, and the discovery part is finished utilizing multi-layered encourage forward artificial neural networks with backspread methodology. This calculation is executed utilizing MATLAB programming. The learning procedure of neurons is utilized to prepare the info face pictures with 1000 cycles to minimize the lapse. In this framework, face distinguish assignment is finished with enhanced precision and achievement rate actually for uproarious face pictures [9].

Face recognition (FR) is a very challenging problem because of the variations in expression, pose and illumination. For most of the present face recognition approaches, the search results have been satisfactory, yet they included a lot of images which were not related to the target images [10]. A new technique has been introduced by us which utilizes an image-based method for human face recognition purpose. This clever framework lives up to expectations by wiping out un-helpful information from face pictures by picture pressure, with the assistance of Sober edge detection (SED) and by contrasting this and the 2-D discrete cosine change (2D-DCT) system for the enhanced proficiency and velocity. SED is a famous edge detection method and has been considered in this approach [11]. The function, an edge is existent in the image toolbox. In this edge function, the Sobel method utilizes the derivative approximation to search edges. Thus, it returns edges at the specific points where the gradient is a maximum of the considered image [12]. For the face recognition system, the developed algorithm creates an image-based approach, using the 2D-DCT or SED for the image compression purpose and the self-organizing map (SOM) neural network for face recognition purposes, which is simulated in MATLAB [13].

3 Proposed Model

The proposed model performs multiple tasks cascaded convolutional network (MTCNN) model for face identification. This model has three convolutional systems (P-Net, R-Net, and O-Net) [14] and can outflank many face recognition benchmarks while holding continuous execution as shown in Fig. 1.

The primary activity is to go in a picture of the program. In this model, a picture pyramid is needed to make and distinguish appearances of every single changed size. At the end of the process, various duplicates of a similar picture are needed to make various sizes to scan for various estimated faces inside the picture.

For each scaled duplicate, a 12×12 phase 1 portion is having that will experience all aspects of the picture, filtering for faces. It begins in the upper left corner, a segment of the picture from (0, 0) to (12, 12). This segment of the picture is passed to P-Net, which restores the directions of a jumping box on the off chance that it sees a face.

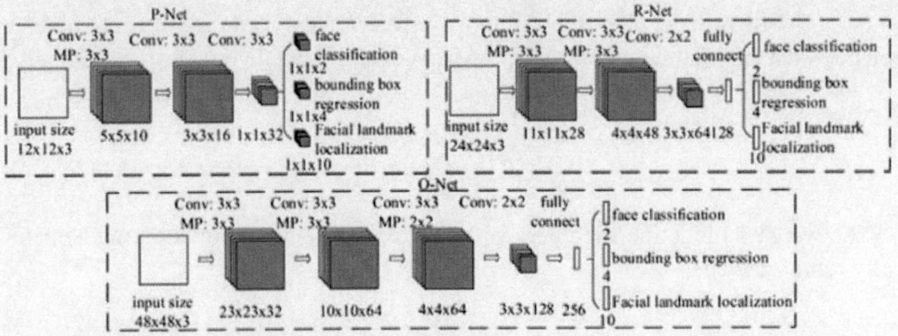

Fig. 1 Architecture of MTCNN model

After the crates with higher certainty have been selected, the arranged framework should be normalized, changing over all the organizing frameworks to that of the real, "un-scaled" picture. Since most portions are in a downsized picture, their directions will be founded on the littler picture.

In any case, there are still a many of bounding boxes left, and a many of them cover. Non-maximum suppression, or NMS, is a strategy that decreases the number of bounding boxes.

In this specific program, NMS is led by first arranging the bounding boxes (and their particular 12×12 bits) by their certainty, or score. In some different models, NMS takes the biggest bounding box rather than the one the system is generally positive about. For each bounding box, a variety of a similar size is made and duplicate the pixel esteems (the picture in the jumping box) to the new cluster. On the off chance that the bounding box is beyond the field of play, the segment of the picture in the jumping box is just duplicated to the new exhibit and fill in everything else with a 0. In the picture over, the new cluster for McCartney's face would have pixel esteems in the left half of the crate, and a few segments of 0s close to the correct edge [14].

R-Net's yield is like that of P-Net: It incorporates the directions of the new, increasingly exact jumping boxes, just as the certainty level of every one of these bounding boxes.

Lastly, the box-contained with lower certainty levels is disposed of and normalize both the bounding box arranges and the facial milestone organizes. At long last, run them through the last NMS. Now, there should just be one jumping enclose for each face the picture [12].

The performance of MTCNN model degrades in the noisy environment. To boost the performance even in a noisy scenario, noise removal methods are deployed at the feeding end of MTCNN model. The algorithm deployed here in this work is non-local means denoising algorithm. The algorithm can be understood as follows.

It is a sort of denoising algorithm implemented at the image processing level. The concept is to replace the noisy pixel value x with a mean value of surrounding pixel y. Mean accountancy $u(x)$ is performed using a weighted function $f(x,y)$ applied in a neighbour region of the target pixel to evaluate the new pixel value which is a smoothen version of neighbourhood intensities. Here, $C(x)$ is a normalized coefficient considered a summation of weighted function over the whole image space Ω.

$$u(x) = \frac{1}{C(x)} \int_{\Omega} v(y) f(x, y) \mathrm{d}y.$$

The weighted function used here is the Gaussian weighted function mathematically represented as,

$$f(p, q) = \mathrm{e}^{-\frac{|B(q)-B(p)|^2}{h^2}}$$

Here, $B(\,)$ represents the local mean in image space and h represents the variance, also working here a primary filtering parameter. Higher the variance, more is the effect of neighbours in decision making.

4 Experimental Setup and Results

The experimental set-up for the proposed model runs on a Window-10, 64-bit machine overclocked with NVIDIA GTX 1060 graphic card (6 GB). The operating system is running over an 8 GB DDR-4 ram. The model is implemented in Python version 3.7.

The original image dataset is cluttered with the noise of increasing variance, and the resultant noisy image is applied to core MTCNN and proposed a unified model for their noise performance. Table 1 depicts a set of test images as well as the results of MTCNN model for face detection, while Table 2 represents the task of face detection performed on input image corrupted by Gaussian noise. It is pertinent from Table 2, that MTCNN model performance degrades in a noisy environment.

To improve the performance, non-local means denoising algorithm is deployed with a Gaussian weighted function. The variance of Gaussian function is kept as 10 with window size 7 and search window 21, respectively. The right column of Table 2 depicts the results showing the outperformance of the deployed model against the core MTCNN model. A comparative plot showing the performance of the proposed method against the core MTCNN model is shown in Fig. 2.

For the ground face count of 9, Fig. 2 shows the number of faces detected in terms of increasing noise. The red plot shows the performance of MTCNN core model, while the blue plot depicts the performance of hybrid MTCNN model with Gaussian denoising feed. It can be observed that due to neighbour-based smoothing, the proposed model has lower performance against MTCNN in lower noise conditions, while in a highly noisy scenario, when facial landmarks are difficult to be identified, the proposed model outperforms.

Table 1 Face detection performed on test images using MTCNN model

Original image	Noiseless_face_detection using MTCNN

Table 2 Results of face detection performed on a denoised image using the proposed method

Noisy_face_detection using MTCNN	Denoised_face_detection using the proposed method

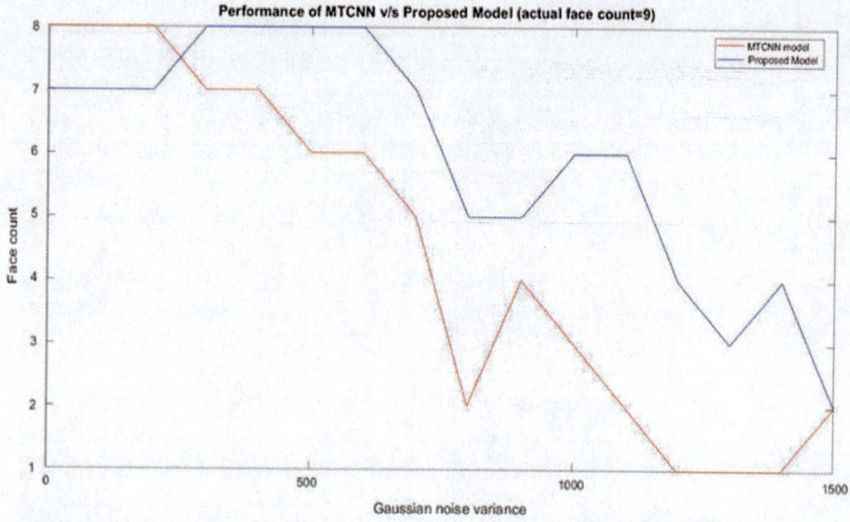

Fig. 2 Comparative plot of performance between MTCNN model and proposed hybrid model

5 Conclusion

Face identification has been a point of major research among scholars. In this work, the performance of core MTCNN model is improved by deploying a non-local means denoising algorithm at the feeding end of the cascaded neural network. The noise performance of deployed methodology is investigated here for Gaussian noise having variance in the range of 300–1500. For the sake of denoising, a Gaussian function is deployed as a weight function having a variance of 10, in the denoising algorithm. It is observed that the addition of the proposed methodology to the core model improved its performance in the noisy scenario. The comparative plot and simulation results confirm our hypothesis to improve the performance of MTCNN using non-local Gaussian denoising method.

References

1. Jose E et al (2019) Face recognition based surveillance system using facenet and mtcnn on jetson tx2. In: 2019 5th International conference on advanced computing & communication systems (ICACCS). IEEE
2. Ma M, Wang J (2018) Multi-view face detection and landmark localization based on MTCNN. In: 2018 Chinese automation congress (CAC). IEEE
3. Xiang J, Zhu G (2017) Joint face detection and facial expression recognition with mtcnn. In: 2017 4th International conference on information science and control engineering (ICISCE). IEEE
4. Rowley HA, Baluja S, Kanade T (1998) Neural network-based face detection. IEEE Trans Pattern Anal Mach Intell 20(1):23–38

5. Hsu R-L, Abdel-Mottaleb M, Jain AK (2002) Face detection in color images. IEEE Trans Pattern Anal Mach Intell 24(5):696–706
6. Gautam D et al (2018) Machine learning-based diagnosis of melanoma using macro images. Int J Numer Methods Biomed Eng 34(5):e2953
7. Yang AP et al (2015) Image denoising based on improved K-SVD and non-local regularization. Comput Eng 41:249–253
8. Viola P, Jones MJ (2004) Robust real-time face detection. Int J Comput Vision 57(2):137–154
9. Hjelmås E, Low BK (2001) Face detection: a survey. Comput Vis Image Underst 83(3):236–274
10. Jones M, Viola P (2003) Fast multi-view face detection. Mitsubishi Electric Research Lab TR-20003-96 3.14
11. Zhang K et al (2016) Joint face detection and alignment using multitask cascaded convolutional networks. IEEE Sig Proces Lett 23(10):1499–1503
12. Shakya S (2020) Analysis of artificial intelligence based image classification techniques. J Innov Image Proces (JIIP) 2(01):44–54
13. Vijayakumar T (2019) Classification of brain cancer type using machine learning. J Artif Intell 1(02):105–113
14. Yang M et al (2013) Non-local means theory based Perona–Malik model for image denoising. Neurocomputing 120:262–267

Chapter 35
Workplace Stress Assessment of Software Employees Using Multi-grade Fuzzy and Importance Performance Analysis

S. Sreedharshini, M. Suresh, and S. Lakshmi Priyadarsini

1 Introduction

Software industry is continuously expanding in India as it is in strong demand globally. India exports software services/projects to more than 95 countries. More than four million direct employees and ten million indirect employees are working in the software industry in India. Every day, the number of employees working in the software industry is increasing. Much fresher is pushed into this industry regardless of their stream of study or interest. Due to the increasing demand for software projects, the employees are forced to work overtime and nightshifts.

When employees are given tasks and targets to be achieved within a short span of time, and if the employee is unable or facing struggles to achieve, it would develop stress. Stress is defined as the state of physical, psychological or social complaints or dysfunctions of individuals feeling unable to bridge a gap on the requirements or expectations placed on them [1]. Workplace stress in the software industry is particularly high and is also increasing at a fast pace. Employees are vulnerable to workplace stress irrespective of age, gender, ethnicity or disability [2].

Research from Capita Employee Benefits reveals that in UK, almost 44% of employees are affected by high stress and quit a job. In 2016, TUC survey states that over 70% of people in UK sites stress as one of the top five hazards. This survey also specifies that stress has increased to 3 points in 2016 when compared to the results

S. Sreedharshini (✉) · M. Suresh
Amrita School of Business, Amrita Vishwa Vidyapeetham, Coimbatore, India
e-mail: dharshudharsh00@gmail.com

M. Suresh
e-mail: drsureshcontact@gmail.com

S. Lakshmi Priyadarsini
Govt Victoria College, Palakkad, Kerala, India
e-mail: lakshpriyadarsini@gmail.com

I. Jeena Jacob et al. (eds.), *Data Intelligence and Cognitive Informatics*,
Algorithms for Intelligent Systems,
https://doi.org/10.1007/978-981-15-8530-2_35

of the survey in 2014. In a recent survey conducted from Korn Ferry—'Workplace stress continues to Mount,' it is said that people are more stressed than five years ago, with 2/3rd saying it costs them to sleep. 13% of the respondents of the survey quit the job due to stress.

This research would concentrate more on evaluating the stress level of the employees in a software company, comparing it with the competitor's company, and will help to identify the factors influencing stress. Those weaker attributes would be addressed by suggesting to improve the work experience of the employees in the workplace. This assessment can be done once a year in the company.

The motivation behind this survey is to assess whether there is stress for the employees of the software company. If there is stress, this model measures the level of the stress using the MGF model and also would identify the weaker attributes that contribute to the employee's stress through IPA. Suggestions for reducing the stress level of the employees are also proposed in the conclusion.

2 Literature Review

Stress is associated with substantial economic consequences, including increased absenteeism, decreased job satisfaction, decrease in worker productivity and physical characteristics of the work environment such as noise, lighting and ventilation [3]. Williams and Cooper [4] measured the occupational stress using OSI scale and developed the outcome of the stress using Pressure Management Indicator (PMI). Lambert and Lambert [5] research on nurses conducted by 17 countries is assessed by identifying and comparing the findings of the research on the role of stress/strain. Kumar and Kumar [6] explored workplace spirituality as a moderator for workplace stress and health, and it is found that stress has a negative impact on health and spirituality has a positive impact on health.

In [7], a new scale to measure job stress is developed by adding the extended set of psychosocial stressor factors and verifies the reliability of the new scale. Chen et al. [8] developed the Ergonomics Workload Stress Index (EWSI) to find the relationship between the stress level and health with the ergonomics of the workplace.

3 Conceptual Model for Assessing Stress Level in the Work Place

The conceptual model is built by collecting the enablers, criteria and attributes through the literature review and expert's opinion. This framework evaluated the stress level of the software industry using three enablers, eight criteria and twenty-seven attributes, and it is depicted in the Table 1. This model mainly concentrated

Table 1 Conceptual model for stress level assessment in workplace

Enablers	Criteria	Attributes
Stressor (S_1)	Intrinsic effort (S_{11})	Error rate (S_{111})
		Less job satisfaction (S_{112})
		Intend to leave a job (S_{113})
	Extrinsic effort (S_{12})	Pressure to work overtime (S_{121})
		Workplace conflicts (S_{122})
		Low pay and benefits (S_{123})
		Deadline pressure (S_{124})
		Job insecurity (S_{125})
	Workplace distraction (S_{13})	Poor work relationship (S_{131})
		Role-related stress (S_{132})
		Project fatigue (S_{133})
Strain (S_2)	Mental health (S_{21})	Depressed (S_{211})
		Anxiety (S_{212})
		Work-life balance (S_{213})
		Lack of concentration (S_{214})
	Physical health (S_{22})	High blood pressure (S_{221})
		Insomnia (S_{222})
		Headache (S_{223})
	Mental demand stress (S_{23})	Overthinking (S_{231})
		Increased responsibility (S_{232})
		Remembering (S_{233})
		Overcommitment (S_{234})
Ergonomics (S_3)	Work environment (S_{31})	Hot temperature (S_{311})
		Lighting discomfort (S_{312})
		Noise (S_{313})
	Work space (S_{32})	Posture discomfort (S_{321})
		Poor layout (S_{322})

on the three main factors of stress, namely stressor [4, 9, 10] strain [2, 11] and ergonomics [3, 8].

Five experts from different software organizations have contributed to the ranking of the factors like enablers, criteria and attributes, called as weightage. Five experts from the ABC Organization have contributed to the ranking of the factors in Table 1 which was called as the rating. These data collections from the 10 different experts were used for the calculation of the stress level assessment for employees in the workplace.

4 Case Study

ABC Organization is a private software company which develops products such as Web sites and applications such as ERP. It is a small-scale organization, which has been started three years ago. ABC Organization has a low volume of employees, where the work is shared by these employees. For them, it would almost take 2–3 weeks to complete a Web site and from 1 month to more than a year to develop an application depending on the project and team size.

Estimation of the project is made, analysis of the time required is identified, and the quotation is sent to the client end. Once it is accepted, the team members are identified and the team is formed. The application is developed and the backend database is built. Test runs are conducted, and finally, the completed project is delivered to the client-side end.

4.1 *Stress Assessment Using Multi-Grade Fuzzy*

The fuzzy logic approach is based on human logic and takes advantage of conceptual knowledge without boundaries. Some of the concepts of fuzzy logic include fuzzy set, linguistic variables, probability distribution and fuzzy if-then rules [12]. MGF overcomes the drawbacks associated with conventional crisp methods.

This framework developed would help the assessment by answering the following questions (Q):

Q1: How to measure the stress levels of the employees in the software industry?
Q2: What are the factors that influence the stress level in employees of software industry?
Q3: How to reduce the stress level in the software industry with the weaker attributes identified?

The weightage of the enablers, criteria and attributes are captured in the following scale (Table 2).

The rating and weightage were collected from the experts in the software industry. The experts were instructed to rate the factors on the scale of 1–10. Here, 0–2 represents the level is very low and so on till, 8.01–10 representing the level 'Extremely

Table 2 Weightage and level of the rating scale

Scores	Level	Grades
0–2	Very low	Stress-free
2.01–4	Fair	Slightly stressful
4.01–6	Moderate	Moderate
6.01–8	High	Stressful
8.01–10	Extreme	Extremely stressful

Table 3 Weightage calculation

Enablers	E_1	E_2	E_3	E_4	E_5	Mean	Weightage (W)
Stressor	8	9	8	5	7	7.4	0.342593
Strain	8	8	6	7	7	7.2	0.333333
Ergonomics	9	8	7	4	7	7	0.324074
						21.6	1

high.' For example, job insecurity, if there is no fear of losing the job, then the ranking would be between 0–4 in accordance to their organization. If there is a fear of losing the job, i.e., there is job insecurity, then the ranking would be between 6–10 accordance with the organization.

The data in the table has been particularly collected for this research purpose from the employees. The rating was collected from the employees of the same organization, five experts had been selected from the case organization, and the data was collected directly from them. The data for the weightage was collected from five experts from five different organizations. The experience of the experts varies from 1 to 15 years to make the data more generalized.

Data was collected from the employees based on their experience; hence, the data would not exhibit much variation when the environment changes, and there is a little chance for minor variation in few attributes, but those variations would not have much impact on the results.

$$\text{Weightage} = \text{Mean}/\sum \text{Mean}$$

The weightage data collected from the experts of different companies was used to calculate the weightage for the factors. The data of the enablers, criteria and attributes was separated. Mean of the factors was calculated, and weightage of the factor was calculated using the above formula. For example, the weightage calculation of the enablers is displayed in Table 3.

Similarly, for criteria (W_i) and attributes (W_{ij}), the respected weightage was calculated with the data collected from the experts of a different organization. The mean normalized weightage of the expert's opinion and the attributes rating are shown in Table 4.

4.1.1 Primary Assessment Calculation

The primary calculation was performed by multiplying the attributes data of the rating collected with the mean of the weightage attributes collected. Matrix multiplication was used to perform this calculation. The formula used is given below:

$$R_{ij} * W_{ij} = L_{ij} \quad [13\text{--}15]$$

Table 4 Rating and weightage by experts

Attributes	E_1	E_2	E_3	E_4	E_5	W_{ij}	W_i	W
S_{111}	9	8	8	9	8	0.35245902	0.28	0.342593
S_{112}	8	7	6	3	8	0.31967213		
S_{113}	6	1	9	9	8	0.32786885		
S_{121}	3	4	7	5	7	0.21717172	0.3	
S_{122}	4	3	9	9	9	0.21717172		
S_{123}	1	3	7	6	3	0.18181818		
S_{124}	6	6	9	9	9	0.22222222		
S_{125}	10	5	9	9	9	0.16161616		
S_{131}	2	10	2	3	6	0.34951456	0.42	
S_{132}	6	7	6	3	4	0.3592233		
S_{133}	1	7	4	7	3	0.29126214		
S_{211}	6	2	6	6	6	0.2208589	0.357798	0.333333
S_{212}	6	8	9	9	9	0.28220859		
S_{213}	6	5	6	7	7	0.23312883		
S_{214}	4	3	8	9	6	0.26380368		
S_{221}	9	3	8	9	6	0.33070866	0.321101	
S_{222}	10	10	9	3	4	0.33070866		
S_{223}	6	10	8	5	6	0.33858268		
S_{231}	2	10	6	5	2	0.32407407	0.321101	
S_{232}	2	10	7	1	2	0.18518519		
S_{233}	4	10	3	3	2	0.21296296		
S_{234}	2	10	6	2	2	0.27777778		
S_{311}	2	8	1	1	1	0.3	0.548387	0.324074
S_{312}	6	3	6	7	7	0.34615385		
S_{313}	2	1	9	7	6	0.35384615		
S_{321}	4	10	7	5	7	0.48809524	0.451613	
S_{322}	10	3	6	6	6	0.51190476		

4.1.2 Secondary Assessment Calculation

The secondary calculation was performed by multiplying the result from the primary assessment calculation with the mean of the weightage criteria collected. Matrix multiplication was performed using the above formula.

4.1.3 Tertiary Assessment Calculation

Tertiary calculations were performed by multiplying the result from the secondary assessment calculation with the mean of weightage enablers collected. Matrix multiplication was performed using the formula mentioned above. The final result from the tertiary assessment calculation was said to be the measure of the stress level of the ABC Organization.

The result of this calculation performed was found to be 5.800929. By applying this to the weightage and level of the rating scale (Table 2), the level of stress of the employees in the organization could be identified. Since the result was around 5, it came under the category of 4.01–6. Thus, the respective grade was 'Moderate.' Hence, the employees in ABC Organization face a moderate stress level in the workplace.

4.2 Classification of Stress Assessment Attributes Using IPA

IPA analysis is a graphic method, with a two-dimensional coordinate system. The average value of importance and performance of different service/product elements, which were calculated in relation to one another, mainly in the area divided into four quadrants [16]. The mean of the attributes rating was taken as performance and was marked in the X-axis, and the weightage of the attributes was marked in the Y-axis as the weightage [17].

In IPA analysis, the respective mean calculated for both the rating and the weightage was listed for each attribute as performance and weightage. The data was plotted in the scattered plot graph, where the average of both the axes was taken and marked with a line in the graph to section it into four different quadrants of the IPA analysis. The inference of the four quadrants of the IPA analysis is described in Table 5. The attributes classified using IPA analysis are depicted in Fig. 1.

5 Results and Discussions

The current model developed for the assessment of the stress level in the software industry had concluded that the stress level in the ABC Organization is 'moderate.' The case company's employee stress level S was found to be around 5.8 from the final results of the MGF, which lies in the scale 4–6; hence, the stress level is moderate for the employees.

The model has also identified four weaker attributes which influence the stress level of the employees. The four weaker attributes are identified from the IPA, i.e., the attributes lying in the first quadrant in the IPA. These weaker attributes can be controlled by the ABC Organization through necessary actions taking against them. Those actions are given as suggestion (Table 6). If the company improves the

Table 5 Inference of IPA

Quadrants—inference	Attributes in the quadrant
Q1—The performance of the attributes in this quadrant is not appreciable, and it is high weightage area where these attributes should be focused more	S_{313}, S_{121}, S_{312}, S_{311}
Q2—The performance of the attributes in this quadrant is good, and the weightage of these attribute is also high, and hence, this should be maintained for a longer period	S_{322}, S_{122}, S_{214}, S_{321}, S_{112}, S_{113}, S_{223}, S_{222}, S_{124}, S_{212}, S_{111}, S_{221}
Q3—The performance of the attributes in this quadrant is not good, and it is not important to improve it	S_{132}, S_{211}, S_{131}, S_{123}, S_{231}, S_{234}, S_{133}, S_{233}, S_{232}
Q4—The performance of the attributes lying in this quadrant is good, but the weightage is low, and hence, it is not important to maintain it	S_{213}, S_{125}

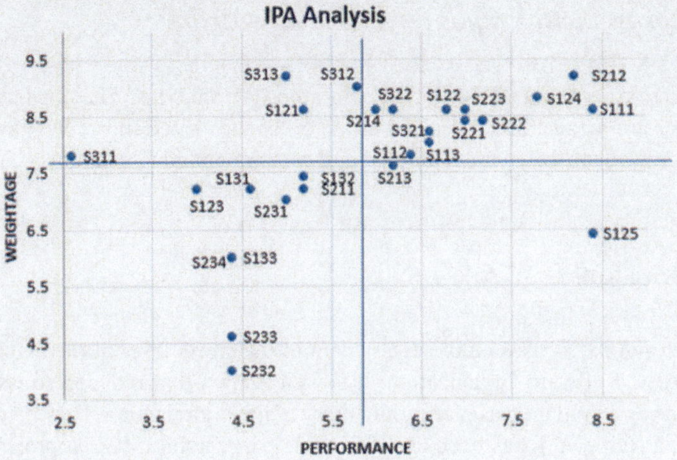

Fig. 1 IPA plot

Table 6 Improvement proposal for weaker attributes

Weaker attributes	Improvement suggestions
Noise (S_{313})	The noise level in the workplace environment is more since the noise is an external factor and could be prevented by placing noise-resistant glasses
Hot temperature (S_{311})	The temperature of the workplace is hot, which induces the stress level of the employees of the organization. Installing air conditioners in accordance with the number of computers and people in the workplace would be one of the solutions for this weaker attribute
Lighting discomfort (S_{312})	The lighting discomfort of the organization could sometime cause eye irritation and headache which reduces the efficiency and increases the stress level; thus, the lighting setup should be changed according to the seating, or the seating arrangement should be changed accordance with the lighting setup
Pressure to work overtime (S_{121})	The continuous pressure to work overtime would cause the employees less interested in their job and induces the stress. Making a standard work time every day would reduce the stress level caused due to the overtime work

experience of the employees in the workplace by altering these attributes, it would reduce the stress level down to 'Very Low.'

5.1 Improvement Proposal for Weaker Attributes

The weaker attributes were identified using the IPA analysis. The attributes in the Quadrant-1 are weaker attributes and are to be concentrated more. Suggestions for improving these attributes are identified and recommended (Table 6).

6 Conclusion

This paper mainly focuses on the assessment of the stress level of the employees in the case ABC Software Organization. It is well known that reduced stress level of the employees would increase the productivity of the organization. Hence, to address this issue, a framework has been created in this paper using MGF approach, which would be measuring the stress level of this organization; in this case, it is found to be 'Moderate.'

In the second part of the paper after measuring the stress level, the factors influencing the stress level has to be identified; thus, IPA analysis is used. Accordance with the results from IPA analysis, the company should mainly focus on the ergonomics factor of the workplace, which is highly lacking. Also, suggestions for improvement of the weaker attributes were provided for reducing the stress level of the organization. This framework is applicable to assess the stress level of employees of any software or software-related industries. In the future, the attributes of the framework can be modified and improved in accordance with the changes faced by the industry.

References

1. Mucci N, Giorgi G, Cupelli V, Gioffrè PA, Rosati MV, Tomei G, Breso-Esteve E, Arcangeli G (2015) Work-related stress assessment in a population of Italian workers. The stress questionnaire. Sci Total Environ 502:673–679
2. Palmer S, Cooper C, Thomas K (2004) A model of work stress. Counselling at work. Winter, 5, 25
3. Thayer JF, Verkuil B, Brosschotj JF, Kevin K, West A, Sterling C et al (2010) Effects of the physical work environment on physiological measures of stress. Eur J Cardiovasc Prev Rehab 17(4):431–439
4. Williams S, Cooper CL (1998) Measuring occupational stress: development of the pressure management indicator. J Occup Health Psychol 3(4):306
5. Lambert VA, Lambert CE (2001) Literature review of role stress/strain on nurses: an international perspective. Nurs Health Sci 3(3):161–172
6. Kumar V, Kumar S (2014) Workplace spirituality as a moderator in relation between stress and health: an exploratory empirical assessment. Int Rev Psychiatry 26(3):344–351
7. Shukla A, Srivastava R (2016) Development of short questionnaire to measure an extended set of role expectation conflict, coworker support and work-life balance: the new job stress scale. Cogent Bus Manage 3(1):1
8. Chen JG, Jung HS, Peacock BJ (1994) A fuzzy sets modelling approach for ergonomic workload stress analysis. Int J Ind Ergon 13(3):189–216

9. Fahlen G, Peter R, Knutsson A (2004) The effort-reward imbalance model of psychosocial stress at the workplace—a comparison of ERI exposure assessment using two estimation methods. Work Stress 18(1):81–88
10. Nübling M, Stößel U, Hasselhorn HM, Michaelis M, Hofmann F (2006) Measuring psychological stress and strain at work—evaluation of the COPSOQ questionnaire in Germany. GMS Psycho-Soc Med 3
11. Padma V, Anand NN, Gurukul SS, Javid SSM, Prasad A, Arun S (2015) Health problems and stress in information technology and business process outsourcing employees. J Pharm Bioallied Sci 7(Suppl 1):S9
12. Vinodh S, Chintha SK (2011) Leanness assessment using multi-grade fuzzy approach. Int J Prod Res 49(2):431–445
13. Vinodh S (2011) Assessment of sustainability using multi-grade fuzzy approach. Clean Technol Environ Policy 13(3):509–515
14. Ganesh J, Suresh M (2016, Dec) Safety practice level assessment using multigrade fuzzy approach: a case of Indian manufacturing company. In: 2016 IEEE International conference on computational intelligence and computing research (ICCIC). IEEE, pp 1–5
15. Sridharan V, Suresh M (2016, Dec) Environmental sustainability assessment using multigrade fuzzy—a case of two Indian colleges. In: 2016 IEEE International conference on computational intelligence and computing research (ICCIC). IEEE, pp 1–4
16. Ormanovic S, Ciric A, Talovic M, Alic H, Jeleskovic E, Causevic D (2017) Importance-performance analysis: different approaches. Acta Kinesiol 11:58–66
17. Dickson D, Ford R, Deng WJ (2008) Fuzzy importance-performance analysis for determining critical service attributes. Int J Serv Ind Manage

Chapter 36
Early Detection of Mild Cognitive Impairment Using 3D Wavelet Transform

B. A. Sujatha Kumari, A. G. Varun Yadiyala, B. J. Aruna, C. Radha, and B. Shwetha

1 Introduction

Alzheimer's disease (AD) is a cumulative disorder which affects memory and other cognitive abilities. It is the common cause of dementia where dementia is designated by downswing in the memory that effects the mental cognitive functions Dementia has different stages. Normal cognitive person have short memory, losing or misplacing things are some symptoms, whereas the person have more memory loss in mild cognitive. During last stage, patients completely depends on others for everyday activities that is, at stage AD. Mild cognitive impairment (MCI) is the stage between the expected cognitive decline of normal ageing and the more serious decline of dementia. Early detection of AD is helpful for the control of the disease. Mild cognitive impairment (MCI) is viewed as early stage of AD. Hence, detecting AD and MCI from normal healthy people is of great importance.

Mild cognitive impairment (MCI) is the stage between the expected normal cognitive decline because of normal ageing and AD. It can involve problems with memory, language, thinking and judgement that are greater than normal age-related changes. Mild cognitive impairment may increase your risk of later developing dementia caused by Alzheimer's disease or other neurological conditions. But some people with mild cognitive impairment never get worse, and a few eventually get better. 3D-DWT is mostly chosen because for feature extraction in hyper-spectral images because the horizontal, vertical and spectral information are processed in parallel ways. 2D wavelets extract the date in two directions which may lead to loss

B. A. Sujatha Kumari · A. G. V. Yadiyala · B. J. Aruna · C. Radha (✉) · B. Shwetha
Department of Electronics and Communication Engineering, JSS Science and Technology University, Mysore, India
e-mail: radharadhac21@gmail.com

B. A. Sujatha Kumari
e-mail: sujathakumari@sjce.ac.in

I. Jeena Jacob et al. (eds.), *Data Intelligence and Cognitive Informatics*, Algorithms for Intelligent Systems,
https://doi.org/10.1007/978-981-15-8530-2_36

of information. The detailed methodology used in the proposed work and related work is discussed in the next section.

2 Related Work

Some of the related work done for the implementation of this project can be briefly analysed below.

Upon pre-processing of the structured MRI using 3D wavelets, the grouping of different stages of the dementia can be studied through three classifiers in [1], they are linear SVM, kernel SVM and kernel SVM trained using PSO with varying time coefficient of acceleration (PSOTVAC). Similarly, 2D dual-tree complex wavelet transform [2] can be used for the feature extraction of the pre-processed MRI images. Tang et al. [2] shows a novel and efficient ML algorithm which is termed as gradient boosting classifier algorithm for early stage grouping of the Alzheimer's. The wavelets obtained from the transform, in this algorithm, are then combined with the biomarkers (like whether the patient is male or female or considering the age of the patient) and is passed onto a proposed classifier algorithm.

Another unique feature extraction technique can be seen in [3], where author proposes a 3D scattering wavelet transform, that can be used to filter the hyper-spectral image volumetric data using a series of wavelet decomposition filters and obtain complex modulus by implementing the local weighted averaging. A feature reduction technique PCA [4] is used for reducing the features of the extracted information. In [5], 3D-DWT is applied on the different regions of brain and further processed to the second stage, and these regions exhibit the changes due to the affect by AD. In the last stage of [5], Fisher's discriminant ratio is used to obtain minimum set of relevant features. Also can get to know about the classification model estimation based on the parameters obtained such as specificity, sensitivity and classification accuracy for evaluating the efficiency of the classifier algorithm.

A new trending solution for the classification is to propose a deep convolutional neural network-based algorithm, which is used in [6] for the detection of AD and other stages that lead to AD using MRI. To learn the feature representations of the MRI data, deep learning techniques can be used. In this study, a four-way classifier is implemented to differentiate between stages such as late-MCI (LMCI), mild cognitive impairment (MCI), Alzheimer's (AD) and cognitively normal persons. This study shows that deep CNN is an effective and better for extraction of features as well as classification of stages. Study [7] shows the use of point-capsule networks and CNN with previously trained 3D auto-encoder, which would improve the predictive performance of the designed system when compared with the deep CNN trained which is initially described. So, according to [7], the three-dimensional capsule network performs better than the 3D-CNN with auto-encoder.

A paper by Hosseini-Asl E, Keynton R, El-Baz A et al. [8] discussed a wavelet transform-based bilateral filtering. Magnetic resonance (MR) images consist of random noise which interrupts the process and results in automatic feature extraction complicated. In this paper, the authors estimated the Rician noise and then applied the 3D wavelets which results in equal decorrelation of 3D images , and since it is 3-dimention, for that purpose, they used 3D bilateral filtering for improving the denoising efficiency, and it also preserves edge features. At last they were evaluated with Neigh shrink threshold algorithm.

Another significant and novel publication by Ozmen et al. [9] proposes an algorithm consisting of decomposition by 3D DWT of MRI images, where the low frequency and high frequency bands are obtained by applying decomposition on the images, weights of each sub-bands were collected and apply denoising on them, then energy values of original image and its sub-band images are calculated image is reconstructed by inverse 3D DWT. Parameters with new weight sub-band performance is evaluated with mean square error and peak signal to noise ratio this method was tested with the 3D healthy brain MRI images which has different noise levels.

For diagnosis of AD, various methods were used, in the paper [10] published by D. Cheng and M. Liu proposed to construct multi-level convolutional neural network to combine and learn the multi-modality features for AD classification using MRI and PET images. Initially, deep CNNs are constructed, then for the high-level features obtained, two CNNs are cascaded for image classification, and generic features from MRI and PET images are learnt automatically for classification.

Vakili S, Khalili M proposed [11] AD diagnosis with deep three-dimensional convolutional neural network, and this can learn universal features apprehend AD bio-markers and adjust to various domain sets of data. Over here three-dimensional convolutional neural network is built on three-dimensional convolutional auto-encoder that is trained before to acquire anatomical shape variations in MRI scans of structural brain, and fully connected layers of the three-dimensional CNN are then tuned for each job particular AD classification.

The previous study on Alzheimer's disease gave an advantage of understanding the structure of MRI and its techniques for identifying and classifying the stages in a brain-related disorder. Pre-processing of 3D MRI [12] is a crucial process of the system for getting more accuracy on the prediction of the stages. 3-D wavelet transform can be used to extract the feature and reduce the quantity of feature by featuring on the region of interest. The review also shows the different algorithm in machine learning for the classification of the extracted features which are compatible with the 3D structure of the MRI. The 3-D wavelet transform was mainly used in all the survey papers for feature extraction and feature reduction of the MRI. The study also gave the detailed view on MRI classification, MRI feature enhancement and techniques on different algorithm for classification.

3 System Architecture and Methodology

The system architecture involves collection of MRI data from ADNI dataset and performing pre-processing steps on the raw data. 3D wavelets are then applied on the processed data. It results in 22 sub-bands, wherein first sub-band is selected which gives the approximate information. The three-level decomposed data is then given as input to the machine learning algorithm which classifies the data as pMCI, sMCI and AD.

The flowchart for the proposed method is shown in Fig. 3. The dataset is collected and pre-processed using SPM tool. Grey matter is collected [13] for the further analysis. 3D wavelets are applied to discard the redundant information. The wavelet coefficients are the converted to vector form and are given as input to classifier algorithm. The algorithm is trained to classify the given data as sMCI, pMCI and AD.

The algorithm of this system is as followed in the steps given below;

1. **Data collection and segregation**
   MRI data is collected from Alzheimer's disease neuroimaging initiative (ADNI) site which are of T1 weighted, 1.5T Baseline. The obtained data contains various age group. But patients aged 65 and above with MCI have high risk of getting dementia, women face high risk of Alzheimer's than men, and hence, obtained data is segregated into normal cognitive, MCI and AD of male and female based on age along with patient id. The datasets supporting the conclusions of this proposed work are available in the Alzheimer's disease neuroimaging initiative (ADNI) repository.
2. **Data segmentation**
   The goal of image segmentation is to divide an image into a set of semantically meaningful, homogeneous and non-overlapping regions of similar attributes such as intensity, depth, colour or texture. MRI image is segmented using SPM tool available in MATLAB. The grey matter is considered for early detection of Alzheimer's disease. Figure 1 shows coronal view, sagittal view and horizontal view of segmented brain MRI image. It is a sample data instance which has been considered in this project. At this pre-processing steps, noises like Riciannoise, Rayleigh noise and Gaussian noises are reduced.

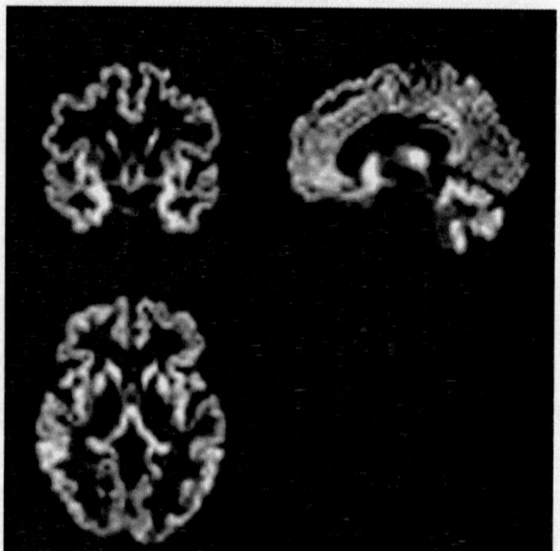

Fig. 1 Segmented grey matter of brain MRI

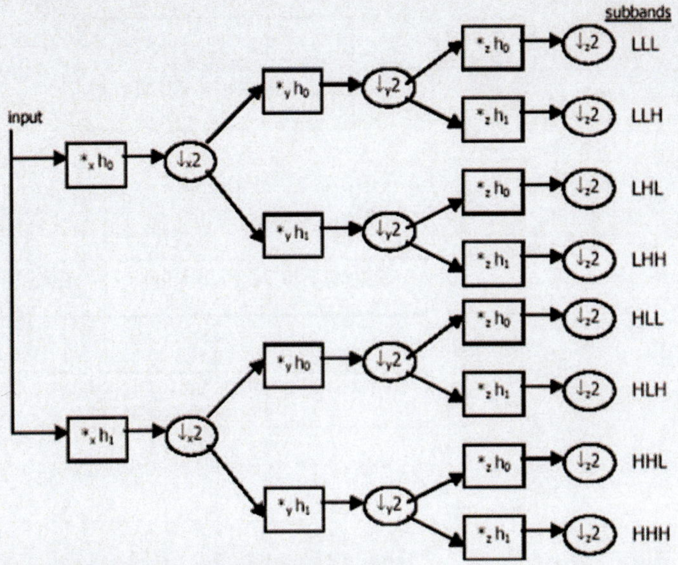

Fig. 2 Three-level decomposition using DWT

3. **Feature selection using 3D-DWT**
The required feature from the MRI is obtained by applying wavelet transform known as DWT, where DWT is a multi-resolution analysis, and it decomposes images into wavelet coefficients and scaling function. In the first level of 3D wavelet decomposition, eight sub-bands [14] are obtained out of which the first sub-band which is a low-pass filter component is selected for the next level decomposition. There are four main components in the wavelet decomposition. Figure 2 shows three level decomposition using wavelets. LL which is low-pass filter component gives approximate information and is used as useful data for the next step. LH is a component which provides horizontal information about the image. HL and HH component gives vertical information and diagonal information, respectively.
4. **Classification Algorithm**
The classification among sMCI (stable MCI), pMCI (progressive MCI) and AD is done to know whether a patient who has diagnosed with MCI will progress to AD or will the patient remain at the MCI stage itself. A hybrid algorithm is developed for efficient classification. Simple KNN algorithm is combined with the efficient SVM algorithm which led to the development of effective hybrid algorithm. KNN algorithm works on the idea that the data points that are nearer to the chosen points are considered in the next iteration. In SVM, the points that are close to the maximum margin hyper-plane are treated as support vectors. When only SVM algorithm is used for classification, it will not provide fully efficient classification since only one point is chose as representative point. That is where the hybrid algorithm becomes more effective. IN KNN-SVM [15] hybrid algorithm, all the

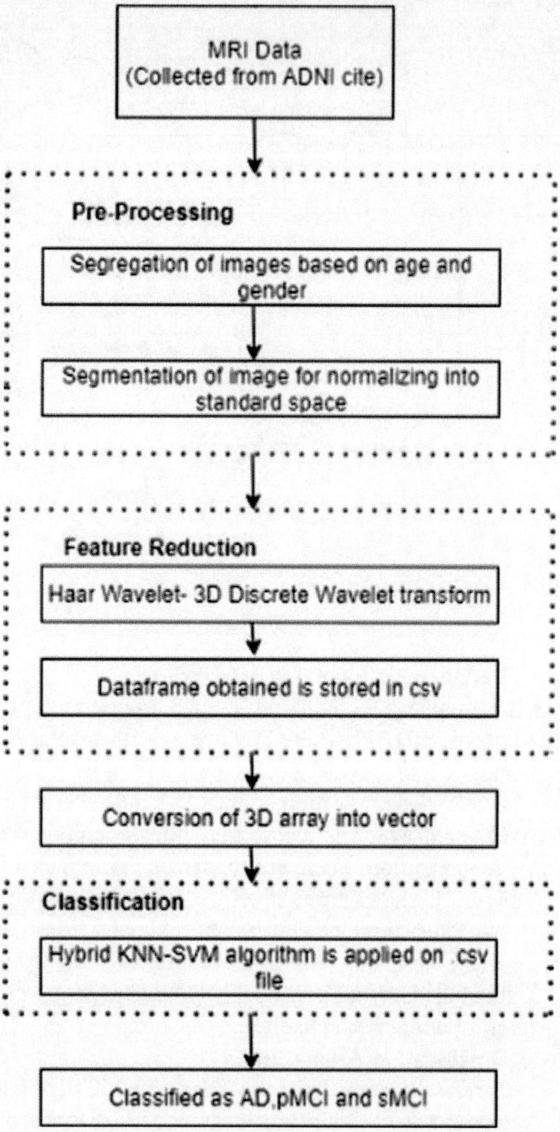

Fig. 3 Flow diagram of algorithm

support vectors in the plane are considered as representative points, and thus, all the information in the plane can be utilized for the classification. In hybrid algorithm, it is trained with the radial basis function, and in the testing scenario, nearest datapoints are considered using K-nearest algorithm. Figure 3 shows Flow diagram of the classification algorithm.

4 Implementation and Results

The implementation of the designed model occurs in the following steps:

1. Pre-processing (segmentation) using SPM12/8 tool from MATLAB
2. Three-level decomposition of pre-processed 3D MRI data using *pywavelets* library of python
3. Classification of the subjects using SVM-KNN classifier.

The patient dataset is collected from ADNI Website. A total of 800 images are collected, and these images are T1 weighted images. Table 2 shows the number of samples collected. The images collected are then classified as male or female. The main aim of this classification is to identify for which age group people this disease is likely to occur. The images are observed, and then, the patients are classified as AD, sMCI (stable MCI) and pMCI(progressiveMCI). A part of the images collected is used for training of the model, and rest of the images are used for testing. Table 1 shows the number of samples used for training the model.

After the data is obtained from ADNI dataset, cleaning of data must be done. A tool called statistical parameter mapping (SPM) is used in order to accomplish the above task. This is done through MATLAB. The image to be segmented is selected as data. Since the grey matter of the brain image is used for classification of the Alzheimer's disease, grey matter in SPM tool is selected.

Features of the pre-processed MRI brain data are extracted using 3D DWT where the pre-processed image is decomposed to three-levels using *pywt* library of Python. Three level DWT is applied and got 22 sub-bands.

$$H_i = \sum_{m=0}^{k-1} x_{2i-m} \cdot s_m(z) \tag{1}$$

$$L_i = \sum_{m=0}^{k-1} x_{2i-m} \cdot t_m(z) \tag{2}$$

Equations (1) and (2) shows low-cut filter and high-cut filter equations that are used to extract base-cut and treble-cut filter components from the image, respectively. 'H_i' indicates the high-pass filter coefficients, and 'L_i' indicates the low-pass filter coefficients, 'x_{2i-m}' represents the pixel element and 's_m and t_m' are the scaling and translation window functions. Harr wavelet transform is used in the proposed work

Table 1 Number of samples collected

Subject type	No. of samples
Cognitively normal (CN)	148
Mild cognitive impairment (MCI)	340
Alzheimer's disease (AD)	312

[16]. In the first level of the 3D decomposition. 8 sub-bands were obtained and the LLL is given as the input for the next level of decomposition since it contains the approximate information of the image. The low pass and high pass information is then extracted again by applying DWT in the second level. The low pass image is then considered and then passed as input to the third level. The coefficients of the low pass image obtained in the third level are considered. Then, the data is stored as a CSV file to process it to the classification model.

4.1 Model Fitting

Towards analysing the performance, the data is trained and tested for the classifiers using SVM-KNN classifier. The stored dataset is modelled in classifier using Python. The implementation of these classifiers is used from Sci-kit library. A hybrid classification algorithm is used which improves the efficiency of the classifier algorithm. The value of *k* is not kept constant, and it is varied according to the number of the datasets used for training. SVM algorithm generally uses an inbuilt function called the kernel, to transform the input data, and then based on these transformations, it finds an optimal boundary or plane in between the possible classes.

Once the reduced set of features is obtained, the next is to fit a classification model so as to accurately classify CN and MCI subjects. Additionally, two more classes of classification have also been done which are sMCI and pMCI, CN and pMCI. The classification between sMCI and pMCI is done to know whether a patient who has diagnosed with MCI will progress to AD (progressive MCI or pMCI) or will the patient remain at the MCI stage itself (stable MCI or sMCI). So, in essence, a three-class classification problem is trying to be solved. Generally, in classifiers, one typically uses stratified *k*-fold cross-validation [17], in which the folds are selected so that each fold contains appropriate proportion of class labels. The ten-fold cross-validation technique is utilized in this project for training and testing of the dataset, where for every 10-fold of training data in the classifier model, onefold of testing data has been given.

4.2 Results

It is found that by using 3D-DWT, the features got reduced by around 75%. This decrease in number of features decreased time complexity for classification to some extent, and also, this affects the classification accuracy of classification algorithm. In fact, in some cases, the classification accuracy did improve by using 3D-DWT. The classification through SVM-KNN classifier gave the highest results compared to all other classification algorithms. The CN and MCI are classified, and their corresponding accuracies are shown in Table 2. It is found that classification accuracy does

Table 2 Accuracy and error rate of the hybrid classifier

Parameters	sMCI	pMCI	AD
Accuracy	0.957	0.932	0.928
Error rate	0.043	0.068	0.072

not significantly change with decrease in features of training set. This is obtained from *sklearn* library, which split data in train/test sets.

Area under ROC curve (or AUC for short) is a performance evaluation metric for binary classification. Usually, AUC is not as sensitive to differences between the class distribution within the training and test samples, so AUC can be used as an alternative to accuracy. Basically, an AUC driven analysis will help in deciding a suitable model for classification, when one may have been trained on a skewed dataset.

AUC and ROC curve is one of the important classifier evaluation parameters for checking any classifier model's performance. ROC is a probability curve, and AUC measures the model's separability of classes. It tells capability of the model for distinguishing between classes. The model is better at predicting 0s as 0s and 1s as 1s, when the AUC is higher. The ROC characteristics for the developed algorithm are shown in Fig. 4. The ROC curve obtained for the graph 4a is 95%. This graph depicts the performance for the classification cognitively normal (CN) versus stable mild cognitive impairment (MCI).

Area under curve obtained for the graph 4b is 93%. This graph depicts the performance for the classification stable mild cognitive impairment (sMCI) versus progressive mild cognitive impairment (pMCI). The AUC curve obtained for the graph 4c is 92%. This graph depicts the performance for the classification progressive mild cognitive impairment (pMCI) versus Alzheimer's disease (AD).

5 Conclusion

The proposed method gives high efficiency compared to 2D analysis, and over here, various stages of AD are identified and classified using hybrid classifier algorithm. Based on the changes in structural brain, various stages of AD are identified. The data obtained by ADNI site is pre-processed, required features are extracted using 3D discrete wavelet transform, and the obtained data-frame of CN versus MCI, sMCI versus pMCI and pMCI versus AD which are the three binary classes later classified using hybrid classifier namely SVN andante result obtained is more accurate as the data from all the three axes of the image was considered.

For future improvements in this work, PET data can be used for AD detection at the early stage using conventional image processing technique. Classification of sMCI versus pMCI can be improvised by devising other methods of classification process ensuring higher classification accuracy. To avoid the pre-processing steps,

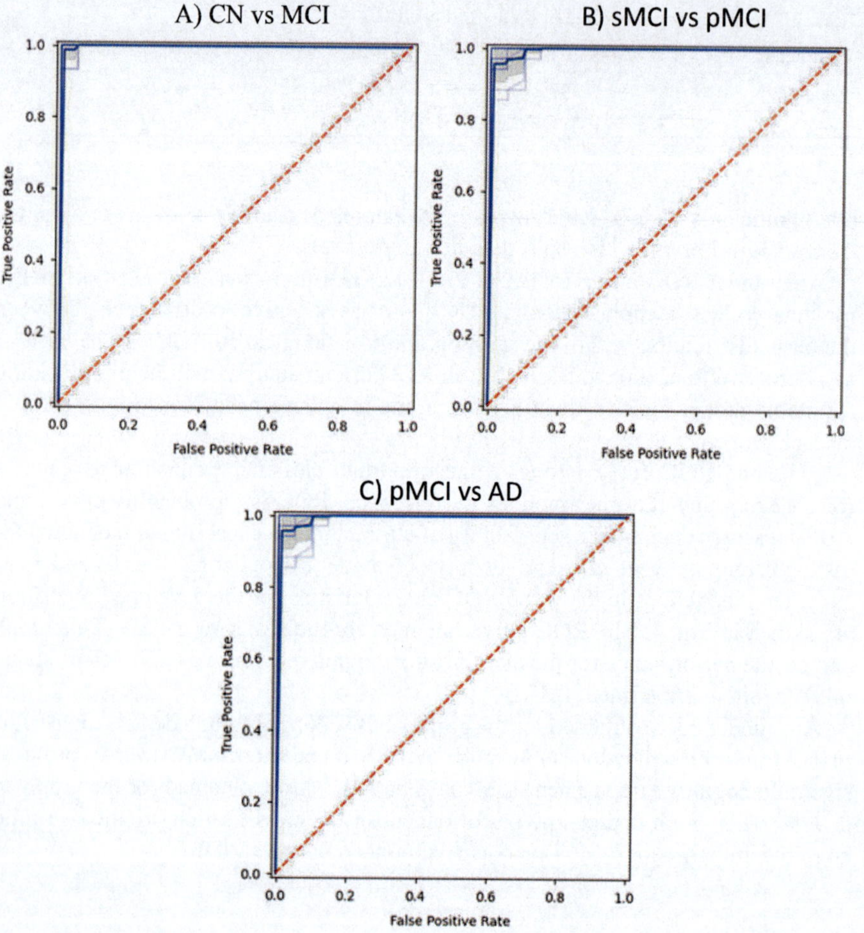

Fig. 4 ROC curves

more sophisticated algorithm such as spherical convolution neural network (CNN) on the 3D MRI can be used.

Acknowledgements We would like to express our gratitude towards our project guide and other faculties who helped us genuinely with all important guidance and support to complete this project successfully. And we sincerely extend our gratitude towards the institution for providing all the facility required to reach our project to successful work. We would also like to thank our friends for giving various suggestions that was very informative and helpful.

References

1. Sujathakumari BA, Abhishek MH, Darshan Singh S, Rakesh DS, Aneesh KN, Mahanand BS (2019) Detection of MCI from MRI using gradient boosting classifier. In: 20191st international conference on advances in information technology. IEEE
2. Tang YY, Lu Y, Yuan H (2015) Hyperspectral image classification based on three-dimensional scattering wavelet transform. IEEE Trans Geosci Remote Sens 53(5):2467–2480
3. Ayaz A, Ahmad MZ, Khurshid K, Kamboh AM (2017) MRI based automated diagnosis of Alzheimer's: fusing 3D wavelet-features with clinical data. In: 2017 39th annual international conference of the IEEE engineering in medicine and biology society (EMBC), Seogwipo, pp 1210–1213
4. Aggarwal N, Rana B, Agrawal RK (2015) 3D discrete wavelet transform for computer aided diagnosis of Alzheimer's disease using T1-weighted brain MRI. Int J Imaging Syst Technol 25:179–190
5. Farooq A, Anwar S, Awais M, Rehman S (2017) A deep CNN based multiclass classification of Alzheimer's disease using MRI. In: 2017 IEEE international conference on imaging systems and techniques (IST), Beijing, pp 1–6
6. Kruthika KR, Rajeswari HD (2019) CBIR system using capsule networks and 3D CNN for Alzheimer's disease diagnosis. Inf Med Unlocked 100227. Available online 22 Aug 2019
7. Wang Y, Che X, Ma S (2012) Nonlinear filtering based on 3D wavelet transform for MRI denoising. EURASIP J Adv Signal Process 2012(Article number: 40)
8. Hosseini-Asl E, Keynton R, El-Baz A (2016) Alzheimer's disease diagnostics by adaptation of 3D convolutional network. In: 2016 IEEE international conference on image processing (ICIP), Phoenix, AZ, pp 126–130
9. Ozmen G, Ozsen S, Yilmaz B (2016) Denoising MR images with weighted 3D discrete wavelet transform. In: 4th international conference on advanced technology & sciences (ICAT)
10. Cheng D, Liu M (2017) CNNs based multi-modality classification for AD diagnosis. In: 2017 10th international congress on image and signal processing, biomedical engineering and informatics (CISP-BMEI), Shanghai, pp 1–5
11. Vakili S, Khalili M (2015) A joint 3D DWT and SPIHT based algorithm for 3D MRI image compression. In: 2015 5th international conference on computer and knowledge engineering (ICCKE), Mashhad, pp 36–41
12. Shakya S (2019) Virtual restoration of damaged archeological artifacts obtained from expeditions using 3D visualization. J Innovative Image Process (JIIP) 1(02):102–110
13. Neffati S, Taouali O (2017) An MR brain images classification technique via the Gaussian radial basis kernel and SVM. In: 18th international conference on sciences and techniques of automatic control computer engineering—STA'2017, Monastir, Tunisia, Dec 21–23, 2017
14. Ozmen G, Ozsen S (2017) A new denoising method for fMRI based on weighted three-dimensional wavelet transform. In: New trends in data preprocessing methods for signal and image classification
15. Khan A, Usman M (2017) Early diagnosis of Alzheimer's disease using machine learning techniques. In: International conference of the IEEE engineering in medicine and biology society (EMBC), Seogwipo
16. Manoharan S (2019) Study on Hermitian graph wavelets in feature detection. J Soft Comput Paradigm (JSCP) 1(01):24–32
17. Zhang Y, Wang S, Phillips P, Dong Z, Ji G, Yang J (2015) Detection of Alzheimer's disease and mild cognitive impairment based on structural volumetric MR images using 3D DWT and WTA KSVM trained by PSOT-VAC. Biomed Signal Process Control 21:58–73

Chapter 37
Spatial Data Mining of Agricultural Land Area Using Multi-spectral Remote-Sensed Images

Parminder Kaur Birdi, Karbhari Kale, and Varsha Ajith

1 Introduction

Satellite sensors data is available for larger areas making it suitable for agricultural land monitoring. This data is acquired at frequent intervals which is useful to understand changes occurring in agricultural land areas. Due to the availability of satellite images covering large-scale areas, it is possible to design efficient agriculture land use monitoring systems by applying different spatial data mining methods [1]. The traditional data mining methods have been majorly focusing on transactional and relational databases. Remote-sensed data (spatial or geo-referenced) collected by various sensors record spatial relations present in the data. These systems record energy reflected at different wavelength ranges present in the electromagnetic spectrum. Satellite images have a huge amount of spatial information of different land covers recorded from the earth's surface. The traditional agricultural monitoring systems are dependent on extensive human efforts, and they usually cover smaller areas. This problem can be overcome using remotely sensed systems as the larger area is covered and human efforts and cost is also lowered. The temporal and spatial resolutions of the input data need to be considered appropriately.

P. K. Birdi (✉) · V. Ajith
Jawaharlal Nehru Engineering College, MGM University, N-6, CIDCO, Aurangabad, Maharashtra 431003, India
e-mail: dhingra.param@gmail.com

K. Kale
Department of CS and IT, Dr. Babasaheb Ambedkar Marathwada University, Aurangabad, Maharashtra 431003, India
e-mail: kvkale91@gmail.com

I. Jeena Jacob et al. (eds.), *Data Intelligence and Cognitive Informatics*, Algorithms for Intelligent Systems,
https://doi.org/10.1007/978-981-15-8530-2_37

Remote-sensed data are used from multi-spectral (MS), hyper-spectral (HS), and panchromatic sensors. Remote-sensed data from AVIRIS, MODIS, IRS-LISS-III, IV, SPOT-4,5, Landsat-7,8, ASTER, GF-2, IKONOS, and Sentinel-2 sensors are mostly used for agricultural land monitoring and crop growth analysis systems. It has been observed that spatial resolutions typically range from 250 to 5.8 m [2, 3].

Spatial data mining (SDM) has developed from the field of data mining and statistics. A large number of tasks including clustering, classification, association rule mining and data visualization is part of spatial data mining. SDM has a number of steps, wherein the selection of appropriate remote sensed images can be started with, followed by preprocessing and normalization of images, application of the spatial mining method for the classification task, visualization, and performance evaluation of results produced. Agricultural monitoring systems have been designed using spatial data mining methods [4]. There are two approaches to remote-sensed image classification; supervised and unsupervised. Maximum likelihood classifier (MLC), support vector machines (SVM), neural networks (NN), and decision tree (DT) are most commonly used supervised methods. DT and NN methods have been observed to generate higher classification accuracy in comparison with the traditional method, MLC. The performance of classifier depends on the training dataset created using handcrafted features. Finding an optimal decision tree design is a compute-intensive task, as the decision tree should have minimum overall classification error, appropriate depth of the tree, and an optimal subset of features to form decision rules.

In this study, a decision tree is constructed using rule-based modeling technique based on the foundation that different land use and land covers (LULC) have different spectral responses. The rules are formed using spectral signature values for the target land use and land covers. To improve classification accuracy and reduce time complexity, NDVI is also used in rule mapping. Rule-based mapping is done at pixel level. The spectral reflectance values and NDVI values of the training dataset are used as input.

2 Literature Review

Elodie et al. [5] used remote-sensed images from SPOT-5 sensor for mapping of cropland. Object-based supervised classification method is designed where expert rules are framed in the first step, followed by extraction of patterns from the dataset, and then, patterns are used to build classification rules. Naive Bayes, support vector machine, random forest, and decision tree methods are used for image classification. Remote-sensed images from Sentinel-2 sensor is also used for experiments. Overall accuracies for all methods are computed, and SVM is found to attain the highest accuracy of 84% as compared to other methods. It has also been concluded that data mining methods are suitable to process huge amounts of data [5]. Schultz et al. [6] used temporal images from Landsat-8 to classify study scene into different crop classes. An automated supervised technique is designed using random forest (RF) classifier, where features for segmentation step are automatically selected reducing human efforts. The proposed method produced a satisfactory overall accuracy for classification [6]. Many studies show that decision trees achieve higher classification accuracies as compared to classifiers like MLC, neural network as the time required to train a DT is less. Decision tree classifier has been applied for cropland monitoring and mapping. Some other studies illustrate the improvement in classification accuracies when spectral signatures are integrated with derived features including texture values and different vegetation indices [2, 3, 7]. In another study by Lebourgeois et al. [8] remote-sensed data from SPOT-5, Landsat-8, and PLEIADES sensors are used for agriculture land mapping. In this study, random forest classifier is applied to produce land cover maps at different levels into two classes and further into 25 crop subclasses. Random forest classifier is optimized by reducing variable count and this, in turn, resulted in less computing time [8]. Gervais et al. [9] carried out experiments using Landsat-5 and MODIS datasets to study the influence of urbanization on the growth life cycle of plants. NDVI values were computed and used to extract three seasons of the vegetation, starting, end, and length of the season. The performance of the experiments was found to be satisfactory [9].

From different research studies, it has been observed that the selection of appropriate features plays a major role in attaining higher classification accuracies using decision tree classifier.

3 Study Area

For conducting experiments, study scene is selected by considering the presence of different land covers with an abundance of sugarcane crop at different growth stages and availability of remote-sensed images. Navin Kaygaon is the study scene, close to Aurangabad district of Maharashtra state and can be located at 19° 20′ 1.4″ to 20° 15′ 19.65″ North and 74° 30′ 43.61″ to 75° 5′ 67″ East [10]. Figure 1 shows the study area acquired from LISS-IV sensor.

The land covers present are waterbody, barefarm, road, settlement, and sugarcane crop at different growth stages.

4 Data Used

Remote sensing data is collected using satellite sensors, and the appropriate sensor is selected based on its spatial, spectral, and temporal resolution. The spectral reflectance values recorded by the sensor can be related to its phenological growth cycle of sugarcane crop. Different stages of the life cycle of a crop are its sowing time, budding, ripening, and harvesting. Spatial resolution, i.e., the size of a pixel, and the number of pixels present in a typical a sugarcane farm also affects the classification accuracy. Tables 1 and 2 show detailed specifications remote-sensed data used for different experiments.

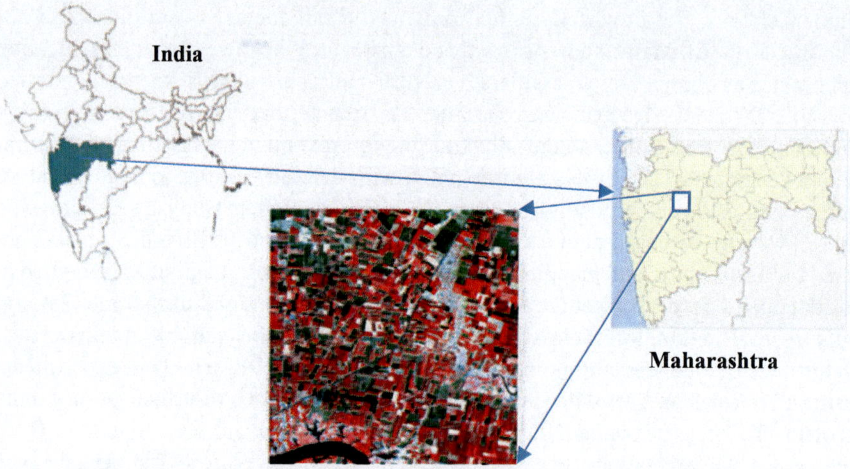

Fig. 1 Satellite image of study area represented by LISS-IV data

Table 1 Landsat-8 OLI sensor data specifications

Mode	Band	Spectral region	Spectral resolution (μm)	Spatial resolution (m)
Panchromatic	Band 8	Panchromatic	0.50–0.68	15
Multispectral	Band 1	Visible	0.43–0.45	30
	Band 2	Visible	0.450–0.51	
	Band 3	Visible	0.53–0.59	
	Band 4	Red	0.64–0.67	
	Band 5	Near-infrared (NIR)	0.85–0.88	
	Band 6	SWIR 1	1.57–1.65	
	Band 7	SWIR 2	2.11–2.29	

Table 2 LISS-IV data specifications

Mode	Band	Spectral region	Spectral resolution (μm)	Spatial resolution (m)
Panchromatic	Single	Panchromatic	0.50–0.75	5.8
Multispectral	Band 2	Visible	0.52–0.59	5.8
	Band 3	Visible	0.62–0.68	
	Band 4	Near-infrared (NIR)	0.77–0.86	

In situ data is also collected for precise remote-sensed image interpretation. For information related to crops cultivated, the growth cycle of crops was collected with the help of local farmers who owned farms of the study area. GPS enabled handheld devices were used to record geographical coordinates of a total of 150 sugarcane farms which ranged in size from half an acre to sixteen acres.

5 Proposed Methodology and Experimentation

This section describes the proposed research methodology used to classify multi-spectral remote-sensed images. Spatial data mining method decision tree has been used as a classifier here [11]. For experiments, multistage decision tree has been designed by applying rules created using band spectral reflectance values and NDVI. Figure 2 shows the proposed flow diagram of the proposed methodology. Since dataset specifications are different, separate decision trees are developed for LISS-IV and Landsat-8 images. According to Tucker et al. [12] and Schmidt et al. [13], physical characteristics of crops change with different growth stages and can be better interpreted using vegetation indices (VI) [12, 13]. The most recommended vegetation index, NDVI, has been used in experiments. NDVI represented by Eq. 1 is calculated as the ratio between red and NIR bands and spectral reflectance values [14, 15].

$$\text{NDVI} = \frac{R_{\text{NIR}} - R_{\text{Red}}}{R_{\text{NIR}} + R_{\text{Red}}} \tag{1}$$

5.1 *LISS-IV and Landsat-8 Dataset Pre-processing*

Since satellite images are not directly usable due to radiometric, geometric, and atmospheric errors present in them. LISS-IV dataset is acquired from IIRS and images that

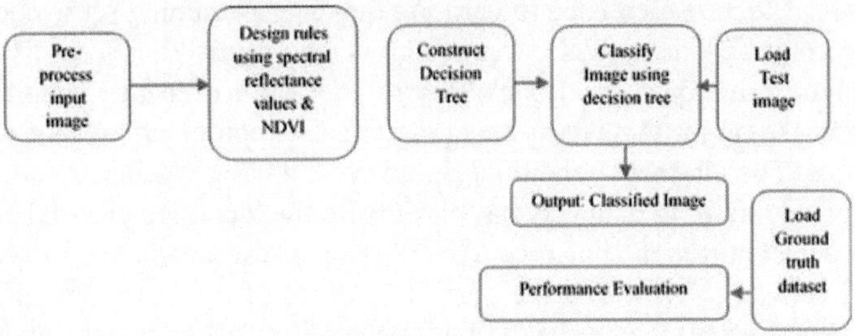

Fig. 2 Flow diagram of the proposed methodology using decision tree classifier

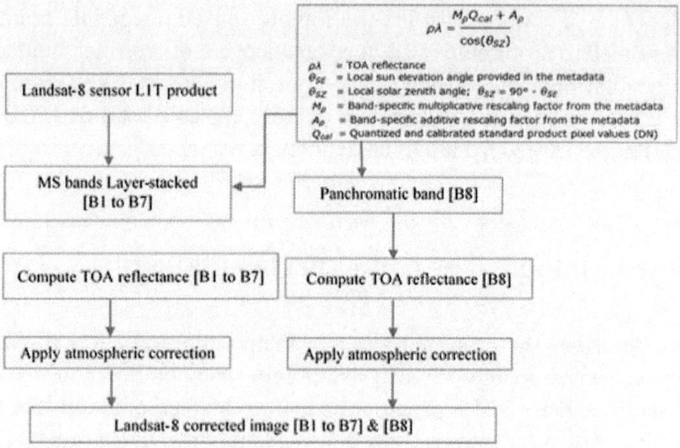

Fig. 3 Landsat-8 dataset preprocessing

are geometrically and radiometrically corrected. So no preprocessing is performed on them. The study area is cropped and layer stacked for further use. Landsat-8 images are preprocessed after acquiring them in L 1T format. The steps performed to preprocess them are shown in Fig. 3. The preprocessing is carried out using ENVI software by applying quick atmospheric correction method on all the images.

5.2 Experimentation

An exponential number of decision trees can be constructed from a given set of attributes or features, where some of the trees can be more accurate than others. Finding an optimal decision tree is computationally infeasible as the search space is exponential in size. Various decision trees were built using trial and error. A widely used decision tree algorithm C4.5 is applied to generate decision trees since reaching an optimum tree is time consuming. A greedy algorithmic model is used to overcome this, where at every level a locally optimal decision is made to construct the decision tree. The set of features are used to partition the data. A training set is constructed by investigating the existing data, where a class label is assigned to every pixel. A recursive algorithm is developed by making a combination of feature values from the training data. This algorithm works fine if every feature combination has a class label assigned to it. The alternate condition can arise for a case, where for some feature combinations no training data gets associated with the identified class labels.

Once a decision tree is constructed, classifying a test data is fast as worst-case complexity is $O(d)$, where d is the maximum depth of the tree. Maximum depth of the decision tree reached a depth of 16, generating 65 leaf nodes and a total of 128 nodes for Landsat-8 image with four classes and seven attributes. The total time taken is 9 min and 19 s with an accuracy of only 50% on test data. The minimum depth as indicated by Landsat-8 decision tree is three and four for LISS-IV image.

A minimum of 50 training pixel values (for every target class) are considered from all spectral bands. The mean values are computed along with standard deviation and further used to frame the rules for decision making. Four land covers as target classes could be selected for Landsat-8 image due to fewer number of training pixels for road and barefarm, whereas six land covers were selected for LISS-IV image classification.

Table 3 shows minimum (min), maximum (max), mean, and standard deviation (SD) values obtained from regions of interest marked as a training dataset for Landsat-8 image. Similarly, Table 5 shows spectral reflectance distribution in all bands for the target classes of LISS-IV image. The two most separable group of classes are processed first, and the subtlest class pair will be processed last. This helps in reducing cumulative error. To ensure better accuracy, error generated at every decision node needs to be reduced by making use of optimal or sub-optimal features for portioning of the data, as given in Table 4.

The highest class separability is shown by water from Crop-Hv and Crop-Gr with a value of 2.0. So, for this design at level 1, water is separated from other land covers using Band 7 (spectral reflectance values are less than 0.048). It has no overlap with other class range. For the next decision rule, class 'Others' show the next best TD value w.r.t to Crop-hv and Crop-Gr. The optimal decision tree for Landsat-8 image is shown in Fig. 4a.

The best features selected are considered for constructing final decision trees for LISS-IV image as shown in Fig. 4b.

6 Results and Interpretation

The performance of decision tree classifier has been evaluated using a confusion matrix. The confusion matrix shows the relationship between the predicted values and actual (ground truth) values. The confusion matrix is further used to calculate overall classification accuracy which is defined as the ratio between correctly predicted values and a total number of predicted values. Kappa coefficient is also computed to measure the agreement between predicted values and actual values, where a zero coefficient value means no agreement and one means complete agreement. With the increase in kappa value, classification accuracy becomes better. The same images have also been classified using maximum likelihood classifier (MLC) for comparison of results, and classified images are shown in Figs. 5 and 6.

Table 3 Spectral reflectance distribution in all bands for Landsat-8 data

	Crop-Hv			Crop-Gr			Water			Others		
	Min	Max	Mean ± SD	Min	Max	Mean ± SD	Min	Max	Mean ± SD	Min	Max	Mean ± SD
B1	0.131	0.145	0.139 ± 0.003	0.132	0.138	0.132 ± 0.001	0.126	0.136	0.129 ± 0.001	0.128	0.228	0.146 ± 0.011
B2	0.112	0.132	0.123 ± 0.004	0.113	0.125	0.116 ± 0.001	0.105	0.117	0.109 ± 0.002	0.108	0.226	0.132 ± 0.015
B3	0.097	0.123	0.110 ± 0.005	0.102	0.116	0.107 ± 0.003	0.081	0.102	0.088 ± 0.004	0.085	0.235	0.122 ± 0.029
B4	0.086	0.137	0.113 ± 0.008	0.081	0.107	0.090 ± 0.005	0.063	0.085	0.071 ± 0.004	0.082	0.257	**0.130 ± 0.029**
B5	0.151	0.329	0.185 ± 0.038	0.279	0.404	0.341 ± 0.022	0.059	0.095	0.068 ± 0.005	0.124	0.372	0.211 ± 0.058
B6	0.141	0.203	0.166 ± 0.012	0.141	0.194	0.151 ± 0.010	0.035	0.062	0.041 ± 0.004	0.105	0.376	0.190 ± 0.045
B7	0.083	0.164	0.126 ± 0.013	0.067	0.105	0.081 ± 0.007	0.023	**0.044**	**0.027 ± 0.003**	0.089	0.329	0.141 ± 0.035

Bold values represent attributes used to form rules for decision tree for Landsat-8

Table 4 Best features for class pairs of Landsat-8 data used in optimal DT

Class pairs	No. of features selected	Best feature selected
Crop-Hv + Crop-Gr	NDVI, B5	NDVI
Crop-Hv + Water	B6, B7	B7
Crop-Gr + Water	B6, B7	B7
Crop-Hv + Others	B5,B4	B4
Crop-Gr + Others	B7, B4	B4
Water + Others	B5, B6, B7	B7

The results show that DT classifier has higher classification accuracy as compared to MLC as optimal features are selected to form classification rules while constructing a decision tree. The first classification rule consists of spectral features which have the highest separability, measured using transformed-divergence matrix. Further classification rules also follow the same selection criterion. Another factor which resulted in improvement of accuracy is the inclusion of NDVI values for classification of sugarcane crop at harvest and grown stage. Table 6 shows the classification results obtained on both the datasets. Since the spatial resolution of LISS-IV dataset is 5.8 m, user's and producer's accuracy (UA, PA) for six classes is found to be better than Landsat-8 dataset as per results obtained by Kaur and Kale [16].

7 Conclusions and Future Scope

Traditionally, agricultural land monitoring is being done using manual methods which are time inefficient and tedious. The work done using manual surveys is also error prone and results in inaccurate data recording. Satellite images are found to be appropriate for this task as they cover larger areas, and availability of periodic information makes them beneficial for doing the task of agricultural land monitoring automated. While performing spatial data mining methods such as decision trees, the accuracy of supervised classification methods depends on correct labeling of training samples and also a suitable amount of training samples. This study shows that multi-spectral images from Landsat-8 and LISS-IV sensors are suitable to carry out this task of agricultural land monitoring using a decision tree. As new techniques such as deep learning have also been used for similar work, the task can be performed using deep convolutional neural networks for more accurate results.

Table 5 Spectral reflectance distribution in all bands for LISS-IV data

	Crop-Hv			Crop-Gr			Water			Settlement			Road			Barefarm		
	Min	Max	Mean ± SD	Min	Max	Mean ± SD	Min	Max	Mean ± SD	Min	Max	Mean ± SD	Min	Max	Mean ± SD	Min	Max	Mean ± SD
B1	0	168	46.58 ± 25	0	205	45.86 ± 19.9	0	155	1.7 ± 9.15	24	255	221 ± 38.05	15	177	60.7 ± 20.8	0	255	160 ± 38.4
B2	0	151	**33.60 ± 19**	**23**	255	**165 ± 40.6**	0	255	0.8 ± 7.94	53	255	**155 ± 33.7**	1	40	**20.24 ± 10.4**	31	255	151.36 ± 48
B3	0	176	64.40 ± 22.5	0	255	26.0 ± 17.2	0	133	**1.17 ± 6.82**	29	255	208 ± 43.5	12	126	58.42 ± 16.6	28	255	164.5 ± 37.2

Bold values represent attributes used to form rules for decision tree for LISS-IV

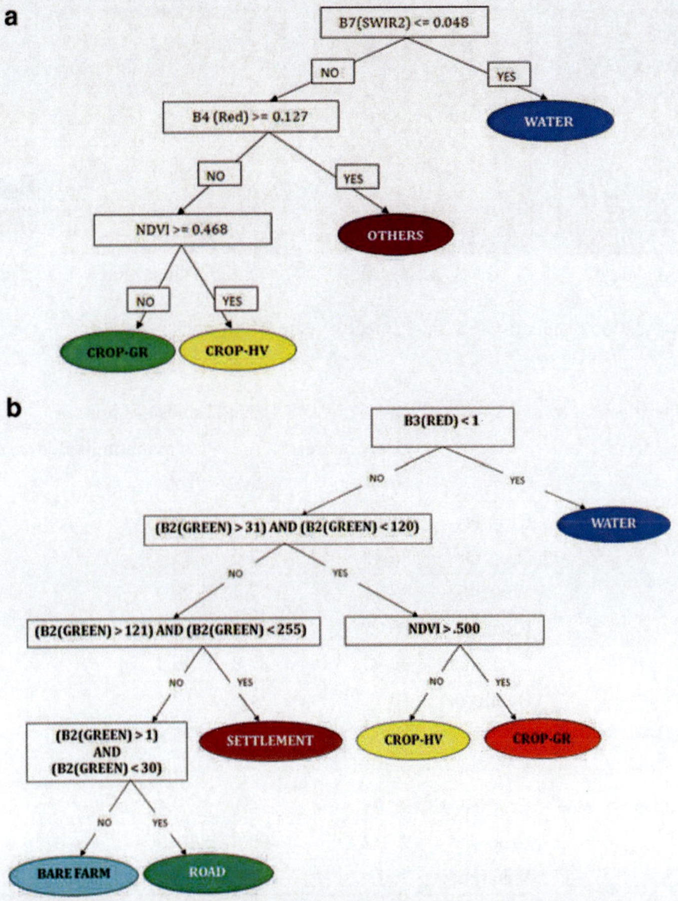

Fig. 4 **a** Optimal decision tree for Landsat-8 image. **b** Optimal decision tree for LISS-IV image

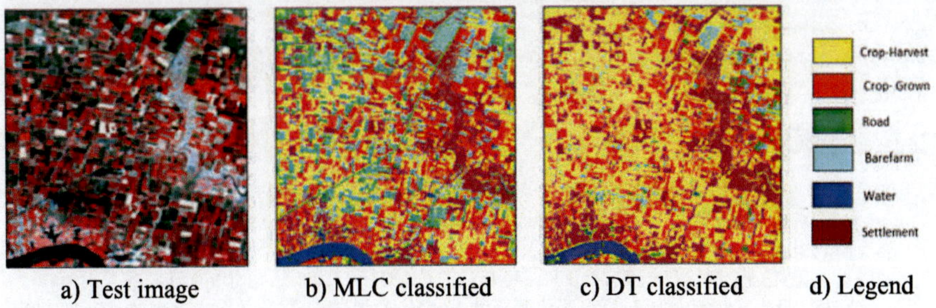

Fig. 5 Classification results for LISS-IV image of date 27th April'17

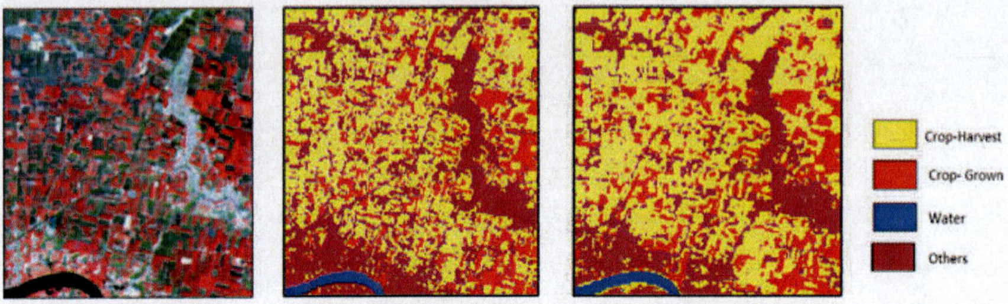

a) Test image b) MLC classified c) DT classified d) Legend

Fig. 6 Classification results for Landsat-8 image of date 24th May'17

Table 6 Classification results for study images of LISS-IV and Landsat-8 images

Study image details		Decision tree		Maximum likelihood classifier	
	Class	UA%	PA%	UA%	PA%
LISS-IV Image of date 27th Apr'17	Crop-Hv	**91.92**	71.74	68.85	55.08
	Crop-Gr	**96.49**	87.4	85.33	98.81
	Waterbody	88.32	97.71	98.3	96.68
	Road	25	20	14.74	88.32
	Settlement	99.95	82.8	96.31	8.99
	Barefarm	10	8	12.4	80.59
	OA	79.45%		75.71%	
	k	0.716		0.692	
Landsat-8 image of date 24th May'17	Crop-Hv	**92.08**	74	79.41	79.41
	Crop-Gr	**97.22**	93.22	93.06	97
	Waterbody	100	100	100	100
	Others	76.55	96.52	88.24	86.54
	OA	88.97%		89.88%	
	k	0.86		0.86	

Bold values represent highest UA values observed

References

1. Openshaw S (1999) Geographical data mining: key design issues. In: Proceedings of geocomputation '99
2. Verma AK, Garg PK, Hari Prasad K, Dadhwal V (2016) Classification of LISS-IV imagery using decision tree methods. In: The international archives of the photogrammetry, remote sensing and spatial information sciences, vol XLI-B8
3. Peña JM, Gutiérrez PA, Hervás-Martínez C, Six J, Plant RE, López-Granados F (2014) Object-based image classification of summer crops with machine learning methods. Remote Sens 6:5019–5041
4. Tso B, Mather PM (2009) Classification methods for remotely sensed data. CRC Press, Taylor & Fransic Group, Boca Raton, pp 3–15

5. Elodie V, Valentine L, Agnes B, Dino I, Maguelonne T, Stephane D, Fidiniaina R (2015) Identifying cropped areas in small growers agricultural regions using data mining for food security. Accessed on https://agritrop.cirad.fr/574632/1/document_574632.pdf
6. Schultz B, Immitzer M, Formaggio A, Sanches I, Barreto Luiz A, Atzberger C (2015) Self-guided segmentation and classification of multi-temporal Landsat-8 images for crop type mapping in southeastern Brazil. Remote Sens 7:14482–14508
7. Huang J, Wang H, Dai Q, Han D (2014) Analysis of NDVI data for crop identification and yield estimation. IEEE J Sel Top Appl Earth Observations Remote Sens 7(11):4374–4384
8. Lebourgeois V, Dupuy S, Vintrou É, Ameline M, Butler S, Bégué A (2017) A combined random forest and OBIA classification scheme for mapping smallholder agriculture at different nomenclature levels using multisource data (Simulated sentinel-2 time series, VHRS and DEM). Remote Sens 9:259
9. Gervais N, Buyantuev A, Gao F (2017) Modeling the effects of the urban built-up environment on plant phenology using fused satellite data. Remote Sens 9(1):99
10. https://www.onefivenine.com/india/villages/Aurangabad-District/Gangapur/Kaygaon. Accessed on 2nd Mar 2017
11. Hunt EB, Marin J, Stone PJ (1966) Experiments in induction. Academic, New York
12. Tucker CJ, Holben BN, Elgin JH, McMurtrey JE (1980) Relationship of spectral data to grain yield variation. Photogramm Eng Remote Sens 46:657–666
13. Schmidt M, Pringle M, Devadas R, Denham R, Tindall D (2016) A framework for large-area mapping of past and present cropping activity using seasonal Landsat images and time series metrics. Remote Sens 8:312
14. Singha M, Wu B, Zhang M (2016) An object-based paddy rice classification using multi-spectral data and crop phenology in Assam, Northeast India. Remote Sens 8:479
15. Pereira RM, Casaroli D, Vellame LM, Junior JA, Evangelista AW (2016) Sugarcane leaf area estimate obtained from corrected normalized difference vegetation index (NDVI). Pesq Agropec Trop 46(2):140–148. www.agro.ufg.br/pat
16. Kaur P, Kale KV (2017) Identification of growth stage of sugarcane crop using decision tree for Landsat-8 data. In: Proceedings of 38th Asian conference on remote sensing

Chapter 38
Person Identification Using Histogram of Gradient and Support Vector Machine on GEI

P. Nithyakani and M. Ferni Ukrit

1 Introduction

Humans have unique features such as iris, retina, fingerprint, palm, sweat, odour and gait. Feature obtaining methods are categorized as person involvement and non-involvement [1]. Without the involvement of a person, face and gait features can be obtained by various methodologies like the histogram of a gradient, cascading technique, hidden Markov model, PCA, LDA, etc. Gait recognition can be performed with low-resolution video since it is non-invasive. Variations in viewpoint, clothing, accessories, walking speed and injury in the leg may affect the performance of recognition.

The human brain has the ability to perform all biological activities in the meantime of interval. The visual system of humans is well trained to analyse the things which have been viewed through the retina in all human movements [2]. Gait cycle of a person has various unique characteristics that enable it to identify a person among the group. Medical studies related to the gait analysis suggested that there are 24 unique gait features to differentiate age, gender, individual, etc. Gait analysis is performed by capturing the video from the viewpoint [3]. The foreground pre-processing and feature extraction methods are integrated into the raw data. The pose estimation methods face problem in accuracy because of internal and external human factors such as changes in the body segment. Generally, the sequence captured is converted to angular vectors represented in the sparse matrix which identifies the individual with their uniqueness in gait feature.

P. Nithyakani (✉) · M. Ferni Ukrit
School of Computing, SRMIST, Chengalpattu, India
e-mail: nithyakp@srmist.edu.in

M. Ferni Ukrit
e-mail: ferniukm@srmist.edu.in

I. Jeena Jacob et al. (eds.), *Data Intelligence and Cognitive Informatics*,
Algorithms for Intelligent Systems,
https://doi.org/10.1007/978-981-15-8530-2_38

2 Background and Related Work

Gait recognition is categorized as model-based and model-free methods [1]. The model-based approach is based on the motion activity structure and angle inclination of bones and spine. Gait features can be obtained from the sensor devices like smartphone, accelerometer and gyroscope fixed in any part of the body. The model-free approach is based on the holistic appearance of a person movement, walking sequence and its silhouette images [4]. Here, gait feature is by the series of background subtraction, grayscale conversion, gait energy image and various machine learning techniques. Gait energy image is widely used as a model-free approach in recognizing human gait due to its remarkable nature [5, 6]. Gait energy image is further processed to develop gait entropy image, gait probability image, gait Gaussian image, etc.

Histogram of gradient is the object detector descriptor used in computer vision applications, human detection, character pattern recognition and face recognition [5]. Histogram of oriented gradient provides robustness in illumination and spatial variation environment by extracting the information of edges. The input image is partitioned into cell blocks for normalizing the illumination and the contrast with speeded up robust feature [7]. Histogram of the gradient is calculated for each cell in the blocks and its resolution varies with respect to the pixels and the cell size [6]. Point spread function is estimated with histogram of oriented gradients for better feature classification. The estimation methodology is given in Fig. 1.

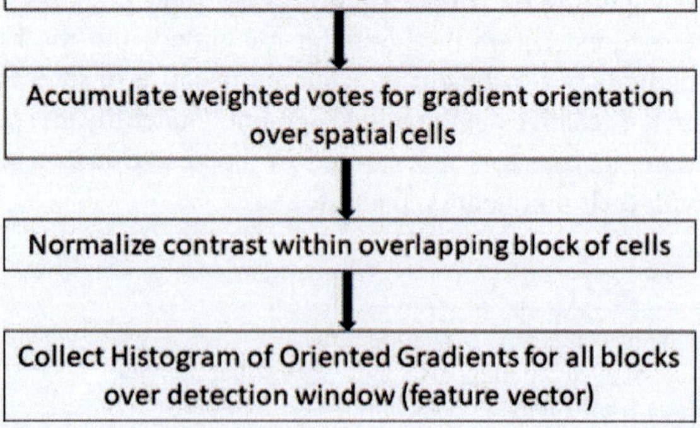

Fig. 1 Estimate of histogram of oriented gradient

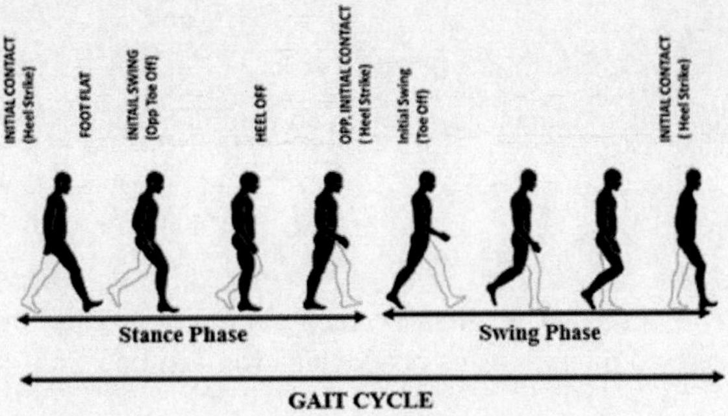

Fig. 2 Gait cycle of an individual walking sequence

3 Overview of the Proposed Method

Video captured (indoor) is converted into frames. Each RGB (red, green, blue) frame is transformed into grayscale and hue saturation intensity (HSI) image which has uniform colour space [8]. Grayscale image improves the speed of processing whereas HSI image differentiates the pixel of shadow to manage the illumination. Background subtraction isolates the image by segmenting the region of interest from the background and forms the silhouette image. There are two ways to obtain a silhouette image. One method provides equal weightage to every sequential image which forms "the motion Energy Image", the other provides declining weightage to the sequential images to form "the Motion History Image". Pre-processing of the image is done for removing noise by convolving with the Gaussian filter. After processing, histogram of gradient is calculated from the processed gait energy image [9].

3.1 Gait Cycle Identification

Gait cycle is identified as two phases: stance and swing. During the stance phase, the heel strikes the ground and bears the bodyweight while swing phase the heel swing to the displacement of another cycle. The complete gait cycle is represented in Fig. 2.

3.2 Histogram of Oriented Gradient

Histogram of oriented gradients is used for extraction of feature and classifying the region of interest by filtering vertically and horizontally [10, 11]. Horizontal filter, h_x, is applied with the value of $[-1, 0, 1]$ and vertical filter, h_y, is applied with the value of $[-1, 0, 1]T$. Horizontal filtered image, G_x, and vertical filtered image, G_y, are formed by

$$G_x = I * h_x \text{ and } G_y = I * h_y$$

The filtered image is derived and combined using the formula

$$G(i, j) = \sqrt{g_x(i, j)^2 g_y(i, y)^2}$$

The dominant gradient angle, θ, is calculated by the inverse of tangent over the vertical filtered image pixel which is given by

$$\theta(i, j) = \tan^{-1}\left[\frac{g_y(i, j)}{g_x(i, j)}\right]$$

Histogram of oriented gradient is obtained with respect to the calculated values of orientation and magnitudes of each pixel which are represented in histogram channels or bins. Magnitudes get accumulated into bins with an angle ranging from 0 to 180° or 0 to 360° [10]. The additional number of bins can be added to increase the performance in detecting the gait pattern of an individual. The gradient is utilized to calculate the weight of each pixel. Finally, all the pixels with magnitude and gradient are concatenated to for the histogram of an oriented gradient.

In general, illuminations, brightness, background and foreground contrast affect the gradient strength of the frame obtained from the captured video. Normalization technique over the histogram of a gradient with the vector v, combining with constant ε, overcomes this problem and the methods used are described in the equation

$$l2\text{-norm} = v \rightarrow \frac{v}{\sqrt{v_2^2 + \varepsilon^2}}$$

$$l1\text{-norm} = v \rightarrow \frac{v}{v_1 + \varepsilon}$$

$$l1\text{-sqrt} = v \rightarrow v/v_1 + \varepsilon$$

Figure 3 depicts the procedure of histogram of oriented gradient. Normalized histogram of the oriented gradient is used for classifying and identification by support vector machine.

3.3 Support Vector Machine

Support vector machine (SVM) is a machine learning technique that is originated from Vapnik's statistical theory and is used to solve binary classification problems. In Vapnik's theory [12], a convex quadratic optimization problem is formulated as

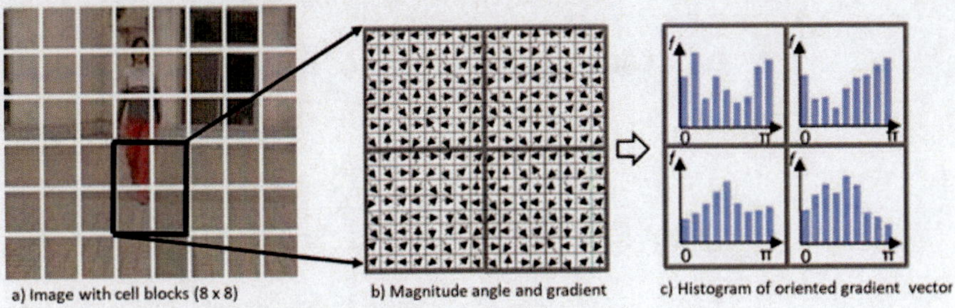

Fig. 3 Histogram of oriented gradient of a gait image

a learning process with a global optimum error surface. Support vector machine is implemented by mapping the input vectors x into a high-dimensional feature space Z through kernel transformation and prior nonlinear mapping. Optimal hyperplane separates the given data sample m which is in the form $<x_i, y_i>$ where x_i is an n-dimensional vector and y_i is the class label which is given by,

$$y_i = f(x_i) = \omega^{\mathrm{T}}\Phi(x_i) + b$$

where w is the "normal vector" perpendicular to the hyperplane, b is the offset parameter, Φ is the function of nonlinear transformation and T represents the matrix transpose. Figure 4 demonstrates the linear kernel SVM with the optimal hyperplane as $wTx + b = 0$. The feature, x, was classified with satisfying the condition $wTx + b \geq +1$ and $wTx + b \geq -1$. Optimal hyperplane separates the training vectors to provide the error probability ratio of support vectors and training samples in number [12].

The optimal hyperplane is constructed in the higher-dimensional feature space, $\Psi(x) \in$ Rd. This induces the application of linear SVM and defines that the data is linearly separable. Weight vector and bias feature the optimal hyperplane. The following derivations solve the optimization problem

$$\min_{w,b,\xi} = \frac{1}{2}w^2 + C\sum_{i=1}^{L}\xi_i$$

$$\text{Subjected to } y_i(w\Psi(x) + b) \geq 1 - \xi_i\xi_i \geq 0$$

where y_i, non-separable slack variables which can be switched between the slack variable penalty and margin width, is controlled by a parameter. Reframing the optimization problem is given by

$$\max_\alpha = \sum_{i=1}^{L}\alpha_i - \frac{1}{2}\sum_{i=1}^{L}\alpha_i\alpha_j y_i y_j \Psi(x_i)\Psi(x_j)$$

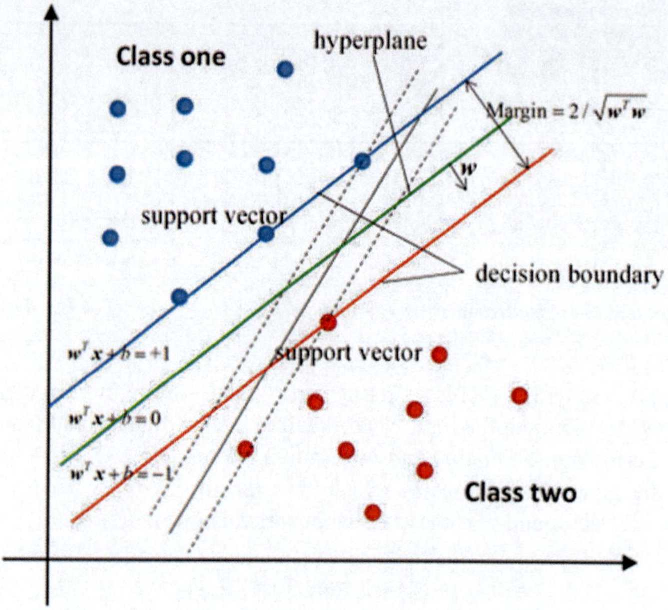

Fig. 4 Linear kernel SVM with optimal hyperplane

$$\text{Subjected to} \sum_{i=1}^{L} \alpha_i y_i = 0 \quad 0 \le \alpha_i \le C, i = 1, 2, \ldots, L,$$

$$\text{where } w = \sum_{i=1}^{L} \alpha_i y_i \Psi(x_i)$$

Lagrange multipliers are used to solve the convex quadratic optimization problem. Inner product calculations are done with mapping function $\Psi(x)$ and kernel function for any two points x_i and x_j. Nonlinear classifier arises from linear through the substitution of kernel evaluation and solves the optimization problem found in inner product values. A cubic kernel is given by

$$k\left(x_i - x_j\right) = \left(x_i - x_j\right)^3$$

4 Implementation

CASIA A data set is used to implement the proposed method with all three sequences (parallel, 45° and 90°). Figure 5 demonstrates the obtained histogram of gradient

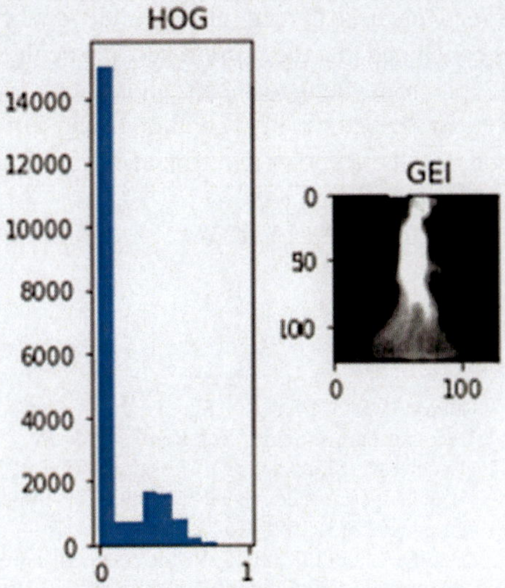

Fig. 5 Histogram of gradient of an gait energy image (GEI)

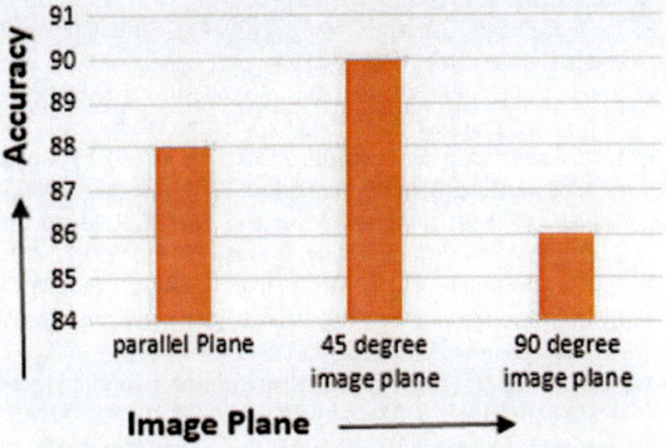

Fig. 6 GEI on different planes and its accuracy with SVM

graph for gait energy image. Support vector machine classifier is used to recognize the individual by comparing the values of a histogram of gradient obtained from the gait energy image. The recognition accuracy of each individual is calculated for parallel, 45° and 90° image plane with 88%, 90% and 86%, respectively. The recognition accuracy is presented in Fig. 6.

5 Conclusion

An automated efficient gait classifier is implemented based on histogram of the oriented gradient, support vector machine. It represents that the unique gait features can be extracted from a histogram of gradients value of gait energy image and can be effectively used to recognize the individual with SVM classifier. With these results, it is concluded that the support vector machine recognizes the different gait patterns of people from young to old age and thus paving the way for developing the recognition system for pathological conditions and surveillance applications. Future work is to compare the accuracy and efficiency of various machine learning algorithms in human gait recognition.

References

1. Nixon MS, Carter JN, Shutler JD, Grant MG (2002) New advances in automatic gait recognition. Inf Secur Tech Rep 7(4):23–35
2. Wagner M, Slijepcevic D, Horsak B, Rind A, Zeppelzauer M, Aigner W (2019) KAVAGait: knowledge-assisted visual analytics for clinical gait analysis. IEEE Trans Visual Comput Graphics 25(3):1528–1542
3. Raj JS, Ananthi JV (2019) Recurrent neural networks and nonlinear prediction in support vector machines. J Soft Comput Paradigm (JSCP) 1(01):33–40
4. Roy A, Sural S, Mukherjee J (2012) Gait recognition using pose kinematics and pose energy image. Signal Process 92:780–792
5. Pang Y, Zhang K, Yuan Y, Wang K (2014) Distributed object detection with linear SVMs. IEEE Trans Cybern 44(11):2122–2133
6. Sun L, Liu G, Liu Y (2013) Multiple pedestrians tracking algorithm by incorporating histogram of oriented gradient detections. IET Image Process 7(7):653–659
7. Li Y, Yang C, Zhang L, Xia R, Fan L, Xie W (2018) A novel SURF based on a unified model of appearance and motion-variation. IEEE Access 6:31065–31076
8. Kumar TS (2019) A novel method for HDR video encoding, compression and quality evaluation. J Innovative Image Process (JIIP) 1(02):71–80
9. Yao L, Kusakunniran W, Wu Q, Zhang J, Tang Z, Yang W (2019) Robust gait recognition using hybrid descriptors based on skeleton gait energy image. Pattern Recogn Lett. https://doi.org/10.1016/j.patrec.2019.05.012
10. Shu C, Ding X, Fang C (2011) Histogram of the oriented gradient for face recognition. Tsinghua Sci Technol 16:216–224. https://doi.org/10.1016/S1007-0214(11)70032-3
11. Sangeetha D, Deepa P (2017) A low-cost and high-performance architecture for robust human detection using histogram of edge oriented gradients. Microprocess Microsyst 53:106–119
12. Vapnik VN (1995) The nature of statistical learning theory. Springer, New York

Chapter 39
Inventory Optimization for Cognitive Demand Scheduler Using Data Analytics

Priyanka Singh, Soma Ghosh, Manish Saraf, and Rahul Nayak

1 Introduction

1.1 What is Cognitive Demand Planner or Scheduler?

While procuring and producing products, lots of stakeholders are involved at various stages like for knowing the exact required quantity of any raw material, the stores are involved, for forwarding the order list to the vendors, buyers are involved, for supplying materials, suppliers are involved, vendors are involved who provide us with the raw materials and finally end-users (customers) are involved to whom finished goods are sold. All the data that are related to each of this process should be available to all the stakeholders in real time which will help the business leaders to take a vital decision in favour of the organization. Hence, cognitive demand scheduler (CDS) is a web-based tool that provides a digital platform to enable end-to-end new age and complete visibility of information from planning to final consumption (Fig. 1).

P. Singh (✉) · S. Ghosh
College of Engineering, Pune, India
e-mail: ps21priyanka@gmail.com

S. Ghosh
e-mail: Gsn.comp@coep.ac.in

M. Saraf · R. Nayak
Reliance Industries Limited, Navi Mumbai, India
e-mail: manish.saraf@ril.com

R. Nayak
e-mail: rahul.nayak@ril.com

I. Jeena Jacob et al. (eds.), *Data Intelligence and Cognitive Informatics*,
Algorithms for Intelligent Systems,
https://doi.org/10.1007/978-981-15-8530-2_39

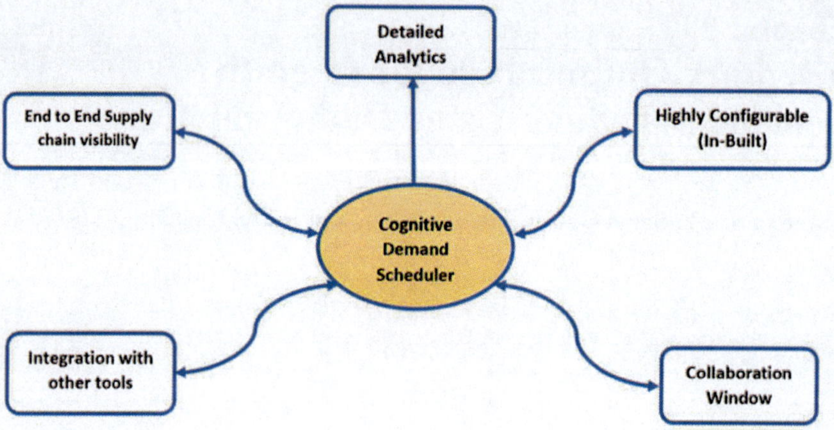

Fig. 1 What makes CDS better?

Features of CDS

- In detail complete supply chain information visibility
- Detailed data analytics coupled with business rules
- Users can collaborate with each other for smoother conduct of operations.

1.2 Role of Inventory Management in the CDS

Inventory plays an important role in CDS as this tool is made to increase the business productivity of an organization that is into selling, buying, trading, planning and production. Hence, for production and consumption, inventory, where the company holds its stock plays a crucial role as inventory, helps to meet future demands. If inventory can be managed that means the core of the supply chain is being managed, since the stock of raw materials or the end products is stored in warehouse were having the inventory, all the planning data which are related to products like various materials, raw materials, stock required for producing the products are also related to inventory. Inventory has various data that are related to planning and production; hence, these are very dynamic in nature so it is important to analyse inventory in real time to have the descriptive statistics of the data which will help the business head to take a better decision. With the help of prescriptive statistics or machine learning, the future demands can be predicted or the date on which particular material's stock will arrive.

1.3 Role of Data Analytics in Inventory Optimization

Data analytics in the field of science is used to draw insights from the raw data. It is used to analyse the raw data to give useful information which will help in taking better business decisions and improves the productivity of the organization. There is a huge amount of data that is involved in supply chain and if descriptive analysis of this data is done it will help us to know the crucial information related to stock such as when the particular stock going to be replenished, what is the total amount of stock available if the consumption of the product is increasing or decreasing down the line and many more such insights. One of the field of data science is data visualization which enables the users to understand bigger and complex data in an easy, intuitive and graphical format. Hence in this Project we have used this technique to show the summary of planning data in enterprise, Descriptive statistics of the inventory data and all the crucial Key performance indicators are collectively represented on the dashboard, that provides an aid to the business leaders to take vital decision for the betterment of an enterprise [1]. Thus, crucial business decisions can be taken in favour of an organization. One of the branches of data science includes predictive analytics and the use of machine learning to predict the future consumption pattern to avoid any pitfalls or to maintain the optimal level of stock to meet the future demands.

1.4 Different Stages Involved in the Procurement of Material in an Inventory

1. Creation of purchase requisition (PR): Purchase requisition is raised by the stores after checking the availability of stock of raw materials to the buyers. Once it is created, all line items are released and when the final line item releases, purchase requisition final release date is being considered as the end of the purchase requisition creation stage.
2. Release of purchase order (PO): Once the requirement is sent to the buyers, they prepare the final list of materials and send the schedule line date for each line item. The final schedule is released for each material and this is called purchase order final release date.
3. Acknowledgement from the vendor: This is the confirmation from the vendor side acknowledging the amount of stock that is available and that they are ready to dispatch the stock. They do it by sending the ASN receipt to the buyers.
4. Dispatch of items by the vendor: After the schedule lines are sent to the vendor along with quantities and date of a schedule of each line item, they create shipping consignment number (SCN) against each schedule line and send the required quantity sometimes before or after the creation of SCN date.

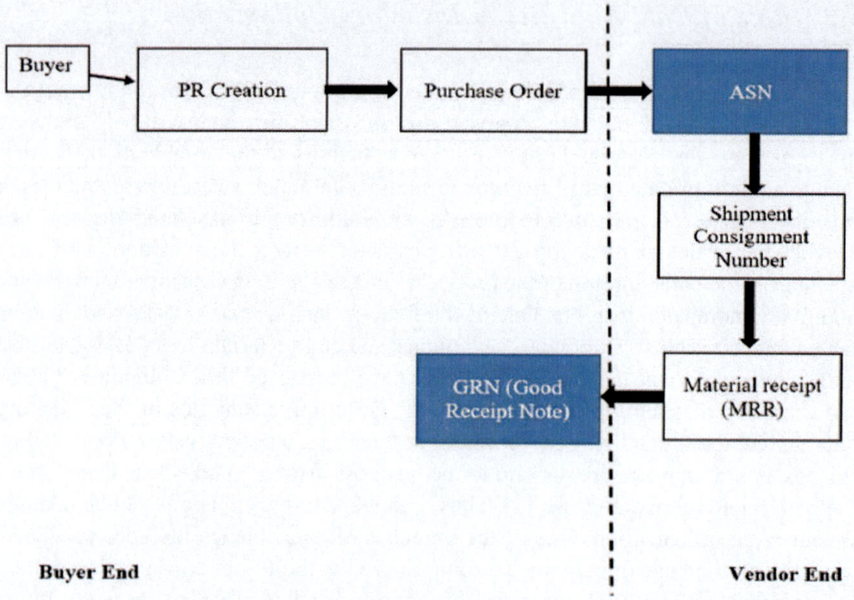

Fig. 2 Processes involved in the procurement process

5. Material received receipt (MRR) [2]: This is the document which is created when the consignment reaches the store. Basically, it is the record which is entered at the entry gate of each store to confirm the arrival of stock at the stores.
6. Good receipt number (GRN): This is the final stage of procurement. This document is created when the product reaches the store warehouse (Fig. 2).

2 Literature Review

2.1 *Summary Table for the Survey Conducted in Relation to Identifying KPIs and Use of Data Visualization Tools in Different Organizations*

See Table 1.

3 Proposed Architecture Diagram

In the proposed model as mentioned Fig. 3, the SAP source data are ingested into the HANA database using SAP Landscape Transformation (SLT) technique which is the

Table 1 Summary table for the literature survey

Title of paper	Publication and year	Summary
Improving business decision making based on KPI management system [3]	IEEE, 2017	Designed a tool called BIM stands for Business indicator management to analyse and monitor the growth and progress of the business. 12 KPI, like collection efficiency, customer satisfaction, an increase of sales, quality of index, satisfaction of employees, claim rate, etc., is identified. BIM also generates real-time graphs as well as the insights from the information that are stored in the database directly. With the help of BIM, 87% customer satisfaction was achieved which was 76% without BIM
Key performance indicators (KPIs) for the evaluation of the demand response in the Technical University of Cluj-Napoca buildings [4]	IEEE, 2019	KPI in this research work was used to evaluate demand response in the technical institute building. KPI cost—economic gain, energy saving and consumption and user engagement in terms of acceptance have been identified. For occupants, KPI like user engagement has been identified to measure the participation, comfort and convenience of the consumers. KPI like climate policies has been identified to find the rate of CO_2 emission or check for the greenhouse effect

(continued)

Table 1 (continued)

Title of paper	Publication and year	Summary
Performance measurement and KPIs for remanufacturing [5]	Springer, 2015	In this research work, several KPIs have been identified for the remanufacturers like cycle times, lead time, the overall effectiveness of the equipment, count of concessions, work in progress, materials that are used, energy consumption that is required, what is the number of recycled materials that are used, greenhouse gas emissions, amount of waste, what is the amount of water that is used, etc.
Improving inventory management performance using a process-oriented measurement framework [6]	Springer, 2010	In this work, KPIs are identified for measuring the performance of inventory management by analysing the IM business process and with the help of the business intelligence approach. In this paper, an IM framework was created by collecting all things together to provide an overview of the IM processes within different organizations. The proposed framework was tested with the help of KPIs along with the bits of advice of experts
A KPI-based condition monitoring system for the beer brewing process [7]	IEEE, 2019	In this paper, authors have used KPIs to monitor the condition for the beer brewing company. Finally, a dashboard is made to represent all the KPI related to brewing like different types of brew, what is the plant condition, what is the availability of a particular brew to the mangers for better understanding

(continued)

Table 1 (continued)

Title of paper	Publication and year	Summary
Interactive data visualization of chatter conditions in a cold rolling mill [8]	2018	visual analytics was used to understand the factors and the conditions that create chatter in the rolling process of steel in the industries. The user can dynamically monitor the conditions of each chatter onset with the help of a web-based interactive dashboard
A survey of scholarly data visualization [9]	IEEE, 2018	In this research work, authors have used scholarly information for the visualization to depict the relations between different authors, papers, citations and to show the scholarly networks. In this paper, the authors have discussed the existing tools, techniques as well as systems that occurs for analysing and visualizing a huge volume of scholarly data
Big data visualization: tools and challenges [10]	2nd international conference on contemporary computing and informatics, 2016	In the survey paper, authors have discussed the importance of big data and the challenges that are involved in it, they even discussed different visualization tools. Some of the visualization tools that are surveyed in this paper were: Tableau, Gephi, Plotly, Excel, Microsoft Power BI, etc.

(continued)

Table 1 (continued)

Title of paper	Publication and year	Summary
Data visualization [11]	International journal of engineering research and advanced technology, 2016	In research work, the data visualization concept has been discussed. Different types of charts like a line graph, bar chart, pie chart and scatter plot is explored
Data analysis and visualization of sales data [12]	World conference on futuristic trends in research and innovation for social welfare, 2016	In the survey paper, sales data are being analysed and visualized. This paper focused on a system that will graphically represent the sales data which will provide a holistic view of the business and that will help to make better decisions to promote the efficient generation of revenues, better management of business operation and tracking the flow of tasks

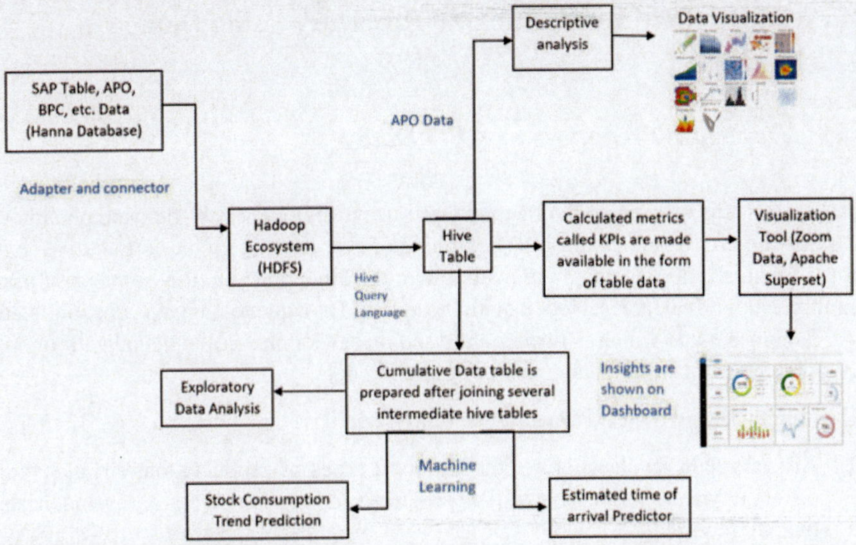

Fig. 3 Proposed architecture diagram

ETL tool to load the real-time data from SAP source system to HANA database. Then with the help of UDIF that is unified data ingestion framework, data from HANA database are ingested in Hive tables. For performing descriptive statistics over APO data, multiple intermediate tables are joined which will be considered as a final data model and then this table is analysed on different metrics with the help of PySpark or HQL. This final model is sent to the visualization tool. For building data model, for using machine learning on such huge production data, PySpark is used as it is scalable and help in providing less latency [13].

Firstly, for building KPIs, a collaborative discussion with the stakeholders is conducted so as to decide the requirement, their perspective, formulae and logics needed to build KPI. Then the developer would collect all these formulae and logic, will model the data in such a way that all the variables involved in the KPI formula is available in table. The data is modelled with the help of Hive query language in our project as our base tables containing the variables of formula is present in Hadoop, finally a cumulative data model is prepared containing all the KPI logics, and then it is fed into the visualization tool for visualizing the KPIs in real time. Then, these tables sent to visualization tools and KPIs are made available to business leaders in the form of dashboards.

After data and visual analytics, machine learning algorithm for predictive analysis is also performed. The data are gathered for last three years and then these data models are prepared and features are engineered and then these data are fit into a machine learning predictor module to predict the stock consumption trend and for predicting the lead time of any material from the day it is being ordered to the day it will be received.

4 Methodology for Optimizing Inventory Using Data Analytics

4.1 Descriptive Analysis of APO Data

APO stands for Advanced Planner and Optimizer. It is part of SAP demand planning. It is a 3-month plan and done at the beginning of every month for the consecutive next three months. The APO data are available in the Hive table but it is segregated into multiple individual tables. Hence, with the help of Hive query language, consolidated APO table is made which is further analysed on several metrics in visualization tool Apache Superset. APO data consist of planning data.

Few metrics on which APO data are analysed:

1. At each plant level, what are the different types of products that are planned, different types of grades that will be produced and different types of raw materials that are required.
2. What are the different raw materials required by each grade and what are their respective quantities?
3. What are the different grades are produced under different products?
4. What is the total count of different products, grades and raw materials across each plant site?
5. What are the five most and least produced grades, consumed raw materials?

4.2 Identify Key Performance Indicators and Visualizing It on the Dashboard

Key performance indicators (KPIs) are the metrics [14] that are used to evaluate how well your system performs. It enables the real-time scenario that is prevailing within the supply chain network. These KPIs are represented graphically and visually on the dashboard. A dashboard contains all the KPI related to the inventory which will help the business leaders to monitor the real time scenario which will further help them to take better decision in favor of the organisation. These KPIs are calculated by taking the different columns present in different hive table. For example For calculating KPI: "*days_to_stock_out*", we have used the formula, total stock divided by consumption rate. Well the *total stock* is present in hive table A having total stock as one of the attributes, where as the value *consumption rate* is coming out from the hive table B. Therefore, we have to use join concepts of hive query language to join several hive tables just to calculate each KPI. As different Hive tables have different attributes that are needed for calculating each KPI, data modelling using Hive query language is used to inculcate different attributes important for the calculation of metric in a single data model. By doing it, all 33 identified KPIs will be inculcated into five data models only which helps us to save memory and saves latency while reading each table. These data models which have the field for each calculated metrics are used as the data source in visualization tool and then on top of each data source KPI is represented on the charts and the dashboard [1] is presented to the business leaders and all the stakeholders to take a precise decision to enhance the productivity of the organization.

Some of the KPIs identified in this research work:

- APO and BPC forecast error to measure the percentage of error while forecasting in APO and BPC planning
- Days to stock out for knowing the exact days of material in which their stock will be replenished
- Overstock indicator to determine which of the materials are overly stocked
- Different numbers of purchasing groups
- The expiry date of each of the raw material to know when the particular stock is getting expired
- Average order size for each of the material
- Different lead times like PR-PO, PO-SCN, SCN-GRN and PR-GRN (full life cycle).

4.3 Prediction of Stock Consumption and Estimated Time of Arrival of Stock Using a Machine Learning Algorithm

Stock availability plays a vital role in any organization to meet its future demand. Hence, it is very important to maintain the optimal amount of stock to provide aid to the end-users to meet their needs. A data model was made using Hive query language which includes fields like material number, plant code, date field which shows that on a particular date what amount of stock was used, stock value is there and its unit of measurement was there. This data model contains the data for the last three years and the stock consumption wanted to be predicted for the coming next three months.

In traditional systems, lead time, which is the difference of days between the date on which order was placed to the date on which order was received, was calculated by taking the average of the past lead times. Machine learning was used in this project to accomplish this task of predicting lead time. For example, as soon as purchase requisition is created, the system would be able to predict the date on which stock will reach to store that is when the GRN will be created.

The algorithm used in this research work was random forest regressor.

Random Forest

The random forest supports classification as well as regression and for both, it uses an ensemble learning method. It is a supervised learning algorithm. The random forest does not support boosting technique, rather it is a bagging technique. The trees in random forests run in parallel. All the trees are independent of each other as there is no interaction between the trees while building the forest. It operates by constructing a multitude of decision trees at training time and outputting the class that is the mode of the classes (classification) or mean prediction (regression) of the individual trees.

Steps of implementation:

1. Import all the necessary libraries related to sklearn, ensemble, NumPy and pandas
2. Data pre-processing was done:
   - For stock consumption prediction, materials which are procured for more than 30 times in 36 months were considered and for estimated time of arrival (ETA) predictor, lead times are added as target variable to make it supervised learning approach. Sum of a stock is calculated by grouping it into month-year-wise to get the aggregate stock for each material on different month-year pair
   - Missing values are deleted as it is both for stock consumption use case all null values are dropped, whereas for ETA, null values are only dropped if it reaches a particular threshold value
   - Outliers are detected and treated using inter-quartile method and skewness
   - Log transformation was done for normalizing the skewness
   - Scaling was done
3. Feature engineering was done to select the best features using ANOVA testing, F-score, P-value, Pearson's correlation and Kendall correlation
4. Test and training splitting: For stock, last 3-month data into test and remaining data into training and for ETA, last 6 months data for test and remaining into training
5. Categorical and numerical fields are treated using label encoding
6. Model was fit
7. Accuracy calculated for training as well as test data
8. Hypertuning was done to tune the hyperparameters to yield better accuracy.

5 Results

5.1 Descriptive Analysis of APO Data

Figure 4 depicts dashboard containing descriptive statistics of APO data on the metrics: the sum of the quantity that was planned for each product type across plant

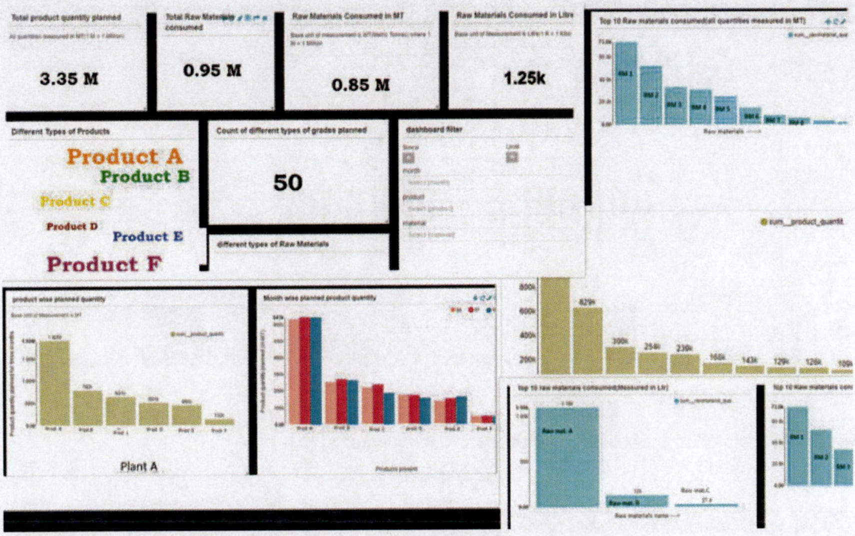

Fig. 4 APO data dashboard [15, 16]

A; the three months planned quantity for different products in plant A; and the planned top 10 materials that are going to be consumed in plant A. There are some raw materials which are liquid in nature hence their base unit of measurement was ltr else everything is measured in metric tonnes (MT) and top 10 planned materials quantity.

5.2 Dashboard for KPI Visualization

Figure 5 represents the dashboard on Zoom data [15] visualization tool for different key performance indicators that are identified for inventory management. The below figure depicts the KPI related to purchasing like average order size, coverage of scheduling agreement, a quantity of material that is ordered, etc. For each material name and its respective plant code, this material is present.

Figure 6 represents the two sections of the dashboard on Zoom data visualization tool for different key performance indicators where the top section consists of the KPI related to planning forecast that is done via APO tool of SAP. The forecast error for APO and BPC is calculated against actual production and the value if represented on the dashboard to tell the business unit how well the planning is.

The second section of the dashboard reveals the key performance indicators that are related to the stock presents in inventory.

For example, days to stock out to represent the number of days that are remaining for the stock of any material to be exhausted completely, how many materials are

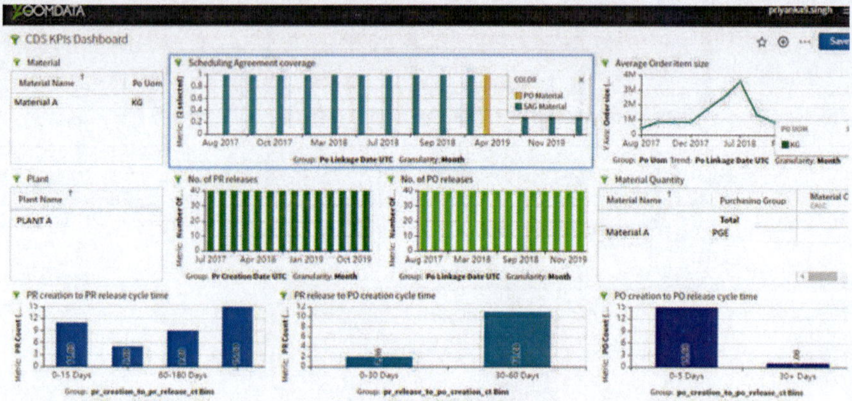

Fig. 5 KPI related to purchasing and procuring in inventory [15]

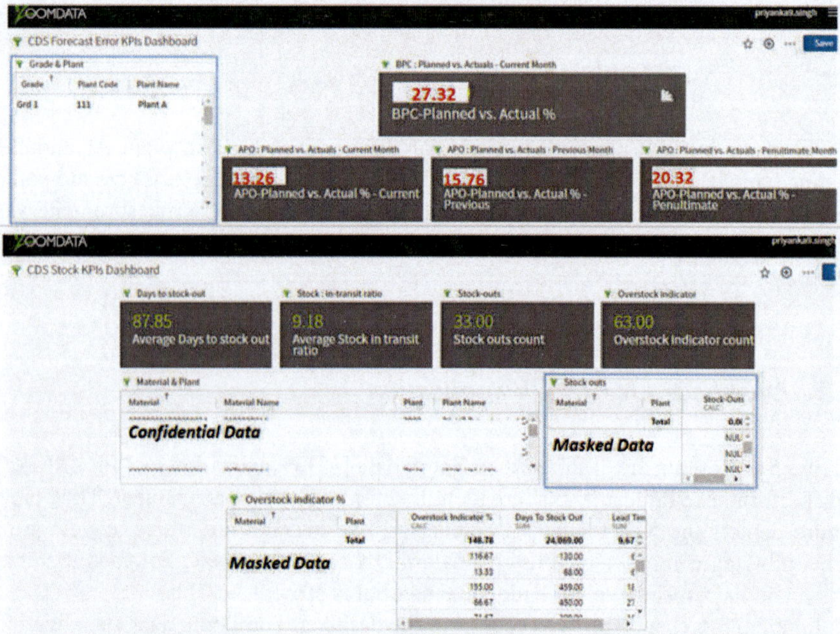

Fig. 6 KPI dashboard [15] for stock and forecast error for APO and BPC

listed as overstock as these materials are in abundance so do not need to order them, How many material's stocks is over to show this stock out indicator was used.

Table 2 Table of accuracy calculated for prediction model using machine learning algorithm and random forest regressor

Data set name	Stock consumption (%)	Number of deliveries (%)	PR-PO lead time (%)	PO-SCN lead time (%)	ETA hit rate (PR-GRN) (%)
Training	85.34	99.9	99.92	64.3	78.13
Testing	82.67	98.9	55.54	54.67	67.7

5.3 Accuracy Table for Stock Consumption and Lead Time Prediction Using a Machine Learning Algorithm

Accuracy was calculated with the help of mean absolute percentage error for both pieces of training as well as test data. The mean absolute percentage error (MAPE), also known as mean absolute percentage deviation (MAPD), is a measure of prediction accuracy of a forecasting method in statistics. It is calculated as the average of the unsigned percentage error, as shown in the example below:

$$\text{MAPE} = 100 \times \frac{1}{n} \times \sum_{i=1}^{n} \left| \frac{\text{Obs}_i - \text{Model}_i}{\text{Obs}_i} \right|$$

Finally, accuracy is calculated as: 100-MAPE (Table 2).

6 Conclusion

Inventory is very important for an industry that is involved in trading, sale and production. As supply chain within an organization is hugely influenced by its inventory and stock so it is vital to optimizing it to meet future demands in an efficient and reducing the inventory cost at the same time. The inventory-related data are highly dynamic in nature so it should be monitored very precisely. Hence, for this, key performance metrics are defined and these are shown on dashboards in real time in this project to make business leaders aware of the current scenario and take better decisions in favour of the organization. Industries use SAP APO for demand planning so it is crucial to know how efficient and how the planning was done for each site, to achieve this, descriptive statistics for the APO data were done and finally visualized on Apache Superset in this research work. Predictive analysis using machine learning algorithm and the random forest was done to predict the future consumption trend of the materials in stock to maintain an optimal amount of stock without overstocking or understocking. The accuracy achieved was 82.67%. The lead time was predicted for better scheduling and planning, the accuracy achieved was 67.7%.

7 Future Scope

In this research work, random forest algorithm was used to predict the lead time or estimated time of arrival of any materials so the next step of this research work would be the use of neural networks and also productionizing the machine learning model in real time.

References

1. Rachad S, El Idrissi Larabi Z, Nsiri B, Bensassi B (2017) Inventory management in closed loop structure using KPIs. Int J Appl Eng Res 12(15):4864–4869. ISSN 0973-4562
2. Reliance Industries Limited, Navi Mumbai
3. de Andrade PRM, Sadaoui S (2017) Improving business decision making based on KPI management system. In: Proceedings of IEEE international conference on systems, man, and cybernetics (SMC), pp 1280–1285
4. Cretu M, Ceclan A, Czumbil L, Şteţ D, Bârgăuan B, Micu DD (2019) Key performance indicators (KPIs) for the evaluation of the demand response in the Technical University of Cluj-Napoca buildings. In: IEEE 8th international conference on modern power systems (MPS), pp 1–4
5. Graham I, GoodallP, Peng Y, Palmer C, West A, Conway P, Mascolo JE, Dettmer FU (2015) Performance measurement and KPIs for remanufacturing. J Remanufacturing 5(10):1–17. ISSN 2210-4690
6. van Heck G, van den Berg J, Davarynejad M, van Duin R, Roskott B (2010) Improving inventory management performance using a process-oriented measurement framework. In: Quintela Varajão JE, Cruz-Cunha MM, Putnik GD, Trigo A (eds) Enterprise information systems. CENTERIS 2010. Communications in computer and information science, vol 109. Springer, Berlin
7. Pasic F, Wohlers B, Dziwok S, Becker M, Heinrich M (2019) A KPI-based condition monitoring system for the beer brewing process. In: 2019 24th IEEE international conference on emerging technologies and factory automation (ETFA), Zaragoza, Spain, pp 1469–1472
8. Pérez D, Díaz Blanco I, Cuadrado A, Rendueles JL, García D (2018) Interactive data visualization of chatter conditions in a cold rolling mill. Comput Ind 103:86–96
9. Liu J, Tang T, Wang W, Xu B, Kong X, Xia F (2018) A survey of scholarly data visualization. IEEE Access 6:19205–19221
10. Ali SM, Gupta N, Nayak GK, Lenka RK (2016) Big data visualization: tools and challenges. In: 2016 2nd international conference on contemporary computing and informatics (IC3I), Noida, pp 656–660
11. Sadiku MNO, Shadare AE, Musa SM, Akujuobi CM (2016) Data visualization. Int J Eng Res Adv Technol (IJERAT) 02(12). ISSN: 2454-6135
12. Singh K, Wajgi R (2016) Data analysis and visualization of sales data. In: 2016 world conference on futuristic trends in research and innovation for social welfare (Startup Conclave), Coimbatore, pp 1–6
13. Ranganathan G (2020) Real time anomaly detection techniques using pyspark frame work. J Artif Intell 2(01):20–30
14. Van Looy A, Shafagatova A (2016) Business process performance measurement: a structured literature review of indicators, measures and metrics. SpringerPlus 5:1797
15. https://www.zoomdata.com/
16. https://superset.apache.org/

Chapter 40
DRIVE SAFE: Lane Deviation Detection and Alert System Using Image Processing Techniques

H. C. Arun, Jisha John, and Aswathy Ravikumar

1 Introduction

About 60–65% of the road traffic accidents are due to the negligence from the part of the drivers; in the year 2017, about 90% lost their lives in India [1]. To enhance the safety of the driver and other passengers and thereby reducing the amount of accidents that occur on road, Advanced Driver Assistance System (ADAS) [2] is used. As the name suggests, the ADAS assists the driver by becoming the third eye. The main feature of ADAS is to increase road safety so as to decrease traffic jams, in turn, increase transportation efficiency. The detection of lane border in images of road scenes is obtained from a camera mounted on the vehicle. The main problems that occur on the lane border detection techniques are shadows which are cast by trees, buildings. It should process the painted and the unpainted roads in a similar fashion; it should also handle the curved roads as well as the straight roads. The lane detection technique can be used for self-driving cars [3]. The key aspects of lane detection are to understand the road length, the number lanes, and their position, whether there is any merging or splitting of lanes, etc., in different road conditions. Most of the accidents that occur are mainly due to vehicle deviating from lane due to physical ailments of the driver, and hence, timely alerts are given can save lives

H. C. Arun · J. John (✉) · A. Ravikumar
Department of Computer Science and Engineering, Mar Baselios College of Engineering and Technology, Thiruvananthapuram, Kerala 695015, India
e-mail: jisha.json@gmail.com

H. C. Arun
e-mail: arunhc2780@gmail.com

A. Ravikumar
e-mail: aswathy.ravikumar@mbcet.ac.in

I. Jeena Jacob et al. (eds.), *Data Intelligence and Cognitive Informatics*,
Algorithms for Intelligent Systems,
https://doi.org/10.1007/978-981-15-8530-2_40

to a great extend. The proposed method aims at the detection of lane deviation and generation of an alert via a hardware module.

2 Related Work

2.1 RANSAC Technique [4]

The full form of RANSAC is "RANdomSAmple Consensus." It is used to estimate whether the dataset is having any outliers by using some mathematical model. RANSAC is a non-deterministic algorithm which produces a result with some probability. It may also contain inliers due to noise, "inliers" are referred to data values which are erroneous, and "outliers" are data which doesn't fit with the proposed model. The two main cause of unfitting of data which results in outliers is incorrect hypothesis and measurements. For a given set of inliers, RANSAC explains how to avoid outliers. This is done with the help of the bird's eye view of the image, called the Inverse Perspective Mapping (IPM) [5]. After that, this image is filtered with selective Gaussian spatial filters which are optimized to detect vertical lines. Once the threshold is applied on the filtered image and thereby maintaining the maximum values only and discarding others, Hough transform [6] is used to detect straight lanes; after that line fitting is done by using the RANSAC, and then, RANSAC spline fitting is used to distinguish between straight lanes with curved ones and thereby detecting curved lanes. The final step is cleaning and localization of detected splines.

2.2 Vanishing Point [7]

In an image coordinate system, all the parallel lines will converge at a vanishing point (Vp). Vp can be estimated using the extrinsic and intrinsic parameters of the camera, and to check whether the lines cross Vp, Hough transform is used. For that, at first, the vanishing point is detected. Then, the disparity map is generated using the pixel difference between the images taken using two cameras. A v-disparity map is then used to create a histogram of disparities. The vertical profile of the road is extracted from the v-disparity map.

2.3 River Flow Model [8]

Lane flow is similar to the river flow, i.e., if there is a path for the river to flow it will use that path; here, the high-intensity value pixels are linked up to all possible high edge values. Kalman filter predicts the future lane position by using the trajectory of previous observations. Here, first the horizontal localization is done this to differentiate road region from the sky region. Then, lane tracking is done by using the river flow concept. Once the lane line is detected, then the tracking is done by using the Kalman filters.

2.4 *Active Line Model [9]*

Active Line Model (ALM) and Hough transform are used for the detection of the lane. The ALM combines with the Hough transform to get better results. The Hough transform is used to extract the vanishing point of the road and is used to estimate the beginning of lane boundaries on the road. The prominent lane line is detected using ALM technology finally.

3 Methodology

All the existing methods focus only on lane detection while the proposed system gives alert to the driver if lane deviation is detected. The main objectives of this work are:

- To alert the driver, if the driver is inattentive or drowsy.
- To identify the lane lines and alerts the driver, if the car goes wrong sidetrack.

Figure 1 shows the overall outline of the proposed lane detection and alert system.

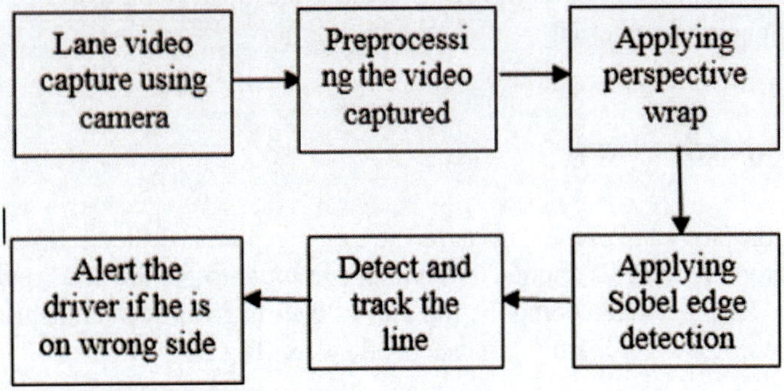

Fig. 1 Lane detection and alert system

3.1 Data Collection

The video of the road is captured using the camera device. The camera is mounted on the rear side of the mirror inside the vehicle, so as to get a clear view of the road lanes.

3.2 Preprocessing

Once the video of the lane is captured, the next step is to preprocess [10] the captured video. Preprocessing of the video has two steps:

- The first step is to convert the captured RGB input into grayscale. This is to reduce the computation cost that is because the RGB input has got three color channels whereas the grayscale has only one-layer image ranging from 0–255. This makes the processing of the three-layer RGB image that takes at least four times longer than processing the grayscale counterpart. It may be negligible if there are very few images in the database to process. If there are thousands of images, then the significant increase in speed can be observed. The complexity of code will also increase if playing around with RGB input. Another important factor is the signal to noise ratio. For lane detection, one of the important factors is to identify the lane lines; in order to do that Sobel edge detector is used which won't work good along with color input.
- The second step in preprocessing is to remove the noise using the Gaussian blur. Gaussian blur or smoothening is very effective in removing the noise. It replaces the pixel with a weighted average of surrounding pixels. The weights come from the Gaussian probability distribution, so the nearest pixels are more influential. It is rotationally symmetric.

3.3 Perspective Warp

Detecting the curvature of the lane using the camera space is difficult. So perspective transformation [11, 12] is applied to the image in order to get the bird's eye view of the image. This is done keeping the top part of the image as such and shrinking the bottom rows. It has got mainly three steps; they are (Fig. 2):

1. Represent the image in a shifted coordinate system

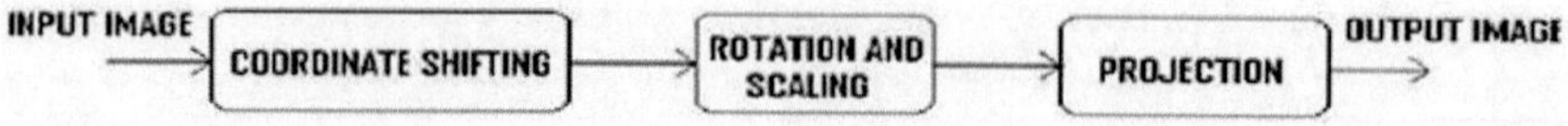

Fig. 2 Perspective warp flowchart

The image is represented in 2D x, y plane. For a grayscale image of resolution 100×100, there are about 10,000-pixel values with pixel intensities ranging from 0–255. The starting point is at (0, 0), and the ending point is at (99, 99) for a 100×100 image.

$$(0,0)(0,1)(0,2) - - - - (0,98)\ (0,99)$$

$$(1,0)(1,1)(1,2) - - - - (1,98)\ (1,99)$$

$$(99,0)\ (99,1)\ (99,2) - - - - (99,98)(99,99)$$

To shift the coordinate, the center point of the above mentioned image is taken as (0, 0) with starting and endpoints at (−50, −50) and (49, 49), respectively.

$$(-50,-50)(-50,-49)(-50,-48) - - - - (-50,-48)\ (-50,49)$$

$$(-49,-50)(-49,-49)(-49,-48) - - - - (-49,-48)\ (-49,49)$$

$$(49,-50)\,(49,-50)(49,-48) - - - - (49,48)\ (49,49)$$

2. Rotate image
   To rotate the image, first change to 3D plane by introducing a third parameter z which is initially set to zero. To rotate the image, the (x, y, z) is multiplied with the R_z matrix to get (p, q, r).

$$R_z = \begin{bmatrix} \cos(\beta) & 0 & -\sin(\beta) & 0 \\ 0 & 1 & 0 & 0 \\ \sin(\beta) & 0 & \cos(\beta) & 0 \\ 0 & 0 & 0 & 1 \end{bmatrix}$$

   Here, $\beta = 45°$.
   After rotation, the height of the image will get reduced in order to retain that use matrix S_H.

$$S_H = \begin{bmatrix} Sx & 0 & 0 & 0 \\ 0 & Sy & 0 & 0 \\ 0 & 0 & 1 & 0 \\ 0 & 0 & 0 & 1 \end{bmatrix}$$

3. Projecting the image on 2D plane

To project the image on to 2D plane using the formulas:

$$u = \frac{\text{fp}}{f - r} + \left(\frac{\text{im}_{\text{size}}}{2}\right)$$

$$v = \frac{\text{fq}}{f - r} + \left(\frac{\text{im}_{\text{size}}}{2}\right)$$

where f = focal length which can have an arbitrary value, and im_size is the size of the image.

3.4 *Sobel Edge Detection*

The Sobel operator [13] is an edge detector which finds a region where there is a sharp change in intensity between adjacent pixels or sharp change in color between adjacent pixels. The Sobel operator uses a 3 × 3 kernel is used to scan for each of the *x*- and *y*-axis. The Sobel operator is separate for *x*- and *y*-direction. The gradient for *x*-direction has negative values on the left-hand side of the matrix and the right-hand side got positive values; the center pixels are preserved whereas for the *y*-direction gradient the negative numbers reside at the bottom of the matrix the top of the matrix contains the positive values, and the middle row pixels are preserved. The Sobel operator finds the difference in gradient between the adjacent pixels of the given image by placing the gradient matrix above it. If upon calculating the gradient matrix the value is positive, then it marks an edge. The Sobel operator is the example of a discrete differentiation operator. Sobel operator performs an approximation operation on the gradient of an image intensity function. The Sobel operator is the combination of differentiation function and Gaussian smoothing. The Sobel operator finds more edges than the Prewitt operator and is computationally faster than the canny edge detector which has an additional overhead of noise reduction step embedded with it. The most important advantage of the Sobel edge detector is its simplicity (Fig. 3).

Given below is the example of applying a Sobel operator on a pixel. Here, this pixel is taken as an edge since it produces a positive value upon applying the Sobel operator. If the output value is negative, then it is discarded (Fig. 4).

Example of applying Sobel operator on an image (Fig. 5).

Fig. 3 Sobel filter[13]

-1	0	+1
-2	0	+2
-1	0	+1

x filter

+1	+2	+1
0	0	0
-1	-2	-1

y filter

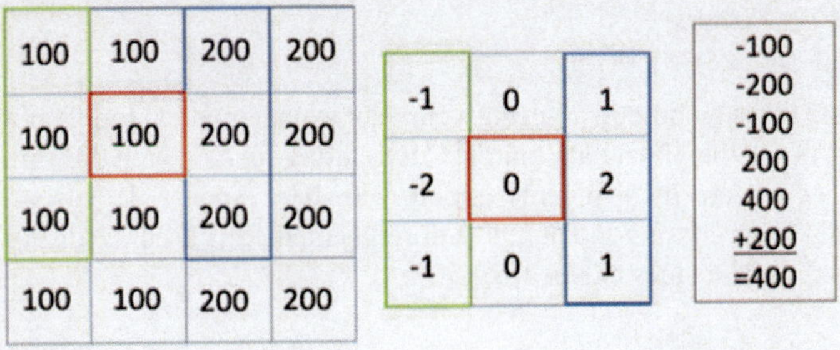

Fig. 4 Sobel operator on a pixel [13]

Fig. 5 Sobel output [13]

3.5 Sliding Window Protocol [14]

The sliding window search is used to track the lane lines and also to differentiate among the right and left lane boundaries. Sliding window protocol is mostly used for object detection. It finds where exactly in the input the object resides. A window of fixed width and height is allowed to scan through the input. The sliding window has got three arguments. First is the input which is followed by the step size and finally the window size. The step size indicates that how many pixels have to be skipped on x- and y-axis while scanning normally it won't set to 1 since no need to scan through each and every pixels which will increase the computational time and thereby the computational cost. The window size indicates the width and height of the window which is used for scanning in terms of pixels.

3.6 *Curve Fitting*

Curve fitting is the technique of fitting a curve or a mathematical function which will be the best fit for the given data points. Curve fitting finds the relationship between predictors which are independent variable with the response dependent variable. Here, for lane curve fitting the numpy's polyfit() method is used. By using numpy's polifit() method, the lanes can be tracked.

3.7 *Alert*

An alert is given if the vehicle crosses the intersection lane and moves on to the wrong side. The alert can be manually switched off only after the driver comes back to the right track.

4 Hardware and Software Setup with Alert Module

The operating system used is Windows 8.1. The front end of the project is written using Python 3.6 by using the PyCharm IDE. The processor used is Intel i7 4th generation. It needs at least 8 GB RAM and 10 GB HDD space.

Figure 6 shows the hardware setup for lane detection. The hardware setup consists of an Arduino (3), Buzzer (1), camera, switch (4), and some jumper wires (2). When the driver drives along the wrong side of the road, the Arduino triggers the alarm

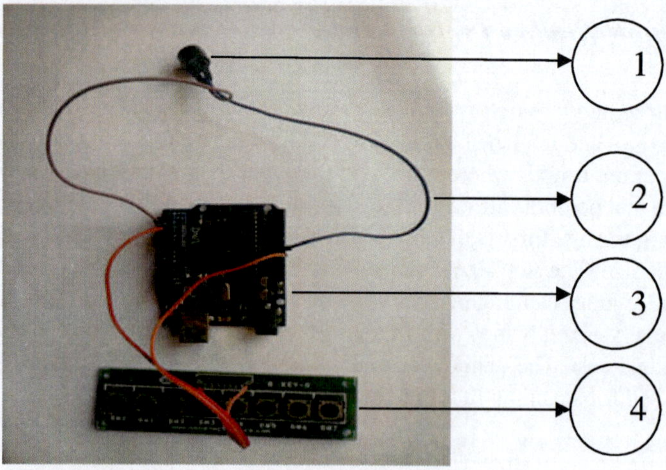

Fig. 6 Hardware module of lane detection

which can be stopped only after he comes back to the right lane of the road by using the switch.

5 Result and Analysis

The proposed system has been tested on many traffic videos and is found to give good results. A series of videos have been collected under traffic and without traffic, and the system shows good detection rate in all situations.

Figure 7 shows no alert situation that is the driver is driving inside the allocated road lane lines, so there is no need any alert. The Euclidean distance is used to check whether the driver crosses the lane line. If he crosses the lane line, then the Euclidean distance will be below a predetermined threshold and an alert is given.

Figure 8 shows the alert situation when the driver crosses the lane line. Here, the driver overtakes the bus and accidentally cuts the middle lane line which is dangerous because there are vehicles which are using the other side to go to the opposite direction, which will lead to an accident. This can be avoided by triggering an alarm when the driver cuts the lane line. This is what exactly the proposed system does. The blue and green dots with the red line in between are used to represent the calculation of the Euclidean distance.

Fig. 7 No alert situation

Fig. 8 Alert situation

6 Conclusion and Future Scope

In this paper, different technologies used for real-time lane detection have been discussed. The literature survey shows that real-time lane detection has scope for future research, as it can be used various road traffic applications. The lane detection technique proposed here can be used as alert systems which can reduce the accidents occur on road. The lane detection technique proposed here has got mainly five modules like preprocessing, perspective transform, Sobel edge detection, Sliding window search, curve fitting, and finally alerting. It can be integrated with a drowsiness detection module to increase the overall security. The lane detection has poor performance on a rainy day but by upgrading the camera this problem can be eliminated.

References

1. https://timesofindia.indiatimes.com/india/90-deaths-on-roads-due-to-rash-driving-ncrb/articleshow/61898677.cms
2. Jimenez F et al (2016) Advanced driver assistance system forroad environments to improve safety and efficiency. ScienceDirect Transp Res Procedia 14:2245–2254
3. Manoharan S (2019) An improved safety algorithm for artificial intelligence enabled processors in self driving cars. J Artif Intell 1(02):95–104
4. Du H et al (2017) The fast lane detection using RANSAC algorithm. Science Direct
5. Muad AM et al (2004) Implementation of inverse perspective mapping algorithm for the development of an automatic lane tracking system. IEEE

6. Koresh MHJD (2019) Computer vision based traffic sign sensing for smart transport. J Innovative Image Process (JIIP) 1(01):11–19
7. Umar et al (2017) Multiple lane detection algorithm based on novel dense vanishing point estimation. IEEE Trans Intell Trans Sys 18(3)
8. Lim KH et al (2012) River flow lane detection and Kalman filtering-based lane tracking. Hindawi Publishing Corporation, New York
9. Hauizhong et al (2010) Research on real time lane line detection technology based on machine vision. In: international symposium on intelligence information processing and trusted computing. IEEE
10. https://towardsdatascience.com/image-pre-processing-c1aec0be3edf/
11. Lin CC et al (2012) A vision based top view transformation for vehicle parking assistant. Sensor. ISSN 1424-8220
12. Venkatesh M et al (2012) Transformation techniques. Int J Sci Eng Res 3(5)
13. https://medium.com/datadriveninvestor/understanding-edge-detection-sobel-operator-2aada303b900/
14. https://www.pyimagesearch.com/2015/03/23/sliding-windows-for-object-detection-with-python-and-opencv/

Chapter 41
Foodborne Illness Investigation Technique Using Machine Learning

Ahmad Alkinj, Shabina Ghafir, M. Afshar Alam, and Md Tabrez Nafis

1 Introduction

Consistently in our life, many megabytes are included as a tweet, remark, posts in Internet-based life locales, photographs, recordings, net surfing, making an account in any Web-based life Webpage and so forth, gives us a knowledge about the plenitude of information. This information is either helpful for one or perhaps pointless for other; however, for an information researcher, it gives part of data escaped the ordinary human eyes or typical PC calculation. In the second 50% of the twentieth century, the artificial intelligence has come into the scene where it makes a machine to act like human, and one of the well-known subfields of artificial intelligence is machine learning, where learning of calculation encourages the machine to take in an undertaking from its experience. With this experience, it causes an amazing calculation to discover the example and use it for future forecasts. The Tom M. Mitchell gives the more point-by-point meaning of machine learning in 1997 that is [1] "A PC program is said to gain as a matter of fact E regarding some class of assignments T and execution measure P, if its presentation at undertakings in T, as estimated by P, improves with experience E."

This review centers around the utilization of AI calculations in different ailment datasets to help them for foreseeing the infection heretofore which will be both advantageous for the patient and the specialists. One of this examination has been done even to survey the understudy assessment investigation utilizing its coursework and class execution which will help to learn the procedure and training industry for bringing novel and imaginative ability. The reasonable ICT is the prerequisite of time as it gives productive, less force utilization gadget with more execution like the one understudy participation framework utilizing the versatile biometric sensors a case of

A. Alkinj (✉) · S. Ghafir · M. A. Alam · M. T. Nafis
Department of Computer Science & Engineering, Jamia Hamdard, New Delhi, India
e-mail: Alkinj1986@gmail.com

I. Jeena Jacob et al. (eds.), *Data Intelligence and Cognitive Informatics*,
Algorithms for Intelligent Systems,
https://doi.org/10.1007/978-981-15-8530-2_41

portable crowdsensing to manufacture a framework for computerized participation in instructive framework even can be stretched out to working environments [2, 3].

1.1 Machine Learning Techniques

The pioneer of manmade reasoning, Arthur Samuel, has just said about AI as a method for programming gives PC "the capacity to learn without unequivocally customized." The AI types are classified into three kinds, for example, regulated learning, solo learning, and fortification learning.

1.1.1 Supervised Learning

The supervised learning [4] issue assists with deriving a capacity from named preparing information. The preparation information comprises preparing models. Right now, preparing models have pair comprising of info object ordinarily, which is a vector, and ideal yield esteem called as a supervisory sign. There are different calculations used to get familiar with the mapping capacity between the information and yield objects. There are two gatherings in directed learning as arrangement and relapse. Arrangement issue is the point at which the yield variable is classification, for example, "white" or "dark" and "malady" or "no infection" [5, 6]. Though, relapse issue is when yield variable is genuine worth, for example, "rupees," "weight," and "temperature."

Be that as it may, this calculation utilized different methodologies like GLM (summed up straight model), SVR (bolster vector relapse), GPR (Gaussian procedure relapse), neural systems, gullible Bayes, bolster vector machine, GBM (Angle Boosting Machine), and so on [7]. As appeared in Fig. 1.

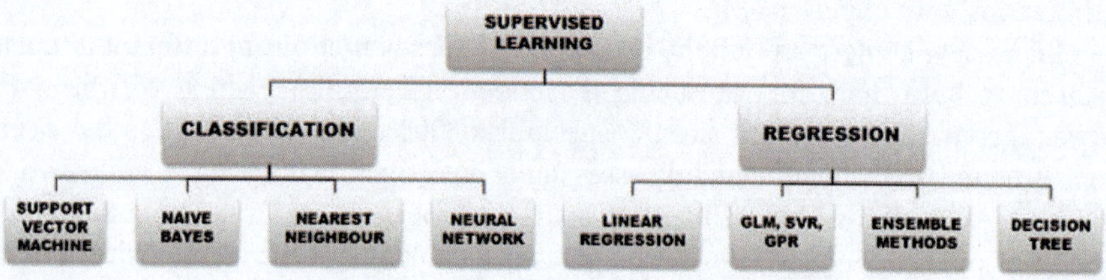

Fig. 1 Supervised Learning algorithm

1.1.2 Unsupervised Learning

The solo learning issue is construing a capacity to portray the unlabeled information, right nowPublications used classification method on various, or arrangement is excluded from the perception. At the point when the gave students model are unlabeled, there is no precision in the yield of important calculation, which is one of the ways of recognizing unaided gaining from administered and fortification adapting individually. One of the primary instances of unaided learning is the issue of thickness estimation in insights, in spite of the fact that there are different other issue and arrangements including abridging and clarifying the key highlights of the information [8, 4].

Further gathering of solo learning is bunching and affiliation. Bunching is about inborn groupings of the information, and such gathering can be statures of understudy in class or school. Though affiliation rules are going to set up intriguing connection among factors and dataset, most popular case of affiliation rule Daniel Forces says [9, 10]. "In 1992, Thomas Blischok, administrator of a retail counseling bunch at Teradata, and his staff arranged an examination of 1.2 million market bushels from around 25 Osco Medication stores. Database inquiries were created to distinguish affinities. The examination 'discovered that somewhere in the range of 5:00 and 7:00 p.m. that shoppers purchased brew and diapers.' Osco chiefs did not abuse the lager and diapers relationship by drawing the items nearer together on the racks." broadly known as market crate examination. Anyway, some different acclaimed calculations under bunching and affiliation rules are K-implies, K-medoids, Gaussian blend, shrouded Markov model, progressive grouping, PCA (head segment investigation), apriori calculation, aprioriDP, FP development calculation, and so forth as appeared in Fig. 2.

Some true utilizations of solo learning are NASA remote detecting, smaller than normal UAV, nano-camera creation innovation, [1] content outline, cosmology extraction, directed showcasing, break elicitation, recommender framework, and so on.

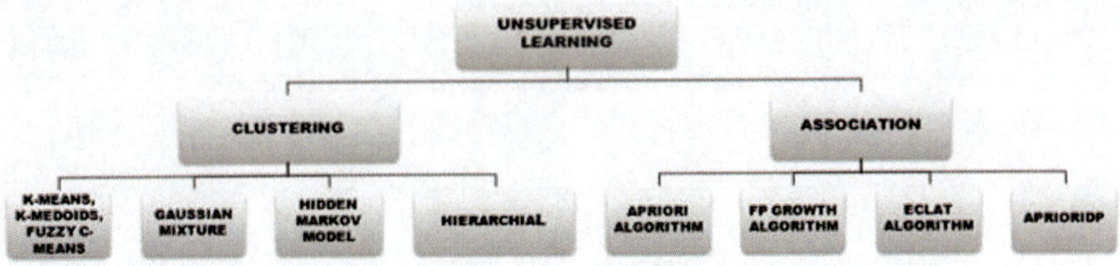

Fig. 2 Unsupervised learning algorithm

1.1.3 Reinforcement Learning

The fortification learning is one for the most part well-known examination in the field of computerized reasoning, and this calculation is a piece of AI which is especially not quite the same as managed and unaided learning. The administered learning gives criticism immediately whether the work has been done is correct or wrong comparably the case which is with unaided learning, then again fortification taking intakes in everything from the experience where it perceives stuff what it has realized in past with no awkward coding [3]. All the fortification learning specialists have expressed objectives that can detect parts of their surroundings and can pick activities to impact their surroundings. The supported learning has different methodologies of the well-known calculations to take a shot at which are transient contrasts, Q-learning, Markov choice procedure, Monte-Carlo strategy, and so on, as appeared in Fig. 3.

2 Related Works

Ghozzi et al. [7] in the proposed work investigate the utilization of AI calculations in the study of disease transmission, deciding episodes from observation information. Novel calculation [7] farringtonOutbreak with farringtonFlexible; and second is Gee utilizing R bundle. This sort of dataset execution is well in semi-directed learning a fortification learning.

Pannala et al. [5] proposed that strategy for this paper [5] shows process opinion examination of café audit where the dataset was gathered from Bangladeshi survey site Priyo Survey (Beta). Pre-preparing toolbox like NLTK with scikit-learn library for grouping language. Calculations are utilized, for example, multinomial credulous Bayes (MNB), bolster vector machine (SVM), K-closest neighbor (KNN), and semantic relapse (LR). Precision got from these calculations is MNB with 77%, SVM with 71.42%, KNN with 74.60%, and LR as 77.79%. Blended sentiments are hard to discover that this paper recommended multiclass classifier to adapt to this issue of a hazy area in future.

Pannala et al. [5] proposed to inquire about a shot [5] at perspective-based assessment examination (ABSA) that is utilized alongside AI calculations SVM (bolster

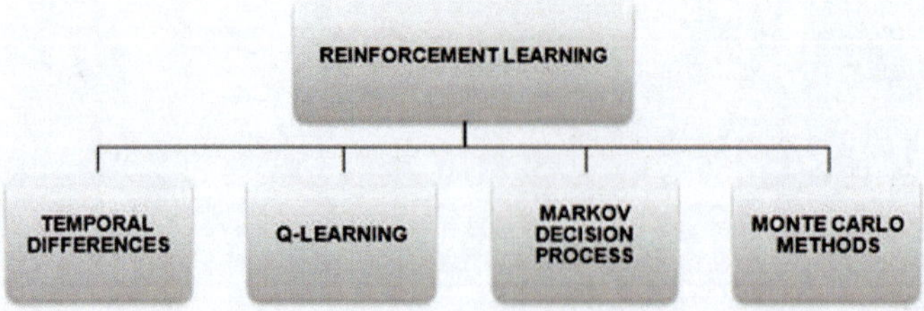

Fig. 3 Reinforcement learning algorithm

Table 1 Publications used classification method on various disease dataset

Publication	Method	Disease	Type of data	Accuracy(%)	Best technique
Ghozzi, S., Zacher, B. et al., 2019	farringtonOutbreak farringtonFlexible hmmOutbreak	Infectious disease	Clinical data	80.89	hmmOutbreak
N. U. Pannala, C. P. Nawarathna, et al., 2016	Multinomial Naïve Bayes (MNB), support vector machine (SVM), K-nearest neighbor, (KNN) and linguistic regression (LR)	Food disease	Web site review data	77, 71.42, 74.60, 77.79	LR
N. U. Pannala, C. P. Nawarathna, et al., 2016	Aspect-based sentiment analysis (ABSA), support vector machine (SVM), logistics regression	Food disease	Restaurant review dataset	66.12, 84.46	Logistic regression
Krishna, A., Akhilesh, V., et al., 2019	Support vector machine (SVM)	Food disease	Restaurant review dataset	94.56	SVM

vector machine) and coordinations relapse. Python language and NLTK library were utilized for examination. Eatery survey dataset is utilized. Exactness got in SVM was 66.12%, and in coordinations relapse 84.46%. The multiclass arrangement is an issue still to be settled.

Krishna et al. [11] proposed and explored paper [11] that this paper additionally manages conclusion investigation online nourishment audit. SVM (bolster vector machine classifier is utilized) with most elevated precision is of 94.56%. The issue with this paper was that this calculation like SVM will not be performed well with the multiclass dataset (Table 1).

3 Problem Statement

The labeled dataset in structured formats is one of the challenges faced by any analyst. However, there is no such method or rule developed which can give a define solution. No free lunch theorem is applied in machine learning which clearly states that there is no shortcut for a problem.

Unlikely there are various industries which produce big data like healthcare, government survey, marketing segmentation, online shopping, etc. Some of the data obtained are really important for future analysis. For data mining or applying machine learning prediction algorithms, it is always a challenge to have a right pick. Supervised learning algorithms deal with label data, and with unlabeled data, it may give good accuracy but cannot guarantee that the accurately classified data are appropriate.

The challenges faced in unsupervised learning algorithms:

- With unlabeled dataset, the training and testing dataset is inaccurate.
- The state-of-the-art algorithms only work well with labeled data.
- Unlabeled dataset if having only numeric data with distinct value, supervised learning will not perform well.
- Labeling the unlabeled data is a costly and time-consuming task. Most of the labeling is generally done manually.
- Preprocessing takes a lot of time

These classification algorithms are even optimized using the nature-inspired algorithm namely particle swarm optimization for attribute reduction which can easily give a proper solution. To an unlabeled dataset, clustering algorithms give a structure to the dataset.

Directly using an unsupervised learning algorithm will give a quick view of the performance of clustering algorithms. These performances can be compared using accuracy, precision, recall, and cross-validating. The best performance model can be further taken with that dataset which can be further used for any kind of analysis and prediction.

4 Present Investigation

4.1 Stages of Supervised Learning of Data Analysis for Foodborne Illness

Stage 1: Data gathering and preparation

This step is extremely vital as a result of the standard and amount of information that you simply gather can directly confirm, however, smart your prophetic model are often.

Stage 2: Cleaning of the dataset

Loading the dataset or making one's own dataset for modeling and prediction.

Stage 3: Visualization of a dataset is important to know attribute performance

There are numerous models that specialists and information researchers have made throughout the years. Some are exceptionally appropriate for picture information, others for arrangements (like content, or music), some for numerical information, others for content-based information.

Stage 4: Attribute reduction using a pandas data frame library.

Before applying any grouping strategy on your information, it is essential to assess whether the informational collections contain significant bunches (i.e., non-arbitrary structures) or not. On the off chance that yes, at that point what number of groups are there. This procedure is characterized as the evaluating of grouping propensity or the attainability of the bunching investigation.

Stage 5: Label encoding for the categorical attribute

Deciding the ideal number of bunches in an informational collection is a key issue in a parceling grouping, for example, k-implies grouping, which requires the client to determine the quantity of groups k to be produced.

Stage 6: Splitting dataset in train and test

The term cluster validation is utilized to plan the methodology of assessing the integrity of grouping calculation comes about. This is essential to abstain from discovering designs in arbitrary information, and in addition, in the circumstance where you need to think about two bunching calculations.

Stage 7: Apply classification algorithm and use AdaBoost with random forest classification

Which utilizes the inward data of the bunching procedure to assess the decency of a bunching structure without reference to outer data. It can be likewise utilized for assessing the number of groups and the fitting bunching calculation with no outside information.

Stage 8: Comparison of a model on the accuracy, cross-validation, AUC, and ROC score

Which comprises in looking at the aftereffects of a group investigation to a remotely referred to come about, for example, remotely gave class names. It quantifies the degree to which group marks coordinate remotely provided class names.

4.2 Supervised Algorithms Used

For this dissertation, four clustering algorithms have been used to make an evaluation namely:

1. Logistic regression (LR)
2. Random forest
3. Ensemble method AdaBoost
4. AdaBoost with random forest

4.2.1 Logistic Regression (LR)

Calculated relapse is an administered learning characterization calculation used to foresee the likelihood of an objective variable.

Logistic Regression Assumptions

Prior to jumping into the execution of calculated relapse, the accompanying suspicions should be known about the equivalent.

- In the event of double-calculated relapse, the objective factors must be paired consistently, and the ideal result is spoken to by the factor level 1.
- There ought not to be any multicollinearity in the model, which implies that the autonomous factors must be free of one another.
- Significant factors for our model should be remembered.
- A huge example size ought to be picked for calculated relapse.

4.2.2 Random Forest Classifier

An arbitrary timberland classifier is a meta estimator that fits various choice tree classifiers on different sub-tests of the dataset and utilizations averaging to improve the prescient precision and power over-fitting. The sub-test size is consistently equivalent to the first information test size, yet the examples are drawn with substitution if bootstrap = True (default). Irregular backwoods, similar to its name infers, comprise of countless individual choice trees that work as a troupe. Every individual tree in the irregular woodland lets out a class expectation and the class with the most votes turns into our model's forecast (see figure beneath) [12].

The central idea driving arbitrary woods is a basic, however, ground-breaking one—the insight of groups. In information science talk, the explanation that the irregular timberland model works so well is: "An enormous number of generally uncorrelated models (trees) working as a board of trustees will beat any of the individual constituent models."

- There should be some real sign in our highlights with the goal that models constructed utilizing those highlights which show improvement over irregular speculating.
- The forecasts (and along these lines the mistakes) made by the individual trees need to have low connections with one another.

4.2.3 Ensemble Method Using AdaBoost

The objective of outfit strategies is to join the forecasts of a few base estimators worked with a provided learning calculation to improve generalizability/heartiness over a solitary estimator. Two groups of gathering techniques are generally recognized:

- In averaging strategies, the driving rule is to assemble a few estimators autonomously and afterwards to average their expectations. Overall, the joined estimator is generally better than any of the single base estimator since its difference is diminished.
- Models: Sacking techniques, Backwoods of randomized trees, …
- On the other hand, in boosting techniques, base estimators are assembled successively and one attempts to decrease the inclination of the consolidated estimator. The inspiration is to join a few feeble models to create an incredible group.
- Models: AdaBoost, Slope Tree Boosting.

4.2.4 AdaBoost

The module sklearn.ensemble incorporates the famous boosting calculation AdaBoost, presented in 1995 by Freund and Schapire. The center guideline of AdaBoost is to fit a grouping of frail students (i.e., models that are just somewhat superior to arbitrary speculating, for example, little choice trees) on more than once adjusted renditions of the information. The forecasts from every one of them are then joined through a weighted greater part vote (or total) to create the last expectation. The information alterations at each supposed boosting cycle comprise of applying loads, to every one of the preparation tests. At first, those loads are good to go to, with the goal that the initial step prepares a frail student on the first information. For each progressive cycle, the example loads are independently changed and the learning calculation is reapplied to the reweighted information. At a given advance, those preparation models that were inaccurately anticipated by the supported model initiated at the past advance have their loads expanded, though the loads are diminished for those that were anticipated effectively. As emphases continue, models that are hard to anticipate get ever-expanding impact. Each consequent powerless student is along these lines compelled to focus on the models that are missed by the past ones in the succession [13].

AdaBoost can be utilized both for arrangement and relapse issues:

- For multiclass arrangement, AdaBoostClassifier executes AdaBoost-SAMME and AdaBoost-SAMME.R.
- For relapse, AdaBoostRegressor actualizes AdaBoost.R2.

5 Proposed Solution

5.1 Experimental Setup

The problem statement is pretty straight. To provide the solution for a labeled and huge dataset model, training is done using supervised learning and ensemble method. To develop any kind of method a proper system, environment and tools are important.

For the current problem statement due to lack of good resource and time, factor with the availability of resources had to be compromised to develop a proper solution.

5.1.1 Requirement Specification

The open-source Anaconda constrictor singular version (officially Anaconda constrictor dispersion) is the simplest method to perform Python/R information science and AI on Linux, Windows, and Macintosh operating system X. With more than 19 million clients around the world, it is the business standard for creating, testing, and preparing on a solitary machine, empowering singular information researchers to:

- Rapidly download 7500+ Python/R information science bundles
- Oversee libraries, conditions, and situations with Conda
- Create and train AI and profound learning models with scikit-learn, TensorFlow, and Theano
- Examine information with adaptability and execution with Dask, NumPy, pandas, and Numba
- Envision results with Matplotlib, Bokeh, Datashader, and Holoviews

ML Tool: Anaconda navigator, Jupyter notebook (.ipynb).

5.1.2 Dataset

Next time you take a chomp, think about this: around one out of six (or 48 million) individuals in the US become ill from eating polluted nourishment every year. In excess of 250 pathogens and poisons have been known to cause foodborne ailment, and practically every one of them can cause an episode.

A foodborne malady flare-up happens when at least two individuals get a similar sickness from the equivalent polluted nourishment or drink. While most food-borne diseases are not part of a perceived episode, flare-ups give significant data on how germs spread, which nourishments cause sickness, and how to forestall contamination.

This dataset gives information on foodborne infection episodes answered to CDC from 1998 through 2015. Information fields incorporate year, state (flare-ups happening in more than one state are recorded as "multistate"), an area where the nourishment was readied, revealed nourishment vehicle, and defiled fixing, etiology (the pathogen, poison, or compound that caused the ailments), status (regardless of whether the etiology was affirmed or suspected), all-out diseases, hospitalizations, and fatalities. In numerous flare-up examinations, a particular nourishment vehicle is not recognized; for these episodes, the nourishment vehicle variable is clear. Are foodborne illness episodes expanding or diminishing? What contaminant has been answerable for the most ailments, hospitalizations, and passings? What area for nourishment planning represents the most serious danger of foodborne sickness?

	# Year	A Month	A State	A Location	A Food	A Ingredie...	A Species	A Serotyp...	A Status	# Illnesses	# Hospita...	# Fatalities
1	1998	January	California	Restaurant						20	0	0
2	1998	January	California		Custard					112	0	0
3	1998	January	California	Restaurant						35	0	0
4	1998	January	California	Restaurant	Fish, Ahi		Scombroid toxin		Confirmed	4	0	0
5	1998	January	California	Private Home/Residence	Lasagna, Unspecified; Eggs, Other		Salmonella enterica	Enteritidis	Confirmed	26	3	0
6	1998	January	California	Restaurant			Shigella boydii		Confirmed	25	3	0
7	1998	January	California	Restaurant						6	0	0
8	1998	January	California	Restaurant	Stuffing, Unspecified; Sandwich, Turkey		Salmonella enterica	Enteritidis	Confirmed	4	3	0
9	1998	January	Colorado	Restaurant	Salad, Unspecified		Virus		Suspected	21		
10	1998	January	Colorado	Restaurant	Peppers		Virus		Suspected	3		0

Fig. 4 View of dataset

Attribute contained in the dataset:

- Year
- Month
- State
- Location
- Food
- Ingredient
- Species
- Serotype/Genotype
- Status (Fig. 4)

5.2 *Procedure and Techniques*

Get-together information once has our gear and liquor, it is the ideal opportunity for our first genuine advance of AI: gathering information. This progression is significant in light of the fact that the quality and amount of information that you assemble will straightforwardly decide how great your prescient model can be. Right now, gathered information will be the shading and the liquor substance of each drink. This will yield a table of shading, alcohol%, and whether it is lager or wine. This will be our preparation information (Fig. 5).

Stage 1: Data gathering and preparation

This dataset gives information on foodborne illness episodes answered to CDC from 1998 through 2015. Information fields incorporate year, state (flare-ups happening in more than one state are recorded as "multistate"), an area where the nourishment was readied, revealed nourishment vehicle and tainted fixing, etiology (the pathogen, poison, or compound that caused the ailments), status (regardless of whether the

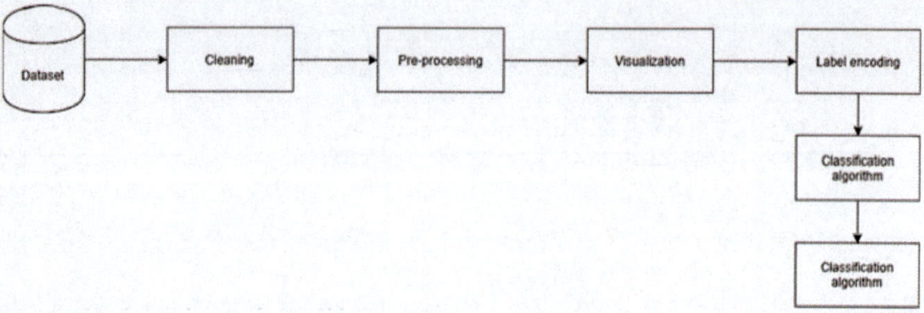

Fig. 5 Pipeline of stages followed for classifying the model

etiology was affirmed or suspected), absolute sicknesses, hospitalizations, and fatalities. In numerous flare-up examinations, a particular nourishment vehicle is not recognized; for these flare-ups, the nourishment vehicle variable is clear. Dataset has nine qualities in a particular year, month, state, area, fixings, species, serotype/genotype, status, diseases, hospitalizations, and fatalities with 19,117 columns.

Are foodborne sickness episodes expanding or diminishing? What contaminant has been liable for the most sicknesses, hospitalizations, and passings? What area for nourishment planning represents the most serious danger of foodborne sickness?

Stage 2: Cleaning of dataset

Describe the dataset using value counts, info, head, tail-like parameter. Then, the column will be dropped which have no use. In our case, ingredients and serotype/genotype do not serve any purpose. The drop() function is used to remove those columns (Figs. 6 and 7).

Then, visualizing the null values using isnull function. The attribute with false means that they are not null, and if it is true, then it is null. So, the true value will be removed using isnull().

Stage 3: Visualization of dataset is important to know attribute performance

After visualizing the dataset with matplotlib and seaborn libraries, various insights are concluded, finding the relation between species, fatalities, and state attributes.

Conclusion:

- Multistate experience 33 death confirmed cases where Parasite, heavy metal, and histamine-like species were the reason
- Followed by state like Pennsylvania, Texas, California, and Rhode Island have less than 10 death cases
- Species like Bacillus cereus, Clostridium Perfringen Bacterium, Scombid, Toxin, mycotoxins, escherichia coli, enterotoxigenic (Figs. 8 and 9).

Stage 4: Label encoding for the categorical attribute

Encoding the categorical data.

```
<class 'pandas.core.frame.DataFrame'>
RangeIndex: 19119 entries, 0 to 19118
Data columns (total 12 columns):
Year                    19119 non-null int64
Month                   19119 non-null object
State                   19119 non-null object
Location                16953 non-null object
Food                    10156 non-null object
Ingredient              1876 non-null object
Species                 12500 non-null object
Serotype/Genotype       3907 non-null object
Status                  12500 non-null object
Illnesses               19119 non-null int64
Hospitalizations        15494 non-null float64
Fatalities              15518 non-null float64
dtypes: float64(2), int64(2), object(8)
memory usage: 1.8+ MB
```

Fig. 6 Getting the information about the dataset

```
df_original.isnull() # identifying missing values per
column
```

	Year	Month	State	Location	Food	Ingredient	Species
0	False	False	False	False	True	True	True
1	False	False	False	True	False	True	True
2	False	False	False	False	True	True	True
3	False	False	False	False	False	True	False
4	False	False	False	False	False	True	False
5	False	False	False	False	True	True	False

Fig. 7 Finding the null values in the dataset

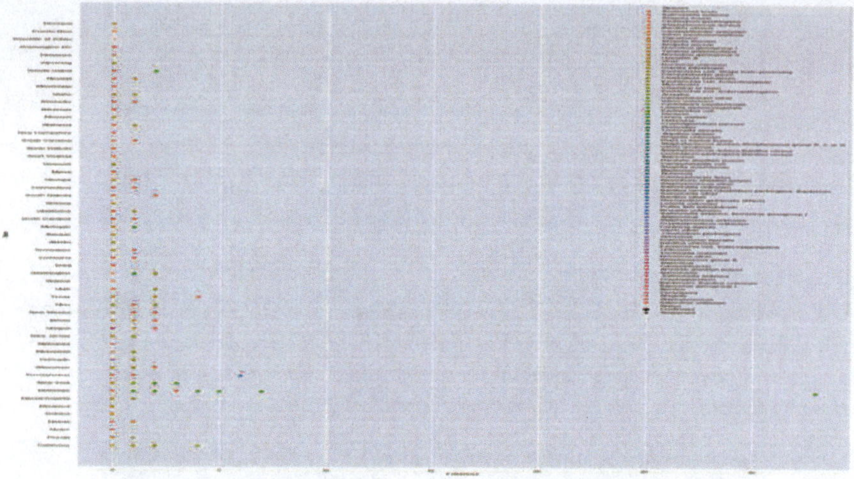

Fig. 8 Visualizing the species causing fatalities which having the confirmed status

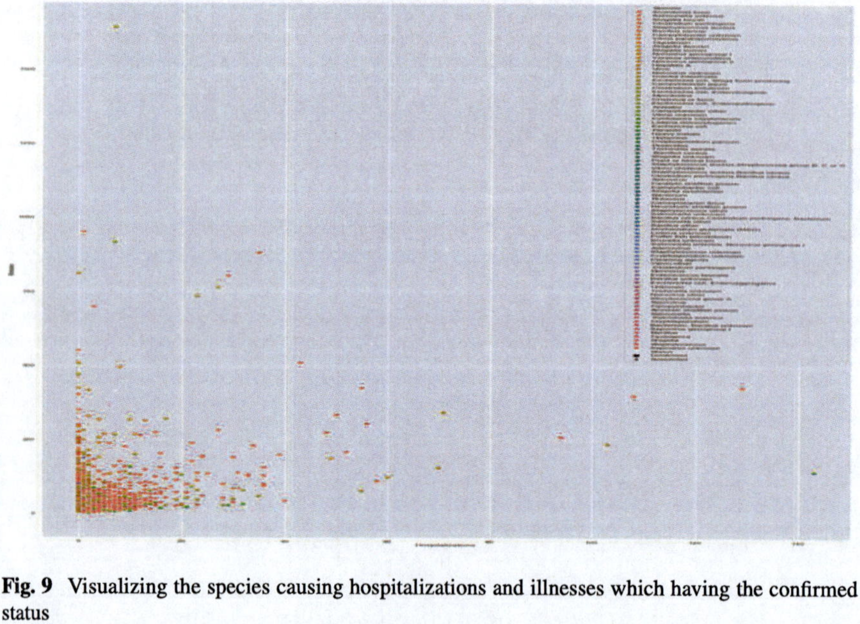

Fig. 9 Visualizing the species causing hospitalizations and illnesses which having the confirmed status

Stage 5: Splitting dataset in train and test

The dataset is created in 80–20 split. 80% of the data is kept for training, and the rest 20% is for testing.

Stage 6: Apply classification algorithm and use AdaBoost with random forest classification

First logistic regression is applied, accuracy is calculated, and found the true positive, true negative, false positive, false negative. Afterwards, recall and precision are found.

Then, using random forest, found the same and simply using an ensemble method of AdaBoost. After finding accuracy cross-validation is done using kfold of sklearn.model_selection with cv = 10.

The random forest was found performing better in this dataset, and later, this model is used with AdaBoost ensemble method. In conclusion, found ensemble with random forest, with better accuracy, and cross-validation score.

Stage 7: Comparison of a model on the accuracy, cross-validation, AUC, and ROC score

6 Discussions

In this classification algorithm, it is a tedious task. First logistic regression is applied, accuracy is calculated, and found the true positive, true negative, false positive, and false negative. Afterwards, recall and precision are found.

Then, using random forest, found the same simply using an ensemble method of AdaBoost. After finding accuracy, the cross-validation is done using kfold of sklearn.model_selection with cv = 10.

The random forest was performing better in this dataset, and later, this model is used with AdaBoost ensemble method. In conclusion, found an ensemble with random forest, with better accuracy, and cross-validation score.

Now, the sensitivity and specificity of the following models will be seen (Tables 2 and 3; Fig. 10).

Table 2 Comparison of sensitivity and specificity between different model for case 1

Model	Sensitivity (TP) (%)	Specificity (FP) (%)
Logistic regression	15.06	98.54
Random forest classification	37.82	94.57
AdaBoost	57.21	87.86

Table 3 Comparison of sensitivity and specificity between different model for case 2

Model	Sensitivity (TP) (%)	Specificity (FP) (%)
Ensemble random forest	63.14	88.34

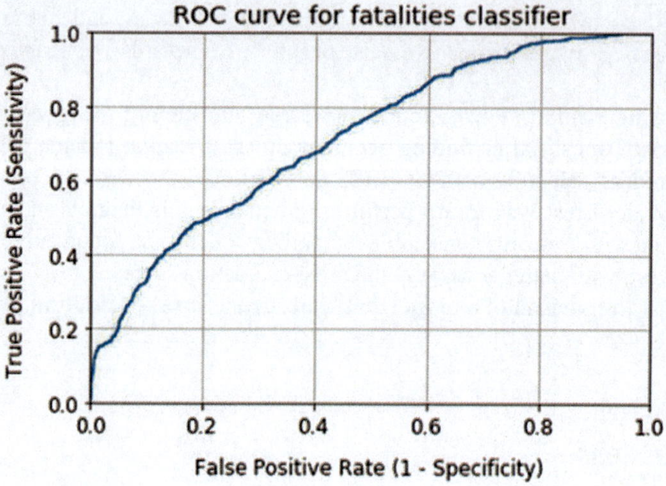

Fig. 10 Logistic regression of TPR-FPR ROC curve

7 Summary and Conclusions

From the above work, it can be concluded that there is no thumb rule to define which classification is better for which kind of dataset. Many papers have been written proving the state-of-the-art algorithms, the random forest is unbeatable and have extremely good results, and this experiment does not prove to go with all the justification is thought. The use of any model solely depends upon the type of data and availability of resource to proof the best model for any kind of dataset.

Also, more the numeric dataset is found with un-normalized value tougher is the processing and labeling. The use of training and testing split in the unlabeled dataset was a challenge, but in this case, it was giving a better outcome cluster compared with the cross-validation. Processing speed for sklearn libraries needs high-speed RAM more than 8 GB. A categorical dataset is hard to shape using NumPy libraries. Complicated instances make analysis time consuming and unavailability of labeled dataset.

8 Future Scopes

The unavailability of a fast processing system, working on Python script, and using scikit-learn packages for the model prediction was not feasible for the current prospect and time frame.

In the current project, one of the datasets just has been used among the record of one dataset. Due to lack of resources and training and testing, this huge dataset is really a challenge which can be better done with attribute optimization algorithms like metaheuristic algorithms. The future scope is unsupervised learning technique hybrid with nature-inspired algorithms that can bring revolutionary changes. Nature-inspired algorithms like Firefly algorithms, Ant colony optimization, genetic algorithms, etc., can help in dimensionality reduction with an increase in performance with a discrete dataset and can be implemented in Tensorflow-Keras for better prediction.

References

1. Mitchell T (1997) Machine learning. McGraw Hill 0-07-042807-7; 2
2. https://www.andor.com/learning-academy/raman-spectroscopy-an-introduction-to-raman-spectroscopy
3. Sutton RS, Barto AG (1998) Reinforcement learning: an introduction, vol. 1, no. 1. MIT Press, Cambridge
4. Joseph SIT (2019) Survey of data mining algorithm's for intelligent computing system. J Trends Comput Sci Smart Technol (TCSST) 1(01):14–24
5. Pannala NU, Nawarathna CP, Jayakody JTK, Rupasinghe L, Krishnadeva K (2016) Supervised learning based approach to aspect based sentiment analysis. In: 2016 IEEE international conference on Computer and Information Technology (CIT), Nadi, pp 662–666
6. Negi PS, Wazir S, Tabrez Nafis M (2020) Discovering frequent high average utility itemset without transaction insertion. In: Karrupusamy P, Chen J, Shi Y (eds) Sustainable communication networks and application. ICSCN 2019. Lecture Notes on Data Engineering and Communications Technologies, vol 39. Springer, Cham
7. Ghozzi S, Zacher B, Ullrich A (2019) Supervised learning for automated infectious-disease-outbreak detection. Online J Public Health Inf 11(1)
8. Raj JS (2019) A comprehensive survey on the computational intelligence techniques and its applications. J ISMAC 1(03):147–159
9. https://www.webmd.com/diabetes/tc/diabetic-nephropathy-topic-overview#1
10. De'Ath G (2007) Boosted trees for ecological modeling and prediction. Ecology 88(1):243–251 (PMID: 17489472)
11. Krishna A, Akhilesh V, Aich A, Hegde C (2019) Sentiment analysis of restaurant reviews using machine learning techniques. Lecture Notes in Electrical Engineering, pp 687–696
12. Wang A, Wan G, Cheng Z, Li S (2011) Incremental learning extremely random forest classifier for online learning. J Softw 22(9):2059–2074
13. Lei L, Wang X (2013) Improved AdaBoost ensemble approach based on loss function. J Comput Appl 32(10):2916–2919

Chapter 42
A Primer on Word Embedding

Satvika, Vikas Thada, and Jaswinder Singh

1 Introduction

The modern world is called the "data age" because data is the new tool or weapon for achieving laurels in the business or corporate era. The correct and updated data (or information) is the most prominent thing required by the commercial houses either small or big to gain a candid advantage over its competitors. Nowadays, more and more companies are investing capital (human resource, money and time) in data collection, processing and analytics. This process involves enormous data harvesting or warehousing from numerous sources whether social media or traditional methods and then processing this whole data so that data collected from various sources is in a common format. It is quite vital as unstructured data is estimated to be approximately 70% of the total data accumulated, which must be brought to the same format to apply mathematical or statistical functions. These functions help in analyzing the data and discovering the hidden patterns or information. By doing so, the companies get a real insight into customer behavior, buying pattern and potential customer identification for peculiar products and services, which is mandatory in the contemporary world to boost the productivity and revenue. There are myriad mechanical or algorithmic paradigms available for customers' opinion mining such as rule-based approach, machine learning and deep learning, which try to learn the exact sentiments behind customers' reviews and predict them accurately [1].

Satvika (✉) · V. Thada
CSE Department, Amity University, Gurugram, Haryana, India
e-mail: satvika16oct@gmail.com

V. Thada
e-mail: vthada@ggn.amity.edu

J. Singh
CSE Department, GJU Hisar, Haryana, India
e-mail: jaswinder_singh_2k@rediffmail.com

I. Jeena Jacob et al. (eds.), *Data Intelligence and Cognitive Informatics*,
Algorithms for Intelligent Systems,
https://doi.org/10.1007/978-981-15-8530-2_42

Rule-based approaches are based on defining a bunch of rules for identifying the intensity and emotions that a word expresses [2]. These rules must be logical and intuitive for a particular domain under consideration and should also take into account the sarcasm and satire. The chief concern in rule-based approaches is their static behavior, i.e., the rules are static and do not change rapidly with time, which make them obsolete soon. However, it must be considered that data is too dynamic these days and hidden patterns contained in data change quickly also [3]. Another imperative hindrance in rule-based systems is the huge amount of hard work and testing involved, which makes it sluggish. Also, rule-based systems give excellent results in a narrow domain but behave poorly when it comes to generalization [4]. Due to these quandaries, the rule-based approach got obsolescent and machine learning algorithms surfaced which is used widely today for data analytics.

Machine learning is the state-of-the-art branch of artificial intelligence, which uses statistical models and algorithms to attain a precise task using inferences and patterns present in the data itself rather than explicit directions from humans [5]. The machines (or computers in this case) learn by gaining knowledge from past know-hows and then try to practice it in the unknown but related domains. There are copious categories of machine learning available chiefly supervised, unsupervised, reinforcement, semi-supervised, self-supervised, multi-instance, inductive, deductive, transductive learning, etc., which are used specifically for particular tasks.

The machine learning approach varies its underlying algorithm according to the varying data and hence more dynamic than rule-based approach [6]. The usual problems solved by machine learning are classification, clustering, dimensionality reduction, anomaly detection, etc., and some of the noteworthy machine learning algorithms to handle these hitches are Naïve Bayes, support vector machine, k-nearest neighbor, decision trees, etc. Overall, it can be said that machine learning is very adaptive which learns patterns from data and therefore can be applied to a plethora of applications like natural language processing, spam detection, sentiment analysis, recommendation systems, genre (multi-class) classification, stock prediction, etc. [7] (Fig. 1).

Fig. 1 How machine learning works [8]

Last but not least is the deep learning methodology, which is nothing but a group of algorithms and models that entail manifold processing layers. Deep learning is the newest branch of artificial intelligence which uses artificial neural networks (similar to neural network present in the human brain) for the process of learning and predicting [9]. Therefore, it can be defined as the science of making machines intelligent by making them learn by example, just like humans do. It is usually an unsupervised kind of learning which can absorb information from either structured or even unstructured data; hence, it is practically more useful than its counterparts.

There are plenty of applications of deep learning like natural language processing, self-driving cars, health care, financial fraud detection, fake news detection, virtual assistants (like Siri, Cortana and Jarvis), etc. The next section discusses the need for word embeddings. Section 3 details the numerous types of word embeddings. Section 4 gives an insight into the traditional word embeddings, and Sect. 5 deliberates the pre-trained word embeddings. Finally, the next section specifies the conclusion for the article.

2 Word Embedding and Its Need

Machine learning, as well as deep learning, basically uses the neural networks and its myriad variations like convolution neural networks, recurrent neural networks, etc. to handle the perplex problems especially in the domain of text processing. Nevertheless, the key point here is that the neural networks cannot directly use natural languages like English or Hindi because neural nets do not understand the words or phrases; instead, they use numerical data. This is where word embedding comes into play, i.e., represent the textual information from languages into statistics.

In short, word embeddings can be explained as the numerical representation of a document texts' semantic meaning [10]. These vectors provide the relation between various words or phrases of the document, i.e., the words having similar meaning have closer vector values, which establish their closeness in the linguistics. For example, the word "male" is proximate to "King" and "Boy" and quite far from "Queen" and "Princess", as shown in Fig. 2.

It is essential to understand that word embedding is not the mere translation of texts into numbers; rather, it conveys the semiology meanings of the words. The distance between vectors represents the similarity between the myriad words [12]. In brief, it can be concluded that computers are incapable of understanding natural language's words and that's why word embeddings are required. Also, encoding phrases into numeric form can make mathematical functions, especially matrix operations successfully operate for NLP tasks.

Word embeddings can be categorized broadly into two categories, namely frequency-based and prediction-based. As the name suggests, frequency-based embeddings take into account the frequency or number of times a word occurs in the document to find its relevance. It is also called count-based embeddings, and its principal point is to give weights according to the occurrence and also the context

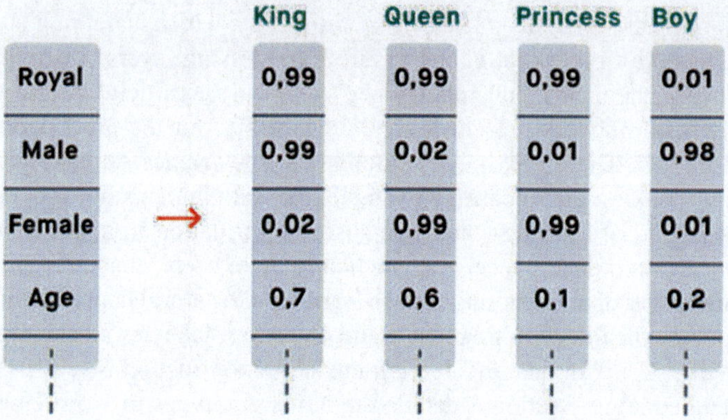

Fig. 2 Example of word embeddings [11]

[13]. Frequency-based word embeddings can again be classified into count vector, TF–IDF vector and co-occurrence. The prediction-based embeddings fundamentally consist of two major approaches: continuous bag-of-words (CBOW) and skip-gram that come together to form the Word2Vec which is the prominent state-of-the-art model of word embedding. Each of the above will be discussed in detail in the following sections.

3 Count Vectorization

Count vectorizing or one-hot encoding (OHE) is the most basic word embedding that works on a simple binary principle and produces a high-dimensional sparse matrix. Firstly, it creates a bag-of-words (vocabulary) from all the given text or corpus, which is defined as the assemblage of similar text. Secondly, it counts the occurrence of each word in the document existing in the corpus. The final output of count vectorizing is a sparse matrix with dimensions D*T, where D denotes the number of documents and T signifies the number of dissimilar words in the vocabulary [14]. This can be understood by the following example:

Document 1 (D1): "The dog ate the cat".

Document 2 (D2): "The lion can eat a cat".

As displayed above, there are two documents, i.e., $D = 2$ and total eight different words, thus making $T = 8$ in the corpus: {"the", "dog", "ate", "cat", "lion", "can", "eat", "a"}. Hence, count vectorizing encoding for the above corpus should be (Table 1):

Table 1 Example of count vectorization

	the	dog	ate	cat	lion	can	eat	a
D1	2	1	1	1	0	0	0	0
D2	1	0	0	1	1	1	1	1

The matrix generated is self-explanatory: If a word appears in a corresponding document, it is given the value "1", or else given "0". If a word appears multiple times, it can be given that frequency, just like "the" word is coming two times in the D1; hence, value of "the" is marked as 2 for D1. As the words "lion", "can", "eat" and "a" are absent in D1, hence they are assigned "0" for D1. Similarly, the words "dog" and "ate" are not given in D2 and marked "0". Each cell in the matrix corresponds to one document and one precise word in the corpus.

Usually, a corpus contains 1000s of sentences with 10s of words, thus producing a lot of words which are not occurring in most of the documents. Thus, the topmost pickle in this embedding is the occurrence of cells with loads of 0s and is hence called the sparsity matrix with high dimensions. Nonetheless, these dimensions or features can be diminished so that visualization is possible because of lower dimensions [15]. One way of doing so is to eliminate commonly occurring words also called "stopwords" from the corpus. This whole process of generating table is called tokenization, i.e., identifying all the individual words present in the corpus.

4 TF–IDF Vectorization

The full abbreviation of TF–IDF is "term frequency–inverse document frequency". The phrase "term frequency" refers to the incidence of a term in a document divided by the entire amount of documents, while the term "inverse document frequency" denotes the logarithmic value (to the base 10) of a whole number of documents divided upon the magnitude of documents a specific term is appearing in (Fig. 3).

The product of TF and IDF values for a word is known as its TF–IDF weight; its higher value signifies rarity of a term and vice-versa [16]. TF–IDF vectorization is the embedding which removes the problem in count vectorizing which simply records the occurrence of a particular word in the document irrespective of its significance [18]. Actually, the commonly used words like "a", "an", "the", "this", "that", "is", "am", "are", etc. are most likely to appear in the documents, and their frequency does not mean they are strongly imperative; instead, they are least important. Hence, TF–IDF tries to eliminate these frequent words by assigning them lower weights and try to include more noteworthy terms by empowering them with more weight values [19]. For example, when a 1000-word document on cryptography comprises the word "cipher" 145 times, the TF value for this word is 145/1000, i.e., 0.145. Also consider if the term "cipher" is contained in 40 documents out of total 1000 documents in the corpus, then IDF value will be calculated by formula log10 (1000/40) → log10(25)

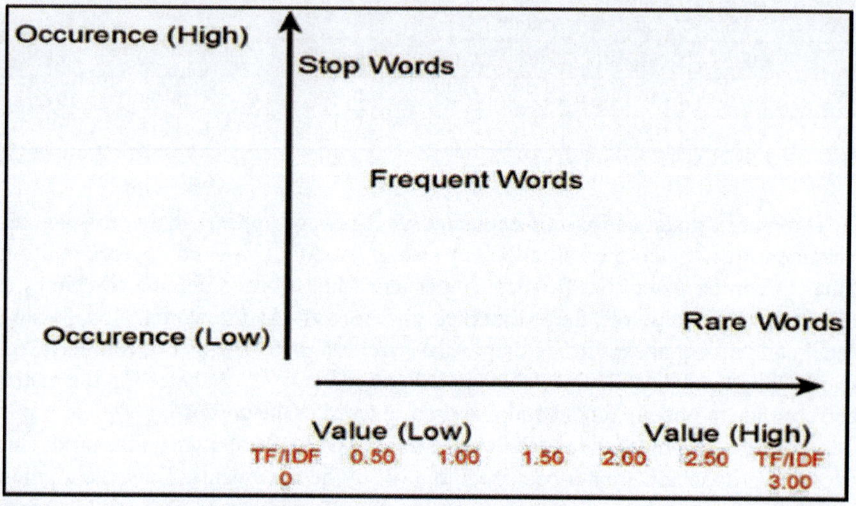

Fig. 3 TF–IDF values for words according to their frequency [17]

→ 1.398. Having calculated both TF and IDF values, the final TF–IDF value can be deliberated by their multiplication:

TF–IDF for word "cipher" is 0.145 * 1.398, i.e., 0.203.

The advantage of TF–IDF can be taken by running a TF–IDF calculation on every word of the corpus and choosing the terms with higher TF–IDF values. Correspondingly, these selected terms can be again be aligned according to their search volumes on the web and lastly selecting the terms with greater search volumes as they make more sense according to the users. One downside of TF–IDF vectorization is that it also produces a high-dimensional representation that may not confine the actual semantic relationship between words.

5 Co-occurrence Vectorization

As the name suggests, it is a matrix that suggests how some words tend to occur jointly and are probably used in a similar context. It is a colossal matrix (even bigger than the one-hot encoding matrix) that is comparable in size to the whole corpus and uses positive whole numbers to mark the presence of co-occurrence of two words and integer value "0" to mark the absence of co-incidence. If some terms are coming multiple times together, the presence is denoted by that frequency [20]. To check the co-occurrence of words, a context window size must be decided, which indicates how many words before and after must be scanned for a particular term. To explain this, let us take an example:

Table 2 Example of co-occurrence matrix

	He	is	the	best	dancer	very	famous	smart
He	0	3	1	1	1	1	1	1
is	3	0	1	1	1	1	2	1
the	1	1	0	1	1	0	0	0
best	1	1	1	0	1	0	0	0
dancer	1	1	1	1	0	0	0	0
very	1	1	0	0	0	0	1	0
famous	1	2	0	0	0	1	0	0
smart	1	1	0	0	0	0	0	0

Corpus: "He is the best dancer. He is very famous. He is smart". This corpus has three sentences, and let us assume the size of the context window is two, i.e., each word will be scanned two words both before and after a given word. The co-occurrence matrix for the above corpus should be the following (Table 2):

The explanation is quite easy; first of all, every word in the given corpus does not occur with itself. Hence, the cell at the intersection of word with itself is marked "0", just like "He" does not occur with "He" in any sentence of vocabulary. The occurrence of "He" and "is" is coming in all three sentences and hence marked "3" for the intersection of "He-is" and also for "is-He". Similarly, the occurrence of "is-famous" is stated as "2" coz in second sentence, "famous" term is coming in a context window of "is" of second sentence as per forwarding propagation and also comes in a context window of "is" of third sentence according to the backward scanning. Also, check some words do not happen together like the pair of "famous-smart" as the context window is only two words long; hence, they are marked "0" for both "famous-smart" and "smart-famous" (Fig. 4).

Although the biggest benefit of co-occurrence vectorization over count vectorization and TF–IDF is that it preserves the semantic relationship between terms and is also comparatively faster [21]. However, the focal drawback is its enormous size, but the modern tools like Hadoop can handle this by factorizing the matrix.

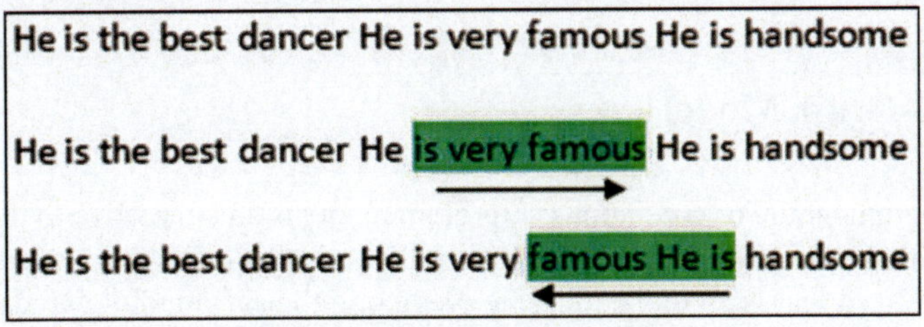

Fig. 4 Co-occurrence for "is-famous" word pair

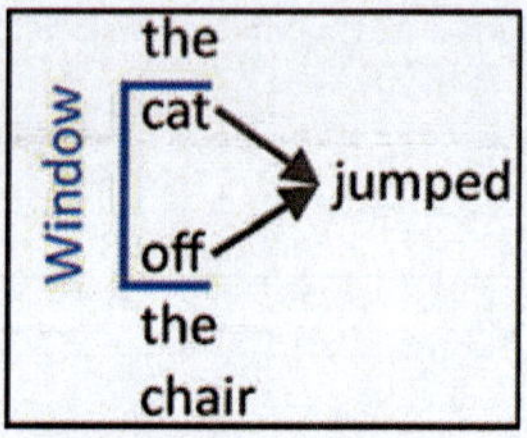

Fig. 5 Prediction using CBOW model [23]

6 Continuous Bag-of-Words (CBOW) Model

Bag-of-words (BOW) can be defined as the means of mining features from text to employ in machine learning or deep learning algorithms. BOW basically involves two entities: a dictionary of recognized words and a way to quantify the presence of these words. CBOW model is one of the fundamental approaches based on BOW and used for word embeddings. It is generally recognized as a learning model that predicts a term by its context and that context maybe a single or multiple words (usual case). Henceforth for every word, "n" words before and after it is considered to check the semantic association among words. It is based on neural networks with hidden layer/s and works on a simple norm that milieu of a word/phrase can be branded by the neighboring words [22]. The dimension of the hidden layer and the output layer should remain the same, but the input layer dimensions can be altered along with the activation function of the hidden layer. Figure 5 shows an example of CBOW model.

As depicted in the above example, the surrounding words, "the", "cat", "off", "the" and "chair", are able to predict the central word "jumped" to complete the sentence. Here, the term "window" refers to the context window that determines how many words afore and later a given term. There are several advantages of CBOW like chiefly diminutive memory usage and being probabilistic in nature comparative to the traditional frequency-based embeddings. The main disadvantages of CBOW are sometimes average prediction for a word. It happens due to the hidden layer's output which is the average of word vectors related to context terms at the input layer [24]. For example, "Windows" can be both a graphical operating system by Microsoft or opening in a wall.

7 Skip-Gram Model

The skip-gram model or continuous skip-gram model is an alternative to the CBOW model with a small twist. Instead of using the surrounding words to unearth the middle word, it learns an embedding by predicting the surrounding words with the help of a given central word. It can be said that skip-gram architecture is an exact reverse of CBOW but keeps the topology same. Nonetheless, it can be imagined that skip-gram is analogous to the one-context architecture of CBOW [25] (Fig. 6).

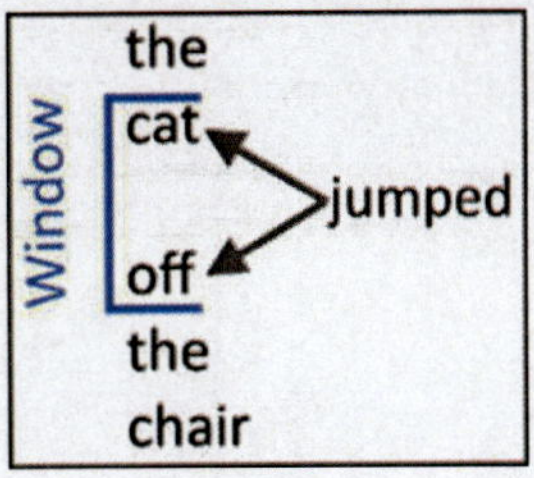

Fig. 6 Prediction using skip-gram model [23]

As shown above, skip-gram approach predicts the neighboring words using the current word, i.e., each central word is given input to a classifier with constant projection layer and predicts the nearby words within a particular range previous to and next to the current word. The training purpose is to learn word vector illustration that is superior at forecasting the close by words. The most important plus of using skip-gram is capturing multiple semantics for each word, like foretelling both graphical operating system by Microsoft and opening in a wall for "Windows". Another chief boon of skip-gram model is its association with negative sampling, which outperforms its other competitors [26].

8 Word2Vec

Word2Vec can be defined as Google's magic wand for word embedding, which has made the process easy and simplified. Word2Vec is based on the amalgamation of CBOW and skip-gram models (the shallow neural network architectures) to map word(s) to the target term(s). It can also be stated as the successor of the neural probabilistic model, which acquires embedding by attaining classification or modeling. Word2Vec was published in a research paper by Tomas Mikolov, Kai Chen, Greg Corrado and Jeffrey Dean (all Google employees) in September 2013 [27]. It is word embedding powerhouse by Google that trained on circa 100 billion words taken from Google News Data with nearly 300 dimensions. Now, Word2Vec can be said as the state-of-the-art word embedding standard as it produces generalized results and that too with lesser dimensions. It is based on a simple working principle that using trivial two-layer neural network for the training of rebuilding linguistic backgrounds of the word [28]. The input of Word2Vec is usually an enormous text corpus and produces "vector space" of huge dimensions, which means assigning a corresponding vector in the space to each exclusive word. At that moment, words of a corpus that share mutual context are located adjacent to each other in vector space and are hence called word vectors [29]. The metric used here is "cosine similarity" or "Euclidean distance" which discovers the likeness between myriad words or documents; the smaller the cosine angle, closer the documents. Cosine similarity can be defined as the amount of cosine of the angle between two non-zero vectors in three-dimensional space [30] (Fig. 7).

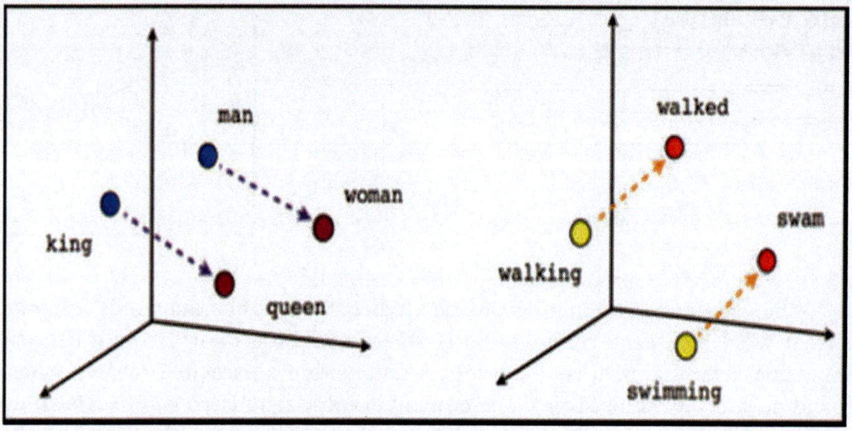

Fig. 7 Example of words in vector space [31]

As Word2Vec works on two alike yet opposite architectures, which both have their own pros and cons. On the one hand, CBOW is way quicker and produces enhanced word presentations for more recurrent terms. On the other hand, skip-gram is better suited for rare words and can also work fine with trivial vocabulary [32]. A major parameter of Word2Vec is sub-sampling, i.e., eliminating high-frequency words above a definite threshold may be subsampled to upsurge the training swiftness. Another imperative factor is dimensionality whose ideal value lies between 100 and 1000 because word embedding quality escalates with increased dimensionality. However, the most prominent parameter for Word2Vec is its training algorithm: hierarchical softmax and/or negative sampling being the most eminent ones. The hierarchical softmax technique uses a Huffman tree to diminish calculation and boost the conditional log-likelihood, thus making it work superior for the rare words. Besides, the negative sampling uses the sampled negative instances and minimalizes their log-likelihood for attaining better results for commonly used words and vectors with lower dimensions [33].

As stated above, negative sampling with continuous skip-gram is a far superior method for embedding, but how does it work? The answer is quite simple: Instead of just using positive samples, add some negative samples too. As the ultimate aim of every embedding algorithm is to better predict the context words according to a particular word, hence each model takes input and output words and its output score forecasts whether these words are neighbors or not in the vector space. If the words are neighbor, the target is set as "1", or else it is given the value "0". A desirable model may generate 100% correctness by giving all 1s but will be unable to learn and generalize. Hence, the negative samples must be added to the dataset for 0s, i.e., randomly designated words that are not neighbors as the output terms. The negative sampling increases efficiency by slightly decreasing the speed, thus maximizing the result. This can be easily understood by an example: If a bag contains 6 red balls, 12 white balls and 2 black balls, then simple sampling has a probability of picking red balls is 6/20 or 0.3, the black ball is 0.1, and the white ball is 0.6. The main shortcoming here is that high-frequency data will be picked (because of higher probability), which typically is of lesser value. Hence, adjusted sampling is done, i.e., reducing probability by 3/4th rule, thereby reducing white ball probability to 0.45, and probabilities of red balls and black balls will be enhanced. The appropriate number of negative samples varies between 5 and 20, although 2 to 5 negative samples are also sufficient [34].

9 GloVe

GloVe stands for Global Vectors for word representations, which can be stated as the extension of Word2Vec, but yielding even better results. It was developed by Jeffrey Pennington, Richard Socher and Christopher D. Manning in 2014 as an unsupervised learning algorithm for gaining vector illustration for words at Stanford [35]. The chief addition to the arsenal here is the inclusion of global statistics and exclusion of window feature for local context to generate improved embeddings. The training here is done on the non-zero values of the global word co-occurrence matrix that depicts how rarely or frequently two terms co-occur. The representations also show fascinating linear substructures of word vector space.

These pre-trained word vectors can be downloaded online, and it comes in different shapes and sizes. The smallest one contains 6 billion tokens with 50, 100, 200 or 300 dimensions, which is 822 MB in size. The subsequent file contains 42 billion tokens with 300 dimensions of 1.75 GB size. The largest file is trained on 840 billion tokens with 300 dimensions and of size 2.03 GB size. An additional feather in its cap is a pre-trained file for Twitter data that contains 2 billion tokens of 25, 50, 100 and 200 dimensions with 1.42 GB size.

10 fastText

fastText by Facebook is the free and open-source yet lightweight word embedding library to create supervised or unsupervised algorithms that are generally used for text representation and classification. fastText is based on two research papers written by Tomas Mikolov et al. in 2016, and now, it offers three models with 1 or 2 million word vectors and 300 dimensions [36]. The maximum model size of fastText is closely 8.2 GB in size.

The chief features of fastText are its pre-trained models for nearly 294 languages, working with general hardware with ease and even compatibility to work on cell phones [37]. However, a striking characteristic of fastText is its ability to yield vectors for even the unknown vocabulary, misspelled words and two words merged into one another. This is achieved by splitting entire terms into a bag-of-n-gram letters, and these subwords can be merged independently to create new words, as shown in

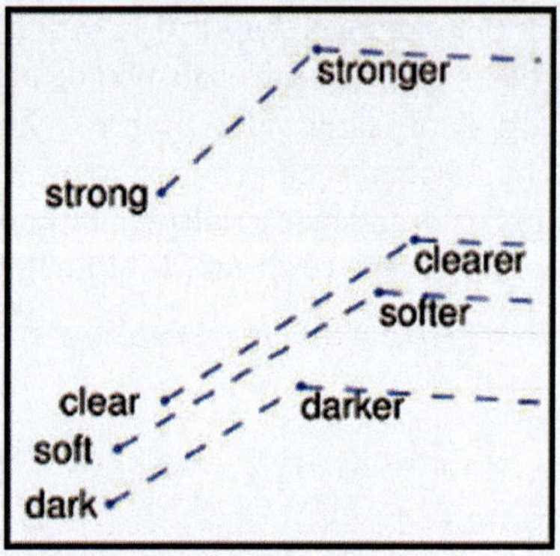

Fig. 8 GloVe showing comparative-superlative terms [35]

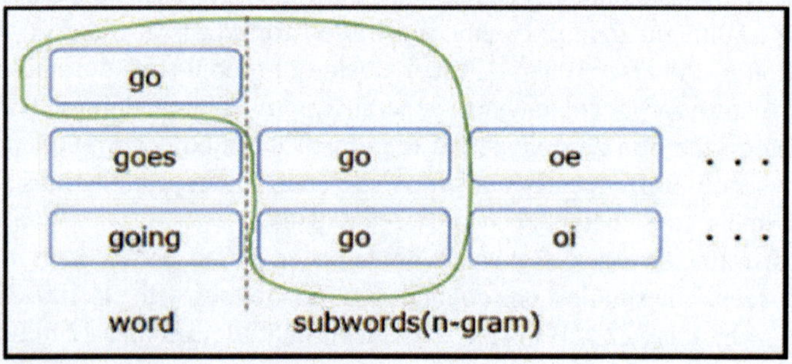

Fig. 9 Learning word embedding through subwords [17]

Fig. 8. The best results of fastText can be achieved by using skip-gram architecture in conjunction with negative sampling. As explained in the previous section, negative samples are random samples from the corpora and the sub-words will be used as positive samples. Also, fastText does not support GPU till now and works only on the CPU [38] (Fig. 9).

11 Poincare Embeddings

Poincare embeddings by Facebook is a novel research idea given by Maximilian Nickel and Douwe Kiela in 2017 that focuses on finding the hierarchical word representations [39]. To trace any term's hierarchical properties, Poincare uses a hyperbolic geometry, where the distance is used to express likeness and associations. The hyperbolic space is preferred over the Euclidean distance because it is an unremitting form of a tree and hence superior to comprehend the organizational property of learning. Due to the above elements, the embeddings generated by hyperbolic space are more

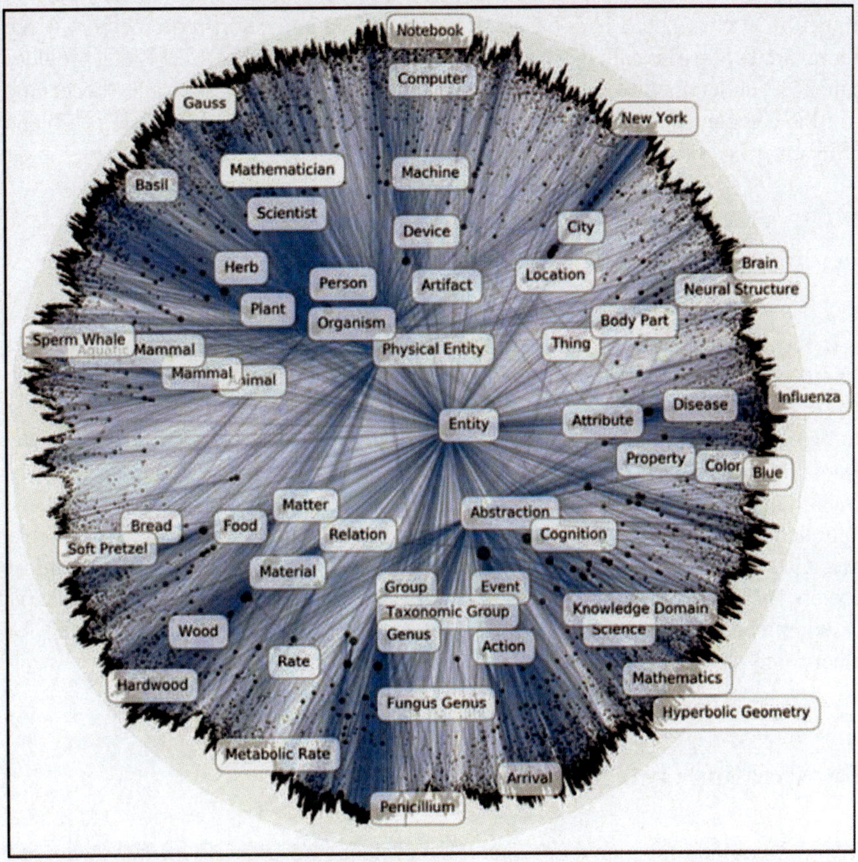

Fig. 10 Poincare embeddings' WordNet [40]

generalized and better for illustration. The key aspect here is that reduced dimensions are used here to create the hierarchical information, which was proved when they generated WordNet [40]. On the downside, Poincare embeddings are less proficient for modeling the perplex patterns due to their dimension limitation (Fig. 10).

12 ELMo

Embeddings for language model (ELMo) is designed to overcome the problems encountered by the Word2Vec and GloVe embeddings. It is defined as the deep contextualized term representation that looks at the whole sentence and then tries to gain knowledge about its syntax and semantics afore allocating embeddings to each word [41]. Similar to fastText, ELMo also concentrates on character level, thus exploiting on subwords and gaining the ability to even identifying the previously unidentified words. ELMo is based on bi-directional long short-term memory (deep neural networks) also called deep bi-directional language model (biLM) that provides enhanced understanding of language and modeling that can envisage succeeding word in a sentence. A major disadvantage of ELMo is that its final yield is a blend of its inner layers' output [42].

13 Probabilistic fastText

It is the most novel type of word embedding introduced by Ben Athiwaratkun, Andrew Gordon Wilson and Amina Anandkumar in June 2018 [43]. It is named after fastText and works similarly too but with additional responsibilities like capturing ambiguous information, getting subwords organization and most importantly the words that are spelled the same but have diverse meanings. For example, the word "fair" could mean a festival or unbiased or even the appearance of a person. Hence to understand in which context any particular term is being, probabilistic fastText uses Gaussian mixture models which are used for word representation instead of vectors [44]. Moreover, probabilistic fastText outclasses the traditional dictionary-based embeddings and fastText in performance both for English language and for other foreign languages.

14 Creating Own Embeddings

It is also possible for a novel or expert data scientist to generate his/her own word embeddings in case he/she is not satisfied with the pre-trained embeddings available like GloVe, ELMo, fastText, etc. The procedure is quite simple, which starts with building a vocabulary list of all the distinctive words and their frequency. Then, a dictionary should be prepared, where each distinct word will be matched with a unique ID. Then, a model must be defined by specifying myriad parameters like the number of dimensions, training algorithm and least count of words to consider while model training and window (utmost space between the target word and its close by words). An imperative point of consideration here is that the default number of all these parameters is also set. Subsequently, this model can be saved either in ASCII or in binary format for future use, and finally, it can supply this data in the neural networks to produce word embeddings [45].

15 Conclusion

This research paper discusses the most vigorous subject matter in the domain of sentiment analysis and that is word embeddings. In comparison with the conventional embeddings available earlier, some novel embedding methods have surfaced and taken the world with a storm. This has caused havoc for the beginners or new-fangled scholars who are trying to learn and finding which out of a plethora of word embedding to use in which situation. This paper is intended to simplify the chaos and present each available embedding technique in a simple way possible describing in detail the working principle, benefits and detriments and also able to work on numerous natural languages. Finally, it can be concluded that no word embedding method can be confirmed as the best method of all because their performances vary with the dataset; thus, one behaves well with a dataset and may yield poor results for another.

References

1. Neethu MS, Rajasree R (2013) Sentiment analysis in twitter using machine learning techniques. In: 2013 Fourth International Conference on Computing, Communications and Networking Technologies (ICCCNT). IEEE, pp 1–5
2. Yin Y, Wei F, Dong L, Xu K, Zhang M, Zhou M (2016) Unsupervised word and dependency path embeddings for aspect term extraction. arXiv preprint arXiv:1605.07843
3. Collomb A, Costea C, Joyeux D, Hasan O, Brunie L (2014) A study and comparison of sentiment analysis methods for reputation evaluation. Rapport de recherche RR-LIRIS-2014-002
4. Tan LK-W, Na J-C, Theng Y-L, Chang K (2012) Phrase-level sentiment polarity classification using rule-based typed dependencies and additional complex phrases consideration. J Comput Sci Technol 27(3):650–666
5. Shalev-Shwartz S, Ben-David S (2014) Understanding machine learning: from theory to algorithms. Cambridge University Press
6. Zou WY, Socher R, Cer D, Manning CD (2013) Bilingual word embeddings for phrase-based machine translation. In: Proceedings of the 2013 conference on empirical methods in natural language processing, pp 1393–1398
7. Das S, Dey A, Pal A, Roy N (2015) Applications of artificial intelligence in machine learning: review and prospect. Int J Comput Appl 115(9)
8. Molnar C (2020) Interpretable machine learning. Lulu.com
9. Deng L, Liu Y (eds) (2018) Deep learning in natural language processing. Springer, Berlin
10. Li Y, Yang T (2018) Word embedding for understanding natural language: a survey. In: Guide to big data applications, pp 83–104. Springer, Cham
11. Text Embedding Beta (2020) https://peltarion.com/knowledge-center/documentation/modeling-view/build-an-ai-model/blocks/text-embedding
12. Zhao Y, Huang S, Dai X, Zhang J, Chen J (2014) Learning word embeddings from dependency relations. In: 2014 international conference on Asian language processing (ialp), IEEE, pp 123–127
13. Kusner M, Sun Y, Kolkin N, Weinberger K (2015) From word embeddings to document distances. In: International conference on machine learning, pp 957–966
14. Tang D, Wei F, Qin B, Yang N, Liu T, Zhou M (2015) Sentiment embeddings with applications to sentiment analysis. IEEE Trans Knowl Data Eng 28(2):496–509
15. Chen Y, Perozzi B, Al-Rfou R, Skiena S (2013) The expressive power of word embeddings. arXiv preprint arXiv:1301.3226
16. De Boom C, Van Canneyt S, Demeester T, Dhoedt B (2016) Representation learning for very short texts using weighted word embedding aggregation. Pattern Recogn Lett 80:150–156

17. Heidenreich H (2018) Introduction to word embeddings. https://hunterheidenreich.com/blog/intro-to-word-embeddings/
18. Paik JH (2013) A novel TF-IDF weighting scheme for effective ranking. In: Proceedings of the 36th international ACM SIGIR conference on Research and development in information retrieval, pp 343–352
19. Gebre BG, Zampieri M, Wittenburg P, Heskes T (2013) Improving native language identification with tf-idf weighting. In: The 8th NAACL workshop on ınnovative use of NLP for Building Educational Applications (BEA8), pp 216–223
20. Gebejes A, Huertas R (2013) Texture characterization based on grey-level co-occurrence matrix. Databases 9:10
21. Liang D, Altosaar J, Charlin L, Blei DM (2016) Factorization meets the item embedding: regularizing matrix factorization with item co-occurrence. In: Proceedings of the 10th ACM conference on recommender systems, pp 59–66
22. Luo Q, Xu W, Guo J (2014) A study on the CBOW model's overfitting and stability. In: Proceedings of the 5th ınternational workshop on web-scale knowledge representation retrieval & reasoning, pp 9–12
23. Word Embeddings for Natural Language Processing (2016) https://deep-solutions.net/blog/WordEmbeddings.html
24. Guo G, Ouyang S, Yuan F, Wang X (2018) Approximating word ranking and negative sampling for word embedding. In: International Joint Conferences on Artificial Intelligence Organization
25. Yang L, Chen X, Liu Z, Sun M (2017) Improving word representations with document labels. IEEE/ACM Trans Audio Speech Language Process 25(4):863–870
26. Goldberg Y, Levy O (2014) word2vec explained: deriving Mikolov et al.'s negative-sampling word-embedding method. arXiv preprint arXiv:1402.3722
27. Mikolov T, Sutskever I, Chen K, Corrado GS, Dean J (2013) Distributed representations of words and phrases and their compositionality. In: Advances in neural information processing systems, pp 3111–3119
28. Yoshioka K, Dozono H (2018) The classification of the documents based on word2vec and 2-layer self organizing maps. Int J Mach Learn Comput 8(3):252–255
29. Mikolov T, Yih W, Zweig G (2013) Linguistic regularities in continuous space word representations. In: Proceedings of the 2013 conference of the north american chapter of the association for computational linguistics: human language technologies, pp 746–751
30. Sidorov G, Gelbukh A, Gómez-Adorno H, Pinto D (2014) Soft similarity and soft cosine measure: similarity of features in vector space model. Computación Y Sistemas 18(3):491–504
31. The Synthesis Project: Materials Word Embeddings (2020) https://www.synthesisproject.org/resources
32. Levy O, Goldberg Y (2014) Dependency-based word embeddings. In: Proceedings of the 52nd Annual Meeting of the Association for Computational Linguistics (Volume 2: Short Papers), pp 302–308
33. Rong X (2014) word2vec parameter learning explained. arXiv preprint arXiv:1411.2738
34. Wu C-Y, Manmatha R, Smola AJ, Krahenbuhl P (2017)Sampling matters in deep embedding learning. In: Proceedings of the IEEE ınternational conference on computer vision, pp 2840–2848
35. Pennington J, Socher R, Manning CD (2014) Glove: global vectors for word representation. In: Proceedings of the 2014 conference on empirical methods in natural language processing (EMNLP), pp 1532–1543
36. Joulin A, Grave E, Bojanowski P, Douze M, Jégou H, Mikolov T (2016) Fasttext. zip: compressing text classification models. arXiv preprint arXiv:1612.03651
37. Santos I, Nedjah N, de Macedo Mourelle L (2017) Sentiment analysis using convolutional neural network with fastText embeddings. In: 2017 IEEE Latin American Conference on Computational Intelligence (LA-CCI), pp 1–5. IEEE
38. Moffat A, Zobel J (1996) Self-indexing inverted files for fast text retrieval. ACM Trans Inf Syst (TOIS) 14(4):349–379

39. Nickel M, Kiela D (2017) Poincaré embeddings for learning hierarchical representations. In: Advances in neural information processing systems, pp 6338–6347
40. Saedi C, Branco A, Rodrigues J, Silva J (2018) Wordnet embeddings. In: Proceedings of the third workshop on representation learning for NLP, pp 122–131
41. Jha V (2018) Insight into hierarchical representations through poincaré embedding. https://www.techleer.com/articles/478-insight-into-hierarchical-representations-through-poincare-embedding/
42. Chami I, Wolf A, Sala F, Ré C (2019) Low-dimensional knowledge graph embeddings via hyperbolic rotations. In: Graph representation learning NeurIPS 2019 Workshop
43. Athiwaratkun B, Wilson AG, Anandkumar A (2018) Probabilistic fasttext for multi-sense word embeddings. arXiv preprint arXiv:1806.02901
44. Choi J, Lee S-W (2020) Improving FastText with inverse document frequency of subwords. Pattern Recogn Lett
45. Brownlee J (2017) Deep learning for natural language processing: develop deep learning models for your natural language problems. Mach Learn Masterysss

Chapter 43
Automatic Cataract Detection Using Haar Cascade Classifier

Jaspreet Kaur, Prerit Sinha, Rahul Shukla, and Vikas Tiwari

1 Introduction

A cataract is caused when normal transparency is lost in the lens of humans because of opacity. The lens is joined by nerves and cells of tissues and is formed of protein and water. The formation of protein clumps, which together blurs the lens and blocks the light to the retina. This type of cataract is mainly caused due to the aging process and in diabetes. The drugs taken in diabetes may also lead to earlier cataract formation in human eyes. Cataracts are mainly of three types based on the location of opacity in the lens—posterior sub-capsular cataract, cortical cataract, and nuclear cataract [1].

Of all the studies, it is observed that cataract is mainly caused due to old age. According to a population-based survey in Chinese-Singaporean community, almost 35% of the cataract affected people were found in people over 40 years old. Further, 7% people aged 40, and 37.5% in 50s, 80.4% in 60s, and 93.3% in 70s were found to show cataract [2, 3].

Due to the rapid increase in population and increasing pollutants in our daily lives, the blindness caused due to age-related cataracts is expected to grow rapidly. There are different risk factors for cataract researched over time, although the proven preventive strategies are very few. So, the timely diagnosis and proper surgical treatment for cataracts can be essential to prevent vision impairment and improvement in daily life standards [4].

J. Kaur · P. Sinha (✉) · R. Shukla · V. Tiwari
Galgotias College of Engineering and Technology(AKTU), Noida, India
e-mail: sinhaprerit@gmail.com

J. Kaur
e-mail: jas.kaur@galgotiacollege.edu

I. Jeena Jacob et al. (eds.), *Data Intelligence and Cognitive Informatics*,
Algorithms for Intelligent Systems,
https://doi.org/10.1007/978-981-15-8530-2_43

There are various other computerized methods used for cataract detection like using SVM and K-NN. In this paper, a system has been developed more concise and easily integrable to an application using Haar cascades.

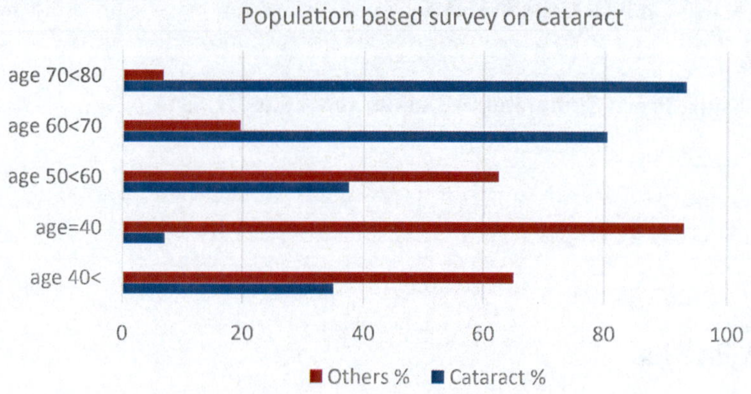

2 Related Work

S.no	References	Analysis made	Limitations
1	Lopez, H. I. M., Garcia, J. C. S., & Mendez, J. A. D. (2016). Cataract detection techniques: A review. IEEE Latin America Transactions, *14*(7), 3074–3079	The methods have been proposed in recent years for the detection and diagnosis of cataracts is discussed	The individual method is not explained
2	Li, H., Lim, J. H., Liu, J., Wong, D. W. K., Tan, N. M., Lu, S., … & Wong, T. Y. (2009, October). Computerized systems for cataract grading. In 2009 2nd International Conference on Biomedical Engineering and Informatics (pp. 1–4). IEEE	The computer-aided methods for cataract differentiation and grading using retroillumination and slit-lamp image are used for determining the opacity of cortex and nucleus in the lenses	The retroillumination method has low accuracy
3	Hnoohom, N, & Jitpattanakul, A. (2017, November). Comparison of ensemble learning algorithms for cataract detection from fundus images. In the 2017 21st International Computer Science and Engineering Conference *(ICSEC)* (pp. 1–5). IEEE	Here, the features extracted from the fundus images are determined using the texture-based and sketch-based methods. It includes using two different machine learning algorithms	The fundus image is costly to obtain

(continued)

(continued)

S.no	References	Analysis made	Limitations
4	Patil, D., Nair, A., Bhat, N., Chavan, R., & Jadhav, D. (2016, December). Analysis and study of cataract detection techniques. In 2016 International Conference on Global Trends in Signal Processing, Information Computing and Communication *(ICGTSPICC)* (pp. 516–519). IEEE	Various cataract detection techniques in the stage of pre-processing, feature extraction, and classifier contraction are discussed with their accuracy percentage	The method discussed is costly and isn't portable
5	Qian, X., Patton, E. W., Swaney, J., Xing, Q., & Zeng, T. (2018, October). Machine Learning on Cataracts Classification Using SqueezeNet. In *2018* 4th International Conference on Universal Village *(UV)* (pp. 1–3). IEEE	The squeeze net model that grades cataract from nuclear, cortical, posterior cataract is implemented. Manual cropping using an image data generator is used to gain a greater percentage of accuracy	Not user friendly, as it doesn't have proper GUI and isn't automated
6	Harini, V., &Bhanumathi, V. (2016, April). Automatic cataract classification system. In 2016 International Conference on Communication and Signal Processing *(ICCSP)* (pp. 0815–0819). IEEE	A combination of two methods in the wavelet domain and spatial domain combined successfully. SVM and RBFN classifier is used for classification	The algorithm doesn't classify more than one different eye disorders. Neuro-fuzzy images can't be classified
7	Bhadra, A. A., Jain, M., &Shidnal, S. (2016, March). Automated detection of eye diseases. In 2016 International Conference on Wireless Communications, Signal Processing and Networking (WiSPNET) (pp. 1341–1345). IEEE	Pattern recognition is applied for the recognition of images combined with BGR color property. It uses the OpenCV library and is tested for several diseased eye images	The dataset used is less, and it has an accuracy of 92%
8	Rana, J., & Galib, S. M. (2017, December). Cataract detection using a smartphone. In 2017 3rd International Conference on Electrical Information and Communication Technology (EICT) (pp. 1–4). IEEE	This is a detailed study of early cataract detection using smartphones	It is not reliable because the classifier is not well trained

(continued)

(continued)

S.no	References	Analysis made	Limitations
9	Li, H., Lim, J. H., Liu, J., &Wong, T. Y. (2007, August). Towards automatic grading of nuclear cataract. In *2007* 29th Annual International Conference of the IEEE Engineering in Medicine and Biology Society (pp. 4961–4964). IEEE	The difference between the clinical grading of cataract systems and computerized grading of cataract systems is discussed. This study is done on the 40 test images	Not user friendly nor automated and slit-lamp images are required
10	Lakra, A., Tripathi, P., Keshari, R., Vatsa, M., & Singh, R. (2018, August). SegDenseNet: Iris Segmentation for Pre-and-Post Cataract Surgery. In 2018 24th International Conference on Pattern Recognition (ICPR) (pp. 3150–3155). IEEE	The SegDenseNet is based on DenseNets. These are based on deep learning algorithms. This is an iris segmentation algorithm detecting the iris for past surgeries if any	When the image of the iris is at an off-angle, few triangle samples and bubbles are observed in the wide off-angle
11	Patwari, M. A. U., Arif, M. D., Chowdhury, M. N., Arefin, A., & Imam, M. I. (2011, May). Detection, categorization, and assessment of eye cataracts using digital image processing. *The first international conference on interdisciplinary research and development* (pp. 22-1)	The error measured in the algorithm is observed due to the gray level variation	Not user friendly as the system is not automated and has no GUI
12	Daud, M. M., Zaki, W. M. D. W., Hussain, A., Ahmad, W. S. H. M. W., &Mutalib, H. A. (2018, July). Connected Ocular Screening System Using Smartphones. In *2018 1st IEEE International Conference on Knowledge Innovation and Invention (ICKII)* (pp. 1–4). IEEE	It is a survey for ocular disease detection using in recent studies using mobile applications	No detailed information, the overview is only present
13	Fan, W., Shen, R., Zhang, Q., Yang, J. J., & Li, J. (2015, October). Principal component analysis based cataract grading and classification. In *2015 17th International Conference on E-health Networking, Application & Services (HealthCom)* (pp. 459–462). IEEE	The sketch features and wavelets extracted from the fundus images are studied using the PCA method to the dimensionality of the above extracted two features	The cataract detection speed is increased, but accuracy is decreased

(continued)

(continued)

S.no	References	Analysis made	Limitations
14	Sigit, R., Kom, M., Satmoko, M. B., Basuki, D. K., & Si, S. (2018, September). Classification of Cataract Slit-Lamp Image Based on Machine Learning. In *2018* International Seminar on Application for Technology of Information and Communication (pp. 597–602). IEEE	The slit-lamp images are preprocessed using the canny method and median filter method. The uniformity and mean intensity are extracted using the pixel scanning method	The classification algorithms have 3.33% of errors in 30 datasets
15	Kaur, M., Kaur, J., & Kaur, R. (2015, October). Low-cost cataract detection system using a smartphone. In *2015* International Conference on Green Computing and Internet of Things *(ICGCIoT)* (pp. 1607–1609). IEEE	The commercially available smartphone is modified using the external microscope for disease detection in eye related to the retina	The working model description is unavailable

3 Data Augmentation Using Augmentor

Data augmentation is done using a python package called augmentor. It is designed to aid augmentation by applying the different features applied by the user for the creation of image data for complicated machine learning processes. Being a basic augmentation tool, it performs the important preprocessing steps required for a program or system. Data augmentation is a process by which a list of the graphic dataset can be extended without varying the specific details in the image. Image augmentation uses various methods such as flip, rotate, crop, shears, and many more. The rotate function here rotates each image by a certain angle to make a combination with the other alterations [5]. Similarly, the flip left_right function flips the images for creating a similar required image. Zoom random function zooms any random image to create further altercation and increasing variation. It has many processes, and it is performed using the pipeline methodology.

For training classifier model, a significant amount of data is needed, and for such data collection, one should have a vast amount of resources and money at their disposal. By using augmentors, a significant amount of resources is created in very limited time and without any specific resources [6].

For creating the final dataset for training our classifier, 40 healthy eye images are taken and 35 cataract affected images of the eye (Figs. 1 and 2).

After feeding these images to augmentor, 1000 positive images (cataract affected eye images) and 1000 negative images (normal eye images) are generated and stored in the separate directory (Fig. 3).

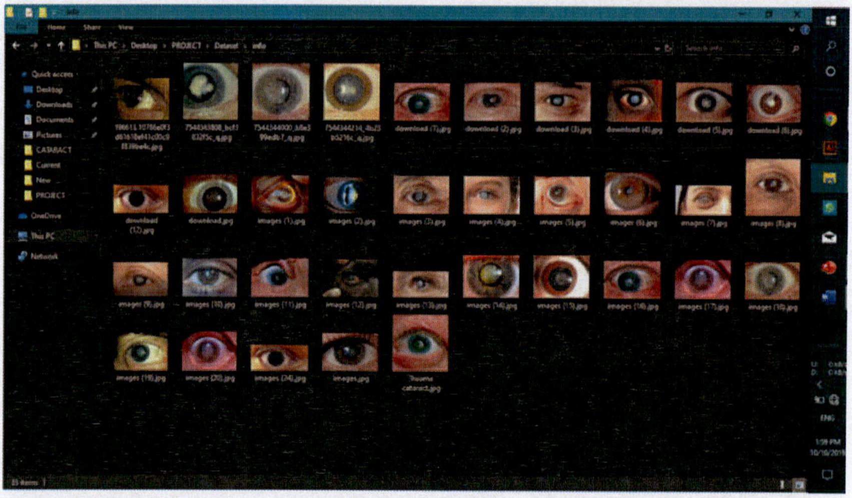

Fig. 1 35 cataract affected eye images

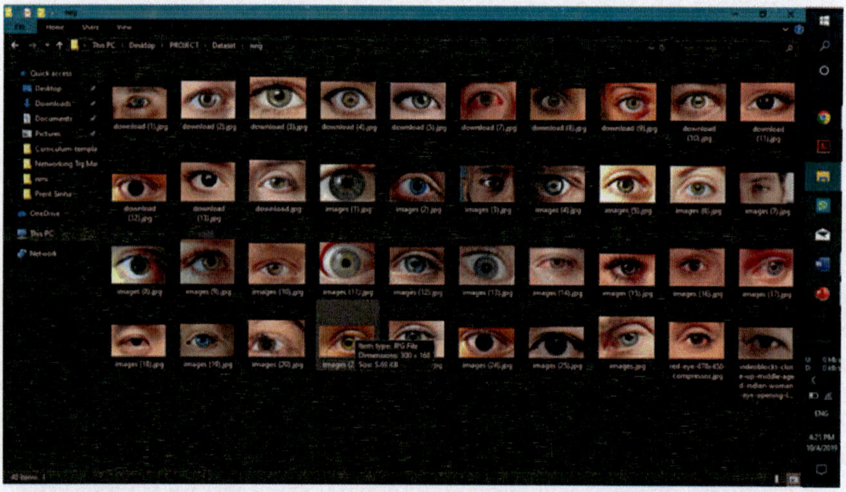

Fig. 2 40 cataract non-affected eye images/healthy eye images

4 Create Haar Cascade File Using OpenCV and Python

Haar cascade classifier is an advanced object detection algorithm introduced by Michael Jones and Paul Viola in their research paper "Rapid Object Detection Using a Boosted Cascade of Simple Features" in the year 2001. This is based on a machine learning algorithm in which a function based on cascade is fed a bunch of negative and positive images of an object for which the classifier has to be developed. This

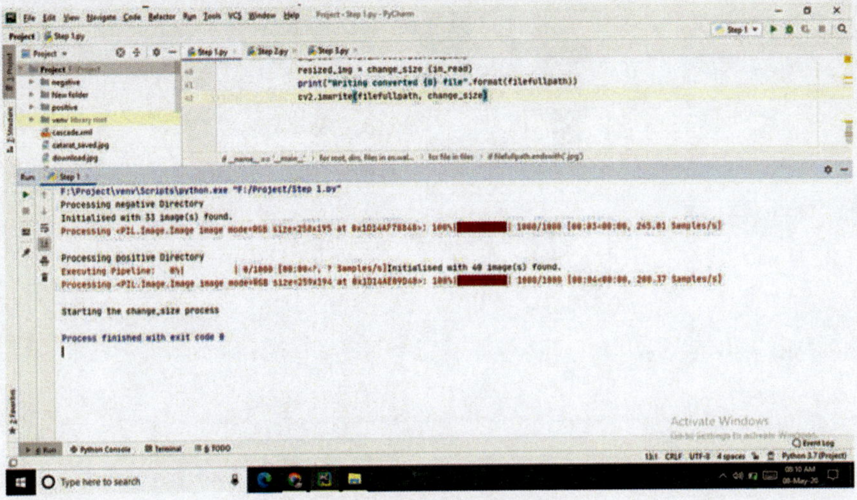

Fig. 3 Final 1000 positive and negative images generated each

results in a trained systematic cascade file in XML format which then determines the object and detects it as well [7].

4.1 *Haar Cascade Model*

The algorithm for Haar cascade classifiers has four stages:

- Haar feature selection,
- Creating integral images,
- AdaBoost training,
- Cascading classifiers.

Haar feature selection consisting of methods like rectangular-Haar-like feature is the area sum of pixels inside the rectangle. These are the integral sums of the areas similar to convolutional sums of the areas.[7, 8].

Rectangular-Haar-like feature is the difference between the area sum of pixels inside the rectangle. These are the integral sums of the areas similar to convolutional sums of the areas. The method described above is known as a two-rectangle feature. Similarly, the four-rectangle feature has two rectangles each for an integral sum of the pixels inside them (Figs. 4 and 5).

Creating integral images—The difference between the integral images in the image creates an approximate value for each feature determined. For example, in the image

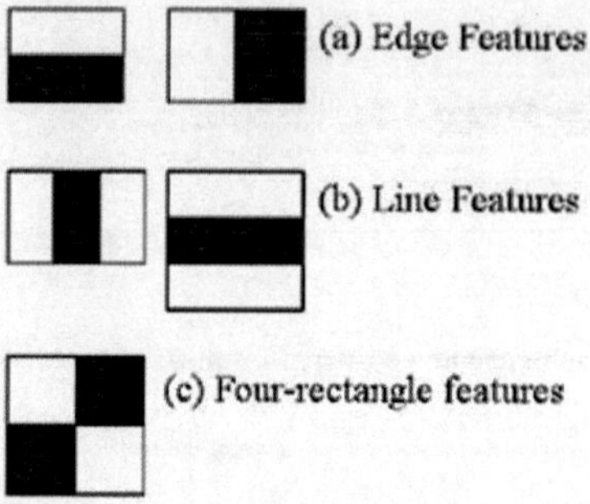

Fig. 4 Haar-like features

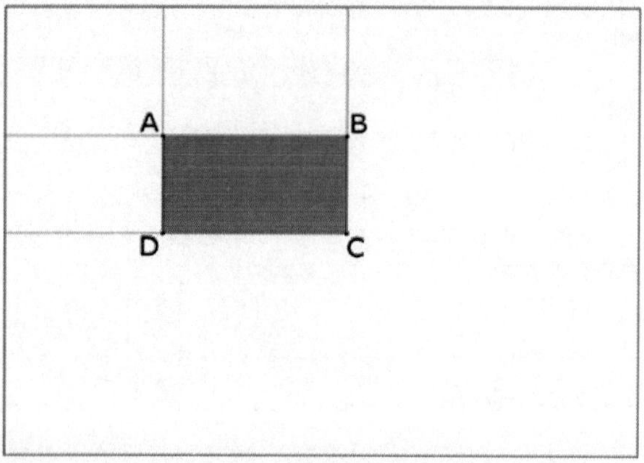

Fig. 5 Integral sum of the shaded region in a rectangle (rectangular feature)

below, the first feature is focused in the area around the eyes. It is observed that the feature near the nose and cheek is less dark than near the eyes.

The integral image between cheek and eyes is clear therefore a good feature. Similarly, thousands of features are determined, and the most relevant one is selected and applied in the property. This is further carried out in AdaBoost training (Fig. 6).

AdaBoost training simply divides the features determined as "weak" classifiers and transforms it into "strong" classifiers via linear combination. Thus, the system learns the difference between the desired object and the undesired one. Here, all of the Haar features determined are said to be a weak classifier that has insufficient accuracy, and strong classifiers are the opposite (Fig. 7).

Now, when these steps get lined up, then a cascade classifier is generated.

Cascade classifier has a collection of weak learners in different stages which are called decision stumps. Decision stumps are weak learners who have insufficient accuracy in their detected features. These weak learners are used for training by a technique called boosting. Boosting is training some information on an average

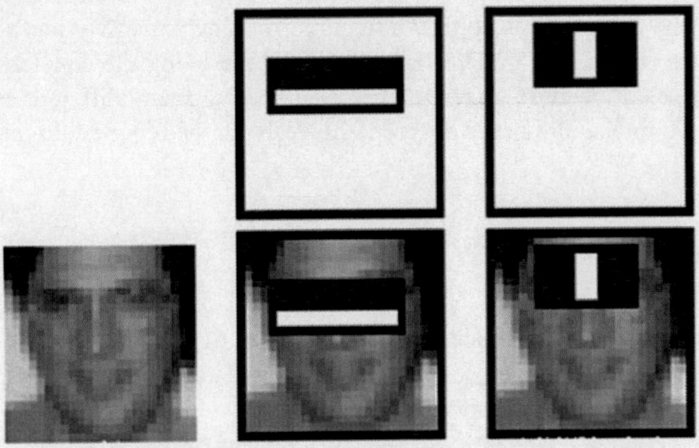

Fig. 6 Integral sum of the images (selection of features)

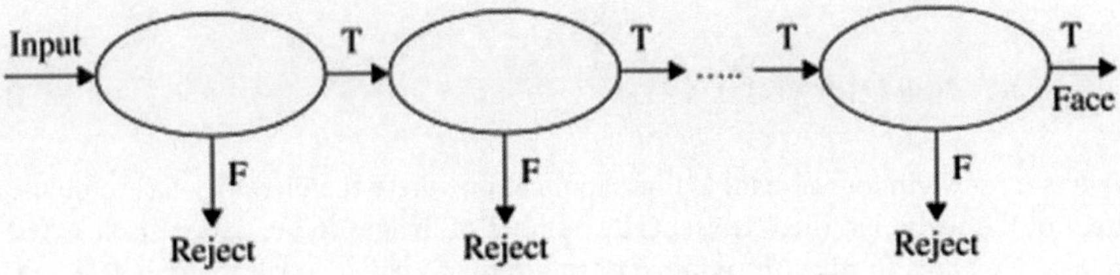

Fig. 7 Different stages for image recognition via strong classifier or cascade classifier

decision of weak learners by boosting. This results in a highly accurate classifier by calculating the average of results gained from simple classifiers or weak learners.

5 Train Haar Cascade File

In our case, 1000 positive images of cataract infected and non-infected eye images were fed into the cascade trainer. These are fed via the directory location of the listed files obtained during augmentation.

5.1 Negative Samples

Required non-cataract (not containing cataracts) eye images are taken as negative samples. It should be noted that negative samples of sample pictures also are referred to as background samples or background pictures and can be used to repose changeably during this document. Delineate pictures could also be of various sizes [9].

However, every image ought to fit the required training window dimension (which corresponds to the model dimensions, most of these being the equal size of your object), as a result of these pictures are accustomed to many different samples for given negative image into many different images with the required dimensions.

5.2 Positive Samples

They are employed for the cascading methods to outline the outlook of the model while making attempts to seek out cataract pictures. The application is supported by two ways that generating a positive sample dataset.

•You will generate a bunch of positives from one positive cataract picture.

•You will provide each of the positive pictures yourself and solely use the tool to chop them out, resized them, and place within the OpenCV required binary format.

5.3 Cascade Trainer GUI

This is a four-window cascade trainer application where the directory has to be the input of the well-classified dataset. The number of images in the dataset is selected in the dialog box. Further, the number of stages for classifier construction is selected. The number of stages determines the accuracy of the classifier. The desired boost type is selected according to the required efficiency (Fig. 8).

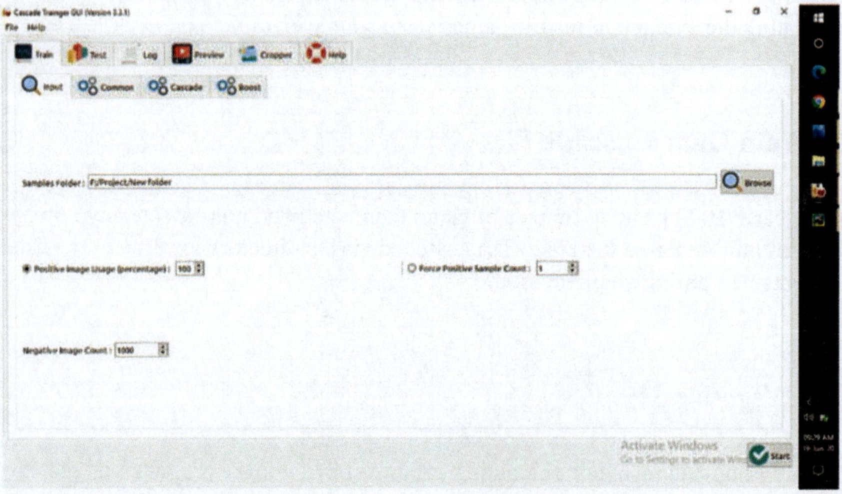

Fig. 8 Cascade generator

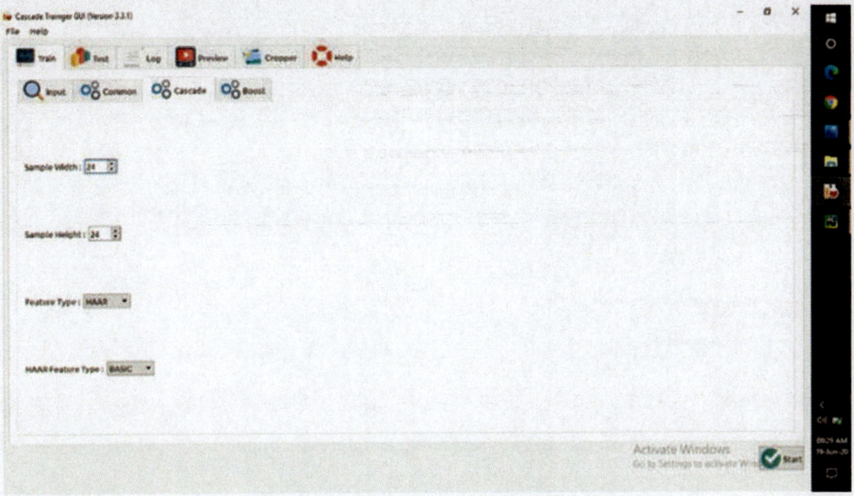

Fig. 9 Selection of feature type and Haar feature type in cascade trainer

At the last stage, the feature type is selected for cascade, i.e., Haar and its feature as BASIC. The sample width and length are also mentioned.

This wizard guides through all the essential steps required and finally generates an XML extension file in the desired directory. The cascade classifier construction undergoes 20 stages, every stage consisting of a group of weak learners. They are referred to as stumps because they are easy classifiers. Boosting technique is employed at every stage which is pretrained. The average selections made by the weak learners which provide the weighted correct classifier are obtained by boosting. The different stages are classified as positive or negative depending on the stages outlined by classifier labels in the sliding window (Fig. 9).

The match of the required result prompts an output positive, and the negative is the result of no match of the desired feature or object. When the result is negative, the detector moves to the next window for the same procedure. Similarly, if found positive, the detector is passed onto the subsequent stage [10, 11]. At last, when the stage classifier classifies the stage positive, the detector reports a found object. These weak classifiers then combined through linear combination generate a strong classifier. This resulting strong classifier is known as the Haar cascade classifier file or XML file.

6 Detection Using Cascade Classifier

The training method provided by OpenCV is a cascade classifier that is used to call the XML file in the code. OpenCV has different methods for calling the XML file, object detection, and image input. cv::CascadeClassifier creates the file

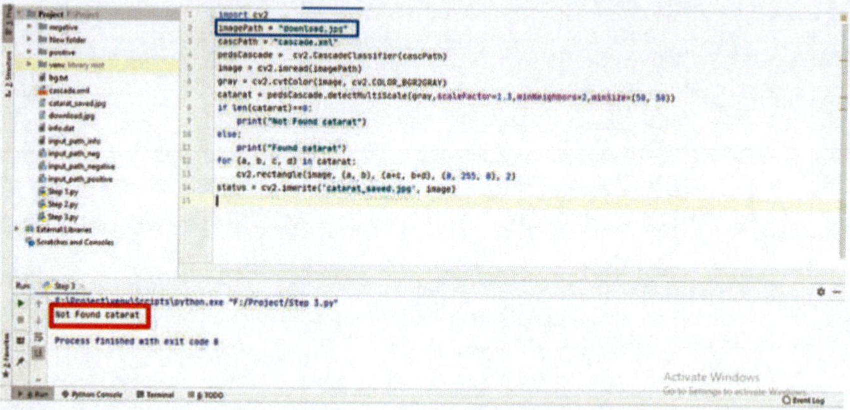

Fig. 10 XML file integration into python script and testing of the input data

cv::CascadeClassifier::load which is used for reading/loading the XML file. The XML file loaded is a pretrained model, which is discussed in the above topics. cv::CascadeClassifier::detectMultiScale detects the desired region in the image. The generated cascade file is further integrated into the python script. In the same script, the test image is received and comparison or detection of the desired feature (cataract) takes place. Here, the test image was a random image downloaded from the Internet. The image was cataract free as the code run result shows "cataract not found". The software used was "Spyder" or "Pycharm" for integrating the script to the XML file. The OpenCV library helps in integrating it into the python script using the above-mentioned functions. Similar testing of other images has been discussed in the results section (Fig. 10).

7 Results

The generated cascade classifier is transformed into a concise XML file. The XML file is further fed to the python script using OpenCV. A random image is chosen from the Internet or obtained from a patient which is fed to the script. Thus, the resulting data is categorized or determined if it is cataract affected or not. If yes, the output is "cataract found", and if not, the output is "cataract not found". This script can be further integrated into a program and then to an application for ease of usage and access (Table 1).

The lens localization accuracy is 95.9%, and lens structure detection is 95% which have been observed using the SVM regression, whereas using machine learning algorithms, the accuracy was found out to be 97.5%. Here, the accuracy depends on the quantity of data fed and the formation of more weak classifier stages. The more weak classifiers are generated in the training the more accurate is the system.

Table 1 Comparison of results from the tested images in terms of accuracy, severity, and error

S.no	Evaluator	Cataract type	Severity %	Error %	Accuracy %
1	Haar	Normal	0	0	100
2	Haar	Normal	0	0	100
3	Haar	Normal	0	0	100
4	Haar	NC	28	−6.67	96.67
5	Haar	NC	37	−7.50	97.50
6	Haar	NC	48	−4.00	96.00
7	Haar	NC	63	5.00	95.00
8	Haar	NC	84	5.00	96.30
9	Haar	NC	100	0.00	95.00
10	Haar	CC	95	5.56	97.20
11	Haar	CC	23	−6.00	97.00
12	Haar	CC	47	−6.00	95.00
13	Haar	CC	73	4.29	93.00
14	Haar	CC	100	0.00	94.00
15	Haar	CC	36	−10	92.00

A total number of 15 test images were taken and tested. The images were of different cataract types. This table describes the result of the tested image in three terms. Severity means the percentage of a cataract affected in the eyes. The cataract types discussed are cortical cataract (CC) and nuclear cataract (NC), and normal is sub-capsular posterior cataract. The accuracy of this system overall reaches a maximum of up to 97.5%.

In some cases such while using K-NN, it shows the maximum accuracy up to 99% but has less efficiency and is not that easily integrable and easy to modify or bring further changes.

8 Conclusion and Future Scope

In this paper, the formulation and application of the automatic cataract detection systems have been briefly discussed. Algorithms based on machine learning are being used, and all the libraries used were of OpenCV. The classifier feature used is Haar, and the boosting type is GAB. The data acquisition is done from the open-source Web sites, and the data is further expanded by a technique called image augmentation. Further formulation of the cascade classifier is done on the python platform. This is a step towards a concise mechanism for cataract detection, as in this technological world, everything wanted to be concise and easily accessible. **Future Scope** of this system is as much as the usage of the OpenCV and python. The development in these sectors is never-ending and more feasible. This system can be further applied in detecting other diseases related to eyes like glaucoma. It can be further developed to grade the different types of cataract. This system prominent work is to be integrated into an application and introducing it to the more accessible platform for public usage.

Acknowledgements We would like to thank the anonymous authors for helping with all the queries and review comments. We would also like to thank our HOD, Project Coordinator, and Class Coordinator for their wise guidance.

References

1. Lopez HIM, Garcia JCS, Mendez JAD (2016) Cataract detection techniques: a review. IEEE Latin Am Trans 14(7), 3074–3079. Wong TY, Loon SC, Saw SM (2006) The epidemiology of age-related eye diseases in Asia. Br J Ophthalmol 90:506–511
2. Li H, Lim JH, Liu J, Wong DWK, Tan NM, Lu S, Wong TY et al (2009) Computerized systems for cataract grading. In: 2009 2nd ınternational conference on biomedical engineering and ınformatics, pp 1–4. IEEE
3. Hnoohom N, Jitpattanakul A (2017) Comparison of ensemble learning algorithms for cataract detection from fundus images. In: 2017 21st International Computer Science and Engineering Conference (ICSEC), pp 1–5. IEEE.
4. Patil D, Nair A, Bhat N, Chavan R, Jadhav D (2016) Analysis and study of cataract detection techniques. In: 2016 International Conference on Global Trends in Signal Processing, Information Computing and Communication (ICGTSPICC), pp 516–519. IEEE
5. Qian X, Patton EW, Swaney J, Xing Q, Zeng T (2018) Machine learning on cataracts classification using SqueezeNet. In: 2018 4th ınternational conference on Universal Village (UV), pp 1–3. IEEE
6. Harini V, Bhanumathi V (2016) Automatic cataract classification system. In: 2016 International Conference on Communication and Signal Processing (ICCSP), pp 0815–0819. IEEE
7. Viola P, Jones M (2001) Rapid object detection using a boosted cascade of simple features. In: Proceedings of the 2001 IEEE computer society conference on computer vision and pattern recognition. CVPR 2001, vol 1, pp I-I. IEEE
8. Rana J, Galib SM (2017) Cataract detection using smartphone. In: 2017 3rd ınternational conference on Electrical Information and Communication Technology (EICT), pp 1–4. IEEE
9. Li H, Lim JH, Liu J, Wong TY (2007) Towards automatic grading of nuclear cataract. In: 2007 29th annual ınternational conference of the IEEE Engineering in Medicine and Biology Society, pp 4961–4964. IEEE
10. Lakra A, Tripathi P, Keshari R, Vatsa M, Singh R (2018) SegDenseNet: ıris segmentation for pre-and-post cataract surgery. In: 2018 24th International Conference on Pattern Recognition (ICPR), pp 3150–3155. IEEE
11. Patwari MAU, Arif MD, Chowdhury MN, Arefin A, Imam MI (2011) Detection, categorization, and assessment of eye cataracts using digital image processing. In: The first international conference on interdisciplinary research and development, p 22-1

Chapter 44
VG2—DNA-Based One-Time Pad Image Cipher

Akhil Kaushik, Vikas Thada, and Jaswinder Singh

1 Introduction

The whole world is today under lockdown because of COVID-19, but the information has always been under lockdown since the medieval age. Anyone who has got the correct information and at the right time could do wonders whether in business or personal life. However, when the corporate world is considered, the information acts as a lifeline to not only win against the competitors by gaining a momentous advantage over them but also to gain the faith of customers. Nevertheless, the information age presently is not limited to the textual data and today loads of data are usually present in the multimedia form. Consequently, the traditional ciphers have lost their meaning today and prove unworthy of dealing with the modern day eavesdropping and hacks. Hence, these conventional cryptographic algorithms must be amalgamated with some novel techniques especially DNA cipher.

DNA crypts arrived in the mid-90s and have been in the limelight ever since. Leonard Adleman implemented this novel approach to solve the famous Hamiltonian Path Problem in 1994, followed by Lipton, Boneh, Dunworth, etc., who solved another enigmatic mathematical problem using DNA computations [1]. Needless to say, the research area became popular among fellow researchers and even today,

A. Kaushik (✉) · V. Thada
CSE Department, Amity University, Gurugram, Haryana, India
e-mail: akhilkaushik05@gmail.com

V. Thada
e-mail: vthada@ggn.amity.edu

J. Singh
CSE Department, GJU Hisar, Haryana, India
e-mail: jaswinder_singh_2k@rediffmail.com

I. Jeena Jacob et al. (eds.), *Data Intelligence and Cognitive Informatics*,
Algorithms for Intelligent Systems,
https://doi.org/10.1007/978-981-15-8530-2_44

loads of young aspirants are using this brilliant approach for seeking answers to NP-hard and NP-complete issues. Since the last few decades, DNA cryptography and steganography have gained huge popularity due to several reasons especially while the security of digital images. Some experts state that it is because DNA ciphers are based on biology rather than the modular arithmetic; hence, the traditional cryptanalyst approaches are unable to break these perplex modern warfare tools [2]. However, the ciphers developed so far are still considered in the infancy stage and there is huge room for enhancement.

In this paper, Sect. 2 highlights the biological background of DNA. Section 3 deliberates the mechanism of one-time pads. Section 4 deals with the basics of symmetric encryption. Section 5 confers the proposed message encryption algorithm. The proposed method with experimental results is listed in Sect. 6. Subsequent Sect. 7 discusses the security and time analysis of proposed cipher. Finally, the conclusions are drawn in Sect. 8 and the future work is also discussed.

2 Biological Background

Primarily, the Deoxyribonucleic Acid (DNA) is defined as the genetic carrier that carries out the hereditary information among genes of all the living organisms from one generation to the subsequent ones. DNA is entailed of two stretched strands of nucleotides, each containing either of the four bases: Adenine, Guanine, Cytosine, and Thymine. These stretched strands of nucleotides form a double-helix structure that was identified first by Watson and Crick in 1953 [3] (Fig. 1).

The nitrogenous bases are attached to each other via a hydrogen bond, where the complementary rule allows Adenine to bond with Thymine using a double bond(=), and Cytosine to combine with Guanine through a triple bond(≡). The main duty of a DNA molecule is to store the genetic information for long durations and carry it for forwarding to the next generation of species [5]. The DNA molecules also have instructions which are obligatory to construct other cell components like RNA molecules and proteins.

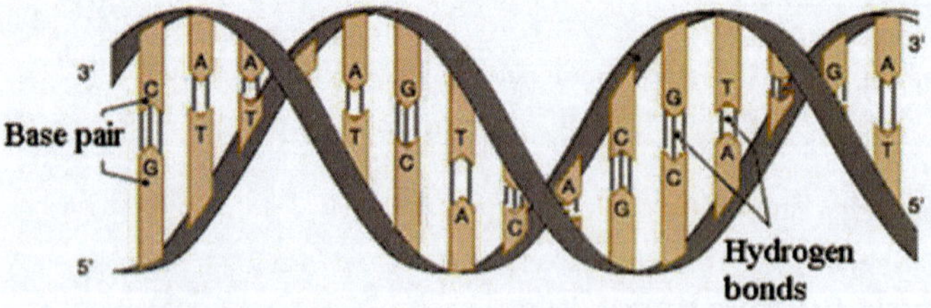

Fig. 1 Structure of DNA [4]

3 One-Time Pads (OTP)

The one-time pad is a historical concept which introduced the timeless breed of provably unbreakable ciphers. It was first described by Frank Miller in 1882 but actually patented by Gilbert Vernam in 1919 and is based on the idea of using a random key for encryption [6]. This key must possess four key properties: single usage only, length more than or equal to the length of plaintext, truly random and kept completely undisclosed. Such a key is then paired (XORed) with the plaintext and the corresponding cipher text is obtained. As the key is used one time only, it is referred to as a one-time pad. The word "pad" was added due to the paper pads, which were used for keys distribution. Once, the message is coded with the OTP, the OTP must be destroyed for safety reasons [7] (Fig. 2).

The OTP became popular in the two world wars where the Americans, British, and Germans used them extensively for the secure transmission of messages during wartime. Eventually, the final contributions were made by Information theorists Vladimir Kotelnikov and Claude Shannon. The former was a soviet citizen whose report remained secret (although submitted in 1941) and latter's work was published in 1949 in Bell Labs Technical Journal [9]. The most striking feature of OTP is that it is used in pairs, i.e., both communicating parties must share the OTP prior to communiqué either through a trusted channel or face-to-face. The additional copies of OTP also increase the likelihood of getting caught and revealing the secrecy of the whole system to spies or eavesdroppers. Conversely, the key distribution problem can be handled easily by quantum cryptography, which ensures no one can eavesdrop and acquire knowledge about the keys [10]. There are other quandaries too that

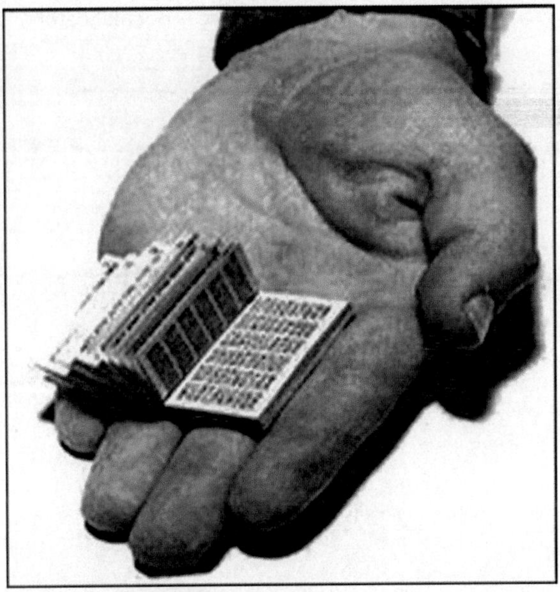

Fig. 2 A Russian One-Time Pad captured by MI5 [8]

are associated with OTPs, like generating truly random numbers. Although a variety of programming languages offer random number generation, but they are pseudo-random in particular. Hence, their usage is discouraged for cryptographic purposes, however the physical phenomenon like cosmic emissions are the only recognized ways to produce truly random numbers [11]. Though, the one-way hash functions like MD5 can offer much needed help in spawning randomness.

4 Symmetric Encryption

Symmetric or secret-key encryption is the conventional type of encoding which basically ruled the world of secrecy from the 1940s to 1970s. It is also known as private-key or single-key cipher due to the fact that it uses a solitary key for encryption of plaintext on one side and decryption into plaintext on the other side. Another key aspect of symmetric enciphering is that the decryption process is reverse and identical to the encryption process. It is based on the postulation that even if the eavesdropper gains some plaintexts and their corresponding ciphertexts, he/she is still unable to deduce the secret key or the algorithm behind it. Hence, a robust and strong enciphering algorithm is the foremost and primary prerequisite for symmetric encoding [12] (Fig. 3).

Another requirement is the secure transfer of key between the communicating parties, because if the eavesdropper detects the private key, the whole communication is compromised. As shown in the image above, the ciphertext can be sent over the insecure channel, but the secret key must be sent secretly through a secure channel. This eventually means that only the encryption key must be kept secret, but the secrecy of the encryption algorithm is not mandatory. It is a big reason why numerous

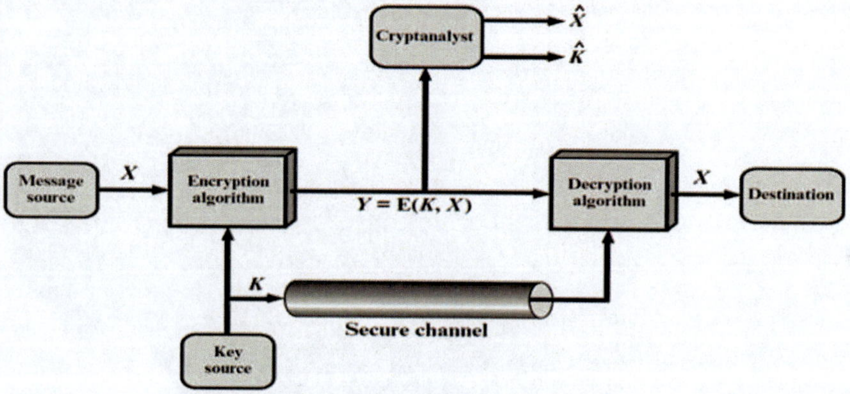

Fig. 3 Symmetric encryption model [13]

symmetric ciphers were implemented in hardware and their low-cost chips were manufactured. This also highlights one major feature of symmetric cipher, i.e., key distribution problem, which must be handled carefully for robust message passing between users.

Another quandary of private key cryptography is the key management. As a single key is used to converse between a pair of users, the upsurge in communicating parties for a user means an exponential increase in the number of keys and that has to be maintained carefully [14]. For example: if a user interconnects with 10 users, he needs to maintain 10 keys, i.e., one for each recipient and if all 10 users interact with each other, they need (10*9)/2 or 45 keys. Now, imagine the situation of interaction among 100,000 users. It is obligatory as the chances of using the wrong key for interaction are manifold.

5 Proposed DNA Encryption Algorithm

The proposed DNA enciphering algorithm here is based on one-time pad encoding primarily and is implied for digital images here. The encoding depends on the two keys: primary and secondary key, each of different length. The primary key is applied on the pixel level and its length depends upon the bit representation of a pixel, i.e., 24 bits (8 bits each for Red, Green, and Blue color). This key is deduced from the random DNA sequences selected from publicly accessible DNA databases like Genbank, FASTA, ASN.1, etc. [15]. Usually, codons (triplet of 3 nucleotide bases) form the primary key to match the binary representation of a pixel.

However, the secondary key is applied beforehand the primary key on a block level. The secondary key composes of 24-bits and is generated as a combination of English characters (A-Z, a-z), digits (0–9), and special characters like ~,! #, $, %, ^, &, @, etc. As the whole image is alienated into 9 blocks, there is a requirement of 9 randomly produced secondary keys. As the secondary keys are randomly formed; hence, they can also be called as secondary session keys (as shown in Fig. 8).

5.1 Message Encryption

The process of encrypting the message in the form of a digital image is described through the following diagram:

Following are the steps of the encryption procedure of the proposed algorithm:

i. Generation of OTP as the primary key from publicly available DNA sequences. It will act as the session primary key.
ii. Generation of the secondary session key from the secondary key as shown in Fig. 4.
iii. Select the digital image to be enciphered.

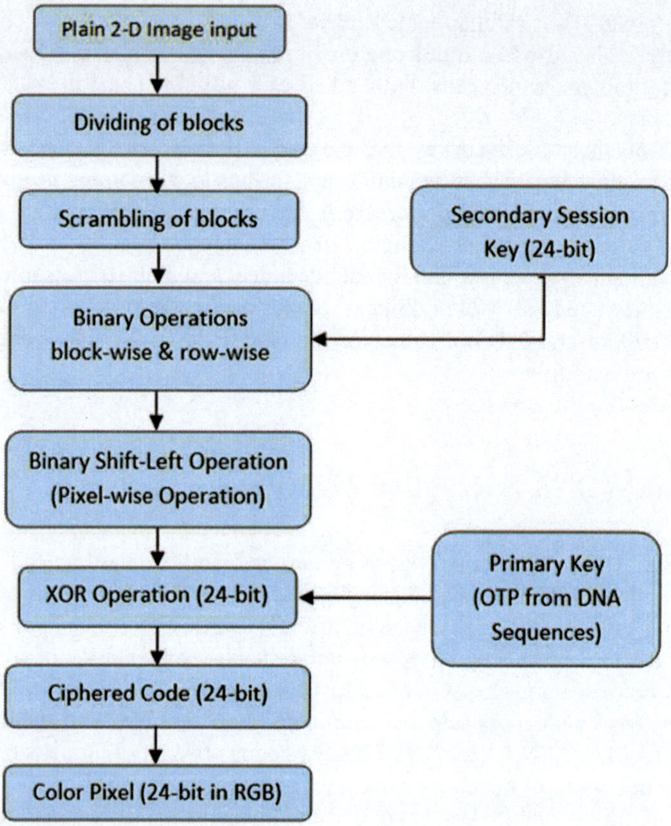

Fig. 4 Encryption procedure of VG2 cipher

iv. Divide the image into 9 blocks of equal size.
v. Shuffle the blocks randomly, so that the image is distorted.
vi. Now, for the first 3 blocks, select red component of 1st block, green color of 2nd block, blue hue of the 3rd block, and transform them into binary.
vii. Apply binary operations (AND, XNOR, and OR) on the output of step vi using the secondary session key.
viii. Repeat the steps vi and vii for blocks in 2nd and 3rd row also, but with different order of these binary operations.
ix. After the block-level encoding, RGB contents of a pixel is converted into binary and shift-left operation is applied 10 times.
x. Then, the primary key (OTP) produced from the DNA sequences are applied on each pixel for further renovation.
xi. Finally, the output of step x can be either converted back into an image.

5.2 Message Recovery

As the VG2 Cipher is based on symmetric encryption, the process of decryption is exactly reverse of the encryption procedure. Before starting the deciphering over the image, the primary, and secondary key must be received using a secure transmission. The DNA sequence used to generate the primary key could be conveyed over the secure telephone or in physical contact. Another alternative can be the usage of secure key transfer mechanisms like the Diffie-Hellman key exchange protocol. The decryption procedure can be explained through the following figure (Fig. 5):

The steps of decryption are explained as follow:

i. The ciphered image is received over the insecure channel.
ii. The primary key is identified through the same DNA sequences available publicly.
iii. The session key is also received through secure transmission and the secondary session key is calculated.
iv. The primary key is applied to the pixel level at the ciphered image.

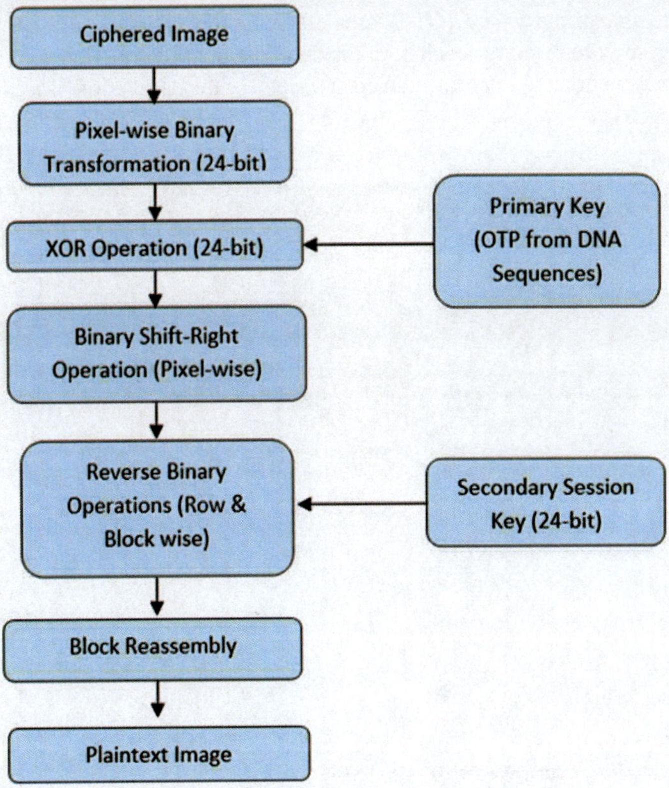

Fig. 5 Decryption process of VG2 Cipher

v. Afterwards, a binary shift-right operation is applied 10 times on pixel-level binary code.
vi. The secondary session key is applied to do the reverse binary operations on blocks row-wise.
vii. The blocks are descrambled and unified into a single image.
viii. The plaintext in the image form is obtained.

6 Experimental Results

The OTP-based encoding algorithm—VG2 cipher is implemented in the Python language using Jupyter Notebook on a machine with Intel Core i5—10th generation processor with 8 GB RAM Hewlett Packard built. Here, is an example of VG2 cipher implementation that shows the screenshots of plain-image and the produced ciphered image (Fig. 6).

The image is divided into 9 blocks and then scrambled to add distortion. The sample image produced after this step is (Fig. 7).

The image is then subjected to the randomly generated secondary key and consequently to the primary key (OTP) created from the DNA databases. One such secondary key produced for each block is demonstrated below (Fig. 8).

The final enciphered image produced at the end of the encryption process is shown below (Fig. 9).

Fig. 6 The input image for VG2 cipher [16]

Fig. 7 The intermediate image after block shuffling

```
"d'x", '"p0', 'cry', '"G"', ',-Q', '17|', '!H4', '*Gy', 'OAZ'
```

Fig. 8 Sample of secondary key produced randomly

7 Performance Evaluation

Cryptography is the binding technique that combines privacy and confidentiality to ensure message security between trusted users but with the untrusted communication channel. For any cryptographic algorithm, the topmost striking characteristics are security and speed [17]. The equilibrium between both these features must be maintained so that its performance supersedes the counterparts. The following sub-sections discuss both of the parameters one by one:

7.1 Security Analysis

The VG2 cipher proposed in the paper has laid the foundation of robustness both in the algorithm and the security keys. The encryption algorithm has perplex steps and encodes both at the grid level and pixel level of the image, which adds another feather of mystique.

Fig. 9 Final output image of VG2 Cipher

The message transmission is usually done through cables, radio channels, fiber-optics, etc. and if the encryption keys are also sent through these channels, there are copious chances of keys falling in the hands of attackers. However, the use of DNA cryptography brings in the layer of biological refuge. This extra-biological layer is enormous in size thus providing a colossal advantage over traditional techniques. Moreover, the primary key used in VG2 cipher is a One-Time Pad (OTP) which is shaped from the random DNA sequences available from open DNA databases like NCBI as shown in Fig. 10.

As the primary key is acting as OTP key (random and used one time only), it makes the attacker's hit and trial outbreak nearly impossible. Moreover, the ciphered image can be converted into DNA sequences also. The biological quandaries of DNA generated OTP and ciphertext in the form of DNA sequences will create enormous overhead for the cryptanalyst, where he needs to crack both algorithm and keys to deduce the plaintext.

Another point to consider here is that the secondary key is also fashioned randomly, thus producing a different secondary key for 9 blocks and they also change with every session. The security is also enhanced due to the fact that the proposed cipher is not fully dependent on either of the keys—primary or secondary. Thus, the practice of using multiple keys and that too random generated with every session increases the robustness and safekeeping of the proposed cipher.

```
CGAACTGGACAGCGCTTCAACGGAACGGATCTACGTTACAGCCTGCATAATGAAAACGGAG
TTGCCGACGACGAAAGCGACTTTGGGTTCTGTCTGTTGTCATTGGCGGAAAACTTCCGTTC
AGGAGGCGGACACTGATTGACACGGTTTAGCAGAAGGTTTGAGGAATAGGTTAAATTGAGT
GGTTTAATAACGGTATGTCTGGGATTAAAGTGTAGTATAGTGTGATTATCGGAGACGGTTT
TAAGACACGAGTTCCCAAAATCAAGCGGGGTCATTACAACGGTTATTCCTGGTAGTTTAGG
TGTACAATGTCCTGAAGAATATTTAAGAAAAAAGCACCCCTCATCGCCTAGAATTACCTAC
TACGGTCGACCATACCTTCGATTATCGCGGCCACTCTCGCATTAGTCGGCAGAGGTGGTTG
TGTTGCGATAGCCCAGTATAATATTCTAAGGCGTTACCCTGATGAATATCCAACGGAATTG
CTATAGGCCTTGAACGCTACACGGACGATACGAAATTATGTATGGACCGGGTCATCAAAAG
GTTATACCCTTGTAGTTAACATGTAGCCCGGCCCTATTAGTACAGTAGTGCCTTGAATGGC
ATTCTCTTTATTAAGTTTTCTCTACAGCTAAACGATCAAGTGCACTTCCACAGAGCGCGGT
GGAGATTCATTCACTCGGCAGCTCTGTAATAGGGACTAAAAAAGTGATGATAATCATGAGT
GCCGCGTTATGGTGGTGTCGGAACAGAGCGGTCTTACGGCCAGTCGTATGCCTTCTCGAGT
TCCGTCCAGTTAAGCGTGACAGTCCCAGTGTACCCACAAACCGTGATGGCTGTGCTTGGAG
TCAATCGCAAGTAGGATGGTCTCCAGACACCGGGGCACCAGTTTTCACGCCGAAAGCATAA
ACGACGAGCAGATATGAAAGTGTTAGAACTGGACGTGCCGTTTCTCTGCGAAGAACACCTC
GAGCTGTAGCGTTGTTGCGCTGCC
```

Fig. 10 Random DNA sequences generated as OTP Key

Table 1 Timing analysis of VG2 cipher

Input size (in pixels)	Encryption time (in seconds)	Decryption time (in seconds)	Total time for execution (in seconds)
480 * 480	2.72	2.6	5.32
720 * 720	5.15	4.95	10.1
960 * 960	8.6	8.5	17.1

7.2 *Timing Analysis*

The proposed cipher is designed to produce faster encoding and decoding so that the performance of VG2 cipher is not compromised. The timing for enciphering and deciphering of the recommended encoding system is given in Table 1:

8 Conclusion

This paper talks a novel encrypting algorithm—VG2 cipher which is useful for image cryptography. The proposed algorithm is based on symmetric encryption, DNA cryptography, and one-time pad key concept. The novel algorithm takes advantage of biological methods and their randomness is ensured due to its wide obtainability from DNA databases. The primary key is extracted from these DNA sequences and an additional secondary key is also used to add superfluous security layer that is also produced randomly. Furthermore, the encryption is done at two levels—grid level and pixel level; thus, making it more multifarious. Also, the same size of plain-image and ciphered image proves the least distortion in the images. The experiments reveal that the encoding and decoding timings of VG2 cipher are also quite low; thus, enhancing the performance. Security and performance analysis of the proposed cipher demonstrate ultimate offence against the attackers and still at an excellent pace. The future work includes hardware realization of encoding algorithm and extending it to audio and video files encryption.

References

1. Kaundal AK, Verma AK (2015) Extending Feistel structure to DNA cryptography J Discrete Math Sci Cryptogr 18(4):349–362
2. Anwar T, Paul S, Singh SK (2014) Message transmission based on DNA cryptography. Int J Bio-Sci Bio-Technol 6(5): 215–222
3. Anirban Ghosh Manju Bansal 2003 A glossary of DNA structures from A to Z Acta Crystallographica Section D: Biol Crystallography 59(4):620–626
4. Borda M, Tornea O (2010) DNA secret writing techniques. In: 2010 8th ınternational conference on communications. IEEE, pp 451–456
5. Grosse I, Herzel H, Buldyrev SV, Eugene Stanley H (2000) Species independence of mutual information in coding and noncoding DNA. Phys Rev E 61(5): 5624
6. Sherwood A (2013) A brief history of one time pads. https://learn.adafruit.com/raspberry-pi-thermal-printer-one-time-pads/a-crash-course-in-one-time-pads
7. Zheng G, Fang G, Shankaran R, Orgun MA (2015) Encryption for implantable medical devices using modified one-time pads. IEEE, Access 3:825–836
8. Ranum MJ (1995) One-Time-Pad (Vernam's Cipher) frequently asked questions. https://www.ranum.com/security/computer_security/papers/otp-faq/
9. Balanici D, Tomsa V, Borda M, Malutan R (2015) Full duplex OTP cryptosystem based on DNA Key for text transmissions. In: International conference for information technology and communications. Springer, Cham, pp 39–48
10. Upadhyay G, NeneMJ (2016) One time pad generation using quantum superposition states. In: 2016 IEEE International Conference on Recent Trends in Electronics, Information & Communication Technology (RTEICT). IEEE, pp 1882–1886
11. O Atsumi S Hayashida R Maruta 2006 Unified OTP cryptosystem with authentication and secrecy Front Artif Intell Appl 147 287
12. Singh G (2013) A study of encryption algorithms (RSA, DES, 3DES and AES) for information security. Int J Comput Appl 67(19)
13. Stallings W (2016) Cryptography and network security: principles and practice, 7th edn. Pearson Education India, February 2016
14. Ghani A, Mansoor K, Mehmood S, Chaudhry SA, Ur Rahman A, Saqib MN (2019) Security and key management in IoT-based wireless sensor networks: an authentication protocol using symmetric key. Int J Commun Syst 32(16):e4139
15. Database resources of the national center for biotechnology information. Nucleic Acids Res 46(D1):D8–D13
16. Methar A (2020) India Gate Image, May 30, 2020. https://www.pexels.com/photo/brown-concrete-india-gate-789750
17. Wu Y, Wu MK, Shu HJ (2018) Research on optimal combination of secondary hybrid encryption algorithm based on K-means clustering algorithm. In: International conference on ıntelligent ınformation hiding and multimedia signal processing. Springer, Cham, pp 134–141

Chapter 45
Hand Gesture Recognition Control for Computers Using Arduino

J. S. Vimali, Senduru Srinivasulu, J. Jabez, and S. Gowri

1 Introduction

Hands-free operation of electronic items is new to this world. Some types of sensors are surrounded by all of us. These can vary from different shape and sizes. These can be of multiple-use or made for any specific purpose, likewise, they have their own uses of data. Such example of sensors is light sensors, heart sensor, flow and level sensor, tilt sensor, color sensor, IR sensor, accelerometer, proximity sensor, smoke, gas and alcohol sensor, and an ultrasound sensor. These sensors collect different types of data and give it to their processor for their own uses.

Sensors are used with different types of devices for their own specific tasks assigned to them. One of the best examples of sensors is its own mobile phone. It is fitted with different types of sensors to control mobile phone features like screen brightness, photo depths, and many other features. These features make the environment more interactive.

The world mainly uses 13 types of sensors. These 13 types of sensors can be used in different ways or with a different configurations to get different types of work done. Each sensor can differ in their shape, size and cost. These sensors can be fitted with any real-time devices such as vehicles, crops, medical field, or for weather forecasting. Sensors can be else called as an input device. These input devices are used to give some input commands to the processer. There can be different types of processors to control these sensors. The project is aimed to design a system in which

J. S. Vimali (✉) · S. Srinivasulu · J. Jabez · S. Gowri
Department of Information Technology, Sathyabama Institute of Science and Technology, Chennai, India
e-mail: vimalijsmtech@gmail.com

S. Srinivasulu
e-mail: sendurusrinivas@gmail.com

I. Jeena Jacob et al. (eds.), *Data Intelligence and Cognitive Informatics*, Algorithms for Intelligent Systems,
https://doi.org/10.1007/978-981-15-8530-2_45

user hand gesture can be used as input and process it with the help of Arduino and perform the triggered actions over the operating system applications.

This project deals with a similar type of real-time usage of the ultrasound sensor. The sensor is used to calculate the change in hand gesture, which will be further used to calculate the change in distance. This data will be used to perform the triggered action over the operating system. There are much application which has inputs to perform some tasks such as scroll up, next, pause, rewind, and scrolls down, etc., functions.

2 Related Work

Godwin Ponraj et al. worked on sensor fusion of leap motion controller and flex sensors using kalman filter for human finger tracking. It contains a wearable glove which can be worn over hands to provide gestures. It needs to be charged on every time interval and requires High maintenance.

Patsakorn Rittitum et al. developed digital scrum board using leap motion. It is a very costly system. It consists of a leap motion sensor and a wearable device.

Leigh Ellen Potter and others also worked on the leap motion sensor. It was named as "The Leap Motion controller: A view on sign language." It is mainly designed to understand Australian signs. It is very costly and contains less accuracy.

Lucy Dodakian designed free-hand interaction with leap motion controller for stroke rehabilitation. The system ranges to very high cost. It consists of a leap motion sensor and a wearable device.

Meenakshi Panwar had designed hand gesture recognition based on shape parameters. It uses infrared sensors to detect gestures. The recognition was very slow and less accurate.

The previous existing systems are hand gesture control system [1, 2] using leap motion devices. These systems are very costly and use various other sensors. These systems are not affordable by common people. It was basically designed for Australian people's. Hand gesture recognition based on shape parameters detects sharp gestures only. It follows all the basic principles for gesture detection. It uses an infrared sensor to detect the movement and detection is not accurate.

The signal could be an image of physical behavior or enthusiastic expression. It incorporates body motion and hand signal. It falls into two categories: inactive motion [3, 4] and energetic motion [5, 6]. For the previous, the pose of the body or the signal of the hand signifies a sign. For the last mentioned, the development of the body or the hand passes on a few messages. Motion can be utilized as a device of communication between a computer and human [7, 8]. It is significantly diverse from the conventional equipment-based strategies and can finish human–computer interaction through signal acknowledgment. Motion acknowledgment decides the client expectation through the acknowledgment of the signal or movemessnt of the body or body parts. Within the past decades, numerous analysts have strived to move forward the hand signal acknowledgment innovation. Hand signal acknowledgment has extraordinary esteem in numerous applications such as sign dialect acknowledgment [9, 10], increased reality (virtual reality) [11, 12], sign dialect mediators for the crippled, and robot control.

3 Proposed Work

This article proposes a simple hand motion control based on Arduino. This system helps us to control various functions of a computer using hand signals. As a replacement of using a keyboard, mouse or joystick, our hand motion can be used to direct few computer tasks such as playing/pausing a clip, moving left/right in a picture slide show, scrolling up/down on a Web page, and more. It is cheaper than the existing system as it contains two Ultrasonic sensors and an ARDUINO board which are available at very low cost.

As shown in Fig. 1, the different hand gestures of user are going to be given as input to the sensors fixed in the particular computer. The ultrasonic sensor receives the inputs and converts those into different operations need to perform on the computer.

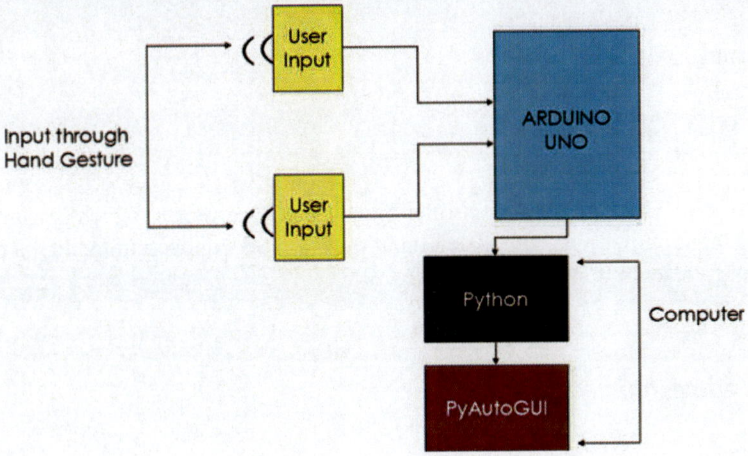

Fig. 1 Block diagram

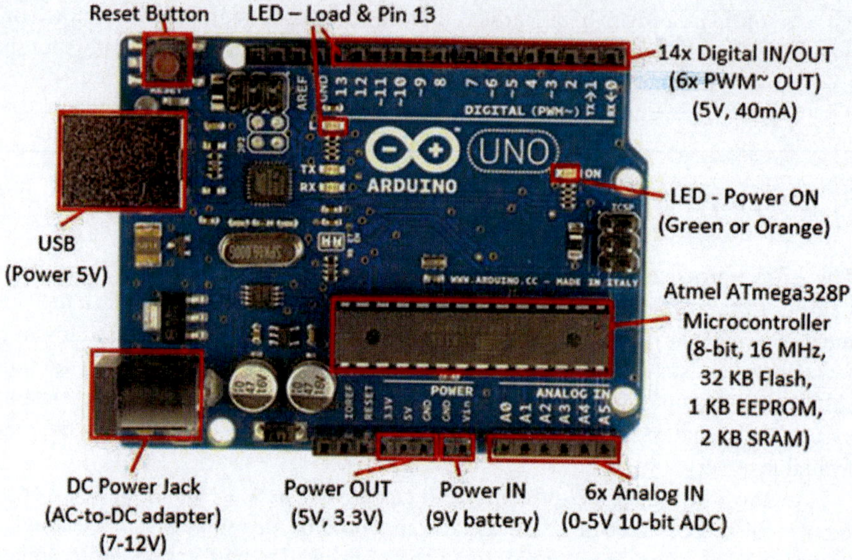

Fig. 2 Aurdino UNO board

4 Background

4.1 Arduino Uno

The Arduino Uno is a microcontroller based on AT mega 328, which is having 14 digital input/output pins, six analog input pins, a USB connection facility, a power jack and a reset button facility (Fig. 2).

4.2 Ultrasonic Sensor

Ultrasonic sensors measure the distance by using ultrasonic waves by which the sensor head emits an ultrasonic wave, i.e., the transmitter head, and the receiver absorbs the wave from the reflected target back. Ultrasonic sensors measure the distance between the emission and the receiving time of the target (Figs. 3 and 4).

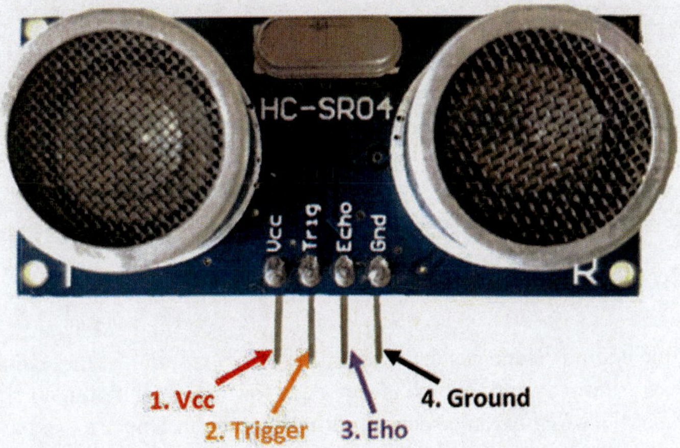

Fig. 3 Ultrasonic sensor

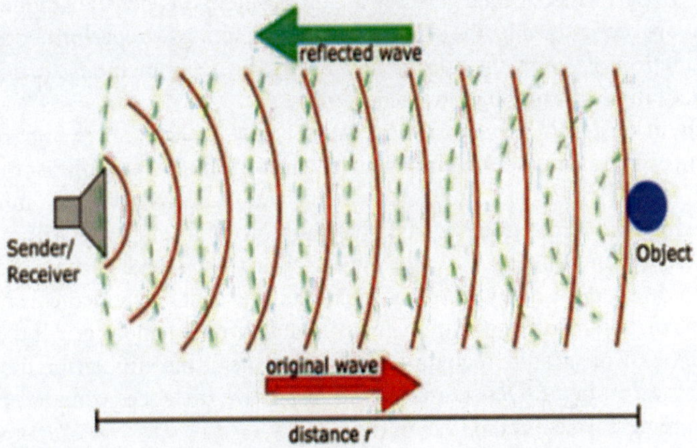

Fig. 4 Signal transmission in ultrasonic sensor

4.3 PyAutoGUI

PyAutoGUI is used to control the hardware like a mouse, keyboard, etc., using programs. GUI automation for human beings can be achieved through a cross-platform Python module provided by PyAutoGUI. It is with sensible defaults and is possibly designed to be very simple. It can simulate various functions of a keyboard as well as a mouse. Pressing hot keys combination is one among them.

4.4 PySerial

Access to the serial ports of microcontrollers can be encapsulated by pySerial. It works as a backend for Python. Python properties can be used to access serial port settings. All the files are purely made on Python.

5 Methodology

The principle behind "Hand gesture recognition using Arduino" is the Arduino-based hand gesture control system, which is very easy for implementation, understanding, and processing. Two ultrasonic sensors with the Arduino Uno are used as the main components to implement the project and to get the desired output.

In this paper, a technique is being introduced, which is based on determining distances by the ultrasonic sensors. According to the specific distances, particular functions are performed. Some gestures are proposed in this new technique and then the actions are recognized by the sensors and then functions are performed according to the instructions given by the user. Here, a few mainstream methods are used, based on hand action recognition by ultrasonic sensors.

The circuit design is very simple but the setup of the sensors is important. The positioning of Ultrasonic sensors is done in such a way that the two ultrasonic sensors are placed at either ends of a computer screen. The respective distance information is collected by the Arduino UNO, using Python programming and a library called PyAutoGUI which converts the data into click action. Hand gesture recognition system can be used for many different functionalities, such as video control, music player control, Microsoft application control, and chrome handling. All these interactions are real-time gesture recognition. By using the ultrasonic sensor, the distance of hand acts as an input. Depending on the respective distances of hand, particular functions are performed in the system (Fig. 5).

5.1 Input

This step is the first and most important step which includes the process of providing hand gestures to the receiver. The receiver here refers to the two Ultrasonic sensors attached on the top of the screen side by side. There are various gestures which can be provided for the automation. Some of them can be as follows:

- Gesture 1: By keeping both the hands at a distance away and in front of the sensor, the video can be played/paused.
- Gesture 2: Video can be forwarded one step by keeping the hand at a particular far distance from the right sensor.

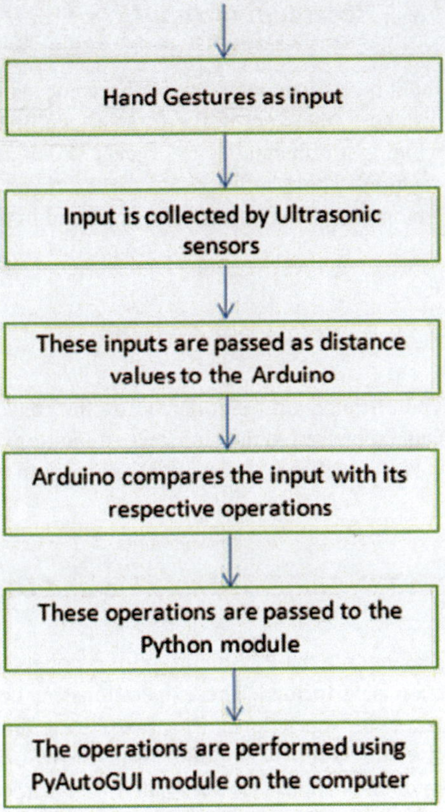

Fig. 5 Work flow process

- Gesture 3: Video can be rewind one step by keeping the hand at a particular far distance from the left sensor.
- Gesture 4: Video can be fast-forwarded by moving the hand from a particular near distance towards the right sensor and vice versa can cause rewind of the video.
- Gesture 5: Volume can be raised by moving the hand from a particular near distance toward the left sensor and vice versa can cause a reduction in the volume of the video.
- Gesture 6: Hand swipe in front of the right sensor can cause moving to the next tab.
- Gesture 7: Hand swipe in front of the left sensor can cause moving to the previous tab.
- Gesture 8: Hand swipe from the left sensor to right can change between tasks.

5.2 Reception of Input

Input is received by the two Ultrasonic sensors in the form of distance values. These distance values are in the form of centimeters. The sensors emit Ultrasonic waves which get reflected by our hands which are present in the front of sensors for the gestures. These reflected waves are received by the sensors and the distance from which the waves reflect back is passed to the Arduino.

5.3 Functions of Arduino

The distance values collected by the sensors are very important. These values are then compared to the respective operations. The operation with which the incoming values matches is collected by the python program as input through pySerial.

5.4 PyAutoGUI Function and Output

The input to the Python program is collected through the pySerial and specific operation is performed. These operations can be controlling the keyboard, mouse cursor, etc., with the help of PyAutoGUI module. According to the specified operation given as input by Arduino through pySerial, particular hotkeys combination of a keyboard can also be automated. The output can be obtained on a real-time basis on the computer screen.

These outputs can be as following:

- Tab switch in a Web browser
- Web page scrolling
- Task switching
- Changing the slides of the presentation
- Play/pause the video
- Change in volume
- Video forward
- Video rewind

6 Results

6.1 Controlling the Video

There are various possible operations which can be performed by the hand gestures using Arduino. Some of them are performed in the following result.

6.2 *Controlling Web Browser*

There are various operations which can be controlled by gestures. Some of them are performed and the result is displayed below.

7 Conclusion

Classification of the text provides the table with automation and simplification. It is amazing to see how marketers, product managers, designers, academics, and engineers will all be able to make use of this technology. Technology's entire idea is to make life easier. Classification of broad textual information helps standardize the system, simplifies and improves user experience. Text mining will be of increasing use in the future as a lot of everyday data in the form of text is generated. These type of gestures can be used to perform basic operations on computer applications. It is an open-source project where the gestures can be changed according to the user's comfort. Hand gesture control can be used in virtual reality applications. Augmented reality is the boom nowadays and hand gesture can play an important role. Reading sign language can be an application of hand gesture.

References

1. Rautaray SS, Agrawal A (2015) Vision based hand gesture recognition for human computer interaction: a survey. Artif Intell Rev 43(1):1–54
2. Jinda-Apiraksa A, Pongstiensak W, Kondo T (2010) A simple shape-based approach to hand gesture recognition. ECTI-CON2010: the 2010 ECTI International Confernce on Electrical Engineering/Electronics, Computer Telecommunications and Information Technology, IEEE
3. Jinda-Apiraksa A, Pongstiensak W, Kondo T (2010) Shape-based finger pattern recognition using compactness and radial distance. In: The 3rd International Conference on Embedded Systems and Intelligent Technology (ICESIT 2010), Chiang Mai, Thailand, February
4. Rokade R, Doye D, Kokare M (2009) Hand gesture recognition by thinning method. In: 2009 International Conference On Digital Image Processing. IEEE, pp 284–287, March
5. Gowri S, Divya G (2015) Automation of garden tools monitored using mobile application. In: International confernce on innovation information in computing technologies. IEEE, pp 1–6, February
6. Li G, Tang H, Sun Y, Kong J, Guozhang Jiang Du, Jiang BT, Shuang Xu, Liu H (2019) Hand gesture recognition based on convolution neural network. Cluster Comput 22(2):2719–2729
7. Li C, Xie C, Zhang B, Chen C, Han J (2018) Deep Fisher discriminant learning for mobile hand gesture recognition. Pattern Recogn 77:276–288
8. Vimali JS, Srinivasulu S, Gowri S (2019) IoT based bank security system. Int J Recent Technol Eng 2324–2327
9. Vinoth Krishnan V, Sai Anand M, Vimali JS (2016) Response mechanism for crisis times from social networks. Int J Pharm Technol (IJPT) 8(2):11958–11966
10. Balamurali R, Vimali JS (2015) Certificate and message authentication acceleration in VANET. IN: Proceedings 2015—IEEE International Conference on Innovation, Information in Computing Technologies, ICIICT 2015
11. Duraipandian M, Vinothkanna MR (2019) Cloud based Internet of Things for smart connected objects. J ISMAC 1(02):111–119
12. Raj JS, Ananthi JV (2019) Automation using Iot in greenhouse environment. J Inf Technol 1(01):38–47

Chapter 46
Infrared Small Target Detection Based on Fractional Directional Derivative and Phase Fourier Spectrum Transform

Sur Singh Rawat, Sashi Kant Verma, and Yatindra Kumar

1 Introduction

Methods for small target detection infrared imaging technique are seen in applications such as warning, infrared search and track system (IRST). The target object generally has less number of pixels in the infrared image. As there is low signal to noise ratio and due to this the target remains in the noisy background due to this effect the detection task become much tough [1–3]. Many algorithms on small target detection have been there which can be divided into two types: sequential detection methods and the single frame detection methods. The sequential detection methods such as dynamic programming algorithms [4, 5] make use of the spatial as well as the temporal information to estimate the candidate targets. As the velocity of the target as compared to the imaging sensor is very high so this degrades the performance of the sequential detection methods. Single-frame detection methods are seemed to be more fast and robust. The typical single-frame detection methods are two-dimensional least mean square (TDLMS), max-mean and max-median filter [6], morphological filters [7–9] also enhance the target but, these methods need the proper design of filter mask to meet the target features; otherwise, they may fail to distinguish the target from

S. S. Rawat (✉) · S. K. Verma
Department of Computer Science and Engineering, JSS Academy of Technical Education, Noida, UP, India
e-mail: sur.rawat@jssaten.ac.in

S. K. Verma
e-mail: skverma.gbpec@rediffmail.com

S. K. Verma
Department of Computer Science and Engineering, GBPIET, Pauri Garhwal, Uttrakhand, India

Y. Kumar
Department of Electrical Engineering, GBPIET, Pauri Garhwal, Uttrakhand, India
e-mail: kyatindra@gmail.com

I. Jeena Jacob et al. (eds.), *Data Intelligence and Cognitive Informatics*, Algorithms for Intelligent Systems,
https://doi.org/10.1007/978-981-15-8530-2_46

the clutter background. Gao et al. were the first who tried the patch image-based approach to infrared small target detection problem and proposed a method, infrared patch image-model (IPI) [1], and this method has exploited the property of non-local self-correlation for background. Patch image and also it assumes that the background patches generally come from a single or the mixture of low-rank subspace clusters. To recover the multi-subspace structure while preserving the local features, recently, robust subspace discovery by block Diagonal adaptive locality-constrained representation (rBDLR)-based method was proposed by Zhao zhang et al. [10]. The IPI model has the l1-norm sparsity problem as well as the constant weighting parameter issue which estimates the background inaccurately, and as a result, there is miss classification of target object. Dai et al.[11] proposed a method which was based on structural information of the background image. This method although has shown good performance over the other methods but an extra burden was created in calculation of weights of the columns. Dai et al. [12] proposed a non-negative IPI method based on partial sum of minimization of singular values(PSMSV) (NIPPS) so as to approximate the background and preserves the singular values. NIPPS has a problem of choosing a proper energy constraint ratio as well as the rank of the matrix. Gao et al. [13], reweighted IPI (ReWIPI) model was based on work discussed in[14] to constrain the background patch image and preserve the background edge information. Problem with this approach was the improper tuning of weighting parameter which could affect the computation of singular value decomposition (SVD) of matrix. To, further, improve the existing IPI-based methods, non-convex rank approximation minimization (NRAM) [15] was proposed. To bring some inner smoothness to the background, TV regularization term was applied in total variation regularization and principal component pursuit (TV-PCP) [16]. Human visual system (HVS) [17] was recently introduced to small target detection, which assumes target as the most salient object. Among the saliency-based methods, Laplacian of Gaussian (LOG)[18, 19], difference of Gaussian (DOG) [20], second-order directional derivative (SODD)[21], local contrast measure (LCM) [22], multi-scale patch-based contrast measure(MPCM) [23],improved difference of Gabor's (IDOGb) [24], local difference measure (LDM) [25], weighted local difference measure (WLDM) [26], spectral residual [27], phase spectrum [28].

Recently, filters based on fractional order has been used for image enhancement in many applications of medical image processing, image denoising, edge detection, and image singularity detection, etc. In order to enhance the edges and preserve the smooth area in the image, the Grumwald–Letnikov (G–L)-based fractional order filter is used which preserve the low-frequency and the high-frequency information of image.

In this paper, we proposed fractional directional derivative (FDD)-based method [29, 30] to improve the above existing filter-based infrared small target detection methods. The contribution of this paper is summarized as follows:

1. An infrared small target detection method based on fractional directional derivative is proposed.
2. Our method enhances the target as well as suppresses the background accurately.

The rest of the paper is framed as follows. In Sect. 2, the methodology of our method is described in detail. In Sect. 3, the experiment of the proposed method with the single infrared image, sequences of infrared images, and also the synthetic image is incorporated, and also its results are then compared with the other baseline methods. Finally, we have a discussion and conclusion which are presented in Sect. 4.

2 Methodology

2.1 Fractional Directional Derivative

In order to refer the fractional order calculus theory, we can divide them into three classical definition of fractional calculus: Riemann–Liouville (R–L) definition, Caputo definition, and the Grumwald–Letnikov (G–L) definition [30]. Among all the three definitions, we prefer G-L definition-based fractional order as the most suitable and easiest definition. α-order Grumwald–Letnikov (G–L) definition-based fractional differential of a function can be defined as

$$\begin{aligned} D^{\alpha}_{G-L}s(x) &= \frac{d^{\alpha}}{[d(x-a)^{\alpha}]}s(x)\Big|_{G-L} \\ &= \lim_{N\to\infty}\left\{\frac{\left(\frac{(x-a)}{N}\right)^{-\alpha}}{\Gamma(-\alpha)}\sum_{k=0}^{N-1}\frac{\Gamma(k-\alpha)}{\Gamma(k+1)}\times s\left(x-k\left(\frac{(x-a)}{N}\right)\right)\right\} \end{aligned} \tag{1}$$

where $[a, x]$ is the signal duration of $s(x)$, and α is any real number which include fraction also. D^{α}_{G-L} represents the Grumwald–Letnikov (G–L) fractional differential operator. The fractional differential of Eq. (1) is easy to calculate as it requires the discrete samples of $s\left(x-k\left(\frac{(x-a)}{N}\right)\right)$ only.

Without loss of generality, for $a = 0$, $[0, x]$ is the duration of signal $s(x)$ which can be divided into N equal shares. As the image processing is the direct processing of discrete pixels. The digital image processing of no causal pixels can be represented as

$$\begin{cases} s_0 \equiv s(x) \\ s_0 \equiv s\left(x+\frac{x}{N}\right) \\ s_{-k} \equiv s\left(x+\frac{kx}{N}\right) \\ s_{-N} \equiv s(2x) \end{cases} \tag{2}$$

For large value of N, we can rewrite Eq. (1) as below to create a fractional differential mask to process the image

$$\frac{d^{\alpha}}{[dx^{\alpha}]}S(x)\bigg|_{G-L} \cong \frac{\left(\frac{x}{N}\right)^{-\alpha}}{\Gamma(-\alpha)}\sum_{k=0}^{N-1}\frac{\Gamma(k-\alpha)}{\Gamma(k+1)}s\left(x+\frac{\alpha x}{2N}-\frac{kx}{N}\right) \tag{3}$$

Let the point $s\left(x+\frac{\alpha x}{2N}-\frac{kx}{N}\right)$ in Eq. (3) be the node and any node between the three neighboring nodes $s(x+(x\,/\,N)-(k\,x\,/\,N)), s(x-(k\,x\,/\,N)), s(x-(x/N)-(kx/N))$ can be expressed using the Lagrange three-point interpolation expression.

$$\begin{aligned} s(\varepsilon) \cong{} & (\varepsilon - x + (x/N))\left(\varepsilon - x + \frac{x}{N} + k\frac{x}{N}\right)s(x + (x/N) - (kx/N)) \\ & - \left(\varepsilon - x - \frac{x}{N} + k\frac{x}{N}\right)\left(\varepsilon - x + \frac{x}{N} + k\frac{x}{N}\right)s(x - (kx/N)) \\ & + \left(\varepsilon - x - \frac{x}{N} + k\frac{x}{N}\right)\left(\varepsilon - x + k\frac{x}{N}\right)Xs(x - (x/N) - (kx/N)) \end{aligned} \tag{4}$$

Assuming $\varepsilon = x + \left(\frac{\alpha x}{2N}\right) - \left(k\frac{x}{N}\right)$ we can solve for $s(x + (\alpha x/2N) - (kx/N))$ as

$$\begin{aligned} s\left(x + \frac{\alpha x}{2N} - k\frac{x}{N}\right) \cong{} & \left(\frac{\alpha}{4} + \frac{\alpha^2}{8}\right)s\left(x + \left(\frac{\alpha x}{N}\right) - \left(k\frac{x}{N}\right)\right) \\ & + \left(1 - \frac{\alpha^2}{4}\right)s\left(x - \left(k\frac{x}{N}\right)\right) \\ & + \left(\frac{\alpha^2}{4} - \frac{\alpha}{4}\right)s\left(x - \left(\frac{x}{N}\right) - \left(k\frac{x}{N}\right)\right) \\ ={} & \left(\frac{\alpha}{4} + \frac{\alpha^2}{8}\right)s_{k-1} + \left(1 - \frac{\alpha^2}{4}\right)s_k + \left(\frac{\alpha^2}{4} - \frac{\alpha}{4}\right)S_{k+1}. \end{aligned} \tag{5}$$

Equation (3) can be rewrite as below using Eq. (5) to create fractional differential mask.

$$\begin{aligned} \frac{\mathrm{d}^{\alpha}}{[\mathrm{d}x^{\alpha}]}s(x)\bigg|_{G-L} & \cong \frac{\left(\frac{x}{N}\right)^{-\alpha}}{\Gamma(-\alpha)}\sum_{k=0}^{N-1}\frac{\Gamma(k-\alpha)}{\Gamma(k+1)} \\ X\left[s_k + \frac{\alpha}{4}(s_{k-1} - s_{k+1}) + \frac{\alpha^2}{8}(s_{k-1} - 2s_k + s_{k+1})\right]. \end{aligned} \tag{6}$$

Taking the top $n+2$ terms from Eq. (6) to approximate the fractional derivative. The fractional mask in eight direction can be calculated, and their coefficients are given below.

$$\begin{cases} a_{s-1} = \frac{\alpha}{4} + \frac{\alpha^2}{8} \\ a_{s0} = 1 - \frac{\alpha^2}{2} - \frac{\alpha^3}{8} \\ a_{s1} = -5\frac{\alpha}{4} + 5\frac{\alpha^3}{8} + \frac{\alpha^4}{16} \\ a_{sk} = \frac{1}{\Gamma(-\alpha)} \left[\begin{array}{c} \frac{\Gamma(k-\alpha+1)}{(k+1)!} \cdot \left(\frac{\alpha}{4} + \frac{\alpha^2}{8}\right) + \frac{\Gamma(k-\alpha)}{(k)!} \cdot \left(1 - \frac{\alpha^2}{4}\right) \\ + \left(\frac{\Gamma(k-\alpha-1)}{(k-1)!}\right) \cdot \left(-\frac{\alpha}{4} + \frac{\alpha^2}{8}\right) \end{array} \right. \end{cases} \tag{7}$$

From Eq. (7), coefficient of a mask can be calculated, for example, if $\alpha = 0.6$ and $n = 5$, then the values of coefficients are $a_{s-1} = 0.195$, $a_{s0} = 0.793$, $a_{s1} = -0.5169$, $a_{s2} = -0.0271$, $a_{s3} = -0.0449$. The sum of the coefficients of the mask is not zero which is different from the integral mask. Processing of 2D image signal $s(x, y)$ needs the operation on the discrete pixels so we are going to process the image by convoluting with the above constructed mask in all the eight directions such as positive and negative x and y co-ordinates, left downward, left upward direction, and right downward and right upward direction. These masks of size $(2m + 1) \times (2m + 1)$ for $k \to n = 2m + 1$ in the eight directions are denoted by P_x^+, P_y^+, P_x^-, P_y^-, P_{LDD}, P_{LUD}, P_{RDD} and P_{RUD}. The resultant image signals after convolution are given below.

$$S_x^\alpha - (x, y) = \sum_{i=-2m}^{1} \sum_{j=-m}^{m} P_x^-(i, j) s(x + i, y + j) \tag{8}$$

$$S_x^\alpha + (x, y) = \sum_{i=-1}^{2m} \sum_{j=-m}^{m} P_x^+(i, j) s(x + i, y + j) \tag{9}$$

$$S_{y^-}^\alpha (x, y) = \sum_{i=-m}^{m} \sum_{j=-2m}^{1} P_y^-(i, j) s(x + i, y + j) \tag{10}$$

$$S_{y^+}^\alpha (x, y) = \sum_{i=-m}^{m} \sum_{j=-1}^{2m} P_x^+(i, j) s(x + i, y + j) \tag{11}$$

$$S_{P_{\text{LDD}}}^\alpha (x, y) = \sum_{i=-m}^{m} \sum_{j=-m}^{1} P_{\text{LDD}}(i, j) s(x + i, y + j) \tag{12}$$

$$S_{P_{\text{LUD}}}^\alpha (x, y) = \sum_{i=-2m}^{1} \sum_{j=-2m}^{1} P_{\text{LUD}}(i, j) s(x + i, y + j) \tag{13}$$

$$S_{P_{\text{RDD}}}^\alpha (x, y) = \sum_{i=-1}^{2m} \sum_{j=-1}^{2m} P_{\text{RDD}}(i, j) s(x + i, y + j) \tag{14}$$

$$S^{\alpha}_{P_{\mathrm{RUD}}}(x, y) = \sum_{i=-2m}^{1} \sum_{j=-1}^{2m} P_{\mathrm{RUD}}(i, j)s(x + i, y + j) \quad (15)$$

2.2 Phase Fourier Transform

The phase Fourier transform (PFT) [28] of an image signal gives the phase information in an image and tells the location of an object present in an image. Phase spectrum information in an infrared image will detect the target as it the most salient information. Following steps are needed to calculate the phase Fourier transform of an image.

Let (x, y) is an image

$$f(x, y) = F(s(x, y)) \quad (16)$$

$$p(x, y) = P(f(x, y)) \quad (17)$$

$$SM(x, y) = g(x, y) * \left\| F^{-1}\left[e^{i \cdot p(x,y)}\right] \right\|^2 \quad (18)$$

where F and $^{-1}$ are the Fourier, and its inverse Fourier transform $P(f(.))$ is the phase spectrum of an image and $g(x, y)$ is the 2-D Gaussian filter at ($\sigma = 8$). (x, y) is the saliency map of an image.

2.3 Approach to Infrared Target Detection Based on Fractional Directional Derivative Mask

The proposed method is shown in Fig. 1. When an infrared image is convoluted with the fractional derivative mask, then the sum of fractional derivate coefficient in all the eight direction is not zero which means that the low-frequency information is preserve, and the high-frequency information is improved unlike the case of the integral mask where the sum of mask coefficient is zero. Secondly, the PFT is used to locate the salient object in the image. The following steps are adopted for the above mentioned approach

1. Infrared image is convoluted with the fractional derivative mask in all the eight directions.
2. The output of the step 1 is taken as input for the PFT which computes the phase information and gives the saliency map which helps to locate the salient object in the image.

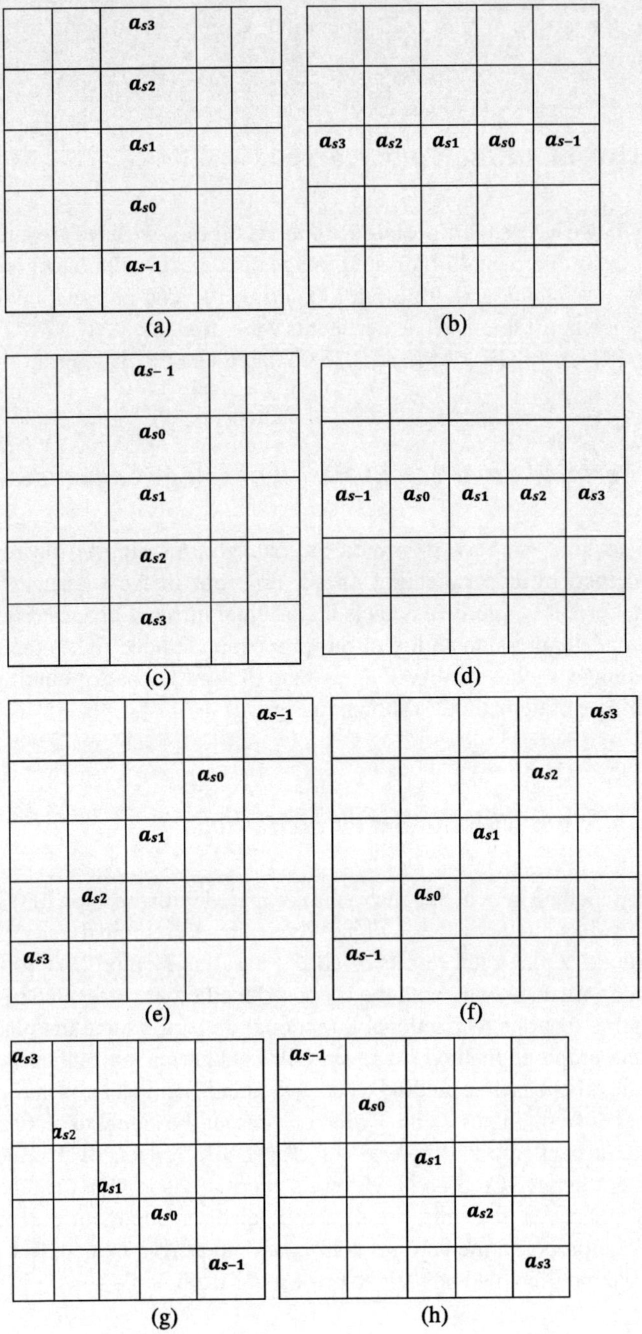

Fig. 1 Fractional differential mask on eight directions

3. Finally, the adaptive threshold operation is applied to get the target more accurately.

3 Experimental Result and Analysis

In this section, we have experimental section, and for this, we have prepared a dataset which includes more than 450 infrared images under different background scenes such as, sea, sky, cloud, and ground for experiments. The information about these images are shown in Table 1. All experiments were implemented on MATLAB 2015 software on PC with 4 GB RAM and 2.20 GHz Intel Core 2 Duo processor (Fig. 2).

3.1 Background Suppression Result of Image Sequences

In this sub-section, we have presented the background suppression result of the proposed method by experimenting on six different image sequences under the complex background. Figure 3a depicts the original infrared images; Fig. 3b shows the background suppression ability of our proposed method. To see the location of target in an image, we have utilize 3D gray map to show the background suppression result of our proposed method as shown in Fig. 3c.

3.2 Comparison with the Baseline Methods

The proposed method here in this section is compared with the three (03) state-of-art methods to evaluate its robustness. Figure 4a depicts the original image sequences. The background suppression results of all the baseline methods like top hat, max-mean, and max-median along with the proposed method are presented in Fig. 4b–d. We can observe that due to fractional directional derivatives and the phase Fourier transform, the proposed method suppresses the background and enhances the target better than the other baseline method where a simple filtering approach is performed. We can observe from Tables 2 and 3 which represent the signal to clutter ratio gain (SCRG) and the background suppression factor (BSF) respectably that the proposed method has got better SCRG and BSF result in comparison with the other baseline methods. Similarly, as the time complexity is also an important parameter in an algorithm. The proposed method has computational efficiency of **0.051 s** which is better than the baseline methods as can be observe from Table 4.

Table 1 Detail information of representative images

Infrared real sequences Nos	Target type	No. of Frames	Image Size	Background characteristics	Target information
Seq. 1	A small ship	30	256 × 200	Sea–sky backgrounds is blurred	• Target size changing • Target in small size • Large imaging distance
Seq. 2	An airplane	250	256 × 200	Clouds with low local contrast and heavy background	• Target size changing • Target small in size • Large imaging distance
Seq. 3	An airplane	250	256 × 200	With changing background	• Target size changing • Large imaging distance • With low SCR
Seq. 4	A helicopter	100	241 × 200	Background changing with time	• Small target size • Large imaging distance • With low SCR
Seq. 5	A ship	100	320 × 240	Changing background	• A one to two target • Small in size
Seq. 6	A person, moving back and forth in forest	250	280 × 228	Background with heavy clouds	• Small in size • Long imaging distance

4 Conclusion

An algorithm based on the fractional directional derivative filter and the PFT is presented in this paper. We first used a fractional directional derivative filter to enhance the infrared image containing the target; then, PFT is applied to obtain a saliency map of the candidate target, and then, adaptive threshold operation is applied to get the final target. The experimental result suggests that the proposed algorithm can suppress the background noise effectively and also detect the target accurately with the low false rate.

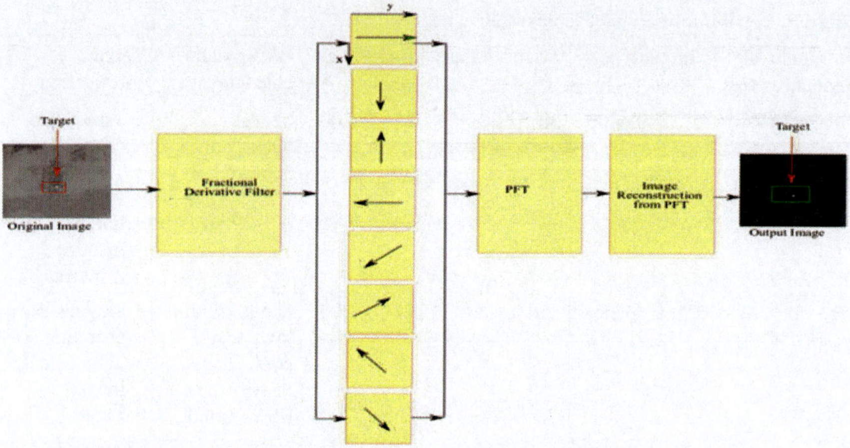

Fig. 2 Proposed method for infrared small target detection based on FDD and PFT

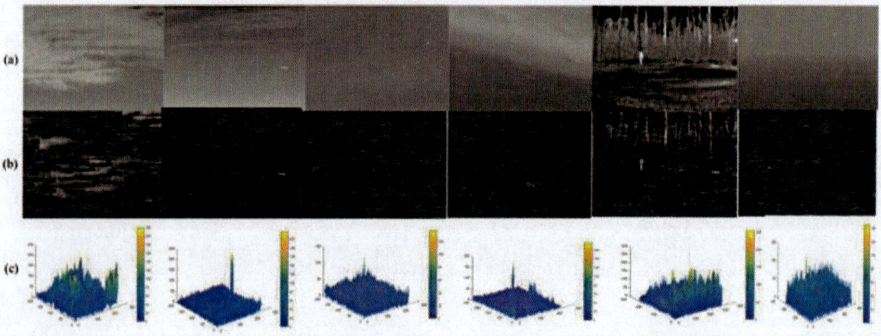

Fig. 3 3D map of original infrared images and the result of the proposed method. **a** Original representative images, **b** Detection result of Fig. 3(**a**), **c** 3D map of proposed method respectably

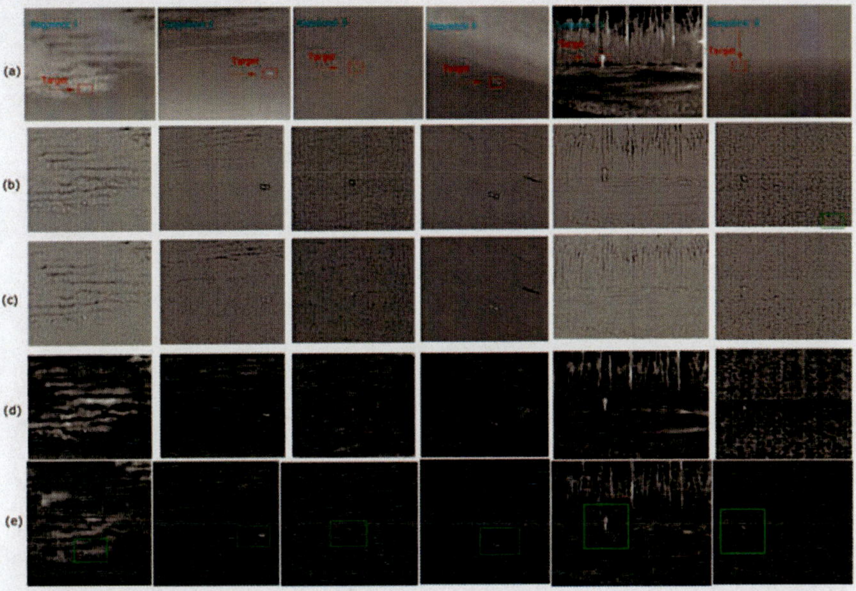

Fig. 4 Result of various methods, **a** original image, **b** Max-mean, **c** Max-median **d** Top hat, and **e** result of proposed method

Table 2 SCRG values of the proposed method

Methods	Image sequences					
	A	B	C	D	E	F
Max-mean	1.16	5.73	7.38	12.74	1.32	12.13
Max-median	1.06	1.71	5.41	1.59	0.26	6.75
Top hat	0.99	1.58	6.46	13.06	0.09	6.92
Proposed method	2.43	10.21	13.89	33.26	1.16	25.5

Table 3 BSF values of the proposed method

Methods	Image sequences					
	A	B	C	D	E	F
Max-mean	0.49	2.34	0.51	3.93	1.34	0.92
Max-median	1.29	3.9	0.75	7.25	2.12	1.2
Top hat	1.38	3.39	0.86	10.11	2.43	1.26
Proposed method	5.04	3.96	0.99	14.04	3.82	1.32

Table 4 Computation time of the proposed method and baseline method

Methods	Times (s)
Max-mean	0.841
Max-median	0.914
Top hat	0.715
Proposed method	**0.051**

Acknowledgements The authors would like to thank all the valuable reviewers.

References

1. Gao C, Meng D, Yang Y, Wang Y, Zhou X, Hauptmann AG (2013) Infrared patch-image model for small target detection in a single image. IEEE Trans Image Process 22(12):4996–5009
2. Shao X, Fan H, Lu G, Xu J (2012) An improved infrared dim and small target detection algorithm based on the contrast mechanism of human visual system. Infrared Phys Technol 55(5):403–408
3. Chen CP, Li H, Wei Y, Xia T, Tang YY (2014) A local contrast method for small infrared target detection. IEEE Trans Geosci Remote Sens 52(1):574–581
4. Barniv Y (1985) Dynamic programming solution for detecting dim moving targets. IEEE Trans Aerospace Electronic Syst 1:144–156
5. Barniv Y, Kella O (1987) Dynamic programming solution for detecting dim moving targets part II: analysis, vol 6, pp 776–788
6. Deshpande SD, Meng HE, Venkateswarlu R, Chan P (1999) Maxmean and maxmedian filters for detection of small targets. In: SPIE's International Symposium on Optical Science, Engineering, and Instrumentation, pp 74–83
7. Bai X, Zhou F (2010) Analysis of new top-hat transformation and the application for infrared dim small target detection. Pattern Recogn 43(6):2145–2156
8. Fortin R, Rivest J (1996) Detection of dim targets in digital infrared imagery by morphological image processing. Optical Eng 35(7):1886–1893
9. Bai X (2015) Morphological infrared image enhancement based on multi-scale sequential toggle operator using opening and closing as primitives. Infrared Phys Technol 68:143–151
10. Zhang Z, Ren J, Li S, Hong R, Zha Z, Wang M (2019) Robust subspace discovery by blockdiagonal adaptive locality-constrained representation. In: Proceedings of the 27th ACM international conference on multimedia. ACM, France, pp 1569–1577
11. Dai Y, Wu Y, Song Y (2016) Infrared small target and background separation via column-wise weighted robust principal component analysis. Infrared Phys Technol 77:421–430
12. Dai Y, Wu Y, Song Y, Gao J (2017) Non-negative infrared patch-image model: robust target background separation via partial sum minimization of singular values. Infrared Phys Technol 81:182–194
13. Guo J, Wu Y, Dai Y (2017) Small target detection based on reweighted infrared patch-image model. IET Image Process 12(1):70–79
14. Gu S, Xie Q, Meng D, Zuo W, Feng X, Zhang L (2017) Weighted nuclear norm minimization and its applications to low level vision. Int J Comput Vision 121(2):183–208
15. Zhang L, Peng L, Zhang T, Cao S, Peng Z (2018) Infrared small target detection via non-convex rank approximation minimization joint l2, 1 norm. Remote Sensing 10(11):1–34
16. Wang X, Zhenming P, Dehui K, Zhang P, He Y (2017) Infrared dim target detection based on total variation regularization and principal component pursuit. Image Vision Comput 63:1–9

17. Itti L, Koch C, Niebur E (1998) A model of saliency-based visual attention for rapid scene analysis. IEEE Trans Pattern Anal Mach Intell 20(11):1254–1259
18. Kim S (2011) Min-local-LoG filter for detecting small targets in cluttered background. Electronics Lett 47(2):105–106
19. Kim S, Yang Y, Lee J, Park Y (2009) Small target detection utilizing robust methods of the human visual system for IRST 30(9):994–1011
20. Wang X, Lv G, Xu L (2012) Infrared dim target detection based on visual attention. Infrared Phys Technol 55(6):513–521
21. Qi S, Ma J, Tao C, Yang C, Tian J (2013) A robust directional saliency-based method for infrared small-target detection under various complex backgrounds. IEEE Geosci Remote Sensing Lett 10(3):495–499
22. Chen CP, Li H, Wei Y, Xia T, Tang YY (2014) A local contrast method for small infrared target detection. IEEE Trans Geosci Remote Sensing 52(1):574–581
23. Wei Y, You X, Li H (2016) Multiscale patch-based contrast measure for small infrared target detection. Pattern Recogn 58:216–226
24. Han J, Ma Y, Huang J, Mei X, Ma J (2016) An infrared small target detecting algorithm based on human visual system. IEEE Geosci Remote Sensing Lett 13(3):452–456
25. Deng H, Sun X, Liu M, Ye C, Zhou X (2017) Entropy-based window selection for detecting dim and small infrared targets. Pattern Recogn 61:66–77
26. Deng H, Sun X, Liu M, Ye C, Zhou X (2016) Small infrared target detection based on weighted local difference measure. IEEE Trans Geosci Remote Sensing 54(7):4204–4214
27. Hou X, Zhang L (2007) Saliency detection: a spectral residual approach. In: IEEE conference on computer vision and pattern recognition. IEEE, pp 1–8
28. Guo C, Ma Q, Zhang L (2008) Spatio-temporal saliency detection using phase spectrum of quaternion fourier transform. In: 2008 IEEE conference on computer vision and pattern recognition. IEEE, pp 1–8
29. GuanJ, ou J, Lai Z, Lai Y (2018) Medical image enhancement method based on the fractional order derivative and the directional derivative, Int J Pattern Recognition Artif Intell 32(3), pp 1857001–18570022
30. Pu Y-F, Zhou J-L, Yuan X, Fractional differential mask: a fractional differential-based approach for multiscale texture enhancement. IEEE Trans Image Process 19(2):491–511

Chapter 47
Detection of Pre-term Delivery by the Analysis of Fetal ECG Signals

P. Sridharan and V. Dhileep

1 Introduction

Internationally, around 140 million births happen each year. Most of these are vaginal births among pregnant women with no recognized risk factors for complications, either for themselves or their infants, at the onset of labor. Around half of all stillbirths and one-fourth of neonatal deaths result from intricacies during labor and childbirth. When the fetal heart rate turns out to be under 99 or more than 180 beats/min, it shows that the fetus has an irregular heart rhythm, implying the baby is under states of hypoxia or ischemia or is influenced by other variables, which can prompt birth defects. In this way, improving the nature of care around the hour of birth has been recognized as the most effective methodology for diminishing stillbirths, maternal, and infant deaths compared with antenatal or postpartum care strategies. Monitoring biomedical signals, through estimation, measurement, assessment, and classification of signal properties is one of the essential tools for examining the progression of diseased states.

For ECG based understanding of fetal distress, various techniques have been implemented dependent on time and/or frequency domains. Analysis of ECG signal to shed light on information regarding the condition of heart that is contained in the signal, using digital signal processing techniques like FFT, STFT, WT, etc., [1]. The frequency spectrum of ECG reflects heart rate potentially retaining more information since the spectrum of ECG signal is affected by waveform components, their time positions within the cardiac cycle as well as by regularity of heart period [2]. An alter implementation of signal processing is where autoregressive modeling

P. Sridharan (✉) · V. Dhileep
Department of Biomedical Engineering, Rajalakshmi Engineering College, Chennai, India
e-mail: sridharan89@hotmail.com

V. Dhileep
e-mail: dhileepvijayan@gmail.com

I. Jeena Jacob et al. (eds.), *Data Intelligence and Cognitive Informatics*, Algorithms for Intelligent Systems,
https://doi.org/10.1007/978-981-15-8530-2_47

(AR) is applied to a modeled ECG signal, and the AR coefficients have been used for classification into arrhythmias such as Atrial Premature Contraction, Premature Ventricular Contraction, Ventricular Tachycardia, and Ventricular Fibrillation [3]. A methodological approach for the cardiotocography (CTG) monitoring, based on a multi-parametric fHR analysis is carried out in [4]. fHR characteristics include fHR mean and variance, parameters based on the estimation of the power spectral density (PSD). This is a step toward the early detection of fetal risk in pregnancy. Signal analysis techniques such as the periodogram power spectrum and spectrogram time-frequency analysis are used to characterize ECG variations, such as normal and different types of heart blocks [5]. Applying PCA to an ensemble of heartbeats and ECG signal, the associated pattern of principal components reflects the degree of morphologic beat-to-beat variability, ST-T changes, etc., which can be used to detect cardiac abnormalities [6]. The abdominal signals (ADS) recorded from the maternal abdomen can be used to obtain fHR on a beat-to-beat basis, allowing the fECG morphology analysis, and also is said to be suitable for long-term screening of the fetal health condition, done using ICA and ESC algorithms [7]. A method to distinguish between ECG signals from the normal sinus rhythm (NSR) database and those from the Arrhythmia database is done in [8], using their power spectral density (PSD) through fast-Fourier transform.

Assessment of five different human emotions (happiness, disgust, fear, sadness, and neutral) using heart rate variability (HRV) signal parameters derived from an electrocardiogram (ECG), by employing discrete wavelet analysis (DWT) is shown in [9]. The k-nearest neighbor (KNN) and linear discriminant analysis (LDA) were used to map the statistical features into corresponding emotions. Defining the electrophysiological differences between IUGR and normal fetuses utilizing NIEA techniques, using parameters taken from the ECG signal—P wave duration, PR interval QRS duration QT interval T/QRS ratio was explored in [10]. DWT is also used in the frame classification of ECG signals that can be employed in signal compression and recognition of arrhythmias. This is achieved by forming a template set which is used in the classification of ECG signals [11]. There are also instances of automatic diagnosis of the classification of Anteroseptal Myocardial Infarction using neural networks with complete trees, based on the features extracted from ECG signals during the training phases, which has an accuracy of 92.5% [12]. Morphological changes in the shape of the ECG signal can reflect the heart's condition. These changes can be considered as the features of the ECG signal, which can be used in its classification. Pattern recognition and feature extraction algorithms are utilized in [13] to achieve this objective.

As of late, the improvement of dynamical framework examination has prompted the presentation of a lot of signal processing techniques focused on the extraction of parameters from acquired biomedical signals. The essential prerequisite for empowering automated investigation is the exact and robust recording of the fHR. The investigation of a fetal HR (fHR) signal speaks to a noninvasive principal tool for checking fetal conditions in the antepartum period. In actuality, a significant segment of ominous fetal results appears to be because of events happening before the labor onset. This paper is to try using digital signal processing (DSP) techniques in the examination of fetal ECG signals. DSP Techniques such as extraction of statistical properties, periodogram, power spectrum, and auto regression modeling, principal component analysis are performed as a combination to differentiate between normal fECG signal and abnormal fECG. Using these algorithms, the signals are classified as:

- Normal fECG—indicating normal delivery and no pre-term complication.
- Abnormal fECG—indicating a pre-term complication and the possibility of early delivery.

2 Methodology

The research work begins with ECG signal data acquisition, followed by pre-processing, parameter and feature extraction and analysis, and finally classification and scoring.

Figure 1 describes the methodology adopted in the analysis of fECG signals. Screening of fECG signals represents the choice of normal signals—Synthetic fECG C1 and C3, and the choice of abnormal signals—ARR and NR Signals obtained; Pre-processing to remove Power-line interference and to perform baseline correction; Parameters that are computed for the collected fECG signals include Statistical Parameters, Frequency Domain plots, AR Modeling, and Principal Component Analysis.

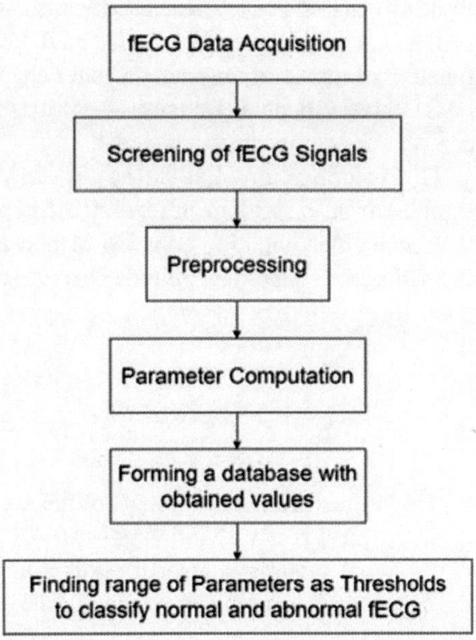

Fig. 1 Fetal ECG processing and analysis flow chart of proposed methodology

2.1 *Data Chosen for Analysis/Screening*

20 signals from the FECGSYNDB were chosen—10 from Case 1 (Fetal movement + noise) and 10 from Case 3 (Uterine contraction + noise) categories. These were considered to be the normal fetal ECG signals. For these signals, a 10-second portion is chosen randomly, as it being more consistent and being the least deviant from the norms. 26 signals from the NIFEA database were chosen from the Arrhythmia/Normal section. These were considered to be abnormal fECG signals. For these signals, a 10-second segment is chosen based on annotations indicating maximum and/or observable deviations from the norm. These 10-second segments of fECG are then used in analysis preceded by pre-processing. The chosen fECG signal segments were non-stationary, containing moderate linear drifts or increasingly complex trends. It has been found that the majority of information in fECG is contained within the frequency band of 7–75 Hz. Thus, to eliminate the drift and noise, a Chebyshev low pass filter of order 1 was designed.

2.2 *Methods Employed in Analysis of Pre-processed FECG Signals*

The pre-processed signal segments were analyzed for various parameters associated with fECG signal:

A. Statistical Parameters—namely Mean, Median, Standard Deviation, RMS, Peak-to-Peak, R Peak. These parameters can help to describe the signal's properties and the nature of the said random signal. Fetal ECG is a random signal and also is a function of time and thus can be better characterized by its statistical properties.

B. Power Spectral Density—describes the distribution of power into frequency components composing that signal.

C. Periodogram—Let $x_w(n) = w(n)x(n)$ denote a windowed segment of samples from a random process x(n), where the window function *w* (classically the rectangular window) contains M non-zero samples. Then, the periodogram is defined as the squared magnitude Discrete-time Fourier Transform of x_w (n) divided by *M*

$$P_{x,M}(w) = \frac{1}{M}\left|\sum_{n=0}^{M-1} x_w(n)e^{-jwn}\right|^2 \tag{1}$$

This value reflects an estimate of the spectral density of the fECG signal.

D. Autoregressive Modeling—An AR model of order four is designed for the fECG signal, which is adapted for extracting features from ECG signals, thus enabling the discrimination of the ECG signals. The general AR model is given by the equation

$$y_t = \sum_{i=1}^{p} \beta_i y_{t-i} + \epsilon_t \tag{2}$$

Where β denotes the AR coefficient, p is the order and ε_t denotes the error term.

Algorithms used to find the AR coefficients of fECG signals are Yule-Walker Algorithm, Burg's Method and Covariance Method.

E. Principal Component Analysis—When the signal is recurrent, like the ECG signal, the analysis is based on samples extracted from the same segment location of different periods of the signal. The parameters that show significant variation in values for each case, and that can be easily incorporated for computation are Latent and Mu. The principal component variances are the latent values and the estimated mean of each variable is Mu. Thus, for this paper, those two parameters in PCA are taken into consideration for analysis and classification.

2.3 Compilation of Obtained Values

Following the multi-parametric analysis, the values obtained for the aforementioned properties are investigated to highlight the difference in their trends and thus used to distinguish the normal and abnormal fECG signals, which is explained in detail in Sect. 3.

3 Results and Discussion

The extracted features and parameters from the various fECG signals are used to discriminate between normal and abnormal fECG. The objective is to show how each method differentiates the ECG waveforms and identifies the predicaments that might lead to pre-term pregnancy.

3.1 Illustration

The figures shown below are representative of the processes done to each of the four classes of chosen fECG signal.

Parameter Computation and Methods of Analysis employed in the investigation of fECG signals

A. ARR Signal—ARR08

See Fig. 2; Tables 1 and 2

The obtained value for the periodogram estimate is 0.074393 (Tables 3 and 4).

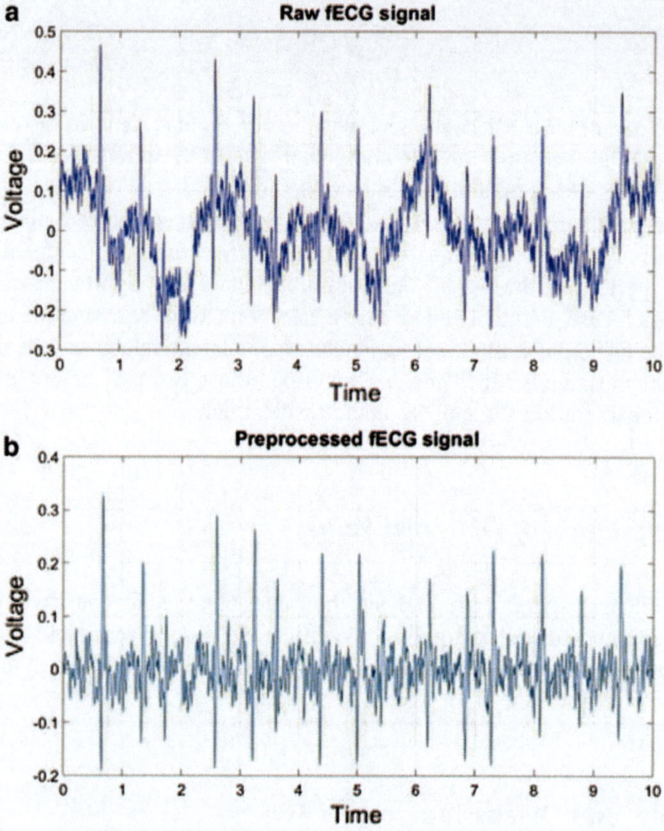

Fig. 2 **a** Raw ARR08 fECG, **b** Pre-processed ARR08 fECG

Table 1 Values obtained for ARR08 Signal's statistical parameters

Parameter	Value
Mean	0.000113
Median	−0.00225
Standard deviation	0.05146
RMS	0.05146
Peak-to-Peak	0.5184
R peak	0.3275

Table 2 Peak values obtained for ARR08 signal's power spectral density estimation

Parameter	Value
PSD1	1051.023
PSD2	0.00042
PSD3	0.000313

Table 3 Values obtained for ARR08 signal's AR coefficients

Algorithm	Coefficient value
AR Yule-Walker	1
	−3.34818
	4.610205
	−3.13076
	0.882674
AR Burg	1
	−3.34848
	4.611011
	−3.13152
	0.882922
AR covariance	1
	−3.34846
	4.611005
	−3.13154
	0.882937

Table 4 Values obtained for ARR08 Signal's PCA Parameters

Parameter	Value
PCA Latent	0.002648
PCA Mu	0.000113

B. NR Signal—NR04

See Fig. 3; Tables 5 and 6.

The obtained value for the periodogram estimate is 1.618555609 (Tables 7 and 8).

C. Syn fECG C1 Signal—Sub01

See Fig. 4; Tables 9 and 10.

The obtained value for the periodogram estimate is 0.027190585 (Tables 11 and 12).

D. Syn fECG C3 Signal—Sub08

See Fig. 5; Tables 13 and 14.

The obtained value for the periodogram estimate is 0.009853093 (Tables 15 and 16).

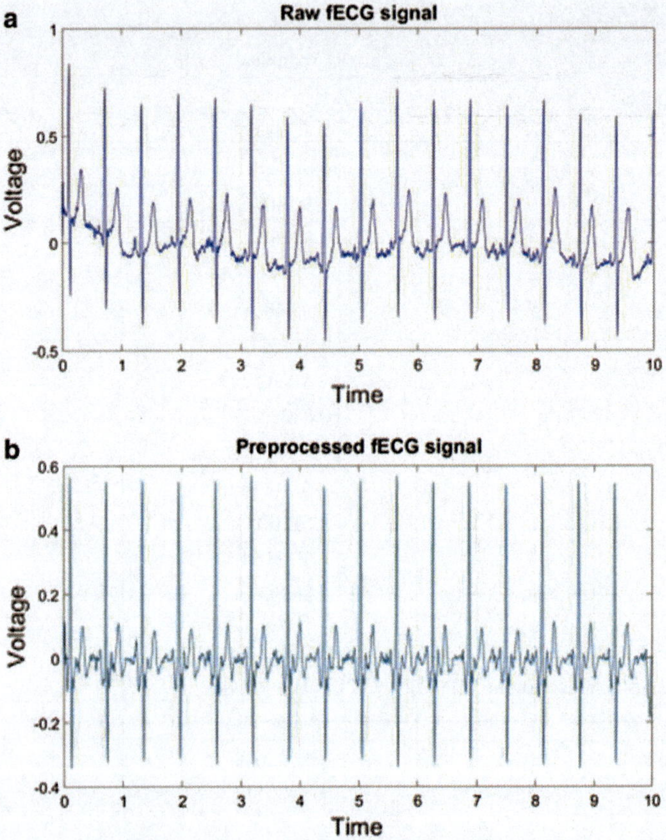

Fig. 3 **a** Raw NR04 fECG, **b** Pre-processed NR04 fECG

Table 5 Values obtained for NR04 signal's statistical parameters

Parameter	Value
Mean	−0.00069
Median	−0.00762
Standard deviation	0.1003
RMS	0.1003
Peak-to-Peak	0.9166
R peak	0.5717

Table 6 Peak values obtained for NR04 signal's power spectral density estimation

Parameter	Value
PSD1	46809.41
PSD2	0.004681
PSD3	0.002382

Table 7 Values obtained for NR04 signal's AR coefficients

Algorithm	Coefficient value
AR Yule-Walker	1
	−3.78891
	5.479768
	−3.58804
	0.898059
AR Burg	1
	−3.78921
	5.480652
	−3.58891
	0.898348
AR covariance	1
	−3.78904
	5.480129
	−3.58838
	0.898167

Table 8 Values obtained for NR04 signal's PCA parameters

Parameter	Value
PCA Latent	0.010059658
PCA Mu	−0.000694474

3.2 *Classification*

The parameters are analyzed to find the range of values to initially classify the fECG signals as normal (Syn fECG C1 and C3) and abnormal signals (ARR and NR).

3.3 *Quantitative Analysis*

The values of these 15 parameters mentioned in Table 17 are analyzed for threshold values and ranges to indicate normal and abnormal fECG signals. After the threshold value is found, a scoring scheme is implemented to indicate severe cases of abnormality and cases far from the said abnormality.

Scoring scheme

- +1 is given if the obtained value crosses the threshold
- 0 is given if the obtained value is equal or less than the threshold

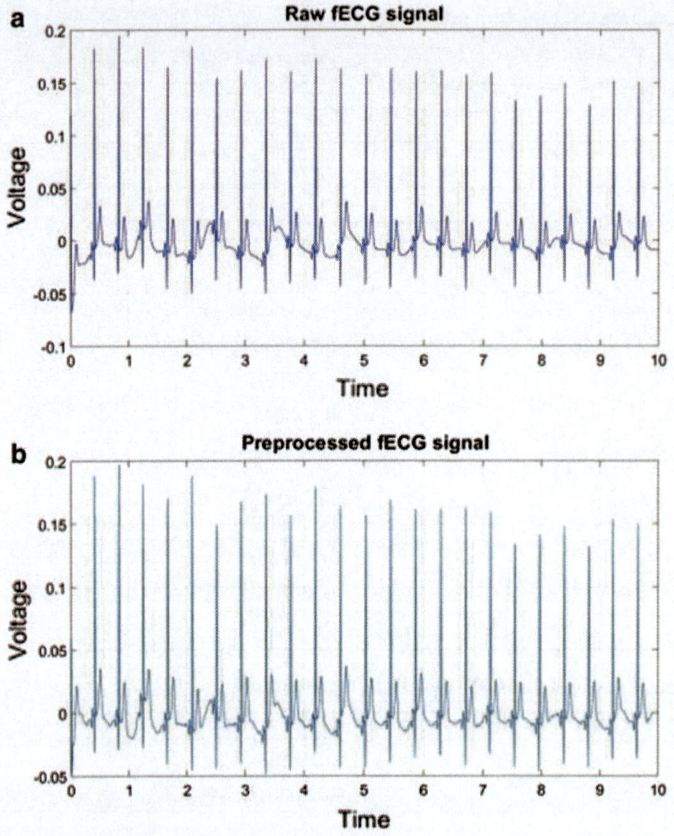

Fig. 4 **a** Raw Sub01 fECG, **b** Pre-processed Sub01 fECG

Table 9 Values obtained for Sub01 signal's statistical parameters

Parameter	Value
Mean	9.44E-05
Median	−0.00532
Standard deviation	0.02294
RMS	0.02294
Peak-to-Peak	0.2461
R peak	0.1974

Table 10 Peak values obtained for Sub01 signal's power spectral density estimation

Parameter	Value
PSD1	187.5991351
PSD2	0.000300159
PSD3	0.000131762

Table 11 Values obtained for Sub01 signal's AR coefficients

Algorithm	Coefficient value
AR Yule-Walker	1
	−1.350646526
	1.202768462
	−0.791178038
	0.27958337
AR Burg	1
	−1.354993699
	1.210069984
	−0.796855224
	0.281021597
AR covariance	1
	−1.354166777
	1.204212075
	−0.791600347
	0.279760539

Table 12 Values obtained for Sub01 signal's PCA parameters

Parameter	Value
PCA Latent	0.000526
PCA Mu	9.44E-05

The score for all 15 parameters is then tallied, based on the initial classification of normality the severity of each case is deduced.

The parameters after undergoing thorough analysis and scoring tend to shed a light on the fetal health status. They can provide an indication of unfavorable fetal outcomes that seem to occur before the labor onset (Table 18).

According to the scoring, Subjects ARR02, ARR08, ARR10, and ARR12 have obtained scores > 7.5, which is a clear indication of their deviation from the threshold values and thus depict high-risk pregnancy; while others with scores < 7.5 are not emergent (Table 19).

According to scoring, Subjects NR01, NR04, NR06, NR08, NR10, NR11, NR12, and NR13 have obtained scores > 7.5, which is a clear indication of their deviation from the threshold values and thus depict high-risk pregnancy; while the others with scores < 7.5 are not emergent (Table 20).

According to the scoring scheme, Sub04, Sub05, and Sub07 have obtained scores < 7.5, which is a clear indication of their deviation from the threshold values depict a possible risk, while the others with scores > 7.5 are indicative of normal pregnancy.

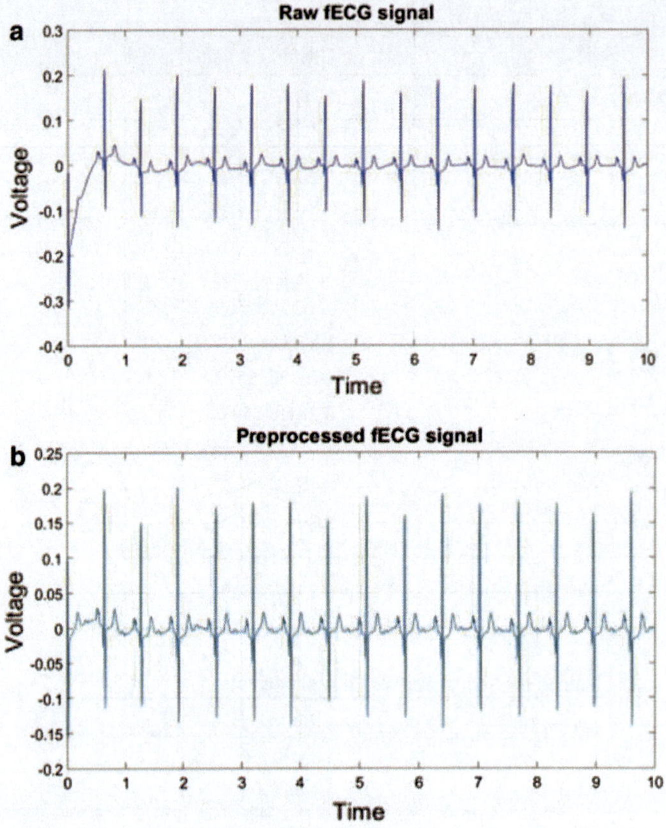

Fig. 5 **a** Raw Sub08 fECG, **b** Pre-processed Sub08 fECG

Table 13 Values obtained for Sub08 signal's statistical parameters

Parameter	Value
Mean	0.0004739
Median	−0.003416
Standard deviation	0.02839
RMS	0.02839
Peak-to-Peak	0.3567
R peak	0.2017

Table 14 Peak values obtained for Sub08 signal's power spectral density estimation

Parameter	Value
PSD1	64.60572552
PSD2	0.000103369
PSD3	4.70861E-05

Table 15 Values obtained for Sub08 signal's AR coefficients

Algorithm	Coefficient value
AR Yule-Walker	1
	−1.335605955
	0.977803619
	−0.388345816
	0.174215101
AR Burg	1
	−1.340283766
	0.993172498
	−0.408948367
	0.187336446
AR covariance	1
	−1.369545635
	1.025119653
	−0.420145349
	0.181540052

Table 16 Values obtained for Sub08 signal's PCA parameters

Parameter	Value
PCA Latent	0.000805851
PCA Mu	0.000473905

The technical and methodological instruments utilized to date for clinical diagnosis don't extract solid quantitative indices connecting fECG and fHR signal patterns with pathophysiological fetal states. If a woman is to develop any condition that may harm the infant in her uterus, then it isn't always identified on traditional monitoring techniques for fECG and fHR, CTG, or Ultrasound. Along these lines, there is a need to provide an investigation of said recorded waveforms to distinguish dangerous conditions that may develop sooner rather than later. This is because each portion of the fECG waveform conveys different sorts of data for the clinician analyzing fetus' condition, alongside the fHR properties that could be especially valuable in checking the fetal prosperity. The fECG and fHR signals could give reliable signs on fetal status, where the abnormal condition produces perceptible variations of the signal qualities, both in the time and in the frequency domain. By investigating these parameter estimates, the physician can detect fetal distress and can aid in the decision of continuing the pregnancy to full term.

Table 17 Average parameter values for each case

Type	ARR	NR	Syn fECG C1	Syn fECG C3
Mean	−0.00047	5.19E-05	−8.9848E-05	−4.4896E-05
Median	−0.00183	−0.00742	−0.0077032	−0.0070534
Standard deviation	0.182388	0.134606	0.04397	0.038867
RMS	0.182378	0.134605	0.04397	0.038867
Peak-to-Peak	1.599408	1.228979	0.48594	0.42474
R peak	0.929392	0.875279	0.3711	0.31697
PSD1	85132.8	68176.53	316.256987	224.1755211
PSD2	0.016688	0.006818	0.00050575	0.000358672
PSD3	0.012179	0.00426	0.00019876	0.000158504
Periodogram	2.761424	2.412651	0.0587484	0.042584754
AR Yule-Walker	1	1	1	1
	−3.74016	−3.83883	−1.68509184	−1.69212027
	5.392671	5.605527	1.6482347	1.672420119
	−3.55318	−3.69146	−0.93811831	−0.96183543
	0.903379	0.925316	0.29374866	0.304466967
AR Burg	1	1	1	1
	−3.74067	−3.83987	−1.68863018	−1.69437393
	5.394056	5.6086	1.65575419	1.678312569
	−3.55445	−3.69452	−0.94613712	−0.96858372
	0.903772	0.92634195	0.29775125	0.308117216
AR Covariance	1	1	1	1
	−3.74033	−3.83951	−1.68974036	−1.6974611
	5.393181	5.607452	1.65389539	1.679742235
	−3.55371	−3.69329	−0.94250627	−0.96678668
	0.903563	0.925899	0.29549591	0.3057367
AR PSD Yule-Walker AR PSD Burg AR PSD covariance	Peaks occurring between 0 – 0.1 NF and steep descent beyond 0.1 NF		Peaks occurring between 0 – 0.5 NF and steep descent beyond 0.5 NF	
PCA Latent	0.043616	0.021085	0.00240846	0.001898797
PCA Mu	−0.00047	5.19E-05	−8.9802E-05	−4.4896E-05

4 Conclusion

Fetal ECG and fHR signals are considered the best diagnostic device for analyzing the status of the fetus in utero. The fetal heart rate may change as the fetus reacts to conditions in the uterus. An abnormal fetal heart rate may show that the fetus isn't getting enough oxygen or that there are other issues. These current examination tools base their expertise on scoring systems that attempt to recreate the time domain

Table 18 Scoring of ARR signals

Subject	Mean	Median	SD	RMS	Peak-to-Peak	R Peak	Periodogram	PSD1	PSD2	PSD3	AR Yule-Walker	AR Burg	AR Covariance	PCA Latent	PCA Mu	Score
ARR01		√					√		√	√				√		5/15
ARR02		√	√	√	√	√	√	√	√	√				√		10/15
ARR03		√														1/15
ARR04		√							√	√						3/15
ARR05			√	√	√	√	√	√	√	√				√		9/15
ARR06	√	√									√	√	√		√	6/15
ARR07						√					√	√	√			4/15
ARR08		√	√	√	√	√	√	√	√	√	√	√	√	√		13/15
ARR09						√					√	√	√			4/15
ARR10			√	√	√	√	√	√	√	√				√		9/15
ARR11		√	√	√	√	√	√	√	√	√				√		10/15
ARR12		√	√	√	√	√	√	√	√	√				√		10/15

Table 19 Scoring of NR signals

Subject	Mean	Median	SD	RMS	Peak-to-Peak	R Peak	Periodogram	PSD1	PSD2	PSD3	AR Yule-Walker	AR Burg	AR Covariance	PCA Latent	PCA Mu	Score
NR01		√	√	√	√	√	√	√	√	√				√	√	11/15
NR02		√													√	2/15
NR03																0/15
NR04	√	√	√	√	√	√	√	√	√	√	√	√	√	√	√	15/15
NR05	√	√													√	3/15
NR06	√		√	√	√	√	√	√	√	√				√	√	11/15
NR07																0/15
NR08		√	√	√	√	√	√	√	√	√				√		10/15
NR09		√									√	√	√			4/15
NR10	√		√	√	√	√	√	√	√	√				√	√	11/15
NR11	√	√					√	√	√	√				√	√	8/15
NR12			√	√	√	√	√	√	√	√	√	√	√	√		12/15
NR13			√	√	√	√	√	√	√	√				√		9/15
NR14							√									1/15

Table 20 Scoring of Syn fECG C1 signals

Subject	Mean	Median	SD	RMS	Peak-to-Peak	R Peak	Periodogram	PSD1	PSD2	PSD3	AR Yule-Walker	AR Burg	AR Covariance	PCA Latent	PCA Mu	Score
Sub01			√	√	√	√	√	√	√	√	√	√	√	√		12/15
Sub02	√		√	√	√	√	√	√	√	√				√	√	11/15
Sub03	√	√	√	√	√	√	√	√	√	√				√	√	12/15
Sub04		√									√	√	√			4/15
Sub05	√	√									√	√	√		√	6/15
Sub06			√	√	√	√	√	√	√	√				√		9/15
Sub07	√	√									√	√	√		√	6/15
Sub08			√	√	√	√	√	√	√	√	√	√	√	√		12/15
Sub09			√	√	√	√	√	√	√	√				√		9/15
Sub10	√	√			√		√	√	√	√				√	√	9/15

criteria utilized by the clinicians, although quantitatively, to conclude if the fetus is in a "distress" condition. With the estimations done in this investigation of the fECG waveforms, obviously, at times there is an articulated deviation from the normal limit threshold, obtained for each type of signal, which shows clearly that there is a chance of risk in pregnancy. The combination of different indices is essentially different between normal and pathological fetuses, while traditional parameters alone are not. Along these lines, it tends to be reasoned that the combination of statistical properties, principal component analysis gives a powerful method for studying the fECG signals for early identification of any abnormal states of the baby. These strategies can be utilized to build a real-time fECG monitoring and analyzing system which helps obstetricians to distinguish fetal distress in the womb of pregnant women and take necessary actions to help their patients. One of the major challenges presented in this field is the different methods employed in the acquisition of the fECG signals and the lack of benchmark standards for various signal properties and their values. A future extension of this paper could involve qualitative and quantitative analysis of said parameters in depth; to not only detect impending risk to the fetus, but also the severity of the said risk, which can help to provide an accurate conclusion of the risk in pregnancy. Also, this work could speak to a step toward an early recognition of fetus at risk during pregnancy. This could likewise be reached out utilizing neural networks to upgrade the dependability of fetal well-being.

References

1. Rajni IK (2013) Electrocardiogram signal analysis—an overview, December 2013
2. Jozef S, Jozef M (2007) Spectral properties of ECG signal, May 2007
3. Srinivasan N, Ge DF, Krishnan SM (2002) Autoregressive modeling and classification of cardiac arrhythmias, October 2002
4. Signorini MG, Magenes G (2003) Linear and nonlinear parameters for the analysis of fetal heart rate signal from cardiotocographic recordings, March 2003
5. Saad NM, Abdullah AR, Low YF (2006) Detection of heart blocks in ECG signals by spectrum and time-frequency analysis, June 2006
6. Castells F, Laguna P, Sornmo L, Bollmann A, Roig JM (2006) Principal component analysis in ECG signal processing, November 2006
7. Taralunga D, Ungureanu M, Strungaru R, Wolf W (2011) Performance comparison of four ICA algorithms applied for fECG extraction from transabdominal recordings, April 2011
8. Segismundo MJ, Villanueva MMF, Zaldua A, Lee-Ramos CM (2012) Spectrogram analysis of electrocardiogram with normal sinus rhythm, arrhythmia, and atrial fibrillation, July 2012
9. Murugappan SM, Zheng BS (2013) Frequency band analysis of Electrocardiogram (ECG) signals for human emotional state classification using Discrete Wavelet Transform (DWT), January 2013

10. Smith V, Nair A, Warty R, Sursas JR, da Silva Costa F, Wallace EM (2019) A systematic review on the utility of non-invasive electrophysiological assessment in evaluating for intrauterine growth restriction, February 2019
11. Huang B, Kinsner W (2002) ECG frame classification using dynamic time warping, February 2002
12. Kumaravel N, Sridhar KS, Nithiyanandam N (1996) Automatic diagnosis of heart diseases using neural network, January 1996
13. Pietka E (1990) Feature extraction in computerized approach to the ECG analysis, May 1990

Chapter 48
Sentiment Analysis of Twitter Data Using Techniques in Deep Learning

S. Gowri, J. Jabez, J. S. Vimali, A. Sivasangari, and Senduru Srinivasulu

1 Introduction

Overcurrent centuries, Twitter has become more popular for sharing the recent activities surrounded by us. For any user, it is very important to know the most trending subjects on Twitter by analysing all the tweets each day. Along with that, since data redundancy occurs, it is significant to find relevant data and the user data is found by sentiment analysis. It is actually slow to discover the relevant data from an enormous number of tweets. Twitter has become crucial in social media for sharing the data in microblogging. In Twitter, the data is very much secured and it is highly confidential. There is a huge collection of data and a lot of trending topics. The main advantage of using Twitter to share the data is microblogging. The term microblogging refers to share the data in microblogs. Twitter is well known for microblogging. This project is designed to separate the data which is either good or bad. By using deep learning algorithms, the relevant data is separated [1]. The main theme of this project is to whether the tweet is either is good or bad. To know whether a particular topic is good or bad, every data cannot be read and comments by the user. There are a lot of comments and suggestions on each topic. Sentiment analysis is used to separate the raw data and the relevant data. Mawkishness examination is the procedure of recognising the thoughts whether a specific topic, and the textual data is optimistic, undesirable or unbiased.

Mawkishness examination is the best method for analysing the theme or a product whether it is an optimistic, undesirable or unbiased. To do this project, a programming language is needed it is 'Python'. Nowadays, Python is a fast-growing language. Each task can be done using Python which makes the work flexible. Many modules are

S. Gowri (✉) · J. Jabez · J. S. Vimali · A. Sivasangari · S. Srinivasulu
Department of Information Technology, Sathyabama Institute of Science and Technology, Chennai, India
e-mail: gowriamritha2003@gmail.com

I. Jeena Jacob et al. (eds.), *Data Intelligence and Cognitive Informatics*, Algorithms for Intelligent Systems,
https://doi.org/10.1007/978-981-15-8530-2_48

needed to be imported in Python code. There is a matplotlib module which shows the bar graph of current scenario news. Every module has its functionality. The data need to be read by Python using a read-csv module in which it is stored in the Excel sheet. The panda's module needs to use for maintaining data as tabular rows and columns. When Twitter data is used for analysing, every word from the Excel column will be taken and analyse using sentiment analysis. The Excel sheet will be working as comma-separated values and it helps extract data for data processing. Python is a high-end language and it is useful for large problem-solving. Python gives an optimisation solution for any problem. The Python was written with 'C' language. In this programme, dictionaries are used for searching whether a particular word has meaning or not.

If yes, then using sentiment analysis it will be processed and be marked as a good or bad word. If no, then that particular word will be discarded and if the words are repeated then, the analysis will remove the duplicates. Whether a particular topic was good or bad will be decided by the users who are raising their voice on Twitter. Based on the comments mentioned in Twitter, the sentiment analysis will judge whether the topic is good news or bad news. Not only Twitter about news in newspapers can be known too via sentiment analysis. Nowadays, everyone is reading e-news. With the help of e-news data, the data can be analysed and can graphically display the news that on which news topic is trending, whether that trending topic is good news or bad news.

So, it may be social media or newspaper, the data is important to analyse using sentiment analysis. Every citizen in this society needs to know about the news what's going around. So, this project is one of the solutions for those who need to know about news whether it is positive, negative or neutral. Finally, to say about this project that it is very helpful for knowing about current trending topics. This project is one of the solutions to express news graphically. The graphical representation shows positive words, negative words and shows the person names on whom the news was going on.

2 Existing System

In the project, the task is to implement a system that implements a trending topic analysis system and that analysis the given topic by performing subject adaptive mawkishness arrangement. Subject adaptive sentimentality arrangement is used to analyse the tweets which are either positive, negative or neutral. This adaptive sentiment classification method performs deep analysis and multi-sequential summarisation. The ultimate task is to analyses the data which is either positive, negative or neutral by performing adaptive sentiment classification. The technique that is used in this project is the adaptive mawkishness arrangement.

This adaptive mawkishness arrangement performs each topic or product analyses in a detailed manner. Examining system delivers a cavernous examination of every subject through adaptive mawkishness arrangement. Multi-tweet consecutive summarisation is used in the existing system [2].

There are some advantages to the existing system. Implementing a trending topic analysis framework that would analyse the given topic by performing subject adaptive mawkishness classification analysis system provides an in-depth analysis by theme adaptive mawkishness organisation and multi-tweet sequential summarisation, which goals to provide a chronological sequence. There are some drawbacks in the existing system. The drawbacks which may lead some misleading of data and product. Sometimes the analysis may not be done properly. Now, let us discuss the drawbacks that are involved in the existing system. The existing work is about halal products which are a danger to the citizens of our country [3, 4]. So, to stop the halal products, the solution was raised and to support it and take a step further this Twitter data analysis helps [5]. The adaptive sentiment analysis classification will classify the data into two groups that are positive and negative [6]. The arrangement acts a most important role in the analysis of the information using deep learning algorithms and multi-tweet summarisations [7]. The collected works will be arranged in chronological order in which it makes algorithms easy to identify the words in dictionaries. The newly tweets should be considered for analysing. The meaningful words need to be collected and stored for classification.

Implementing deep learning prototypical that can categorise Netflix evaluations as optimistic or undesirable. The prototype will yield a complete evaluation as an input (word after word) and deliver proportion evaluations for examining if the evaluation carries an optimistic or undesirable mawkishness [8].

The tweets data are dynamically uploaded in the database. So, via API's the database of Twitter need to be connected with and collect the tweet data for data cleaning and data processing [9]. The stored data need to be analysed and to be displayed in graphical format. The collected data should be useful for citizens purpose. The main theme of this project is for social responsibility citizens across the world. This project is very useful for the citizens who are interested to know about their daily news. This adaptive sentiment classification performs each topic or product analyses in a detailed manner. Investigation system delivers a cavernous examination for each theme through adaptive mawkishness arrangement. Multi-tweet sequential summarisation is useful for the existing system [10].

3 Proposed System

The proposed system through foreground dynamic topic modelling. This allows an end user to wholly examine a trending to a superior level of a feature. Preprocessing stage, the idealised system also knobs non-English tweet conversion that is indispensable to avert neglecting of community estimation about the trending subject. It as stream-based approach and semantic-based approach. Redundancy check is performed to remove duplicates from the selected tweets followed by a threshold check to ensure a fair mixture of user's opinions. The proposed system is a foreground dynamic topic modelling. This proposed system enables us to check every trending topic at a deep level of analysation, this method uses to check the non-English tweets on everyone's opinion, and this method is a steam-based and semantic-based approach.

3.1 Recurrent Neural Network (RNN)

A recurrent neural network (RNN) is a class of counterfeit neural systems where associations among hubs structure a coordinated diagram laterally a fleeting succession. This permits it to display fleeting unique conduct. Gotten from feedforward neural systems, RNNs can utilise their interior condition (memory) to procedure variable-length arrangements of information sources. This makes them pertinent to assignments, for example, unsegmented, associated penmanship acknowledgement or discourse acknowledgement.

The articulation "irregular neural framework" is used carelessly to suggest two extensive classes of systems with an equivalent general structure, where one is restricted drive and the other is an unbounded inspiration. The two classes of systems show transient exceptional lead. A constrained drive discontinuous framework is an organised non-cyclic outline that can be unrolled and superseded with a cautiously feedforward neural framework, while a limitless inspiration monotonous framework is a planned cyclic chart that cannot be unrolled.

Both restricted drive and endless inspiration dreary systems can have additional taken care of states, and the limit can be under direct control by the neural framework. The limit can in like manner be superseded by another framework or graph if that unites time delays or has analysis circles. Such controlled states are suggested as gated state or gated memory and are a bit of long transient memory systems (LSTMs) and gated dreary units. This is similarly called feedback neural network (Fig. 1).

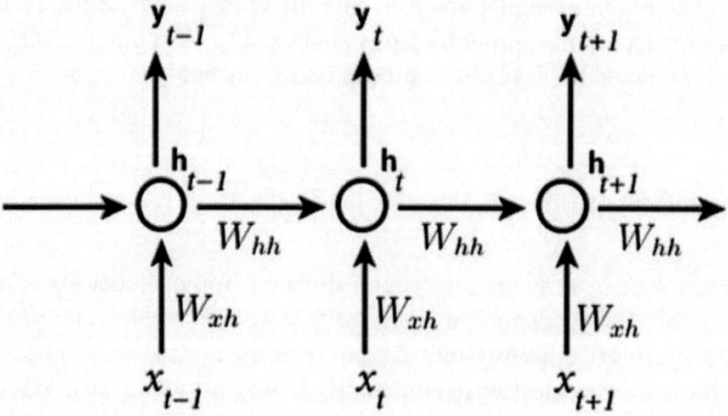

Fig. 1 Recurrent neural network

$x(t-1)$, $x(t)$, $x(t+1)$ are sequential inputs that depend on each other (such as words in a sentence). $y(t-1)$, $y(t)$, $y(t+1)$ are the outputs. Unique for RNN is the fact that the calculation of the currently hidden state $h(t)$ of the neurons for the input $x(t)$ depends on the previously hidden state $h(t-1)$ for the previous input $x(t-1)$. W_{xh} and W_{hh} are weight matrices that connect the input $x(t)$ with the hidden layer $h(t)$, and $h(t)$ with $h(t-1)$, respectively. Along these lines, a repeat is acquainted with the neural system which can be considered as a memory on the past sources of info. In principle, along these lines' "vanilla" RNNs can utilise data in subjectively long arrangements, yet practically speaking, they are restricted to thinking back just a couple of steps.

3.2 Long Short-Term Memory LSTM

Long momentary memory (LSTM) is a fake recurrent neural network (RNN) engineering used in the field of significant learning. Not in the least like standard feedforward neural systems, LSTM has input affiliations. It cannot simply strategy single data centres, (for instance, pictures), yet what's more entire groupings of data, (for instance, talk or video). For example, LSTM is material to endeavours, for instance, unsegmented, related handwriting acknowledgement, talk acknowledgement and inconsistency recognisable proof in organising traffic or IDSs (interference area structures).

A common LSTM unit is made out of a cell, a data entryway, a yield gateway and a neglect entryway. The phone reviews regard over optional time ranges and the three gateways control the movement of information into and out of the telephone. LSTM systems are fitting to requesting, planning and making front throws reliant on the time course of action data since there can be slacks of dark range between noteworthy events in a period game plan. LSTMs were made to deal with the vanishing tendency issue that can be experienced while planning traditional RNNs. The relative absence of care towards gap length is a touch of slack of LSTM over RNNs, hid Markov models and other gathering learning methods in different applications (Fig. 2).

3.3 Implementation of the Proposed Methodology

The proposed system also removes the duplicates from the selected tweets by a threshold check to ensure a fair mixture of user's opinion. Sentiment analysis of Twitter using big data helped us to analysis a huge number of datasets the various providers can be further compared and judge which one is best. To measure the performance with the existing system, initially implemented with MapReduce architecture. In the next phase, spark frame will be implemented and the performance will be analysed.

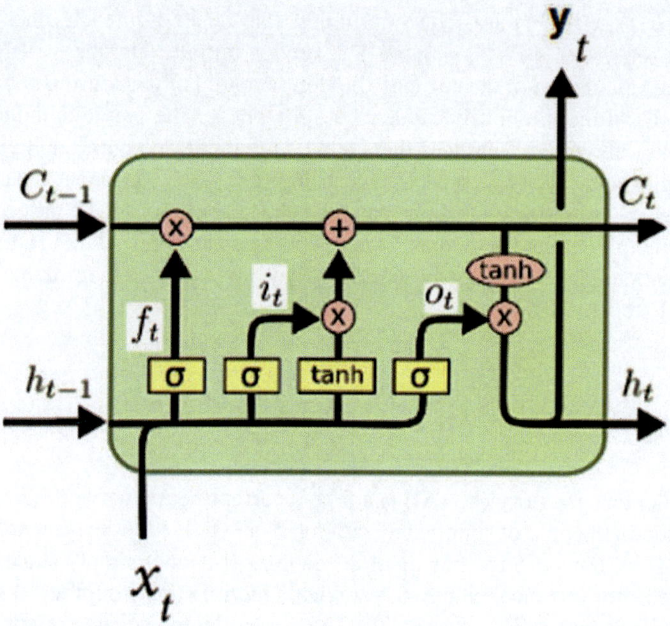

Fig. 2 Long short-term memory

3.4 Algorithm

The implementation for the system is done with the combination of the following snippets of code.

3.4.1 Pre-processing

```
# remove twitter handles (@user)
combi['tidy_tweet'] = np.vectorize(remove_pattern)(combi['tweet'], "@[\w]*")

# remove special characters, numbers, punctuations
combi['tidy_tweet'] = combi['tidy_tweet'].str.replace("[^a-zA-Z#]", " ")
```

3.4.2 Extraction of Hashtags

```
# extracting hashtags from non racist/sexist tweets
HT_regular = hashtag_extract(combi['tidy_tweet'][combi['label'] == 0])

# extracting hashtags from racist/sexist tweets
HT_negative = hashtag_extract(combi['tidy_tweet'][combi['label'] == 1])

# unnesting list
HT_regular = sum(HT_regular,[])
HT_negative = sum(HT_negative,[])
```

3.4.3 Frequently Used Hashtags Collection

```
# selecting top 10 most frequent hashtags
d = d.nlargest(columns="Count", n = 10)
plt.figure(figsize=(16,5))
ax = sns.barplot(data=d, x= "Hashtag", y = "Count")
ax.set(ylabel = 'Count')
plt.show()
```

3.4.4 Outcomes from the Input Context

```
# bag-of-words feature matrix
bow = bow_vectorizer.fit_transform(combi['tidy_tweet'])

# TF-IDF feature matrix
tfidf = tfidf_vectorizer.fit_transform(combi['tidy_tweet'])

# splitting data into training and validation set
xtrain_bow, xvalid_bow, ytrain, yvalid = train_test_split(train_bow, train['label'],
                    random_state=42, test_size=0.3)
```

Hashtags, a simple automated method can be provided for what people think. Thus, collecting information from networks and analysing it using big data techniques has behind the traditional database approach. The duplicates are removed from the selected data that is either can be data or data from the product to analyse it is positive, negative or neutral. The proposed system is very much useful for analysing the trending topics which are generated by second. The data is analysed at a deep level. The data is analysed at a greater level that tells whether it is positive, negative or neutral. The data can be of anything like data from Twitter, news, Facebook posts, Instagram posts or any social media news [11, 12]. The proposed system uses a method called foreground dynamic topic modelling. This method checks for the removal of duplicates (Fig. 3).

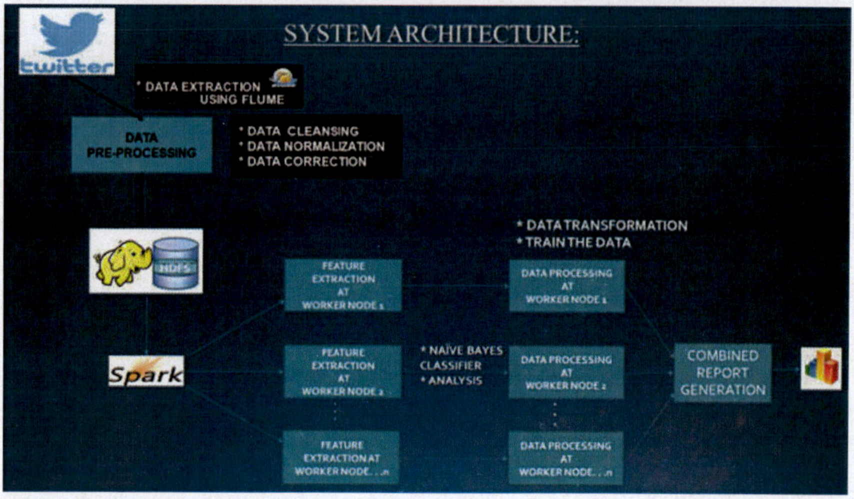

Fig. 3 Overview of system architecture

4 Results and Discussion

The result and discussion for this project are to collect the tweets, data or some product from Twitter. The main theme is analysing the data to prove whether the data or product is positive, negative or neutral. There are some methods to analyse the data or product whether it is positive, negative or neutral. There are some major issues in the related work. The most important thing is to select the particular language for the project and the thing is to collect the entire data or data related to the product. The data is collected in the form of Excel sheets. These Excel sheets will help us to analyse the data whether the data or product data is optimistic, undesirable or unbiased. This analysation is done in the dictionary words. The words which are present in the dictionary are only get analysed. The words which are not present in the dictionary are not analysed and these words will move into the neutral part. And the most important thing is to build the graphical representation of the words that are collected from the Twitter or newspaper. The data can be anything which is either tweets, Facebook posts, trending news or halal products [13, 14].

Sentiment analysis is the best method for analysing the subject or a product whether it is an optimistic, undesirable or unbiased. To do this project, a programming language is needed it is 'Python'. Nowadays, Python is a fast-growing language. Every task can be done using Python which makes the work flexible. Many modules are needed to be imported in Python code. There is a matplotlib module which shows the bar graph of current scenario news. Each module has its functionality. The data need to be read by Python using the read-csv module, in which it is stored in an Excel sheet. There are a lot of comments and suggestions on each topic. Sentiment analysis is used to separate the raw data and the relevant data. Sentiment examination is the procedure of recognising the thoughts whether a specific subject, the product is optimistic, undesirable or unbiased (Figs. 4, 5, 6, 7 and 8).

Fig. 4 Normal words

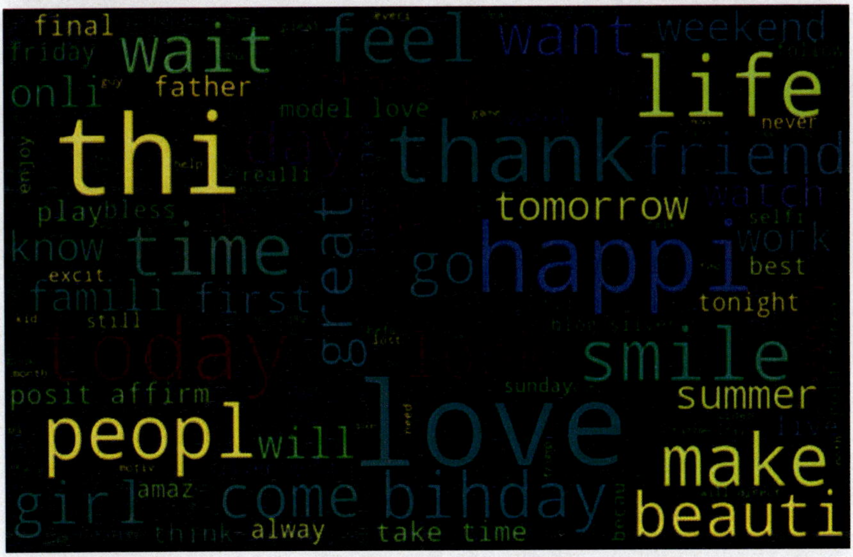

Fig. 5 Positive words

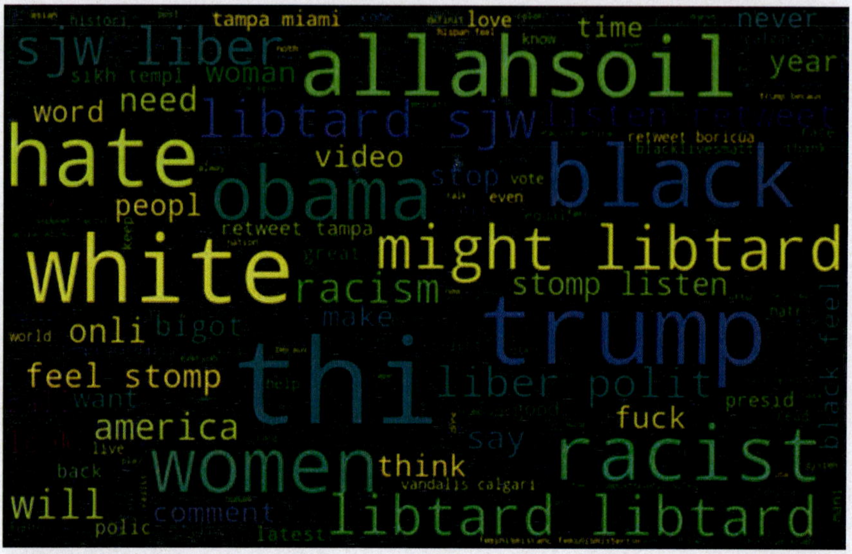

Fig. 6 Negative words

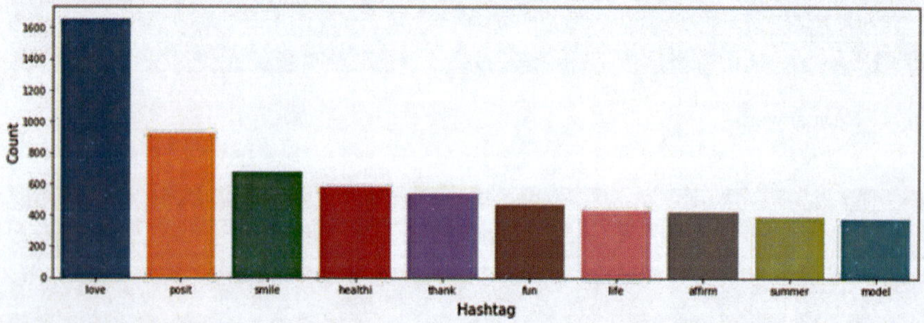

Fig. 7 Emotions

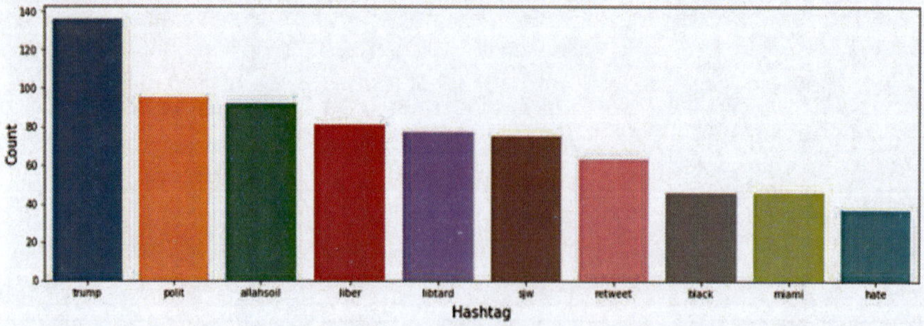

Fig. 8 Hashtags

5 Conclusion

Sentiment analysis of Twitter using big data helped us to analysis a huge number of datasets. The variety of providers can be further compared and judge which one is best. Data has been analysed properly. This project creates interest in every citizen to know about current trending news. Every user has their perception about any trending topic, but this project provides correct details about news and conveys the proper message to the users who use this project. Finally, I conclude that this project is very useful to the citizens, all among the world.

References

1. Ioffe S, Szegedy C (2015) Batch normalization: accelerating deep network training by reducing internal covariate shift. In: Proceedings of the 32nd international conference on machine learning, vol 37, Lille, France, Feb 2015
2. Feizollah A, Anuar NB, Salleh R, Amalina F, Ma'arof RUR, Shamshirband S (2013) A study of machine learning classifiers for anomaly based mobile botnet detection. Malaysian J Comput Sci 26(4):251–265
3. Battour M, Ismail MN, Battor M (2010) Toward a halal tourism market. Tourism Anal 15(4):461–470
4. Aoun I, Tournois L (2015) Building holistic brands: an exploratory study of Halal cosmetics. J Islamic Marketing 6(1):109–132
5. Battour M, Ismail MN (2016) Halal tourism: concepts, practices, challenges and future. Tourism Manage Percept 19:150–154
6. Song J, Lee S, Kim J (2016) Inference attack on browsing history of Twitter users using public click analytics and twitter metadata. IEEE Trans Dependable Secure Comput 13(3) (May–June 1 2016)
7. Goodfellow I, Bengio Y, Courville A (2016) Deep Learning. MIT Press, Cambridge, MA, USA
8. Oppermann A (2019) Sentiment analysis with deep learning of Netflix reviews. Available: https://towardsdatascience.com/sentiment-analysis-with-deep-learning-62d4d0166ef6
9. Srilekha K, Sudha RV (2018) Using public click analytics and Twitter metadata occurring inference attack on browsing antiquity of twitter users. Int J Eng Res Technol (IJERT) ISSN: 2278–0181, NCICCT - 2018
10. Jabez J, Gowri S, Vigneshwari S, Albert MJ, Srinivasulu S (2019) Anomaly detection by using CFS subset and neural network with WEKA tools. Smart innovation systems and technologies
11. Dharani K, Rajesh M, Senthil Prakash T (2016) Deduction attack on browsing history in Twitter using public click analytic and metadata. Int J Comput Res Dev (IJCRD) 1(2): 2456–3137
12. Dataset. https://www.trackmyhashtag.com/twitter-dataset
13. Kamin D (2019) The rise of Halal tourism. Available: https://www.nytimes.com/2019/01/18/travel/the-rise-of-halal-tourism.html
14. Belopilskaya Y. How Halal Tourism is reshaping the global tourism industry. Available: https://hospitalityinsights.ehl.edu/halal-tourism-global-industry

Chapter 49
Part-of-Speech Tagging in Mizo Language: A Preliminary Study

Morrel V. L. Nunsanga, Partha Pakray, Mika Lalngaihtuaha, and L. Lolit Kumar Singh

1 Introduction

Natural Language Processing (NLP) is a branch of artificial intelligence that deals with natural languages. It is an approach to the process, explore, examine and understand a large amount of text data. It helps resolve ambiguity in many languages. It provides a mechanism to help a computer to understand, interpret and process human languages and helps to resolve ambiguity in many languages. Part-of-speech tagging is one of the basic preprocessing modules for any language processing. It is the process of defining and marking up a word in a corpus (text) using corresponding part of speech such as noun, pronoun, adjective or a lexical class maker, based on its context in the sentence and definition in the corpus. It indicates designated parts of speech for each and every word in a sentence. So, it provides information about how the particular word is being used in the sentence or within the scope of a phrase. Assigning part of speech is sometimes complicated, as a specific word may suggest various implications depending on the sentence where the word is being utilized.

M. V. L. Nunsanga (✉) · M. Lalngaihtuaha · L. Lolit Kumar Singh
Mizoram University, Mizoram, India
e-mail: morrelhmar@mzu.edu.in

M. Lalngaihtuaha
e-mail: pein2991@gmail.com

L. Lolit Kumar Singh
e-mail: llksingh@yahoo.co.in

P. Pakray
NIT, Silchar, India
e-mail: parthapakray@gmail.com

I. Jeena Jacob et al. (eds.), *Data Intelligence and Cognitive Informatics*,
Algorithms for Intelligent Systems,
https://doi.org/10.1007/978-981-15-8530-2_49

The nomenclature 'Mizo' signifies both the name of a tribe and the language. The Mizo tribe is an ethnic group, settled mainly in a mountainous region Mizoram and believed to be migrated from China. Other than the mainland Mizoram, many of the Mizos reside in Myanmar, Tripura, Assam, Manipur and Nagaland. The term Mizo is actually an umbrella term to denote the various clans, such as Hmar, Pawi, Paite, Lakher, Ralte. Mizo as a language, it belongs to the Tibeto-Burman group of languages [1] now spoken by around a hundred thousands of individuals mostly from the territory of Mizoram and Chin State in Myanmar. The Mizo language known as *Mizo ṭawng* or *Zo ṭawng* is also called as *Duhlian ṭawng*. The Mizo language, as in its present form, is generally based on *Lusei (Lushai* or *Lushei)* dialect but is evolved gradually with significant influence from its other Mizo sub-tribes such as Pawi, Hmar, Paite, Lai. The Mizo alphabet, consisting of 19 consonants and 6 vowels, was initially introduced by the pioneer Welsh Missionaries in the year 1984 and is based on roman script.

Regarding language processing, Mizo language is in its infant stage and is less-resource language. Due to its unique features and morphological structures, there are issues and challenges from the computational point of view in Mizo languages, and it is highly required to put research efforts for the development of the language processing. In this research work, the morphological structures of the Mizo language are studied and proposed the abstract representation of the morphological features of the Mizo language.

2 Related Works

The works of POS tagging started way back in the year 1960s [2–6]. Since then, various researchers have come up with different techniques and methods to improve the process of POS tagging for different languages in terms of accuracy and efficiency. Most of the earlier tagging systems were based on the rule-based method or stochastic methods.

TAGGIT [6] was one of the pioneer works on rule-based POS tagging. It was used initially for tagging of the Brown Corpus and correctly tagged 77% of the corpus; the rest was manually tagged over a period of several years. Eric Brill [7] has developed one of the most widely used rule-based POS taggers with the accuracy rate of 95–99%. The ENGTWOL [8] was other prominent rule-based taggers for English, proposed during 1995 by Karlsson et al.

The stochastic-based model has been used in the tagging system since the nineteen sixties [2, 4, 5] were among the first taggers of this kind. Several researchers employed this method and improved the performance of the tagging system. One of the notable pioneering works related to the annotated corpus was created in the English Penn Treebank project [9]. A significant amount of works has been done for English language and many other European languages [10–14]. Thorstens Brants [13] proposed efficient statistical parts of speech tagger named Trigrams'n'Tag (TNT). The tagger used the second-order Markov models for part-of-speech tagging. He claimed that the tagger not only yields better accuracy but was the fastest of its kind in training and tagging.

Muaidi [15] presented a POS tagger in Arabic languages based on neural networks. The Levenberg–Marquardt learning algorithm was utilized to train the network. The performance of the developed tagger was tested on the two datasets: training dataset (consisting of 19,848 words) and testing dataset (consisting of 4962 words). The result of 98.83% accuracy was achieved with the training dataset, and 90.21% accuracy was achieved with the testing dataset. They have proved that the Levenberg–Marquardt learning neural network is more effective than the traditional back propagation learning algorithm for tagging Arabic words.

A considerable amount of research works on POS tagging for different Indian languages have already been done. Sarkar et al. [16] presented a trigram HMM-based POS tagger for Indian languages such as Hindi, Bengali, Marathi, Telugu. The performance comparison result showed that the trigram POS tagger performed better than the bigram POS tagger for all four languages. The experiment proved that better performance was obtained for handling unknown words when the trigram tagger uses analysis of prefix, suffix and word-type than the trigram tagger that uses analysis of suffix only.

Joshi [17] have proposed Hidden Markov Model-based POS tagger for Hindi. IL POS tagset was used for the development of this tagger. They claimed that they have developed a test corpus of 500 sentences (11,720 words), and they obtained that number of correct POS tags assigned by the system was 10,798, and the accuracy was 92.13%. Much more research works on POS tagging in the Hindi language in different approaches are also available [].

Sanjeev et al. [18] proposed a POS tagger for Panjabi language using a Bigram Hidden Markov model. The HMM model was implemented using Viterbi algorithm. The model was tested on the corpus containing 26,479 words and achieved 90.11% accuracy.

Gupta et al. [19] developed an automatic POS tagger for Urdu by using HMM and CRF-based approaches. 7000 sentences were made as training data for machine learning and obtained 83.37%; 81.03% accuracy was obtained on the test dataset using CRF and HMM-based approaches, respectively.

Patra et al. [20] developed POS tagger for Kokborok using rule-based methods and statistical methods. In the system, 26 tagsets were used to tag 42,537 words in the corpus. In the statistical methods, the two machine learning classifiers and Support Vector Machines (SVM) and Conditional Random Field (CRF) were employed. Using rule-based method, the accuracy of 69% and 81.67% and 84.46% accuracies were obtained from CRF-based and SVM-based methods, respectively.

Paul et al. [21] proposed POS tagger for Nepali language using one of the most popular statistical models, i.e. Hidden Markov Model (HMM). They used the corpora collected from Technology Development for Indian Languages (TDIL) and the BIS tagset of 42 tags. 96% accuracy for known words was achieved with this tagger.

Singh et al. [22] presented POS tagger for Manipuri language and claimed that the proposed tagger was the first POS tagger for Manipuri language. They claimed that 3784 Manipuri sentences of 10,917 unique words were given as input to the tagger engine and obtained 65% single tagged correct words, 4% multiple tagged correct words, 23% unknown words tagged; wrong tagged words was 8%.

Sunita et al. [23] discussed POS tagging for Khasi language using the HMM approach. 7500 words and 312 words have been used for the training dataset and testing dataset, respectively, and obtained 76.70% accuracy.

Pakray et al. [24] has initiated the works on POS tagging for Mizo language. They laid the groundwork towards the development of the Mizo POS system in which they proposed 24 tagsets for Mizo language, by following the Penn Treebank tagset system. In their works, a Mizo-to-English dictionary was also developed which contains English words and its corresponding meaning in Mizo language, synonyms to that word and its POS tag for each synonym.

3 General Methodology of POS Tagging

The general methodology of POS tagging system is discussed in this section

3.1 Tokenization

It is the basic task of splitting or dividing sentences into pieces called tokens and removing unwanted symbols or characters from the corpus. It is the common foremost task of natural language processing.

3.2 Designing a Tagset

Tagset is a set of lexical symbols or labels, which have been used to describe the grammatical nature of tokens in the corpus. As the morphological structures of different languages are different, there can be no standard tagsets that is used by ever researcher. It is required to design a good tagset to represent morphological features of that particular language.

Table 1 Words, tones and POS class

Words	Tones	Example	POS Class
Sa (meat)	Short falling tone	***Sa** ka lei* (I purchase meat)	Noun
Sa (hot)	Long high tone.	*Tui **sa*** (Hot water)	Adjective
Sa(to build)	Short low tone	*In ka **sa*** (I build a house)	Verb
Lei(bridge)	Long low tone	***Lei** sei* (Long bridge)	Noun
Lei (tongue)	Long high tone	*Ka **lei*** (My tongue)	Noun
Lei(to buy)	Short rising tone	*Ka **lei*** (I buy/purchase)	Verb
Lei (soil)	Short rising tone	***Lei** ka lei* (I purchase soil)	Noun

3.3 *Building the Corpus or Dataset*

Corpus is a collection of a large and structured set of texts (including transcription of speech or signs) constructed with a specific purpose. One of the first basic requirements in the development of computational linguistics is having a good corpus. Corpus plays an essential role in the development of language processing and is the driving force for the supervised learning-based NLP systems. It provides a testbed for statistical linguistics analysis.

3.4 *Developing the Tagger*

Generally, the existing part of speech taggers can be grouped into three fundamental classifications, such as rule-based techniques, stochastic-based strategies and hybrid methods. Rules-based techniques use grammatical rules of the language and required contextual information in the tagging system to assign POS tags. It requires in-depth knowledge of the language to develop a proper set of rules. Stochastic-based POS tagging, on the other hand, incorporate with frequency, probability or statistics. It is based on the probability of tagging occurrence, and the system needs to be trained properly to tag corpus. The hybrid methods, of course, combine the above two methods to develop a model for the tagging systems. The efficiency and accuracy of these defined methods could be different for different languages (Table 1).

4 Discussion on Some Unique Features of Mizo Language

In this section, some unique features of the Mizo language are being discussed.

Mizo language is a tonal language. Meaning of a word could be different depending on the tones of uttering and the context in which the word is being used. According to Fanai [25], tones of Mizo language can be basically classified into four lexical tones such as low, high, falling and rising. Some linguists further add sub-divisions viz. short and long for all these basic four classifications, thus differentiating into eight different tones. How a single word indicates can vary as its tone differs are shown below:

Mizo language is free word order. Generally, Mizo language follows Object–Subject–Verb, as in *Chaw*|Object *ka*|Subject *ei*|Verb (I eat food). But other linguists mentioned that Mizo language is a 'free word order' [25]. That means that subject, object and verb in a sentence could be arranged in any order to make the sentence meaningful. For instance, consider the following sentences:

Mawiin lehkhabu a chhiar — Subject – Object – Verb
Lehkhabu Mawiin a chhiar — Object – Subject – Verb
A chhiar lehkhabu Mawiin — Verb – Object – Subject
Lehkhabu a chhiar Mawiin — Object – Verb – Object

All these sentences indicate that 'Mawii reads book'.

Gender of a person can easily identify: Another unique feature of Mizo language is that name of a person can usually be identified with the last letter of that person's name. A person's name that ends with a letter 'a' is male, and a name that ends with a letter 'i' is female. For example, *Lallawmsangi* is a female, and *Lallawmsanga* is a male. However, some exceptional cases do not follow a normal Mizo naming pattern.

Mizo language has double adverb. A double adverb is another unique feature of the Mizo language. It is a word (but is written two times) that can imaginatively describe the subject and express the character of the subject. For example, *Lui a luang ri **hum hum***', the word '*hum hum*' describes how the river flows with strength/pressure and billowing roar.

5 General Part of Speech in Mizo Language

For the study and research on part of speech in Mizo language, information is taken and produced from accessible material such as various Mizo Grammar books written by different people [26, 27]. Basic parts of speech in Mizo language are shown in Table 2.

6 Proposed POS Tagsets for Mizo Language

After thorough study and analysis of the morphological structure of Mizo *Ţawng* (Mizo language), tagsets consisting of 37 tags have been proposed covering all the grammatical information of tokens in a corpus as provided in Table 3.

Table 2 Part of Speech in Mizo Language

POS Name	Description	Example
Common noun	Generic name of things	*pangpar* (flower), *zawhte* (cat)
Proper noun	Individual name of things	*Roa* (person's name)
Abstract noun	An intangible concept such as an emotion, a feeling, a quality or an idea	*beiseina*(hope), *hlauhna*(Fear)
Personal pronoun	Simple substitute for the proper name of a person or people	*Kan* (we)
Possessive Pronoun	A pronoun that replaces a noun and shows ownership	*Ka* (my)
Relative pronoun	A pronoun used to connect a clause or phrase to a noun or pronoun	*hi,chu,kha*
Demonstrative pronoun	A pronoun that points towards the noun it replaces, indicating its position	*Saw saw, khi khi.*
Interrogative pronoun	A pronoun that is used to ask a question	*tu* (who),*eng* (what)
Verb base form	A word that is used to describe an action, state or occurrence	*thlawk* (fly), *ziak*(write)
Nounal verb	When a noun is used as a verb in a sentence it becomes a nounal verb	*Rilru puitling* (matured mind)
Double verb	The verb word is written two times to exaggerate	*Ziak ziak* (always write)
Adverb	A word that modifies or qualifies an adjective, verb or other adverb	*Tlan chak* (ran fast)
Nounal adverb	When noun is used as an adverb	*Sawi hmanlai* (saying old things)
Double adverb	When adverb is used two times in a sentence they become double adverb	*Mai mai,tak tak*
Demonstrative adverb	An adverb used to indicate the location of an event or situation	*Lo kal rawh* (come here)
Interrogative adverb	An adverb that used to ask questions related to reason, manner, time or place	*Khawiah nge* (where is it)
Adjective base form	Words that describe the qualities or states of being of nouns or pronouns	*Mawi* (beautiful), *Lian* (big)
Nounal adjective	When some noun is used as an adjective in a sentence, it becomes a nounal adjective	*vawk in* (pig house)

(continued)

Table 2 (continued)

POS Name	Description	Example
Double adjective	The adjective word is written two times to exaggerate	*Ţhaţha*(good one, s)
Comparative adjective	An adjective used to compare two objects	*zawk* (which)
Superlative adjective	Adjectives that allow the comparison of nouns to the highest or lowest limits of the quality is compared	*ber* (most)
Interrogative adjective	An interrogative adjective asks a question and describes a noun	*Khawii khuange i nih* (which town are you from)
Postposition	Words that precede or follow noun phrases	*atan*(for),*aţang* (from)
Coordinating conjunction	It is a part of speech that connects words, phrases or clause	*chuan, leh* (and)
Interjection	Expresses a spontaneous feeling or reaction	*Awi!, Karei!*
Particles	Words that have no meaning on its own	*alawm,rawh,asin*

7 Conclusion and Future Works

In this paper, the general parts of speech in Mizo have been discussed, and a list of tags consisting of 37 tagsets is proposed to tag and describes the grammatical information of words in a corpus. Efforts were made to study different tagging system, which helps to examine the detailed morphological features of the Mizo language. This work also provides a significant part-of-speech tagging in Mizo language and the unique features of the Mizo language. To create a dataset annotated with this proposed tagsets in the future, and further, it is also planned to develop a HMM-based tagger for Mizo language.

Table 3 POS tagsets design for Mizo Language

Sl. no.	Tag	Description	Example
1	CMN	Common noun	*Mipui* (crowd),*pangpar* (flower),*tlang* (mountain)
2	PPN	Proper noun	*Lali,Mawia* (Person's name)
3	ABN	Abstract noun	*Hmangaihna* (Love),*Huaisenna* (Bravery)
4	PSP	Personal pronoun	*Ka* (I), *Kan* (We),*In* (You),*An* (They)
5	POP	Possessive pronoun	*Ka* (My), *Kan* (Our), *In* (Your), *An* (Their)
6	RLP	Relative pronoun	*Kha* (that), *khing* (those)
7	MP	Demonstrative pronoun	*Hei hi la rawh* (Take this), *hei hi* (This)
8	IP	Interrogative pronoun	*Tunge* (Who),*Khawi zawk*(Which),*hei, heng*
9	VB	Verb base form	*Ding* (Stand),*Kal* (walk),*Ziak* (write)
10	NVB	Nounal verb	*Ka va Bazar ang e* (I'm going to Bazar)
11	DVB	Double verb	*Ţawng ţawngsuh.* (Don't always speak)
12	NRB	Nounal adverb	*Nupui a nei naupang hle.* (He married a very young wife)
13	RB	Adverb base form	*A mawi lutuk* (It is very beautiful)
14	DRB	Double adverb	*Ziak Char Char* (Write Continuously)
15	MRB	Demonstrative adverb	*Rawn,han,zu,va,lo, Lo kal rawh* (Come here)
16	IRB	Interrogative adverb	*Arsa hi engtinnge ka siam dawn?* (How am I going to prepare this chicken?)
17	JJ	Adjective base form	*Aizawl chu khawpui mawi tak a ni* (Aizawl is a very beautiful city)
18	MJJ	Demonstrative adjective	*Heng mi lemte hi ka ziah vek a ni* (All these pictures are sketched by me)
19	DJJ	Double adjective	*Tha tha* (good ones),*mawi mawi* (Beautiful ones)
20	NJJ	Nounal adjective	*Bawngsa* (beef, i.e. meat of cow)
21	CJJ	Comparative adjective	*Liana aiin ka sang zawk* (I am taller than Liana)
22	SJJ	Superlative adjective	*Te ber* (smallest), sang ber (tallest)
23	IJJ	Interrogative adjective	*Eng chaw nge i siam* (what food do you prepare?)
24	PPT	Postposition	*President atan an thlang* (He was elected for the President)
25	CC	Coordinating conjunction	*Lala leh Mawiate an zai mek*(Lala and Mawia are singing), & (and)
26	UH	Interjection	*Karei!*(How awful!)*Eheu!* (wow!)

(continued)

Table 3 (continued)

Sl. no.	Tag	Description	Example
27	PT	Particles	*I kaltur a ni* (You must go)
28	SYM	Symbol	[]-!@#$%^*()
29	SH	Slash	/\
30	,	Comma	,
31	.	Fullstop	.
32	:	Colon	:
33	;	Semi colon	;
34	?	Question mark	?
35	QM	Quotation mark	",", ', '
36	CD	Cardinal number	*1,2,3,pakhat,pahnih,sangkhat,za*
37	NG	Negation	*Mawi lo* (not beautiful),*Lian lo* (not big)

Acknowledgements The authors would like to acknowledge the Centre for Natural Language Processing, NIT, Silchar, and the Depart of Information Technology, Mizoram University for supporting this work.

References

1. Lorrain JH (1975) Dictionary of the Lushai languages (Reprinted) The Asiatic Society
2. Klein Sheldon F, Simmons R (1963) A computational approach to grammatical coding of English words. J ACM 10:334–347
3. Harris Z (1963) String analysis of the language structure. Linguistic Soc Am 39(3):473–478. https://doi.org/10.2307/411129
4. Stolz WS, Tannenbaum PH, Carstensen TV (1965) A stochastic approach to the grammatical coding of English. Commun ACM 399–405
5. Bahl LR, Mercer RL (1976) Part of speech assignment by a statistical decision algorithm. In: Proceedings of IEEE international symposium on information theory, pp 88–89
6. Greene BB, Rubin GM (1971) Automatic grammatical tagging of English. Technical Report, Department of Linguistics, Brown University
7. Brill E (1992) A simple rule-based part of speech tagger. In: Proceedings of the third conference on applied natural language processing, pp 152–155, March 1992
8. Karlsson F, Voutilainen A, Heikkila J, Anttila A (1995) Constraint grammar: a language-independent system for parsing unrestricted text. Mouton de Gruyter, Berlin
9. Marcus MP, Marcinkiewicz MA, Santorini B (1993) Building a large annotated corpus of English: the Penn Treebank, Comput Linguistics 19(2):313–330
10. Garside R (1987) The claws word-tagging system. The computational analysis of English: a corpus-based approach, pp 30–41
11. Solorio T, Liu Y (2008) Part-of-speech tagging for English-Spanish code-switched text. In: Proceedings of the conference on empirical methods in natural language processing, Honolulu, Hawaii, 25–27 Oct 2008
12. Chanod JP, Tapanainen P (1995) Tagging French: comparing a statistical and a constraint-based method. In: Proceedings of the seventh conference on European Chapter of the Association for Computational Linguistics, Dublin, Ireland,, pp 149–156

13. Brants T (2000) TnT-A statistical part of speech tagger. In: Proceedings of the sixth conference on applied natural language processing. Seattle, Washington, pp 224–231
14. Altunyurt L, Orhan Z (2006) Part-of-speech tagger for Turkish. Technical report, Department of Computer Engineering, June 2006
15. Hasan M (2014) Levenberg-marquardt learning neural network for part-of-speech tagging of arabic sentences. WSEAS Trans Comput 13:300–309
16. Sarkar K, Gayen V (2013) A trigram HMM-based POS Tagger for Indian Languages. In: Satapathy S, Udgata S, Biswal B (eds) Proceedings of the international conference on Frontiers of Intelligent Computing: Theory and Applications (FICTA). Advances in intelligent systems and computing, vol 199. Springer, Heidelberg
17. Nisheeth J, Darbari H, Mathur I (2013) HMM based POS tagger for Hindi. In: Proceedings of international conference on Artificial Intelligence, Soft Computing (AISC-2013)
18. Sharma SK, Lehal GS (2011) Using hidden Markov model to improve the accuracy of Punjabi POS tagger. In: 2011 IEEE International Conference on Computer Science and Automation Engineering
19. Gupta V, Joshi N, Mathur I (2016) POS tagger for Urdu using Stochastic approaches. In: Proceedings of the second international conference on information and communication technology for competitive strategies. ACM
20. Patra BG, Debbarma K, Das D, Bandyopadhyay S (2012) Part of Speech (POS) Tagger for Kokborok. In: Proceedings of COLING 2012: Posters, Mumbai, pp 923–932
21. Paul A, Purkayastha BS, Sarkar S (2015) Hidden Markov model based part of speech tagging for Nepali language. In: Proceedings of the international symposium on advanced computing and communication (ISACC), Silchar, pp 149–156
22. Singh TD, Bandyapadhyay S (2008) Morphology Driven Manipuri POS Tagger. In: Proceedings of IJCNLP-08, Hyderabad, India, pp 91–98
23. Warjri S, Pakray P, Maji AK, Lyngdoh S (2019) Identification of POS Tag for Khasi Language based on Hidden Markov Model POS Tagger. In: 20th international conference on computational linguistics and intelligent text processing, La Rochelle, France, 7–13 Apr 2019
24. Pakray P, Pal A, Majumder G, Gelbukh A (2016) Resource building and POS tagging for Mizo language. IEEE CS, pp 3–7, ISBN 978-1-5090-0323-5
25. Fanai LT (2015) Tones in Mizo language. J Human Soc Sci 1(1). ISSN(P): 2395-7352, eSSN: 2581-6780
26. Lalzarzova K (2016) MizoTawng grammar & composition (Revised & Enlarged), 1st edn. K. Sangzawna, Aizawl, Mizoram
27. Thangzikpuia PC (2019) MizoTawng Grammar (Based on its usage and unique features), 1st edn. P.C. Thangzikpuia, Aizawl, Mizoram

Chapter 50
Phishing Website Prediction: A Comparison of Machine Learning Techniques

Anjaneya Awasthi and Noopur Goel

1 Introduction

As the people are becoming more and more reliant on information technology and the services provided by its data manipulation devices and the network, people somehow are getting exposed to many threats present on the web. This reliance has made the people relaxed as well as careless too. This means as soon as the users are becoming less security concerned and aware, they are giving the chance to the hackers and other opportunists who are waiting for them to be less security aware. The attackers can attack using many techniques such as DDOS [2] and phishing attack, etc. So, the society needs the tools that make the security arrangements for their data and eventually for the people whom this data belong to. These less aware, relaxed people become the soft and easy target for the phishers and hackers and other kinds of cybercriminals who are continuously in search of these target users. As the public Internet has open and free access available to all the people in the world, this quality of the public Internet allows the miscreants do their planning and in most of the cases they successfully execute their plans too. Phishing is one of the most dangerous kinds of threat for our society and knowledge about this threat is only the way for the users to keep themselves safe from such kind of threat and it is possible only when users will be able to get the correct information about the website they are trying to visit. For this purpose, this study may be helpful. In this paper, the dataset used is having more than 2400 websites data with many attributes, and these attributes are used

A. Awasthi (✉) · N. Goel
Department of Computer Applications, VBS Purvanchal University, Jaunpur, UP, India
e-mail: anjaneyaawasthi@gmail.com

N. Goel
e-mail: noopurt11@gmail.com

I. Jeena Jacob et al. (eds.), *Data Intelligence and Cognitive Informatics*,
Algorithms for Intelligent Systems,
https://doi.org/10.1007/978-981-15-8530-2_50

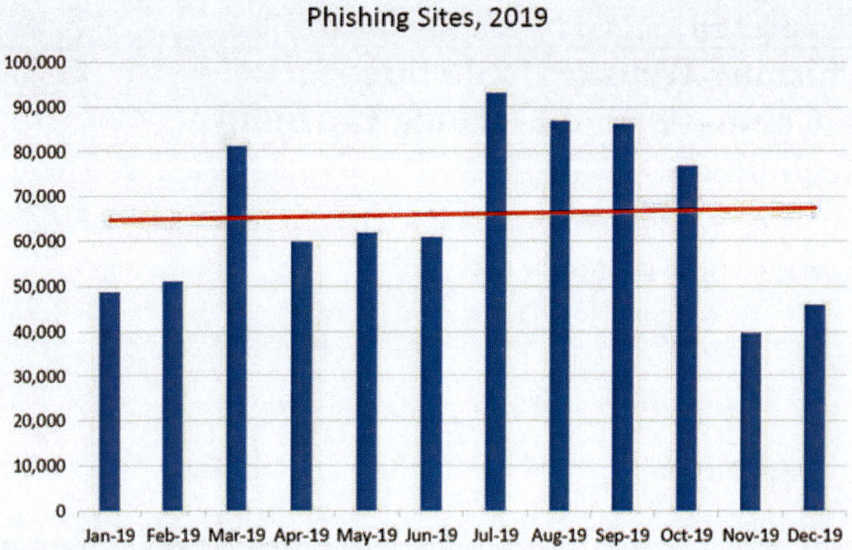

Fig. 1 Phishing sites data year 2019 [3]

as features to get the URL correctly predicted that is it a phishing or non-phishing website. As per the 4th quarter report of the year 2019 released by APWG [3] in February 2020, the total no. of unique phishing websites found in the last three months of 2019 are 76,804, 39,580, and 45,771, respectively. A number of unique phishing emails exchanged are 45,057, 42,424, and 45,072, respectively, and the no. of brands targets were 333, 325, and 341, respectively. Month-wise phishing website data is given in Fig. 1.

These attacks affected almost every sector worldwide including IT and IT-enabled services. For instance, the sector-wise 4th quarter report of these sectors is given below in Fig. 2.

A solution to this problem is the need of the hour and this comparative study is done to aid the developers in the development process of such a solution.

Section 2 presents a literature review containing previous researches by various authors from different parts of the world. Section 3 discusses the methodology used in this paper, i.e. 6 machine learning approaches to predict the phishing websites, all of the algorithms produced accuracy more than 92% in prediction where in case of random forest the achieved prediction score is of 97.56%. Section 4 presents the results and discussion where the results produced by these algorithms are displayed in both the tabular and graphical formats. Section 5 is the conclusion of this paper.

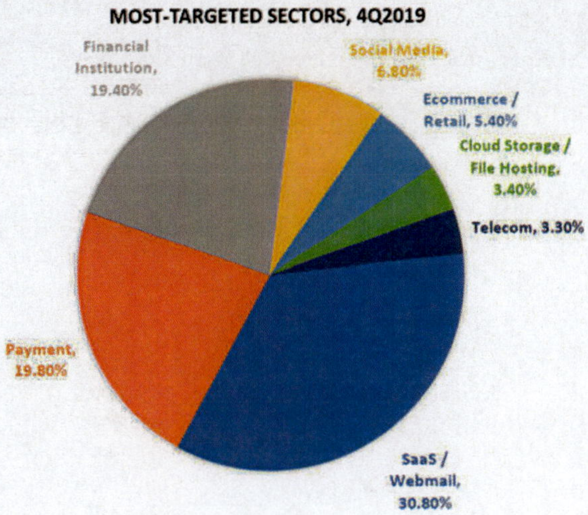

Fig. 2 Sector-wise data of 4th quarter 2019 [3]

2 Literature Review

In this section, the previous studies performed by various researchers are discussed. Some of the studies are feature extraction oriented while some have used the features that are extracted either from URLs or email addresses to detect phishing websites and web pages. Many websites are providing the phishing website data for the research and study purposes worldwide, for instance, Phishtank, APWG, and UCI, etc. The literature survey of previous studies with their predicted accuracy and limitations is given in Table 1.

Many authors have compared, tested, and exploited many of the machine learning algorithms and their functionalities. The six of them found to be the most suitable have also compared and presented a comparison of the results.

3 Methodology

In this paper, 6 machine learning classifier algorithms have been used, namely: logistic regression [6, 15–18], decision tree [6, 17], random forest [6, 16, 18], support vector machine [8], k-nearest neighbour [12], and Gaussian Naïve Bayes [10] to experiment on the dataset taken from UCI machine learning website [1]. Classified using these classifiers and presented the predictions achieved through these algorithms. The results are displayed in percentage using the tabular format as they were produced by these algorithms and also in graphical format plotted these results too for the ease of understanding of the readers.

Table 1 Literature review of previous studies

Sl. No	Author (s)	Proposed year	Classifiers used	Accuracy produced (%)	Limitations
1	Hong et al. [4]	2020	AdaBoost, RF, SVM, 1DConv, LSTM	90	Feature selection not appropriate may be improved
2	Satapathy et al. [5]	2019	NB, ELM, ANN	89.3	Very low accuracy because features not selected from a large dataset
3	Kiruthiga et al. [6]	2019	MS Visual Studio Express, C#	96	No ML classifier used and no features selected
4	Cuzzocrea et al. [7]	2019	J48, HT, RT, RF, LMT, decision stump	N/A	Less features used, feature selection may be improved
5	Mao et al. [8]	2019	SVM, DT, RF, AdaBoost	97.31	Based on CSS only, more features may be considered
6	Kang et al. [9]	2019	RF, C4.5,JRip, PART, SVM, NB	96.17	Association may be incorporated with advanced ensemble techniques
7	Wang et al. [10]	2019	GNB, LR, GBDT, CNN, RNN	96.01	Takes too long to train the classifiers
8	Jain et al. [11]	2018	NB, SVM	91.28	Very less features (5 only) used
9	Ali [12]	2017	BPNN, RBFN, SVM, NB, C4.5, kNN and RF	N/A	Accuracy not given
10	Abdelhamid [13]	2017	C4.5, eDRI, Ridor, OneRule, CR, BN, SMO, AdaBoost	96	Low phishing detection rate as compared to ML
11	Gupta [14]	2016	RF, NNC, BC	88	Very less accuracy
12	Chiew et al. [15]	2015	Logo detection with ML	93.4	Misinterpretation of logo images

3.1 Dataset Analysis

As mentioned earlier, the dataset is taken and used from the UCI Machine Learning Repository; the attributes and their column names available in the .csv file are given below in Table 2.

The dataset is divided into two parts; the first part (80% of the dataset) is used for training of the classifiers and the other part (20% of the dataset) is used for testing purpose. Then, these two are fed as input into all of our six classifiers one by one to classify the data and make the prediction. The process is explained more clearly in Fig. 3.

3.2 Machine Learning Classifiers Used

The three classifiers of machine learning which have been used are described in this section.

3.2.1 Logistic Regression Classifier

The logistic regression is derived from "logistic" and it was later called "logit" introduced by Cox [19] and Theil [20] in 1966 and 1969. Daniel McFadden [21] laid a theoretical foundation for the new logistic regression concept [22]. It predicts the probability in two values only, while a linear regression predicts the values outside the range of (0 to 1).

3.2.2 Decision Tree Classifier

A decision tree [8] is a supervised learning predictive modelling tool that generates a decision tree from root following a top-down approach that involves the partitioning of data. By splitting the data and constructing the decision tree, the dataset is then split into different attributes, and the decision node is chosen based on max data gain, keep dividing dataset by its branches. The algorithm runs recursively on non-leaf branches until data is classified.

3.2.3 Random Forest Classifier

Random forest [16] or random decision trees is an unsupervised learning technique based on ensemble learning. Used for classification and regression, it works by using multiple decision trees for training and giving output prediction in the form of a class.

Table 2 Description of the dataset

S. No	Column (attribute) name	Column description	Data type	Possible values
1	"has_ip",	IP address present in URL	Int	{1,0}
2	"long_url"	Too long length of URL	Int	{1,0,−1}
3	"short_service"	URL shortening service used	Int	{0,1}
4	"has_at"	@ Symbol within the URL	Int	{0,1}
5	"double_slash_redirect"	"//" used for redirection	Int	{1,0}
6	"pref_suf"	Unwanted prefix and suffix	Int	{−1,0,1}
7	"has_sub_domain"	If URL have sub-domain	Int	{−1,0,1}
8	"ssl_state"	URL contain SSLfinal state	Int	{−1,1,0}
9	"long_domain"	Domain registration timeline	Int	{0,1,−1}
10	"favicon"	Favicon present	Int	{0,1}
11	"port"	If port is enabled	Int	{0,1}
12	"https_token"	If HTTPS token within URL	Int	{1,0}
13	"req_url"	If requests another URL	Int	{1,−1}
14	"url_of_anchor"	If contain URL of anchor	Int	{−1,0,1}
15	"tag_links"	Tags having links	Int	{1,−1,0}
16	"SFH"	Incorrect use of server form handler	Int	{−1,1}
17	"submit_to_email"	Submitting data to an email	Int	{1,0}
18	"abnormal_url"	Abnormal looking URL	Int	{1,0}
19	"redirect"	Redirection present	Int	{0,1}
20	"mouseover"	If on_mouseover event is used	Int	{0,1}
21	"right_click"	If RightClick is disabled	Int	{0,1}
22	"popup"	If it calls for a Pop Up Window	Int	{0,1}
23	"iframe"	Iframe ambiguity	Int	{0,1}
24	"domain_Age"	If age of domain is too short	Int	{−1,0,1}
25	"dns_record"	If DNS record not present	Int	{ 1,0}
26	"traffic"	If web traffic is too less	Int	{−1,0,1}
27	"page_rank"	If Page_Rank is not appropriate	Int	{−1,0,1}

(continued)

Table 2 (continued)

S. No	Column (attribute) name	Column description	Data type	Possible values
28	"google_index"	If not found in Google Index	Int	{0,1}
29	"links_to_page"	If links are pointing to external page	Int	{1,0,−1}
30	"stats_report"	If statistical report is not appropriate	Int	{1,0}
31	"target"	Result	Int	{1, −1}

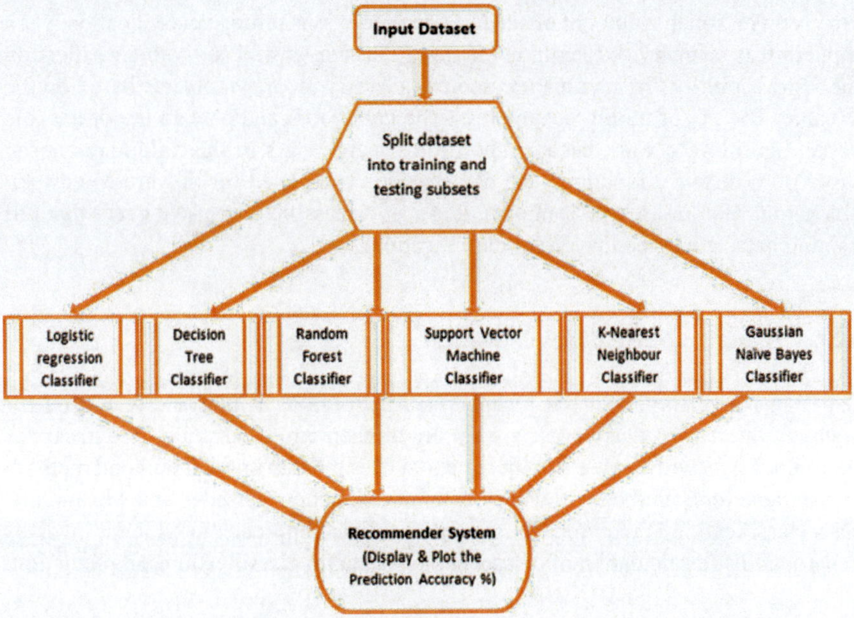

Fig. 3 Phishing website detection and prediction process

May be stated that the random forest is an improved version of a decision tree technique that exploits the capabilities of multiple decision trees with some changes. The extension includes bagging and random feature selection that suggests construction of variance controlled group of decision trees. The random forest algorithm includes bagging technique.

3.2.4 Support Vector Machine Classifier

Support vector machine [23, 24] classifier is a supervised learning tool that is useful for both regression and classification. The main theme of working of SVM is that it is a binary classification algorithm that separates the data points to find a hyperplane in case of many possible inputs. It handles the outliers efficiently and in case of high dimensional spaces works well. It uses decision functions also known as support vectors to perform classification. At least 4 types of kernels are used for classification SVC with linear kernel, linear SVC, SVC with RBF kernel, and SVC with the polynomial kernel.

3.2.5 K-Nearest Neighbour Classifier

K-nearest neighbour [24] is said to be the most simple classifier algorithm and yet it can give a reasonable amount of accuracy in predictions through classification when applied. It is a supervised learning classifier that works well for both classification and regression. It is nonparametric and can classify unknown objects based on the K-value. The no. of training examples is the value of K and play an important role in the algorithm to work because the nearest neighbours of this value are chosen based on a distance function. The distance functions used for this are Minowski, Manhattan, Hamming, and Euclidian. It is a slow classifier because it needs that full training dataset is to be loaded for classification.

3.2.6 Gaussian Naïve Bayes Classifier

Gaussian Naïve Bayes [24] is a supervised learning algorithm and it is used for predictive modelling that works based on Bayes theorem. Ideally, it is used in case of continuous data with the use of binning and it uses the maximum likelihood method. It is suitable for both the cases when binary classification is needed and when multi-class classification is required. It uses a probabilistic approach in which it involves the probabilistic calculation of dataset and test data for classification and prediction.

4 Results and Discussion

In this section, the details of data and the proposed approach are discussed in which the 6 machine learning algorithms are applied on the dataset in Python for classification and prediction of the phishing websites from the dataset, and the pre-processed dataset has been taken from UCI Machine Learning Repository: phishing websites dataset, [1]. Here, the detailed description of data is being presented, i.e. how much instances of phishing websites and how much is of non-phishing websites. The first value (−1) indicates the value of an attribute in result attribute which contains the details of the distribution values as (−1) value indicates that the website is phishing website, and (1) value indicates that the website is a non-phishing website, the total no. of instances is 2456 and the total no. of attributes is 31 (30 features and 1 result).

Total no. of instances: 2456.

Total no. of attributes: 31.

Total no. of phishing websites: 1362.

Total no. of non-phishing websites: 1094.

As per the dataset, the number of phishing website data instances is 1362, while the number of non-phishing website instances is 1094. That sums up to our total data instances, i.e. 2456. The next is to show this distribution diagrammatically, and this is shown this data in Fig. 4.

Dataset is presented using exploratory data analysis (EDA) which is an approach for data analysis that incorporates many graphical techniques are used. A histogram is used to examine the underlying structure and extraction of important variables, and to develop a model and to determine optimal factors. To represent the statistical summary of the data and to identify best features, as shown in Fig. 5.

Further, this data is input into all 6 classifiers mentioned earlier. The confusion matrix is used to present the summarised prediction results of a problem where the classifiers are used for the problem. Incorrect and correct prediction summary is broken into classes and their values are given. It shows the confusion of the classification model used while making predictions. It also gives errors and their types as well. The constituent true positive and true negative parameters are obtained from another source as the confusion matrix. The confusion matrix values true positive and true negative are obtained from the classifier's expectation and also referred to as prediction and true and false terms are obtained from external observations.

Definition of the terms:

- Positive (P): when an observation is found positive.
- Negative (N): when an observation is not found positive.
- True positive (TP): When the observation is found positive and also is predicted to be positive.
- False negative (FN): when an observation is found positive, but it is being predicted as negative.

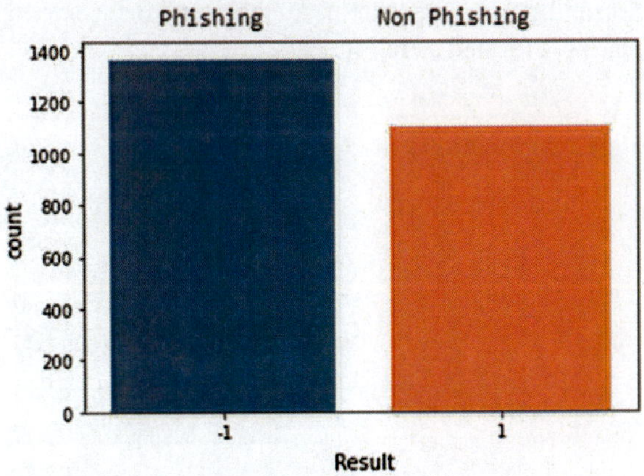

Fig. 4 Frequency of phishing and non-phishing data in a dataset

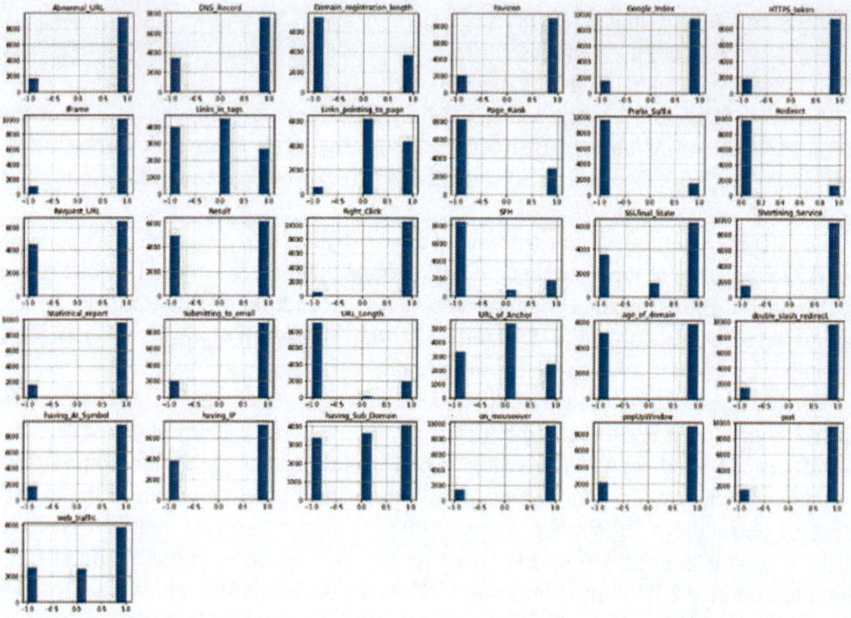

Fig. 5 Histogram for the dataset

- True negative (TN): when an observation is found negative, and it is also being predicted as negative.
- False positive (FP): when an observation is found negative, but it is being predicted as positive.

The constituent true positive and true negative parameters are obtained from other sources as the confusion matrix. The confusion matrix values true positive and true negative are obtained from the classifier's expectation and also referred to as prediction and true and false terms are obtained from external observations. The confusion matrix is formed as in Fig. 6.

Fig. 6 Confusion matrix

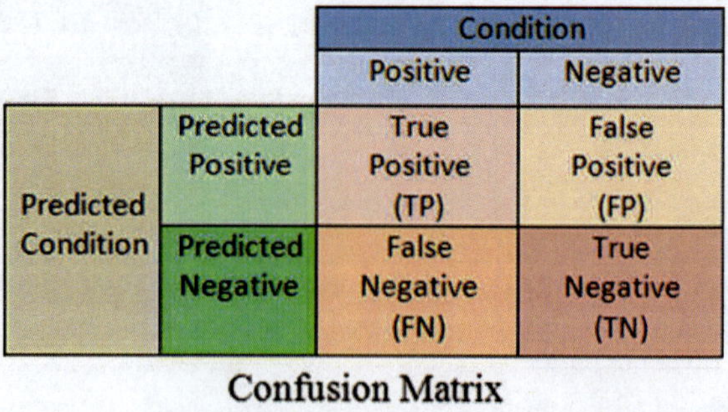

Table 3 Confusion matrix metric formula table

Type of metric	Formula	
Accuracy	$\text{ACC} = \frac{\text{tp+tn}}{\text{tp+fp+tn+fn}}$	(1)
Recall	$\text{Recall} = \frac{\text{tp}}{\text{tp+fn}}$	(2)
Precision	$\text{Precision} = \frac{\text{tp}}{\text{tp+fp}}$	(3)
F1-score	$\text{F} = 2 \cdot \frac{\text{precision} \cdot \text{recall}}{\text{precision+recall}}$	(4)
Specificity	$\text{Specificity} = \frac{\text{tn}}{\text{tn+fp}}$	(5)

A classification report contains many constituent variable parameters that show the different parameters used for the calculations of an accuracy score. Accuracy score is used to determine how good is the model in predicting the goodness of our model used for the observation. So the ability of our model has equal weights on positive and negative classes. Recall value that is also referred to as hit rate, true positive rate (TPR), or sensitivity, is calculated by using a formula as given below in Table 2. True negative rate (TNR), or selectivity, is also known as specificity is calculated by using a formula as given below in Table 2. Positive predictive value (PPV) also known as precision is calculated by using a formula as given below in Table 2. F1-score or balance f-score, also known as traditional f-measure, is the harmonic mean of sensitivity and precision as described in the description of the formula used to get these terms calculated is given below in Table 3.

All the classifiers, first, the logistic regression algorithm is applied, then next, the decision tree algorithm has been applied, and next, the random forest classifier algorithm is applied, then support vector machines classifier, then k-nearest neighbour, and lastly the Gaussian Naïve Bayes have been applied on this dataset has been chosen. For this, the data have been split as specified earlier and the results obtained in the form of the confusion matrix, and classification report and root mean squared error are given below in Table 4.

The results achieved by using these 6 classifiers are displayed in the form of a bar chart below in Fig. 7.

The results achieved using these 6 algorithms are encouraging and give an understanding to the readers for them to work further with more machine learning techniques or they may choose to work using the best performing algorithm, as it may give more accurate and refined results if different combinations of the algorithms and techniques are tried and tested, and the comparison of the result is given in Table 5 for better understanding and clarity.

Various previous researches in the area of phishing websites prediction have been selected and analysed and presented the comparative results achieved by 6 algorithms in the table. There are several other techniques in machine learning which may be of help.

Table 4 Computational table for all classifiers

Classifier used	Confusion matrix	Phishing and non-phishing	Precision	Recall	F1-score	Support	Root mean squared error
Logistic regression	[[265 8] [19 200]]	−1 1	0.93 0.96	0.97 0.91	0.95 0.94	273 219	0.4685
Decision tree	[[260 13] [13 206]]	−1 1	0.95 0.94	0.95 0.94	0.95 0.94	273 219	0.4597
Random forest	[[270 3] [9 210]]	−1 1	0.97 0.99	0.99 0.96	0.98 0.97	273 219	0.3123
Support vector machines	[[268 5] [12 207]]	−1 1	0.96 0.98	0.98 0.95	0.97 0.96	273 219	0.3717
K-nearest neighbour	[[263 10] [13 206]]	−1 1	0.95 0.95	0.96 0.94	0.96 0.95	273 219	0.4324
Gaussian Naïve Bayes	[[255 18] [23 196]]	−1 1	0.92 0.92	0.93 0.89	0.93 0.91	273 219	0.5773

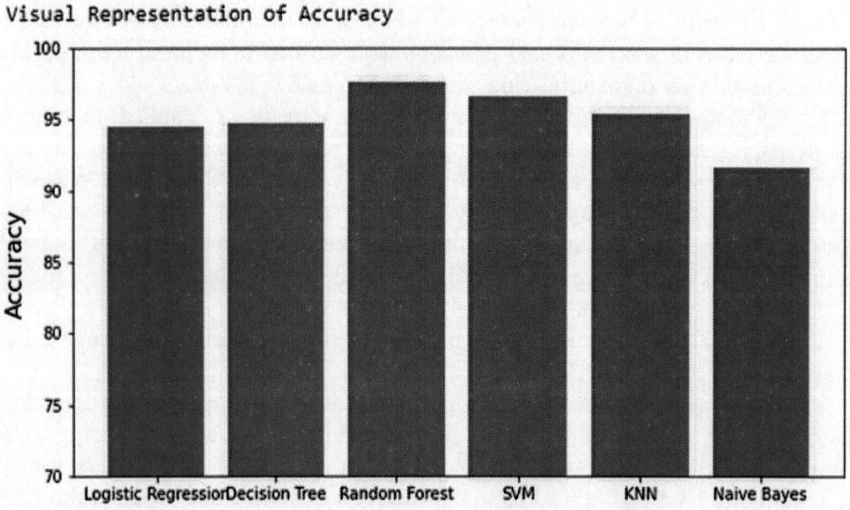

Fig. 7 Performance comparison of classifiers

5 Conclusion

The experiment has been conducted to research on a very important security aspect of phishing websites detection and prediction while understanding its importance in the life of a person who is somehow associated with any of the data stores or networks.

Table 5 Comparison of results (accuracy)

Classifier used	Accuracy achieved (%)
Logistic regression classifier	95.12
Decision tree classifier	94.71
Random forest classifier	97.56
Support vector machines Classifier	96.54
K-nearest neighbour classifier	95.32
Gaussian Naïve Bayes classifier	91.66

It is very difficult for a person nowadays to keep away from technology and the data exchange among people so it is needless to say that it has become the need of the people. To get people safe from the threats like phishing, it is much needed to develop tools that they may be helpful for the users getting properly informed about these traps. The testing is conducted and results are displayed comparatively to help for further study and research in this area. In the present study, the random forest classifier has outperformed the rest of the classifiers. In this area of research trying to work, furthermore, to contribute to making the systems more accurate and help the community fight this threat.

References

1. Dua D and Graff C (2019) UCI machine learning repository [http://archive.ics.uci.edu/ml]. Irvine, CA: University of California, School of Information and Computer Science
2. Pasumponpandian A, Smys S (2019) Ddos attack detection in telecommunication network using machine learning. J Ubiquitous Comput Commun Technol 01(01):33–44. https://doi.org/10.36548/jucct.2019.1.004
3. APWG|Phishing activity trends reports. https://apwg.org/trendsreports/. Last accessed 2020/05/05
4. Hong J et al (2020) Phishing URL detection with lexical features and blacklisted domains. In: Adaptive autonomous secure cyber systems. Springer International Publishing, pp 253–267. https://doi.org/10.1007/978-3-030-33432-1_12
5. Satapathy SK et al (2019) Classification of features for detecting Phishing web sites based on machine learning techniques 8:424–430
6. Kiruthiga R, Akila D (2019) Phishing websites detection using machine learning. 2:111–114. https://doi.org/10.35940/ijrte.B1018.0982S1119
7. Cuzzocrea A et al (2019) A machine-learning framework for supporting intelligent web-phishing detection and analysis. ACM Int Conf Proc Ser i:19–21. https://doi.org/10.1145/3331076.3331087
8. Mao J (2019) Phishing page detection via learning classifiers from page layout feature
9. Leng K et al (2019) A new hybrid ensemble feature selection framework for machine learning-based phishing detection system. Inf Sci (Ny) 484:153–166. https://doi.org/10.1016/j.ins.2019.01.064
10. Wang W et al (2019) PDRCNN : precise phishing detection with recurrent convolutional neural networks
11. Jain AK, Gupta BB (2018) PHISH-SAFE : URL features-based phishing detection system using machine learning. Springer, Singapore. https://doi.org/10.1007/978-981-10-8536-9

12. Ali W (2017) Phishing website detection based on supervised machine learning with wrapper features selection 8(9):72–78
13. Abdelhamid N et al (2017) Phishing detection: a recent intelligent machine learning comparison based on models content and features. In: 2017 IEEE International conference on ıntelligence and security ınformatics: security and big data, ISI 2017. Institute of Electrical and Electronics Engineers Inc, pp 72–77. https://doi.org/10.1109/ISI.2017.8004877
14. Comparıson of classıficatıon algorıthms to detect Phıshıng web pages usıng feature selectıon and 4 (2016). https://doi.org/10.5281/zenodo.61181
15. Chiew KL et al (2015) Utilisation of website logo for phishing detection. Comput Secur. https://doi.org/10.1016/j.cose.2015.07.006
16. Abdelhamid N, Abdel-jaber H (2017) Learning comparison based on models content and features, pp 72–77 (2017)
17. Kumar A, Gupta JBB (2018) A machine learning based approach for phishing detection using hyperlinks information number of unique Phishing sites detected. J Ambient Intell Humaniz Comput 0. https://doi.org/10.1007/s12652-018-0798-z
18. Srinivasa R et al (2019) Two level filtering mechanism to detect phishing sites using lightweight visual similarity approach. J Ambient Intell Humaniz Comput 0123456789. https://doi.org/10.1007/s12652-019-01637-z
19. Cox DR (1966) research papers in probability and statistics (Festschrift for J. Neyman). Wiley, London
20. Theil H (1969) A multinomial extension of the linear logit model. Int Econ Rev (Philadelphia) 10(3):251. https://doi.org/10.2307/2525642
21. McFadden D (1973) Frontiers in econometrics. Academic Press, New York
22. Cramer JS (2005) The origins of logistic regression. SSRN Electron J 167–178. https://doi.org/10.2139/ssrn.360300
23. Gudikandula P (2019) Recurrent neural networks and LSTM explained. 01(01):33–40
24. Master Machine Learning Algorithms. https://machinelearningmastery.com/master-machine-learning-algorithms/. Last accessed 2020/06/16

Chapter 51
Comparison of Computer Vision Techniques for Drowsiness Detection While Driving

Deepanshu Yadav, Divya Mohan, and Amrita Jyoti

1 Introduction

Accidents occur almost daily on the road, on the tracks, and sometimes also in the air. About 40% of road accidents occur due to drained drivers drowsing off while driving, and this is said in a study by Central Road Research Institute (CRRI) on the 300-km expressway connecting Agra and Lucknow [1]. In the years 2016 and 2017, the police department of states/UTs recorded casualties of 1,50,785 persons and 1,47,913 persons, respectively. This means that on average, around 400 people lose their lives on Indian roads [2].

There are several promising techniques to detect drowsiness in drivers, and keeping an eye on the drivers all the time is an effective one. A computer vision system helps in developing methods that attempt to mimic the behavior and functionality of human vision. Computer vision is one of the emerging fields which is focused on helping computers to see. It is a specialization of artificial intelligence and machine learning and hence thereby has the potential to solve a large number of real-life problems. There are many computer vision algorithms present to detect drowsiness in the driver. These algorithms first perform automated extraction eyes from images of a driver's face and then process it to determine drowsiness in the driver's eye. But the question arises which algorithm bypasses another in terms of efficiency and time complexity.

D. Yadav (✉) · D. Mohan · A. Jyoti
CSE Department, ABES Engineering College, Ghaziabad, India
e-mail: deepanshu.16bcs1077@abes.ac.in

D. Mohan
e-mail: divya.16bcs1173@abes.ac.in

A. Jyoti
e-mail: amrita.jyoti@abes.ac.in

I. Jeena Jacob et al. (eds.), *Data Intelligence and Cognitive Informatics*,
Algorithms for Intelligent Systems,
https://doi.org/10.1007/978-981-15-8530-2_51

To understand the working of this project, one should be familiar with the concept of neural networks, machine learning, or artificial intelligence in general.

1.1 Objective

The objective of this is to implement and compare two most promising and different approaches available in the market to detect if a person (driver, to be specific) is in a paired or an impaired state while driving. The sole reason for this research-based project is to compare the outcomes of each solution on various parameters so that a visible and quantifiable difference could have been between both, and so, then people could use any as per their requirement.

2 Related Work

Various approaches and products have been made for this problem, and a selected few have been considered while moving on with this project.

In this paper [3], the authors have proposed and developed "THE COPILOT" which is a Second Generation PERCLOS Monitor, which is funded by and the Federal Motor Carrier Administration and National Highway Traffic Safety Administration of USA. PERCLOS is defined as the ratio/percentage of the time duration for which the subject's eyes are closed over a specified period. The Copilot consists of a digital camera integrated with a low-cost digital signal processor (DSP) and uses infrared light to detect the state of eyes (open or close) in both day and night time. It is a very low-cost drowsiness monitor aimed for commercial use.

In this paper [4], candidates have used the Viola–Jones algorithm to detect the facial expression and as well as the location of the eyes. Eyeblink duration and blink frequency are the criteria for detecting drivers' levels of drowsiness by eye tracking, and PERCLOS was used to confirm the results. The main focus was to determine the relative levels of drowsiness among the subjects and not the total driving fatigue levels as investigating issues such as fatigue entail a lot of effort and are highly complex. A software was developed that could receive color frames from the camera placed in front of the driver and calculate coordination of facial details. Finally, by comparing and fissuring, the information obtained from interpreting assessment criteria about the alertness or sleepiness together with information resulted from image processing, and the software for detection of the levels of drowsiness was developed and promoted.

In this paper [5], the method uses the OpenCV Python-based engine to measure for how long the driver's eyes were closed and to detect the drowsiness of the driver. If the eyes were closed for some threshold time duration, it was assumed that they are starting to sleep, so play an alarm to alert them and save them from any potential accidents.

The camera Logitech C920 was used for this project because it is compatible with Raspberry Pi devices. The eye aspect ratio, on the other hand, is a more convenient solution that has a very simple calculation which is based on the ratio of distances between facial landmarks of the eyes. This method is fast, efficient, and easy to implement for eye blink detection. The SciPy package was used to compute the Euclidean distance between facial landmark points in the EAR calculation.

In this paper [6], various methods for detection of drowsiness while driving have been looked upon and discussed. There are three different techniques currently in consideration that are being used by researchers to detect drowsiness which is:

(A) Image processing-based techniques
(B) Artificial neural network-based techniques
(C) Electroencephalograph(EEG)-based techniques.

Upon looking in all methods, Jay D. Fuletra and Dulari Bosamiya have concluded that EEG is not feasible as no one can tie electrodes to themselves while driving; image processing although sometimes produces false alarms and is the most feasible of all due to easy implementation and consideration of various surrounding changes in images/live video feed.

3 Proposed Methodology

3.1 Approach-1. (Dlib + EAR)

See Fig. 1.

The following steps show the working of this approach.

Step 1 Input image frames from a car desk camera.
Step 2 Use Dlib to search for the eyes and mark the coordinates of the eye.
Step 3 Use eye coordinates to calculate the EAR.
Step 4 Count the consecutive frames with EAR close to zero.
Step 5 Trigger alarm if the count reaches a certain threshold.

Dlib is a modern C++ toolkit developed by Davis E. King. It contains machine learning algorithms and is open-source software released under a Boost Software

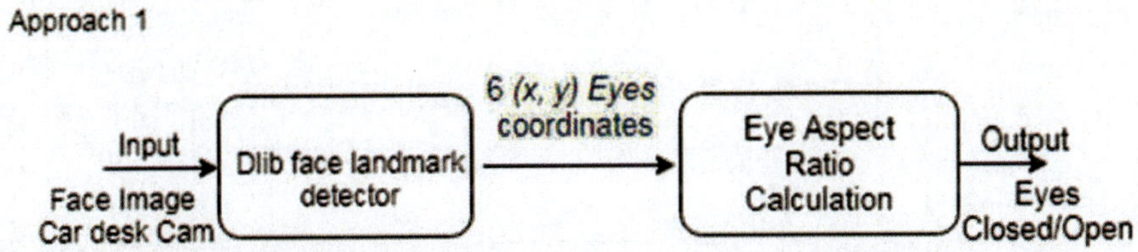

Fig. 1 Block diagram approach 1

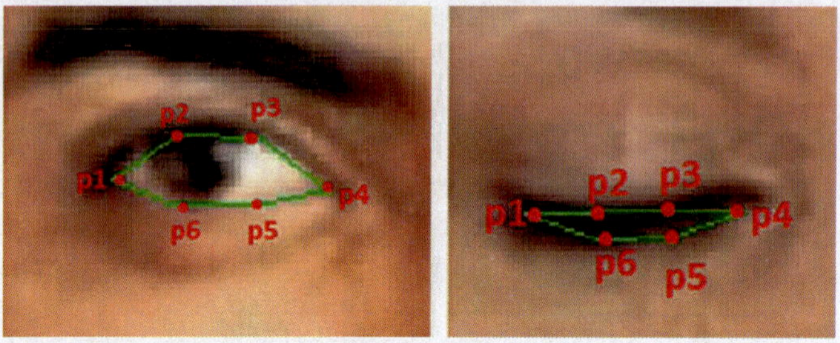

Fig. 2 6 (x, y) coordinates associated with the eye

License [7]. Dlib library includes a facial landmark detector, which is an implementation of the One Millisecond Face Alignment with an Ensemble of Regression Trees [8] paper by Kazemi and Sullivan [8]. This facial landmark detector receives image frames from the video of the driver's face by the car desk camera and returns 6 (x, y)-coordinates representing the borderline of each eye.

The height and width of these eye coordinates have a relation between them, given by Čech and Soukupová in their paper released in 2016, "Real-Time Eye Blink Detection using Facial Landmarks" [9], and following is an equation to make this relation estimable called the eye aspect ratio(EAR):

$$\text{EAR} = \frac{\|p2 - p6\| + \|p3 - p5\|}{2\|p1 - p4\|} \tag{1}$$

In the EAR equation $p1, p2, p3, p4, p5, p6$ are 6 (x, y) coordinates associated with the eye as shown in Fig. 2. When an eye is open, then EAR is mostly constant and is approaching zero while closing an eye. A counter variable stores frame counts when EAR is close to zero and resets if EAR is not close to zero; if EAR remains close to zero for certain consecutive frames, an alarm is triggered to notify the driver.

3.2 Approach 2: (Haar-Cascade + CNN)

See Fig. 3.

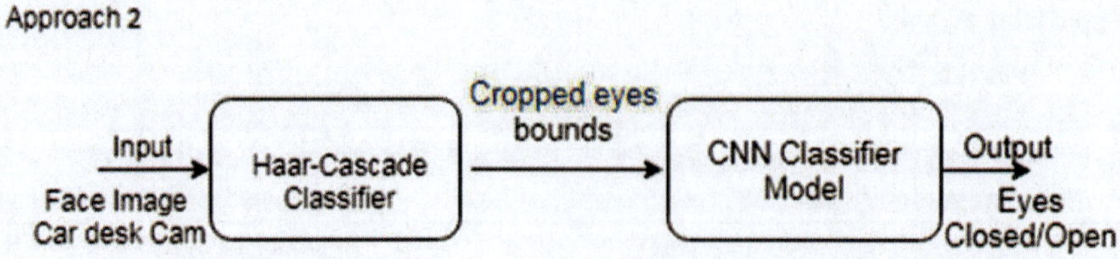

Fig. 3 Block diagram approach 2

Haar feature-based cascade classifiers are another effective object detection method based on the machine learning algorithms, which was proposed by Michael Jones and Paul Viola in their paper, “Rapid Object Detection using a Boosted Cascade of Simple Features” in 2001. In this approach, the cascade function is trained using a lot of negative and positive images. Positive images contain the object which is to be detected and negative images do not.

In this approach, OpenCV (Haar-cascade classifier) is used for collecting images from a camera and feeding them into a deep learning model [9] to determine whether the person’s eyes are “Closed” or “Open.” The following steps show the working of this approach.

Step 1 Input image frames from a car desk camera.
Step 2 Use OpenCV to detect eyes and feed them to the classifier.
Step 3 The classifier will label whether eyes are closed or open.
Step 4 Count the consecutive frames with closed eyes.
Step 5 Trigger alarm if the count reaches a certain threshold.

Keras using convolutional neural networks (CNNs) is used to build the classifier model. CNN is a deep neural network that is very popular for image recognition and performs splendidly for photo classification purposes. For training of classifier, 7000 images of eyes were segregated and labeled as open/close. This dataset was diverse based on contrast and lighting condition. The following are the layers of CNN model architecture used in the approach:

- Convolutional layer; 32 nodes, kernel size 3
- Convolutional layer; 32 nodes, kernel size 3
- Convolutional layer; 64 nodes, kernel size 3
- Fully connected layer; 128 nodes.

3.3 Comparison Scenario

Accuracy and time are the two major bases of our comparison to compare the results of both approaches. Accuracy helps to define the correctness of the approaches, and time helps to determine whether or not alarm will be triggered on time in order to alert drivers.

The comparison has been divided into two steps:

First, the accuracy and time are compared for the first stage of both approaches (i.e., Dlib vs. Haar-cascade). This helps to justify which algorithm could detect eyes efficiently and also, checks if wearing glasses affects the accuracy of any approach, and by how much percentage?

Secondly, the accuracy at the second stage is also compared (i.e., EAR vs. CNN). This helps to determine how well an approach detects the state of eyes as closed or open.

4 Implementation and Results

4.1 Results for Dlib (Approach-1) and Haar-Cascade (Approach-2).

See Fig. 4.

Table 1 and Fig. 5 compare the ability of both approaches to detect eyes from image frames. For comparison, three different types of datasets are taken.

1. Normal face images dataset: collection of 3189 images of the frontal face having dimensions 200 height × 200 height. Faces in images are diverse in contrast, complexion, and ages.
2. Face images with glasses dataset: collection of 250 images of a frontal face wearing glasses having dimensions 200 height × 200 height. Faces in images are diverse in contrast, complexion, and ages.

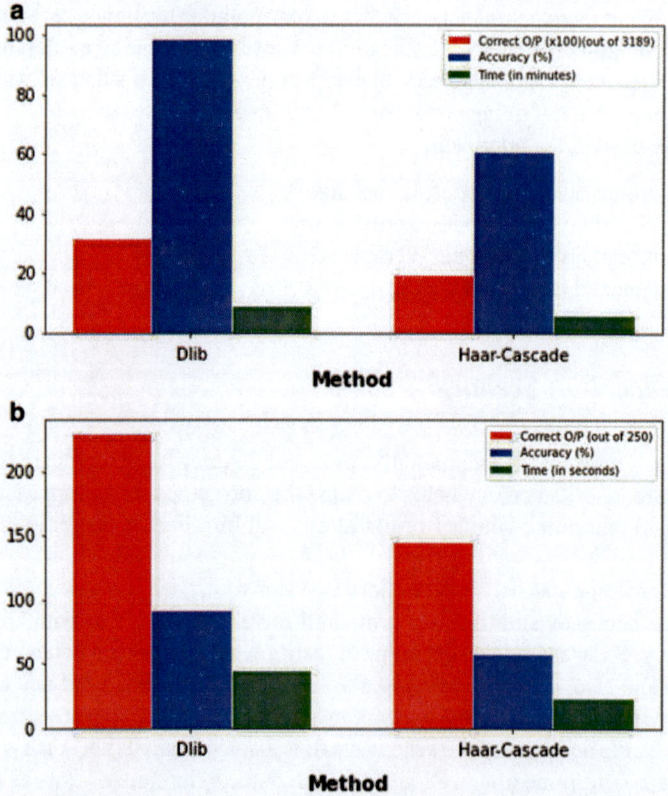

Fig. 4 **a** Dlib versus Haar-cascade for normal face dataset, **b** Dlib versus Haar-cascade for faces wearing glasses dataset

Table 1 Comparison table for Dlib and Haar-cascade

S. No.	Task	Input	Dlip results			Haar-cascade results		
			Correct O/P	Accuracy (%)	Time (min:s)	Correct O/P	Accuracy (%)	Time (min:s)
1	Detect eyes	Normal face images	3143/3189	98.56	9:20	1945/3189	60.99	6:05
2	Detect eyes	Face images with glasses	229/250	91.6	0:45	145/250	58.0	0:24
3	Detect eyes	Driving face cam video		93.44	8:22		68.50	5:23

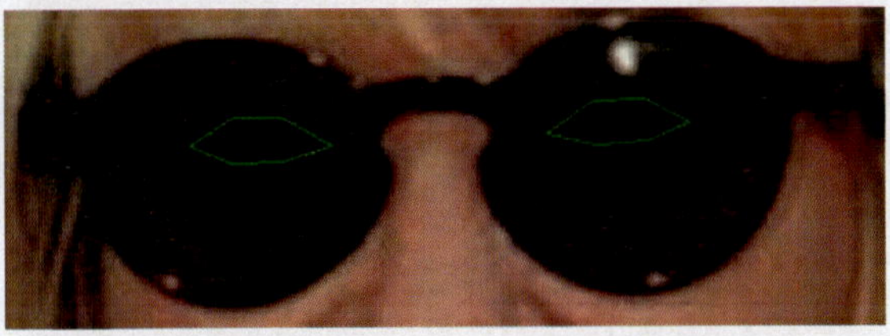

Fig. 5 Example of error result of Dlib. *Note* Results of time in both approaches might change as per system configuration, but the relationship shown in output will remain the same

3. Driving facecam video: The length of the video is 1 min. Frame dimensions are 508 height × 360 widths, and the frame rate is 30 frames per second.

5 Observation from the Results and Above Table:

- The average of the difference between the accuracy of Dlib and Haar-cascade is 32.03%, which denotes Dlib is much more efficient as compared to Haar-cascade to detect eyes from frames.
- The average of the difference between the total runtime of Dlib and Haar-cascade is 2:11 min, which denotes Haar-cascade is much more time-efficient as compared to Dlib to detect eyes from frames.
- In the case of Dlib, the accuracy dropped by 6.95% for the dataset of faces with glasses and a 3% drop in case of Haar-cascade for the same dataset.
- Dlib has a tendency to sometimes assume coordinates of eyes even though eyes are not visible, and this is done by reading other facial landmarks like the nose, eyebrows, etc., as shown in Fig. 5. This ability of this algorithm might prove helpful in other cases, but in our cases, it reduces the accuracy of the algorithm.

Table 2 Comparison table for EAR and CNN

S. No.	Task	Input	Ear results			CNN results		
			Correct O/P	Accuracy	Time (min:s)	Correct O/P	Accuracy	Time (min:s)
1	Eyes closed/open	Images with 100% eye detection	245/258	94.96%	0:32	253/258	98.06%	0:34
2	Eyes closed/open	Normal face images	235/250	94.0%	0:37	103/250	41.20%	0:32

5.1 *Following Are the Results for EAR (Approach-1) and CNN (Approach-2)*

Table 2 and Fig. 6 compare the ability of both approaches to detect whether the eye is open/closed. For comparison, two different types of datasets are taken.

1. Images with 100% eye detection: Selective 258 images from normal face images dataset on which both Dlib and Haar-cascade detected eyes.
2. Normal face images dataset: collection of 250 images of the frontal face having dimensions 200 height × 200 height. Faces in images are diverse in contrast, complexion, and ages.

5.2 *Following Are the Examples of the Output of Eye Aspect Ratio (Approach 1):*

See Fig. 7.

5.3 *Following Are the Examples of the Output of Haar-Cascade (Approach 2):*

See Fig. 8.

5.4 *Observation from the Results and Above Table:*

- For the CNN classifier, true positive (TH) = 253 and false positive (FP) = 5 Therefore, sensitivity = 0.98.

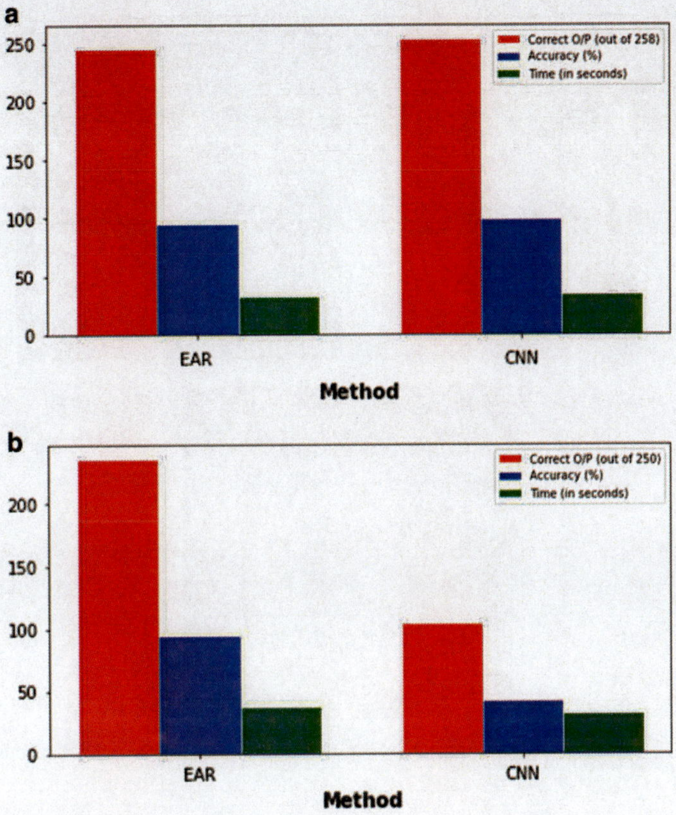

Fig. 6 **a** EAR versus CNN for images with confirmed eye detected dataset, **b** EAR versus CNN for images with normal face dataset

- Point 1 shows that the accuracy of CNN is 2.7% more than EAR when the correct output of Dlib and Haar-cascade is given to EAR and CNN, respectively.
- Point 2 shows that the overall accuracy of approach 2 reduced to 41.20% from 98.06% majorly because of low accuracy of Haar-cascade, whereas only a slight dip in accuracy is observed in approach 1.

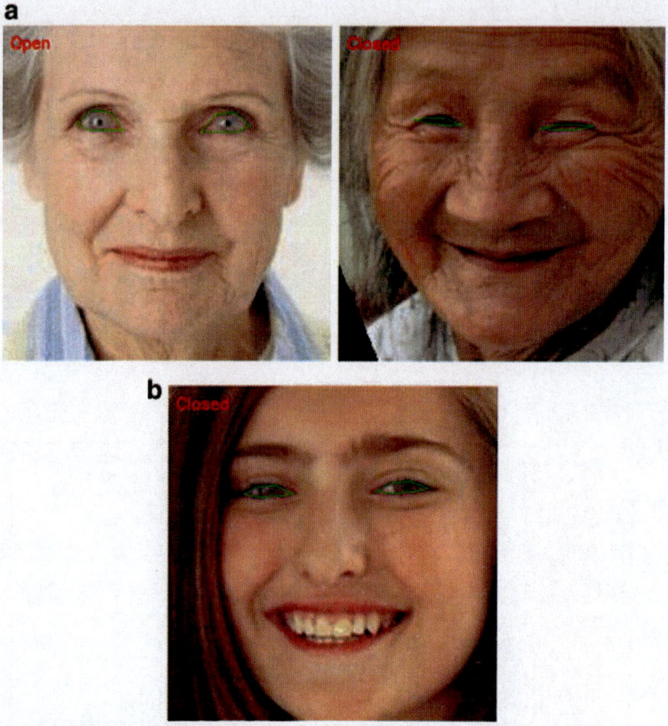

Fig. 7 **a** Correct output, **b** error output

6 Conclusion

The results of Table 1 clearly indicate that Dlib is much more accurate than Haar-cascade to detect eyes from image frames. By implementing Dlib on a high configuration system will reduce the time barrier in order to alert drivers on time. Eye aspect ratio is a simple and yet efficient approach to detect whether eyes are closed/open but produces error results under two scenarios: first, when the driver is smiling which tends to reduce the size of the eye and secondly, the driver having exceptionally small eyes.

The accuracy of both the algorithms reduces in case the driver wears glasses. This is highly dependent on the visibility of the eye to the face camera while wearing glasses.

The overall conclusion from the results is to use Dlib and EAR (approach 1) for more accurate drowsiness detection in drivers, whereas another approach of using Dlib with CNN classifiers could also be explored.

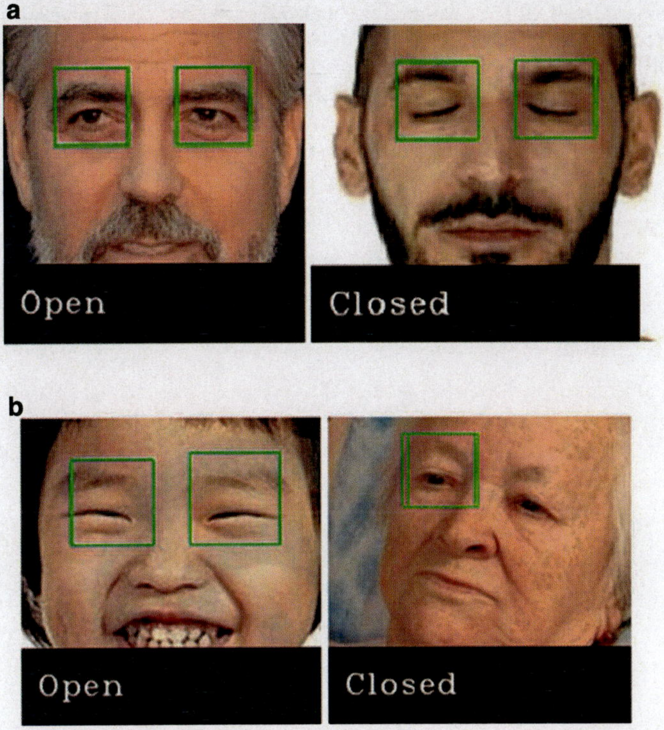

Fig. 8 **a** Correct output, **b** error output

References

1. Retrieved from https://www.financialexpress.com/india-news/40-of-highway-accidents-occur-due-to-drivers-dozing-off/1659901/
2. Retrieved from https://www.motorindiaonline.in/driver-welfare/drowsy-driving-a-safety-challenge/
3. Grace R, Steward S (2001) Drowsy driver monitor and warning system. In: Proceedings of the first international driving symposium on human factors in driver assessment, training, and vehicle design: driving assessment 2001 https://doi.org/10.17077/drivingassessment.1010
4. Adrian Rosebrock, Author at PyImageSearch—Page 21 of 43. (2017, June 19). Retrieved from https://www.pyimagesearch.com/author/adrian/page/21/
5. Poursadeghiyan M, Mazloumi A, Nasl Saraji G, Baneshi MM, Khammar A, Ebrahimi MH (2018) Using image processing in the proposed drowsiness detection system design. Retrieved from https://www.ncbi.nlm.nih.gov/pmc/articles/PMC6174048/
6. Fuletra J, Bhatt D (2013) A survey on driver's drowsiness detection techniques. Int J Recent Innov Trends Comput Commun. 1:816–819
7. Dlib C Library. (n.d.). Retrieved from https://Dlib.net/
8. Kazemi V, Sullivan J (2014) One millisecond face alignment with an ensemble of regression trees. In: 2014 IEEE conference on computer vision and pattern recognition. https://doi.org/10.1109/cvpr.2014.241

9. Soukupova T, Cech J (2016). Real-time eye blink detection using facial landmarks. Retrieved from https://vision.fe.uni-lj.si/cvww2016/proceedings/papers/05.pdf
10. Team DF (2020) Intermediate python project—driver drowsiness detection system with OpenCV & Keras. Retrieved from https://data-flair.training/blogs/python-project-driver-drowsiness-detection-system/

Chapter 52
Machine-Learning Based Fault Diagnosis of Electrical Motors Using Acoustic Signals

Ravi Teja Grandhi and N. Krishna Prakash

1 Introduction

Electrical machinery like induction motor and DC motors has a drastic growth in its production and hence its usage, as the establishment of small scale industries is on the rise globally. These machinery are mostly of a rotating type which has bearings, rotors and stator circuits demanding regular maintenance to avoid the performance degradation in the machinery as well to avoid the loss in production to the industries [1]. The rotating parts of the electrical motors normally wear-out due to lack of regular maintenance or more operating time. To diagnose the fault at the early stages of failure and avoid the breakdown of motors, normally, industries follow maintenance techniques like preventive and predictive maintenance where the former includes time-based monitoring or usage-based monitoring and the latter includes condition-based monitoring of the machinery.

Time-based monitoring is the traditional approach that is to monitor the motors on a regular time to time basis and usage-based maintenance technique is used when the usage of the machinery is not the same all the time. Condition-based monitoring is another predictive approach resulting in maintenance cost reduction over the other techniques as it produces a warning before the failure of the machinery by monitoring vibration of the motor, acoustics of the motor, and motor current. This method involves motor current signature analysis (MCSA) in which stator and motor current are analyzed to diagnose the fault in the machinery [2, 3].

The Usage of faulty motors may result in accidents and loss in production. Often repair and replacement of these motors cost to the original price of the machinery. MCSA is done to identify the faults in the rotating machines. Stator current signals

R. T. Grandhi (✉) · N. Krishna Prakash
Department of Electrical and Electronics Engineering, Amrita School of Engineering, Amrita Vishwa Vidyapeetham, Coimbatore, India
e-mail: raviteja.grandhi0112@gmail.com

I. Jeena Jacob et al. (eds.), *Data Intelligence and Cognitive Informatics*,
Algorithms for Intelligent Systems,
https://doi.org/10.1007/978-981-15-8530-2_52

are processed and analyzed for broken rotor bars, misalignment, and bearing defects [4, 5]. Several acoustic signal-based approaches had been proposed which resulted in better efficiencies in the diagnosis of both electrical and mechanical faults (bearings, broken bars, etc.) in motors [6–10]. Machine-learning-based fault diagnosis methods were proposed to classify the faults [11, 12]. Statistical feature extractions like mean, standard error, minimum, maximum, sum, median, standard deviation, skewness, kurtosis, sample variance, and range from the recorded signal were used to classify the faults. Classifiers like support vector machine, decision tree, Learning Vector Quantization (LVQ) were used for accurate classification [13, 14]. Studies suggest that acoustic signal processing is advantageous over other techniques when the procedure of measuring the machine parameters involves high risk appending the fact that the cost of sensors used to measure the machinery metrics is greatly reduced [15]. This paper proposes the usage of smartphones for recording the acoustic signals and performing the classification using machine learning techniques thereby reducing the significant hardware involved in the literature mentioned.

2 Proposed Approach

The acoustic signals are recorded from a three-phase induction motor with mobile phones. Formation of dataset involves two stages as shown in Fig. 1: (i) filtering and splitting the recorded acoustics into a specific length of two seconds each; (ii)

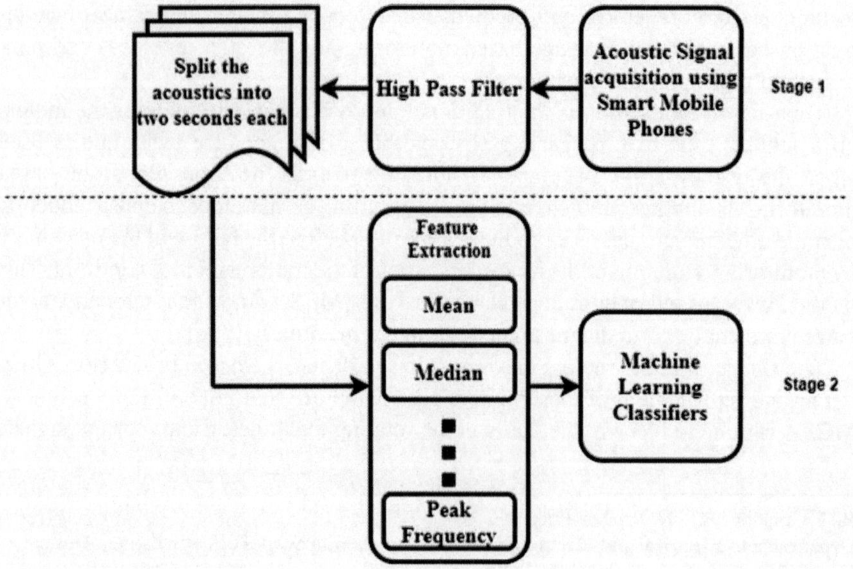

Fig. 1 Proposed approach

Table 1 Time-domain features

Features	Formulae
Mean	$\frac{\sum_{i=1}^{n} Xi}{n}$ X_i, the array of the normalized frames and n is the total number of samples
Median	$X\left(\frac{n+1}{2}\right)$ X is the set of normalized frames
Skewness	$\frac{3(\text{Mean}-\text{Median})}{\text{Standard Deviation}}$
Kurtosis	$E\left[\left(\frac{(X-\mu)^4}{\sigma}\right)\right]$

extraction of the features from the split signals and classifying using machine learning techniques. The acoustic data of one-minute duration from the three-phase induction motors are recorded using the voice recorder application in the mobile phone under no-load and loaded conditions.

Ambience noises usually of lower frequencies in the recorded acoustic signals are eliminated using the high pass filter technique designed for a specific cutoff frequency. Fast Fourier Transform (FFT) is applied to the filtered signal for conversion into frequency domain signals. Once the cutoff frequency is decided, the lower frequencies below that cutoff frequency are nullified and the reconstructed signal is obtained using inverse FFT for further feature extraction. The filtered signals are split into smaller samples with a specific length of two seconds each. To enhance the input signal and to create an accurate dataset for the machine learning algorithms the filtered signals are split into smaller samples of two seconds each.

To convert the time-domain acoustic signals as a structured dataset for training and testing, statistical features are required. Time-domain features are extracted from the split signals which include mean, median, skew, kurtosis, and signal peak-frequency in the frequency domain. Normalizing the amplitude of the signal is done to change the amplitude to meet particular criteria. One such type of normalization is to change the amplitude between -1 and $+1$ is represented in Eq. (1).

$$\text{NormalizedAmplitude} = \frac{X - \min(X)}{\max(X) - \min(X)} \quad (1)$$

where X represents the set of un-normalized amplitude values.

Mathematical representation of time-domain features [13] is tabulated in Table 1.

3 Machine Learning Models

Machine learning (ML) algorithms perform classification or regression on a given set of data. ML algorithms are used to perform fault classification and three models

Decision Tree, K-Nearest Neighbor, and Support Vector Machine are considered for the proposed fault diagnosis.

The Decision tree is a classification model in supervised learning, where supervised learning means that the target class of the samples is known. In this model, the prediction is done using the information gain of an individual feature of the data set based on entropy [1]. The nodes in the decision tree are divided based on the highest information gain. Exclude one by one feature and calculate information gain for each feature in order to get the leaf nodes. The mathematical formula behind entropy is represented in Eq. (2).

$$\text{Entropy} = \sum_{i=1}^{c} -p_i * \log_2(p_i) \tag{2}$$

where p_i is the frequentist probability of a class in the data set, the maximum entropy is n for 2^n classes. The Information Gain (IG) that is calculated from the entropy at each level is mentioned in Eq. (3).

$$\text{IG}(Y, X) = E(Y) - E(Y|X) \tag{3}$$

where the information gain is calculated by subtracting the entropy of Y given X from just entropy of Y which reduces the uncertainty about Y.

K-nearest neighbor, a commonly used classification model uses the Euclidean distance metric represented in Eq. (4) to predict the class of an unknown sample based on the pre-defined value of K [1].

$$\text{Euclideandistance} = \sqrt{\sum_{i=1}^{n} (X_i - Y_i)^2} \tag{4}$$

where X and Y in (4) are the two data points.

Support Vector Machine (SVM) uses statistical learning theory combined optimization techniques to perform both classification and regression [13].

For assessing AI models few metrics can be calculated from the confusion matrix. They are precision, recall, $f1$-score, and Mathew correlation coefficient. $F1$-score is the function of precision and recall whose mathematical equation is described in Eq. (5).

$$F1 - \text{score} = \frac{2 * (\text{precision} * \text{recall})}{\text{precision} + \text{recall}} \tag{5}$$

Mathew correlation coefficient has its advantages over the $f1$ score in finding the best model based on the dataset [16]. The mathematical equation of the Mathew correlation coefficient is given in Eq. (6).

$$\mathrm{MCC} = \frac{((\mathrm{TP} * \mathrm{TN}) - (\mathrm{FP} * \mathrm{FN}))}{\left(\sqrt{(\mathrm{TP} + \mathrm{FP})(\mathrm{TP} + \mathrm{FN})(\mathrm{TN} + \mathrm{FP})(\mathrm{TN} + \mathrm{FN})}\right)} \tag{6}$$

where TP is True Positive, TN is True Negative, FP is False Positive, and FN is False Negative. MCC shall be the model metric to select the best model.

4 Results and Discussion

Acoustic signals are recorded using different model mobile phones at sampling rates 44.1 and 48 kHz for the brand new, to-be-faulty and faulty induction motors.

The recorded acoustic signals before filtering are shown in Figs. 2 and 3, the signals are filtered using the high pass filter by choosing a cutoff frequency of 1 kHz to remove the ambient noise further split into smaller samples and then normalized.

Feature dataset is calculated from the filtered and split acoustic signals which are normalized are shown in Figs. 4 and 5 thereby forming the input data for the ML

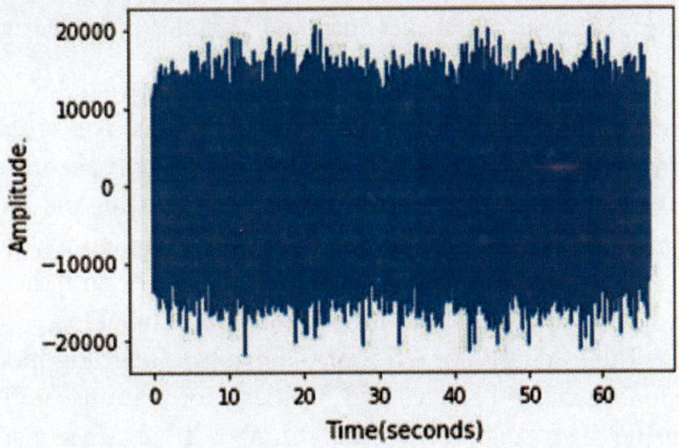

Fig. 2 A pre-filter acoustic signal in time-domain

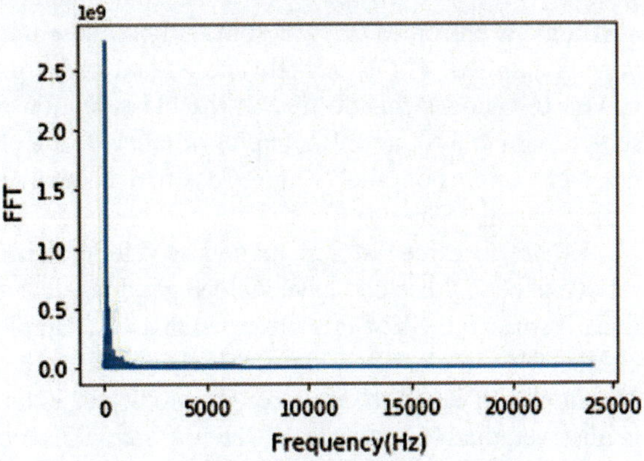

Fig. 3 Frequency components in the acoustic signal before filtering

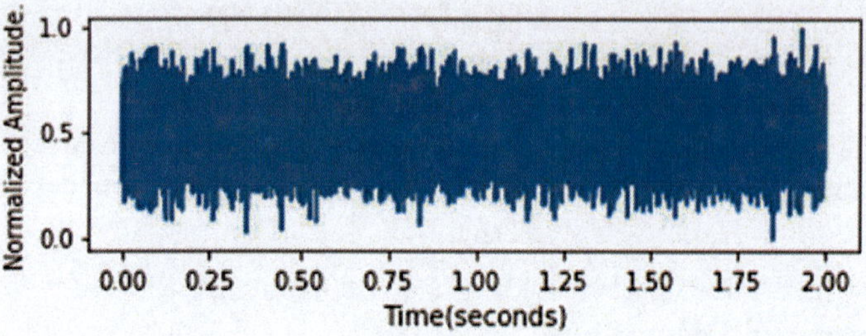

Fig. 4 Acoustic signal after filtering and splitting in the time domain

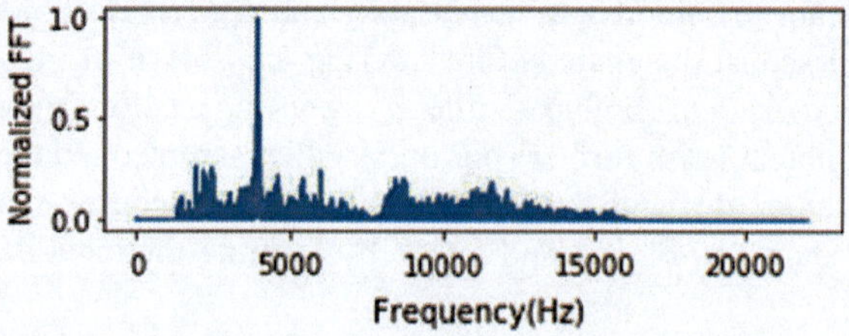

Fig. 5 Acoustic signal after filtering and splitting in the frequency domain

models. The feature dataset for 953 samples is divided as a training set and testing set where 666 samples (70%) of the dataset is used for training the models and 286 samples (30%) of the dataset is used for testing the models. The samples of the test data for an induction motor that is aged and considered to be faulty are labeled as label '0' whereas samples of the test data for an induction motor which is currently working in proper condition is labeled as label '1'.

The evaluation metric for the machine learning models includes accuracy, precision, recall, $F1$ score, and Mathew correlation coefficient. $F1$ score is calculated using the precision and recall of label '0' data and also for label '1' in each model.

For the decision tree model, the root node for this data set is the peak-frequency with the entropy of 0.966. Other features that are split into the leaf nodes have the next highest entropies in each split and continue until the entropy of the last leaf node is nullified. The base node and its leaf nodes for this data set are represented in Fig. 6. Using Decision tree as the classification model it is observed that 165 samples of label '0' and 91 samples of label '1' are classified correctly whereas the other 11 samples of label '0' are classified as label '1' and 19 samples of label '1' are classified as label '0'.

As there are three different kernels used for classification using SVM, for the data set extracted in this study have yielded good results with Polynomial kernel than the linear kernel. For SVM it is observed that 152 samples of label '0' and 99 samples of label '1' are classified correctly whereas the other 24 samples of label '0' and 11 samples of label '1' are wrongly predicted. Finally, for K-Nearest Neighbor it is observed that 164 samples of label '0' and 95 samples of label '1' are classified

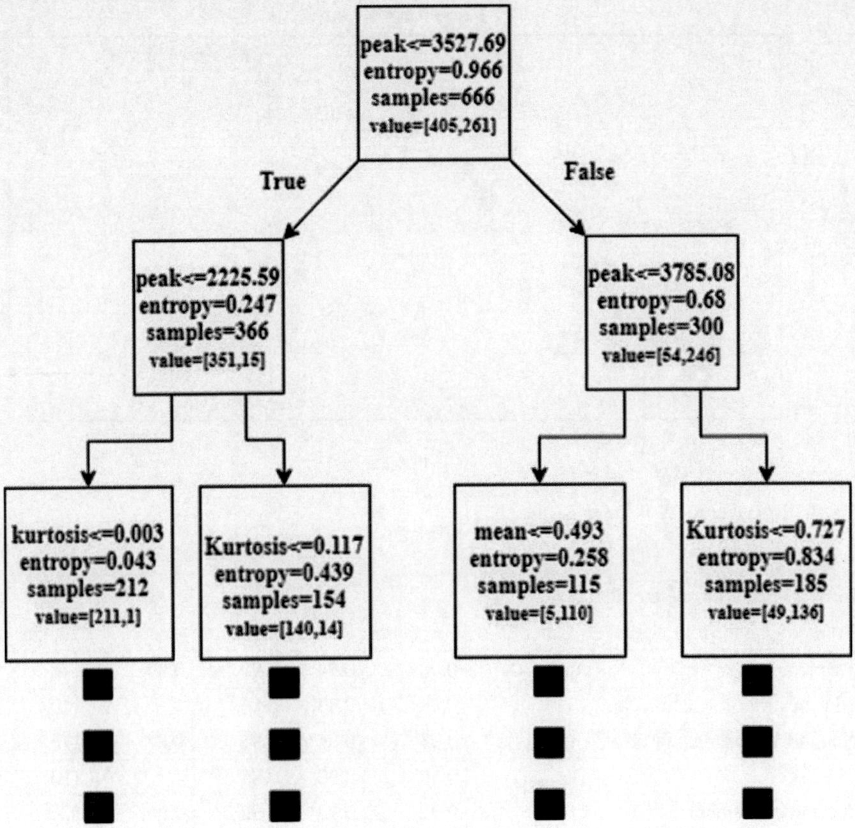

Fig. 6 Base node and leaf node for the extracted dataset

correctly whereas the other 12 samples of label '0' and 15 samples of label '1' are misclassified. The K value used to train the classification model is selected based on the Mean error calculated for K range from 1 to 40 which is depicted in Fig. 7 and the highest accuracy is observed at $K = 4$.

With these observations precision, recall, $F1$ score and accuracy are calculated for the Decision tree, SVM, and K-NN as shown in Table 2.

The data set being imbalanced, meaning the number of samples for label '0' and label '1' are not the same, the discussed metrics may not be sufficient to decide on a suitable model for classification. Therefore, the Mathew correlation coefficient is used to choose the best model. The correlation coefficients are found to be 0.77, 0.75, and 0.80 for the decision tree, SVM, and K-NN respectively. K-Nearest Neighbor classification method had the highest rate of Mathew correlation coefficient in differentiating between faulty motor from the non-faulty motor acoustics among all other algorithms analyzed in this approach.

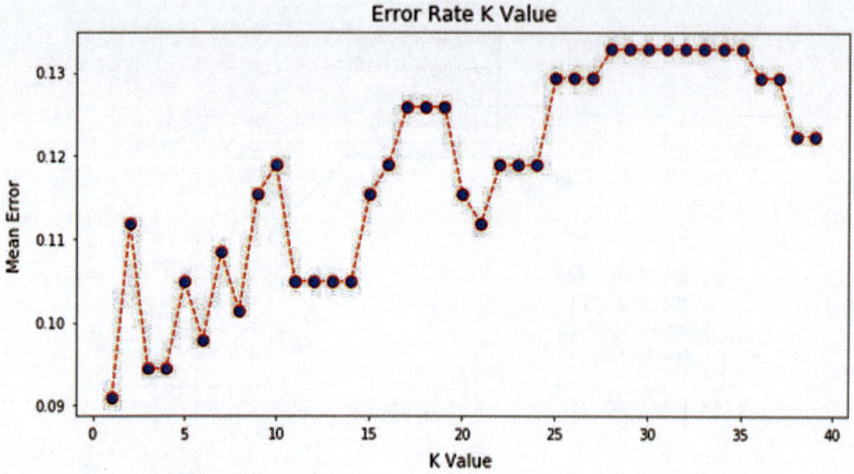

Fig. 7 Error rate for K value in K-means

Table 2 Performance metrics

Predicting algorithm	Label (0/1)	Precision	Recall	f1-score	Accuracy	Support
Decision tree	0	0.90	0.94	0.92	0.89	176
(Entropy of information gain)	1	0.89	0.83	0.86		110
Support vector machine	0	0.93	0.86	0.90	0.87	176
(SVM)	1	0.80	0.90	0.85		110
K-nearest neighbor	0	0.92	0.93	0.92	0.90	176
(K-NN)	1	0.89	0.86	0.88		110

5 Conclusion

The proposed model performs fault diagnosis of electrical motors simply and efficiently using signals recorded by mobile phones. This not only eases the work of fault detection but also helps in eliminating the cost required in setting up the other fault diagnosis equipment. Firstly the recorded signals are filtered for ambient noise and the refined signals are split into smaller samples for which the feature data set is obtained. Now various efficient classification algorithms are applied on this data set to build accurate fault diagnosis classifiers. Out of all algorithms considered K-Nearest Neighbor is found to produce satisfactory results with the highest rate of accuracy and Mathew correlation coefficient of 0.80, close to +1 is observed while differentiating between faulty motor from the non-faulty motor acoustics. This proposed method can be referred for classification of mechanical faults of the motors in specific type like Imbalance of rotor and shaft, bearing faults etc.

References

1. Dineva A, Mosavi A, Gyimesi M, Vajda I, Nabipour N, Rabczuk T (2019) Fault diagnosis of rotating electrical machines using multi-label classification. Appl Sci 9
2. El Hachemi Benbouzid M (2000) A review of induction motors signature analysis as a medium for faults detection. IEEE Trans Ind Electroni 47(5):984–993
3. Vigneshkumar S, Shankar VK, Krishna PN, Supriya, P (2018) Fault detection in gearbox using motor electrical signature analysis. In: 9th international conference on computing, communication and networking technologies, Bangalore, pp 1–6
4. Seera M, Lim CP, Ishak D, Singh H (2012) Fault detection and diagnosis of induction motors using motor current signature analysis and a hybrid FMM–CART mode. IEEE Trans Neural Netw Learn Syst 23(1):97–108
5. Guo H, Liu M (2018) Induction motor faults diagnosis using support vector machine to the motor current signature. In: IEEE Industrial Cyber-Physical Systems (ICPS), St. Petersburg, pp 417–421
6. Germen E, Başaran M, Fidan M (2014) Sound based induction motor fault diagnosis using kohonen self organizing map. Mech Syst Signal Process 46(1):45–58
7. Verma NK, Sevakula RK, Dixit S, Salour A (2016) Intelligent condition based monitoring using acoustic signals for air compressors. IEEE Trans Reliab 65(1):291–309
8. Glowacz A (2016) Diagnostics of rotor damages of three-phase induction motor using acoustic signals and SMOFS-20-EXPANDED. Mech Syst Signal Process 41:507–515
9. Glowacz A (2019) Fault diagnosis of single-phase induction motor based on acoustic signals. Mech Syst Signal Process 117:65–80
10. Rzeszucinski P, Orman M, Pinto CT, Tkaczyk A, Sulowicz M (2018) Bearing health diagnosed with a mobile phone: acoustic signal measurements can be used to test for structural faults in motors. In: IEEE Ind Appl Mag, 24(4):17–23
11. Chopra P, Yadav S (2016) Fault detection and classification by unsupervised feature extraction and dimensionality reduction. In: Complex & intelligent systems, pp 25–33
12. Vununu C, Lee S-H, Kwon K-R (2018) A sound based machine fault diagnosis system using artificial neural networks. In: International conference of artificial intelligence, pp 459–463
13. Zheng H, Zhou L (2012) Rolling element bearing fault diagnosis based on support vector machine. In: 2nd international conference on consumer electronics, communications and networks (CECNet), Yichang, pp 544–547
14. Sreekumar KT, George KK, Santhosh Kumar C, Ramachandran KI (2019) Performance enhancement of the machine-fault diagnosis system using feature mapping, normalisation and decision fusion. In: IET Sci, Measur Technol 13(9):1287–1298
15. Baydar N, Ball A (2003) Detection of gear failures via vibration and acoustic signals using wavelet transform. Mech Syst Signal Process 17(4):787–804
16. Chicco D, Jurman G (2020) The advantages of the Matthews correlation coefficient (MCC) over F1 score and accuracy in binary classification evaluation. BMC Genomics 21

Chapter 53
Study of Users Attitude and Classification of Comments and Likes from Facebook Using RapidMiner

Mohammad Kashif Khan and Nafisur Rahman

1 Introduction

Social networking websites like Instagram, Facebook, Twitter, LinkedIn, etc., are some of the most vastly used social networks. Their opinions, feedbacks, and other information they share can be of vast importance to data scientists/analysts in numerous ways. The study of the emotions and sentiments of the users means the implementation and use of different Natural Language Processing (NLP) [1] techniques, machine translation techniques; analysis of text and computational linguistics helps us in data mining [2] (recognizing), knowledge discovery, and extraction of useful data from the big data. It aims at determining the different social behaviors of users such as attitude, emotions, opinions, judgment. [3]. Data mining, in general terms, is the process of extracting data of importance from a huge amount of data and then bringing out inferences and results by data evaluation and interpretation.

M. K. Khan · N. Rahman (✉)
Department of Computer Science and Engineering, School of Engineering Sciences and Technology, Jamia Hamdard, New Delhi, India
e-mail: nafiis@gmail.com

M. K. Khan
e-mail: mdkashifkhan1990@gmail.com

I. Jeena Jacob et al. (eds.), *Data Intelligence and Cognitive Informatics*,
Algorithms for Intelligent Systems,
https://doi.org/10.1007/978-981-15-8530-2_53

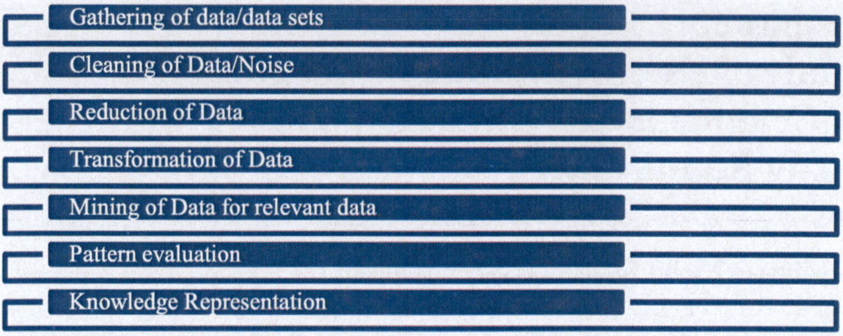

Data mining has the following steps:

Various algorithms and programs are implemented in these steps as per our requirements to extract the desired patterns [4].

Further, the concept of data classification is used. It is defined as the process of arranging and classifying data into a more relevant, organized, and useful manner to make it more useful, efficient, and more productive for our use [5]. The process includes labeling the data to make it easier to find and track. It further removes the duplicate if any, which decreases the used storage and increases the speed of the process [6]. An algorithm or a program that carries out classification and connects the data provided as input to a certain category in the meantime putting a label on it [7]. Text analysis or text mining [8] is the way in which an algorithm or an AI technology is used to convert unstructured data in the form of text into easily understandable, meaningful, and processable data. Generally, NLP is used, information extraction/retrieval techniques, or pre-processing of text for the same.

The main generator of this research area is social networks and bloggers. These online networks are driven by people's choices and accepted or rejected by their say. The websites are now rich with different commercial aspects, knowledge stores, entertainment sources, books and articles, and far more.

In this work, three different classifiers are being used to extract the thoughts and sentiments of the people; they share on Facebook through their posts and classify them into different categories. In addition, compare the results to seek out which classifier gives the simplest end in terms of higher precision and recall ratios and accuracy.

In this paper, naïve Bayes will be used [9], support vector machine [10], and decision tree [11] to perform the experiment. Later, the three results will be compared to each classifier and point out the best one.

An illustration of the above could be projected as below, where datasets are taken, and those are processed by selecting attributes and applying validations done through classifiers, and the results are drawn (Fig. 1).

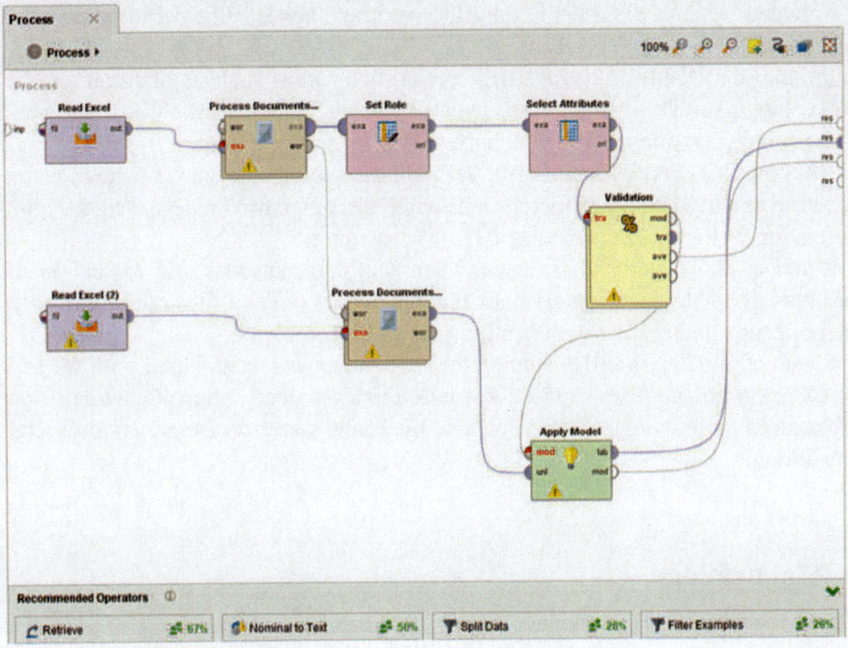

Fig. 1 Main process of RapidMiner

2 Literature Review

A huge amount of research has previously been done in the area of attitude and sentiment inspection of humans. Tim O'Keefe and Irena Koprinska [12], in 2009, suggested a new approach to select features, attributes, and weight and then applied Naïve Bayes and SVM classifier on the data. They selected only 29% of the selected attributes and obtained a classification accuracy of 87.15%.

Duwairi et al. [13], in 2014, performed sentiment analysis on Arabic Tweets. They prepared a database of 25,000+ tweets and inspected and labeled them into neutral, negative, and positive by the use of the tool RapidMiner.

Tripathi et al. [14], in 2015, studied the sentiment analysis tweets in English from Twitter using RapidMiner tool. They used the text mining technique to build sentiment classifier that predicted the mood of the user to be happy, neutral, or sad.

Owais et al. [15], in 2015, suggested a better way for the detection of sarcasm from user-given content. They performed feature extraction of all articles in their training set; later, they assigned some constant weights to all the words after the pre-processing task. They found that the SVM's with BNS feature scaling, when combined together gave the best results.

Pathak and Nafis [16], in 2015, identified influential in Twitter using Google's PageRank Algorithm [17] and considering the dynamism of Retweet or mention functionality to optimized the calculation of the ranking score. More the number of Retweets by multiple users in the follower graph better is the user's influencing capability.

Ibrahim and Salim [18], in 2016, studied behavioral analysis of tweets in the Arabic language. They gathered the tweets by using Twitter microalgae. Later, they applied some classifiers like SVM, naive Bayes, *k*-nearest neighbor.

Bhuvaneswari and Parimala [19], in 2016, tried to classify movie reviews using naïve Bayes and SVM. They matched the outcomes of the two models, included with correlation weight feature. They got 97.05% accuracy.

Singh et al. [20], in 2016, research on sentiment analysis and evaluation of similarity for heterogeneous-domain product reviews. They illustrated sentiment categorization and scaling with similarity among the reviews.

Shujat et al. [21], in 2016, studied the recommender system based on SNA and the CoCos problem. They devised a solution that obtained and exploited data and information by tracking the activities and the usage patterns of the users on social networks.

3 Methodology

RapidMiner is being used as an implementation tool in this paper. RapidMiner is freeware, open-source data science software that provides a unified environment for operations like, predictive analysis, text mining, and machine learning, etc. It is a client–server (master–slave) model, where the server is installed on premises of the user or in cloud infrastructure that are either public or private and used for both home/office and commercial purposes and applications. It is used for learning data analytics, training with different datasets, research in various fields, applying rapid prototyping, development of application and provides support for all the steps of the data mining process which includes gathering and preparation of the data, visualization of results, validation of datasets, and optimization of data. In this paper, the operators of feature selection are used, classification, and validation, etc. In order to read an object from the data repository, the data access operators are used (In this case,retrieve operator is used). For feature selection, optimize selection (Brute Force) is used which gives the most relevant attributes of the given example set by trying all possible combinations.

Through RapidMiner, the datasets are gathered through a plug-in known as the Facebook operator, and process cocuments operator is then filter and transforms the information into useful information, by evaluating the pattern defined in classifiers.

The three most popular classifiers are used for classification. They are:

1. K-NN,
2. Naïve Bayes, and
3. Decision Tree.

A *k*-nearest neighbors (K-NN) classifier stores and processes all the available data and categorizes the new incoming data on the basis of the level of similarity between the data. K-NN classifier has been used in statistical estimations as well for pattern recognition. K-NN is selected based on a distance metric.

This algorithm may be a non-parametric method that is used for many classification and regression techniques. In all the cases, the input holds of the k closest training sets within the metric space. The output then depends on whether k-NN is employed for classification or regression.

In k-NN classification, the object output may be defined as a class. An object is then viewed by its neighbors on the basis of votes, which is then assigned to the category of the most common to its k nearest neighbors. If $k = 1$, then the single nearest neighbor is assigned.

In k-NN regression, the object output is the value for the property. This value is the same as the values of its k-nearest neighbors.

The algorithm of the k-NN is a machine learning which is merely based on instance approximated for every one computation which is performed during the classification process. The most beneficial technique is often to assign votes to the contribution of its neighbors in classification and regression, so as that the nearer neighbors contribute more to the standard than the more distant ones.

To understand the k-NN classification and k-NN regression is to take a group of objects and categorizes or the object property value, respectively. No explicit training steps are required for the training set for the algorithm.

There are three choices to measure the distance, but the most popular choice is Euclidean.

Euclidean distance is an ordinary distance of straight-line projected between two points in Euclidean space, where this space becomes a metric space.

In Cartesian coordinates, if $a = (a_1, a_2, \ldots, a_n)$ and $b = (b_1, b_2, \ldots, b_n)$ are two points in Euclidean n-space, then the Euclidean distance (d) from a to b or from b to a is given by the Pythagorean expression:

$$D(a, b) = \sqrt{\sum_{i=1}^{n} (b_i - a_i)^2}$$

where a and b are query points, and d is the length of the line segment connecting ab.

Naive Bayes classifiers are for models of data that assign and provide the unique label of classes to all input occurrences. It is based on Bayes' Theorem and contains a collection of algorithms for classification. Several algorithms share a common principle, i.e., independent pairs are being classified with respect to each other. A naive Bayes classifier is extremely expandable and needs a variety of linear variables (features/predictors) in a learning problem. Naive Bayes is a probability model which works on conditions.

A decision tree classifier is used as the second algorithm. It utilizes a tree formation to organize a series of test questions and conditions. Decision tree classifier is the most extensively used technique of classification. To solve the classification problem, it applies a straightforward idea. In decision tree, internal nodes, and root contain the state which is to be tested while all the end vertex are assigned a label, Negative (No), or Positive (Yes).

Figure 2 depicts the flow of the main process designed in RapidMiner. Retrieve operator is used to reading an object from the data repository. Optimize selection (Brute Force) operator is used to select the most relevant attributes by trying all possible combinations.

Figure 3 depicts the use of a cross-validation operator, to access the analytical performance of the used learning operator. They are generally utilized to evaluate how precisely a prototype will accomplish in the real run. It can also be programmed to return the labeled data if desired by the user.

Figures 4, 5, and 6 depict the sub-processes within the cross-validation operator. *K*-nearest neighbor classifier, naive Bayes classifier, and decision tree classifier are

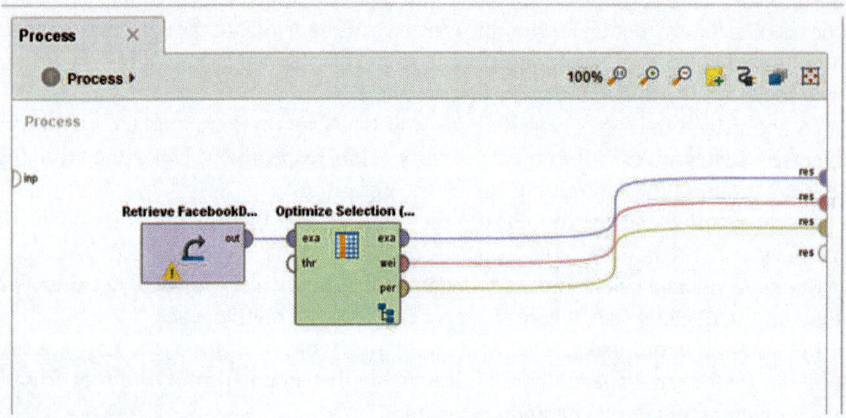

Fig. 2 Starting main process

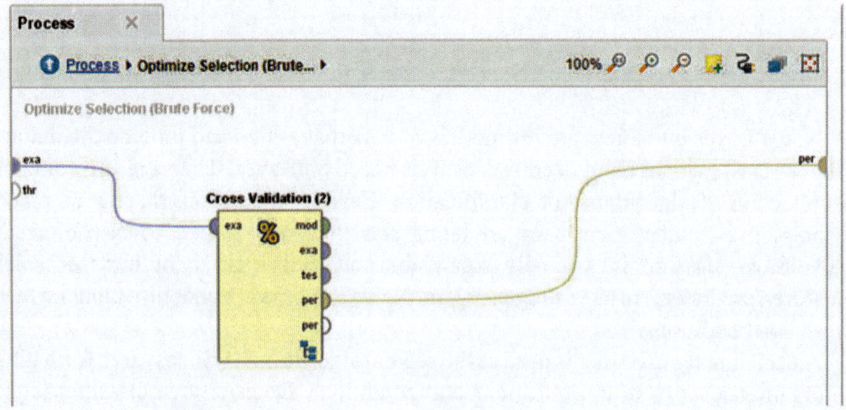

Fig. 3 Optimize feature selection operator

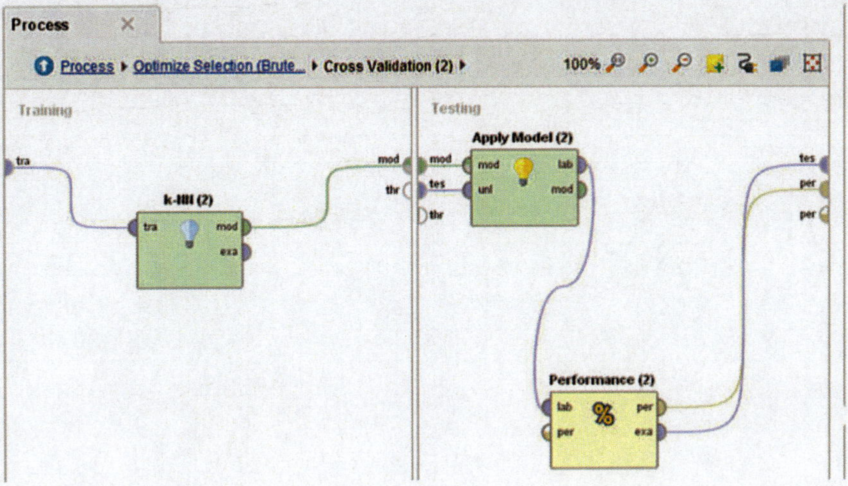

Fig. 4 *K*-NN classifier

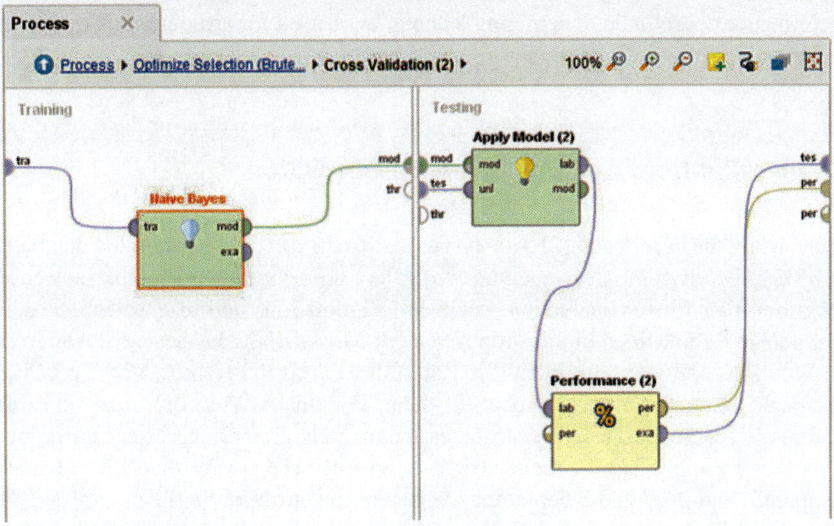

Fig. 5 Naïve Bayes classifier

being used, respectively. The cross-validation operator consists of two sub-processes, namely training and testing. As all be known, the training set applies as an input to the classifiers which gives the model as the output. Then, the model may be implemented on already learned or trained models on example sets which give labeled data as the output. Then, this labeled data is passed as an input to the performance operator. The

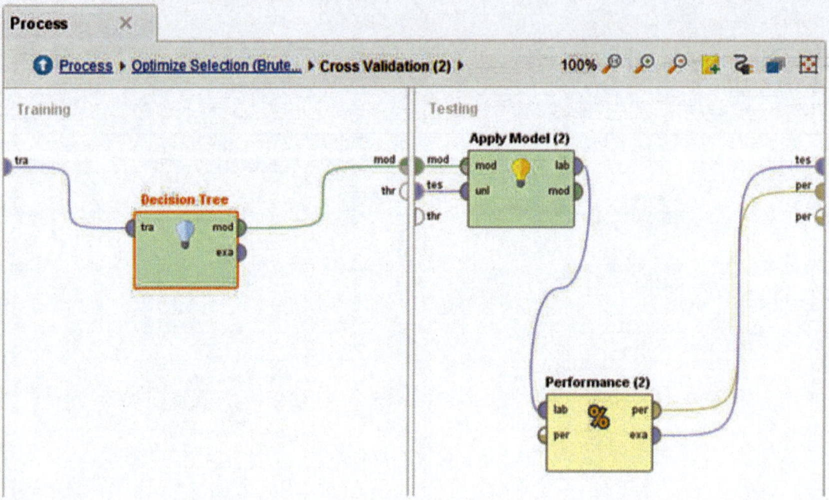

Fig. 6 Decision tree classifier

performance operator, in data mining techniques, is used for evaluation of analytical performance.

4 Investigations and Performance Inspection

In this work, the Facebook [22] dataset is used; this dataset can be generated manually or by the use of some data analytic tool. They are refined with feature selection operators, like forward selection, backward elimination, optimize selection, etc., available in RapidMiner. In this proposed work, almost 1000 datasets are taken from the Facebook operator plug-in built-in RapidMiner and fetched best, worst, positive, and negative comments and processed them. For this dataset, optimize selection features select the most relevant attributes and try all combinations; after that various classifiers are applied on the reduced dataset, and their precision and recall values are compared. Figures 7 and 8 depict the weights of the attributes in the case of *K*-NN, Figs. 9 and 10 depict the naïve Bayes, and Figs. 11, and 12 depict the decision tree for the optimize selection (Brute Force) approach and graphical representation.

In this section, the information that's utilized in this work is also described; the posts are collected from Facebook [22] social networking site which is in the text tongue. The posts are collected from Facebook. Then processed the text mining through operators available in RapidMiner on the information before applying to the classifiers for training and testing. For training purposes, the posts are classified into two sorts of labels- positive and negative. These labels are going to then train the classifier and support this learning to predict the label of the testing dataset. The

AttributeWeights (Optimize Selection (Brute Force))

attribute	weight
ID	0
Name	1
Date Joi...	1
Activity	0
No of Fol...	1
Profile Pi...	0
No of Fri...	1
Likes Co...	0
Comme...	0
Shares ...	0
Tag Count	0
Pokes C...	1

Fig. 7 Attributes weight for *K*-NN classifier

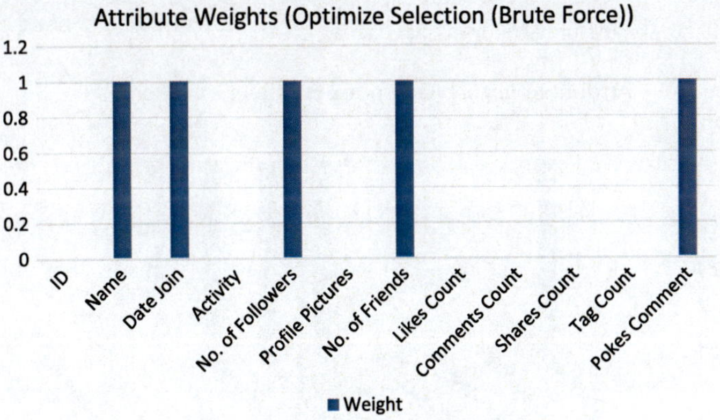

Fig. 8 Graphical attributes representation for *K*-NN classifier

AttributeWeights (Optimize Selection (Brute Force))

attribute	weight
ID	0
Name	0
Date Joi...	1
Activity	0
No of Fol...	0
Profile Pi...	0
No of Fri...	1
Likes Co...	1
Comme...	1
Shares ...	1
Tag Count	0
Pokes C...	0

Fig. 9 Attributes weight for naive Bayes classifier

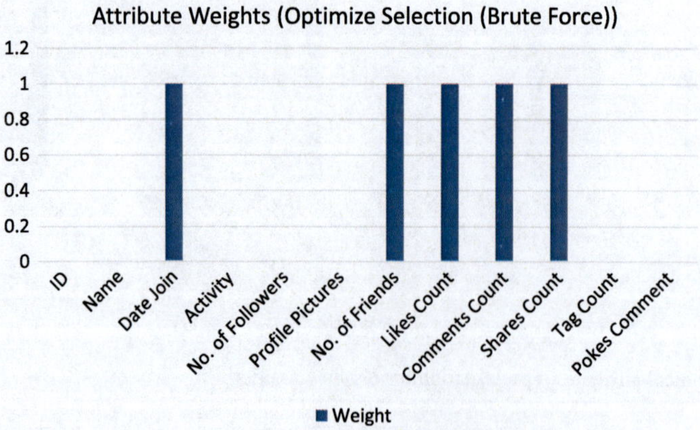

Fig. 10 Graphical attributes representation for naive Bayes classifier

AttributeWeights (Optimize Selection (Brute Force))

attribute	weight
ID	0
Name	0
Date Joi...	1
Activity	1
No of Fol...	0
Profile Pi...	0
No of Fri...	0
Likes Co...	1
Comme...	0
Shares ...	1
Tag Count	0
Pokes C...	1

Fig. 11 Attributes weight for decision tree

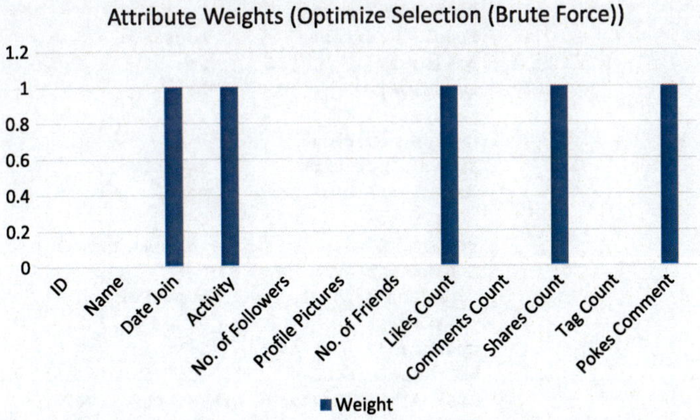

Fig. 12 Graphical attributes representation for decision tree

dataset provided for the testing and supported the training provided to the classifier. Three different classifiers are being used to predict the label. The prediction results finally done by the classifiers are shown in below figures.

Figures 13, 14, and 15 depict the confusion matrix for the K-NN, naïve Bayes, and decision tree classifier—the template will do that for you.

A comparison graphical figure is as below (Fig. 16):

Fig. 13 K-NN performance vector

PerformanceVector

```
PerformanceVector:
accuracy: 50.72%
ConfusionMatrix:
True:    positive         negative
positive:         230      239
negative:         2        18
precision: 90.00% (positive class: negative)
ConfusionMatrix:
True:    positive         negative
positive:         230      239
negative:         2        18
recall: 7.00% (positive class: negative)
ConfusionMatrix:
True:    positive         negative
positive:         230      239
negative:         2        18
AUC (optimistic): 0.992 (positive class: negative)
AUC: 0.500 (positive class: negative)
AUC (pessimistic): 0.069 (positive class: negative)
```

Fig. 14 Naïve Bayes performance vector

PerformanceVector

```
PerformanceVector:
accuracy: 64.42%
ConfusionMatrix:
True:    positive         negative
positive:         152      94
negative:         80       163
precision: 67.08% (positive class: negative)
ConfusionMatrix:
True:    positive         negative
positive:         152      94
negative:         80       163
recall: 63.42% (positive class: negative)
ConfusionMatrix:
True:    positive         negative
positive:         152      94
negative:         80       163
AUC (optimistic): 0.815 (positive class: negative)
AUC: 0.544 (positive class: negative)
AUC (pessimistic): 0.485 (positive class: negative)
```

PerformanceVector

```
PerformanceVector:
accuracy: 84.66%
ConfusionMatrix:
True:   positive        negative
positive:       224     67
negative:       8       190
precision: 95.96% (positive class: negative)
ConfusionMatrix:
True:   positive        negative
positive:       224     67
negative:       8       190
recall: 73.93% (positive class: negative)
ConfusionMatrix:
True:   positive        negative
positive:       224     67
negative:       8       190
AUC (optimistic): 0.984 (positive class: negative)
AUC: 0.850 (positive class: negative)
AUC (pessimistic): 0.723 (positive class: negative)
```

Fig. 15 Decision tree performance vector

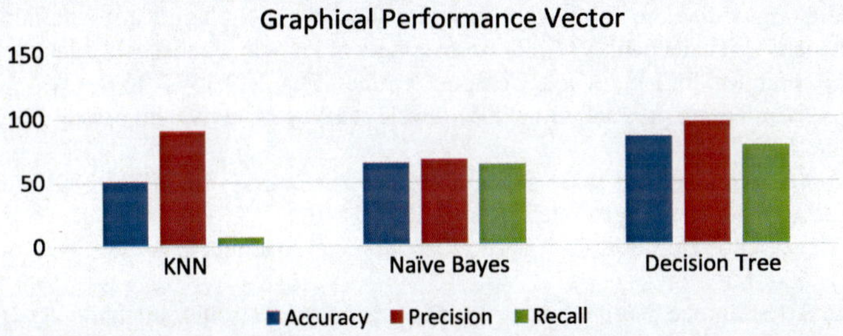

Fig. 16 Graphical representation of performance vector of all classifier

5 Conclusion and Future Work

In the current paper, the attitude classification of social media users is performed by the use of different techniques of data mining. Three different classifiers are made used— K-NN, naive Bayes, and decision tree. The result shows that the accuracy of *K*-NN was 77.50%, naive Bayes was 80%, and the decision tree was 78%. In our work, found that naive Bayes was the classifier with the most accurate results, and it was recommended to be used with the dataset from social networking websites. Decision tree was the next best performer.

This work can be extended by increasing the size of the dataset. The number of labels and change the labels can also be increased based on the needs and the type of study. A more massive and sophisticated dataset is intended to use. Other languages, apart from English, can also be included in a single study. The inclusion of special characters and non-letter characters would be a challenge. Better results might be getting if the smileys, emoticon, stickers, etc., could be incorporated because they are being used massively in social networks to represent different expressions and feelings.

References

1. Young T, Hazarika D, Poria S, Cambria E (2018) Recent trends in deep learning based natural language processing [Review Article]. IEEE Comput Intell Mag 13(3):55–75. Available: https://doi.org/10.1109/mci.2018.2840738
2. Wu X, Zhu X, Wu G-Q, Ding W (2014) Data mining with big data. IEEE Trans Knowl Data Eng, 26(1):97–107. Available: https://doi.org/10.1109/tkde.2013.109
3. Pang B, Lee, L (2008) Opinion mining and sentiment analysis. Found Trends® Inf Retrieval 2(12):1–135 Available: https://doi.org/10.1561/1500000011
4. Ramageri B (2010) Data mining techniques and applications. Indian J Comput Sci Eng 1(4):301–305
5. Chen Y, Lin Z, Zhao X, Wang G, Gu Y (2014) Deep learning-based classification of hyperspectral data. IEEE J Sel Topics Appl Earth Observ Remote Sens 7(6):2094–2107. Available: https://doi.org/10.1109/jstars.2014.2329330
6. https://digitalguardian.com/blog/what-data-classification-data-classification-definition
7. Priyadharsini C, Thanamani A (2014) An overview of knowledge discovery database and data mining techniques. Int J Innov Res Comput Commun Eng 2(1):1571–1578
8. Hotho A, Nürnberger A, Paaß G (2005) A brief survey of text mining. Ldv Forum, 20(1). Available: 10.1.1.447.4161
9. Rish I (2001) An empirical study of the naive Bayes classifier. In: IJCAI 2001 workshop on empirical methods in artificial intelligence 3(22):41–46
10. Meyer D, Leisch F, Hornik K (2003) The support vector machine under test. Neurocomputing 55(1–2):169–186
11. Safavian S, Landgrebe D (1991) A survey of decision tree classifier methodology. EEE Trans Syst, Man, Cybern 21(3):660–674
12. O'Keefe T, Koprinska I (2009) Feature selection and weighting in sentiment analysis. In: Proceeding of 14th Australasian document computing symposium, Dec 2009, Sydney, Australia
13. Duwairi R, Marji R, Sha'ban N, Rushaidat S (2014) Sentiment analysis in Arabic tweets. In: 2014 5th international conference on information and communication systems (ICICS), 2014. Available: https://doi.org/10.1109/iacs.2014.6841964
14. Tripathi P, Vishwakarma SK, Lala A, Sentiment analysis of English Tweet Using Rapidminer. In: International conference on computational intelligence and communication networks, 2015, pp 668–672
15. Owais S, Nafis T, Khanna S (2015) An improved method for detection of satire from user-generated content. Int J Comput Sci Inf Technol 6(3):2084–2088
16. Nafis M, Pathak AK (2015) To find influential's in Twitter based on information propagation. Int J Comput Appl 118(13):22–25. Available: https://doi.org/10.5120/20805-3507
17. https://www.link-assistant.com/news/google-page-rank-2019.html
18. Ibrahim MA, Salim N (2016) Sentiment analysis of Arabic tweets: with special reference restaurant Tweets. IJCST 4(3):173–179
19. Bhuvaneswari K, Parimala D (2016) Correlation based feature selection for movie review sentiment classification. Int J Adv Res Comput Commun Eng 5(7):325–332. Available: https://doi.org/10.17148/IJARCCE.2016.5765

20. Singh M, Nafis T, Mani N (2016) Sentiment analysis and similarity evaluation for heterogeneous-domain product reviews. Int J Comput Appl 144(2):16–19. Available: https://doi.org/10.5120/ijca2016910112
21. A. Shujat, M. Tabrez, and Sharma V (2016) A solution to CoCoS problem in recommender system based on SNA. Int J Comput Appl 144(3):5–10. Available: https://doi.org/10.5120/ijca2016910148
22. https://facecebook.com

Chapter 54
Novel State Disturbance Based Multi-level Inverter with Sliding Mode Control

Telma Johnson and V. Shijoh

1 Introduction

Nowadays, the renewable energy systems are becoming more effective and suitable for work along with the smart grid application. The renewable energies are from wind, solar, biogas, etc. The exact use of this renewable system will reduce the overall pollution and increases the natural way of producing energy for various applications [1]. The developing of smart grids, as well as the renewable energy, a DC to AC inverter, has been extensively used for various kinds of the renewable energy system. Hence, the inverters become a major role for various purposes of the renewable energy conversion system. From renewable energy, there is a power conversion that has been taken place, and this same energy produced is used for manual and other purposes of energy usage. In this power conversion process, they have been widely used in smart grid, servomotors, distributional system and electric vehicle system, etc. These uses of renewable energy enhance a pollution controlled and more efficient method of using renewable energy in an energy system [2].

Taking the distributional generation (DG) system, the output voltage is the main factor affecting the end side and also the connected loads are considered for the better distributional performance of the system. This output voltage and connected loads are linearly dependent. That is when there is an increment in load, it will affect the output voltage performance and also the output current of the system. For the precise regulation manner of an inverter, there must be obtained the connected devices for finding the load variation [3]. The control performance is a serious issue faced in

T. Johnson · V. Shijoh (✉)
Electrical & Electronics Engineering, Jyothi Engineering College, Cheruthuruthy, Thrissur, Kerala, India
e-mail: shijohv@jecc.ac.in

T. Johnson
e-mail: telmajohnson1997@gmail.com

I. Jeena Jacob et al. (eds.), *Data Intelligence and Cognitive Informatics*, Algorithms for Intelligent Systems,
https://doi.org/10.1007/978-981-15-8530-2_54

a distributional or an electric vehicle system. Taking the same, the DG and electric vehicles have a no load to full load deflection. This deflection will cause the output voltage regulation and performance dreadfully. When the output voltage goes wrong, it will also affect the output current of the entire system. Hence, the measurement of current and voltage becomes a main factor for the system. Hence, for measuring the output current, a wide range of current band sensors is becoming an essential component. But it may cause the system more bulky and costly. This is the main disadvantage of a traditional DC to AC inverter system. And another disadvantage is that there is a misleading of output voltage due to change in connecting load at the input side, it may also increase the total harmonic distortion (THD), and the spiking increment of THD may also cause the system efficiency. Hence, increment of total harmonic distortion causes another defect factor [4].

For smooth and easy operation and getting output here implementing a seven-level multilevel inverter in between the DC/DC converter and load. This multi-level topology is used for reducing the power component and hence to minimize the overall size and weight. This multi-level inverter can generate output voltage with extremely low distortion [5]. The working of an inverter, generating pulse is essential. From the several modulation techniques, taking sinusoidal pulse width modulation (SPWM), this modulation technique is used because it has lesser total harmonic distortion and the developing of a pulse is easy as compared with the other modulation techniques [6]. During this study, the SMC is enough for recognizing the matched uncertainty. This SMC is implemented here for the field of power converters against matched external disturbances caused by the DC offset.

The overall idea of this paper is to produce a high-voltage gain with less distortional factor from a low-voltage without using current sensors. Because of using the current sensors, it costs higher and adding this sensor will enlarge the overall size and weight of the entire system. This proposal leads the way of introducing the system without a current sensor, and the way of measuring the current in this system is introduced by using novel state disturbance observer (NSDO) [7].

Part 1 contains the introduction zone of the entire proposing system. Part 2 describes the concept of the mentioned proposed system. The simulation and its results conclusion and future scope are detailed in remains. TheDC–DC converter, inverter and the controlling system are designed and simulated in MATLAB/Simulink platform to validate the design.

2 Concept of Proposed System

In this paper, it is discussing about how to reduce the total harmonic distortion and voltage regulation at the output side and to generate a high voltage from low-voltage input by sensing the output current instead of using current sensors. By using a multi-level topology here, it will give a fine output with less harmonic content. The load disturbance may occur by using various loads in a distributional system. It will affect the output at the end, and the harmonics also interrupts the overall efficiency. So

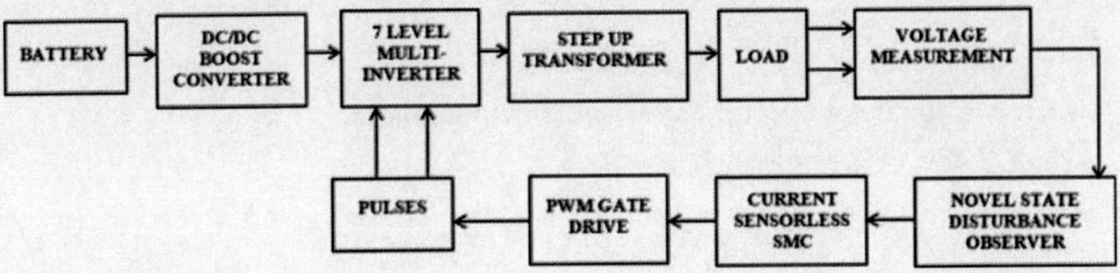

Fig. 1 Block diagram of the proposed system

avoiding the effect of mismatched load disturbance and harmonics, the SMC scheme is developed here. In this study, the NSDO is the one used to reconstruct the inductor current in the presence of variation occurred during loaded condition [8].

By avoiding the use of current sensors, the output side is connected with a voltage measurement system and it measures the total voltage at the load side and after measuring the voltage from the load side this voltage is given into a disturbance observer. Hence, through the disturbance estimation and feed-forward compensation, the total harmonics presented along with the output voltage can be removed, this NSDO is used here for rejecting the higher-order harmonics and caused by the nonlinear load, and this system with NSDO provides the current sensorless running mode. The sliding mode surface provides the controlling panel of current against the load voltage and the gate drive propagates the pulse to the inverter corresponds to the output of the NSDO [9].

Figure 1 shows the basic block diagram of the proposed system. Generally, a DG system-based renewable energy source consists of mainly three parts: input power, conversion of the power unit and load. In this, the main component is the energy conversion system unit [10]. Here battery source used as an input of the proposed system is shown, and it consists of a DC–DC boost converter at the input side which is followed by a multi-level inverter. By using this multi-level inverter, various level voltages can be applied, and the results can occur with low distortion.

The boost converter used here will help to control the output from the input side which will reduce the voltage of the inverter. Because of the single switch in the DC–DC converter, the double frequency components are raised by the switching of the inverter. By replacing the microcontroller by a PID controller with an observer-based system, the inverter topology is a seven-level based by replacing the normal feedback PID controller with an observer-based system. The state inductor current is developed by the NSDO approach [11]. The control system provides a reference current wave. After each cycle of load variation and the disturbance occurred during the variation is checked with the reference wave at the control system [12].

General state space equation for a system with input control vector u can be written as

$$\mathring{x} = Ax + Bu \tag{1}$$

$$y = Cx \tag{2}$$

where y is the output vector and C is the output matrix.
Hence, the observed state be

$$\breve{\hat{x}} = A\hat{x} + Bu + K\left(y - C\hat{x}\right) \tag{3}$$

where K is the gain matrix (Fig. 2).
For a sliding mode control scheme considering the uncertain system as

$$x_2 = \frac{\mathrm{d}x_1}{\mathrm{d}t} \tag{4}$$

$$u + M = m\frac{\mathrm{d}x_2}{\mathrm{d}t} \tag{5}$$

where x_1 represents position, x_2—velocity, m—moment of inertia and M—uncertainty in moment of inertia. Corresponding sliding surface is defined as

$$s = \mathring{e} + ke \tag{6}$$

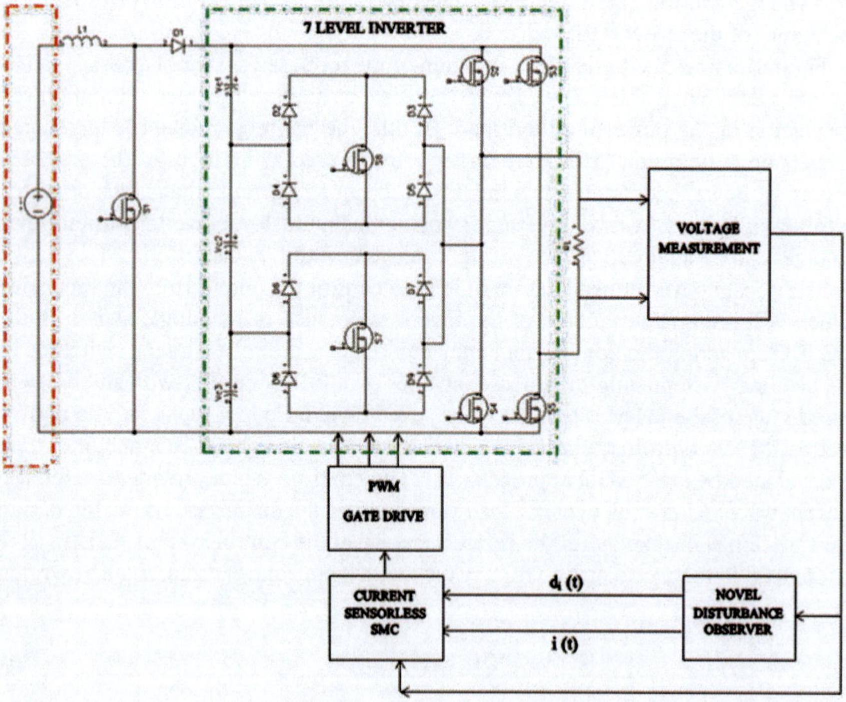

Fig. 2 Circuit diagram of the proposed system

$$s = x_1 - \mathring{x}_d + kx_1 - kx_d \tag{7}$$

$$s = x_2 - \ddot{x}_d + k\mathring{e} \tag{8}$$

$$s = \frac{u}{m} + \frac{M}{m} - x\ddot{d} + kx_2 - kx\mathring{d} \ldots \tag{9}$$

where $e = \times 1 - x_d$ is the position tracking error. x_d is a reference or ideal signal, and k is the gain parameter. Thus, sliding mode control signal obtained for adaptive sliding mode control is

$$u = -M + m\,(\ddot{x}_d - kx_2 + k\mathring{x}_d + \mathring{s}) \tag{10}$$

$$u = -\,(M + m\,(-\ddot{x}_d + kx_2 - k\mathring{x}_d + \alpha \sin(s))) \tag{11}$$

2.1 Simulation Results

The proposed system is simulated in MATLAB/Simulink. Here both the converting and inverting stages are explained. From the source battery, it feeds the DC–DC boost converter and follows to the multi-level inverter system. The controller used here is 8 bit ATMEGA 16, and at the boost converter IRF250 MOSFET is used. For the theoretical analysis, the simulation part is explained as follows (Fig. 3).

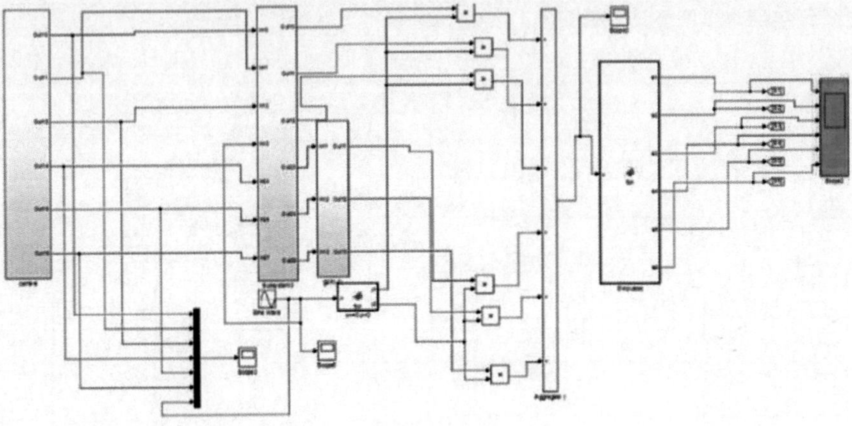

Fig. 3 Simulink model of the proposed system

The first stage consists of a boost DC–DC converter with input as a battery with a feedback loop to maintain the output voltage fixed. Then next stage is an inverter which inverts the DC power coming from the converter to AC output. The peak of the AC output voltage is taken at every 15 kHz of sample time and is fed back to the control system to maintain the output constant. Figure 4 is the switching pulses for the inverter switches are developed using an observer-based SMC with unipolar PWM technique [12].

The modulation index obtained from the SMC in Fig. 5, the output is given to the unipolar PWM technique to generate the PWM output and is given to the switches. The total simulation of the system is done, and the simulation results are as follows. The system is simulated for a 12–240 V boost converter. The output of the converter is given to the inverter section designed for an RL load. The load is then replaced with a nonlinear load having a value equal to 70% of the initial value chosen in

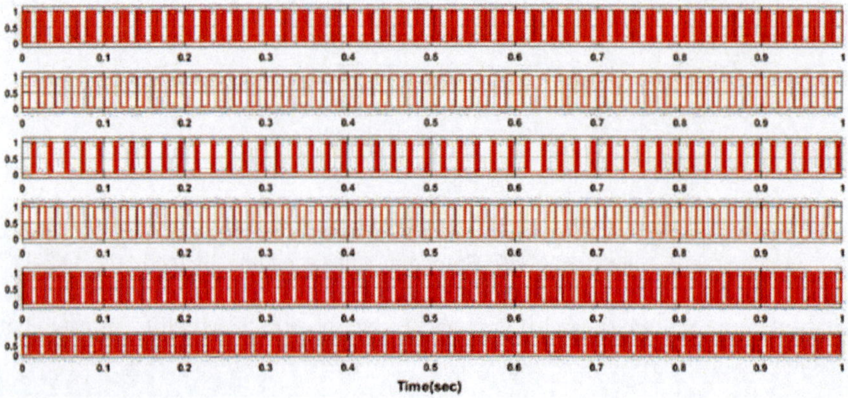

Fig. 4 Switching pulses for inverter

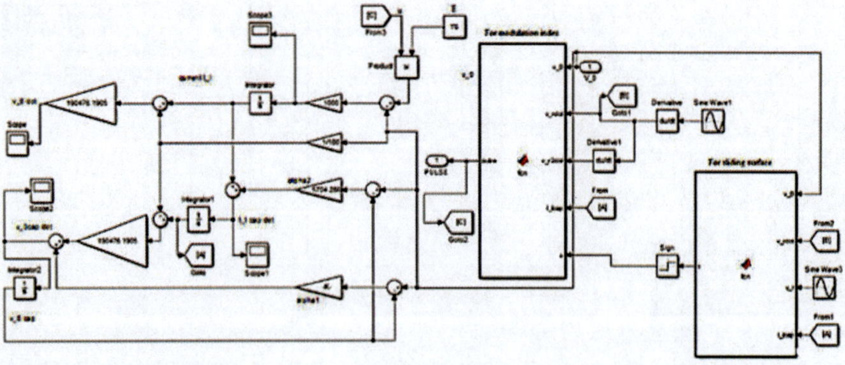

Fig. 5 Observer-based SMC

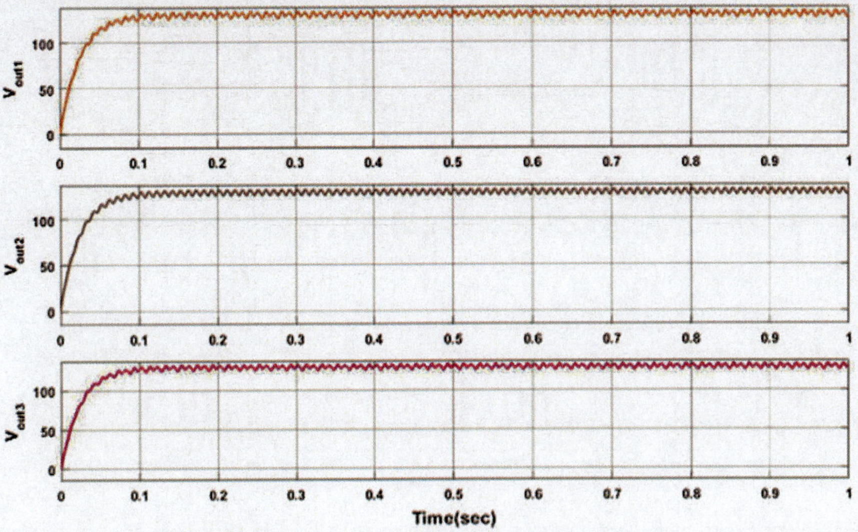

Fig. 6 Output voltage waveform

order to show the variation. The boost converter switching is done with a switching frequencyof50 kHz, and the PID parameters are determined by trial and error method.

Figure 6 shows the output voltage load variation. The load is changed to 70% of the initial value at 0.05th second of the simulation. Because of the reduction of load current, the modulation index generated with the basis of developed inductor current also reduces its value for a half cycle which will reduce the voltage. After that cycle modulation index reaches its previous value to maintain the output voltage constant. The simulation requirements are tabulated in Table 1.

The load variation voltage is plotted in voltage to the time axis. The voltage varies with increment in the time graph and is more sinusoidal with the increased amplitude level of voltage. The three scope output voltages are shown inFig. 6.

Figure 7 is the output of seven-level multi-level inverters. Using higher level multi-inverter gives smooth and accurate output result. The THD can be minimized or reduced to a particular level by adding multi-level inverting topology. In Fig. 8, seven-level inverting output voltages are shown. From this figure, it is clear that

Table 1 Simulation parameters

Sl. No	Simulation parameters	Range
1	Input voltage	12 V
2	Capacitor	1000μF
3	MOSFET	IRF250
4	Diode	IN4007
5	Output voltage	250 V

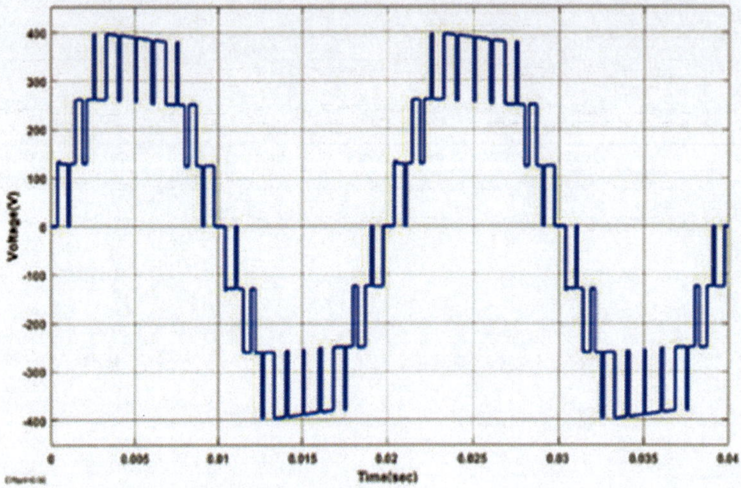

Fig. 7 Seven-level multi-level inverter output voltage

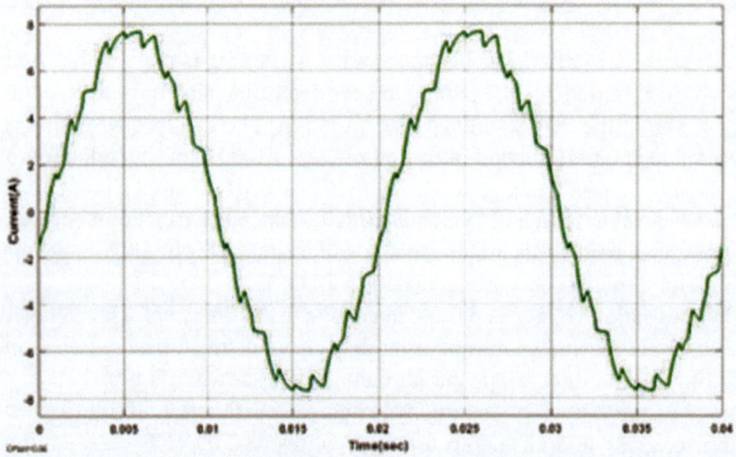

Fig. 8 Inductor current and observed inductor current from the simulation

the output voltage becomes more accurate with less distortion when it is used by a multi-level inverter system.

Figure 8 The inductor current obtained from the simulation diagram and the inductor current developed by the observer system are similar even though there is some variation initially. So that the current developed is on the basis of the SMC scheme, and any variations in the SMC arise because of the developed inductor current. The system is working properly to reach the original value of the inductor current in the power circuit

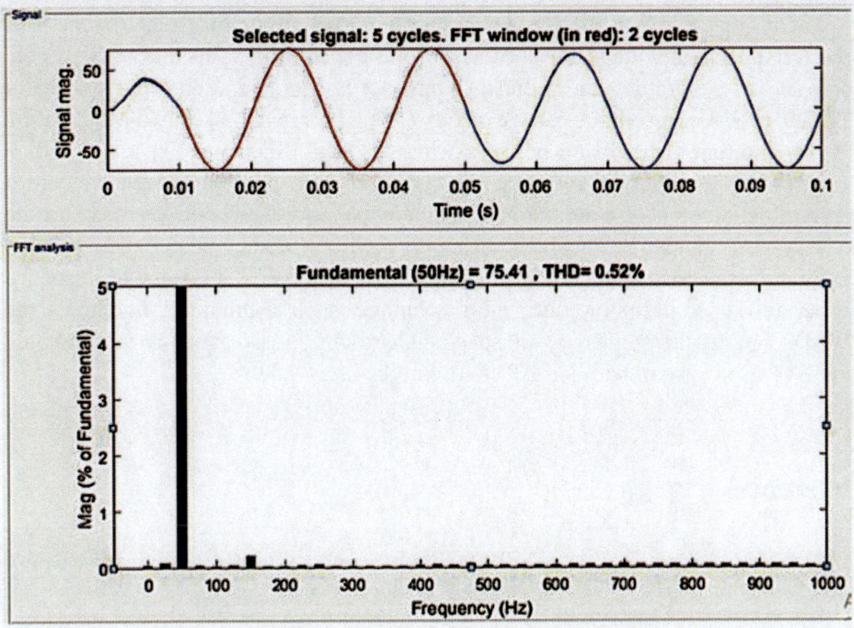

Fig. 9 THD analysis of the system

Figure 9 shows the THD analysis of the circuit developed and is showing a THD value of 0.52 percentages with a magnitude of 240 V peak value. The THD is less than the permissible value which is 5 percentages. Considering the case of an inverter system with a normal PID-based control scheme, THD after reaching the system into steady state and taking a number of cycles after that gives a value of 4 percentages.

3 Conclusion and Future Scope

In this paper, the main objective is to reduce the THD and develop a state disturbance observer and from this observer developing a feedback current without using the current sensors. The voltage at the output is measured, and this is used to generate gate pulse without current sensors. Through this, the overall size can be minimized, and the entire system efficiency can be improved. A seven-level multi-inverter is implemented and simulated via MATLAB/Simulink. This seven-level topology helps to reduce the THD. The observations have concluded with results that the THD of the system reduced 0.5% in simulation; whereas, for the PID controller the THD is more than 4%. The feedback current is developed through the observer system. Then the current is given to the SMC scheme to generate the modulation signal, and the whole system is implemented in a single microcontroller. There is no feeding back of the analog signals in the PC-SIMULINK which made the system complex in previous studies. Considering the variation in load through the disturbance state based on the output voltage, the current sensor is unnecessary. This makes the system compact and economic. The proposed converter is operated at an input voltage of 12 V and produces an output voltage of 240 V. The THD obtained from the simulation is 0.5% and that of hardware is 3.5%; whereas, for a PID controller in simulation THD is 4%. The input converter can be made to a high gain interleaved system for the required amount of gain and low rated components. The control circuit for the boost system can also be developed using the observer-based system to maintain the output voltage constant. Considering the case of harmonics, it is easy to introduce further developed harmonic mitigation techniques such as providing notches to the PWM at a particular angle to avoid specified harmonics. Also, it can be incorporated with PV-based system with MPPT control at the input side.

References

1. Li S,Yang J, Chen WH, et al (2014) Disturbance observer-based control: methods and applications, 1st edn. CRC Press, Boca Raton, FL., USA
2. Cho H, Yoo SJ, Kwak S (2015) State observer based sensor less control using Lyapunov's method for boost converters. IET Power Electron 8(1):1119
3. Biel D, Fossas E, Guinjoan F (2009) Sliding mode control in electro-mechanical systems, 2nd edn. CRC Press, Boca Raton, FL,USA
4. Calais M, Agelidis VG (1998) Multilevel converters for single-phase grid connected photovoltaic systems—an overview. Proc IEEE Int Symp Ind Electron 1:224–229
5. Wang Z, Li1 S, Yang J, Li1 Q (2018) Current sensorless sliding mode control for direct current alternating current inverter with load variations via a USDO approach. IET Power Electron 08(12):1755–4535
6. Trancy Ni L, Patterson DJ, Hudgins JL (2012) High power current sensorless bidirectional 16-phase interleaved DCDC converter or hybrid vehicle application. IEEE Trans Power Electron 27(3):11411151
7. Rodríguez J, Lai JS, Peng FZ (2002) Multilevel inverters: a survey of topologies, controls, and applications. IEEE Trans Ind Electron 49(4):724–738
8. Ahmed KH, Massoud AM , Finney SJ, Williams BW (2009) sensorless current control of 3 phase inverter based distributed generation. IEEE Trans Power Deliv 24(2):1971
9. Shtessel Y, Edwards C, Feidman L et al (2014) Sliding mode control and observation,1st edn. Springer, New York, USA
10. Smys S, Ranganathan G (2019) Robot assisted sensing, control and manufacture in automobile industry. J ISMAC 1(03):180–187
11. Umanand (2012) Power electronics essential applications. Wiley, India Pvt. Ltd.
12. Baig Z, Zeadally S (2019) Cyber-security risk assessment framework for critical infrastructures. Intell Autom Soft Comput 25:121–129

Chapter 55
Survey on Graphical Password Authentication System

Shikhar Singh Patel, Akarsh Jaiswal, Yash Arora, and Bharti Sharma

1 Introduction

In this generation of information technology, the most growing part is the data itself. Trillions of bytes are being exchanged in the form of information passing, message exchange and are stored as a resource or asset on a daily basis. To make it secure, the most general method chosen is securing it with a password. Now, passwords are of different types or forms, and the most common among them is alphanumeric passwords and is because there are more than a trillion permutations possible. In general, some kind of personal relativity is used in our password so that it can be remembered easily. It generates another risk of leakage of personal information via our password in case someone gets access to the password or steals it. Many times, only numbers are used as our password or PIN which in general is easy to predict by someone who is in our touch for a long time. Usage of short and meaningful passwords also has its pros and cons, on one hand, it is easy to remember, but on the other hand, it is easy to predict.

To overcome these problems, various methods had been in trial, but one successful till date is 'graphics as a password'. Graphical passwords are no different than alphanumeric passwords, and it just uses pixels or patterns as input instead of characters and numbers. The major difference is its flexibility. It comes with a wide range of options, and one can easily pick the original among trillions. People can draw their own patterns as well and can set it as their passwords. Another option that is highly popular nowadays is using yourself as your password, i.e., the use of fingerprints, retina, face, etc. Biometrics is quite well known for decades now, but a disadvantage with this is setting up devices is very cost-inefficient. The same problem comes with

S. S. Patel (✉) · A. Jaiswal · Y. Arora · B. Sharma
Department of Computer Application, National Institute of Technology Kurukshetra, Kurukshetra, India
e-mail: 1shikhar1@gmail.com

I. Jeena Jacob et al. (eds.), *Data Intelligence and Cognitive Informatics*, Algorithms for Intelligent Systems,
https://doi.org/10.1007/978-981-15-8530-2_55

a retina scanner as well. Graphical passwords have provided an option to overcome the difficulty of remembering alphanumeric passwords as well as reduce the risk of password to be stolen, as people generally use small and simpler passwords so that they can remember and use frequently. At the same time, people use the same password for different accounts which makes all but one account unsafe in most cases. Often images, patterns, or any other graphical passkeys are easy to be remembered and recognized in front of many random images. Another advantage of using graphical passwords is to remove the problem of leakage of personal information which generally been used in our passwords. On the other hand, text and numeric provide a limited possibility to renew, and attackers easily recognize the pattern of setting up the passwords.

2 Overview of the Authentication Methods

These are those three major authentication areas which are currently being used:

- Token/artificial identity-based authentication
- Biometric/personal identity-based authentication
- Knowledge/info identity-based authentication.

Token-based authentication is also called smart authentication, as it takes minimum time to validate whether the requesting one is authentic or not. Smart cards, key cards, bank cards are being widely used these days. Most of those token-based authentication comes under knowledge-based authentication as well. For example, ATM cards come with a PIN. Biometric uses personal attributes as an unlocking method such as fingerprints, retina, and facial recognition. These attributes are known to be unique to any individual, whereas facial recognition sometimes fails to perform that optimum. The major challenge behind using these methods is their cost. These are soft-hard devices where the whole system is needed to set up to provide security to our information, which is cost-inefficient but ensures the highest security. Knowledge-based authentication is further divided into two minor fields, i.e., text-based and picture-based. Picture-based passwords use either recall technique or recognition technique.

3 Related Work

3.1 Recall-Based Techniques

It states the method of reproducing the previously inserted or selected information as a password while being at the registering stage.

Draw-A-Secret (DAS) Technique

The Draw-a-Secret (DAS) technique was proposed by Mayer et al. in [1], where the user will draw a picture using a mouse on a rectangular 2D grid. The coordinates of the grid occupied by the image are stored in a graphical order. During the authentication, the user needs to re-draw the same picture on the grid. If the drawn picture reaches the same coordinate sequence on the 2D grid, the user is authenticated.

Background DAS (BDAS) Technique

Here, the wallpaper will be added as an enhancement to the original DAS. Therefore, the background image here is a clue. The password is a free-form graphic created by the user on the grid based on the wallpaper of his or her choice. Wallpaper is used to draw passwords. People may make drawing passwords more complicated and difficult to predict. In order to authenticate, the user must recreate the same graphic on the grid using the background image (Fig. 1).

Syukri Signature Technique

This technique was proposed by Syukri, Okamoto and Mombo in 1998 [2]. In this system, the user can draw his or her signature with the help of the mouse. It includes two stages, namely registration, and verification. At the registration stage, the user is asked to draw a signature with the help of the mouse. This information is stored in a database. At the verification stage, the entry accepts the signature, performs

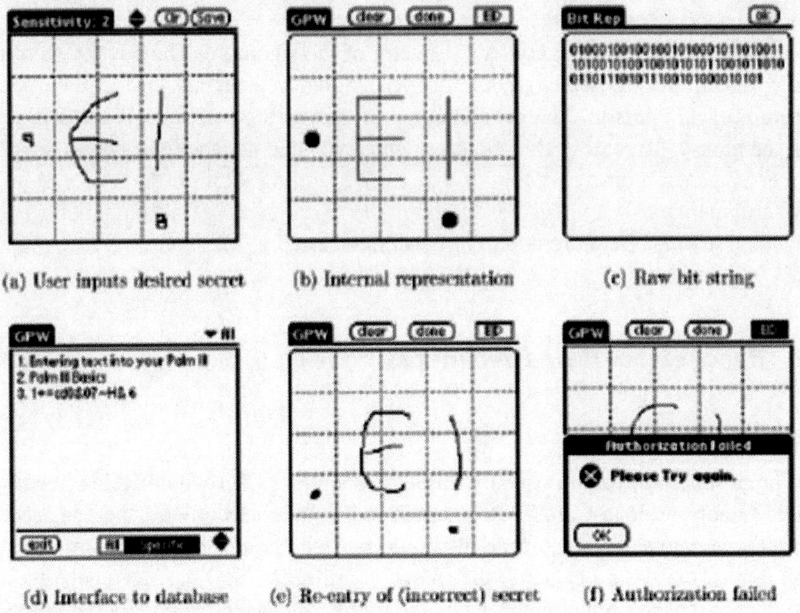

Fig. 1 Background DAS (BDAS) technique

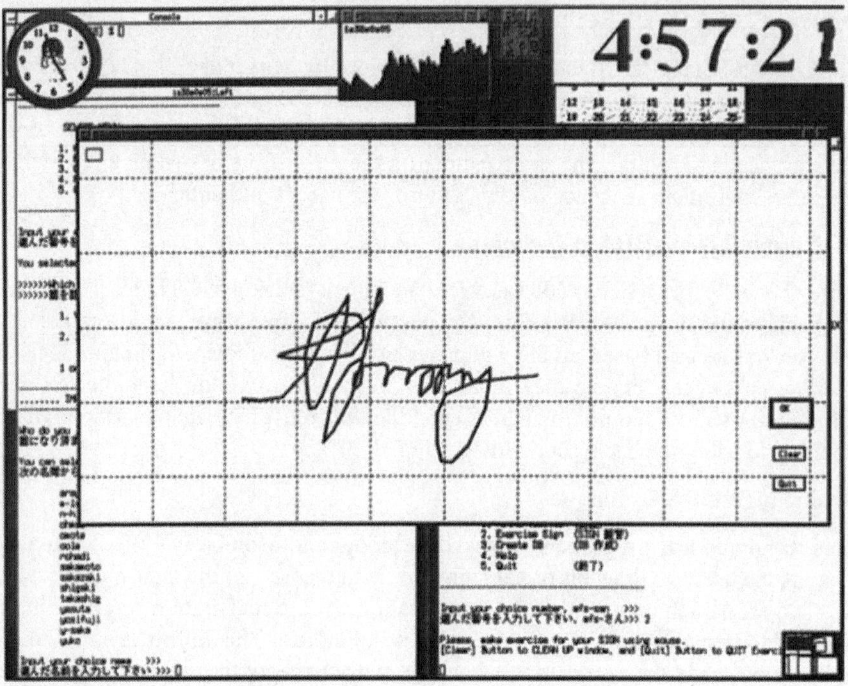

Fig. 2 Syukri signature technique

normalization, and refines another parameter of the signature. The advantage of the Syukri algorithm is no need to remember your signature, and it is hard to make a fake signature of any person. The disadvantage of this method is to draw the signature with the mouse. To remove this inconvenience, you can use the light pen as an input device. Since such devices are not widely used, adding them to the new system as new equipment can be expensive. Although the researchers of this study believe that this technique may be more useful in small devices (Fig. 2).

3.2 Recognition-Based Technique

PassPoint Technique

Pass point scheme was proposed by Susan Wiedenbeck, Jim Waters, Alex Brodskiy, Nasir Memon and Jean-Camille Birget in 2005 [3]. In this technique, the image could be any type, but at the same time, it should be rich enough to have many possible click points. On the other hand, the image is not secret and has no other function than to help the user remember the clicked points. The user selects several points in the image in a specific order. During the login time, the user must click next to the

Fig. 3 Pass point technique

selected click points, within the allowable distance (adjustable), for example, within 0.20 cm from the real click point. The disadvantage of this technique is the login time is longer than the alphanumeric method (Fig. 3).

Pass face Technique

Here, the face is used as a password [4]. This is based on the assumption that people can easily remember people's faces. Use a grid of nine human faces. Among the nine faces, one is known to the user, and the others are spots. The user must identify the familiar face in nine faces. And proceeding until all four faces registered in the recording phase is identified (Fig. 4).

Blonder Technique

The scientist blonder describes graphical password authentication. In this, the user used the default predetermined image. The user clicked on all those tap regions while creating the password in a particular order. For the authentication, the user must click on that approximate area of those tap region in the predefined sequence [5].

Sobrado and Birget Technique

In this method, during registration, numerous icons (pass objects) are displayed to the user, and the user has to select some icons. During authentication, users have to identify those preselected icons and have to click inside the triangle bound by the pass objects [6] (Fig. 5).

Pass-Matrix Authentication

In this method [7], firstly, the user needs to register his information like user id, password, mobile number, and email-id. After that, three images assigned randomly to the user. In that image, the user needs to choose the coordinate squares of the images as the password. These coordinates details will be stored in the database. During the login time, the user has to enter their user id and password, and if the user id and password are valid, a one time password will be sent to the user's mobile number and email-id. This OTP contains the coordinate point of all three images.

Fig. 4 Pass face technique

Fig. 5 Sobrado and Birget technique

Persuasive Cued Click Point

In this method [8], during the registration time, the user enters his or her details like name, phone number, email-id, etc., and for password setup, users click on the points of the images and then submit all the details after that a unique id generated. During the login time, the user has to choose the same click point that has been selected during registration time.

Captcha as Graphical Password

This authentication scheme contains a captcha-based graphical password that is proposed in [9]. This scheme works based on objects such as flowers, vehicles, animals, and character strings. In this scheme, every time all objects like character string, flowers, and animals are randomly organized. Then, it is a challenge for the user to choose the correct user.

Text-Based Shoulder Surfing and Key Logger Resistant Graphical Password

In this scheme [10], there are 72-character set and a circle divided into eight sectors, all the 72-character set are randomly and equally distributed among these eight sectors. During the registration period, the user has to set a password and choose a color. The length of the password should be between 8 and 15 characters. During the login time, the system displays that circle. The circle can be rotated clockwise or anticlockwise. The user needs to rotate the circle containing the *i*-th pass character of his or her password and then click on the confirm button. Repeat this process until confirming all the characters of the password.

4 Comparison Between Different Authentication Systems

Authentication system	Security	User friendly	Availability	Cost
Token-based	Less	Yes	All-time	Less
Biometric-based	High	No	Not all time	High
Knowledge-based	Very high	Yes	All-time	Less

5 Various Types of Attack Techniques

Name	Description
Brute force attack	This is a very basic and easiest attack technique. In the brute force attack, the attackers check all the possible password combinations one by one until they crack the password
Dictionary attack	A dictionary attack is very similar to a brute force attack. In this attack, the attacker checks only common words that most users choose. So, this attack is more feasible than the brute force attack for attackers
Shoulder surfing	In this attack, the attacker obtains information about their password by watching over the victim's shoulder
Key logger	In key logger attack, the attacker installs a program on the victim's computer. When the user enters his or her password, the program records all the keystrokes and sends it to the attacker

6 Advantage and Disadvantage of Various Graphical Password Authentication Systems

References	Scheme	Advantage	Disadvantage
[5]	Blonder	Security is high	In this scheme, passwords are hard to remember
[2]	Syukri signature technique	The security of this scheme is very high because anyone cannot make a fake signature easily	Sometimes, there is a problem in signature matching
[4]	Pass face technique	The face is easy to remember, so this scheme is better than others	Password space is very small
[6]	Sobrado and Birget scheme	Shoulder surfing resistant	If the number of images is large, then it is hard to remember
[11]	Scalable shoulder surfing resistant textual-graphical password scheme	The security is very high, and it is resistant to shoulder surfing	Login process takes a too long time
[7]	Image segmentation	This scheme has a very large password space	Registration and login process takes a too long time

(continued)

(continued)

References	Scheme	Advantage	Disadvantage
[8]	Persuasive cued click point	In this scheme, security is very high	–
[9]	Graphical password (captcha)	This scheme is best for smartphone devices, and security is also very high	This scheme also does not have any major drawback
[10]	Text-based shoulder surfing and key logger resistant	Resistant to shoulder surfing and key logger	–

7 Conclusion

This paper reviews various schemes that come under graphical password authentication techniques. Here, the pros and cons have been discussed with the appropriate examples in the previous sections. The major thing to look at here is how to deal with those loopholes like the quality of the image, password generation, and hash functions associated with it, magnifications, space, etc. These are not actually the drawbacks, but the areas that just need to be taken care of to get appropriate results. The main use of graphical passwords is that it is easy to memorize graphical passwords such as images or patterns instead of random texts or numbers. It also brings some practical issues like quality of the image, color contrast, and the interleaving pixels, i.e., the resolution of the image. For generating a large set of graphical passwords from a single image, it must be large enough. No traditional methods of attacks can break graphical passwords (such as brute force searches, dictionary attacks, or spyware). Often, graphics technology using the current password is still immature. To achieve higher maturity and practicality, cryptographic graphics technology requires more research and the use of research.

References

1. Mayer IJA, Monrose F, Rubin MKRAD (1999) March. The design and analysis of graphical passwords. In: Proceedings of the 8th USENIX security symposium
2. Syukri AF, Okamoto E, Mambo M (1998) A user identification system using signature written with mouse. In: Australasian conference on information security and privacy. Springer, Berlin, Heidelberg, pp 403–414
3. Wiedenbeck S, Waters J, Birget JC, Brodskiy A, Memon N (2005) PassPoints: design and longitudinal evaluation of a graphical password system. Int J Hum Comput Stud 63(1–2):102–127
4. R. U. Corporation (2005) How the passface system works

5. Blonder G (1996) Graphical passwords. United States Paten 5559961, Lucent Technologies. Inc., Murray Hill
6. Sobrado L, Birget JC (2002) Graphical passwords. The Rutgers Scholar, an Electronic Bulletin for Undergraduate Research 4.
7. Tabrez S, Sai DJ (2017) Pass-matrix authentication a solution to shoulder surfing attacks with the assistance of graphical password authentication system. In: 2017 international conference on intelligent computing and control systems (ICICCS). IEEE, pp 776–781
8. Kaja, S, Gupta D (2017) Graphical password scheme using persuasive cued click points. In: 2017 International conference on smart technologies for smart nation (SmartTechCon). IEEE pp 639–643.
9. Kolekar VK Vaidya MB (2015) Click and session based—captcha as graphical password authentication schemes for smart phone and web. In: 2015 International conference on information processing (ICIP). IEEE, pp 669–674
10. Jitendra N, Vinay NS, Ram PS, Sidhardha PN, Deepthi D (2020) Text-based shoulder surfing and key logger resistant graphical password. Eng Sci11(3):214–223
11. S3PAS: A scalable shoulder-surfing resistant textual-graphical password authentication technique, Huanyu Zhao and Xiaolin Li Scalable Software Systems Laboratory Department of Computer Science Oklahoma State University, Stillwater, OK 74078, USA.

Chapter 56
Predictive Analysis for Early Detection of Alzheimer's Disease

B. A. Sujathakumari, M. Charitha Shetty, H. M. Lakshitha, P. Jain Mehulkumar, and S. Suma

1 Introduction

AD is an irremediable and progressive disorder which is said to cause by abnormal accumulation of amyloid beta and neuro-fibrillary tangles in brain which causes problems with respect to memory loss and behaviour [1]. Beta-amyloid and hyper-phosphorylated are the two main proteins that causes AD [2]. Sometimes, genes that are linked to AD might also cause the dementia. AD is seen to be the common in aged group of people with almost affecting 6% of the population over 65 years of age, and affecting more than 47.5 million people worldwide. Mild cognitive impairment (MCI) is recognised as an early stage of the disease. To predict AD at earlier stage, research in mild cognitive impairment (MCI) is mainly focused by which the progression of MCI subjects to AD can be delayed [3]. As no treatments can neither reverse nor stop the AD progression, predicting the AD at the early stage has become a very important which provide patient to access different treatment options and opportunity to participate in clinical trials.

B. A. Sujathakumari · M. Charitha Shetty (✉) · H. M. Lakshitha · P. J. Mehulkumar · S. Suma
ECE Department, Sri Jayachamarajendrandra College of Engineering, JSS Science and Technology University, Mysuru, Karnataka, India
e-mail: charithashetty935@gmail.com

B. A. Sujathakumari
e-mail: sujathakumari@sjce.ac.in
URL: https://sjce.ac.in/ec-faculty/sujathakumari-b-a

H. M. Lakshitha
e-mail: hmlakshitha@gmail.com

P. J. Mehulkumar
e-mail: mehuljain3038@gmail.com

S. Suma
e-mail: sumaspu@gmail.com

I. Jeena Jacob et al. (eds.), *Data Intelligence and Cognitive Informatics*, Algorithms for Intelligent Systems,
https://doi.org/10.1007/978-981-15-8530-2_56

For human brain imaging, MRI has proven to be the most efficient method. MRI is extensively used for the detection of MCI. In our work, the MRI data is obtained from ADNI database. Along with MRI, ADNI also PET scans, cognitive assessments scores, genetic information, blood biomarkers and information about CSF biomarkers which helps in distinguishing MCI progression to AD [4].

2 Related Works

Following papers were referenced to go ahead and build an algorithm for the early detection of AD.

In paper [1], a deep learning approach and multimodal recurrent neural network approach have been applied to predict from MCI progression towards AD. Along with longitudinal CSF biomarkers, cross-sectional neuro-imaging and cognitive-related biomarkers are considered. Necessary dataset is taken from ADNI website. From results in the prediction model for MCI to AD conversion was observed that single modality of data yielded up to 75% accuracy (AUC = 0.83), while longitudinal multi-modal data resulted in 81% accuracy (AUC=0.86). The approach was found to predict subjects who are at risk of developing AD.

In paper [5], longitudinal 3D-T1 weighted MRI's were obtained from ADNI. Study considered 79 normal controls (CN), 50 preclinical Alzheimer's, and 274 MCI/AD. Patient data at baseline and periodic visits are considered. Jacobian determinant maps have been obtained from the results of SPM12 toolbox from which structural changes in brain are statistically analysed. Longitudinal voxel-based classifier was used. The classifier resulted in AUC = 0.87 when tested using receiver operating characteristic curve analysis.

In paper [6], the relationship between data points of different time intervals is analysed using the random forest classifier. The input was obtained from TADPOLE grand challenge which included demographic data and genetics-related data and has a vector which has information of historic time series of patient visits. The normalised ventricles volume has been concentrated for the study. From results, AUC of 0.82 and a classification accuracy of 0.73 were obtained in comparison with a SVM classifier that resulted in AUC of 0.62.

The authors in paper [7] investigate on how to predict progression to AD. Data obtained from National Alzheimer's Coordinating Center (NACC) has been used. To deal with uneven time interval data, the approach of long short-term recurrent neural network (RNN) is used in this paper which depends on enhanced many-to-one RNN structure. The results of proposed approach showed that it could predict AD progressions of patients on their next visits with 99% of accuracy. Thus, it confirms that the approach using RNN can predict progression by taking into account the temporal lobe region and also periodic or historic visits of patients.

In paper [8], structural MRI scans are used to find the likelihood of automatic classification of Alzheimer's disease using different dimensionality reduction techniques. Further study is done by considering neurodegeneration as detected by spike in CSF tau, brain atrophy and loss of neuron stated from MRI; memory loss and general cognitive decline obtained from cognitive assessment and the amyloid beta imaging as measured in cerebrospinal fluid (CSF) and PET amyloid imaging. After dimensionality reduction, standard machine learning techniques like decision tree, rule-based techniques and neural networks are applied on the reduced dataset. ML techniques like random forests, SVM, and recently developed deep neural networks have resulted in a very good performance in classification [9].

In paper [10], the diagnostic results of 248 MCI patients from 2 to 5 years of follow-up time was compared and its effects of advancement on entorhinal cortex and hippocampal volumes were observed. For 5 year timeline, the model gave higher accuracy (AUC = 73% at 2 years vs AUC = 84% at 5 years) when the combination of hippocampal and entorhinal cortex was considered. Even the specificity results (56 versus 71%) were better. The performance of the ML models proposed was comparable to other ML models for the 5 year follow-up time considered.

As discussed in the papers, different algorithms used different datasets for analysis based on the requirement. It was observed from one of the approaches that the cognitive assessment and multi-modal approach showed highest accuracy considering demographic data of the patients [1]. For time-varying data, random forest classifier showed better accuracy when compared to SVM classifier. Amyloid beta and tau species biomarkers were considered to be major contributors for predicting Alzheimer's disease [6]. Concentrating on the advancement of disease on hippocampal region and entorhinal cortex volumes gave good accuracy [10]. Summarising these observations and discussions, it can be concluded that with the combination of brain imaging data, genetic information and demographic data, neuropsychological test results, cerebrospinal fluid biomarkers like amyloid beta and tau species as input to the efficient classifiers like random forest classifiers can result in very high predictive accuracy.

3 Software and Materials Requirement

The following are the software and materials required for the work:

Python: This project uses Python language as it is a great interpreted, object-oriented and interactive programming language apart from being open-source programming language. It eases the usage as it has a lot of code libraries and it is simpler compared to other programming languages. Python along with R is gaining popularity in the domain of analytic. This research work has used Python under Anaconda open-source distribution.

MATLAB: Matrix laboratory (MATLAB) is designed to process with a programming language that expresses matrix and array mathematics directly. It is used for manipulating matrices, implementing the algorithms, interfacing with other programming languages like C, C++, Java, Python, etc., and to plot functions and data. For our project, MATLAB is needed for the pre-processing step which can be done by the toolbox called SPM8.
Statistical parametric mapping (SPM): It is used to test hypotheses about functional imaging data [11]. These ideas have been integrated into a toolbox called SPM which can be made use with the help of MATLAB. It can be used for the analysis of fMRI, PET, MEG, etc. Input images which were considered in our project were realigned, registered to standard template, spatially normalised, and smoothed.
MRIcro: It is used to view medical images on Windows and Linux computers. MRIcro is a free-standing programme, it includes tools that support SPM.
ADNI dataset: ADNI was started by many medical institutes and nonprofit organisations. Its primary goal has been to test if MRI, clinical data, bio-markers, PET's, and neuropsychological assessment can be amalgamised to check the progression of MCI and early AD [12]. The raw image data along with biomarker data and clinical info are taken from ADNI dataset for our project.

4 Proposed Work

The proposed work of the project is explained in this section. The first step is data acquisition through the ADNI website. The data selection is based on the category of age and not based on gender separation. Data is separated based on patient ID for 3 year follow-up line for further process.

The dataset collected is then segmented to remove unwanted region of the brain mainly skull stripping. When segmentation procedure is carried out, registration of the brain MRI with the standard MNI template is also done. The MRI can be used as 3D itself or 2D slices can be considered for the analysis. Extracted 2D slice is considered for the work. The feature reduction is performed to remove the redundant data and also makes the computation easy. The feature extraction and reduction is carried out using multilevel 2D DWT in the proposed work. The machine learning classifier considered for the purposed of prediction and classification is linear regression and random forest classifier.

The block diagram of the proposed work is shown in Fig. 1. After the selection, segmentation, feature extraction, and the next step are training the classifier. The training set is used for training the algorithm along with CSF biomarkers and clinical data. The testing set validates the classifier model. The accuracy result from the classifier how best the data has been classified at an early stage and how the addition of CSF biomarkers and clinical data has showed an impact on accuracy for early detection.

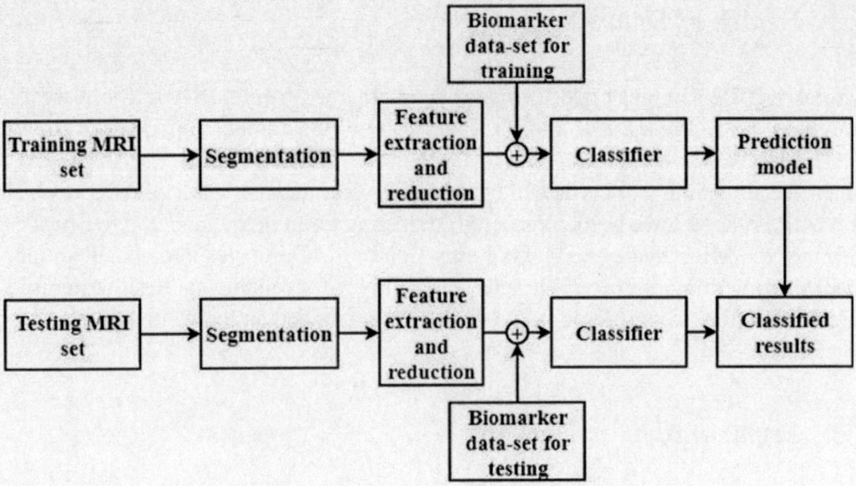

Fig. 1 Block diagram of proposed work

5 Implementation

The implementation part of the work is explained in this section.

5.1 Data Acquisition

The initial step of the work is data collection from standard ADNI dataset. Particularly, data is acquired from ADNI-1 where patients 5 years data can be obtained. ADNI consists of data and samples such as MRI scans, blood biomarkers, CSF biomarkers, cognitive test results, and PET scans.

The study data and resources from the North American ADNI study are available through the ADNI website which includes data of AD patients, MCI subjects, and CN. The data considered for the study in the project mainly includes:

1. CN: Subjects from 1.5 T 3 year data.
2. MCI: Subjects from 1.5 T 3 year data.

The data obtained from the ADNI will be three-dimensional (3D). The three different views of an MRI image can be viewed with the help of MRIcro software which helps in visualisation of the NIfTI and ANALYSE format files.

5.2 Details of Dataset

A total of 2182 samples of MRI images were obtained from ADNI 1. These are T1 weighted, 1.5T, 3 years MRI images. Table 1 shows the number of patients in ADNI 1 dataset.

Since, the aim of our work is detection at an early stage and prediction of AD, only CN and MCI datasets have been considered for testing training the model. After further refining, the MRI images and taking only CN and MCI images into consideration, the following number of total samples and samples for training and testing (tenfold cross-validation is considered in the work) were obtained as shown in Table 2.

5.3 Segmentation

Segmentation is an important step after the separation of patients MRI-based on their ID. It is a method for cleaning up of the unwanted part of the MRI scan mainly the skull part [13–15]. Segmentation is nothing but feature selection method which removes irrelevant data from the MRI image. Segmentation in our work is carried out in SPM 8 toolbox of MATLAB 2014a. The segmented MRI image consists of grey matter, white matter, and CSF. Grey matter of segmented image is considered for further processing in the work since it contains maximum features.

The segmented grey matter of CN and MCI subjects are shown in Fig. 2a and b, respectively. After the selection of segmented grey matter image for further processing, the selection of slice becomes important. The middle slice of the segmented MRI data is said to have maximum and necessary features. Feature extraction is done from the middle slice.

Table 1 Number of samples from ADNI dataset

Subjects	No. of samples
CN	750
MCI	981
AD	541

Table 2 Number of CN and MCI samples taken for classification

Subjects	Total samples	Samples for testing	Samples for training
CN	205	20	185
MCI	343	34	309

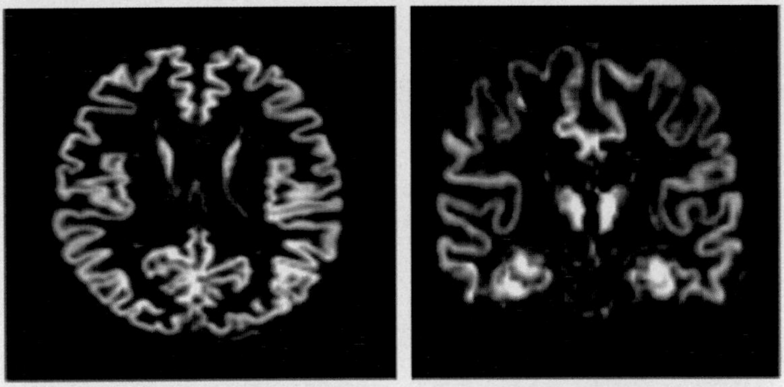

(a) Segmented grey matter of CN subject

(b) Segmented grey matter of MCI subject

Fig. 2 Segmented images

5.4 *Feature Extraction and Reduction*

Once the unnecessary segments of the brain MRI are removed using segmentation, feature extraction technique is applied on the segmented images for the extraction of necessary features which is the basis for classification and prediction [16]. The extracted features through the feature extraction technique will be in the form of matrix which are then reduced to lesser dimensional matrix without changing the properties of overall extracted features. In this work, multilevel two-dimensional discrete wavelet transform (DWT) is carried out for feature extraction and reduction. The low pass filter (LPF) and high pass filter (HPF) coefficients are considered as 2D-wavelets.

A two-level 2D DWT decomposition is shown in Fig. 3 where at first level of decomposition, 4 subbands are obtained and at second level, 7 subbands are obtained, out of which pair of diagonal, horizontal, and vertical coefficients are obtained and one approximation coefficient is obtained. Set of diagonal, horizontal, and vertical coefficients are together said to be detailed coefficients and are not considered in the work. The approximation coefficients (LPF coefficients) are most important com-

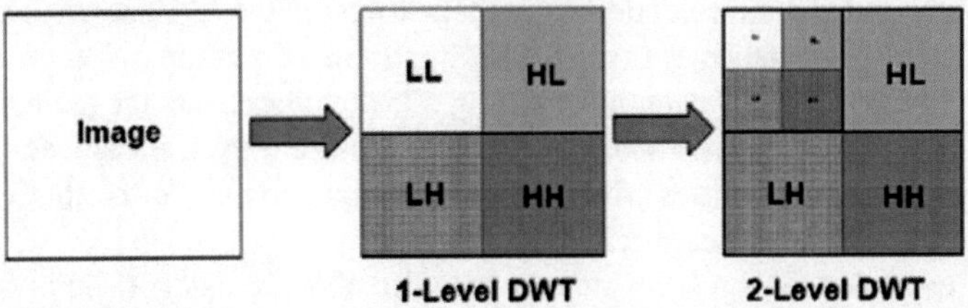

Fig. 3 Two-Level 2D DWT decomposition process

pared to the detailed coefficients (HPF coefficients) and is further considered in the work.

The total number of features of the brain MRI was about 9919. Out of these many features, after performing 2D DWT, the number of features was reduced to 2530 (approximation coefficients). Min-max scaling is also included for maintaining the feature value between 0 and 1. The reduced features are the wavelets of 2D DWT which are stored in .csv file for the next process.

5.5 Biomarker and Clinical Data

Biomarkers measures the movements of a cell. They are very much needed for early warning of abnormality in the health systems. Thus, biomarkers help in the early prediction of any cognitive disorder or impairment in the nervous system.

Many biomarkers are studied for their potential to detect early stages of AD. Our work includes consideration of tau and beta-amyloid levels in CSF. Research suggests that AD in early stages can be detected via CSF levels changes of tau and beta-amyloid, proteins which are strongly linked to Alzheimer's [17].

The other clinical data are considered along with CSF biomarkers for the prediction of the disease in the work. The clinical data includes MMSE score, age, gender, education, and genetic data like apolipoprotein-4. All the above clinical data along with CSF biomarkers are added along with the wavelets to obtain better results compared to using only 2D DWT wavelets for the proposed work. The combined data is given to the classifiers for the prediction.

5.6 Classification

Classification in context of the work refers to separation of the subjects according to the stages [18, 19]. CN vs MCI subjects are classified and classification accuracy along with other metrics are computed in the work. Machine learning (ML) classifiers—logistic regression and random forest classifiers are chosen for computing the results since one is binary classifier and the other is said to be the best classifier which yields more accuracy.

The reason behind the selection of random forest is that it is faster to train than other decision trees, hence, working with hundreds of features is easy. Speed of prediction is significantly faster than training. On the other hand, the main reason for choosing logistic regression model is because it is not only a classification model, but also provides probabilities which makes it unique compared to other classifiers which only provide final classification.

Logistic Regression: It is a 'statistical learning' technique categorised under supervised ML methods. It uses a linear regression equation as a building block and attempts to maximise the quality of output on a training set to produce discrete binary outputs. This algorithm does not need too much of computational resources and is easy to manage. It produces output which is well calibrated to predict probabilities. *Random Forest*: It consists of large number of single decision trees which group as the name suggests and is said to be a very stable algorithm. Each individual tree in gives out a class prediction and the most voted class becomes our model's prediction. In this project, random forest classifier is chosen because it creates numerous trees based on the data and combines the output of these trees. It also handles nonlinear parameters and missing values efficiently and automatically.

6 Results

For analysing the performance, the two classifiers were trained and tested with the data using tenfold stratified cross-validation method. The tenfold divides the entire dataset into 10 sets within 1 set is used for testing and remaining 9 of them for training. This process is repeated for about 10 times and average of all 10 repetitions is reported. This ensures that every data present in the dataset has been used for testing the model. Once the reduced set of features is obtained, the next is to fit a classification model so as to accurately classify CN and MCI subjects. Different classification algorithms—random forest and logistic regression are used for classification. Implementation of these classifiers is used from Scikit library.

6.1 Comparison of Accuracy

It is found that classification accuracy does not significantly change with decrease in features of training set. This is obtained from sklearn library, which split data in train/test sets. The final accuracy was calculated as the average accuracy of tenfold validation. It is found that by using DWT, the features got reduced by around 75%. This decrease in number of features decreased time complexity for classification. However, this did not affect the classification accuracy of classification algorithm. In fact, in some cases, the classification accuracy did improve by using 2D DWT. The comparison of accuracy for CN vs MCI is shown in Table 3.

Table 3 Comparison of accuracies for classifiers used

Model	Accuracy (wavelets) (%)	Accuracy (wavelets + biomarkers) (%)
Logistic regression	77.18	94.34
Random forest	84.30	91.7

Classification accuracy of 77.18% was obtained with only 2D DWT wavelet coefficients. But the accuracy was increased to 94.34% when biomarker and clinical data was added using logistic regression. In the random forest classifier, the accuracy was 84.30% with only DWT wavelet coefficients, but the overall accuracy increased to about 95.7% when the biomarker and clinical data was added. Thus, the random forest showing best performance for classification and prediction.

6.2 *Comparison of Other Performance Metrics*

Some of the other performance or classification metrics like specificity, sensitivity, precision, error rate, false positive rate and area under the curve (AUC) has been calculated using the confusion matrix obtained from the classifier which are shown in Figs. 4 and 5. Other performance matrices are tabulated in Tables 4 and 5.

The performance metrics have been tabulated in Table 4 considering only DWT wavelets coefficient for evaluation of classifier. Table 5 shows the performance metrics considering both DWT wavelets coefficient along with biomarkers and clinical data for evaluation. The AUC curves obtained with and without using biomarker data using logistic regression and random forest classifier are shown in Figs. 4 and 5. The curves show that random forest classifier is the best in terms of performance compared to the binary classifier-logistic regression.

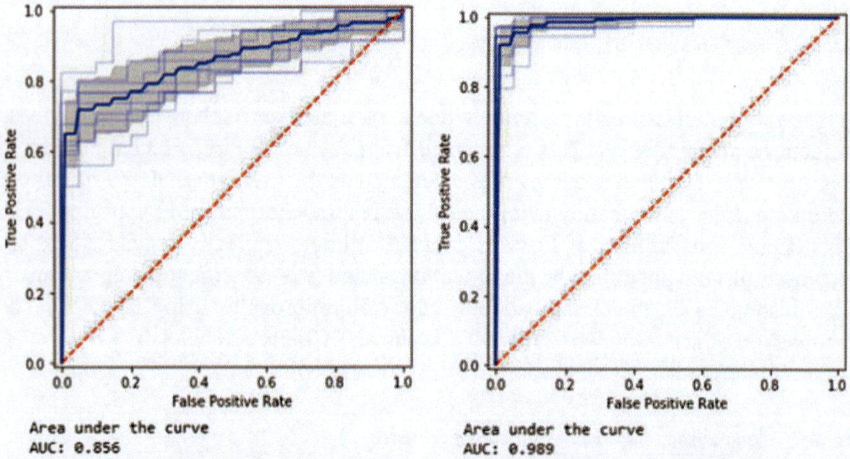

(a) Only DWT data

(b) Biomarkers and DWT wavelets combined

Fig. 4 AUC curve for logistic regression

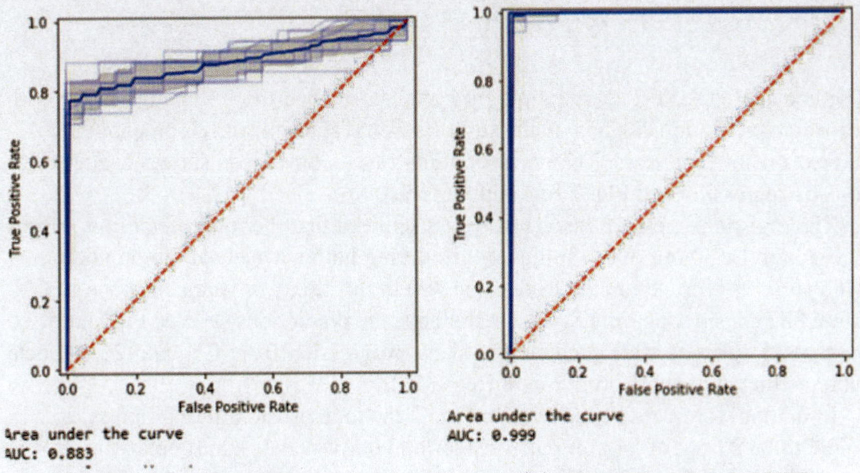

(a) Only DWT data

(b) Biomarkers and DWT wavelets combined

Fig. 5 AUC curve for random forest

Table 4 Metric values for classifier evaluation considering only DWT wavelet coefficients

Parameters	Logistic regression	Random forest
Sensitivity (%)	77.2	75.5
Specificity (%)	77.0	99
AUC	0.856	0.99
Precision	0.849	0.992
Error rate	0.228	0.156

Table 5 Metric values for classifier evaluation considering both DWT wavelet coefficients along with biomarker and clinical data

Parameters	Logistic regression	Random forest
Sensitivity (%)	93.1	86.8
Specificity (%)	96.3	100
AUC	0.989	0.999
Precision	0.978	1.0
Error rate	0.05	0.082

6.3 Statistical Plot of Predictive Analysis

The considered 1.5T 3 years follow-up line data showed the following results with respect to prediction that how many subjects would show the development in the AD in next one or two years. The corresponding results have been shown in Fig. 6 for logistic regression and Fig. 7 for random forest.

The logistic regression results in Fig. 6, showed that the total prediction in progression to be 30 out of 54. In input considering both wavelets and biomarkers, 29 MCI subjects were predicted to develop AD in the next few years but none in CN. But while considering only DWT coefficients, the predictions showed that out of 20 subjects, 5 subjects were predicted to show progression from CN, and 26 subjects were predicted to show progression from MCI.

In random forest results showed in Fig. 7, the total prediction in progression was found to be 28 out of 54 in input considering both wavelets and biomarkers. It was found that out of 28 total predictions, 27 MCI subjects were predicted to develop AD in next few years but none in CN. But while considering only DWT coefficients, the predictions showed that out of total 23 subjects, 22 subjects were predicted to show progression from MCI but none in CN.

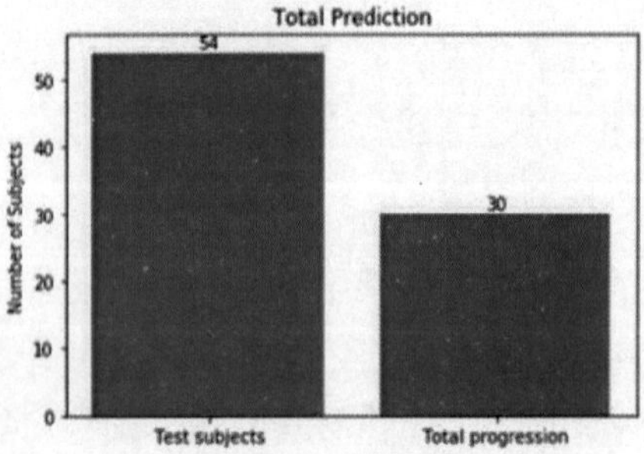

(a) Total Predictions

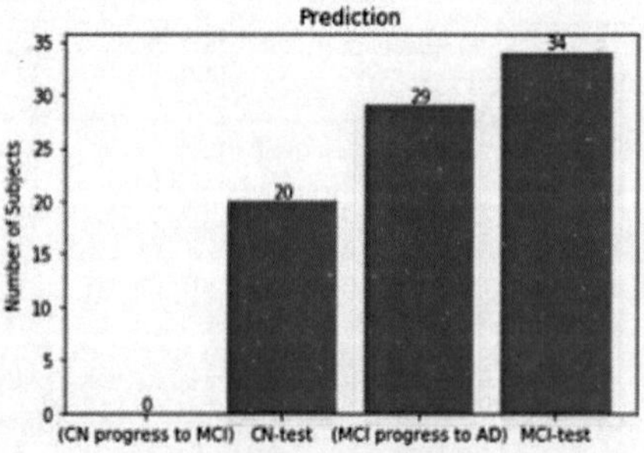

(b) Predictions with wavelet and biomarkers

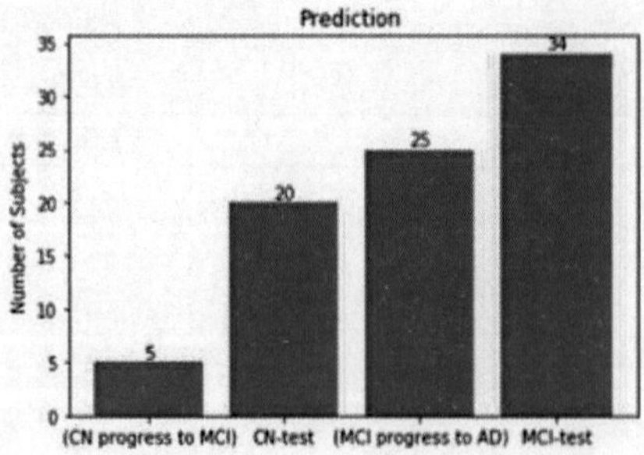

(c) Predictions with wavelets only

Fig. 6 Logistic regression chart of predictions

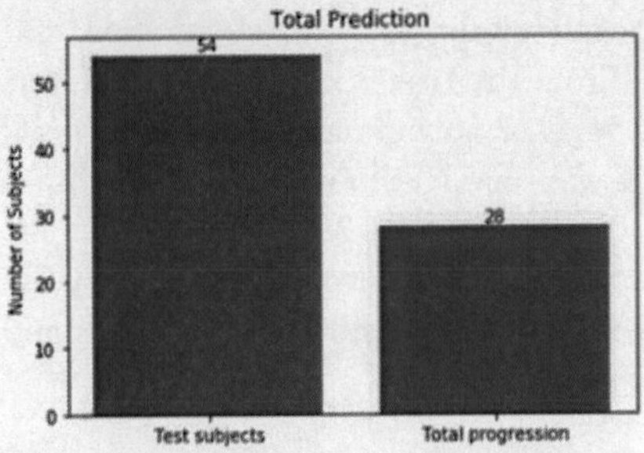

(a) Total Predictions considering both DWT and biomarkers

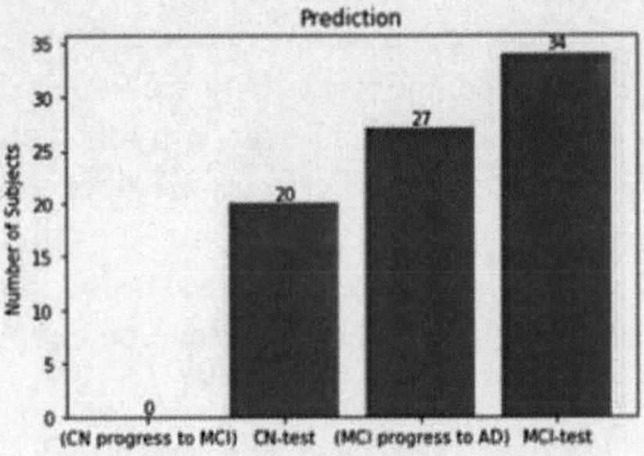

(b) Predictions with wavelets and biomarkers for CN and MCI seperately

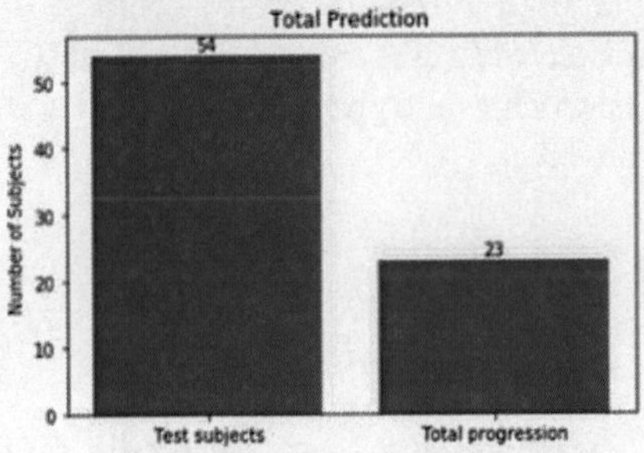

(c) Total Predictions with wavelets only

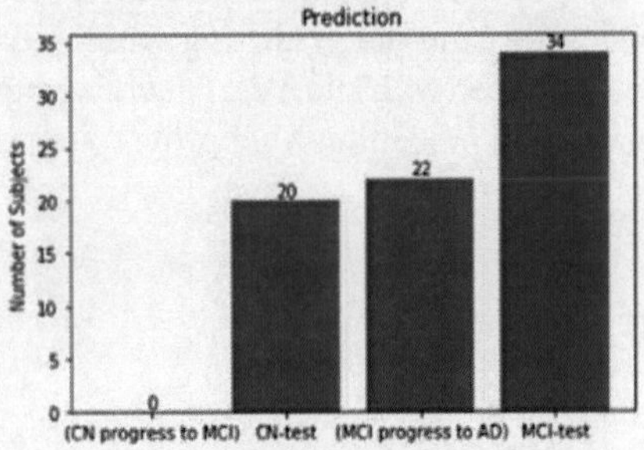

(d) Predictions with wavelets only for CN and MCI seperately

Fig. 7 Random forest chart of predictions

Thus, it was observed that random forest showed the best prediction since it gave more accurate predictions in number of patients who would develop progressions in next few years than the logistic regression and also the classifier with high AUC is said to give better prediction results and hence random forest showed more accurate predictions. And it was seen that there was no improvisation in results for prediction in logistic regression even after the addition of biomarker and clinical data, i.e. the number of predicted subjects who would show progression remained the same with or without biomarker.

7 Conclusion and Future Scope

In this project, a combination of pre-processed longitudinal structural brain imaging data, phenotype data, genetics-related information, neuro-psychological assessments, and CSF biomarkers has been considered as an input for the classifier algorithm. Using this dataset, a high classification performance was obtained. The early stage longitudinal data, i.e. CN and MCI considered for study were classified, and thus the prediction accuracy was analysed. From the results of classifiers, it was clearly observed that the random forest classifier was more efficient in terms of AUC compared to logistic regression algorithm. It was also observed in both the classifiers that the addition of biomarker clinical data to the MRI-2D DWT coefficients increased the accuracy of the classification compared to the accuracy from only 2D DWT coefficients. Hence, it can be concluded that the inclusion of biomarkers and demographics along with MRI data can prove to increase accuracy and precision of the prediction algorithm.

In the near future, the model would be developed for no or little dimensionality reduction as some amount of data is believed to be lost when the dimension of the 3D scans is reduced with leading to a lower accuracy in the prediction of progression. Instead of 1.5T MRI scans, 3T MRI would be used to produce more accuracy. In order to expand the predicting window, combined measures from sMCI and pMCI would be used with latest deep learning models which would give better classification accuracy in predicting MCI progression towards AD.

References

1. Lee G, Nho K (2019) Predicting Alzheimer's disease progression using multi-modal deep learning approach. J Sci Rep
2. Bellevillea S, Fouqueta C (2014) Detecting early preclinical Alzheimer's disease via cognition, neuropsychiatry, and neuroimaging - qualitative review and recommendations for testing. J Alzheimer's Dis 375–382
3. Salvatore C, Cerasa A, Castiglioni I (2018) MRI characterizes the progressive course of AD and predicts conversion to Alzheimer's dementia 24 months before probable diagnosis. Front Aging Neurosci 10 Article 135 (2018)
4. Moscoso A, Silva-Rodriguez J, Aldrey J, Cortes J (2019) Prediction of Alzhiemer's disease with MRI beyond the short term-implication for the design of predictive models. Elsevier's
5. Petrone PM, Casamitjana A, Falcon C, Artigues M (2019) Prediction of amyloid pathology in cognitively unimpaired individuals using voxel-wise analysis of longitudinal structural brain MRI. J Alzheimer's Res Ther
6. Moore PJ, Lyons TJ, Gallacher J (2019) Random forest prediction of Alzheimer's disease using pairwise selection from time series data. J PLOS ONE
7. Wang T, Qiu RG, Yu M (2018) Predictive modeling of the progression of Alzheimer's disease with recurrent neural networks. J Sci Rep
8. Arvesen E (2015) Automatic classification of Alzheimer's disease from structural MRI. Master's thesis in Computer Science-Journal of AD
9. Khan A, Usman M (2015) Early diagnosis of Alzheimer's deisease using machine learning techniques—a review paper. In: Proceedings of the 7th international joint conference on knowledge discovery, knowledge engineering and knowledge management
10. Moscoso S-R, Aldrey C, Fernández-Ferreiro G-L, Ruibal, A (2019) Prediction of Alzheimer's disease dementia with MRI beyond the short term-Implications for the design of predictive models. NeuroImage: Clinical 23:101837
11. http://radinfo2004.tripod.com
12. Yang Z, Zhuang X, Sreenivasan K, Mishra V, Cordes D (2019) Robust motion regression of resting-state data using a convolutional neural network model. Front Neurosci 13 Article 169, pp 1–14

13. Garg RK, Kulshreshtha A (2017) A review of MRI image processing techniques employing segmentation and classification. Int J Comput Sci Trends Technol (IJCST) 5(2):
14. Sa BK, Sa AS, Lala K, Antonya A, Kurupa AM (2017) Alzheimer's detection based on segmentation of MRI image. Sci Dir 474–481
15. Ridgwaya GR, Camaraa O, Scahillb RI, Cruma WR, Whitcherc B, Foxb NC, Hilla DLG (2017) Longitudinal voxel-based morphometry with unified segmentation—evaluation on simulated Alzheimer's disease. IEEE
16. Segovia F, Gorriz JM, Ramýrez J, Salas-Gonzalez D, lvarez IA, Lopez M, Chaves R (2011) A comparative study of feature extraction methods for the diagnosis of Alzheimer's disease using the ADNI database. Elsevier, pp 64–71
17. https://www.pacificneuroscienceinstitute.org/brain-health/diagnostics-procedures/blood-and-csf-biomarker-tests-for-dementia-detection
18. Kruthika KR, Rajeswari HDM (2018) Multistage classifier-based approach for Alzheimer's disease prediction and retrieval. Elsevier's
19. Stalin Babu G, Tirumala Rao SN, Rajeswara Rao R (2019) Exploring of classification methods for early detection of Alzheimer's disease. Int J Eng Adv Technol (IJEAT) 8(6):5206–5211

Chapter 57
Feature Extraction from Ensemble of Deep CNN Model for Image Retrieval Application

Vijayakumar Bhandi and K. A. Sumithra Devi

1 Introduction

Multimedia technology and storage devices have undergone multifold transformation in last two decades. This expansion in digital world has made image analysis techniques like image retrieval still an interesting research problem. Various image retrieval methods have been designed and used effectively in real-world applications. The most important method is content-based image retrieval method (CBIR). This method is built to leverage content similarity between images.

Handcrafted Features: CBIR applications have been built to use various image features depending on the application area. However, widely used features are color property, texture assembly, and shape form. Such descriptors are carefully designed and have proven to work efficiently in most of the CBIR applications. These features are generally called as handcrafted features. Color property of the image does not vary with variation in scale and alignment. Hence, it is a widely used prominent feature for image retrieval in traditional CBIR systems. Color property of the image is very well-captured by color histogram method. Texture property can capture repeating arrangement and disparity in intensity of the image. Gray-level co-occurrence matrix method (GLCM) is widely used for statistically measuring texture feature. Various practical image retrieval applications are created using GLCM statistical features like homogeneity parameter, dissimilarity measure, correlation property, energy feature, and angular second moment (ASM) character. Hu moment invariants is one of the

V. Bhandi (✉)
Jain (Deemed-to-be University), Bangalore, India
e-mail: vijay.bhandi@gmail.com

K. A. Sumithra Devi
Dayananda Sagar Academy of Technology and Management, Bangalore, India
e-mail: sumithraka@gmail.com

I. Jeena Jacob et al. (eds.), *Data Intelligence and Cognitive Informatics*,
Algorithms for Intelligent Systems,
https://doi.org/10.1007/978-981-15-8530-2_57

important features used in shape analysis. Hu moments are invariant to translation orientation, rotational alignment, and scale deviation of the objects.

Drawbacks of Handcrafted Features: Although traditional handcrafted features perform decently in image retrieval, they have some major drawbacks. There are various handcrafted features available and choice of features greatly influences the retrieval results. Selection of improper features brings down the performance of the system. Also, a set of features which performs well in one domain may not yield good results in another domain. Hence, there is a need to design systems that can compute effective image features automatically and remove the dependency on prior knowledge of the target application area.

Convolutional Neural Networks for Image Retrieval: Machine learning enables systems to learn important insights automatically from input data through training. Neural networks approach of machine learning is inspired by biological neural networks and designed to replicate the way human brains learn. Neural networks can identify complex patterns and handle a greater number of patterns at a given time. Neural networks can be applied to solve problems in any domain, since they are general function approximations. Convolutional neural network (CNN) is widely used to analyze visual imagery. The name is inspired by the convolution operator. Convolution learns rich insights about input image by deriving numerous simple data squares from it. Deep CNNs are capable of learning very powerful and complex features from input image data. Hence, features acquired by deep CNN can serve as input to image processing systems including CBIR. Although deep CNNs perform exceptionally well in learning image features, they have a major drawback. Developing a best performing deep CNN system from scratch requires large datasets, enormous computing environments and execution time. Hence, features learnt by well-established deep CNNs possess huge potential to help in this situation. Many such proven CNN models are distributed with model weights for quick implementation. They are pre-trained on enormous datasets and attained high accuracy levels. Many recent image classification research works have made use of such deep CNN models to transfer the knowledge learnt by them. Well-established deep CNN models like ResNet, VGGNet, AlexNet, and GoogleNet are extensively employed in such studies.

This work intends to repurpose a deep CNN (pre-trained) model as a feature generator. Figure 1 summarizes the proposed transfer learning approach applied to CBIR application.

This research work has created a novel image retrieval methodology to employ VGG16 deep CNN model architecture, retrain it on target experimental dataset with a custom-built classifier, and create ensemble with another popular deep CNN model ResNet50. Our proposed framework extracts features out of an ensemble of two models (VGG16 retrained and ResNet50 pre-trained). The new image retrieval technique is subjected to extensive experimental study on a phytomorphology dataset (Swedish leaf images collection).

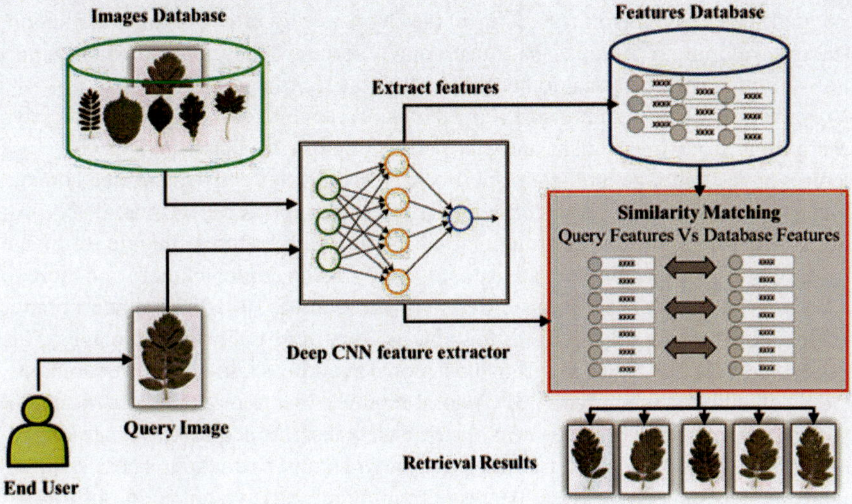

Fig. 1 Transfer learning for CBIR application

2 Related Work

Earlier image retrieval methods were text-based and depended mainly on human annotations for search operation. Such methods involve huge efforts and cost. It is greatly influenced by human perception and bias [1]. Content-based image retrieval method (CBIR) was designed to apply visual content descriptors for effective search operation [2]. The most practical CBIR systems make use of visible descriptors like color destitution, texture arrangement, shape forms [1, 2]. Such visual descriptors can be combined to create fusion features to represent an image efficiently [1, 3]. Color feature is stable and very robust. It can be implemented easily and needs less storage as compared to other properties. Color histogram technique is best-suited to express color distribution in an image [1–3]. Hue saturation value (HSV) space is more closer to human impression about colors. It performs better against red green blue (RGB) space [1]. Texture property is another salient descriptor used in CBIR. Texture feature is designed to measure the surface distribution at granular level. Gray-level co-occurrence matrix method (GLCM) collects texture distribution compactly [1]. GLCM statistical features like homogeneity parameter, dissimilarity measure, correlation property, energy feature, and angular second moment (ASM) character are very useful in image texture analysis [2]. Shape feature is another conspicuous property that can help to distinguish image content visually. Hu-invariant moments are a statistical measure for shape feature. Hu-invariant moments feature has been widely used in image recognition as a shape feature [4].

In the field of image retrieval, the identification and extraction of features are paramount steps [5]. Low-level image descriptors designed for earlier image retrieval systems are not very effective in depicting the contents. These are meticulously designed and need right understanding of the target area for effective implementation. Hence, developing techniques that can produce expressive features without an expert intervention is very essential to boost the performance of a CBIR system.

Convolutional neural networks (CNNs) are capable of acquiring expressive descriptions from image data and can perform automatic classification [6]. Such features have greater generalization ability. CNNs utilize effective dispersed connections, weight distribution, and dimensional subsetting approaches. Hence, CNNs produce elementary network construction that can generate features suitable for image processing [6]. Due to all these advantages, CNNs being deployed more and more to address various issues pertaining to digital image analysis [6–10]. To make a brand-new model very efficient, one needs to train it on very large collection of images. This is a considerable hindrance in setting up a model to address issues in newer domains, where collections are evolving [3]. Also, it requires vast computation environment and longer execution [11]. However, the extensive insights acquired by state-of-the-art deep CNN models can be transferred through feature extraction. These state-of-the-art models have undergone extensive training on massive collection of images such as ImageNet [11]. A new network can be created with weights shared by a pre-trained model, and this approach will help to develop cost-effective applications [11]. Well-established deep CNN models such as ResNet, VGGNet, AlexNet, and GoogleNet are engaged in digital image studies [11]. VGG16 model has delivered noticeable results in different image processing implementations [12–16]. Residual neural network or ResNet is another eminent pre-trained model. ResNet are real deep models and consist of more than 50 layers [17]. ResNet model delivered remarkable results in classifying images [18–20].

Plants are really vital for the well-being of this planet and all living things. Identification of plants can be done through phytomorphology or leaves analysis. Manual analysis and classification of leaves are difficult and time-consuming task. That is due to the existence of wide variety of leaf structures and composition. Hence, there is a great need for image analysis-based approach in pyhotmorphology area. The automatic leaf image processing techniques can help in disease recognition at minimum cost and error without much expertise [21]. Various works have been done on the classification of leaves using image classification techniques [22]. Plant leaf classification using handcrafted features like texture and shape descriptors is built [23]. CBIR using various shape features were built for leaf image search [24, 25]. Manual identification of appropriate properties for distinguishing different kinds of crop diseases is a complex and time-consuming task. Machine learning techniques can help here to automate different tasks like classification, detection, and search. A convolutional neural network-based framework is built to automatically classify rice leaf diseases [21]. Convolutional autoencoder approach is implemented to extract features for classifying leaves [26]. Pre-trained CNNs are suitable for automatic leaf disease classification [21]. No significant works available with respect to image retrieval using CNNs for leaf images.

3 Proposed Methodology

Our proposed CBIR method has two modules: offline indexing stage and online query stage as outlined in Fig. 2. Offline indexing module is a one-off feature extraction activity. Images are preprocessed to convert into required color space and resize to required size. Then each image from the given database is processed to compute features (Handcrafted or deep CNN). Resulting feature sets are stored in an indexing database for image retrieval. In the online query module, the search operation to identify similar images is carried out. Query image is processed to compute required features [Handcrafted or deep CNN features]. These features are matched with the features saved from offline indexing module. Similarity distance method is used to identify matching images depending on the number of images to be retrieved. Three CBIR frameworks are developed in this work: first a baseline CBIR using handcrafted features, second framework using features from pre-trained CNN models, and third is the proposed CBIR which uses features from ensemble of retrained and pre-trained deep CNNs.

3.1 CBIR Using Handcrafted Features

This is a basic CBIR framework created to replicate a traditional CBIR system. Here, the proposed work uses color histogram, texture, and shape descriptors for image similarity search. The retrieval results from this method are compared with the proposed method to measure the improvement. Section 4.1 explains the handcrafted features applied here.

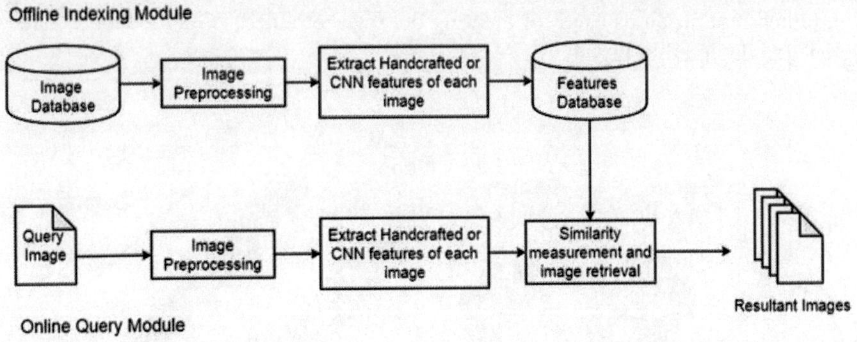

Fig. 2 Modules in CBIR method

3.2 CBIR Using Features from Pre-trained Deep CNN

This method is designed to use pre-trained deep CNN model as a feature extractor. Deep CNN models are made up of various layers that incrementally compute valuable features from an input image. Figure 3 depicts such a generic structure.

This method intends to repurpose two well-known models: ResNet50 and VGG16 as feature generators. VGG16 is developed by Visual Geometry Group (University of Oxford) [16]. It consists of 16 layers and trained effectively on natural images collection (ImageNet). The collection categorized into 21000+ subcategories and contains 14+ million images in total. Fully connected layers stacked at the top of the model perform the final classification. The proposed method intends to extract 4096 features from the fully connected (FC2) layer. ResNet50 is another eminent pre-trained model. It is created based on residual networks [16] which made creation and training of extremely deep CNN a reality. ResNet50 consists of 50 layers stacked in multiple convolution and identity blocks. The proposed work is intended to extract 2048 features from average pooling layer of ResNet50. Image retrieval is performed using features extracted from each of the models (VGG16 and ResNet50) separately. Retrieval performance of this method will be compared against the proposed CBIR framework to gauge the improvement.

3.3 Proposed CBIR Using Features from Ensemble of Deep CNN

This is our proposed methodology. Our aim is to improve the retrieval precision by training one of the deep CNN on target image collection and combine with learning from another pre-trained model. VGG16 pre-trained architecture is used for creating a retrained model. Steps used for creating our own retrained VGG model are listed below and depicted in Fig. 4.

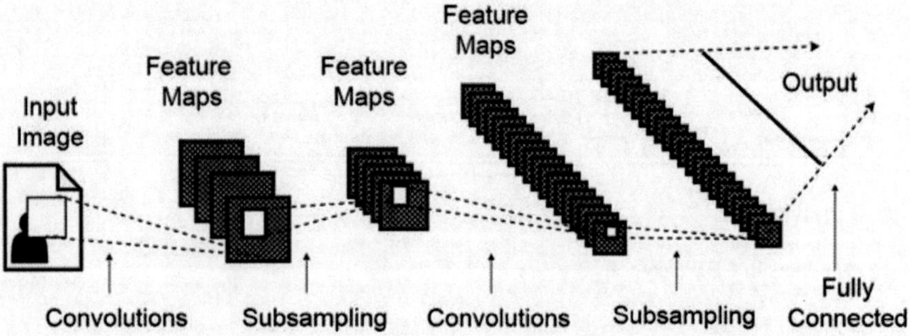

Fig. 3 Multilayered architecture of a deep CNN

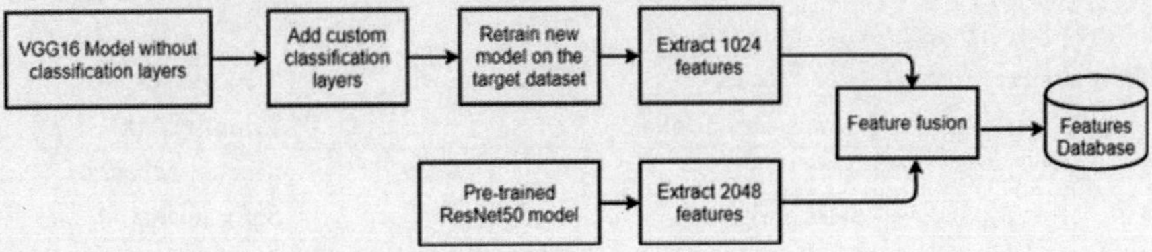

Fig. 4 Feature extraction in proposed CBIR

Create a new model by copying all hidden and classification layers from VGG16 model. Remove classier layers and last hidden layer. Freeze the hidden layers in new model. Add our own custom classifier and dropout layers. Retrain the new model with images from target leaf dataset. The features from this retrained VGG model are clubbed with features from a ResNet50 pre-trained model. The features obtained can be plugged into the feature extraction step of general CBIR methodology as in Fig. 2 to create a new CBIR method.

3.4 *Implementation*

Following configuration is used in the implementation of different methodological approaches here: Python (3.6.10), Keras (2.3.1), TensorFlow (1.14) backend on 64 bit Windows 8.1 with 8 GB RAM & Intel core (i5) processor.

4 Experimental Work

New CBIR methodology is validated by image retrieval tests on multi-class Swedish leaf images dataset, which was used in a well-known image classification study earlier [22]. The dataset consists of 1125 color images of leaves. The leaves are picked from 15 types of trees. 75 images per tree type correspond to a class in the collection. Each image consists of a single leaf or strand scan. Table 1 notes the class names of the collection.

Figure 5 shows the sample images from the collection. The dataset is best-suited for multi-class image analysis work and brings in the variety with respect to size, appearance, and distribution.

Query image picked from every class is provided as input in our experiments. As search users, the proposed work intends to retrieve images from the same class as the input image. This work treats the images retrieved from the same class to be relevant results. Hence, results fetched from other classes will be noted as irrelevant images. Experiments are conducted by changing the number of images returned in each run.

Table 1 Swedish leaf image dataset

Class number	Tree species	Class number	Tree species
1	Ulmus carpinifolia	9	Ulmus glabra
2	Acer	10	Sorbus aucuparia
3	Salix aurita	11	Salix sinerea
4	Quercus	12	Populus
5	Alnus incana	13	Tilia
6	Betula pubescens	14	Sorbus intermedia
7	Salix alba 'Sericea'	15	Fagus silvatica
8	Populus tremula		

Fig. 5 Sample images from leaf dataset

Precision rate is the standard measure to evaluate the performance of image retrieval systems. Precision is measured as in Eq. (1)

$$\text{Precision} = \frac{\text{Number of relevant images retrieved}}{\text{Total number of images retrieved}} \quad (1)$$

The image retrieval experiment is carried out in three steps as explained below sections. Same query images have been used in all steps.

4.1 *Image Retrieval Using Handcrafted Features*

A CBIR framework is developed for image retrieval by using fusion of selected handcrafted features. Handcrafted features applied in our baseline framework are documented in Table 2. Color property is computed by global color histogram distribution. Here histogram is determined over HSV space with bin values [Hue-8,

Table 2 List of handcrafted features applied in baseline CBIR

Category	Handcrafted features
Color features	Global color histogram (GCH)
Texture features	Gray-level co-occurrence matrix (GLCM): Homogeneity, Correlation, dissimilarity, energy and angular second moment (ASM)
Shape features	Hu Moments

Saturation-12, and Value-3]. GLCM method is implemented to determine above-listed texture features. Hu moments are applied to capture shape form.

A fusion vector is formed by combining all the computed features. During offline indexing activity, the vectors of fusion are saved to a database for future use. During search operation, similar feature vector is generated for each input image. This vector is compared against every entry in the database, and most similar ones are marked for retrieval. The similarity is determined by cosine distances metric. The experiment is repeated for different retrieval sizes (5, 10, 15, 20, etc.). Average precision rates for various fetch sizes and for different classes are measured. Tables 3 and 4 present the results from our experimental work for this CBIR technique with handcrafted features.

4.2 Image Retrieval Using Features Extracted from a Pre-trained Deep CNN Model

The proposed CBIR method extracts features from two well-known pre-trained models: VGG16 and ResNet50. This is to gauge the retrieval performance of features extracted from pre-trained models without being retrained. FC2 layer of VGG16 model generates 4096 units, whereas GlobalAveragePooling2 layer of ResNet50 model generates 2048 units. Each image is processed through these models, and features generated are stored in database. While searching images, the database features are compared with pre-trained features generated for input image. The similarity is determined by cosine distances metric. Tables 3 and 4 present the results from our experimental work for this CBIR technique with deep CNN features.

4.3 Image Retrieval Using Proposed CBIR Method

The proposed methods have created a fusion of features from retrained VGG model and pre-trained ResNet50 model. This is developed based on our earlier work [10]. A custom VGG16 model is created by adding our own custom classification layer

Table 3 CBIR Average precision rate per fetch size

Fetch size (across classes)	Handcrafted features (%)	VGG16 (%)	ResNet50 (%)	VGG16-ResNet50-Ensemble (%)
5	78.67	97.33	100.0	98.67
10	66.67	98.00	98.67	99.33
15	60.87	97.33	97.73	99.13
20	56.67	95.67	97.00	98.00
25	53.33	94.40	95.73	98.13
30	49.73	92.40	95.00	97.40
35	47.60	91.87	93.27	95.80
40	45.40	90.20	92.33	94.93
45	42.93	89.40	90.80	94.53
50	40.93	88.40	88.40	93.87
60	38.20	85.73	83.80	92.13
70	36.27	82.00	78.40	90.27
80	33.73	74.60	72.07	84.07
90	31.40	68.40	66.87	76.40
100	29.13	63.20	62.60	69.87

over state-of-the-art VGG16 model. Our custom classification model consists of dense layer with 1024 units using ReLU activation. A dropout is added followed by softmax activation for handling categorical probabilities associated with multi-class data. The resultant model is retrained on the target leaf dataset. To train the model, 10 images are selected from each class. Validation done with 4 images per class. The proposed work has used simple 10 epochs and RMSprop optimizer with very low learning rate for retraining. ResNet50 pre-trained model is used without any retraining to generate required features. A fusion vector of features is produced by combining 1024 features from retrained VGG model and 2048 pre-trained ResNet50 features. The feature vectors are saved to database for every image processed from the dataset. Similarly, fusion features are extracted for each ‘ image during image search operation. Features selection is used to determine the most suitable features required for image retrieval and to lower the computation cost. Univariate chi-squared statistical test selection method is used to select the best 500 features for comparison. The similarity is determined by cosine distances metric. Tables 3 and 4 present the results from our experimental work for this new CBIR technique.

Table 4 Average precision rate for image retrieval per image class

Image class	Handcrafted features (%)	VGG16 (%)	ResNet50 (%)	VGG16-ResNet50-Ensemble (%)
1	65.93	77.13	83.87	85.80
2	82.07	95.53	92.73	96.80
3	12.60	52.00	72.93	72.53
4	10.60	91.53	81.60	94.47
5	73.00	84.53	80.13	92.47
6	31.80	88.93	82.40	92.93
7	54.40	95.27	96.53	95.93
8	36.07	96.20	96.20	96.80
9	32.20	77.40	67.60	75.47
10	56.27	96.60	96.40	96.80
11	47.93	83.07	90.47	95.47
12	57.80	94.27	91.87	96.80
13	50.53	86.07	92.87	96.73
14	42.80	95.53	94.20	96.80
15	57.53	94.87	92.87	96.73

5 Results and Discussion

From the proposed experiment, it is noted that the CBIR which was created as a baseline by applying handcrafted features has attained overall average retrieval precision of 47.44%. Features generated by VGG16 (Pre-trained) model resulted in 87.26% average precision, whereas features generated from ResNet50 (pre-trained) model gives 87.51% average precision. New CBIR technique that used features fusion from ensemble of custom VGG16 (retrained) and state-of-the-art ResNet50 (pre-trained), attained overall average retrieval precision rate of 92.17%.

Figure 6 depicts the improvement in precision as recorded across different retrieval sizes. A clear precision improvement is observed across all fetch sizes for the proposed CBIR (VGG16-ResNet50-Ensemble).

Figure 7 depicts the improvement in precision rate as recorded across different classes. The improvement in precision rate is observed across all image classes of the dataset. Also improvement in precision from the proposed methodology as compared to standalone pre-trained CNN models is also seen. From all these experiments, it is evident that the new CBIR method (VGG16-ResNet50-Ensemble) that used fusion features from ensemble of retrained VGG16 and pre-trained ResNet50 deep CNN models outperform traditional CBIR (handcrafted features) and standalone deep CNN models (VGG16 and ResNet50).

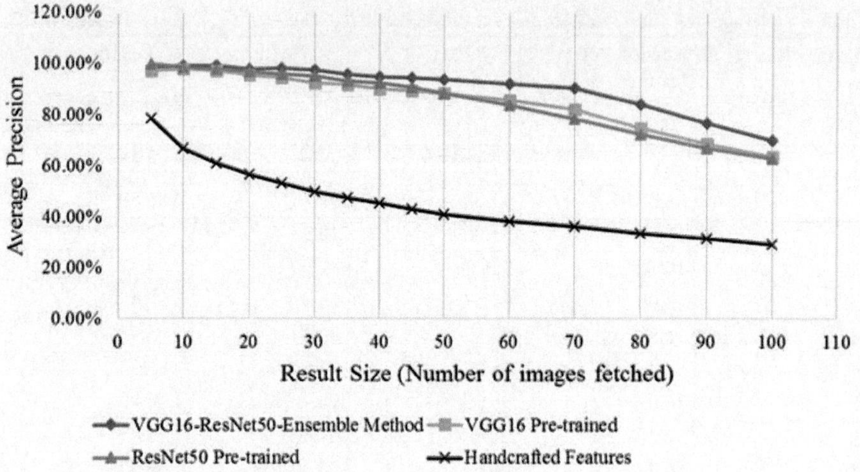

Fig. 6 Average precision rate per retrieval size

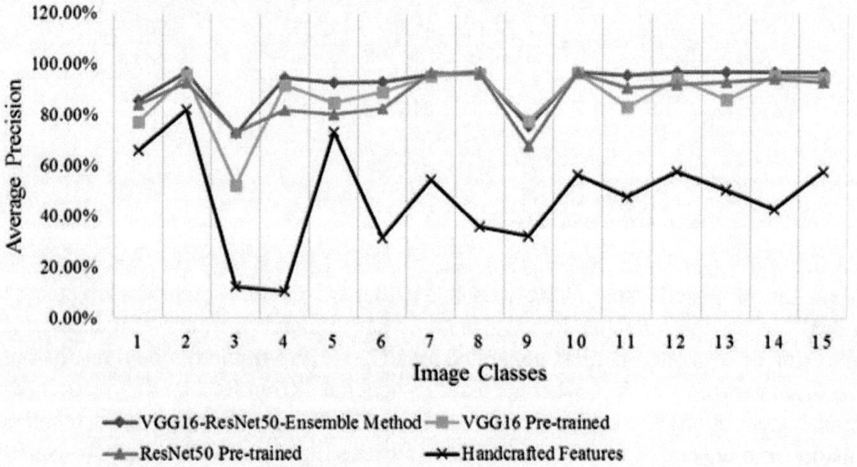

Fig. 7 Average precision rate per each class

6 Conclusion

This research work has proposed a new methodology to create fusion of features generated from retrained VGG16 (with custom classifier) and ResNet50 (pre-trained) deep CNN models. The new technique was applied to retrieve images from Swedish leaf images dataset. New framework (VGG16-ResNet50-Ensemble) outperformed baseline CBIR by achieving average precision of 92.17% versus 47.44% reported by baseline CBIR (handcrafted features). The proposed framework also performed better when compared to standalone deep CNN models. A clear precision improvement is observed across all fetch sizes and classes of the data collection. Empirical results depict that fusion of features from ensemble of retrained and pre-trained deep CNN models is very useful in image search and retrieval tasks. These features outperform handcrafted features and standalone pre-trained deep CNN models.

References

1. Yue J, Li Z, Liu L, Fu Z (2011) Content-based image retrieval using color and texture fused features. Math Comput Modell 54(3–4):1121–1127
2. Vimina ER, Poulose Jacob K (2012) Image retrieval using local color and texture features. In: Zhang T (eds) Mechanical engineering and technology. Advances in intelligent and soft computing, vol 125. Springer, Heidelberg. https://doi.org/10.1007/978-3-642-27329-2-105
3. Oujaoura M (2013) Color, texture and shape descriptor fusion with Bayesian network classifier for automatic image annotation. Int J Adv Comput Sci Appl 12(4):22–29
4. Diao Z, Zheng A, Wu Y (2011) Shape feature extraction of wheat leaf disease based on invariant moment theory. In: 5th computer and computing technologies in agriculture (CCTA). Beijing, China, pp 168–173. https://doi.org/10.1007/978-3-642-27278-3-18f
5. He Y, Xueliang L, Richang H (2016) Image classification via fusing the latent deep CNN feature. In: Proceedings of the international conference on internet multimedia computing and service. ICIMCS, Xian, China, pp 110–113 https://doi.org/10.1145/3007669.3007706
6. Song J, Gao S, Zhu Y, Ma C (2019) A survey of remote sensing image classification based on CNNs, In: Big earth data, pp 232–254. https://doi.org/10.1080/20964471.2019.1657720
7. Zhang Z, Wang Y, Liu Q, Li L, Wang P (2016) A CNN based functional zone classification method for aerial images. In: IEEE international geoscience and remote sensing symposium, IGARSS, Beijing. China, pp 5449–5452. https://doi.org/10.1109/IGARSS.2016.7730419
8. Jang H, Yang H, Jeong D, Lee H (2015) Object classification using CNN for video traffic detection system. In: 21st Korea-Japan joint workshop on frontiers of computer Vision, FCV, Mokpo, pp 1–4. https://doi.org/10.1109/FCV.2015.7103755
9. Gao XW, Hui R (2016) A deep learning based approach to classification of CT brain images. SAI Computing Conference. SAI, London, pp 28–31. https://doi.org/10.1109/SAI.2016.7555958
10. Bhandi V, Sumithra Devi KA (2019) Image retrieval by fusion of features from pre-trained deep convolutional neural networks. In: 1st international conference on advanced technologies in intelligent control, environment, computing and communication engineering, ICATIECE, Bangalore, India, pp 35–40. https://doi.org/10.1109/ICATIECE45860.2019.9063814
11. Alom MZ, Taha TM, Yakopcic C, Westberg S, Sidike P, Nasrin MS, Hasan M, Van Essen BC, Awwal AA, Asari VK (2019) A state-of-the-art survey on deep learning theory and architectures. Electronics 2019(8):292
12. Yang B, Cao J, Ni R, Zhang Y (2018) Facial expression recognition using weighted mixture deep neural network based on double-channel facial images. IEEE Access 6:4630–4640
13. Shaha M, Pawar M (2018) Transfer learning for image classification. In: Second international conference on electronics, communication and aerospace technology, ICECA, Coimbatore, India. https://doi.org/10.1109/ICECA.2018.8474839
14. Selimović A, Meden B, Peer P, Hladnik A (2018) Analysis of content-aware image compression with VGG16. In: IEEE international work conference on bioinspired intelligence, IWOBI, San Carlos, Costa Rica. https://doi.org/10.1109/IWOBI.2018.8464207
15. Gopalakrishnan K, Khaitan SK, Choudhary A, Agrawal A (2017) Deep convolutional neural networks with transfer learning for computer vision-based data-driven pavement distress detection. In: Construction and building materials, pp 322–330
16. Simonyan K, Zisserman A (2015) Very deep convolutional networks for large-scale image recognition

17. He K, Zhang X, Ren S, Sun J (2016) Deep residual learning for image recognition. In: 2016 IEEE conference on computer vision and pattern recognition (CVPR), Las Vegas, NV, pp 770–778. https://doi.org/10.1109/CVPR.2016.90
18. Xie H, He Y, Lei H, Han T, Yu Z, Lei B (2018) Deeply supervised residual network for HEp-2 cell classification. In: 2018 24th international conference on pattern recognition. ICPR, Beijing, China, pp 699–703. https://doi.org/10.1109/ICPR.2018.8545751
19. Jannesari M et al (2018) Breast cancer histopathological image classification: a deep learning approach. In: 2018 IEEE international conference on bioinformatics and biomedicine. BIBM, Madrid, Spain, pp 2405–2412. https://doi.org/10.1109/BIBM.2018.8621307
20. Ma W, Zhou R, Zhao Y, Xia Y, Fenster A, Ding M (2019) Plaque recognition of carotid ultrasound images based on deep residual network. In: 2019 IEEE 8th joint international information technology and artificial intelligence conference, ITAIC, Chongqing, China, pp 931–934. https://doi.org/10.1109/ITAIC.2019.8785825
21. Justine B, Samuel F, Jérőme T, Pierre-Luc S (2019) Convolutional neural networks for the automatic identification of plant diseases, In: Frontiers in plant science, vol 10, p 941. https://doi.org/10.3389/fpls.2019.00941
22. Oskar JOS (2001) Computer vision classifcation of leaves from swedish trees. Master's Thesis, Linkoping University, Sweden. http://www.cvl.isy.liu.se/en/research/datasets/swedish-leaf/
23. Beghin T, Cope JS, Remagnino P, Barman S (2010) Shape and texture based plant leaf classification. In: Blanc-Talon J, Bone D, Philips W, Popescu D, Scheunders P (eds) Advanced concepts for intelligent vision systems. ACIVS, lecture notes in computer science, vol 6475. Springer, Berlin, Heidelberg, Germany. https://doi.org/10.1007/978-3-642-17691-3-32
24. Caballero C, Aranda MC (2010) Plant species identification using leaf image retrieval, In: Proceedings of the ACM international conference on image and video retrieval, CIVR, association for computing machinery, pp 327–334. New York, NY, USA. https://doi.org/10.1145/1816041.1816089
25. Mouine S, Yahiaoui I, Verroust-Blondet A (2012) Advanced shape context for plant species identification using leaf image retrieval. In: 2nd ACM international conference on multimedia retrieval, ICMR, Hong Kong, China. https://doi.org/10.1145/2324796.2324853
26. Paco Ramos MM, Paco Ramos VM, Fabian AL, Osco Mamani EF (2019) A feature extraction method based on convolutional autoencoder for plant leaves classification. In: Orjuela-Cañón A, Figueroa-García J, Arias-Londoño J (eds) Applications of computational intelligence. ColCACI. Communications in computer and information science, vol 1096. Springer, Cham. https://doi.org/10.1007/978-3-030-36211-9-12

Chapter 58
Performance Analysis on Machine Learning Algorithms with Deep Learning Model for Crop Yield Prediction

Supreetha A. Shetty, T. Padmashree, B. M. Sagar, and N. K. Cauvery

1 Introduction

Agriculture is a significant contributor to India's GDP (16.9–17.9%) and employs millions of people. Few people consider farming for basic production and few take up as a business. Agriculture includes crop production, livestock, supplying raw materials to industries, etc. This work is limited to the scope of crop production. Traditionally, the farmer would predict the outcome based on the previous year outcome and experience. In recent years due to the unforeseen climate changes and various environmental factors, the traditional way of prediction is not efficient.

With the technology innovation, weather forecasting has significantly improved and the deployment of wireless sensor networks has helped to understand other environmental factors like soil type, water level, precipitation and so on. Machine learning and deep learning techniques are playing a vital role in prediction in various fields. The researchers are conducting studies on applying these techniques in agriculture for various purposes like automation, precision agriculture, to identify the plant diseases, forecasting the yield, etc. Most of the studies focus on a particular region and crop. There are limited studies carried out for crops in Karnataka region. Hence, this work aims at leveraging deep learning model to estimate yield for crops like Rice, Jowar, Maize and Ragi grown in Karnataka.

S. A. Shetty (✉) · T. Padmashree · B. M. Sagar · N. K. Cauvery
List of features considered, Bangalore, India
e-mail: supreethaas.sit19@rvce.edu.in

T. Padmashree
e-mail: padmashreet@rvce.edu.in

B. M. Sagar
e-mail: sagarbm@rvce.edu.in

N. K. Cauvery
e-mail: cauverynk@rvce.edu.in

I. Jeena Jacob et al. (eds.), *Data Intelligence and Cognitive Informatics*,
Algorithms for Intelligent Systems,
https://doi.org/10.1007/978-981-15-8530-2_58

The paper is organized as follows: Sect. 2 outlines the literature survey, Sect. 3 briefs about methodology followed by experimental results, conclusion, future work and references.

2 Literature Survey

Techniques such as Random forest (RF), decision tree classification, Multi-layer Perceptron (MLP), Convolutional Neural Networks (CNN), Recurrent Neural Networks (RNN) have shown promising results in predicting accurate results. The accuracy of models depends on how the features are selected and how well the model has been trained.

Fukuda et al. [1] proposed a random forest algorithm to predict Mango yield by considering the water supply from both irrigation technique and rainfall. Different models of RF were built and compared with each other, where each model considered either only Irrigation or only Rainfall or both Irrigation and Rainfall. This study helped to understand on which water source is effective in the mango yield process.

Romero et al. [2] performed a study on identifying the features by finding the association rules and by designing different classifiers to predict the yield of durum wheat. Association rule mining outperformed other classifiers. This experiment was conducted in the Buenos Aires province.

M5-Prime regression trees were found to be the best fit for predicting yield by Alberto et al. when different ML techniques were tested on 10 different crop datasets in Mexico. Factors like Planting area, Irrigation water depth, Solar radiation, rainfall, maximum temperature, minimum temperature and number of days required for cultivating crop is considered for training the model [3].

The productivity of sugarcane was estimated by using Random forest models which were trained using climate data collected starting from the year before harvest until the year of harvest. The model was trained and tested to predict yield for two terms of the year the sugarcane was produced. It was observed that the model performed better in the later stage compared to the first one since the feature variables were added up to train the model [4].

Random Forest and Gradient Boosting techniques were applied to classify the sugarcane yield as low, medium and high whose accuracy was 71.83% and 71.64%, respectively [5]. Various factors like cane type, water level, previous year actual yield and predicted yield, soil type and fertilizer information were considered. In addition to the abovementioned two techniques authors also applied other classification techniques like Naive-Bayes, SVM, KNN and Ada-Boost and found that these didn't give accurate results.

Random forest (RF) resulted in high accuracy and prediction when analyzed on 3 different crops namely Wheat, Maize and Potato. It was tested by keeping multiple linear regression as a benchmark. Variables like temperature (mean, maximum, minimum), day length, Nitrogen application rate were ranked by RF based on its importance on crop yield [6].

Remote sensing data, Climate data and previous year yield data were used for training Support Vector Machine, Random Forest, Extremely Randomized Trees and Deep learning models. The validation comparison of these models is presented by authors [7] which also included testing of these models on different seasons to verify their performance.

The rice yield prediction was carried out by Mariappan et al. [8] in Tamil Nadu region by considering the water supply, nitrogen content, fertilizer used, season the crop was grown. The data of 10 years were collected to train the model to predict the yield.

An RNN with long short-term memory (LSTM) cells was proposed to predict the soybean crop yields in Argentina by training the model using MODIS satellite imagery and achieved satisfactory results in predicting Brazil soybean harvests. The authors [9] claim that the training model with freely available satellite images is better than collecting other features manually.

Remotely sensed images were used to train a convolutional neural network (CNN) to predict rice grain yield of Southern China. This outperformed the traditional vegetation-index based method [10]. As discussed in [9], the satellite data is easily available and cost-effective, but this is challenging to estimate the yield of crops grown during monsoon. One such crop is Paddy which is a staple food of most parts of India. So, the authors [11] have considered soil and weather data to predict the paddy yield. This experiment was carried out for a particular district and was extended to predict the yield at finer level i.e., at Taluk level.

A hybrid CNN-RNN framework along with random forest was proposed by Saeed Khaki et al. [12]. It predicts corn and soybean yield considering past yield performance, management, weather, and soil properties.

A study on five different AI models to predict the corn and soybean yields were carried out by Kim et al. [7]. Features such as advanced vegetation index, minimum temperature, maximum temperature and soil moisture were considered.

3 Methodology

Tasks involved in predicting the yield is shown in Fig. 1.

3.1 Data Gathering

The data gathering includes collecting the various features that affect the crop yield. The past yield data is collected from state government agriculture website [13] and weather data like temperature, humidity and pressure are collected from Meteorological website [14]. The details of features and target variable are given in Table 1.

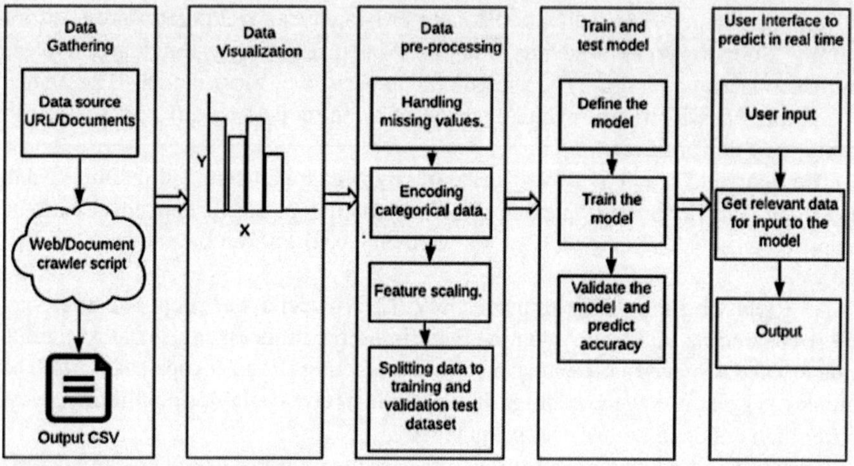

Fig. 1 Steps involved in crop prediction

Table 1 List of features considered

S. No.	Input/output variable	Description
1	Area	Area of land in terms of hectares
2	Year	Year for which data is collected (2010–2016)
3	District	30 districts of Karnataka
4	Season	Cropping seasons—rabi, kharif and summer
5	Crops grown	Jowar, maize, ragi and rice
6	Production	Production of crop in tonnes
7	Maximum temperature	Maximum temperature in degree celsius
8	Minimum temperature	Minimum temperature in degree celsius
9	Average temperature	Average temperature in degree celsius
10	Maximum pressure	Maximum pressure in milli bar (mbar)
11	Minimum pressure	Minimum pressure in milli bar (mbar)
12	Average pressure	Average pressure in milli bar (mbar)
13	Maximum humidity	Maximum humidity in terms of percentage of water present (%)
14	Minimum humidity	Minimum humidity in terms of percentage of water present (%)
15	Average humidity	Average humidity in terms of percentage of water present (%)
16	Yield	Yield in kilogram per hectare

3.2 Data Visualization

Since the data is collected from various sources and the format differs, it must be integrated into one CSV file in order to use it for training the model. The data visualization step deals with understanding the distribution of data in the dataset.

Histogram of data distribution of the features is given in Fig. 2. From the visualization, it is observed that the Area and Production are skewed right where most of the values lie on the left side. The average humidity, pressure and temperature, maximum temperature, minimum humidity data are normally distributed. Most of the entries in the dataset have average humidity of 60–65%, the average pressure of 1008 mbar and average temperature lies between 27 and 29 °C. Histogram of maximum and minimum pressure shows the presence of an outlier.

The effect of each variable on yield is presented using box-and-whisker plot for categorical variables and scatter plot for continuous variables as depicted in Figs. 3 and 4 respectively.

The box-and-whisker visualization shows that Maize crop and Chamarajanagar district have a higher average yield. There is a little variance in average yield between different seasons and across years. The boxplot of district variable gives information that 3 districts have the lowest average yield around 500 kg/ha and most of the districts have an average yield above 1500 kg/ha.

The scatter plot of variables shows that there is no linear relationship between the features and the target variable. However, it gives the idea of the presence of outliers.

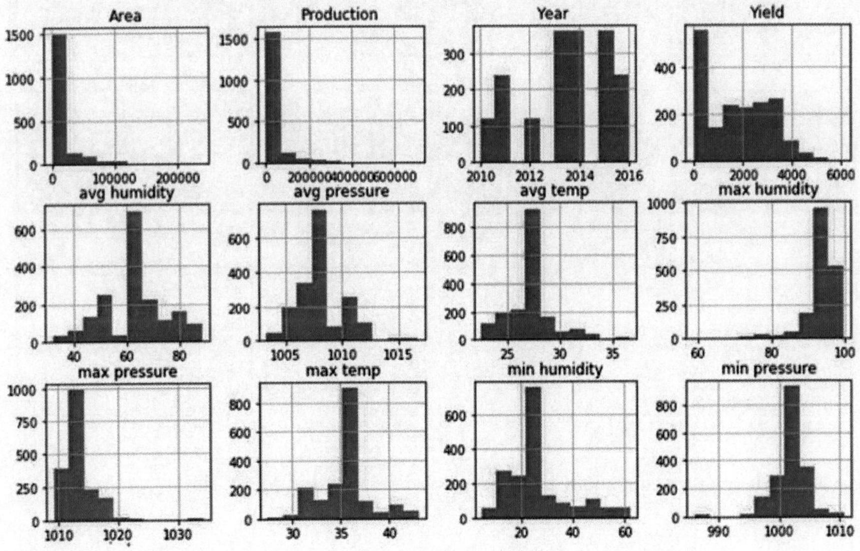

Fig. 2 Data distribution

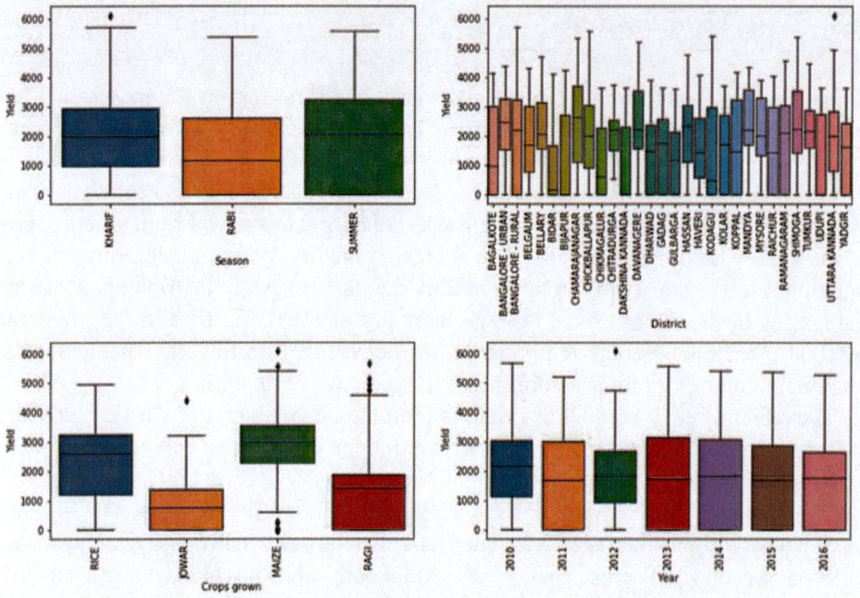

Fig. 3 Box-and-Whisker plot for categorical variables and their relationship with Yield

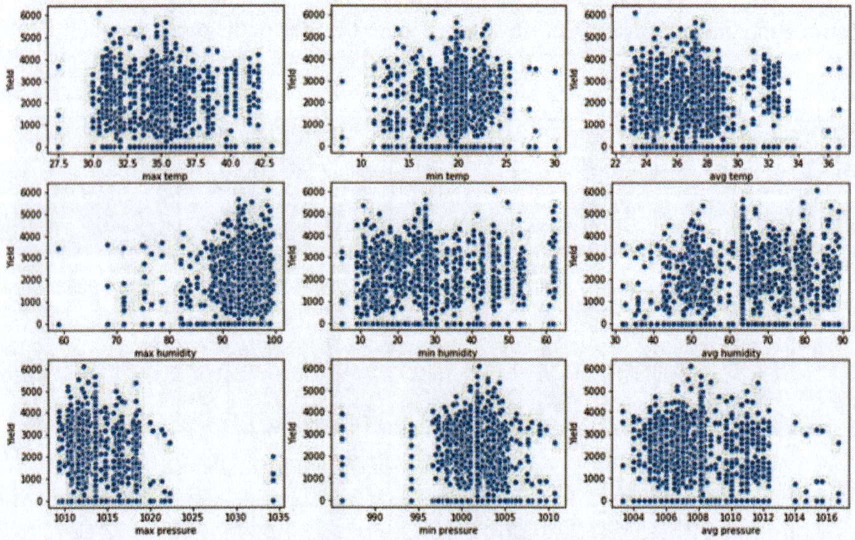

Fig. 4 Scatter plot for continuous variables and their relationship with Yield

3.3 Data Preprocessing

This step includes cleaning of collected data i.e., handling of missing data, feature selection, encoding the categorical data, feature scaling and splitting the data in to train and test data.

- **Handling of missing data**: Missing values of features can be handled by deleting the rows or by assigning a constant value for missing values or by replacing with a mean value of that feature. Since these data are important to train the model, the missing values are replaced with the mean value.
- **Feature selection**: The dataset contains 15 features—District, Season (Rabi/Kharif/Summer), Crops grown (Jowar/Rice/Maize/Rice), Area, Production, Year, Maximum Temperature, Minimum Temperature, Average Temperature, Maximum Pressure, Minimum Pressure, Minimum Pressure, Maximum Humidity, Minimum Humidity and Average Humidity and 1 target variable Yield. Out of these 15 features (or input variables), District, Season, Year, Area and Production doesn't impact the yield. Hence these 5 features are dropped and the rest of the 10 features are selected for further processing.
- **Feature scaling**: As it is observed in data distribution, each feature varies in different scales. Hence feature scaling technique is used to transform the features to range between 0 and 1.
- **Encoding the categorical data**: The categorical values must be encoded to numbers as the string values are not interpreted by models. Two popular ways of encoding are integer encoding and one hot encoding. In this step, one hot encoding technique is applied to the categorical data where the binary vector is created for each category.
- **Splitting the data into train and test data**: The data is split in 80:20 ratio where 80% of data is considered for training the model and 20% for validation.

3.4 Train and Test the Model

This step involves executing the model on the train data and verifying the model using the test dataset. 80% of data is considered for training the model and 20% for validation. The regression models considered for this proposed work are:

- **Multi-layer perceptron model**: Multilayer perceptron model is a deep learning technique with more than one hidden layer. A fully connected multi-layer perceptron model with 1 input layer, 3 hidden layers and 1 output layer (Fig. 5) is built using Keras sequential model. Rectified Linear Unit (ReLU) is used as an activation function in input and hidden layers. Mathematically, ReLU transforms the input to the output in range 0 to z, where z is a positive number. This is the most commonly used activation function in neural networks. Since crop yield prediction is a regression problem, the linear activation function is set at the output.

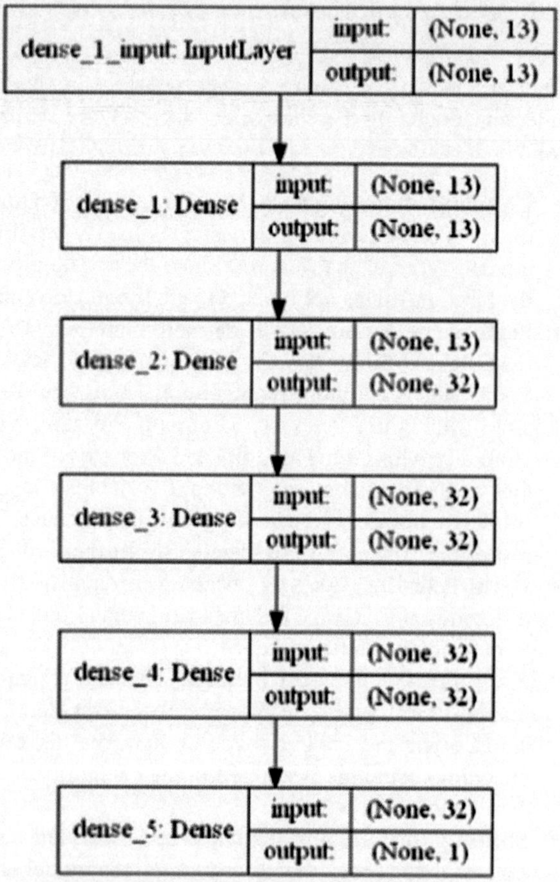

Fig. 5 Multi-layer perceptron model

Keras is used to implement the MLP network. The initial training weights for each layer is added using Keras weight initializer.

- **Random forest regression**: Previous studies have shown that a machine learning technique, Random forest regression model performs better in crop yield prediction. In this study, RandomForestRegressor () function from sklearn library is used with default settings (100 trees). In this technique, the dataset is divided into multiple samples and trees are formed for each sample. Then the predictions from each tree are combined to form the final prediction (Fig. 6).

3.5 *User Interface to Predict in Real-Time*

Once the model is set up, it must be deployed to be used in real-time. Hence a simple web page is developed using Flask through which a user can enter Location, a crop

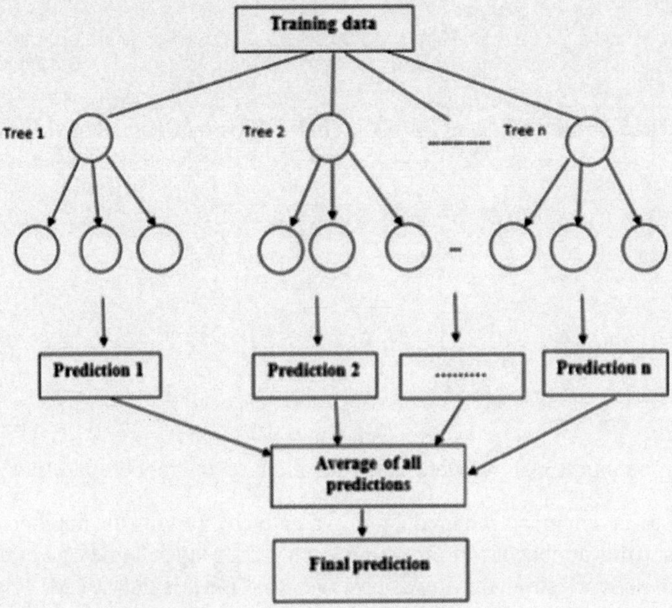

Fig. 6 Random forest regression model

grown and month values, based on which weather information is fetched and model is called for prediction. The predicted value is printed on the screen.

4 Experimental Results

4.1 Metrics for Evaluation

Various metrics to measure the accuracy of regression models are:

- **Mean Absolute Error (MAE)**: This is an average of the absolute difference between the actual values and the predicted values of every sample.

$$\text{MAE} = \sum_{i=1}^{n} |y_i - x_i| \tag{1}$$

- **Mean Squared Error (MSE)**: MSE is the average of the square of the difference between the actual and predicted values of the data.

$$\text{MSE} = \frac{1}{n}\sum_{i=1}^{n}(y_i - x_i)^2 \quad (2)$$

- **Root Mean Squared Error (RMSE)**: This is the square root of the Mean Squared Error.

$$\text{RMSE} = \sqrt{\frac{1}{n}\sum_{i=1}^{n}(y_i - x_i)^2} \quad (3)$$

4.2 Results

The metrics mentioned above were used for validating the models in this experiment.

- The multi-layer perceptron model was evaluated by varying the number of hidden layers and the number of neurons within each hidden layer. Table 2 gives the results for each network structure. It was observed that for this dataset, MLP with fully connected 3 hidden layers with 32 neurons each performed better.
- Random forest algorithm is evaluated by differing the number of trees in the forest. It was observed that Random forest with 50, 100 and 500 trees performed better compared to Random forest with 10 trees. The comparison is shown in Fig. 7.
- The network that performs better in the multi-layer perceptron model is compared against the random forest regression model with 100 tresses and the result is given in Table 3.

Table 2 Result metrics for varying network structure

Network structure	Mean absolute error (MAE)	Mean square error (MSE)	Root mean square error (RMSE)
(13/7/1)	0.1363	0.0364	0.1908
(13/7/4/1)	0.1422	0.0385	0.1964
(13/7/4/2/1)	0.1366	0.0376	0.194
(13/32/1)	0.1338	0.0356	0.1888
(13/32/32/1)	0.1365	0.0375	0.1937
(13/32/32/32/1)	0.1233	0.0344	0.1855
(13/64/32/1)	0.1326	0.0366	0.1915
(13/64/32/16/1)	0.1292	0.0351	0.1875
(13/64/32/16/8/1)	0.1284	0.0361	0.1902

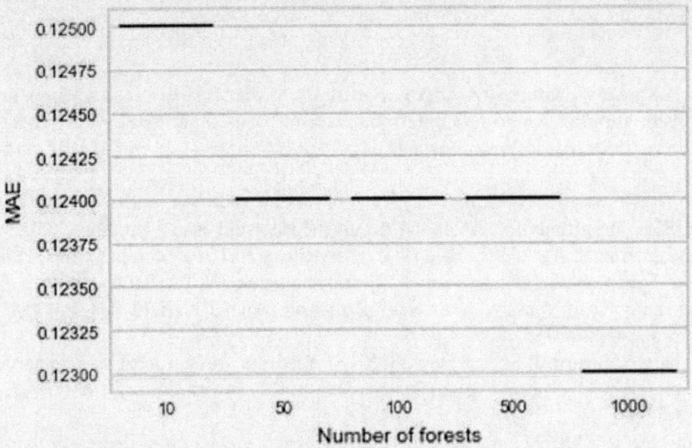

Fig. 7 Comparing the performance of Random forest by varying the number of trees

Table 3 Performance comparison of random forest versus mlp

Model/metrics	Mean absolute error (MAE)	Mean square error (MSE)	Root mean square error (RMSE)
Multi-layer perceptron	0.123	0.034	0.185
Random forest regression	0.124	0.029	0.171

5 Conclusion

The purpose of this work was to leverage the deep learning technique to predict the yield of major crops grown in Karnataka region. The data collected included yield data of 4 crops namely Jowar, Rice, Maize and Ragi for Kharif, Rabi and summer seasons along with maximum, minimum and average values of temperature, humidity and pressure for 7 years. The user-friendly system developed with a trained model to accurately predict the yield would help the user to estimate the yield before-hand and grow the crop accordingly. It was observed that the datasets fit well in both the multi-layer perceptron model and Random forest regression model.

6 Future Work

This study can be further improved to include more features like soil properties, rainfall data, wind speed, satellite images to build a robust model. It can also be extended by training the model with more crops.

References

1. Fukuda S, Spreer W, Yasunaga E, Yuge K, Sardsud V, Müller J (2013) Random forests modelling for the estimation of mango (mangifera indica l. cv. chok anan) fruit yields under different irrigation regimes. Agric Water Manage 116:142–150. https://doi.org/10.1016/j.agwat.2012.07.003
2. Romero JR, Roncallo PF, Akkiraju PC, Ponzoni I, Echenique VC, Carballido JA (2013) Using classification algorithms for predicting durum wheat yield in the province of buenos aires. Comput Electron In Agric. 96:173–179. https://doi.org/10.1016/j.compag.2013.05.006
3. Gonzalez-Sanchez A, Frausto-Solıs J Ojedabustamante W Predictive ability of machine learning methods for massive crop yield prediction. Span J Agricul Res. https://doi.org/10.5424/sjar/2014122-4439.
4. Everingham Y, Sexton J, Skocaj D et al (2016) Accurate prediction of sugarcane yield using a random forest algorithm. Agron Sustain Dev 36:27. https://doi.org/10.1007/s13593-016-0364-z
5. Charoen-Ung P, Mittrapiyanuruk P (2018) Sugarcane yield grade prediction using random forest and gradient boosting tree techniques. In: 2018 15th ınternational joint conference on computer science and software engineering (JCSSE), Nakhonpathom pp 1–6
6. Jeong JH, Resop JP, Mueller ND, Fleisher DH, Yun K, Butler EE et al (2016) Random forests for global and regional crop yield predictions. PLoS ONE 11:e0156571. https://doi.org/10.1371/journal.pone.0156571
7. Kim N, Lee Y-W (2016) Machine learning approaches to corn yield estimation using satellite images and climate data: a case of iowa state. J Korean Soc Surveying Geodesy Photogrammetry Cartography 34(4):383–390. https://doi.org/10.7848/ksgpc.2016.34.4.383
8. Mariappan K, Ben Das JA (2017) A paradigm for rice yield prediction in Tamilnadu. In: 2017 IEEE Technological ınnovations in ICT for agriculture and rural development (TIAR), Chennai, p 1821
9. Wang AX, Tran C, Desai N, Lobell D, Ermon S (2018) Deep transfer learning for crop yield prediction with remote sensing data In: Proceedings of the 1st ACM SIGCAS conference on computing and sustainable societies. ACM, San Jose CA, USA, p 50
10. Yang Q, Shi L, Han J, Zha Y, Zhu P (2019) Deep convolutionalneuralnetworks for rice grain yield estimation at the ripening stage using uav-based remotely sensed images. Field Crops Res 235:142–153. https://doi.org/10.1016/j.fcr.2019.02.022
11. Guruprasad RB, Saurav K, Randhawa S (2019) Machine learning methodologies for paddy yield estimation in India: a case study. In: IGARSS 2019—2019 IEEE ınternational geoscience and remote sensing symposium, Yokohama, Japan, pp 7254–7257
12. Khaki S, Wang L, Archontoulis SV (2010) A CNN-RNN framework for crop yield prediction. Front Plant Sci 10: n. pag. Crossref. Web.
13. https://kapricom.org/crop_production_statistics.html
14. https://www.timeanddate.com/weather/
15. Vijayakumar T, Vinothkanna R (2020) Mellowness detection of dragon fruit using deep learning strategy. J Innovative Image Proc (JIIP) 2:01:35–43
16. Raj JS, Ananthi JV (2019) Recurrent neural networks and nonlınear predictıon ın support vector machınes. J Soft Comput Paradigm (JSCP) 1(01):33–40

Chapter 59
A Novel Hybrid Cipher Using Cipher Squares for Better Security

Adilakshmi Visali Jayanthi

1 Introduction

In today's world, communication through technology has become a daily activity. The information is sent in many forms like emails, SMS, chat messengers, online forums, calls, etc. and desire that every information sent is received only by the intended receiver securely. When it comes to the organizations, data for them is very crucial and any theft or loss of it would lead to grave consequences. That is the reason why Information Security plays an important role in communication technology. To securely transmit a message, many methods of encryption and decryption have been devised. Similarly, many algorithms are invented to decrypt the message without knowing the actual key. Cryptology has grown into a branch of science. Cryptology is the study of both cryptography and cryptanalysis [1]. Cryptography is the art of securely transmitting the message over communication channels. Cryptography helps us provide Authentication, Non-repudiation, Data Integrity, Confidentiality, and a secured Channel for communication. The ciphers are divided into two types- classical ciphers and modern ciphers. Classical Ciphers are further divided into Substitution Ciphers and Transposition ciphers. Secret key cryptography is a method in which a single cryptographic key is shared between the sender and the receiver. A key is used to control the two cryptographic operations- encryption and decryption. This key which is used to both encrypts and decrypt the message is kept secret by both parties. The randomness of the key is one of the deciding factors of the robust algorithm.

As described in [2], Cryptanalysis, also called code-breaking, is the study of various methods of deciphering the ciphertext without knowing what the key is. Encryption, also called Encipher, is a method in which the information is converted into ciphertext using any cryptographic encryption algorithm. Decryption, also called Decipher, is the conversion of the ciphertext into the original information using a

A. V. Jayanthi (✉)
New Delhi, India

I. Jeena Jacob et al. (eds.), *Data Intelligence and Cognitive Informatics*,
Algorithms for Intelligent Systems,
https://doi.org/10.1007/978-981-15-8530-2_59

Fig. 1 The layout of the proposed 3 * 3 cipher square

1	2	3
8	X	4
7	6	5

Decryption Algorithm. That is the input to the decryption algorithm, ciphertext, is the output generated from the encryption algorithm. An encryption algorithm describes the method in which the encryption must be done. Many methods like the index of coincidence, frequency analysis, etc. are used to decipher the ciphertext.

2 Literature Survey

2.1 Hybrid Ciphers

In [3], the authors described the hybrid ciphers. Hybrid Ciphers are those ciphers that encrypt the text message into both alphabets and symbols i.e., the ciphertext is a combination of both alphabets and symbols. When both alphabets and symbols are used, the frequency of the repetitive alphabets is reduced making cryptanalysis difficult to perform on hybrid ciphers.

2.2 Dynamic Key Generation

In [4], the authors described the Caesar cipher. A Caesar cipher is one of the oldest known ciphers invented by Julius Caesar. He used this technique to send messages to his aides securely. In the paper, authors have described a novel method to dynamically generate the secret key depending on the length of the text to be encrypted. Dynamic key generation induces randomness in the algorithm thereby making it difficult to decipher without the prior knowledge of the key (Fig. 1).

2.3 Zigzag Cipher

This cipher is a variant of a transposition cipher called rail-fence Cipher, which is inspired by Polybius square modeling [5]. Greeks invented a special tool called Scytale, to encipher and decipher messages [6]. The plaintext which is to be encrypted

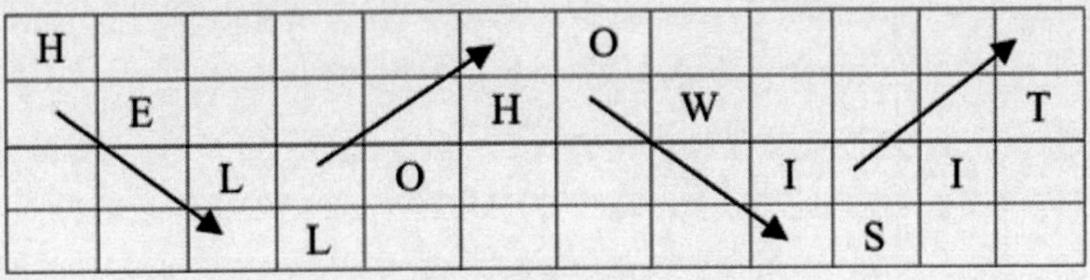

Fig. 2 An example illustrating zigzag cipher with 4 rails

is taken and is written downwards diagonally on successive 'rails' of an imaginary fence, then moving up when one reaches the bottom rail. When one reaches the top rail, the message is written downwards again until the whole plaintext is written out. The message is then read off in rows. A zigzag cipher is well illustrated in Fig. 2.

Here, the key is nothing but the number of rows implemented which is easier to find out by using a brute-force attack. The security of the algorithm can be increased by combining it with another cipher technique [5].

3 Proposed Algorithm

The proposed method can be used for transferring the data securely over various communication channels. This can be used to generate hybrid ciphers, which are the ciphers that contain the combination of both alphabets and symbols in the ciphertext. Hybrid ciphers are the ciphers that contain both alphabets and symbols, due to which the repetitive letters are reduced making it difficult for cryptanalysis. Our proposed method may also be extended for implementing cross-language ciphers with more than two languages. Our proposed method can also be extended to encipher and decipher numerical data like that of account numbers.

In our proposed method, a novel two-dimensional 3 * 3 Cipher Squares is used in order to encrypt and decrypt the messages. In the proposed cipher square, the middle element is X is the letter of the plaintext which is to be encrypted. Whereas the elements in the numbered positions i.e., from 1 to 8 positions are the probable encryptions for the letter 'X.' That is, in our proposed method, any direction can be moved, right, left, up, down, and diagonal to choose the cipher letter. This will lead to choosing a random cipher letter out of these 8 probable elements making the encryption robust. When in the ciphertext, there is no one-to-one mapping among the letters in the ciphertext and plaintext, code-breaking using frequency analysis is much difficult. The layout of a single cipher square is illustrated in the Fig. 1. And the layout of the entire cipher consisting of 26 3 * 3 cipher squares which are used for both encryption and decryption is illustrated in Fig. 3.

1	2	3	9	10	11	16	17	18	24	25	26	32	33	34
8	A	4	16	B	12	23	C	19	31	D	27	39	E	35
7	6	5	15	14	13	22	21	20	30	29	28	38	37	36

40	41	42	48	49	50	56	57	58	64	65	66	72	73	74
47	F	43	55	G	51	63	H	59	71	I	67	79	J	75
46	45	44	54	53	52	62	61	60	70	69	68	78	77	76

80	81	82	88	89	90	96	97	98	104	105	106	112	113	114
87	K	83	95	L	91	103	M	99	111	N	107	119	O	115
86	85	84	94	93	92	102	101	100	110	109	108	118	117	116

120	121	122	128	129	130	136	137	138	144	145	146	152	153	154
127	P	123	135	Q	131	143	R	139	151	S	147	159	T	155
126	125	124	134	133	132	142	141	140	150	149	148	158	157	156

160	161	162	168	169	170	176	177	178	184	185	186	192	193	194	200	201	202
167	U	163	175	V	171	183	W	179	191	X	187	199	Y	195	207	Z	203
166	165	164	174	173	172	182	181	180	190	189	188	198	197	196	206	205	204

Fig. 3 The layout of the 26 3 * 3 cipher squares

3.1 Encryption Algorithm

Step-1: Take the plaintext as *plaintext* and the secret key value as *key*. For each and every letter in the *plaintext*, perform Steps 2 and 3.

Step-2: Calculate the element position to the letter of the plaintext using (1) as *position*. Here,

$$\text{position} = ((\text{text\_position} + \text{key}) \bmod 8) + 1 \tag{1}$$

Randomness is incorporated in the algorithm with the text_position. Here, for the same alphabet which is repeated in the plaintext, every time any one of the eight elements is assigned according to the position value calculated as per Eq. (1). This is well-explained in the example that is given in the following section C.

Step-3: Once the position is calculated, replace the plaintext with the element present at that position.

Step-4: Apply the rail-fence cipher technique with the same given key value. The ciphertext is obtained.

3.2 Decryption Algorithm

Step-1: Take the ciphertext which is the output of the above-proposed encryption algorithm as *ciphertext*. For each and every letter of the *ciphertext,* perform steps 2 and 3.

Step-2: Apply rail-fence decryption technique with the given key value.

Step-3: Upon the obtained result after applying the rail-fence decryption algorithm, if the ASCII value of the character to be decrypted is less than or equal to 119, Decryption Function is

$$\text{plain\_text\_character} = ((\text{ASCII Value})/8) - 3 \tag{2}$$

Else,

$$\text{plain\_text\_character} = ((\text{ASCII Value})/8) - 4 \tag{3}$$

Finally, *PlainText* is obtained. The algorithm is further explained with the below example.

An extra zigzag cipher is used in our algorithm so as to increase the security of our algorithm as it is already proven that a combination of cipher algorithms and techniques has increased the security and made it difficult for code-breaking [4, 7–10].

This method can also be extended to Cross-Language ciphers [3] wherein two languages are used for secure communication between two parties. In [3], the authors described a Cipher that uses two languages English and Hindi. In this cipher that they described, the corresponding ASCII values of the plaintext are converted into the UNICODE values in the Hindi Language for encryption and vice versa for decryption. The algorithm may further be extended to be used for images as in [11]. The working of the proposed algorithm is better described in the example that follows.

3.3 Example

Let us take the example cipher squares to be as given as in the Fig. 4, 26 two-dimensional 3 * 3 cipher squares. These 26 cipher squares are used for 26 letters of the English alphabet. If you want to encrypt and decrypt numbers as well, another 10 more cipher squares can be made corresponding to 0–9 numerals. Let us encrypt and decrypt using the proposed encryption and decryption algorithms.

For example, let the plaintext to be encrypted be 'HELLO HOW IS IT' and the secret key is 3. (Spaces are removed before encryption). The ASCII values are referred to from [12].

	!	"
'	A	#
&	%	$

(	)	*
/	B	+
.	-	,

0	1	2
7	C	3
6	5	4

8	9	:
?	D	;
>	=	<

@	A	B
G	E	C
F	E	D

H	I	J
O	F	K
N	M	L

P	Q	R
W	G	S
V	U	T

X	Y	Z
_	H	[
^	]	\

`	a	b
g	I	c
f	e	d

h	i	j
o	J	k
n	m	l

p	q	r
w	K	s
v	u	t

€		‚
‡	L	ƒ
†	…	„

ˆ	‰	Š
	M	‹
Ž		Œ

	‘	’
—	N	“
–	•	”

˜	™	š
Ÿ	O	›
ž		œ

	¡	¢
§	P	£
¦	¥	¤

¨	©	ª
¯	Q	«
®		¬

°	±	²
·	R	³
¶	µ	´

¸	¹	º
¿	S	»
¾	½	¼

À	Á	Â
Ç	T	Ã
Æ	Å	Ä

È	É	Ê
Ï	U	Ë
Î	Í	Ì

Ð	Ñ	Ò
×	V	Ó
Ö	Õ	Ô

Ø	Ù	Ú
ß	W	Û
Þ	Ý	Ü

à	á	â
ç	X	ã
æ	å	ä

è	é	ê
ï	Y	ë
î	í	ì

ð	ñ	ò
÷	Z	ó
ö	õ	ô

Fig. 4 The layout of example 26 3 * 3 cipher squares

Encipher:

PlainText: HELLOHOWISIT

Key: 3

CipherText:

\˜dE‡YÛ½Ç†š
f

Decipher:

CipherText:

\˜dE‡YÛ½Ç†šf

Key: 3

Plain Text: HELLOHOWISIT

In the above example, the first '*H*' in the plaintext is encrypted as '&#92' whereas the second '*H*' as '&#89' is observed. Similarly, the two L's have been assigned two different values. Thus, the same alphabet of the plaintext is not always mapped to the same alphabet as in the case of traditional ciphers like Caesar cipher, rail-fence cipher, etc, thereby increasing the robustness of the algorithm by inducing randomness. When a reasonable length of ciphertext is known, it can be broken with reasonable confidence using cryptanalysis techniques like frequency analysis. But with the randomness induced using our proposed method, it is difficult to analyze the pattern or to analyze the distribution of letters in the plaintext as the ciphertext obtained has both alphabets and symbols with varied frequencies.

3.4 Comparative Analysis of the Distribution of the Letters in the Example Ciphertext with Existing Cipher Techniques—Caesar and Rail-Fence Ciphertexts

The frequency distribution curve of the plaintext and the obtained ciphertext when the traditional methods of Caesar cipher, rail-fence cipher, and our proposed cipher are used for the example plaintext 'HELLOHOWISIT' using cipher squares are illustrated in the Fig. 5. By looking at the graph in Fig. 5, it is clearly observed that the frequency distribution of the letters in the proposed cipher using the cipher square method is more distributed as compared to the traditional methods. The X-axis consists of the letters of the English Alphabet and the Y-axis consists of the corresponding frequencies of each of the letters in the above example ciphertext. The letters in the ciphertext of our proposed cipher are more distributed across the ASCII values which affect the frequency of their occurrence in the ciphertext. In the traditional methods, the frequency of letters in the plaintext and the cipher text remained the same making it easy for cryptanalysis whereas in the proposed cipher method, there is no correlation between the frequencies of the letters in the plaintext and the ciphertext which makes it difficult for the cryptanalysis.

Along with the above example ciphertext, a better-distributed frequency distribution of letters is found for other examples as well. This cipher technique has given better security than the existing techniques of Caesar and rail-fence ciphers.

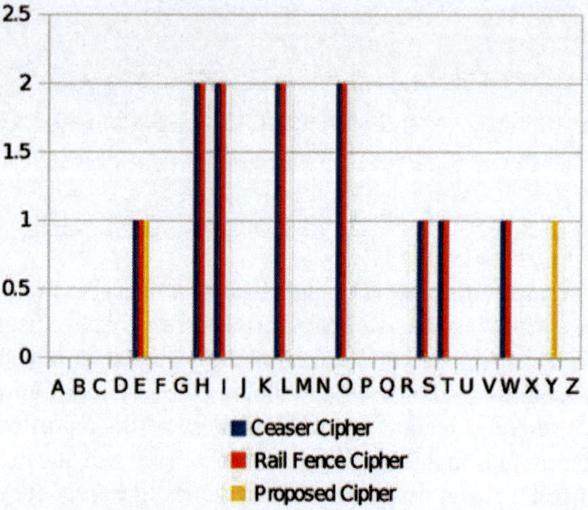

Fig. 5 Graph illustrating the frequency distribution of the letters in the ciphertexts after applying-Caesar cipher, rail-fence cipher and our proposed cipher upon the plaintext 'HELLO HOW IS IT.' The *X*-axis consists of the letters of the English Alphabet and the *Y*-axis consists of the frequency of the occurrence of each of these letters in the above example ciphertext

Implementing the above cipher for Unicode Characters can be quite a challenging task. Also, it has the limitation of the secret key is shared and not being generated completely dynamically.

4 Conclusion

With the proposed method, it becomes difficult for the intruder to decipher the message. It helps in securing the data in security attacks like man-in-the-middle attacks. When this cipher is used, it becomes much more difficult for cryptanalysis as the frequency of repetitive letters is much more decreased. This method can also be extended to cross-language ciphers and for image steganography. Even then, it will be difficult for the intruder to decipher the message without knowing the secret key. Also, a letter is not always mapped to the same letter as in a Caesar cipher, rail-fence cipher, or other traditional ciphers. As every letter may be mapped to any one of the eight possible elements of the cipher square, the complexity of cryptanalysis increases. This method can be combined with other cipher techniques to make much more robust ciphers. Cryptanalysis can be done in detail in further study.

References

1. Stallings W, Tahiliani MP (2014) Cryptography and network security: principles and practice, vol 6. Pearson, London
2. Forouzan BA, Mukhopadhyay D (2011) Cryptography and network security, Special Indian Edition, 2nd edn. TataMcGrawHill Education Private Limited, New Delhi
3. Singh L, Johari R (2015) CLCT: cross-language cipher technique. In: International symposium on security in computing and communication. Springer International Publishing, pp. 217–227
4. Verma P, Gaba GS, Monga H (2016) Modified caesar cipher using rectangular method for enhanced security. J Commun Technol Electron Comput Sci 8(1). https://doi.org/10.22385/jctecs.v8i0.116
5. Utama Siahaan AP (2016) Rail fence cryptography in securing information. Int J Scient Eng Res 7(7):535–538
6. [Online]. Available: http://practicalcryptography.com/ciphers/rail-fence-cipher
7. Bhalaji N (2020) Efficient and secure data utilization in mobile edge computing by data replication. J ISMAC 2(01):1–12
8. Suma V (2019) Security and privacy mechanism using blockchain. J Ubiquitous Comput Commun Technol (UCCT) 1(01):45–54
9. Saini B (2015) Modified caesar cipher and rail fence technique to enhance security. Int J Trend Res Dev 2(5)
10. Lubis F, Simbolon H, Batubara T, Sembiring RW (2017) Combination of caesar cipher modification with transposition cipher. Adv Sci Technol Eng Syst J 2:22–25 https://doi.org/10.25046/aj020504
11. Rachmawati D et al (2019) Combination of columnar transposition cipher caesar cipher and lempel ziv welch algorithm in image security and compression. J Phys Conf Ser 1339:012007
12. ASCII character codes and html, octal, hex and decimal chart conversion [Online]. Available: http://www.asciitable.com

Chapter 60
A Review on Dynamic Virtual Machine Consolidation Approaches for Energy-Efficient Cloud Data Centers

B. Prabha, K. Ramesh, and P. N. Renjith

1 Introduction

The cloud computing model plays a pivotal role in accessing the computational resources for customers by means of the World Wide Web. Cloud computing is a platform where customizable computational resources (e.g., physical servers, storage devices, networking components and applications) and services are shared on the basis of on-demand access. The rapid deployment of these resources is accomplished with nominal management efforts and marginal collaborations with the service providers [1]. The applications and services of cloud computing have been widely grabbed the attention of both academia and industry which in turn increases the entire world data centers (DCs) which greatly demands energy consumption costs [2]. It is also mandatory to deliberate the computing power proficient usage and global warming. Data and communication technology is focusing on a guideline of vitality utilization concern, especially in the distributed computing area, over the whole network [3].

The energy-saving mechanism progress is considered to be an important factor in cloud applications' through energy consumption realization. The data center power consumption is directly relying on cloud computing global agreement which is demanded dramatically [4]. There involves plenty of interconnected servers for providing innumerable cloud services globally which in turn composes numerous

B. Prabha (✉)
IT, Loyola-ICAM College of Engineering and Technology, Chennai, India
e-mail: prabha.bala32@gmail.com

K. Ramesh · P. N. Renjith
CSE, Hindustan Institute of Technology and Science, Chennai, India
e-mail: rameshk.me@gmail.com

P. N. Renjith
e-mail: pnrenjith@hindustanuniv.ac.in

I. Jeena Jacob et al. (eds.), *Data Intelligence and Cognitive Informatics*,
Algorithms for Intelligent Systems,
https://doi.org/10.1007/978-981-15-8530-2_60

interconnected servers reflecting huge energy quantity. Information technology (IT) hardware's and cooling framework regularly devours 88% of the complete power for present day DC which is an unpleasant power breakdown report [5]. The unpredictable traffic, transfer speed and correspondence necessities across applications are the applications deciding the information serious applications in current DCs.

The operational expenses, ownership overall cost and revenue are the factors determining the energy consumption and performance variability [6]. The quality of service (QoS) degradation and application performance varying capacity are the drawbacks of considerably large data centers, hardware failures and software anomalies. Virtual machine (VM) integration is a mechanism for using virtualized data centers for energy and resource efficiency by enveloping as many VMs as conceivable on one physical system. The improved VM placement helps in energy savings by permitting VMs consolidation dynamically through live VM movement and switching idle hosts to lower energy modes [7].

The basis of a peak, off-peak or optimal CPU utilization of VMs is greatly used for diminishing energy consumption through VM consolidation by enclosing binding into the less number of processing servers deprived of QoS as shown in Fig. 1. The

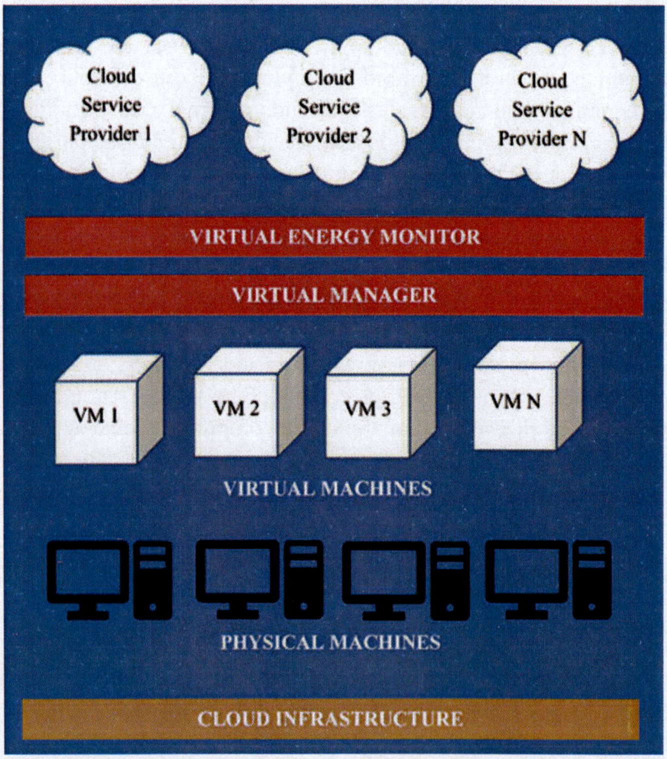

Fig. 1 Cloud energy monitoring mechanism

two aspects of cloud tasks dynamic nature influences consolidation. (1) CPU-process load equivalence through VMs (i.e., the similar CPU utilization traces and the co-occurrence of their peaks) and (2) the exchanging of data across VMs (i.e., data equivalence) [8]. Once their peak utilizations overlap does not happen or proceeds data correlation into consideration, either of the two CPU-load correlations, consolidating VMs are concentrated by utilizing heuristics in various researches.

Integer linear programming (ILP)-based techniques are considered as an unachievable method during execution time because of increased complexity for obtaining an optimal solution. Alike, the limitation of merits for enormous problems since heuristics tends to problem-specific and not as much of sensitive to dynamic environs. Machine learning (ML) and reinforcement learning (RL) are satisfactory methods [9] and are the solutions for handling vital problems with huge state and/or action spaces. The extensive range of measurements assessment (energy, QoS, network, etc.) is accomplished by integrating VM allocation which is the significant task while deploying the ML technique with configurability limitations. Various approaches for mitigating these issues are (i) an optimal solution in energy consolidation is obtained by multi-objective optimization techniques, (ii) the advantages of VM allocation approaches are influenced by hyper-heuristics which is considered as a capable solution. Hyper-heuristics are "heuristics that choose heuristics" [10] and based on the current DC status the method is chosen, with improved trade-offs than the methods used separately. (ii) Metaheuristics approaches used to deal with the issues such as CPU-load interrelationship and merging VMs with their peak usages. These approaches are briefly discussed in this review paper.

The remainder of this paper is structured as follows. Sect. 2 summarizes the background study of the survey. Section 2 divided into four subheadings are briefly discussed. (2.1) Conventional methods for energy efficient, (2.2) linear programming approaches for energy efficient, (2.3) heuristic and metaheuristic methods for energy efficient, (2.4) machine learning methods for energy efficient. Then, Sect. 3 introduces issues from existing methods. At last, a summary of the conclusion is drawn in Sect. 4.

2 Background Study

The enormous core computer resources in a specific environment are termed as a data center or "server farm" which possess unified administration for the vital business for users and enterprises each moment. The cloud computing data center influenced by virtual resources sum corresponds to "resource pool". The power consumption may be the cause for data center energy consumption. Hence, the new energy power generation utilization is a significant factor in energy saving. The high energy costs are due to the huge energy consumption which in turn causes carbon emissions which is against sustainable development. Cloud systems' energy consumption is one of the prevalent research topics in current years. The energy consumption models for cloud systems are built using various efforts. The categorization of earlier approaches based on power-aware dynamic VM consolidation techniques: (i) their proficiencies for considering power and delay overheads associated with VM migration and power mode transitions (PMT) and (ii) their control policy level (centralized and distributed).

The major topics considered for review in this work are:

- Traditional heuristics approaches that were used to address the energy consolidation issues.
- Metaheuristics approaches used to deal with the issues such as CPU-load similarities, combining VMs when their peak usage do not coincide.
- Multi-objective optimization techniques used to attain an optimal solution in energy consolidation.
- Machine learning techniques that were adopted in attaining a better solution to the above-mentioned problems.
- Hyper-heuristics approaches that were adopted to leverage the benefits of VM allocation approaches.

All the above-revealed existing approaches were exploited to solve the energy consolidation problems that in turn guarantee QoS.

In recent times, machine learning methodologies are concentrated for attaining significant results. Nevertheless, now, very potential and novel approaches are highly demanded to handle the dynamic nature of cloud data. This review will also help in motivating the usage of recent deep learning approaches to deal with efficient energy consolidation of cloud data centers. Figure 2 shows the block diagram of energy-efficient cloud data centers for various approaches.

2.1 Review of Conventional Methods for Energy-Efficient Dynamic Consolidation in Data Centers

Beloglazov and Buyya [10] utilized dynamic VM consolidation for identifying host overload by designing Markov host overload detection (MHOD) and OPTimal offline algorithm (OPT). The system management plays a vital role in attaining the QoS goal with respect to OTF parameter, considering workload independent QoS metric.

Kim et al. [11] framed a model for energy-efficient data centers namely Pearson product-moment correlation and first-fit-decreasing heuristic virtual machines (VM) for energy-efficient data centers allotment. The server consolidation and voltage/frequency (v/f) scaling are harnessed together for data centers power management by mitigating the global power consumption, whereas substantial quality of service (QoS) requirements.

Chen and Shen [12] merged complementary VM with spatial/temporal awareness in cloud data centers for developing preliminary virtual machines (VMs) distribution stratagem.

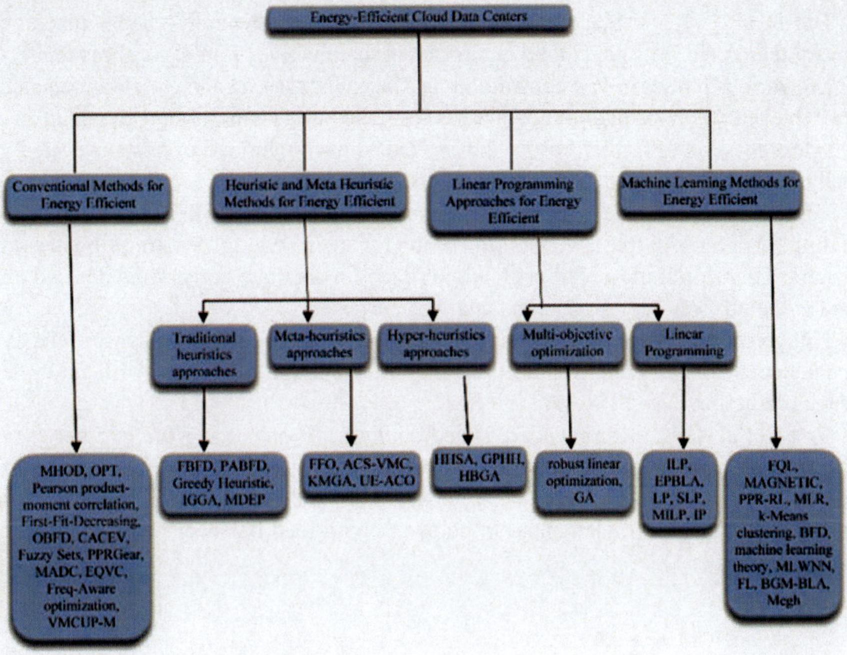

Fig. 2 Block diagram of energy-efficient cloud data centers

Esfandiarpoor et al. [13] developed a method for data center energy improvement named modified best fit decreasing (OBFD). The consolidation of VM in the servers is accomplished to reduce the active racks by structuring a technique named three-structure-aware virtual machine (VM).

Ahvar et al. [14] allocated VMs in a distributed cloud by utilizing a carbon cost emission-efficient VM placement method (CACEV) and fuzzy pairs. The optimization of overall cost and carbon emission is significantly achieved by means of merging several metrics.

Lin et al. [15] suggested a methodology for virtual machines (VMs) peak workload similarity measuring for consolidation VMs and enhancing resource utilization through mathematical quantity-peak similarity.

Ruan and Chen [16] modeled a system namely performance-to-power ratio gear (PPRGear) for virtual machine (VM) assignment for power consumption in the cloud.

Farahnakian et al. [17] developed a scheme for accomplishing dynamic virtual machine (VM) consolidation task via designing multi-agent-based dynamic consolidation (MADC). A local agent is mainly used in this system for decision of overloading of PM utilizing reinforcement learning approach. The supervisor is recognized as a global agent for dynamic optimization of VM placement depending on local agents' decisions.

Liu et al. [18] introduced energy-efficient and QoS dynamic virtual machine consolidation (EQVC) in a cloud environment that involves four algorithms conferring to various phases in VM consolidation. The energy saving and QoS maintenance are achieved by choosing redundant VMs from the hosts formerly they overload and transfer the VMs to further hosts. The avoidance of re-overload of hosts is done by a physical model with adaptive reserved resources.

Arroba et al. [19] developed a system by merging dynamic voltage and frequency scaling (DVFS) with freq-aware optimization for saving energy consumption in data centers. The CloudSim toolkit is greatly utilized for accurate power model based on data gathered from real servers and analysis purpose.

Nguyen et al. [20] proposed an ideology for cloud data centers energy-efficiency enhancement by utilizing virtual machine consolidation algorithm with multiple usage prediction (VMCUP-M).

Rossi et al. [21] recommended a methodology for enhancing the gap between energy utilization and application performance by means of e-eco orchestration of different energy-savings techniques. Energy-efficient cloud orchestrator—e-eco—a management system implementation helps in cloud load balancer deciding during execution (Table 1).

2.2 *Review of Heuristic and Metaheuristic Methods for Energy-Efficient Dynamic Consolidation in Data Centers*

Abdullah et al. [22] utilized various approaches such as fast best fit decreasing (FBFD) algorithm, dynamic utilization rate (DUR) algorithm, VM migration for intelligent VMs allocating into hosts, utilization space, dynamic consolidation of virtual machines (VMs) in cloud data centers, respectively.

Beloglazov and Buyya [23] introduced a stratagem for dynamic consolidation of virtual machines (VMs) established on historical data from the resource usage by VMs analysis namely power-aware best fit decreasing (PABFD) heuristics.

Kansal and Chana [24] offered an energy-aware virtual machine migration for cloud computing through metaheuristic algorithm entitled firefly optimization (FFO). The greatly loaded virtual machine (VM) to the least loaded active node is migrated by utilizing the proposed technique, despite the fact that the performance and energy efficiency of the data centers are conserved. The future work concentrates on energy-aware load balancing technique for cloud computing.

Wu et al. [25] modeled an algorithm for dynamic virtual machine (VM) consolidation in a heterogeneous cloud data center by using greedy heuristic and improved grouping genetic algorithm (IGGA). The greedy heuristic and a swap operation are greatly utilized for consolidation score optimization along with an IGGA.

Table 1 Inference of conventional methods for energy-efficient dynamic consolidation in data centers

S. No	Author name	Method name	Advantages	Disadvantages
1	MHOD and OPT [10]	Markov host overload detection (MHOD) and OPTimal offline algorithm (OPT)	Overload detection in the host	VM manager within the OpenStock platform
2	First-fit-decreasing [11]	Pearson product-moment correlation and first-fit-decreasing	Reduce the global power consumption	Not get an optimal solution
3	Virtual machines (VMs) [12]	Initial virtual machines (VMs) allocation	Reduces the number of PMs used	Not focused to enhance the pattern detection method
4	OBFD [13]	Modified best fit decreasing (OBFD)	VMs servers to minimize no. of active racks	Not support resource allocation
5	CACEV [14]	Cost and carbon emission-efficient VM (CACEV)	Optimize overall cost	Do not have the capability of machine learning
6	Peak similarity [15]	Mathematical quantity-peak similarity	Improve resource usage	Large time consuming
7	PPRGear [16]	Performance-to-power ratio gear (PPRGear)	Reduce energy utilization	little performance degradation
8	MADC [17]	Multi-agent-based dynamic consolidation (MADC)	Reduce the power consumption in a large-scale data center	It is not suitable for hybrid VM consolidation

(continued)

Table 1 (continued)

S. No	Author name	Method name	Advantages	Disadvantages
9	EQVC [18]	Energy-efficient and QoS dynamic virtual machine consolidation (EQVC)	Energy consumption, QoS guarantees	It is supported only single-factorized load in VM
10	Freq-aware optimization [19]	Freq-aware optimization	Accurate power model	It does not support large dataset
11	VMCUP-M [20]	Virtual machine consolidation algorithm with multiple usage prediction (VMCUP-M)	Reduces power consumption in servers	Focused on power consumption
12	E-eco [21]	E-eco	The best trade-off between performance and energy savings	Not suitable for a real-time environment

Xiao et al. [26] suggested minimal data-accessing energy path (MDEP) for scheduling data-intensive workflows which is a novel heuristic for mitigating the energy consumption of intensive data accessing.

Farahnakian et al. [27] obtained a reduction in energy consumption of cloud data centers by using metaheuristic algorithm called ant colony system-based virtual machine consolidation (ACS-VMC).

Haghighi et al. [28] accomplished for resource management by utilizing hybrid k-means clustering with genetic algorithm (KMGA) virtualization system which uses k-means clustering for mapping task and dynamic consolidation method, which is enhanced by a micro-genetic algorithm. The CloudSim 3.0.3 version is utilized for the experimentation with Expert-Choice software tools.

Pang et al. [29] introduced a scheme for reduced energy consumption by utilizing user experience-oriented-based ant colony optimization (UE-ACO). The energy consumption is mitigated by UE-ACO which is the advanced version of ant colony algorithm which fulfills the users' experience for balancing the power consumption and user experience in cloud DC.

Tsai et al. [30] suggested a cloud computing scheme for hyper-heuristic scheduling algorithm (HHSA). The dynamic determination of heuristic for recognizing superior candidate solutions is achieved by deploying diversity detection and improved detection operators. CloudSim (a simulator) and Hadoop (a real system) are chiefly involved for performance analysis, and the perturbation technique to enrich the performance of the proposed algorithm.

Tan et al. [31] introduced rules for virtual machine (VM) selection and creation which is obtained using genetic programming-based hyper-heuristic (GPHH) for online two-level resource allocation problem in container-based clouds and physical machine (PM) selection of first fit is also deployed.

Singh and Dhir [32] recommended a cloud computing energy reduction technique namely hypercube-based genetic algorithm (HBGA) for virtual machine (VM) with quality of service (QoS) constraint (Table 2).

2.3 Review of Linear Programming Approaches for Energy-Efficient Dynamic Consolidation in Data Centers

Nasim et al. [33] presented a robust linear optimization theory for quantification of uncertainty impact in modern data centers. The overbooking ratios such as virtual machine (VM) resource demands, migration-related overhead or the power consumption model of the servers used mainly emphasis the impact of different parameter uncertainties on the energy efficiency.

Ibrahim et al. [34] suggested a scheme for diminishing power consumption in a cloud data center by means of integer linear programming (ILP) dynamic task-scheduling algorithm. The dynamic nature of the cloud environment is achieved

Table 2 Inference of heuristic and metaheuristic methods for energy-efficient dynamic consolidation in data centers

S. No	Scheme	Method name	Advantages	Disadvantages
1	FBFD [22]	Fast best fit decreasing (FBFD)	Improves the energy consumption	It is not suitable for multiple system resources
2	PABFD [23]	Power-aware best fit decreasing (PABFD)	Reduce energy consumption	It is not an investigation of more complex workload models
3	FFO [24]	Firefly optimization (FFO)	Achieve energy efficiency	Not focused energy-aware load balancing
4	IGGA [25]	Greedy heuristic and improved grouping genetic algorithm (IGGA)	Dynamic virtual machine (VM) consolidation	Not focused on user requirements for performing VM consolidation
5	MDEP [26]	Minimal data-accessing energy path (MDEP)	Reduce the energy consumption	Not support VM scheduler in VM hypervisor level
6	ACS-VMC [27]	Ant colony system-based virtual machine consolidation (ACS-VMC)	Reduces energy consumption	It is not implemented in OpenStack cloud platform
7	KMGA [28]	K-Means clustering with genetic algorithm (KMGA)	Reduce the energy consumption of data centers	It is not focused on predict hosts
8	UE-ACO [29]	User experience-oriented-based ant colony optimization (UE-ACO)	Energy consumption while satisfying user experience	Uncertain time to convergence
9	HHSA [30]	Hyper-heuristic scheduling algorithm (HHSA)	Reduce the makespan of task scheduling	It is not finding more effective detection operators
10	GPHH [31]	Genetic programming-based hyper-heuristic (GPHH)	Reduce energy consumption	It is requiring less information about the problem
11	HBGA [32]	Hypercube-based genetic algorithm (HBGA)	Reduction in energy consumption	It is very slow

by an adaptive genetic algorithm (GA) and for providing near-optimal scheduling solution for reducing energy consumption. The validation of stated adaptive GA is achieved by cloud infrastructure simulation and performance, and quality evaluation is studied in this environment.

Rasouli et al. [35] introduced virtual machine (VM) dynamic replacement for diminishing power consumption on the basis of learning automata by an approach named extended-placement by learning automata (EPBLA).

Shidik et al. [36] recommended a fuzzy Markov normal algorithm specifically for dynamic virtual machine (VM) consolidation regarding the enhancement of efficient energy utilization in a cloud data center, which relies on this approach for mathematical programming. The categorization of VM candidate attributes done through Fuzzy logic. Subsequently, the conclusion about the destination category to migrate the VM from overload host has made using Markov normal algorithm.

Liu et al. [37] presented an integer linear programming (ILP) function that diminishes the power usage of cloud computing data center obtaining through schedule the tasks with least quantity of servers, concurrently, sustains the response time constraints of the task. Additionally, established that the required summation of task reply time and active server count to converge such time constraints have enclosed by greedy task-scheduling technique.

Zhang et al. [38] suggested a linear programming (LP) to enable the energy-proficient task scheduling in cloud data centers, which combines a linear processor frequency to perform every process, and these frequencies have been allocated with the time slices by an LP approach.

Anastasopoulos et al. [39] aimed at a non-renewable power utilization of the data centers (DCs), and a stochastic linear programming (SLP) has proposed. Besides, the dimensionality reduction approaches, especially the sample average approximation and Lagrangian relaxation have involved tackling the growing feature of computational complications in SLP designs.

Guérout et al. [40] suggested the methods, namely genetic algorithm (GA) and a mixed integer linear programming (MILP) intended for quality of service (QoS) in cloud processing, which is multiple objective optimizations of four applicable cloud QoS objectives: a GA and a MILP method. The complicacies in VM allocation have grown by the modeling of dynamic voltage and frequency scaling (DVFS) for the conservation of power on physical systems.

Dai et al. [41] recommended an integer programming (IP) to diminish the power utilization of multi-tenant cloud data centers. Initially, the formulation of this strategy consists of virtual machine (VM) scheduling with least energy in data centers accompanying the bandwidth that ensures the IP problem and validates its NP-hardness.

Ferreto et al. [42] presented a linear programming (LP) for efficient power utilization in virtualized cloud DC. The design for heuristics to handle virtual machine (VM) migration that prioritizes the VMs with stable capability. The implemented trails by employing TU-Berlin and Google data center workloads to distinguish location control approach in contrast to current eager-migration-based strategies.

Pahlavan et al. [43] concentrated on the reduction of cloud data centers power consumption by stating an integer linear programming (ILP). The combination of server placement and chassis consolidation for power consumption reduction is done for high-performance computing data centers (Table 3).

Table 3 Inference of linear programming approaches for energy-efficient dynamic consolidation in data centers

S. No	Scheme	Method name	Advantages	Disadvantages
1	Robust linear optimization [33]	Robust linear optimization theory	Higher total energy consumption	Data uncertainty in mathematical programming problems
2	ILP [34]	Integer linear programming (ILP)	Reducing energy consumption	It does not support different providers
3	EPBLA [35]	Extended-placement by learning automata (EPBLA)	Reduce power consumption, energy consumption	The transition from a state is to a single particular next state for each input
4	Fuzzy Markov normal [36]	Fuzzy Markov normal algorithm	Energy consumption, SLA violation	Decision-making in an environment of uncertainty and imprecision for real-world problems is a complex task
5	ILP [37]	Integer linear programming (ILP)	Reduces server energy consumption	Not support for real-world environment
6	LP [38]	Linear programming (LP)	Energy-efficient task scheduling	It is not easy to define a specific objective function
7	SLP [39]	Stochastic linear programming (SLP)	Reduce non-renewable energy consumption	A specific objective function is laid down, it may not be so easy to find out financial and other constraints
8	GA and MILP [40]	Genetic algorithm (GA) and a mixed integer linear programming (MILP)	Reduce energy consumption	It is not supported for real-world environments
9	IP [41]	Integer programming (IP)	Reduce the energy consumption	It is not suitable for larger datasets
10	LP [42]	Linear programming (LP)	Reduces the number of migrations with minimal penalty	Takes large time consuming
11	ILP [43]	Integer linear programming (ILP)	Minimize total power consumption	It is a mathematical optimization or all of the variables are restricted to be integers

2.4 Review of Machine Learning Methods for Energy-Efficient Dynamic Consolidation in Data Centers

Masoumzadeh and Hlavacs [44] suggested a fuzzy Q-learning (FQL) for the effective consolidation of virtual machines (VM) in the cloud processing data center. This method has fragmented into four phases to make decisions, i.e., (1) host overfitting detection, (2) VM selection, (3) host underfitting detection, and (4) VM placement. FQL was known to be an online method for decision-making with respect to improve the selection process of VM in the dynamic VM consolidation protocol.

Haghshenas et al. [45] presented an approach, called a multi-agent machine learning-based approach for energy-efficient dynamic consolidation (MAGNETIC), specifically for computing data centers, which is considered to be a localized-distributed low-overhead failure-aware dynamic virtual machines (VM) consolidation approach that intended to decrease the power usage in wide-ranging data centers. Here, the method initiated the selection of sufficient enough power mode and frequency from every host, whereas choosing runtime on the basis of distributed multi-agent machine learning (ML), and overcomes the corresponding VMs through the clustered heuristic approach.

Pahlevan et al. [46] suggested by taking the factors, like power, quality of service (QoS), network congestion, movement of resources, and adjustability in several circumstances of data center (DC) and defined an allocation method, namely heuristic and a machine learning (ML)-based virtual machines (VM).

Ding et al. [47] intended to balance the performance and energy through efficient consolidation of virtual machines (VM) by proposing a performance-to-power ratio-based reinforcement learning (PPR-RL). The reason behind the proposal is that this is an upbeat methodology, which is capable to fulfill the authors' intention by executing dynamic VM consolidation by taking both forecasted resource utilization and PPR of heterogeneous hosts.

Haghshenas and Mohammadi [48] have recommended a regression-based technique for energy-effective consolidation of VM in data centers, which forecasts this energy usage of VMs and hosts on the basis of their previous performance detail, concerning the exploitation of those forecasts in varied issues of the overall process.

Rybina and Schill [49] suggested the method, namely multiple linear regression (MLR) as regards the forecasting of power utilization of cloud data centers. This method has employed in the server at the time of migrating the VMs on the basis of resource exploitation metrics using the servers' respective MLR.

Dabbagh et al. [50] incorporated the methods of k-means clustering, theoretical Wiener filter and revised best fit decreasing (BFD) in order to propose an energy-conscious resource furnishing system for data centers of cloud, which comprises three key elements, i.e., data clustering, workload prophecy, and power management.

Cai et al. [51] suggested a strategy of machine learning to conserve energy resources in cloud computing as it reduces the redundant computation, which offers the possibility of high energy-efficiency cloud computing. The eradication of unwanted redundant calculation obtained by tallying the input and output of the calculation.

Tang et al. [52] recommended an approach, namely multiple linear regression and wavelet neural network (MLWNN) to extensively decrease the power utilization of servers, network, and cooling systems in a cloud data center. Initially, an energy-systematic cloud data center framework consisting of structural design, function, and the model of power consumption, has to be constructed.

Thein et al. [53] recommended the methods of reinforcement learning (RL) and fuzzy logic (FL) for cloud data centers as regards energy-efficient resource allocation. Proposed a system that will demonstrate efficient progress to achieve extensive energy efficacy in a data center and avoid violation in the service-level agreement (SLA) to leverage the green computing services.

Kumar et al. [54] presented a learning automata (LA) along with game theory in which let LA be the players, in that way, each player has a separate payoff assumed from the consumption and load of energy on the VM. Players communicate with the hypothetical atmosphere to step forward like the applicable choice of VMs and based on the response obtained from the atmosphere, they upgrade their action probability vector.

Tao et al. [55] intended to solve the DM-VM issue by proposing a bucket-code binary graph matching learning algorithm (BGM-BLA), besides the run time reallocation of virtual machine of cloud have scrutinized. Additionally, demonstrated the triple-objective optimization model for DM-VM to manage the power utilization, communication within VMs and migration outlays for the normal function of the platform.

Basu et al. [56] recommended a digital reinforcement learning algorithm, known as Megh for energy and performance-efficient resource management. Artail et al. [57] proposed a speedy cloud framework with performance enhancements for computational cloud, provided by interrogating hardware acceleration into the processing services (Table 4).

3 Issues from Existing Methods

Modern cloud data centers (DCs) are in the position to proactively face the issues of growing resource requirement for computing and handle the difficulties in energy efficacy. Obviously, there necessitated the development of guidelines for providing the resource that must be conscious about the features of VM, such as CPU operations, data conversion, in run time circumstances. (1) The role of conventional methods is inadequate for the environments in wide-ranging DC, as they lacks to prove their flexibility, as well as applicability; (2) enhancement of the energy efficacy in data centers concurrent to the assurance of QoS, accompanying the identification of hardware/software failures-triggered performance variability of servers, is considered to be the two major difficulties in resource management of wide-ranging cloud fundaments, as these concerns impact their proficiency; (3) preserving the energy in cloud

Table 4 Inference of machine learning methods for energy-efficient dynamic consolidation in data centers

S. No	Scheme	Method name	Advantages	Disadvantages
1	FQL [44]	Fuzzy Q-learning (FQL)	Energy-performance trade-off in cloud data centers	It is not implementing more VM selection
2	MAGNETIC [45]	Multi-agent machine learning-based approach for energy-efficient dynamic consolidation (MAGNETIC)	Reduces data center energy consumption	Not support for parallel applications at a time
3	ML [46]	Heuristic and machine learning (ML)	Energy and network traffic for reduced scenarios	High error-susceptibility
4	PPR-RL [47]	Performance-to-power ratio-based reinforcement learning (PPR-RL)	Reduces the energy consumption	It does not support real cloud infrastructure
5	Regression [48]	Regression-based approach	Improves energy efficiency	It is assumed that the cause and effect relationship between variables remains unchanged
6	MLR [49]	Multiple linear regression (MLR)	Increased energy consumption of the servers	It assumes that is a dependent and independent variable which is incorrect many times
7	BFD [50]	k-Means clustering, stochastic Wiener filter and modified best fit decreasing (BFD)	Reduces energy consumption	Euclidean distance for all the data points in the cluster and it will become stable
8	Machine learning [51]	Machine learning theory	Save energy consumption, energy efficiency	This system not focused in the heterogeneous environment
9	MLWNN [52]	Multiple linear regression and wavelet neural network (MLWNN)	Suitable for low workload cloud data center	It is required processors with parallel processing power
10	RL and FL [53]	Reinforcement learning (RL) and fuzzy logic (FL)	Energy-efficient resource allocation	It is not always accurate
11	LA and game theory [54]	Learning automata (LA) and game theory	Energy consumption	Not focused on security problems

(continued)

Table 4 (continued)

S. No	Scheme	Method name	Advantages	Disadvantages
12	BGM-BLA [55]	Binary graph matching-based bucket-code learning algorithm (BGM-BLA)	Energy consumption, communication between VMs	Not suitable for a real-time environment
13	Megh [56]	Megh	Energy-efficient resource management	Not support for large-scale cloud data center
14	Speedy cloud [57]	Speedy cloud	Increased resource utilization	Not focused on scheduling jobs among CPUs and FPGAs

environments enables the eradication of the hotspot belong to overfitted hosts and the shifting of under-fitted hosts to inactive mode via active migrations for VMs; (4) though the conventional techniques process the destinations of VM migration for sustaining the service-level agreements (SLA), yet it is inappropriate with packing VMs on specific physical hosts; (5) energy efficacy owing to the complicacies in real-time application scenarios of cloud computing; (6) the energy utilization quantity of data center certainly increases the expenditures of operation.

4 Conclusion and Future Work

Forecasting the cloud computing data center's dynamic consolidation of energy utilization helps to improve it as well as leads to a constant augmentation. Hence, the cloud applications in the engineering domain turn out to be dependent on energy consumption administration and effective energy utilization of computing. The evolution of huge quantity of models accompanying wide-ranging features has arisen by considering the modeling to predominantly include the technique for confronting the energy consumption. In this way, our preference apparently goes behind a dynamic method for surveying, namely service-level agreement (SLR) to observe the modeling endeavors of the present strategies to exploit it as a base for analyzing the energy consumption of cloud applications. Numerous research and modeling endeavors have inspired by the cloud computing environment, as it deals with the manageable environmental elements and the elements of application performance associated with task processing and data transmission. Recognizing the energy-oriented features helps this survey to ensure that the energy utilization of computing applications during this process for computation and communication are the critical issues of the current researchers. Similarly, the significant factors of the workload are known to be the task size and the data size, as these predominantly utilizes the operation and data transmission of CPU. Fundamentally, the approach of divide-and-conquer has primarily backed to this work with regard to learning the energy utilization in the study of cloud processing. Afterward, the power consumption models reach consecutive and progressive augmentation by the gradual restructuring of vital energy issues as regards to decreasing the complex trade-offs as well as disputes regarding the efficacy of energy. Apparently, the approaches that are necessitated for the approaches must be standard, heuristic, metaheuristic, hyper-heuristics, linear programming, and machine learning, concerning the energy consumption of cloud data centers.

The scheme for heuristic and machine learning (ML) performs well with diminution of energy utilization in the cloud data center's dynamic consolidation process. This survey investigated the standard and traditional approaches using performance-to-power ratio gear (PPRGear), multi-agent-based dynamic consolidation (MADC), and Markov host overload detection (MHOD). The heuristic approach has been scrutinized through several algorithms like ant colony system-based virtual machine consolidation (ACS-VMC), firefly optimization (FFO), hypercube-based genetic algorithm (HBGA), greedy heuristic and improved grouping genetic algorithm (IGGA), hyper-heuristic scheduling algorithm (HHSA), and genetic programming-based hyper-heuristic (GPHH). After that, integer linear programming (ILP), linear programming (LP), and stochastic linear programming (SLP) have taken to account to observe the linear programming method. The analysis of machine learning method includes fuzzy Q-learning (FQL), performance-to-power ratio-based reinforcement learning (PPR-RL), multiple linear regression (MLR), multiple linear regression and wavelet neural network (MLWNN), and regression-based method. Moreover, this survey evidently confers the virtues and drawbacks of the strategies involved in the energy utilization of the processing center in cloud computing. The enhancement for energy efficacy in data centers concurrent to the assurance of QoS, accompanying the identification of hardware/software failures-triggered performance variability of servers, are known to be the two significant difficulties of wide-ranging cloud fundaments' resource management, as these troubles impact their proficiency. To overcome this challenge, this study can be further stretched through developing the switched energy mode option unit, and a reallocation unit can be developed to concurrent handling of available hosts and virtual machine to enable the execution of model-driven simulations, concerning the disclosure of additional information regarding the combinational and factorial impacts on cloud applications' energy consumption.

References

1. Feller E, Ramakrishnan L, Morin C (2015) Performance and energy efficiency of big data applications in cloud environments: A Hadoop case study. J Parallel Distrib Comput 79:80–89
2. Zapater M, Tuncer O, Ayala JL, Moya JM, Vaidyanathan K, Gross K, Coskun AK (2014) Leakage-aware cooling management for improving server energy efficiency. IEEE Trans Parallel Distrib Syst 26(10):2764–2777
3. Hammadi A, Mhamdi L (2014) A survey on architectures and energy efficiency in data center networks. Comput Commun 40:1–21

4. Feng Y, Li B, Li B (2013) Price competition in an oligopoly market with multiple iaas cloud providers. IEEE Trans Comput 63(1):59–73
5. Li X, Garraghan P, Jiang X, Wu Z, Xu J (2017) Holistic virtual machine scheduling in cloud datacenters towards minimizing total energy. IEEE Trans Parall Distrib Syst 29(6):317–1331
6. Zapater M, Fraga D, Malagón P, Banković Z, Moya JM (2015) Self-organizing maps versus growing neural gas in detecting anomalies in data centres. Log J IGPL 23(3):495–505
7. Avgerinou M, Bertoldi P, Castellazzi L (2017) Trends in data centre energy consumption under the european code of conduct for data centre energy efficiency. Energies 10(10):1–18
8. Zhou Z, Hu ZG, Yu JY, Abawajy J, Chowdhury M (2017) Energy-efficient virtual machine consolidation algorithm in cloud data centers. J Central South Univ 24(10):2331–2341
9. Ismaeel S, Karim R, Miri A (2018) Proactive dynamic virtual-machine consolidation for energy conservation in cloud data centres. J Cloud Comput 7(1):1–28
10. Beloglazov A, Buyya R (2013) Managing overloaded hosts for dynamic consolidation of virtual machines in cloud data centers under quality of service constraints. IEEE Trans Paral Distrib Syst 24(7):1366–1379
11. Kim J, Ruggiero M, Atienza D, Lederberger M (2013) Correlation-aware virtual machine allocation for energy-efficient datacenters. Des Automat Test Eur Conf Exhib (DATE), 1345–1350
12. Chen L, Shen H (2014) Consolidating complementary VMs with spatial/temporal-awareness in cloud datacenters, IEEE Conf Comput Commun 1033–1041
13. Esfandiarpoor S, Pahlavan A, Goudarzi M (2015) Structure-aware online virtual machine consolidation for datacenter energy improvement in cloud computing. Comput Electr Eng 42:74–89 (2015)
14. Ahvar E, Ahvar S, Mann ZÁ, Crespi N, Garcia-Alfaro J, Glitho R (2016) CACEV: a cost and carbon emission-efficient virtual machine placement method for green distributed clouds. In: IEEE International Conference on Services Computing (SCC), pp 275–282
15. Lin W, Xu S, Li J, Xu L, Peng Z (2017) Design and theoretical analysis of virtual machine placement algorithm based on peak workload characteristics. Soft Comput 21(5):1301–1314
16. Ruan X, Chen H (2015) Performance-to-power ratio aware virtual machine (VM) allocation in energy-efficient clouds. In: IEEE ınternational conference on cluster computing, pp 264–273
17. Farahnakian F, Pahikkala T, Liljeberg P, Plosila J, Tenhunen H (2014) Multi-agent based architecture for dynamic VM consolidation in cloud data centers. In: IEEE 40th euromıcro conference on software engineering and advanced applications, pp111–118
18. Liu Y, Sun X, Wei W, Jing W (2018) Enhancing energy-efficient and qos dynamic virtual machine consolidation method in cloud environment. IEEE Access 6:31224–31235
19. Arroba P, Moya JM, Ayala JL, Buyya R (2017) Dynamic Voltage and Frequency Scaling-aware dynamic consolidation of virtual machines for energy efficient cloud data centers. Concurrency Comput: Pract Experience 29(10):1–13
20. Nguyen TH, Di Francesco M, Yla-Jaaski A (2017) Virtual machine consolidation with multiple usage prediction for energy-efficient cloud data centers. IEEE Trans Serv Comput
21. Abdullah M, Lu K, Wieder P, Yahyapour R (2017) A heuristic-based approach for dynamic vms consolidation in cloud data centers. Arab J Sci Eng 42(8):3535–3549
22. Beloglazov A, Buyya R (2012) Optimal online deterministic algorithms and adaptive heuristics for energy and performance efficient dynamic consolidation of virtual machines in cloud data centers. Concurrency Comput Practice Experience 24(13):1397–1420
23. Kansal NJ, Chana I (2016) Energy-aware virtual machine migration for cloud computing-a firefly optimization approach. J Grid Comput 14(2):327–345
24. Wu Q, Ishikawa F, Zhu Q, Xia Y (2016) Energy and migration cost-aware dynamic virtual machine consolidation in heterogeneous cloud datacenters. IEEE Trans Serv Comput 1–13
25. Xiao P, Hu ZG, Zhang YP (2013) An energy-aware heuristic scheduling for data-intensive workflows in virtualized datacenters. J Comput Sci Technol 28(6):948–961
26. Farahnakian F, Ashraf A, Pahikkala T, Liljeberg P, Plosila J, Porres I, Tenhunen H (2014) Using ant colony system to consolidate VMs for green cloud computing. IEEE Trans Serv Comput 8(2):187–198

27. Haghighi MA, Maeen M, Haghparast M (2019) An energy-efficient dynamic resource management approach based on clustering and meta-heuristic algorithms in cloud computing ıaas platforms. Wirel Pers Commun 104(4):1367–1391
28. Pang S, Xu K, Wang S, Wang M Wang S (2020) Energy-saving virtual machine placement method for user experience in cloud environment. Math Prob Eng 1–9
29. Tsai CW, Huang WC, Chiang MH, Chiang MC, Yang CS (2014) A hyper-heuristic scheduling algorithm for cloud. IEEE Trans Cloud Comput 2(2):236–250
30. Tan HM, Mei Y (2019) A hybrid genetic programming hyper-heuristic approach for online two-level resource allocation in container-based clouds. IEEE Cong Evol Comput (CEC) 2681–2688
31. Singh N, Dhir V (2019) Hypercube based genetic algorithm for efficient vm migration for energy reduction in cloud computing. Stat Opt Inf Comput 7(2):468–485
32. Nasim R, Zola E, Kassler AJ (2018) Robust optimization for energy-efficient virtual machine consolidation in modern datacenters. Cluster Comput 21(3):1681–1709
33. Ibrahim H, Aburukba RO, El-Fakih K (2018) An integer linear programming model and adaptive genetic algorithm approach to minimize energy consumption of cloud computing data centers. Comput Electr Eng 67:51–565
34. Rasouli N, Razavi R, Faragardi HR (2020) EPBLA: energy-efficient consolidation of virtual machines using learning automata in cloud data centers. Cluster Comput 1–15
35. Shidik GF, Azhari A, Mustofa K (2016) Improvement of energy efficiency at cloud data center based on fuzzy Markov normal algorithm VM selection in dynamic VM consolidation. Int Rev Comput Soft (IRECOS) 11(6):511–520
36. Liu N, Dong Z, Rojas-Cessa R (2013) Task scheduling and server provisioning for energy-efficient cloud-computing data centers. In: IEEE 33rd ınternational conference on distributed computing systems workshops, pp 226–231
37. Zhang Y, Wang Y, Wang H (2016) Energy-efficient task scheduling for DVFS-enabled heterogeneous computing systems using a linear programming approach. In: IEEE 35th ınternational performance computing and communications conference (IPCCC), pp 1–8
38. Anastasopoulos M, Tzanakaki A, Simeonidou D (2016) Stochastic energy efficient cloud service provisioning deploying renewable energy sources. IEEE J Select Areas Commun 34(12):3927–3940
39. Guérout T, Gaoua Y, Artigues C, Da Costa G, Lopez P, Monteil T (2017) Mixed integer linear programming for quality of service optimization in clouds. Fut Gen Comput Syst 71:1–17 (2017)
40. Dai X, Wang JM, Bensaou B (2015) Energy-efficient virtual machines scheduling in multi-tenant data centers. IEEE Trans Cloud Comput 4(2):210–221
41. Ferreto TC, Netto MA, Calheiros RN, De Rose CA (2011) Server consolidation with migration control for virtualized data centers. Fut Gener Comput Syst 27(8):1027–1034
42. Pahlavan MM, Goudarzi M (2014) Power reduction in HPC data centers: a joint server placement and chassis consolidation approach. J Supercomput 70(2):845–879
43. Masoumzadeh SS, Hlavacs H (2013) Integrating VM selection criteria in distributed dynamic VM consolidation using fuzzy Q-Learning. In: Proceedings of the 9th ınternational conference on network and service management (CNSM), pp 332–338
44. Haghshenas K, Pahlevan A, Zapater M, Mohammadi S, Atienza D (2019) MAGNETIC: multi-agent machine learning-based approach for energy efficient dynamic consolidation in data centers. IEEE Trans Serv Comput 1–14
45. Pahlevan XQ, Zapater M, Atienza D (2017) Integrating heuristic and machine-learning methods for efficient virtual machine allocation in data centers. IEEE Trans Comput-aided Des Integr Circ Syst 37(8):1667–1680
46. Ding W, Luo F, Gu C, Lu H, Zhou Q (2020) Performance-to-power ratio aware resource consolidation framework based on reinforcement learning in cloud data centers. IEEE Access 8:15472–15483
47. Haghshenas K, Mohammadi S (2020) Prediction-based underutilized and destination host selection approaches for energy-efficient dynamic VM consolidation in data centers. J Supercomput 1–18

48. Rybina K, Schill A (2016) Estimating energy consumption during live migration of virtual machines. In: IEEE international black sea conference on communications and networking (BlackSeaCom), pp 1–5
49. Dabbagh M, Hamdaoui B, Guizani M, Rayes A (2015) Energy-efficient resource allocation and provisioning framework for cloud data centers. IEEE Trans Network Ser Manage 12(3):377–391
50. Cai XB, Ji YX, Han K (2017) Energy efficiency optimizing based on characteristics of machine learning in cloud computing. ITM Web of Conf 12:1–5
51. Chandy A (2019) Smart resource usage prediction using cloud computing for massive data processing systems. J Inf Technol 1(02):108–118
52. Raj JS (2019) Efficient Information Maintenance Using computational ıntelligence ın the multi-cloud architecture. J Soft Comput Paradigm (JSCP) 1(02):113–124
53. Thein T, Myo MM, Parvin S and Gawanmeh A (2018) Reinforcement learning based methodology for energy-efficient resource allocation in cloud data centers, J King Saud University-Comput Inf Sci,1–26
54. Kumar N, Zeadally S, Chilamkurti N and Vinel A (2015) Performance analysis of Bayesian coalition game-based energy-aware virtual machine migration in vehicular mobile cloud, IEEE Network 29(2):62–69
55. Tao F, Li C, Liao TW and Laili Y (2015) BGM-BLA: a new algorithm for dynamic migration of virtual machines in cloud computing. IEEE Trans Serv Comput 9(6):910–925
56. Basu D,Wang X, Hong Y, Chen H and Bressan S (2019) Learn-as-you-go with megh: Efficient live migration of virtual machines, IEEE Trans Parallel Distrib Syst 30(8):1786–1801
57. Artail H, Saghir M, Sharafeddin M, Hajj H, Kaitoua A, Morcel R and Akkary H (2019) Speedy cloud: Cloud computing with support for hardware acceleration services, IEEE Trans Cloud Comput 7(3):850–865

Chapter 61
A Secure Video Steganography Using Framelet Transform and Singular Value Decomposition

Meenu Suresh and I. Shatheesh Sam

1 Introduction

Currently, because of the evolution of Internet and network technology distribution of enormous multimedia files can be done in a fraction of second. The networks are uncovered to vindictive attacks due to the swift advance in multimedia technology. In the digital world, security and confidentiality of data play the most significant concern. Steganography fulfills these qualities as it is known as concealment of confidential information inside a cover medium. A video acts as the cover medium in video steganography, and the secret information is any image or text [1–3].

In the new era, video steganography can be used for application such as information broadcasting, military services, banking, etc. Videos are predominantly used for hiding confidential data because (i) Larger space to hide, (ii) Concealed information is unnoticeable.

The methods in steganography can feasibly be used in spatial as well as transform domain. The pixels of the cover medium carries the confidential data in the spatial domain. A simplest stenographic technique is the LSB technique in the spatial domain. The secret data is replaced in place of LSB bit of pixel [4]. During the extraction process, data can be retrieved smoothly, and the quality of the image is also not affected [5].

M. Suresh (✉)
Research Scholar, Reg No. 17213112162013, Dept. of Computer Science, Nesamony Memorial Christian College affiliated to Manonmaniam Sundaranar University, Abishekapatti, Tirunelveli, Tamil Nadu 627012, India
e-mail: meenupillai1988@gmail.com

I. Shatheesh Sam
Associate Professor, Dept. of PG Computer Science, Nesamony Memorial Christian College affiliated to Manonmaniam Sundaranar University, Abishekapatti, Tirunelveli, Tamil Nadu 627012, India
e-mail: shatheeshsam@yahoo.com

I. Jeena Jacob et al. (eds.), *Data Intelligence and Cognitive Informatics*,
Algorithms for Intelligent Systems,
https://doi.org/10.1007/978-981-15-8530-2_61

Transformation techniques such as DCT or DWT are initially used to transform the video fame in the transform domain. This helps in maintaining the quality of the video as the data is embedded into the respective transformation coefficients. Sub-dividing the video frames and then transforming it tends to increase the video quality. It also helps in increasing the security with the use of pseudorandom sequence generators. In [6], chaotic systems based on several seed maps were presented. A control sequence was used in the approaching iteration to distinguish which seed map to be used. The proposed system showed enhanced chaoticity. In [7], an encryption scheme based on permutation and substitution which operated on chaos method was proposed. In [8], to increase the keyspace, a pseudorandom sequence generator in its encryption method was proposed. In [9], using logistic and standard maps pseudorandom number sequence was generated to control the encryption operations.

In [10], a data hiding system which provides double-layer security to the secret data that mainly concentrates on the principles of scrambling as well as pseudo-random data hiding is presented. In [11], it is seen that as scene changes were detected, DWT was applied to normalize the cover data as well as the secret data, as the secret data was embedded into the coefficient's LSB value.

The steganography domain also witnesses the use of curvelet transform. It is a directional transform which is multiscale and uses edges to create a sparse representation of objects. In [12], hybrid curvelet transform as well as genetic algorithm-based steganography is proposed. The embedding was done using LSB based, and choices to embed the secret bits were done by Genetic Algorithm (GA) to achieve imperceptibility. In [13], Finite Ridgelet Transform (FRT) along with DWT-based image steganography was proposed. To the secret image, Arnold scrambling was applied.

In [14], the cover was decomposed using DCT, and the AC coefficients were acquired from the same, and using a zigzag format, it was scanned from least to a most significant one. The embedding locations were selected using a chaotic map, and bit by bit entrenching was carried out.

In the present work, framelet transform and singular value decomposition were used to improve the security and quality of the video. To enhance the robustness of the method and to make the hidden data undetectable, video frames were transformed into the frequency domain.

The flow of the paper is stated as follows. Section 2 explains the video steganography system that is proposed. Section 3 covers the experimental results, and the conclusion is in Sect. 4.

2 Proposed Method

This paper combines framelet transform and singular value decomposition to produce a secure and robust video steganography method. The method starts by detecting changes in the scene to discover the keyframes. The frame where the first scene change is discovered which is considered as a keyframe. The keyframe is then converted to HSV color space. Transformed coefficients are got from the keyframes

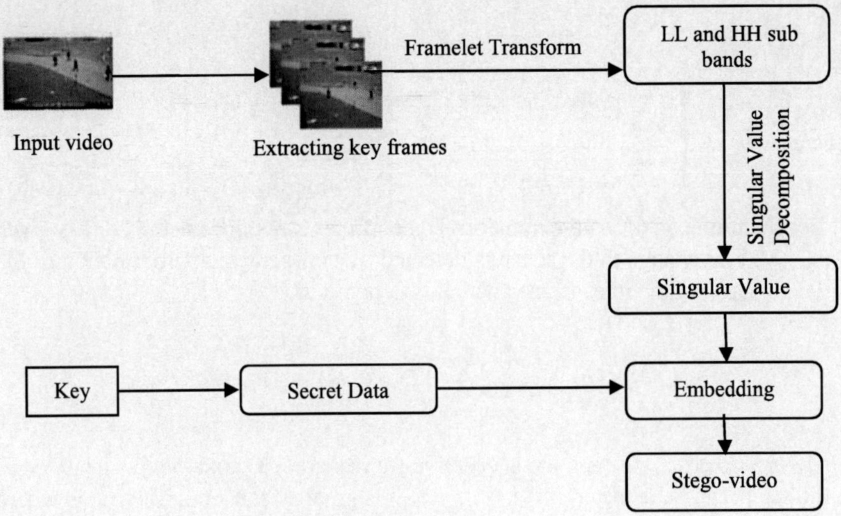

Fig. 1 Schematic representation of embedding process

after applying the framelet transform. Then, singular value decomposition is applied to those transformed coefficients after which the secret data is embedded. Preceding to embedding to increase the security, the secret data is pre-processed using encryption. Figure 1 Illustrates the architecture of the proposed method. The following are the phases of the proposed method,

1. Detecting changes in a scene
2. The transformation from RGB color space to Uncorrelated Color Space (UCS) HSV
3. Encryption of secret data
4. Data embedding
5. Process of extraction.

2.1 Detecting changes in a scene

The dimension of the video frame is denoted by $d \times a$. Scene change detection method is applied to extract the keyframes that are used to hide the secret data. Let frames be denoted by f_r which can be symbolized as $f_r = \{f_{r_0}, f_{r_1}, \ldots, f_{r_n}\}$.

Then, Discrete Cosine Transform (DCT) is used to detect scene change as,

$$\hat{D}(X, Q) = \alpha(X)\alpha(Q)$$

$$\sum_{m=0}^{d-1}\sum_{n=0}^{a-1} f(m, n) \cos\left[\frac{(2m+1)X\pi}{2d}\right] \cdot \cos\left[\frac{(2n+1)Q\pi}{2a}\right]$$

$$m = 0, 1, \ldots, d-1 \text{ and } n = 0, 1, \ldots, a-1 \tag{1}$$

$$\text{where } \alpha(D) = \begin{cases} \sqrt{\frac{1}{d}} \text{ for } D = 0 \\ \sqrt{\frac{2}{d}} \text{ for } D \neq 0 \end{cases} \text{ and } \alpha(V) \begin{cases} \sqrt{\frac{1}{a}} \text{ for } Q = 0 \\ \sqrt{\frac{2}{a}} \text{ for } Q \neq 0 \end{cases}$$

For the frame, $f_r(d, a)$ the transformed coefficients obtained using DCT is given $\hat{D}(X, Q)$. The change in the scene is detected by consecutive approximation of the variance between the inter-frames which is denoted as,

$$v_r\left(fa_{r_i}, fa_{r_{i1}}\right) = \sum_{m=0}^{d-1} \sum_{n=0}^{a-1} \left[\hat{D}_i(X, Q) - \hat{D}_{i+1}(X, Q)\right] \tag{2}$$

where $m = 0, 1, \ldots, r_n$ and the successive frames are fa_{r_i} and $fa_{r_{i1}}$. The changes in the scene can be detected if $v_r\left(fa_{r_i}, fa_{r_{i1}}\right)$ is greater than zero. The frame where the first change in the scene is distinguished which is used to embed the secret data. Based on the size of confidential data for hiding, one or more successive frames can be chosen.

2.2 Transformation from RGB to Uncorrelated Color Space (UCS) HSV

The RGB keyframe thus obtained is converted to the HSV color space for embedding the data as,

$$f_{r\text{UCS}} = \text{RGB2HSV}(f_r) \tag{3}$$

After the application of Eq. 3, the frame is separated into $H_{\text{UCS}}, S_{\text{UCS}}, V_{\text{UCS}}$ for hiding the data. The secret data is pre-processed prior to hiding, thereby increasing the security against attackers. The encrypted data is then embedded into the intensity plane V_{UCS}. After embedding in the plane V_{UCS}, in order to obtain the stego image combined it with the rest of the two planes. Afterward, the stego image is converted back to RGB color space and then combined with the rest of the frames to get the stego video.

2.3 Encryption of secret data

In the following section illustrates the process of encrypting the sensitive contents prior to embedding it into the cover image. Before embedding, the secret data is pre-processed using encryption. For encrypting, two steps are followed, namely key generation and nonlinear diffusion.

2.3.1 Generation of Key

The key is defined with the help of proposed enhanced twisted logistic map which is generated as:

for $q = 1$ to Len step 1

for $c = 1$ to Wei step 1

$s_{u+1} = [\mu - ke_1 - t_u/(1 - t_u) + l_j] \bmod 1$

$t_{u+1} = [\mu + t_j - ke_2 + l_u(1 \times (s_u + 1)^2))] \bmod 1$

$l_{u+1} = [\mu + (s_{u+1} - t_{u+1} \times ke_3)/l_u] \bmod 1$

$S_{q,c} = \lfloor s_{q,c+1} \times 256 \rfloor$

$T_{q,c} = \lfloor t_{q,c+1} \times 256 \rfloor$

$L_{q,c} = \lfloor l_{q,c+1} \times 256 \rfloor$

end

$s_{q+1,1} = s_{q,+c+1}$

$t_{q+1,1} = t_{q,+c+1}$

$l_{q+1,1} = l_{q,+c+1}$

end

where $0 \leq \mu \leq 3.999$ wherein it takes values between 3.455 and 3.999. $|ke_1| > 38.4$, $|ke_2| > 34.6$, $|ke_3| > 31.3$ to increase keyspace. The chaotic keys are $S_{q,c}$, $T_{q,c}$, $L_{q,c}$. The size of ke_i and that of the secret image is the same.

2.3.2 Nonlinear Diffusion

In this, firstly RGB diffusion is obtained by 5 LSB circular shift method. The steps that are followed to XOR all the RGB channels to the circular shift with the help of the first chaotic key is as shown below:

for $q = 1$ to Len

for $c = 1$ to Wei

$\mathrm{HRe}_{q,c} = (\mathrm{Rd}_{q,c} >>> 5) \bmod 256$

$\mathrm{HGr}_{q,c} = (\mathrm{Gn}_{q,c} >>> 5) \bmod 256$

$\mathrm{HBl}_{q,c} = (\mathrm{Be}_{q,c} >>> 5) \bmod 256$

$\mathrm{YRd}_{q,c} = \mathrm{HRe}_{q,c} \oplus \mathrm{X}_{e,f}$

$\mathrm{YGn}_{q,c} = \mathrm{HGr}_{q,c} \oplus \mathrm{X}_{e,f}$

$\mathrm{YBe}_{q,c} = \mathrm{HBl}_{q,c} \oplus \mathrm{X}_{e,f}$

end

end

where the first chaotic key is represented as $\mathrm{X}_{e,f}$. $\mathrm{HRe}_{q,c}$, $\mathrm{HGr}_{q,c}$ and $\mathrm{HBl}_{q,c}$ are the values that are obtained after applying a circular shift. After performing the XOR operation, $\mathrm{YRd}_{q,c}$, $\mathrm{YGn}_{q,c}$ and $\mathrm{YBe}_{q,c}$ are the nonlinear diffusion. Thus, encrypting the secret data makes it potent to withstand any attack.

2.4 Data Embedding

In this section, the actual data embedding mechanism is described. It is mainly based on 2D-framelet transform and singular value decomposition.

Step 1: Obtaining coefficient by applying 2D-framelet transformation to the intensity plane V_{UCS} as,

$$C_o = \Re(V_{\mathrm{UCS}}) \tag{4}$$

where $\Re$ is the framelet transform.

As any change in the chosen sub-bands, HH and LL cannot be identified by HVS, they are preferred.

Step 2: Apply singular value decomposition to the coefficients as,

$$SV_d = \Im(C_O) \tag{5}$$

where $\Im$ is singular value decomposition.

Step 3: Embed the secret data obtained from Sect. 2.3 in singular values using an appropriate embedding factor ϕ as,

$$SV_{d_{\mathrm{new}}} = SV_d + \phi * E_{\mathrm{nc}} \tag{6}$$

where E_{nc} is the encrypted secret data.

Step 4: Acquire reformed coefficients by applying inverse SVD as,

$$\text{SV}'_d = \text{SVD}(\text{SV}_{d_{\text{new}}}) \tag{7}$$

Step 5: Apply inverse framelet transform to get the modified V_{UCS} plane containing the secret data.

$$V'_{\text{UCS}} = \Re^{-1}(C_o) \tag{8}$$

After embedding, combining the plane V_{UCS} with rest of the two planes is done and then converting the stego image back to RGB color space is performed, combined with rest of the frames to get the stego video "S."

2.5 *Process of extraction*

The process of extraction that is the reverse of the process of embedding is summarized below,

Step 1: As discussed in Sect. 2.1, finding the keyframe from the stego video "S" using scene change technique.

Step 2: As explained in Sect. 2.2, the transformation of the keyframe from RGB color space to HSV color space, and selection of intensity plane V_{UCS} is done.

Step 3: Apply framelet transform on V_{UCS} as,

$$C'_o = \Re(V'_{\text{UCS}}) \tag{9}$$

Step 4: Apply singular value decomposition on coefficients as,

$$\text{SV}'_d = \Im(C'_o) \tag{10}$$

Step 5: With the same scaling factor used for embedding recover the secret data.

$$E'_{\text{nc}} = \frac{1}{\phi}(\text{SV}'_{d_{\text{new}}} - \text{SV}_d) \tag{11}$$

3 Experimental Results

The effectiveness of the system proposed is verified using videos taken from [15] and manually generated confidential images. The proposed system was quantified using measures such as Peak Signal to Noise Ratio (PSNR), embedding capacity and is compared with two existing methods, namely Suresh and Shatheesh Sam [16] and Barve et al. [17]. Table 1 shows the properties of the cover videos used.

The embedding capacity is calculated as,

$$\text{Ca} = \frac{\text{Max\_size of data embedded}}{\text{size\_of cover video}} \tag{12}$$

PSNR is calculated as,

$$\text{PSNR} = 10\log\left(\frac{(255)^2}{M}\right)\text{dB} \tag{13}$$

where M is the mean square error.

The performance of the proposed system in comparison with the existing methods is illustrated in Table 2.

Table 1 Properties of videos used

Video	Size	Frame_rate
V1	525 × 340	19
V2	758 × 585	28

Table 2 Analysis of the proposed method with the existing methods

Cover video	Methods	PSNR	Embedding capacity (%)
Video 1	Proposed	38.1356	72.3
	Suresh et al.	33.3728	52.8
	Barve et al.	21.4663	43.9
Video 2	Proposed	38.4786	75.5
	Suresh et al.	33.1355	55.9
	Barve et al.	22.3738	46.8

4 Conclusion

This paper proposed video steganography using framelet transform and singular value decomposition to hide the secret data. With the help of the value attained from inter-frame difference changes in the scene are detected, and from that keyframes are identified. Framelet transform and SVD are applied to the keyframe to entrench the secret data. After embedding, the stego video is transmitted through the network and finally; at the receiver end, the secret data is extracted. The structure aids for better time–frequency localization enhances capacity and security. The proposed method provides an embedding capacity of 75.5% and PSNR of 38 dB. Thus, the visual quality and security obtained using the proposed method are much higher than the existing methods. In future, different wavelet transform, chaotic map along with optimization techniques can be used to enhance secure video steganography.

References

1. Sun W, Lu Z-M, Wen Y-C, Yu F-X, Shen R-J (2013) High performance reversible data hiding for block truncation coding compressed images. SIViP 7:297–306. https://doi.org/10.1007/s11760-011-0238-4
2. Yu C-M, Wu K-C, Wang C-M (2011) A distortion-free data hiding scheme for high dynamic range images. Displays 32:225–236. https://doi.org/10.1016/j.displa.2011.02.004
3. Zeng X-T, Ping L-D, Pan X-Z (2010) A lossless robust data hiding scheme. Pattern Recogn 43:1656–1667. https://doi.org/10.1016/j.patcog.2009.09.016
4. Chan C-K, Cheng LM (2004) Hiding data in images by simple LSB substitution. Pattern Recogn 37:469–474. https://doi.org/10.1016/j.patcog.2003.08.007
5. Khare R, Mishra R, Arya I (2014) Video steganography using LSB technique by neural network. In: 2014 international conference on computational intelligence and communication networks. IEEE, Bhopal, India (2014)
6. Zhou Y, Bao L, Chen CLP (2013) Image encryption using a new parametric switching chaotic system. Signal Process 93:3039–3052. https://doi.org/10.1016/j.sigpro.2013.04.021
7. Belazi A, Khan M, El-Latif AAA, Belghith S (2017) Efficient cryptosystem approaches: S-boxes and permutation–substitution-based encryption. Nonlinear Dyn 87:337–361. https://doi.org/10.1007/s11071-016-3046-0
8. Belmeguenai A, Ahmida Z, Ouchtati S, Djemii R (2017) A novel approach based on stream cipher for selective speech encryption. Int J Speech Technol 20:685–698. https://doi.org/10.1007/s10772-017-9439-8
9. Patidar V, Pareek NK, Sud KK (2009) A new substitution–diffusion based image cipher using chaotic standard and logistic maps. Commun Nonlinear Sci Numer Simul 14:3056–3075. https://doi.org/10.1016/j.cnsns.2008.11.005
10. Parah SA, Sheikh JA, Hafiz AM, Bhat GM (2014) Data hiding in scrambled images: a new double layer security data hiding technique. Comput Electr Eng 40:70–82. https://doi.org/10.1016/j.compeleceng.2013.11.006
11. Ramalingam M, Isa NAM (2016) A data-hiding technique using scene-change detection for video steganography. Comput Electr Eng 54:423–434. https://doi.org/10.1016/j.compeleceng.2015.10.005
12. Muhammad N, Bibi N, Mahmood Z, Kim D-G (2015) Blind data hiding technique using the Fresnelet transform. SpringerPlus 4:832. https://doi.org/10.1186/s40064-015-1534-1

13. Thanki R, Borra S (2018) A color image steganography in hybrid FRT–DWT domain. J Inf Secur Appl 40:92–102. https://doi.org/10.1016/j.jisa.2018.03.004
14. Saidi M, Hermassi H, Rhouma R, Belghith S (2017) A new adaptive image steganography scheme based on DCT and chaotic map. Multimed Tools Appl 76:13493–13510. https://doi.org/10.1007/s11042-016-3722-6
15. Video watermarking using DWT—file exchange. MATLAB Central. https://in.mathworks.com/matlabcentral/fileexchange/52225
16. Suresh M, Shatheesh Sam I (2018) High secure video steganography based on shuffling of data on least significant DCT coefficients. In: 2018 second international conference on intelligent computing and control systems (ICICCS). IEEE, Madurai, India, pp 877–882
17. Barve S, Nagaraj U, Gulabani R (2011) Efficient and secure biometric image stegnography using discrete wavelet transform. Int J Comput Sci Commun Netw 1:96–99

Chapter 62
Collective State Implementation on Particle Swarm Opimization for Feature Selection

Jaswinder Singh, Soham Pathak, and Ritwik Bandyopadhyay

1 Introduction

Feature selection is a process of finding the best set of features such that the accuracy of the model is maintained or increased. Feature selection plays a key role in the machine learning process [1]. However, with feature selection being a non-deterministic polynomial time (NP) hard combinatorial problem, finding the best set of features becomes increasingly hard as the total number of features increases. Exhaustive search and using all the features become impractical [2]. Over the years, techniques like GA, BPSO [3] were made and improved [4] upon for feature selection [5] as they guide toward the best feature set with a quicker convergence of the associated loss function.

Our goal in this paper is to develop a variant of BPSO which provides quicker and better convergence. A new method, namely Collective State Implementation-Binary Particle Swarm Optimization (CSI-BPSO) is proposed in this work. To resolve the limitations of BPSO, the collective state seen in natural swarms in the BPSO algorithm is replicated using the local and global center of masses. The proposed CSI-BPSO was tested with five benchmark datasets. To examine the efficiency of CSI-BPSO, 50 independent runs were performed for each algorithm (CSI-BPSO, BPSO, GA), for 100 iterations involving 10 individuals.

J. Singh (✉) · S. Pathak · R. Bandyopadhyay
KIIT deemed to be University, Bhubaneshwar, India
e-mail: jaswinder9051998@gmail.com

S. Pathak
e-mail: sohamkiit636@gmail.com

R. Bandyopadhyay
e-mail: ritwikbanrg@gmail.com

I. Jeena Jacob et al. (eds.), *Data Intelligence and Cognitive Informatics*,
Algorithms for Intelligent Systems,
https://doi.org/10.1007/978-981-15-8530-2_62

2 Original Binary Particle Swarm

2.1 Existing Work

$$v_i^d(t+1) = wv_i^d(t) + c_1r_1(\text{pbest}_i^d(t) - x_i^d(t) + c_2r_2(\text{gbest}^d(t) - x_i^d(t) \quad (1)$$

where d is the dimension of search space, pbesti is the personal best of the ith individual [6], gbest is the global best, v is the velocity vector, x_i is the position of the ith particle, w is the inertial weight, c_1 and c_2 are acceleration factors, and r_1 and r_2 are random numbers between (0, 1). The velocity is maintained within a specified range of 6 and −6 [7].

A probability value is calculated using the velocity [8], by passing the velocity through the sigmoid function. The position vector (feature set) is calculated by comparing a randomly generated value (rand) between (0, 1) to the probability value as shown in Eqs. (2) and (3).

$$S\left(v_i^d(t+1)\right) = \frac{1}{1 + \exp\left(-v_i^d(t+1)\right)} \quad (2)$$

$$x_i^d(t+1) = 1 \text{ if rand} < S\left(v_i^d(t+1)\right) \text{ else } 0 \quad (3)$$

In BPSO, pbest and gbest play the role of guiding the particle to move toward the global minimum. Iteratively, the pbest and gbest are updated as shown in Eqs. (4) and (5).

$$\text{pbest}_i(t+1) = x_i(t+1) \text{ if } F(x_i(t+1)) < F\left(\text{pbest}_i(t)\right) \text{ otherwise pbest}_i(t) \quad (4)$$

$$\begin{aligned} \text{gbest}(t+1) &= \text{pbest}_i(t+1) \\ &\text{if } F\left(\text{pbest}_i(t+1)\right) < F(\text{gbest}(t)) \text{ otherwise gbest}(t) \end{aligned} \quad (5)$$

$F(.)$ is the fitness function, and t is the number of iterations.

While gbest and pbest are useful in guiding the population toward the minimum in case of BPSO algorithm, they aren't sufficient to redirect the individuals which are far off from the minimum, and hence, a significant portion of the population's contribution is usually negligible.

3 Collective State Impact

3.1 Literature Study

In their study, Tobias Bäuerle, Robert C. Löffler, and Clemens Bechinger provide experimental evidence for the idea of system-independent organization principles of collective states [9] of natural swarms of birds by investigating the stability of swirls formed by light-responsive active colloids which adjust their individual motion to positions and orientations of neighbors. They discuss the change in orientation of the swarm members with respect to their neighbors.

In our research, this concept of change in orientation of particles is taken with respect to the communal and local center of masses, to incorporate the fluid behavior of swarm and mold it to increase the exploration and exploitation power in a binary particle swarm optimization.

3.2 Collective State Implementation—Binary Particle Swarm Optimization

Two radii R_p and R_o are chosen such that $R_o > R_p > 0$. For all individuals in the population center of mass is calculated using individuals within radius R_p and R_o, respectively [10].

Center of mass with respect to R_o (communal) is responsible for increased exploitation power while R_p (local) is responsible for increased exploration power in the CSI-BPSO.

The mass is defined [11] of each individual as shown in Eq. (6) (Fig. 1).

$$M_i = \frac{\text{fitness}_i - \text{Worst fitness in the population}}{\text{Best fitness in the population} - \text{Worst fitness in the population}} \tag{6}$$

Center of mass is calculated for radii R_p and R_o of the individual, respectively, as shown in Eqs. (7), (8), (9), (10).

$$S_p\text{: Subset of all individuals whose euclidean distance with ith individual is less than } R_p \tag{7}$$

$$S_o\text{: Subset of all individuals whose euclidean distance with ith individual is less than } R_o \tag{8}$$

$$\text{C.O.M}_i^d(R_p) = \frac{\sum_{j \in S_p} M_j x_j^d}{\sum_{j \in S_p} x_j^d} \tag{9}$$

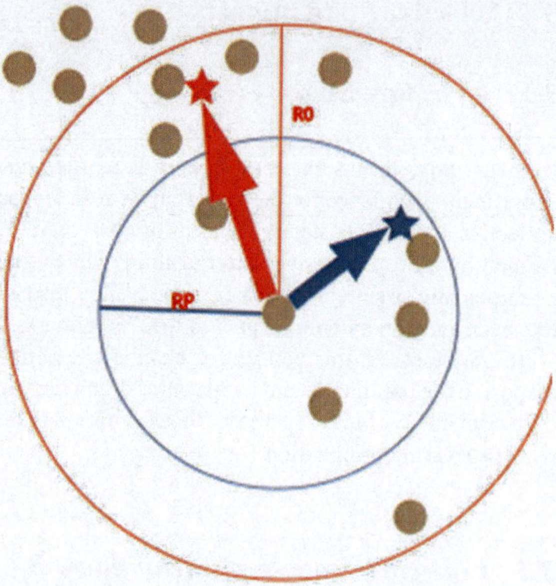

Fig. 1 Representation of local and communal area

$$\text{C.O.M}_i^d(R_o) = \frac{\sum_{j \in S_o} M_j x_j^d}{\sum_{j \in S_o} x_j^d} \tag{10}$$

A cumulative center of mass [12] is calculated using vector addition of the two center of masses multiplied by respective constants (11).

$$\text{C.O.M}_i^d = \text{Constant}_p * \text{C.O.M}_i^d(R_p) + (1 - \text{Constant}_p) * \text{C.O.M}_i^d(R_o) \tag{11}$$

where Constantp is a constant between (0, 1) chosen to balance between the impact of exploration and exploitation. In our research, Constantp is chosen as 0.2.

An individual which is closest to C.O.M_i is chosen from the population as cbesti (cumulative best).

The velocity of the particle is updated as shown in Eq. (12).

$$\begin{aligned} v_i^d(t+1) &= wv_i^d(t) + c_1 r_1(\text{pbest}_i^d(t) - x_i^d(t) \\ &+ c_2 r_2(\text{gbest}^d(t) - x_i^d(t) + c_3 r_3\left(\text{cbest}_i^d(t) - x_i^d(t)\right) \end{aligned} \tag{12}$$

$$\text{pbest}_i(t+1) = x_i(t+1) \text{ if } F(x_i(t+1)) < F\left(\text{pbest}_i(t)\right) \text{ otherwise pbest}_i(t) \tag{13}$$

$$\begin{aligned} \text{gbest}(t+1) &= \text{pbest}_i(t+1) \\ &\text{if } F\left(\text{pbest}_i(t+1)\right) < F(\text{gbest}(t)) \text{ otherwise gbest}(t) \end{aligned} \tag{14}$$

where c_1, c_2, and c_3 are chosen constant and r_1, r_2, and r_3 are randomly generated numbers between (0, 1). The position of the particle is updated similarly (15) (16).

$$S\left(v_i^d(t+1)\right) = \frac{1}{1+\exp\left(-v_i^d(t+1)\right)} \tag{15}$$

$$x_i^d(t+1) = 1 \text{ if rand} < S\left(v_i^d(t+1)\right) \text{ else } 0 \tag{16}$$

Pseudocode of CSI-PSO

- Initialize a population of particles, X_i ($i = 1, 2, \ldots, N$)
- Evaluate the fitness of particles for each species, F(Sn) using the fitness function
- Calculate masses of individuals
- for $t = 1$ to the maximum number of iteration, $T_{\max}$
  - for $n = 1$ to the number of individuals

    Calculate center of mass R_p and R_o
    Calculate cumulative center of mass
    Find individual closest to cumulative center of mass and set it as cbest individual
    Update the velocity of a particle
    Convert the velocity into a probability value
    Update the position of a particle

  - Evaluate the fitness of particle by applying the fitness function
  - Update pbest and gbest
- End if criteria are met.

4 Results

In binary representation, used 0 as the absence of that feature and 1 as the presence. For the purpose of this research, classification tasks are performed using KNN on five different datasets. Each dataset is divided into train and test with a 30% split ratio, and the fitness value is calculated on the test set using the following formula (17)

$$\text{Fitness} = \alpha\frac{|S|}{|T|} + (1-\alpha)\text{Error},$$
$$\text{where Error} = \frac{\text{Number of wrongly classified instances}}{\text{Total number of instances}} \tag{17}$$

Table 1 Dataset information

Dataset name	Number of features	Dataset ID
Cardiac single proton emission computed tomograph	44	D1
Ionosphere	34	D2
Libras movement	91	D3
Musk	168	D4
Spam mail	109	D5

where $|S|$ is the length of feature subset; $|T|$ is the total number of features in each dataset. Error is the classification error rate, and α is the parameter; for the purpose of this research value of 0.01 is chosen for α (Table 1).

4.1 Parameter Setting

50 independent runs are performed for each algorithm for 100 iterations with 10 individuals each. The best fitness scores are recorded at each iteration. Table 2 contains all the parameter values for the three algorithms, respectively.

Table 2 Parameter settings

Parameters	BPSO	CSI-BPSO	GA
Population size	10	10	10
Number of iterations	100	100	100
Number of runs	50	50	50
w	0.9	0.9	–
c_1	2	2	–
c_2	2	2	–
c_3	–	2	–
V_{max}	6	6	–
V_{min}	−6	−6	–
$Constant_p$	–	0.2	–
R_p	–	root((Total features in dataset)/2)	–
R_o		root((Total features in a dataset)/4)	–
Mutation rate	–	–	0.08

4.2 Experiment Results

CSI-BPSO outperforms GA, BPSO in mean and min best fitness across 50 runs, as shown in Table 3. The low standard deviation of the best fitness is indicative of the robustness and consistency of the results across the independent runs. The performance of the algorithms across every iteration can be seen in Figs. 2, 3, 4, 5 and 6.

Table 3 Experiment results

Dataset ID	Feature selection method	Minimum best fitness on the 100th iteration across 50 runs	Maximum best fitness on the 100th iteration across 50 runs	Mean best fitness on the 100th iteration across 50 runs	STD for the 100th iterations across 50 runs	Mean accuracy across 50 runs (%)
D1	Proposed Solution	**0.00182**	**0.12693**	**0.08113**	**0.02889**	**92.08**
	Genetic algorithm	0.16886	0.29375	0.21994	0.03612	78.25
	Particle swarm	0.0442	0.16864	0.11361	0.03018	88.833
D2	Proposed solution	**0.02044**	**0.05023**	**0.03969**	**0.00744**	**96.18**
	Genetic algorithm	0.06979	0.11855	0.09743	0.01255	90.603
	Particle swarm	0.02074	0.05986	0.04557	0.00759	95.660
D3	Proposed solution	**0.21494**	**0.26**	**0.23754**	**0.00994**	**76.38**
	Genetic algorithm	0.26122	0.31656	0.29309	0.01284	70.870
	Particle swarm	**0.21494**	0.26189	0.24263	0.01016	75.925
D4	Proposed solution	**0.03285**	**0.03958**	**0.03612**	**0.00179**	**96.80**
	Genetic algorithm	0.04321	0.05976	0.05284	0.00359	95.168
	Particle swarm	0.03203	0.04357	0.03835	0.0025	96.612
D5	Proposed solution	**0.0327**	**0.04019**	**0.03568**	**0.00157**	**96.86**
	Genetic algorithm	0.04186	0.05905	0.05138	0.00432	95.333
	Particle swarm	0.03335	0.03945	0.03655	0.00158	96.813

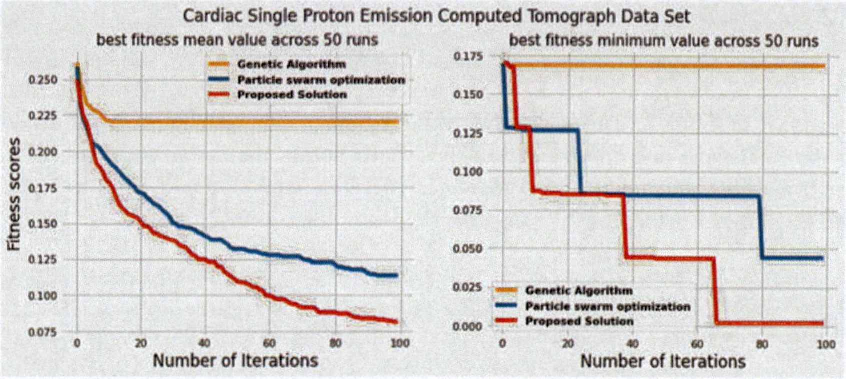

Fig. 2 Cardiac single proton emission computed tomograph dataset results

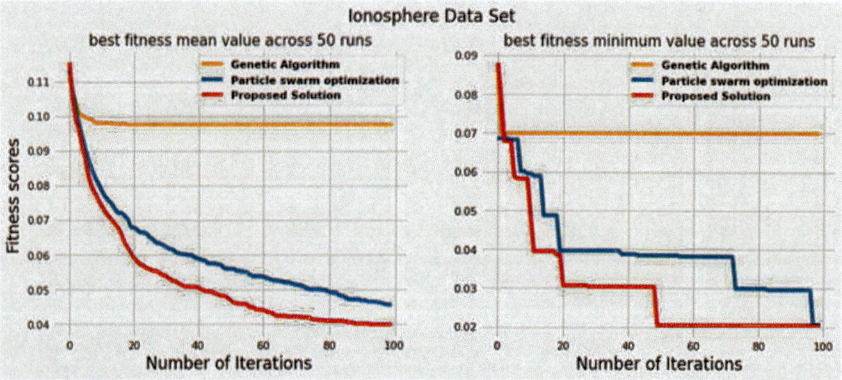

Fig. 3 Ionosphere dataset results

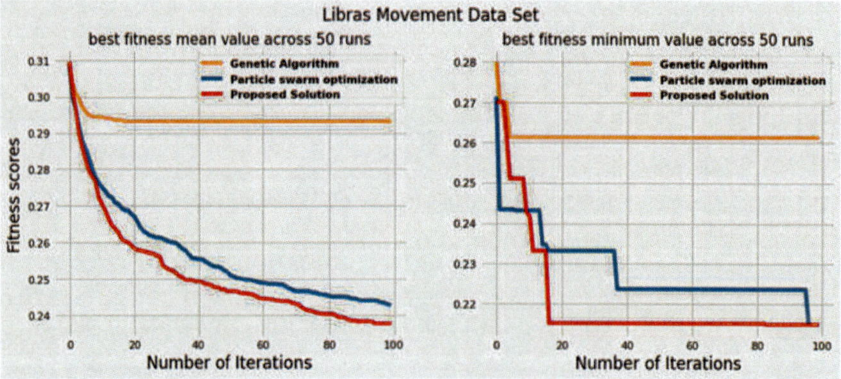

Fig. 4 Libras dataset results

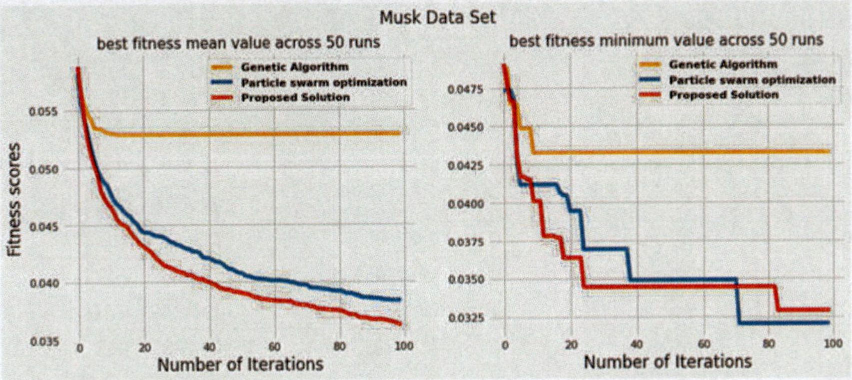

Fig. 5 Musk dataset results

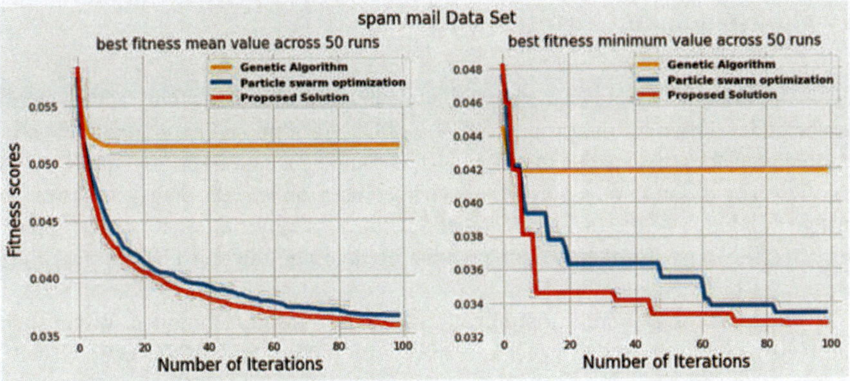

Fig. 6 Spam mail dataset results

A two sample t-test with a 95% confidence level was applied to examine whether the lower loss achieved by the proposed CSI-BPSO was significantly lesser than the other methods. The null hypothesis of the t-test assumes that the loss of the two methods is similar. If the p-value is less than 0.05, the null hypothesis is rejected, and there is a significant difference in loss of two methods. The t-test is performed between CSI-BPSO, GA and CSI-BPSO, BPSO, taking the values of CSI-BPSO as base values. Table 4 contains the results of the performed t-test. The null hypothesis is rejected for each of the datasets, suggesting that better performance CSI-BPSO is statistically significant.

Table 4 *T*-test results

Dataset ID		BPSO	GA
D1	*p*-value	3.9015×10^{-7}	3.5914×10^{-37}
	Statistic value	5.44200998691518	21.004637669290
D2	*p*-value	0.00019521	1.2357×10^{-42}
	Statistic value	3.8714764287812597	27.69498240533
D3	*p*-value	0.0138088069993	2.36304×10^{-41}
	Statistic value	2.50734981345334	23.9469856425
D4	*p*-value	$1.98082462775 \times 10^{-6}$	1.411494×10^{-41}
	Statistic value	5.090843937437118	29.1825954571
D5	*p*-value	0.007274	6.23285×10^{-33}
	Statistic value	2.74128800037458	23.938

5 Conclusion and Future Work

The work done here is an implementation of the proposed algorithm. The program has been developed for the proposed CSI-BPSO algorithm and explained the mathematical nuances that lead us to believe that it could be a better option than BPSO or GA. Further, all the three algorithms have been used on various datasets to provide further validation of the fact that CSI-BPSO can be a better alternative to its counterparts. The experiment was done on existing datasets; the metrics on a real-time system are to be evaluated later.

There has been a surge in optimizers with co-evolution techniques. While such implementations are scarce in feature selection domain. CSI-BPSO coupled with co-evolution strategies can prove to be a better alternative.

References

1. Sivaganesan D (2019) Design and development ai-enabled edge computing for intelligent-iot applications. J Trends Comput Sci Smart Technol (TCSST) 1(02):84–94
2. Zeadally S, Adi E, Baig Z, Khan IA (2020) Harnessing artificial intelligence capabilities to improve cybersecurity. IEEE Access 8:23817–23837. https://doi.org/10.1109/access.2020.2968045
3. Clerc M, Kennedy J (2002) The particle swarm, explosion, stability, and convergence in a multi-dimensional complex space. IEEE Trans Evol Comput 6:58–73
4. Eberhart RC, Kennedy J (1995) A new optimizer using particle swarm theory. In: Proceedings of the 6th international symposium on micro machine and human science, Nagoya, Japan. IEEE Service Center, Piscataway, NJ, pp 39–43
5. Angeline P (1998) Evolutionary optimization versus particle swarm optimization: philosophy and performance differences. In: Proceedings of evolutionary programming VII. Springer, Berlin, pp 601–610

6. Al-kazemi B, Mohan CK (2002) Multi-phase discrete particle swarm optimization. In: Proceedings of the 4th international workshop on frontiers in evolutionary algorithms, Kinsale, Ireland, Jan 2002
7. Afshinmanesh F, Marandi A, Rahimi-Kian A (2005) A novel binary particle swarm optimization method using artificial immune system. In: Proceedings of the international conference on computer as a tool (EUROCON 2005), Belgrade, Serbia, Nov 21–24, pp 217–220
8. Banka H, Dara S (2015) A hamming distance based binary particle swarm optimization (HDBPSO) algorithm for high dimensional feature selection, classification and validation. Pattern Recogn Lett 52:94–100
9. Poli R, Kennedy J, Blackwell T (2007) Particle swarm optimization. Swarm Intell 1:33–57
10. Bäuerle T, Löffler RC, Bechinger C (2020) Formation of stable and responsive collective states in suspensions of active colloids. Nat Commun 11:2547. https://doi.org/10.1038/s41467-020-16161-4
11. Rashedi E, Nezamabadi-Pour H, Saryazdi S (2009) GSA: a gravitational search algorithm. Inf Sci 179(13):2232–2248
12. Mendes R (2004) Population topologies and their influence in particle swarm performance. Doctoral thesis, Escola de Engenharia, Universidade do Minho, Portugal

Chapter 63
Enhanced Dragonfly Algorithm Adapted for Wireless Sensor Network Lifetime Optimization

Miodrag Zivkovic, Tamara Zivkovic, K Venkatachalam, and Nebojsa Bacanin

1 Introduction

One subtype of ad hoc and wireless networks are wireless sensor networks (WSNs). The WSNs are formed out of a large amount of small sensing devices-sensors. Every node is typically a very small and cheap device, with scarce resources, limited processing power, and energy supply. Communication between nodes is performed in a wireless manner. These networks are usually used to monitor specific physical phenomena in a given environment, which is often dangerous and hazardous. When deployed in remote areas, WSNs must operate without human supervision, therefore they have to be independent and capable to self-organize. Any sort of maintenance, configuration, or battery replacement is typically not feasible. The applications of WSNs can be found in many domains, from military and industry to volcano and habitat monitoring, medicine, etc.

Due to limitations in processing power, data storage, memory, and energy constraints, individual sensors can process just a small portion of the total collected data. Sensors are powered by batteries that have limited charge, and very rarely have the

M. Zivkovic · N. Bacanin
Singidunum University, Danijelova 32, 11000 Belgrade, Serbia
e-mail: mzivkovic@singidunum.ac.rs

N. Bacanin
e-mail: nbacanin@singidunum.ac.rs

T. Zivkovic
School of Electrical Engineering, University of Belgrade,
Bulevar Kralja Aleksandra 73, 11000 Belgrade, Serbia
e-mail: zt125040p@student.etf.bg.ac.rs

K. Venkatachalam (✉)
School of Computer Science and Engineering, VIT Bhopal University, Bhopal, India
e-mail: venkatachalam.k@vitbhopal.ac.in

I. Jeena Jacob et al. (eds.), *Data Intelligence and Cognitive Informatics*,
Algorithms for Intelligent Systems,
https://doi.org/10.1007/978-981-15-8530-2_63

option to collect additional energy by themselves (i.e. heat, solar energy, etc.). Since manual replacement of the batteries is not feasible, every sensor must preserve the energy, typically by shutting down components when they are not needed.

Even if the nodes follow a strict energy-saving plan, nodes at the key positions in WSN can be depleted fast if the routing hasn't been carefully selected and optimized. Communication in WSN is a costly operation—energy needed to transfer one bit of information is directly proportional to the square of the distance within destination and the source nodes. If nodes have been depleted fast, this could render the complete network useless. Therefore, to keep the network alive and avoid fast node depletion, it is required to keep the inter-node communication at a minimum level, by selecting and applying proper routing algorithm which is efficient in terms of minimizing the total amount of transmissions.

The nodes are deployed in a random fashion in dense layout—i.e. dropped by a plane flying over the rain forest or war zone. The main goal is that the nodes are as close as possible to the target phenomena, so they can provide valuable measurements and good coverage. After deployment, sensor nodes must be able to operate completely on their own, without any additional resources and with limited energy at their disposal, with the unknown positions of the individual nodes, and to communicate wireless, while trying to keep the network lifetime as long as possible [1].

If the nodes on the crucial routing positions die fast, the communication across the complete network could become impossible. Numerous approaches can be applied to keep the sensors from depletion and prolong the overall network lifetime. One traditional technique that is still popular today is the clustering and hierarchical routing protocol. The algorithm starts by disengaging network into clusters, and every cluster has one sink node, named the cluster head (CH), and member nodes which are assigned to the CH. The CH collects data gathered by other sensors in its cluster and is responsible to transmit the aggregated data to the base station (BS). The data transmission to the BS is considered to be very energy-consuming operation, and has a direct influence on the lifetime of the WSN, as the CHs deplete the energy more rapidly than other sensors in the topology. If the CHs die too fast, it could cut out the complete cluster from the WSN. This can be avoided by optimal selecting and rotating CHs, which will prolong the overall lifespan of the network. The process of forming the clusters and selecting the CH for each cluster is based on the energy level of the nodes. Energy is most important factor in the formation of the clusters and also in the CH election process.

The research presented in this paper has tried to optimize the process of electing the CHs by utilizing swarm intelligence metaheuristics. The proposed work has applied the basic dragonfly algorithm (DA), as well as enhanced DA (EDA), that is proposed in this research, to prolong the overall lifetime of the WSN. It should be noted that the DA has never been utilized to solve this kind of problem before.

Paper is structured in the following way: Section 2 gives swarm algorithms overview and their applications to the WSN domain, Section 3 provides the mathematical model of the clustering problem in WSN, which was used in our simulations, Section 4 present inner workings of the original and enhanced DA algorithms. Experimental results are given in Sect. 5. Final remarks and conclusion is given in the Sect. 6.

2 Swarm Algorithms Applications for WSN Domain Review

The clustering problem in WSN can be categorized in the group of NP-hard problems. According to computational complexity theory, these challenges can't be solved in polynomial time by utilizing traditional deterministic algorithms. However, by applying stochastic algorithms for these types of problems, such as swarm intelligence metaheuristics, satisfying solutions within reasonable computational time could be obtained [2–4].

Swarm intelligence metaheuristics are inspired by the processes found in nature. Through the processes of evolution and natural selection over the billions of years, nature has proven to be extremely successful in solving very difficult problems. All species must ensure the survival of the fittest individuals and the propagation of beneficial traits to the new generation. The survival typically depends on the maximization of the fundamental processes such as feeding and mating. According to the recent literature overview, there are dozens of algorithms that have been inspired by these basic evolutionary processes. These approaches have already proven to be very powerful in solving numerous real-life hard tasks in the past decade.

Metaheuristics which have been inspired by the processes found in nature can be separated into two distinctive groups, evolutionary algorithms (EA) and swarm intelligence algorithms, respectively. The most famous and successful EA approach is the genetic algorithm (GA). The GA has been applied in solving various NP-hard problems in different application domains with very promising results, including image processing optimization [5], machine learning [6], cloud computing [7], internet of things and big data [8] problems, and it was also applied in different challenges that belong to the WSN domain, such as maximum coverage deployment [9]. The second group of algorithms, swarm intelligence, have been inspired by cooperative behavior within the groups of bees, ants, moths, fireflies, dragonflies, bats, fish, wolves, longhorn beetles, etc. and the social behavior they can express when in large groups. These otherwise primitive individuals can show highly coordinated and intelligent behavior when they are found in swarms, and these features were used as an inspiration for numerous swarm intelligence algorithms.

Some of the most prominent swarm algorithms and their applications include: artificial bee colony (ABC) [10, 11], cuckoo search algorithm (CS) [12], monarch butterfly optimization (MBO) [13], grey wolf optimization (GWO) [14], firefly algorithm (FA) [15], elephant herding optimization (EHO) [16–18], moth search algorithm (MS) [19, 20], fireworks algorithm (FWA) [21, 22], tree growth algorithm (TGA) [23], whale optimization algorithm (WOA) [24, 25], bat algorithm (BA) [26] and artificial flora (AF) algorithm [27].

The clustering problem in WSN has been addressed by several swarm intelligence approaches, most of them focusing on the process of CH selection, with some encouraging results [28]. In [29] a PSO-based scheme was proposed to address the hot spot problem in a cluster-based WSN. The clustering process was divided into two phases, with the application of the routing and clustering algorithms which were proven to be energy efficient. During the clustering phase, to prevent the fast depleting of CHs, a smaller amount of member nodes were assigned to their corresponding clusters. A similar approach has been also proposed in [30]. Another PSO based algorithm was shown in [31], which improved the WSN network lifetime by helping in the process of forming clusters and selecting CH. Another promising swarm intelligence approach for this task was presented in [32]. The cluster nodes in the context of WSNs were observed as the flocks of birds, while the algorithm utilized received signal strength indicator (RSSI) with encouraging results.

Besides network lifetime optimization, many other WSN NP-hard challenges were also tackled by using swarm intelligence approaches, for example in solving WSN localization problem, swarm algorithms obtained significant results [2, 17, 33, 34].

3 Clustering Problem Model Formulation

The proposed simulations have utilized the same experimental setup as proposed in [31]. Sensors nodes were randomly deployed in 100×100 m area. Each node in the WSN topology is involved in two processes: data gathering and transmission of the collected data. Every node in the cluster collects data and sends it to the corresponding CH directly (if the CH is within its transmission range) or through some other node that belongs to the same cluster. CH has the role to process all data that have been received from the member nodes within its cluster and to forward the collected data towards BS. The clustering algorithm is used in each round to elect the CHs. After CH has been elected, it will perform data collection, aggregation, and communication with the BS in front of its cluster. The efficiency of the algorithm used for CH selection process depends on the fitness function which was utilized. In [31], new fitness function was proposed for this role, which considers two most critical aspects of the clustering process: first, the energy needed for the communication between cluster members and their corresponding CH, and second, the energy that CH requires for data aggregation and transfer to the BS. In the first process, the energy consumption required for communication between cluster members and their CH can be modeled by the Eqs. (1)–(2).

$$E_1(j) = \sum_{k=1}^{K} \sum_{n_{k_j} \in C_k} \{\frac{E_{k_j} - f(k_j, C_k)}{E_{\max} - E_{\min}}\} \times k_j,\ j = 1, 2, ..., n \tag{1}$$

$$s(C_k, k_j) = \min s(k_j, C_k),\ \forall k = 1, 2, ..., K \tag{2}$$

Second process, the amount of energy required for communication between CH and the BS is given as:

$$E_2(j) = \sum_{k=1}^{K} \{\frac{E_{C_k} - g(C_k, BS)}{E_{\max} - E_{\min}}\} \times C_k \tag{3}$$

Finally, the total amount of energy which is necessary for sending M bits of data from the individual sensor to the BS can be modeled by the following equation:

$$F(j) = E_1(j) + \mu E_2(j) \tag{4}$$

4 Original and Enhanced Dragonfly Algorithm

Dragonfly algorithm (DA) was originally introduced in year 2015 by Mirjalili [35]. This approach was inspired by the behavior exhibited by the dragonflies, a large family of almost 3000 different species that live in all parts of the world. All dragonflies share the same behavior which can be expressed through two main phases in their life cycles, named nymph phase and adult phase respectively.

In the nymph phase, dragonflies live underwater, before morphing into the adult individual. Even in this aquatic larval stage, dragonflies are capable and successful predators. They hunt other water insects and even in some cases smaller fish. The dragonfly's adult phase is much more interesting because of the unique swarm behavior they express. In the adult phase, dragonflies are considered to be formidable predators concerning their size. Dragonflies hunt practically all other small insects such as midges, mosquitoes, butterflies, flies, moths, and even other, smaller dragonflies. Dragonflies form swarms for two main reasons: hunting the prey (also known as static, feeding swarm) and migration (dynamic, migratory swarm).

In a static swarm, dragonflies will patrol over relatively small areas in smaller groups and hunt their prey—other flying insects. Static swarm is characterized by local movement and fast, harsh changes in the flying path during the hunt. However, the dynamic swarm of dragonflies expresses very different behavior. In this case, a large number of dragonflies will form the migration swarm, which is characterized by flying long distances in one direction.

The DA algorithm was inspired by both of these two types of swarm behavior since they can be used for modeling two crucial phases of the optimization process-exploration (diversification) and exploitation (intensification). In the context of the dragonfly swarm behavior, the exploration phase corresponds to the static feeding swarm, where dragonflies join into the smaller sub-swarms and fly over multiple smaller distances. The exploitation phase, in this case, corresponds to the dynamic swarm, where dragonflies join into the large swarms and fly in one direction

The general behavior all swarms tend to follow is based on the three fundamental principles—separation, alignment, and cohesion. Each of the fundamental swarm principles can be modeled mathematically, which is given further in this section.

The first fundamental swarm principle is separation, which states that the individual dragonflies try to avoid collision with the other individuals in the static neighborhood. The separation is mathematically modeled with Eq. 5.

$$S_i = -\sum_{j=1}^{N} X - X_j, \tag{5}$$

where X marks location of the given dragonfly, X_j marks location of the j-th dragonfly in the static neighborhood, while N represents the total number of dragonflies in the neighborhood.

The alignment states that the dragonfly will adjust its flight speed to the speed of other surrounding dragonflies in the neighborhood, and this behavior can be modeled as:

$$A_i = \frac{\sum_{j=1}^{N} V_j}{N}, \tag{6}$$

where V_j labels the flight speed of the dragonfly j in the neighborhood.

The cohesion states that the dragonflies will tend to move towards the center of the mass of the neighborhood. This rule can be modeled mathematically as shown in Eq. 7:

$$C_i = \frac{\sum_{j=1}^{N} X_j}{N} - X \tag{7}$$

Furthermore, it is possible to extract two additional behavioral patterns from the main objective of the swarm of dragonflies—survival. All dragonflies are attracted to the source of food (prey), and try to fly away from the enemies. Attraction to the prey can be modeled by the following expression:

$$F_i = X^+ - X, \tag{8}$$

where X^+ represents the coordinates of prey solution.

Equation that models avoid enemy individuals by the dragonflies is given as:

$$E_i = X^- + X, \tag{9}$$

where X^- labels the coordinates of an adversary.

Taking into account these two additional behavioral patterns, there are five main factors in total which affect the process of updating the positions of dragonflies in the swarm. The overall behavior of all individual dragonflies is the combination of these five patterns.

When updating positions of dragonflies and simulating their movement, two vectors are considered: step and position vectors. Step vector (ΔX) is used to indicate the flight direction of dragonflies (in this case defined in one dimension) and it is modeled by the Eq. 10.

$$\Delta X_{t+1} = (sS_i + aA_i + cC_i + fF_i + eE_i) + w\Delta X_i, \quad (10)$$

where s denotes the weight of separation factor, S_i defines the separation of the i-th dragonfly, a and A_i define the weight of the alignment factor and alignment of the i-th dragonfly, respectively. Parameter c denotes the cohesion weight factor, C_i represents i-th dragonfly cohesion, f defines food factor feight and F_i represents the source of food for the i-th dragonfly. Parameter e is used to establish weight of the enemy factor, and finally, E_i defines the location of i-th individual's enemy. Additionally, inertia weight is denoted as w, while the number of current iteration is marked with t.

The position vector is updated according to the step vector as given in the following expression:

$$X_{t+1} = X_t + \Delta X_{t+1} \quad (11)$$

Modification of the separation, alignment, cohesion, food, and enemy weight factors (given as s, a, c, f, and e in Eq. 10), can achieve different explorative and exploitative behaviors during the optimization process.

The dragonflies move within the search space performing Lévy flight if there is no neighbor, and their positions are updated using following equation:

$$X_{t+1} = X_t + Levy(d) \times X_t, \quad (12)$$

where t denotes current iteration, while d represents the dimension of positions vector. Furthermore, the Lévy flight is determined by using the Eq. 13:

$$Levy(x) = 0.01 \times \frac{r_1 \times \sigma}{|r_2|^{\frac{1}{\beta}}}, \quad (13)$$

where r_1 and r_2 are denote pseudo-random numbers in the interval between 0 and 1, $\beta = 1.5$, and σ is calculated according to the Eq. 14.

$$\sigma = \left(\frac{\Gamma(1+\beta) \times sin(\frac{\pi\beta}{2})}{\Gamma(\frac{1+\beta}{2}) \times \beta \times 2^{\frac{\beta-1}{2}}} \right)^{\frac{1}{\beta}} \quad (14)$$

where Γ is given as $\Gamma(x) = (x-1)!$.

4.1 Proposed Enhanced Dragonfly Algorithm

As in every other swarm algorithm, convergence speed and solutions' quality is largely influenced by the established adjustment between the processes of exploration and exploitation. In some cases, when this balance is shifted more towards exploitation, it may happen in some runs that the search process converges and stagnates in some of the sub-optimal parts of the search space. Similarly, when this balance is shifted in favor of exploration, optimal parts of the space are usually determined, however, the algorithm lacks the power to perform a fine search in these regions, and as a consequence results with worse quality are generated.

This research work has been conducting extensive simulations with the original DA on standard benchmark instances and concluded that the basic version of this metaheuristics properly adjusts the intensification-diversification trade-off. However, it has been concluded that some phases of the search process could be improved. In later iterations, when an algorithm has already located the auspicious part of the search space, by conducting fine-grained exploitation around the current best solutions could lead to better results' quality.

To accomplish this, the proposed work has adopted a uniform crossover operator from the GA, similarly as it was performed in [11]. In later iterations, the lowest fitness individual from the population is substituted by utilizing the uniform crossover operator with random and actual best solution from the population by using following expression:

$$X_{new,j} = \begin{cases} X_{best,j} & \text{if } \phi \geq 0.5 \\ X_{rnd,j} & \text{otherwise,} \end{cases} \tag{15}$$

where $X_{new,j}$ denotes j-th parameter of newly generated solution that is used as a replacement for the worst solution, while $X_{rnd,j}$ and $X_{best,j}$ represent parameter j of the random and best solutions from the population, respectively. Parameter ϕ, that calculated for every argument j represents a pseudo-random number between 0 and 1 and it is used to control the uniform crossover operator.

However, applying uniform crossover operator in early stages of the algorithm's execution may lead to sub-optimal results, hence initial population may be generated far from the optimal domain of the search space. To avoid such a scenario, an additional control parameter—breakpoint (bp) is introduced. The uniform crossover operator mechanism is applied only in case when the following relation is satisfied: $t \geq bp$. In this way, for example if solutions in $t \geq bp$ iterations are far for optimal region, uniform crossover operator will not be triggered and the search will not converge to local optimums.

Our proposed algorithm starts the process of optimization by generating a set of pseudo-random solutions, which are created in the same manner as described in [11]. In each iteration, the positions and steps for each individual from the population are updated by applying Eqs. (10), (11) and (12). In later phases of execution, the worst solution is replaced by hybrids between best and randomly selected individuals in each iteration.

The proposed approach named as enhanced dragonfly algorithm (EDA)—EDA pseudo-code is given in Algorithm 1 (termination criteria is denoted as $MaxIter$). It is noted that the presented pseudo-code is applicable for single-objective optimization problems since the challenge that is tackled in this paper belong to this group.

Algorithm 1 Enhanced dragonfly algorithm

```
Generate initial population of dragonflies X_i, (i = 1, 2, 3, ..., N)
Step vectors initialization ΔX_i, (i = 1, 2, 3, ..., N)
while t ≤ MaxIter do
  Determining objective function values of all individuals (dragonflies)
  Updating food source and enemy locations
  Recalculating weights, s, a, c, f, e and w
  Determine S, A, C, F, and E by utilizing expressions (5)–(9)
  Neighboring radius updating
  if the current dragonfly has at least one neighboring dragonfly then
    Velocity (step) vector update by using Eq.(10)
    Location vector recalculation by employing Eq. (11)
  else
    Recalculate location (position) vector using Eq. (12)
  end if
  Correct newly generated positions to fit within the boundaries of the search space if necessary
  Evaluate fitness for all individuals in the population
  if t ≥ bp then
    Replace worst solution by applying uniform crossover (Eq. 15)
  end if
  t++
end while
```

5 Simulations and Discussion

The experiments conducted within this research have utilized the same simulation setup as in [31], since there is already a goal to provide a precise and objective comparative analysis between our proposed EDA and algorithms presented in this paper. In [31], a famous particle swarm optimization (PSO) algorithm was implemented for this challenge. The simulated area where nodes were placed was a two-dimensional ($2D$) area with a size $100\ U \times 100\ U$. Sensors represented with coordinates (x, y) are randomly deployed over the observed simulation area by using a pseudo-random number generator. The proposed experiments have assumed that the BS is located within the simulated area borders. The clustering model which was used throughout all performed experiments is described in Sect. 3.

Table 1 Summary of model parameters

Parameter	Value
Area	100×100
E_{DA}	5 nJ/bit
Sensors number	100
E_{mp}	0.00013 pJ/bit
Node energy E_0	0.5 J
Msg. length	4000 bits
E_{TX}	50 nJ/bit
P_o	0.1
E_{RX}	50 nJ/bit
E_{fs}	10 pJ/bit

Original DA, EDA, as well as the WSN framework are developed by utilizing JDK (Java Development Kit) 11 and all simulations were performed on Intel® CoreTM i7-8700K CPU with 64GB RAM and Windows 10 O.S.

An overview of the parameters of the used model has been presented in Table 1 [31].

This research work has also implemented the original DA version for the clustering problem, with a goal to give precise performance metrics of the proposed EDA aproach. As stated in Sect. 1, basic DA for this problem was not found in the literature survey. Adaptations of the DA and EDA were named LEACH-DA and LEACH-EDA, respectively.

In all performed simulations, the same basic parameters' values are used for the LEACH-DA and LEACH-EDA: the number of solutions in the population (SN) was set to 20, and $MaxIter$ was set to 4000. The same parameters' values were also used in the LEACH-PSO approach [31]. Other basic DA control parameters' were adjusted as in [35]. By performing empirical simulation, it is concluded that the EDA obtains the best results if the uniform crossover operator is applied in the last 30% iterations, and in compliance with this, the value of *bp* parameter is set to 3200 ($MaxIter \cdot 0.7$).

The proposed LEACH-EDA algorithm was evaluated based on the two important network key performance indicators: network lifetime and the total number of packets that were sent to the BS. Network lifetime was observed through the number of alive and dead nodes. The performance of the proposed LEACH-EDA was compared against basic LEACH, LEACH-PSO, and the original version of the DA algorithm (LEACH-DA). In every conducted experiment, the proposed model has obtained the average results from 30 independent runs. To allow easier comparing of the results, the same number of iterations and runs as in simulations with the basic LEACH and LEACH-PSO is used as presented in [31].

The obtained results of the network lifetime indicator are shown in Table 2. When the basic LEACH algorithm is observed, it can be seen that data collection and

Table 2 Network life time in terms of dead nodes (expressed as the number of iterations after nodes die)

Algorithm	First dead	Half dead	All dead
LEACH	1100	1265	1570
LEACH-PSO	1150	1340	3880
LEACH-DA	1120	1290	2850
LEACH-EDA	1170	1390	3930

transmission will cause that the first sensor node dies after 1100 rounds. After 1265 iterations half nodes are dead, and after 1,570 iterations, all nodes die. When the LEACH-PSO algorithm is observed, the first node will die after 1150 rounds, half of the sensors die after 1340 iterations, and after 3880 iterations, all nodes are dead. When the original DA (LEACH-DA) is observed, the first node dies after 1120 rounds, half of the sensors are dead after 1290 iterations and after 2850 rounds, all sensors die. In the case of the proposed EDA algorithm (LEACH-EDA), it is clearly observed that the first node dies after 1170 rounds, half of the sensors die by the 1390 rounds, and after 3930 iterations, all sensors are dead.

From the obtained results it is observed that the LEACH-DA achieves slightly better performances compared to the basic LEACH algorithm. On the other hand, basic LEACH-DA performs worse than the LEACH-PSO. The proposed LEACH-EDA approach achieves better performances than both basic LEACH and LEACH-PSO algorithms. More precisely, the LEACH-EDA outperforms the LEACH-PSO by 20 rounds (almost 2 %) when the proposed work compares the moment when the first node dies, by 50 rounds (around 3.5 %) when the resarch work compares the moment when half of the nodes are dead, and by 50 rounds (slightly over 1%) when all nodes are dead. Therefore, it is concluded from the analyzed results that by utilizing the proposed LEACH-EDA approach it is possible to increase the lifetime of the WSN, and at the same time to reduce the overall power consumption of the sensor nodes. Moreover, from presented metrics it can be clearly noted that the devised EDA outstandingly boosts the performance of the original DA.

Comparative analysis of the dead nodes parameter in one algorithm run between the LEACH-DA and LEACH-EDA approaches is depicted in Fig. 1 (breakpoints—first dead, half dead, and all dead are denoted with markers).

Comparative analysis based on the second observed parameter, which is the total number of packets sent inside the network, is shown in the Table 3. LEACH, LEACH-PSO, LEACH-DA and LEACH-EDA were compared after 500, 1500, 2500, 3500 and 4000 rounds. Obtained results show that the total number of packets sent by utilizing the original LEACH algorithm is 13,000, while LEACH-PSO and LEACH-DA can achieve the numbers of 20,000 and 16,100 packets, respectively. Our proposed LEACH-EDA approach is capable of sending a total number of 21,000 packets. It can be seen that there is a significant improvement in the total number of packages sent by applying the LEACH-EDA compared to the LEACH and LEACH-DA algorithms

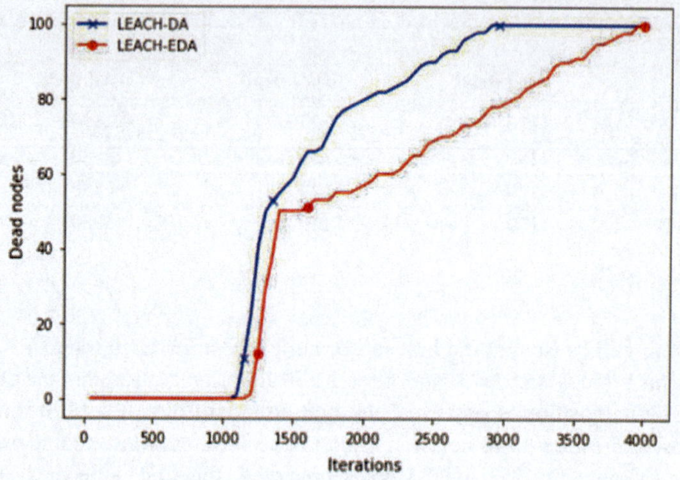

Fig. 1 LEACH-EDA versus LEACH-DA in terms of dead nodes comparative analysis

Table 3 Packets sent to the BS comparative analysis

Algorithm	**Iterations**				
	500	**1500**	**2500**	**3500**	**4000**
LEACH	5000	11,000	13,000	13,000	13,000
LEACH-PSO	7000	13,500	17,600	18,600	20,000
LEACH-DA	5600	12,100	14,500	15,300	16,100
LEACH-EDA	7500	14,000	18,300	19,600	21,000

(25–60 percent), while our simulations have shown that LEACH-EDA performed slightly better than the LEACH-PSO. More precisely, LEACH-EDA outperformed the LEACH-PSO by 5 percent, and compared to the LEACH-DA, proposed LEACH-EDA achieves significantly better results.

Visual representation of results provided in Table 3 is given in Figure 2.

6 Conclusion

The presented research work has applied the proposed enhanced DA (EDA) swarm intelligence metaheuristics for solving the clustering problem in WSNs. The experiments are thoroughly performed in the simulated environment with the sensor nodes randomly distributed over the observed area in each run. Moreover, the results are compared against the original DA, basic LEACH, and LEACH-PSO algorithms.

The scientific implication of this research is twofold: first, it has been managed to improve the original DA metaheuristics, and secondly, our proposed approach has

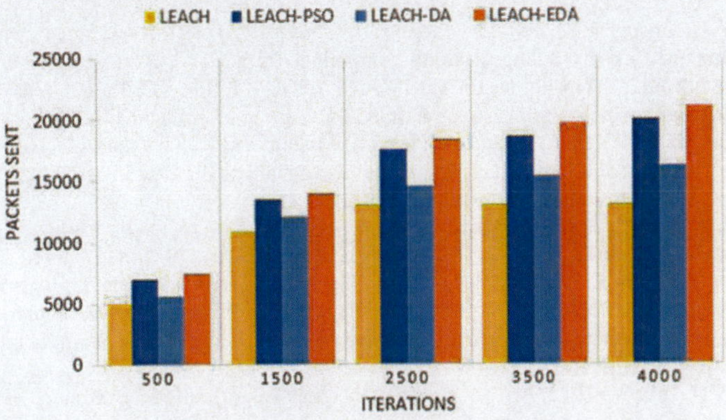

Fig. 2 LEACH-EDA versus LEACH-DA in terms of dead nodes comparative analysis

proven to be a more efficient clustering algorithm that other approaches used for solving this WSNs challenge.

Obtained results indicate that our proposed EDA achieved better performance than some other outstanding metaheuristics. Moreover, the EDA managed to significantly outperform basic DA in all conducted tests. As an overall conclusion, it can be stated that the proposed EDA has a lot of potential in tackling the clustering problem in WSNs. In the future, it has been planned to adapt the EDA for solving other WSNs NP-hard challenges as well.

Acknowledgements The paper is assisted by the Ministry of Education, Science and Technological Development of Republic of Serbia, Grant No. III-44006.

References

1. Zivkovic M, Branovic B, Marković D, Popović R (2012) Energy efficient security architecture for wireless sensor networks. In: 2012 20th Telecommunications Forum (TELFOR), pp 1524–1527
2. Goyal S, Patterh MS (2014) Wireless sensor network localization based on cuckoo search algorithm. Wireless Personal Commun 79:223–234
3. Strumberger I, Bacanin N, Tuba M (2017) Enhanced firefly algorithm for constrained numerical optimization, IEEE congress on evolutionary computation. In : Proceedings of the IEEE International Congress on Evolutionary Computation (CEC 2017), pp 2120–2127
4. Tuba E, Tuba M, Dolicanin E (2017) Adjusted fireworks algorithm applied to retinal image registration. Stud Inf Control 26:33–42
5. Magliani F, Sani L, Cagnoni S, Prati A (2019) Genetic algorithms for the optimization of diffusion parameters in content-based image retrieval. In: Proceedings of the 13th International Conference on Distributed Smart Cameras, ICDSC 2019. ACM, New York, NY, USA, pp 14:1–14:6
6. Goldberg DE (1989) Genetic algorithms in search, optimization and machine learning. Addison-Wesley Longman Publishing

7. Zhan Z-H, Zhang G-Y, Ying-Lin, Gong Y-J, Zhang J (2014) Load balance aware genetic algorithm for task scheduling in cloud computing. In: Dick G, Browne WN, Whigham P, Zhang M, Bui LT, Ishibuchi H, Jin Y, Li X, Shi Y, Singh P, Tan KC, Tang K (eds) Simulated evolution and learning. Springer International Publishing, Cham, pp 644–655
8. Elhoseny M, Abdelaziz A, Salama AS, Riad A, Muhammad K, Sangaiah AK (2018) A hybrid model of internet of things and cloud computing to manage big data in health services applications. Future Generation Comput Syst 86:1383–1394
9. Yourim Y, Yong-Hyuk K (2013) An efficient genetic algorithm for maximum coverage deployment in wireless sensor networks. IEEE Trans Cybern 43(5):1473–1483
10. Tuba M, Bacanin N, Beko M (2015) Multiobjective rfid network planning by artificial bee colony algorithm with genetic operators. In: Tan Y, Shi Y, Buarque F, Gelbukh A, Das S, Engelbrecht A (eds) Advances in swarm and computational intelligence. Springer International Publishing Cham, pp 247–254
11. Bacanin N, Tuba M (2012) Artificial bee colony (ABC) algorithm for constrained optimization improved with genetic operators. Stud Inf Control 21:137–146
12. Bacanin N (2010) Implementation and performance of an object-oriented software system for cuckoo search algorithm. Int J Math Comput Simulation 6:185–193
13. Strumberger I, Tuba M, Bacanin N, Tuba E (2019) Cloudlet scheduling by hybridized monarch butterfly optimization algorithm. J Sensor Actuator Netw 8(3):44
14. Bacanin N, Bezdan T, Tuba E, Strumberger I, Tuba M, Zivkovic M (2019) Task scheduling in cloud computing environment by grey wolf optimizer. In: 2019 27th Telecommunications Forum (TELFOR), pp 1–4
15. Bacanin N, Tuba M (2014) Firefly algorithm for cardinality constrained mean-variance portfolio optimization problem with entropy diversity constraint. In: The Scientific World Journal, special issue Computational Intelligence and Metaheuristic Algorithms with Applications, vol 2014, no Article ID 721521, p 16
16. Wang G-G, Deb S, dos S Coelho L (2015) Elephant herding optimization. In: Proceedings of the 2015 3rd International Symposium on Computational and Business Intelligence (ISCBI), pp 1–5
17. Strumberger I, Minovic M, Tuba M, Bacanin N (2019) Performance of elephant herding optimization and tree growth algorithm adapted for node localization in wireless sensor networks. Sensors 19(11):2515
18. Strumberger I, Bacanin N, Beko M, Tomic S, Tuba M (2017) Static drone placement by elephant herding optimization algorithm. In: Proceedings of the 24th Telecommunications Forum (TELFOR)
19. Wang G-G (2016) Moth search algorithm: a bio-inspired metaheuristic algorithm for global optimization problems. Memetic Computing
20. Strumberger I, Sarac M, Markovic D, Bacanin N (2018) Moth search algorithm for drone placement problem. Int J Comput 3
21. Tuba E, Strumberger I, Bacanin N, Tuba M (2018) Bare bones fireworks algorithm for capacitated p-median problem. In: Tan Y, Shi Y, Tang Q (eds) Advances in swarm intelligence. Springer International Publishing, Cham, pp 283–291
22. Bacanin N, Tuba M (2015) Fireworks algorithm applied to constrained portfolio optimization problem. In: Proceedings of the 2015 IEEE Congress on Evolutionary Computation (CEC 2015)
23. Strumberger I, Tuba E, Zivkovic M, Bacanin N, Beko M, Tuba M (2019) Dynamic search tree growth algorithm for global optimization. In: Camarinha-Matos LM, Almeida R, Oliveira J (eds) Technological innovation for industry and service systems. Springer International Publishing, Cham, pp 143–153
24. Mirjalili S, Lewis A (2016) The whale optimization algorithm. Adv Eng Softw 95:51–67
25. Strumberger I, Bacanin N, Tuba M, Tuba E (2019) Resource scheduling in cloud computing based on a hybridized whale optimization algorithm. Appl Sci 9(22):4893
26. Tuba M, Bacanin N (2015) Hybridized bat algorithm for multi-objective radio frequency identification (rfid) network planning. In: Proceedings of the 2015 IEEE Congress on Evolutionary Computation (CEC 2015)

27. Bacanin N, Tuba E, Bezdan T, Strumberger I, Tuba M (2019) Artificial flora optimization algorithm for task scheduling in cloud computing environment. In: Yin H, Camacho D, Tino P, Tallón-Ballesteros AJ, Menezes R, Allmendinger R (eds) Intelligent Data Engineering and Automated Learning – IDEAL 2019. Springer International Publishing, Cham, pp 437–445
28. Sambo DW, Yenke BO, Förster A, Dayang P (2019) Optimized clustering algorithms for large wireless sensor networks: a review. Sensors 19(2)
29. Azharuddin M, Jana PK (2016) Particle swarm optimization for maximizing lifetime of wireless sensor networks. Comput Electr Eng 51:26–42
30. Wang J, Gao Y, Liu W, Sangaiah AK, Kim H-J (2019) An improved routing schema with special clustering using pso algorithm for heterogeneous wireless sensor network. Sensors 19(3)
31. Yadav A, Kumar S, Vijendra S (2018) Network life time analysis of wsns using particle swarm optimization. Procedia computer science. International conference on computational intelligence and data science, vol 132, pp 805 – 815
32. Jung S-G, Yeom S, Shon M, Kim D, Choo H (2015) Clustering wireless sensor networks based on bird flocking behavior, pp 128–137, 06 2015
33. Strumberger I, Beko M, Tuba M, Minovic M, Bacanin N (2018) Elephant herding optimization algorithm for wireless sensor network localization problem. In: Camarinha-Matos LM, Adu-Kankam KO, Julashokri M (eds) Technological innovation for resilient systems. Springer International Publishing, Cham, pp 175–184
34. Strumberger I, Tuba E, Bacanin N, Beko M, Tuba M (2018) Monarch butterfly optimization algorithm for localization in wireless sensor networks. In : 2018 28th International Conference Radioelektronika (RADIOELEKTRONIKA), pp 1–6
35. Mirjalili S (2015) Dragonfly algorithm: a new meta-heuristic optimization technique for solving single-objective, discrete, and multi-objective problems. In: Neural computing and applications

Chapter 64
Detection of Pneumonia Clouds From Chest X-ray Images

L. Suganthi, K. Nirmala, S. Deepa, K. Nagalakshmi, and M. Santhya

1 Introduction

Among developing countries like India, pneumonia is the leading cause of death. It is an infection of the lungs that may be caused by bacteria, viruses, or fungi. Due to the number of processes with identical signs and symptoms, the clinician assessing the patient with a confirmed or suspected diagnosis of pulmonary infection faces a diagnostic challenge, and the radiographic results of pneumonia do not include a clear aetiological diagnosis [1]. In addition, radiographic manifestations of a given infectious process can differ depending on the patient's immunological status and the pre- or co-existing lung disease.

Several lung diseases are diagnosed by analyzing lung tissue patterns in pulmonary images, and thus, segmentation and analysis are one of Computer Assisted Diagnosis (CAD)'s most significant sections [2, 3]. The CAD image was a remarkable and revolutionary step in detecting lung abnormalities early. It involves automated identification of lung nodules, 3D reconstruction of lung structures, enhancement of the area of interest, clarification and separation. It helps to improve the accuracy and confidence of any lung anomaly detection device in decision-making. CAD involves steps like segmentation, separation, thresholding, neural networking for morphological identification, and feature extraction.

Chest X-ray (CXR) is an effective method for pneumonia diagnosis, and many clinical decisions rely heavily on radiological results from it. This is also fairly inexpensive compared to other imaging treatments, so millions can afford this. Pneumonia is often diagnosed with chest X-rays, and, specifically, by computer-aided pneumonia diagnosis, doctors can diagnose the disease more reliably and

---

L. Suganthi (✉) · K. Nirmala · S. Deepa · K. Nagalakshmi · M. Santhya
Department of Biomedical Engineering, Sri Sivasubramaniya Nagar College of Engineering, Chennai, India
e-mail: suganthi1@gmail.com

I. Jeena Jacob et al. (eds.), *Data Intelligence and Cognitive Informatics*, Algorithms for Intelligent Systems,
https://doi.org/10.1007/978-981-15-8530-2_64

effectively [4–8]. The challenging task for lung segmentation in CXR is to trace the lung boundary accurately. Extraction of the lung area is difficult due to the presence of structures such as ribs, clavicles, mediastinum, and blood vessels in the lungs. Several algorithms were developed to remove the effect of these structures and to only extract lung region to test their abnormality [8–16].

This paper aims at the development of an automated lung segmentation and pneumonia (lung inflammation that appears as a white patch of cloud in the X-ray) detection method to improve the imaging and enhancement of the various pathological conditions of the lungs. Segmentation of the lung region from the chest X-ray accurately using the watershed algorithm and graph cut method. Feature extraction using GLCM feature extraction and selecting the dominant features by feature reduction using ANOVA test and fed to the classifier for classification of X-ray image into an image with pathology and without pathology. Validation of the classifier performance by comparing the original images with ground truth images.

1.1 Methodology

The CXR images are taken from the Kaggle database that includes hundred control images and two hundred diseased images. The proposed methodology dealt, first, with pre-processing images to suppress and remove noise and, at the same time, to enhance image features. Morphological operations have been carried out on the pre-processed image to identify the region of interest. Then, image segmentation was performed using the watershed algorithm and graph cut method; lastly, feature is extracted, and significant features are identified and lastly fed to classifier for identifying the lung with pathology (Fig. 1).

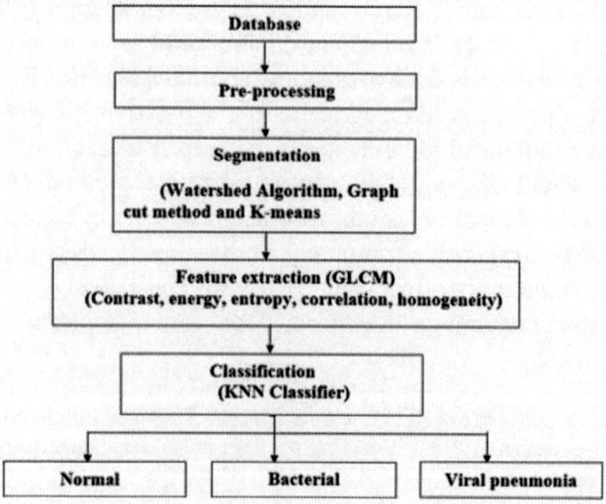

Fig. 1 Workflow of proposed methodology

1.2 Pre-processing

Pre-processing is performed to improve the image data, which suppresses unwanted distortions and also enhances some important image features for further processing. To increase accuracy, a pre-processing sequence was performed before segmentation. Input image contrast is modified first, with a different threshold value. The foreground and background are separated by the values obtained from the histogram, as black and white. Then, morphological operations are applied to the resultant image to remove the unwanted bone parts and other small regions present in the image. By clearing the border, the unwanted region present in the surroundings of the lung including clavicle bone is removed to enhance the region of interest. After the image has been dilated, to smooth the boundary of the lung region is done by filling the gaps and holes with the disk structuring element of radius 3. The small, unnecessary region of less than 300 pixels present in the image is removed by using area open. The image is finally inverted for the segmentation process.

1.3 Segmentation

Watershed segmentation, graph cut method, and K-means clustering have been implemented to segment the region of interest. Each method had limitations and inconsistencies that can be seen in the following results, and the above methods were validated using the Jaccard similarity coefficient. The distance transform is applied to the pre-processed image to find the local minima in the watershed transform. Then, the watershed transform is applied, starting the process of filling from to the local minima's and marking the boundary at the point where the two local minima region intersect. The boundary marked X-ray image will then be overlaid with the original input image. The overlay function will mark the segmented lung region of the original image. In the graph cut method, the rectangle region is defined by assuming the inner region as the foreground and the outer region as the background. The unknown region is separated by the use of K-means clustering. In K-means clustering, a set of K Gaussian is set to the data. And Gaussian distribution parameters, such as mean and variance, are estimated for each cluster and cluster weight. After learning the parameters for each data point, the probability of belonging to each cluster is calculated.

1.4 Feature Extraction and Feature Reduction

The relevant features are extracted from the segmented image to form feature vectors. These feature vectors are then used by classifiers to recognize the input unit with a target output unit. The GLCM features are extracted from the segmented lung image.

The calculation process of GLCM is similar to the histogram calculation. The pixel pair values are selected from the image, and the count value at a corresponding location in the GLCM matrix is incremented by one. This process is repeated for all pixel pairs. Usually, the entire image is divided into many sub-blocks of size 8 × 8, 16 × 16, or 32 × 32. The separate GLCM matrix is needed to be calculated for every sub-blocks of the image. The contrast, correlation, energy, homogeneity, entropy are the GLCM features extracted using (1–4).

$$\text{Correlation} = \sum_{i=0}^{N-1}\sum_{j=0}^{M-1}(G(i,j)\,(i * j) - (\mu x * \mu y) \tag{1}$$

$$\text{Energy} = \sum_{i=0}^{N-1}\sum_{j=0}^{M-1}(G(i,j))^2 \tag{2}$$

$$\text{Homogeneity} = \sum_{i=0}^{N-1}\sum_{j=0}^{M-1}G(i,j)/1 + |i - j| \tag{3}$$

$$\text{Entropy} = \sum_{i=0}^{N-1}\sum_{j=0}^{M-1} -G_{ij*}\log G_{ij} \tag{4}$$

where $G(i, j)$ is the normalized gray level co-occurrence matrix; μ is the mean.

From the segmented lung, the GLCM features such as energy, entropy, correlation, homogeneity, contrast are extracted, and the features of the lung images with and without pathology are tested by the ANOVA test, which can be used to find *p*-values (*p*-value less than 0.05) that mean the characteristics of the respective lungs are different, and (*p*-value greater than 0.05) that mean that the characteristics of the respective lungs are similar, and that features can be eliminated for further classification.

1.5 Classification

The extracted features are fed to the K-nearest neighbor (K-NN) and multiclass SVM classifier for identifying the lungs with and without pathology [17, 18]. Approximately, 22 features are extracted using GLCM, and the dominant features were fed to the K-NN Classifier and SVM classifier. The K-value is initialized, and the distance between the test sample and the training samples are calculated. The more repeated minimum distance class shall be assigned to the test sample.

2 Results and Discussion

The chest X-ray images are enhanced and pre-processed to get the region of interest. The contrast of input image is first adjusted, with specific threshold value the foreground and the background are separated as black and white by the values obtained from the histogram [19, 20]. Then, morphological operations are applied to the resultant image to remove the unwanted bone parts and other small regions present in the image. Figure. 2 shows the output of the pre-processing steps applied on sample control and diseased image (a) Input image (b) Contrast adjustment (c) Thresholding (d) Clear border (e) Dilation (f) Area open. Techniques such as watershed segmentation and graph cut method and K-means clustering were implemented. Each method had restrictions and inconsistencies which can be seen in the following results, and the above methods were validated using the Jaccard similarity coefficient. The segmented output from both methods is shown in Fig. 3.

The segmented results obtained from both the methods are validated using the Jaccard similarity coefficient using Eq. (5). The result of both the methods is compared with the ground truth images (10 images), and the Jaccard similarity coefficient index obtained for the segmentation methods are shown in Table 1. The Jaccard similarity coefficient value can be interpreted, where the higher value derived means the closer the image mask comparable to the ground truth image. Conversely, the lower value means the lesser the image mask and the ground truth image. From Table 1, it can be inferred that all scores are relatively high above 0.7 out of (for exact

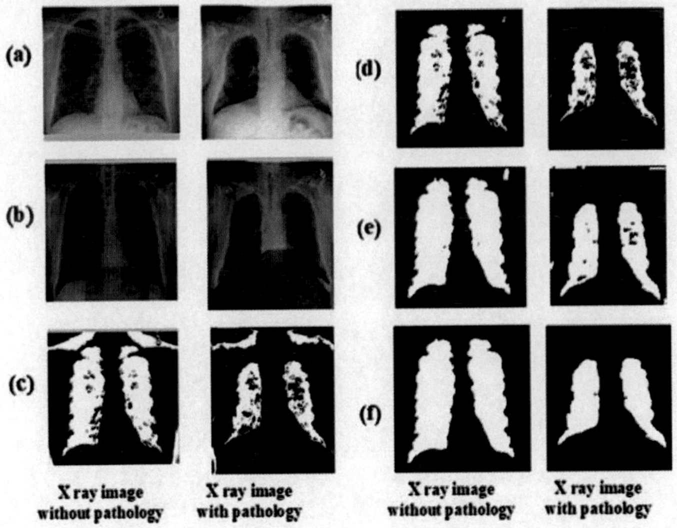

Fig. 2 Output of the pre-processing steps applied on X-ray image without and with pathology. **a** Input image, **b** Contrast adjustment, **c** Thresholding, **d** Clear border, **e** Dilation, **f** Area open

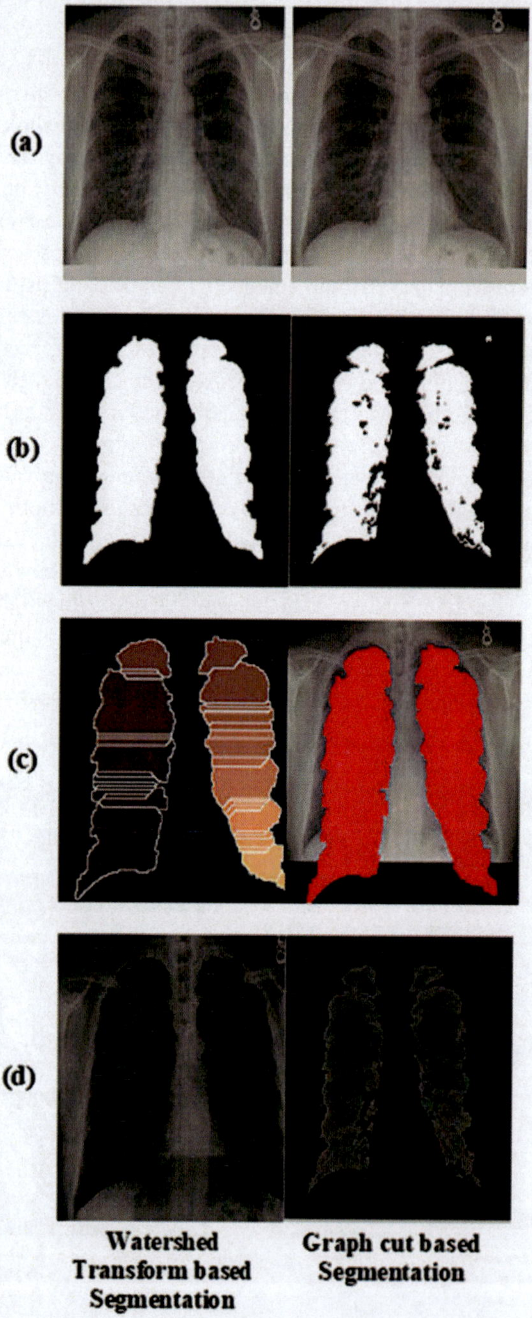

Fig. 3 Output of the segmentation methods based on watershed transform, graph cut applied on a image. **a** Input image, **b** pre-processing, **c** identification of ROI, **d** segmented image

Table 1 Validation of the segmentation process

Image	Jaccard similarity coefficient value	
	Watershed transform segmentation	Graph cut method segmentation
1	0.799	0.645
2	0.685	0.634
3	0.699	0.592
4	0.825	0.742
5	0.689	0.654
6	0.732	0.562
7	0.726	0.623
8	0.734	0.702
9	0.692	0.639
10	0712	0.635
Average	0.731	0.643

similarity). Similarly, the result of method 1 is higher than method 2. Therefore, the watershed algorithm is concluded as good for the extraction of features.

$$J(a,b) = \frac{|a \cup b|}{|a \cap b|} \tag{5}$$

where a, b represent the image before and after segmentation, respectively.

Using the *p*-value of the ANOVA calculation, the correlation feature appears not to be dominant and is therefore neglected. Remaining features such as energy, contrast, homogeneity, entropy are found to be dominant and to be used for further progress (Tables. 2, 3, and 4).

The precision, sensitivity, specificity, negative predictive (NPV) value, and the accuracy of the K-NN and SVM classifier are computed and tabulated in the Tables 5 and 6, respectively. The precision, sensitivity, specificity, negative predictive value, and the accuracy of the classifier are calculated by using the True Positive (TP), True

Table 2 ANOVA (*p*-value) comparison of pathological image features

Features	Bacteria infected lung images	Viral infected lung images
Energy	0.00072	0.000511
Contrast	0.009209	0.003759
Correlation	0.092758	0.161683
Homogeneity	0.016143	0.018548
Entropy	0.00135	0.01564

Table 3 Confusion matrix of the SVM classifier

Total no. of images = 300	Non-pathological	Bacteria	Viral
Non-pathological	80	14	11
Bacteria	15	75	6
Viral	10	8	81

Table 4 Confusion matrix of K-NN classifier

Total no of images = 300	Non-pathological	Bacteria	Viral
Non-pathological	87	8	5
Bacteria	6	84	10
Viral	4	7	89

Table 5 Validation of K-NN classifier

Total no. of images = 300	Non-pathological	Pathological condition	
Non-pathological	TP = 87	FP = 13	Precision = 87%
Pathological condition	FN-10	TN = 190	NPV = 95%
	Sensitivity = 89%	Specificity = 93%	Accuracy = 92%

Table 6 Validation of SVM classifier

Total no. of images = 300	Non-pathological	Pathological condition	
Non-pathological	TP = 80	FP = 27	Precision = 87%
Pathological condition	FN-25	TN = 170	NPV = 95%
	Sensitivity = 76%	Specificity = 86%	Accuracy = 82%

Negative (TN), False Positive (FP), and False Negative (FN) values using (6–10).

$$\text{Sensitivity} = (\text{TP}/(\text{TP} + \text{FN})) * 100 \tag{6}$$

$$\text{Specificity} = (\text{TN}/(\text{TN} + \text{FP})) * 100 \tag{7}$$

$$\text{Accuracy} = ((\text{TP} + \text{TN})/(\text{TP} + \text{TN} + \text{FP} + \text{FN})) * 100 \tag{8}$$

$$\text{Precision} = (\text{TP}/(\text{TP} + \text{FP})) * 100 \tag{9}$$

$$\text{NPV} = (\text{TN}/(\text{TN} + \text{FN})) * 100 \tag{10}$$

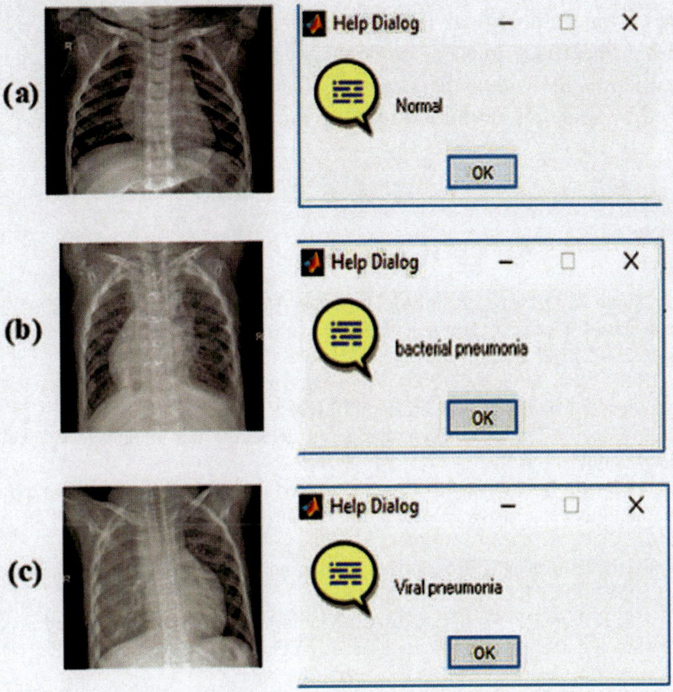

Fig. 4 GUI display of the proposed algorithm for the classification of the three categories of an input image. **a** Control image, **b** Bacterial pneumonia image, **c** Viral pneumonia image

GUI was developed to facilitate the diagnosis of the purpose. When the input image is fed to the system, the proposed algorithm automatically detects the type of infection as shown in Fig. 4.

3 Conclusion

In this paper, the CXR image is pre-processed for better segmentation result of the lung. Two segmentation methods such as watershed algorithm and graph cut method are used, and their results were validated by comparing the Jaccard similarity coefficients between segmented image and ground truth image. Segmented lung from the watershed method is concluded for the extraction of a feature as it gives a good result. Approximately, 22 features are extracted from the lung region using the GLCM feature extraction method. From the extracted features, the reduction feature is applied to obtain dominant features of images without and with pathological condition (Bacteria and Viral). The Analysis of Variance (ANOVA) test is used to reduce the characteristics. The dominant features are then fed to two classifiers, such as K-NN and SVM, for classification. The systems were trained with 300 test and train images (Non-pathological-100, Bacterial-100, Viral-100). The K-NN classifier results with an accuracy of 92%, sensitivity of 89%, and specificity of 93%. The SVM classifier results with an accuracy of 80%. Further, the classifier accuracy can be increased by selecting dominant features and considering more data.

References

1. Awai K, Murao K, Ozawa A, Komi M, Hayakawa H, Hori S, Nishimura Y (2004) Pulmonary nodules at chest CT: effect of computer-aided diagnosis on radiologists' detection performance. Radiology 230(2):347–352
2. Armato III SG, Sensakovic WF (2004) Automated lung segmentation for thoracic CT: impact on computer-aided diagnosis1. Acad Radiol 11(9):1011–1021
3. Marten K, Engelke C (2007) Computer-aided detection and automated CT volumetry of pulmonary nodules. Eur Radiol 17(4):888–901
4. Qin C, Yao D, Shi Y et al (2018) Computer-aided detection in chest radiography based on artificial intelligence: a survey. BioMed Eng OnLine 17:113. https://doi.org/10.1186/s12938-018-0544-y
5. van Ginneken B, Hogeweg L, Prokop M (2009) Computer-aided diagnosis in chest radiography: beyond nodules. Eur J Radiol 72(2):226–230
6. Zakirov AN, Kuleev RF, Timoshenko AS, Vladimirov AV (2015) Advanced approaches to computer-aided detection of thoracic diseases on chest X-rays. Appl Math Sci. 9(88):4361–4369
7. Suzuki K (ed) (2012) Machine learning in computer-aided diagnosis: medical imaging intelligence and analysis. IGI Global
8. Taghizadieh A, Ala A, Rahmani F, Nadi A (2015) Diagnostic accuracy of chest X-ray and ultrasonography in detection of community acquired pneumonia; a Brief Report. Emergency 3(3):114
9. Ticinesi A, Lauretani F, Nouvenne A, Mori G, Chiussi G, Maggio M, Meschi T (2016) Lung ultrasound and chest X-ray for detecting pneumonia in an acute geriatric ward. Medicine 95 (27)
10. Bhuvaneswari P, Therese AB (2016) Detection of cancer in lung with K-NN classification using genetic algorithm. Proc Mater Sci 10:433–440
11. Candemir S, Jaeger S, Palaniappan K, Musco JP, Singh RK, Xue Z, Karargyris A, Antani S, Thoma G, McDonald CJ (2014) Lung segmentation in chest radiographs using anatomical atlases with non-rigid registration. IEEE Trans Med Imaging 33(2):577–590
12. Hua P (2010) Segmentation of lung tissue in CT images with disease and pathology.
13. Kasu NR, Saravanan C (2018) Segmentation on chest radiographs using otsu's and K-means clustering methods. In 2018 International conference on inventive research in computing applications (ICIRCA), pp 210–213. IEEE
14. Mansoor A, Bagci U, Foster B, Ziyue Xu, Papadakis GZ, Folio LR, Udupa JK, Mollura DJ (2015) Segmentation and image analysis of abnormal lungs at CT: current approaches, challenges, and future trends. RadioGraphics 35(4):1056–1076
15. Sharma N, Aggarwal LM (2010) Automated medical image segmentation techniques. J Med Phys/Assoc Med Phys India 35(1):3
16. Candemir S, Antani S (2019) A review on lung boundary detection in chest X-rays. Int J Comput Assist Radiol Surg 14(4):563–576
17. Delimayanti MK (2017) Feature extraction and classification of thorax X-ray image in the assessment of osteoporosis. In 2017 4th International conference on electrical engineering, computer science and informatics (EECSI), pp 1–5. IEEE

18. Qin C, Yao D, Shi Y, Song Z (2018) Computer-aided detection in chest radiography based on artificial intelligence: a survey. Biomed Eng Online 17(1):113. https://doi.org/10.1186/s12938-018-0544-y
19. Bindhu V (2019) Bıomedıcal image analysis using semantic segmentation. J Innovative Image Processing (JIIP) 1(02):91–101
20. Chandy A (2019) A review on iot based medical imaging technology for healthcare applications. J Innovative Image Process (JIIP) 1(01):51–60

Chapter 65
Effectiveness of Blockchain Advancement in Patient Statistical Monitoring Network

M. Malathi, R. S. Krupasree, T. Lalitha Bhuvaneshwari, and S. Gokulraj

1 Introduction

The real, incontrovertible archive of p2p network communications from the blocks which are linked to their transactions and noted in a digital ledger called a blockchain [1]. It is an intelligent peer-to-peer network that uses distributed databases to categorize, distribute, and archive information that is also known as Internet value. In 2009, Nakamoto created the Bitcoin social network and established the "Creation block" [2, 3]. Precise use of blockchain technology would lead to the security and sharing of intellectual property [4]; this technology takes all areas of infotech and communication systems; this practice far takes grown quickly in recent years [5]. Blockchain is classified upon how the data is treated, its nature and the events that the user will take. These include public, private, permission-free consortium. Through this technology, the public can freely access all the details. The blockchain-style consortium allows only the chosen node party to engage in the consensus process distributed [6]. The blockchain goes far beyond storing cryptocurrency transactions like Bitcoin. This breakthrough has driven the enormous value growth of cryptocurrencies and significant investment in blockchain start-ups undertaking capital. The market for blockchain technology is expected to increase by 2021 [7]. Surveys were released about the application of blockchain in various fields [8]. A blockchain may have any information that a cryptographic hash can encrypt. Blockchain technology is regarded as an untrustworthy distributed ledger for recording, storing, sharing, analyzing, and verifying the exchange of medical data between various users like a patient, doctors, etc., [9]. Blockchain has been used to facilitate medical data processing, information sharing, and managing medical history [10] as a support infrastructure for HIE [11]. Blockchain technology will be applied to the health care industry and HIE projects,

M. Malathi (✉) · R. S. Krupasree · T. Lalitha Bhuvaneshwari · S. Gokulraj
Sri Krishna College of Technology, Coimbatore, India
e-mail: m.malathi@skct.edu.in

I. Jeena Jacob et al. (eds.), *Data Intelligence and Cognitive Informatics*,
Algorithms for Intelligent Systems,
https://doi.org/10.1007/978-981-15-8530-2_65

according to previous research. On referencing the research paper [12], Ethereum is the best framework for creating distributed apps using smart contracts and also the best security framework.

2 Blockchain Framework for Health Care

Blockchain applications may be public or private. Everyone can take part in a public rollout, and everyone has the network at their fingertips. All participants are listed in a personal blockchain. Bitcoin is the eminent instance of a public distributed blockchains, a universe leading sector in an extant. Specific blockchains typically chore in a constrained system to additional deprive the prevailing system [13]. Specific applications requiring a large memory space and transfer that the blockchain paradigm can provide for patient medical records and care delivery effectively [14]. According to a blockchain database to track the arrhythmic prominences, stacked denoising autoencoders remained mounted [15]. Stacked denoising autoencoders are a deep network used to store huge amount of information for longer timescales [16]. It is mostly used in systems which are being used continuously for large memory purpose. It has a great act in application to classify images since it is not used in health care application. Progressive blockchain [17] sets a new standard for digitally distributed forms of patient and medical data management. When the request has arrived the execution, phase takes place or "sendTransaction" when an end user makes a call. Though it is fast process, it operates on a read-only pattern. It won't be sent to the network, so you don't need to verify for network interaction. Transactions are made using "sendTrasaction," which is carried out by supervised miners and can have a drastic impact on the chain [17].

3 Ledgers

Distributed ledgers synced to many sites, and it will become the witness for any kind of transactions. Every node in the whole network contains the duplicate of the earlier recordings. So, it can't be hacked easily. For example, the network contains ten nodes, and each node contain the copy of the earlier recording and distributed widely all along with the network; if anyone wants to change the data of one node, it will check the copy of the earlier recording in each node in the network; if the result is positive, he can change the data else he can't. For these security purpose, distributed ledgers are used in blockchain. The focus on database management and indelible records by blockchain has made it a great tool for tracking the distributed supply chain. This would likely focus on falsified drugs because blockchain ledgers would track production deliveries through each phase until they are shipped for end customer processing, clinical transportation, inter-hospital pharmacy, private practice. Ledgers held in the blockchain enable ownership and responsibility to be monitored during transit, limiting liability to protect doctors and pharmacists who prescribe drugs to patients. Bitcoin is typically mistaken for blockchain. Blockchain technology can also be implemented by Bitcoin. The Bitcoin protocol created by Bitcoins is the Bitcoin program's Ledger entries. Nevertheless, the distributive ledgers for cryptocurrencies appear to be minimal. All hierarchical ledger systems, where valuable data are ordered, would be removed. It made the newly opened, distributed ledgers possible. Decentralized distributive ledgers provide more economical and efficient on comparing to centralized ledgers [18] exponentially diminishing value curves operate decentralized distributive ledgers. If any of the nodes want to access, the access signal must be given by all other connected nodes in the system network this makes the framework decentralization [19].

4 Digital Contracts

Smart contracts are the block of code which acquire the characteristics of the fundamentals of blockchain, which contain an unchangeable data history, and the facility to decrease the malfunction. Smart contracts organize their own procedures according to the need of users, mitigate third party conflicts, and reduce operating costs. It records the code, self-evaluates it, and executes it in the blockchain network. In a public blockchain, users are charged to initialize or implement the smart contracts in it. By being public, the functionalities are minimum, and the language used is solidity which is widely used in the distributed applications [20]. Public smart contracts are more suitable for common users. The major demerit of a public smart contract is privacy this can be overcome by permission smart contract. Permissioned smart contracts are mostly used in business corporations [21]. Smart contracts have been used in Ethereum framework. Ethereum developers make the people understand the need for blockchain by introducing the smart contracts in blockchain to the public. When invoked, these smart contracts must be executed by the distributed nodes, which free the top management to concentrate more on big decisions. The key reason for this is the security disadvantage of the existing blockchain technologies which cannot meet their requirements. It creates issues around an organization's security and application protection. Purposeful deceptive activities may occur, for example, in local parts hidden from public scrutiny. So that the blockchain application must be an app that can be the most common blockchain game hosting P2P blockchain in its entirety, and this can be fulfilled by the use of smart contract [22].

5 Existing Security System

Blockchain has been chosen in many fields by its security system. In the past few years, blockchain has been working in so many platforms like v1.o-bitcoin, v2.o-Ethernet, and its currently using the v3.o-EOS platform which supports the development of distributed applications. There are various distributed cloud storages like sia, storej, IPFS, etc. IPFS is an existing distributed file system which replaces the hypertext transfer protocol. IPFS is a content-based address whereas HTTP is a domain-based address. By replacing HTTP to IPFS makes the blockchain more secure and faster. For cryptography MD5, SHA1 and RSA algorithms had been used earlier. MD5 and SHA1 hash functions used in cryptography [23]. MD5 gives the output of 128-bit code. This output can be processed from several stages. At first, it creates a consecutive hexadecimal number, and at each next stage, it interprets the value of the current block and those retrieved from the previous block. MD5 are mostly used for digital signatures [24]. SHA1 which is the series of SHA0 produces an output of 160-bit hash code which consists of 40-hexadecimal value. But both MD5 and SHA1 have high collision rate which is two separate networks with the same probability of solving the same hash are high. RSA involves both encrypted and digital signature algorithm [25]. RSA runs by using both the public key and also the private key. RSA needs to be encrypted the large database at the same time so it becomes slow. To check the security of public keys, it needs a reduce cost [26]. It has now been updated by SHA256 algorithm which is extensively used in blockchain. SHA which means secure hashing function, 256 counts the length of the fixed bit. In SHA256, bitcoin creates a 65-hexadecimal encrypted value. SHA256 helps to inverse the hash order so, it is impossible to decode the original content, and it has the best collision resistant [27]. By analyzing these existing features, to implement blockchain in the health care field is the best solution.

6 Proposed System

Patient-generated data could be received from health apps, wearables, implants, home medical devices, activity monitoring, peripherals, mHealth solutions, etc. At the same time, clinical data and health records could be received from hospital medical devices, EHR/EMR, ambulatory med-devices, connected devices, etc. These received data are then encrypted, digitally signed, and moved toward data lakes and blockchain refer in Fig. 1. A database that holds large quantities of structured, quasi-structured, and non-structured data called a data lake. Here, it stores different types of health data which is structured and non-structured, and the message is encoded and applying a one-way hash code to that document done by cryptographic technique. The cryptographic technique provides the hash code for each block in the network. These hash codes can't be updated if it is created once. The modification of data in a blockchain is not possible without the permission of authenticated user because each block contains

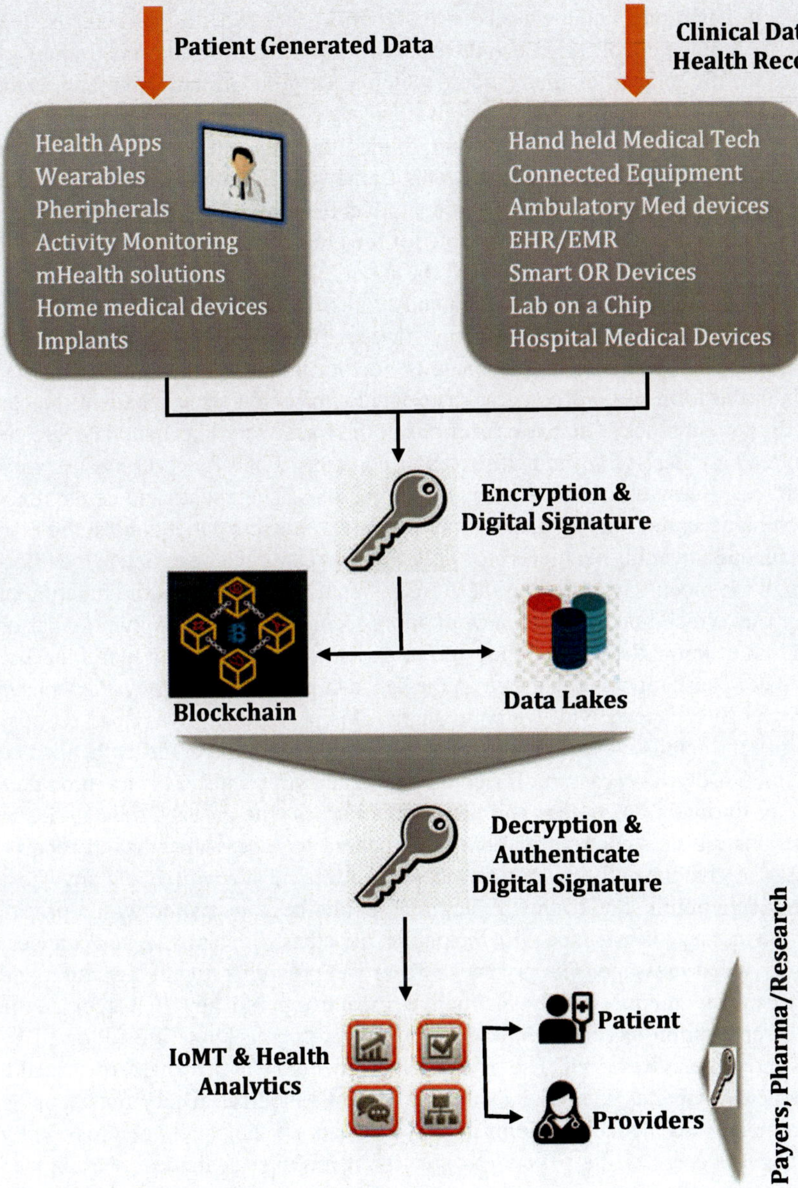

Fig. 1 Blockchain in health care—flow diagram

the data, its hash code, and the hash code of its prior block. So, the data is distributed among the whole network, and also the framework is decentralized. This framework contains the digital credit called Bitcoin which is a cryptocurrency. Ethereum is the framework since it contains its own cryptocurrency called ethers as same as Bitcoin. Ethereum uses pow consensus algorithm. In the Ethereum framework, there will be a pay for each transaction called gas. This framework is the most suitable for the creation of smart contracts which specified its own functions and make itself as invulnerable. This framework makes probable for the creation of distributed apps which has smart contracts, in the GUI format by the use of Ethereum virtual machine. Blocks are arranged in the form of information and address by the use of interplanetary file system. To achieve any kind of transaction, there are certain rules. There are multiple kinds of users in the framework like patients, doctors, nurses, administrates. The user level is in the form of GUI contains distributed apps; if the user is patient, he can select the patient block. The basic functions of the framework like an update, create, delete, read are deal with the help of smart contracts. Each function has assigned grant access. Every user can't access all the functions. The update and delete function can be accessed only by users like doctors and nurses; patients have the read access function over his health records only. Patients can't update or delete the information in the block. By assigning these roles, one user can't access the function of another user which makes the framework more secure. This technology provides us with a link to know the health report of the patients. Each of these reports marked with a date. These reports are always in the hands of patients. Only the patients have power over their reports with which to share. This proves the privacy and security of the patient details. After this process, the data are decrypted and authenticated with the RSA digital signature. Based on health analytics and IoMT, the providers who refer to doctors or nurses and patients can access the decrypted data. Health applications are created by providers for a patient to access their health reports. Patients can give the authorization to access their reports at any time and anywhere and provide a public key. In the cryptograph, public keys are paired with a private key. If the message is assigned with the public, the other with a private key can view and modify that message. The private key user can provide a digital signature, and anyone can view the message by having the respective public key. Hence, twice the time the encryption has been done to ensure the security. These are followed by payers and pharma/Research. The payers can deal with the patients with their health care policies. For that, the payers want the patient's medical history for choosing the best policy. So blockchain helps the payers to access the patient records if only patient grant access to the payers. To give the respective medicines, pharma also wants to know the history of patient health care. So, payers and pharma/research grant with a public key to view the records since they can't modify them. For the above processes, used solidity language which invokes Python and Javascript on a public blockchain platform. Ethereum to create and run smart contracts and leads to the safe distributions of the patient health records. Ethereum blockchain technology helps us to remove intermediates between the public and the attainment of personnel health reports. Thus, this process leads to a great change in the health industry. The efficiency of the framework can be evaluated by analyzing the throughput and latency of the proposed system.

7 Result

To grasp how a system is proposed in a real case scenario various functions of the framework are tested by users in the performance evaluation experiment conducted. With a rising number of transactions, time for execution also increases. Various transactions are carried out for the specific roles covered by the proposed system. When only one person uses the program the algorithm assigns roles, add and looking patient histories which take 19 s, 2 min, and 1 min correspondingly for the specified roles of the proposed system to be executed. The time will increase simultaneously when the user numbers are rising. During the experiment as simulated, the user numbers as indicated in graph Fig. 2 and evaluated the proposed output. These mock-ups are performed on the proposed system which is Ethereum framework and are evaluated at the final stage of the cycle. While performing this experiment, it is observed that as the user numbers and requests increase the device throughput considerably increase steadily. This steady improvement represents the effectiveness of the Ethereum framework.

Latency is the time gap between arising a request from the component, and some other system component generates a response. The discrepancy between both of these aspects is latency. The average latency of the suggested system was measured here using JMeter. When testing the proposed framework's latency, JMeter's is simulated the number of users. In JMeter, millisecond latency is measured. The following graph Fig. 3 offers a description of the system's average latency, together with the proposed framework's throughput. In this experiment, the highest latency recorded

Fig. 2 Number of users versus throughput—graph

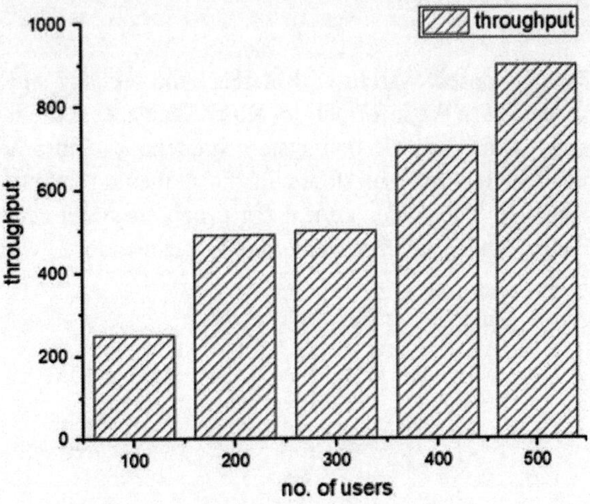

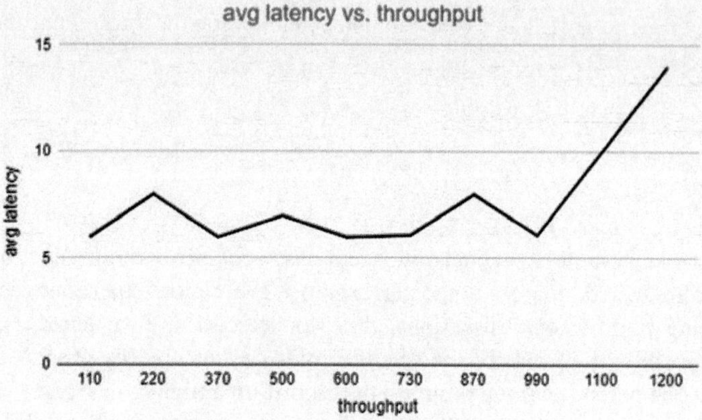

Fig. 3 Average latency versus throughput—graph

is 14ms. The efficiency of this system is also accessed by calculating transaction size and cost. The charge or expense of this system may be measured associated with different transactions. In this Ethereum framework, transaction costs are measured in "ETH" the cryptocurrency of Ethereum, and it has associated with its other units such as wei, gwei. The formula for calculating transaction fees for Ethereum is also discussed here. Transaction cost can be calculated by the product of consumption of gas, and its cost could be described as follows, transaction cost = gas consumption × cost of gas. When the patients' records are stored once, IPFS creates a hash value for the corresponding records. Blockchain stores these hashes when they had been created, so users like doctors, patients can access it whenever they want. By creating these cryptographic hashes, the patients' records are secured in the blockchain [28].

8 Conclusion

This proposed system will increase the security and safety of the transaction and accessing of the information. No one can access the information without the knowledge of the user. If this system implemented in a health care application, all the medical records are stored in the network securely. It could be used for future research. The major aim of the proposed system is that the user can access their record whenever they want, and they can choose the members who can access their reports.

References

1. Krawiec RJ, Housman D, White M, Filipova M, Quarre F, Barr D, Nesbitt A, Fedosova K, Killmeyer J, Israel A, Tsai L (2016) Blockchain: opportunities for health care
2. Swan M (2015) Blockchain: blueprint for a new economy. O'Reilly Media, Inc
3. Nakamoto S (2008) Bitcoin: a peer-to-peer electronic cash system
4. Ishmaev G (2017) Blockchain technology as an institution of property. Metaphilosophy 48(5):666–686
5. Forrest P (2016) Blockchain and non-financial services use cases
6. Zheng Z, Xie S, Dai H, Chen X, Wang H (2017) An overview of blockchain technology: architecture, consensus, and future trends. In: Proceedings of the 2017 IEEE international congress on big data (BigData Congress), Boston, MA, USA, 11–14 Dec 2017, pp 557–564
7. McKinsey & Company (2017) Blockchain technology in the insurance sector. In: Proceedings of the Quarterly Meeting of the Federal Advisory Committee on Insurance (FACI), New York, NY, USA
8. Christidis K, Devetsikiotis M (2016) Blockchains and smart contracts for the internet of things. IEEE Access 4:2292–2303
9. Yue X, Wang H, Jin D, Li M, Jiang W (2016) Healthcare data gateways found healthcare intelligence on blockchain with novel privacy risk control. J Med Syst 40(10):218
10. Liu W, Zhu SS, Mundie T, Krieger U (2017) Advanced block-chain architecture for e-health systems. In: IEEE 19th international conference on e-Health networking, applications and services (Healthcom), Dalian, pp 1–6
11. Mettler M (2016) Blockchain technology in healthcare: the revolution starts here. In: 2016 IEEE 18th international conference on e-Health networking, applications and services (Healthcom). IEEE, Munich, Germany, 14–16 Sept 2016, pp 1–3
12. Stokkink Q, Pouwelse J (2018) Deployment of a blockchain-based self-sovereign identity. In: Proceedings. IEEE international conference on internet things (iThings), pp 1336–1342
13. Mühle A, Grüner A, Gayvoronskaya T, Meinel C (2018) A survey on essential components of a self-sovereign identity. Comput Sci Rev 30:80–86
14. Pirtle C, Ehrenfeld JJ (2018) Med Syst 42: 172 available via https://doi.org/10.1007/s10916-018-1025-3 © Springer Science+Business Media, LLC, part of Springer Nature
15. Vora J, Nayyar A, Tanwar S, Tyagi S, Kumar N, Obaidat MS, Rodrigues JJPC (2018) BHEEM: a blockchain-based framework for securing electronic health records. 2018 IEEE Globecom workshops (GC Wkshps)
16. Juneja A, Marefat M (2018) Leveraging blockchain for retraining deep learning architecture in patient-specific arrhythmia classification. In: IEEE EMBS international conference on biomedical & health informatics (BHI), Las Vegas, NV, USA, pp 393–397
17. Der U, Jähnichen S, Sürmeli J (2047) Self-sovereign identity, opportunities and challenges for the digital revolution. arXiv:1712.01767
18. Davidson S, De Philippi P, Potts J (2016) Economics of blockchain. Available via https://ssrn.com/abstract=2744751. 8 Mar 2016
19. Nofer M, Gombe P, Hinz O, Schiereck D (2017) Blockchain. Bus Inf Syst Eng 59:183–187
20. Dannen C (2017) Introducing ethereum and solidity
21. Blockchain-based Smart Contracts—Applications and Challenges Yining Hu University of New South Wales and Data61-CSIRO Madhusanka Liyanage University College Dublin Ahsan Manzoor Rovio Entertainment Kanchana Thilakarathna University of Sydney Guillaume Jourjon Data61-CSIRO Aruna Seneviratne University of New South Wales, 11 Jun 2019
22. Dannen C.: Introducing ethereum and solidity foundations of cryptocurrency and blockchain programming for beginners. Apress
23. Wang H, Wang Y, Cao Z, Li Z, Xiong G (2019) An overview of blockchain security analysis. In: Yun X et al (eds) Cyber security. CNCERT 2018. Communications in computer and information science, vol 970. Springer, Singapore

24. Rivest R (1992) The MD5 message-digest algorithm. Network Working Group
25. Wang X, Fen D, Lai X, Yu H (2004) Collisions for Hash Functions, in Crypto
26. Zhou X, Tang X (2011) Research and implementation of RSA algorithm for encryption and decryption. In: Proceedings of 2011 6th International Forum on Strategic Technology (IFOST), vol 2. IEEE, pp 1118–1121
27. NIST Brief Comments on Recent Cryptanalytic Attacks on Secure Hashing Functions and Continued Security Provided by SHA-1, NIST, August 2004.
28. Niranjanamurthy K, Kumar S, Saha A, Chahar DD (2016) Comparative study on performance testing with JMeter. Int J Adv Res Comput Commun Eng 5(2):70–76

Chapter 66
Generation of Alternative Clusterings Using Multi-objective Particle Swarm Optimization

Avinash Navlani and V. B. Gupta

1 Introduction

Extracting patterns such as groups, similar items, dissimilar items from a dataset without any extra information or in an unsupervised manner are known as clustering. It offers several use cases like customer segmentation, document clustering, fraud detection, image segmentation, sports, and information retrieval [1]. Clustering is an approach for generating groups of homogeneous sets of items, but it is unsupervised nature unable to offer consistency in results. This lack of consistency is due to multiple representations and perspectives of groups. Alternative clustering is a kind of pattern discovery process that finds all the best possible sets of clusterings from different representations and views by maintaining the lowest intra-cluster distance and highest inter-cluster distance [2].

Saha and Mitra [3] divided alternative clustering into two major categories objective function-based methods and data transformation-based methods while Troung and Battiti [4] divided alternative clustering into two major categories unsupervised and supervised methods. The objective-based method uses a single or multi-objective objective function to generate clustering, and data transformation-based methods to perform projects on data to generate alternative clustering. Unsupervised methods generate clustering without using labels while supervised learning users ask for more alternative clustering. This paper proposed an objective function-based approach to generate alternative clusterings. Objective function comprises two objectives: cluster quality and distinctness. Two objective functions designed one for minimization and another for maximization problem of clustering. Objective function will optimize

A. Navlani (✉) · V. B. Gupta
School of Data Science and Forecasting, Devi Ahilya Vishwavidyalaya, Indore, India
e-mail: avinashnvln8@gmail.com

V. B. Gupta
e-mail: vbgupta.davv@gmail.com

I. Jeena Jacob et al. (eds.), *Data Intelligence and Cognitive Informatics*,
Algorithms for Intelligent Systems,
https://doi.org/10.1007/978-981-15-8530-2_66

using particle swarm optimization. This approach is capable of optimizing the cluster quality and separation both and producing a series of alternative clusterings.

The later portion of the paper is organized as follows: Sect. 2 discusses the background and related work on alternative clustering. In Sect. 3, the objective function is designed and digs down into the proposed method in Sect. 4. Section 5 discusses the experimental results of the proposed method. Finally, concluding summary and a brief description of future work in Sect. 6.

2 Related Work

The main objective of alternative clustering methods is to generate different types of clusterings for understanding new perspectives of unsupervised data. These clusterings can be generated either using objective function or transformations. To generate clusterings multiple objectives such as clustering quality, distinctness can be optimized using the genetic algorithm and also with a variable number of clusters [3]. Another view is presented on generating alternative clustering iteratively based on multiple objectives using genetic algorithms at cluster level rather than object level [4]. These multiple objectives can be optimized using an evolutionary-based approach with an automatic number of cluster determination [5].

Alternative clustering can also be generated using information-theoretic approach by utilizing the hierarchical information-theoretic approach by ensuring maximum mutual information within the cluster and minimizing information utilization between different clusters [6]. Another algorithm on information-theoretic approach is MinCEntropy++, which compares alternative clustering using several information-theoretic measures such as Havrda-Charvat's α structural entropy [7]. Another approach is based on agglomerative hierarchical clustering method COALA with cannot link constraints Bae and Bailey [8].

Multiple objective-based alternative clustering is less explored and needs more attention. The objective functions can be optimized using evolutionary-based methods such as particle swarm optimization (PSO). PSO offers faster convergence, diversity in search space, parallel execution, and it outperforms most of the solutions compared to the genetic algorithm [9]. Multiple objective particle swarm optimization (MOPSO) not only handles multiple objectives at the same time but also offers a wider range of applications [10].

3 Objective Function for Alternative Clustering

3.1 Internal Evaluation Measures

The internal evaluation measure assesses the generated clustering based on the feature column only. Here, the clustering method will not use any label to compare the results, this type of performance evaluation known as internal or intrinsic evaluation measures. The following internal evaluation measures can be utilized to evaluate the performance:

Davies–Bouldin Index
The Davies–Bouldin index (DBI) score used to evaluate clustering quality. Davies–Bouldin index considers intra-cluster similarity and inter-cluster differences. The centroid linkage is used to compute the inter-cluster distance. It is conveyed as the ratio of within-cluster distance to between-cluster distance [11]. DBI score for a cluster analysis should be close to 0. Davies–Bouldin index computational cost in terms of time is linear. Davies–Bouldin index formula can be described as follows:

$$\mathrm{DBI} = \frac{1}{n} \cdot \sum_{i=1}^{n} \max_{c_i \neq c_j} \left(\frac{\sigma_i + \sigma_j}{d(c_i, c_j)} \right) \tag{1}$$

where n is the number of clusters. σ_i is the mean distance (which measures the scatter within the cluster) of all the points in ith cluster from the centroid c_i. $d(c_i, c_j)$ is the separation between cluster centroids c_i and c_j.

Silhouette Coefficient
The Silhouette coefficient quantifies the clustering performance using the pairwise differences between-cluster and within-cluster distances [12]. The higher the Silhouette value betters the clustering. The computational cost in terms of time is quadratic with respect to the number of clustered vectors. Silhouette coefficient formula can be written as follows:

$$\mathrm{SC} = \frac{1}{n} \sum_{i=1}^{n} S_i \tag{2}$$

$$S_i = \frac{b(i) - a(i)}{\max\{a(i), b(i)\}} \tag{3}$$

where $a(i)$ is the average distance of ith to all other vectors, and $b(i)$ is the average distance of ith element to the vectors in other clusters. The score ranges from -1 to $+1$. -1 indicates highly scattered clusters, 0 indicates overlapped clusters, and $+1$ for highly compact clusters.

3.2 External Evaluation Measures

The external evaluation measure quantifies the quality of the generated clustering. It evaluates performance using the available target label. Here, the clustering evaluation method will use any target label to compare the results, this type of performance evaluation known as external or extrinsic evaluation measures. The following external evaluation measures can be utilized to assess the performance:

Fowlkes–Mallows Score
The Fowlkes–Mallows score is utilized to discover the closeness among the clusters and can be explained as the geometric mean of the precision and recall. This is why it is also known as G-measure. Fowlkes–Mallows score's higher value represents more similar clusters. The score ranges from 0 to 1 [13]. The formula of Fowlkes–Mallows score can be described as follows:

$$\mathrm{FM} = \sqrt{\mathrm{Precision} * \mathrm{Recall}} \tag{4}$$

V-Measure Score
V-measure or Validity-Measure is an entropy-based evaluation measure, similar to normalized mutual information which describes the decency of a clustering solution. Its core components are homogeneity and completeness. Mathematically, V-measure is the harmonic mean of homogeneity and completeness scores [14]. The V-measure score ranges from 0 to 1. The higher score of the V-measure represents more similar clusters.

$$V_\beta = (1+\beta)\frac{h \cdot c}{\beta \cdot h + c} \tag{5}$$

where h stands for homogeneity score, c stands for completeness score, and β is the weight of homogeneity score.

Jaccard Index
The Jaccard index is utilized to assess the similarity between the two sample sets. 1 index represents that all the elements in both clusters are common, and 0 represents that all the elements in clusters are distinct [15]. The lower value of the Jaccard index represents more distinct clusters. Jaccard index formula for two sample set can be described as follows:

$$\mathrm{JI} = \frac{|A \cap B|}{|A \cup B|} = \frac{\mathrm{TP}}{\mathrm{TP} + \mathrm{FP} + \mathrm{FN}} \tag{6}$$

Rand Index
Rand index can be explained as the ratio of correctly clustered data items to all the data items. The lower value of the Rand index is preferred because it represents

distinct clusters [16]. Rand index formula can be described as follows:

$$\mathrm{RI} = \frac{\mathrm{TP} + \mathrm{TN}}{\mathrm{TP} + \mathrm{FP} + \mathrm{FN} + \mathrm{TN}} \tag{7}$$

4 Designing the Objective Function

Our proposed algorithm experimented on two objective functions one for minimization and other for a maximization problem. Objective function-based techniques are easy to define, implement, and optimize. The minimization problem objective function uses a combination of internal clustering evaluation measure Davies–Bouldin index, and external clustering evaluation measures the Rand index and Jaccard index. This objective function is similar to [3]'s fitness function. This objective function is explained by the following equation.

$$\text{Objective Function (Min.)} = \lambda * \mathrm{DBI} + \left(\frac{1-\lambda}{2}\right) * \mathrm{RI} + \left(\frac{1-\lambda}{2}\right) * \mathrm{JI} \tag{8}$$

where λ is the trade-off parameter between quality and dissimilarity of clustering.

The maximization problem objective function uses a combination of internal clustering evaluation measure Silhouette coefficient, and external clustering evaluation measures Fowlkes–Mallows and V-Measure score. This objective function is explained by the following equation.

$$\begin{aligned}\text{Objective Function (Max.)} = \lambda * \mathrm{SC} + \left(\frac{1-\lambda}{2}\right) * \mathrm{FMI} \\ + \left(\frac{1-\lambda}{2}\right) * V\text{-measure}\end{aligned} \tag{9}$$

where λ is the trade-off parameter between quality and dissimilarity of clustering.

5 Proposed Method

Our proposed method uses a multi-objective particle swarm optimization (MOPSO) algorithm to optimize the objective function for multiple objectives and generate alternative clusterings with high quality and dissimilarity. PSO is an evolutionary population-based stochastic optimization approach provoked by the social conduct of bird flocks. It is based on how birds search for food in a group? This similar concept is simulated in the computer world for finding the optimal solution in search space named as particle swarm optimization (PSO). PSO is a combination of the cognitive and social components. PSO is faster and more adaptive in nature compared to the genetic algorithm, especially in case of multi-objective optimization problems; PSO strategies are the more intuitive and best fit for finding Pareto solutions. Each particle is adjusted by its velocity based on its momentum and two best values. The first one is the best preceding position known as the personal best, and the second one is the best position in the swarm neighborhood known as global best. Here, personal best and global best are values of the objective function in a particular iteration. Their values keep updating in the iteration. The portion that adjusts the velocity based on personal best position is called a cognitive segment, and the area affected by the best position in the swarm neighborhood is considered as the social segment.

Particle velocity

$$v_{k+1}^{i} = w_k v_k^i + c_1 r_1 \left(p_k^i - x_k^i\right) + c_2 r_2 \left(p_k^g - x_k^i\right) \tag{10}$$

Particle position:

$$x_{k+1}^{i} = x_k^i + v_{k+1}^{i} \tag{11}$$

where x_k^i Particle position, v_k^i particle position, p_k^i best individual particle position, p_k^g best swarm position, w_k constant inertia weight, c_1, c_2 cognitive and social parameters, respectively, r_1, r_2 random numbers between 0 and 1 (Fig. 1).

The proposed algorithm takes essentials inputs such as λ, c_1, c_2, r_1 and r_2 and loads the data. First, it initializes the particles and computes the global fitness. After that, it compares and updates personal best and global best. Then, it performs the clustering and computes the objective function and changes its velocity and position. It keeps running until predefined iterations are completed and saves the top alternative clusterings.

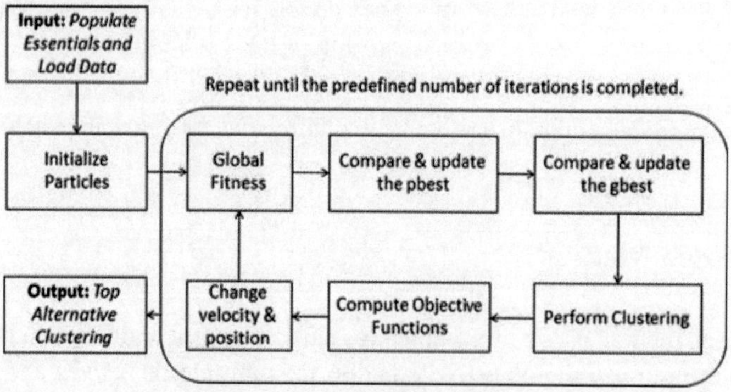

Fig. 1 Flow chart of proposed method

6 Results and Discussion

Our approach has experimented on seven datasets from UCI repository: Iris, Diabetes, Breast Cancer, Liver Patient, Zoo, Red wine, and White wine [17]. Datasets are explained in Table 1.

Evaluation of alternative clusterings is performed in terms of two multi-objective functions. The objective function scores for the top two alternative clusterings for every dataset are mentioned in Table 2.

Our proposed method results can be validated by comparing with k-means clustering results for both the multi-objective functions and shown in Table 3. K-means clustering algorithm is most common, widely accepted, and deployed in industrial

Table 1 Datasets information

Datasets	Number of observations	Number of feature columns	Number of classes
Iris	150	4	3
Diabetes	768	8	2
Breast cancer	699	9	2
Liver patient	579	10	2
Zoo	101	16	7
Red wine	1599	11	6
White wine	4898	11	7

Table 2 Objective function scores for top two clustering

Datasets	Best clustering	Objective function score (Maximization)	Best clustering	Objective function score (Minimization)
Iris	AltClust 1	0.671042	AltClust 1	0.341104
	AltClust 2	0.671042	AltClust 2	0.341104
Diabetes	AltClust 1	0.449485	AltClust 1	0.281903
	AltClust 2	0.449485	AltClust 2	0.361981
Breast cancer	AltClust 1	0.713027	AltClust 1	0.39937
	AltClust 2	0.710942	AltClust 2	0.39937
Liver patient	AltClust 1	0.654944	AltClust 1	0.179912
	AltClust 2	0.620207	AltClust 2	0.259079
Zoo	AltClust 1	0.604965	AltClust 1	0.224478
	AltClust 2	0.576613	AltClust 2	0.226537
Red wine	AltClust 1	0.333631	AltClust 1	0.06246
	AltClust 2	0.306784	AltClust 2	0.066451
White wine	AltClust 1	0.229463	AltClust 1	0.085223
	AltClust 2	0.229463	AltClust 2	0.085223

Table 3 Comparison of the proposed method with *k*-means

Datasets	Objective function (Min.)		Objective function (Max.)	
	Our method	*K*-means	Our method	*K*-means
Iris	0.34	0.54	0.67	0.67
Diabetes	0.28	0.36	0.45	0.45
Breast cancer	0.4	0.4	0.713	0.71
Liver patient	0.16	0.16	0.65	0.62
Zoo	0.22	0.22	0.61	0.61
Red wine	0.062	0.08	0.33	0.31
White wine	0.085	0.12	0.229	0.226

applications. Our proposed method scores for both objective functions work well compared to the *k*-means clustering algorithm.

7 Conclusion

This paper has proposed a multiple objective-based particle swarm optimization method for producing alternative clustering. Multiple objectives-based PSO frames the clustering problem and searches for the optimum solution in parent front solutions. Two types of objective functions maximization and minimization have been used. Both objective functions have two objectives: cluster quality and dissimilarity. Each objective function tries to maximize the cluster quality and minimize the dissimilarity. The results of our implemented method worked well compared to the *k*-means clustering methods in terms of clustering quality and dissimilarity. Our proposed approach is able to produce a series of alternative clusterings. In the future, variable-length alternative clusterings (clusterings with different-different numbers of clusters) can be generated using the proposed method and can be experimented on the large datasets.

References

1. Han J, Kamber M, Pei J (2011) Data mining: concepts and techniques. Morgan Kaufmann, San Francisco
2. Navlani A, Gupta V (2017) Alternative clustering. In: Applying predictive analytics within the service sector. IGI Global, Hershey, pp 1–12
3. Saha M, Mitra P (2014) VLGAAC: variable length genetic algorithm based alternative clustering. In: Loo CK, Yap KS, Wong KW, Teoh A, Huang K (eds) Neural information processing. ICONIP. Lecture notes in computer science, vol 8835. Springer, Cham
4. Truong DT, Battiti R (2012) A cluster-oriented genetic algorithm for alternative clustering. In: IEEE 12th International conference on data mining (ICDM), Brussels

5. Handl J, Knowles J (2007) An evolutionary approach to multiobjective clustering. Evol Comput IEEE Trans 11(1):56–76
6. Dang XH, Bailey J (2010) A hierarchical information-theoretic technique for the discovery of nonlinear alternative clusterings. In: 16th ACM SIGKDD international conference on knowledge discovery and data mining, New York
7. Vinh NX, Epps J (2010) minCEntropy: a novel information-theoretic approach for the generation of alternative clusterings. In: IEEE 10th International conference on data mining (ICDM), Sydney
8. Bae E, Bailey J (2006) COALA: a novel approach for the extraction of an alternate clustering of high quality and high dissimilarity. In: The sixth IEEE international conference on data mining, Hong Kong
9. Saed AAA, Kadir WMNW (2011) Applying particle swarm optimization to software performance prediction an introduction to the approach. In: 5th Malaysian conference in software engineering (MySeC), Johor Bahru
10. Lalwani S, Singhal S, Kumar R, Gupta N (2013) A Comprehensive survey: multi-objective particle swarm optimization (MOPSO) algorithm: variants and applications. Trans Comb 2:89–101
11. Davies DL, Bouldin DW (1979) A cluster separation measure. IEEE Trans Pattern Anal Mac Intell PAMI-1(2):224–227
12. Rousseeuw PJ (1987) Silhouettes: a graphical aid to the interpretation and validation of cluster analysis. J Comput Appl Math 20:53–65
13. Fowlkes EB, Mallows CL (1983) A method for comparing two hierarchical clusterings. J Am Stat Assoc 553–569
14. Rosenberg A, Hirschberg J (2007) V-measure: a conditional entropy-based external cluster evaluation. In: Proceedings of the 2007 joint conference on empirical methods in natural language processing and computational natural language learning (EMNLP-CoNLL), Czech Republic
15. Tan PN, Steinbach M, Kumar V (2005) Introduction to data mining. Pearson, London
16. Rand WM (1971) Objective criteria for the evaluation of clustering methods. J Am Stat Assoc 846–850
17. Dua D, Graff C (2017) UCI machine learning repository. https://archive.ics.uci.edu/ml/

Chapter 67
Region of Interest-Based Encryption of Biomedical Image

Srinivas Vodnala, Shubhankar Majumdar, and Pallab Kumar Nath

1 Introduction

Segmentation of images is the process of taking a digital image and breaks it into several pixel segments. Segmentation of the image aims to simplify and/or alter the representation of an image into something more meaningful and understandable. This is the method of partitioning a digital image with similar attributes into several, distinct regions containing each pixel (sets of pixels, also known as superpixels). The purpose of this is to turn the representation of an image into something more relevant and easy to interpret.

Cryptography is the science of reliably transmitting information against possible third-party adversaries. Encryption of an aspect of cryptography is the mechanism by which information is encoded to remain secret or difficult to access unapproved clients. This assures the preservation of private data, confidential information and can improve the security of communications between client and server applications. Essentially, when the information is encrypted, no matter what unauthorized entity or content cannot be able to access it.

In this paper, the concepts of image segmentation are used to get the region of interest (ROI) of an image, and then, encryption algorithms for encrypting the entire image and ROI and find the respective computational times and compare them.

S. Vodnala (✉) · S. Majumdar
Department of Electronics and Communication Engineering, National Institute of Technology Meghalaya, Shillong, Meghalaya, India
e-mail: t18ec001@nitm.ac.in

S. Majumdar
e-mail: shub@nitm.ac.in

P. K. Nath
NeuRonICS Lab, DESE, IISc. Bangalore, Bangalore, India
e-mail: pkniitkgp@gmail.com

I. Jeena Jacob et al. (eds.), *Data Intelligence and Cognitive Informatics*,
Algorithms for Intelligent Systems,
https://doi.org/10.1007/978-981-15-8530-2_67

In Sect. 2, the related work done in this domain is discussed. In Sect. 3, descriptions of the method followed for extraction of ROI of the USG image, and then, a discussion regarding the encryption algorithms used for encryption is given. Finally, in Sect. 4, summary of the results is produced, and in Sect. 5, conclusions are made.

2 Related Works

Image segmentation is the principal approach in digital image processing; in this Otsu, thresholding is one of the fundamental methods of thresholding in which threshold value is calculated automatically. This method finds its applications in many areas like object detection, pattern recognition, and some of the predominant areas of computer vision as well.

Recognition and retrieval of information from some of the degraded digital documents are often a cumbersome process, but binarization of such documents makes it easy to read it; this kind of binarization is done using Otsu thresholding [1, 2]. It also finds its footprint in finding the object in moving frame, tracking, and recognizing complex signs in an image, extracting the meaning full information from the complex background, helping in differentiating foreground and background [3–5].

Contours is an image segmentation technique, which has very much contribution in medical image segmentation. It helps in segmenting the required region of interest by detecting the shape of the objects from images like computed tomography (CT), bio images in which shape of the objects can be tracked [6, 7]. Some of the compelling applications can also be found in the field of robotics, where it helps robots to find the gestures of people and sorting different objects by their shapes [8, 9]. It also helps to find out to which class a particular object belongs by getting to know the shape of the object [10].

Image is one of the most significant data portrayal styles and broadly utilized in most of the applications. Images are regularly traded between two parties over the uncertain network. In request to shield the images from interference, the image which is shared ought to be encrypted before transmission. For securing the image data, cryptography algorithms are considered. There are two types of cryptographic ciphers, symmetric and asymmetric ciphers. The schematic diagram of symmetric and asymmetric key encryption is shown in Fig. 1.

RSA is one of the most widely used asymmetric key cryptographic algorithms. It is important as it provides privacy, verification, information respectability, and non-repudiation. An RSA process is modular exponentiation that involves repeated multiplications of modulations. RSA operand sizes must be 1024-bit or greater for safety reasons [11]. It is generally used to make sure about system traffic, email, web-based business, and e-business frameworks for applications in advanced marks and encryptions.

The ElGamal encryption system is an asymmetric key encryption algorithm based on the Diffie–Hellman for public key cryptography involving an exchange of key [12]. It is the most broadly utilized other option to RSA because of its computational

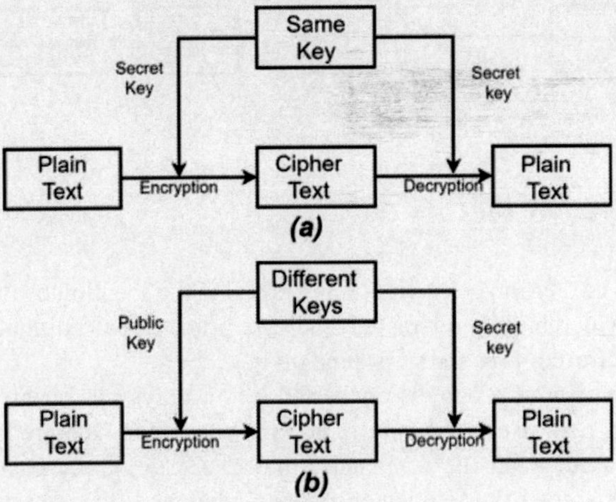

Fig. 1 **a** Symmetric encryption, **b** Asymmetric encryption [15]

efficiency. Effectively versatile to numerous sorts of cryptographic gatherings [13]. It gets the strength from the supposition that the discrete logarithms can't be found in practical time span for a given number, while the reverse activity of proficient power can be computed.

Because of the coming of asset obliged patterns, for example, advanced mobile phones and savvy gadgets, the environment of computing is evolving. Since now a day's everyday life is profoundly interlaced with pervasive networks, the significance of security is growing. A lightweight encryption algorithm is vital for secure correspondence between these sorts of asset obliged devices, and numerous researches have been examining this field. So a cipher of lightweight block known as LEA was proposed, originally focused for proficient usage on microprocessors, as it is quick when executed in software, and besides, it has a little memory impression [14]. To consider ongoing innovation, every necessary estimation utilize 32-bit wide operations. Also, the algorithm is contained not all that so-called complex S-Box like structures but basic addition, rotation, and XOR activities.

3 Methodology

The objective of calculating computation times of the program when the entire USG image of a baby and a part of it, i.e., ROI of that image is encrypted which is being proposed in this paper. Once the computational times are acquired, the conclusions can be drawn from it, regarding which process takes the least time and which is better for encryption.

Fig. 2 Flowchart of the proposed method

3.1 Extraction of ROI

To get the ROI of any image, it would be helpful if the location of the pixel values of that region which must be extracted is known prior. After getting the coordinates of that location further processing is done on it.

First, USG image of a baby is converted into a grayscale image so that no. of dimensions gets decreased, and in turn, no of computations required to process the image is decreased. After the conversion into grayscale, apply Otsu thresholding to the image, this gives the segmented image. Now apply the contour operation to the segmented image and get the coordinates of required contour. With the help of the location of coordinates extract the region of interest (ROI) of the image. The complete flow of the above process is shown in the below flow chart of Fig. 2.

3.2 Otsu Thresholding

Thresholding is one of the image segmentation techniques, which help in simplifying the visual data for further analysis. It is a non-direct activity that changes over a grayscale image into a binary image where the two levels are appointed to pixels that are underneath or over the predetermined limit esteem, i.e., certain threshold value. In simple words, if the value of a pixel is higher than a threshold value, it is allocated with one value (might be 0), else it is allocated with another value (might be 255).

For Otsu Thresholding, a threshold value is not chosen randomly as done in simple thresholding but is determined automatically. A bimodal image (two distinct image values) is considered. The histogram generated contains two peaks. So, a generic condition would be to choose a threshold value that lies in the middle of both the histogram peak values.

Otsu's thresholding technique includes emphasizing through all the possible limit esteems and registering an extent of spread for the pixel levels each side of the threshold, for instance, the pixels that either fall in background or foreground. The point is to find the threshold esteem where the sum of background or foreground spreads is at its minimum. This can be explained by considering foreground and background as two classes and calculating intra-class variance and inter-class variance, with the help of which the required threshold value can be achieved. Significance of variance is that the higher the value of variance, the more dispersed the data is.

As far as the intra-class variance (*I*, for short), the lower the value of "*I*" is, the less scattered the data in each class is (foreground and background). As the outcome, to get the optimal threshold esteem is to locate the minimum value of *I*.

Accordingly, in case of the calculation of inter-class variance (Ir, for short), the value should be maximum to get an optimal threshold value. If the considered value of Ir is minimum. It means that the dispersity between two classes is lowest (which is not a good result) which generates the requirement for the maximum value of "Ir".

This whole discussion can be summarized by the following equations:

Let us assume that in our image, L gray levels (1, 2, 3, …, L) and n_i denote the number of pixels that has at level i. N is the total number of pixels and given by [5]:

$$N = \sum_{i=1}^{L} n_i \tag{1}$$

The probability of occurrence of level i is given by [5]:

$$p_i = \frac{n_i}{N} \tag{2}$$

If our image is considered which has two classes C_1 and C_2 distinguished by threshold X. C_1 consists of pixels with levels (1, 2, …, X), and C_2 consists of pixels with levels ($X + 1$ … L). Let P_1 (X) and P_2 (X) be the cumulative probabilities, U_1 (X) and U_2 (X) denote the mean levels, and σ^2 (X) and σ^2 (X) denote the variances of the classes C_1 and C_2, respectively. These values are given by [5]:

$$P_1 = \sum_{i=1}^{X} p_i \tag{3}$$

$$P_2 = \sum_{i=X+1}^{L} p_i \tag{4}$$

$$U_1(X) = \sum_{i=X+1}^{X} i * (p_i / p_1(X)) \tag{5}$$

$$U_2(X) = \sum_{i=X+1}^{L} i * (p_i / p_2(X)) \tag{6}$$

$$\sigma_1^2(X) = \sum_{i=1}^{X} (i - U_1(X))^{2*} (p_i / P_1(X)) \tag{7}$$

$$\sigma_2^2(X) = \sum_{i=X+1}^{L} (i - U_2(X))^2 * (p_i / P_2(X)) \tag{8}$$

Let U, $\sigma_1^2(X)$, and $\sigma_{\text{Ir}}^2(X)$ represents the mean level of the image, the intra-class variance, and the inter-class variance, respectively, and given by [5]:

$$U = P_1(X).U_1(X) + P_2(X).U_2(X) \tag{9}$$

$$\sigma_I^2(X) = P_1(X).\sigma_I^2(X) + P_2(X).\sigma_I^2(X) \tag{10}$$

$$\sigma_{\text{Ir}}^2(X) = P_1(X).(U_1(X) - U)^2 + P_2(X).(U_2(X) - U)^2 \tag{11}$$

The threshold decided by minimizing the intra-class variance of (10) or minimizing the inter-class variance (11).

3.3 Contours

A contour can be portrayed as an ordered list of edges or by a curve. Instances of curves incorporate line segments and cubic splines. There are a few standards for a decent curve portrayal. Edges must be connected into a portrayal for a locale boundary. This portrayal is known as a contour. Discovering shapes resemble discovering white object from a dark background. In this way, an object to be found ought to be white and background ought to be dark. For that purpose, Otsu thresholding is applied to the respective image, which helps in separating the region of interest from black or dark background and then can able to create an outline in an image which helps to stand out the region of interest; i.e., in this case, it is the outline of a baby from the background. The outline of the image is a fundamental element of the image, conveying most of the data of the image. Contours help in extracting this outline. Contours are characterized as the line joining all the focuses along with the limit of a picture that is having a similar intensity. It comes as a convenient option for analyzing the shape and finding the size of the object of interest by storing the (x, y) coordinates of the respected object of interest.

The whole process of ROI extraction can be given by below pseudo algorithm:

- Read the input image.
- Convert into grayscale.
- Consider the image that has L gray levels and distributes them between two classes C_1 and C_2.
- Calculate the respective cumulative probabilities P_1 (X), P_2 (X) and mean levels U_1 (X) and U_2 (X).
- Calculate intra-class and inter-class variances, i.e., $\sigma_I^2(X)$ and $\sigma_{\text{Ir}}^2(X)$.
- The corresponding threshold value depends on minimizing $\sigma_I^2(X)$ or maximizing of $\sigma_{\text{Ir}}^2(X)$.
- Get the resultant threshold image.
- Apply contour detection which traces the contour of an image along with similar intensities.
- Store the coordinate location of contour (x, y).
- Extract the contour which is ROI.

4 Results and Discussion

First, the methodology explained as in the flow chart of Fig. 2 is performed on the respective biomedical images, and required ROI's are extracted accordingly. After the extraction of a region of interest (ROI) of the required image, next step is to perform encryption and decryption operations on the actual image of a baby and extracted ROI of the image and compare the computational times of both. For this purpose, consider three asymmetric cryptographic algorithms, i.e., RSA, Elgamal, and LEA. Table 1 shows the respective computational times of the algorithms of the entire image and region of interest of that image and ROI extraction times in milliseconds, respectively.

This technique of ROI extraction is performed on biomedical images of baby, brain, lungs, and retina which give satisfactory results. Here, biomedical images of a baby of dimensions of pixels 259 × 188, the brain of dimensions 442 × 442 pixels, lungs of dimensions 256 × 256 pixels, and retina of 300 × 200 pixels are taken as input images, respectively, and of Otsu and detection of contours segmentation is performed to extract the required region of interest. The results of these images after these operations can be seen in Fig. 3.

In Fig. 3, (a) biomedical image of the baby is taken as input with dimensions of 259 × 188 pixels, and after segmenting and extraction of ROI, the image has dimensions

Table 1 Computational times of entire biomedical image of baby, brain, lungs, retina and respected ROI and cumulative times of encryption of ROI and its extraction in milliseconds

Algorithm	Computational times in ms		Computational + ROI extraction (ms)
	Entire image	ROI	
Image of baby			
RSA	2943.99	1880.12	1883.83
Elgamal	919.40	448.41	452.12
LEA	658.89	325.35	329.06
Image of brain			
RSA	7737.34	4579.27	4579.28
Elgamal	721.28	244.60	244.61
LEA	448.29	176.45	176.46
Image of lungs			
RSA	3744.57	2586.52	2586.524
Elgamal	225.58	101.23	101.234
LEA	152.69	86.97	86.974
Image of retina			
RSA	3360.85	2435.86	2435.863
Elgamal	116.63	70.50	70.503
LEA	75.50	58.75	59.293

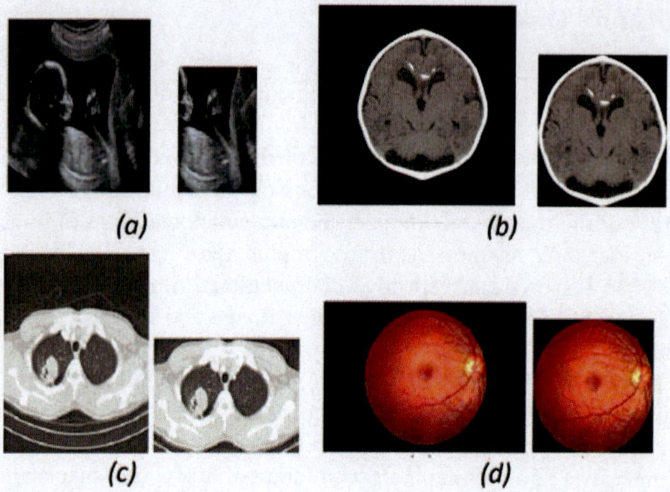

Fig. 3 Result of ROI extraction a image of baby, b image of brain, c image of lungs, d image of the retina

of 141 × 137 pixels, (b) shows the original image of the brain with dimensions of 442 × 442 pixels and its respective ROI of dimensions 292 × 336 pixels, (c) shows the original image of lungs with dimensions of 256 × 256 pixels and its respective ROI of dimensions 256 × 143 pixels and similarly, (d) shows the original image with dimensions of 300 × 200 pixels and its respective ROI of dimensions 181 × 177 pixels.

It can be observed from all the results that the computational times of entire images are higher than their respective ROI's; the reason being that the dimensions and size of the entire image in comparison with the size and dimensions of its ROI is higher in value. Also, time taken for ROI extraction varies for different images because the extracted ROI of a particular image has certain dimensions and sizes which differ from other images ROI's dimensions and size after their respective extractions that is a reason why this change is observed.

Below bar graph in Fig. 4 shows the comparative analysis of computational times of respected algorithms when their operation is performed on the entire image and required region of interest (ROI) of the image of baby, brain, lungs, and retina, respectively.

Figure 5 represents the result of Otsu thresholding of the respective images.

Table 2 PSNR, SSIM, NCC values of the otsu segmented images are calculated and tabulated.

The basic difference between the PSNR and SSIM, NCC is PSNR method which is developed for calculating absolute errors of an image; usually, it is considered higher the PSNR value better the image quality, whereas SSIM is a method for finding the similarity between the images and considers the structural similarity between the

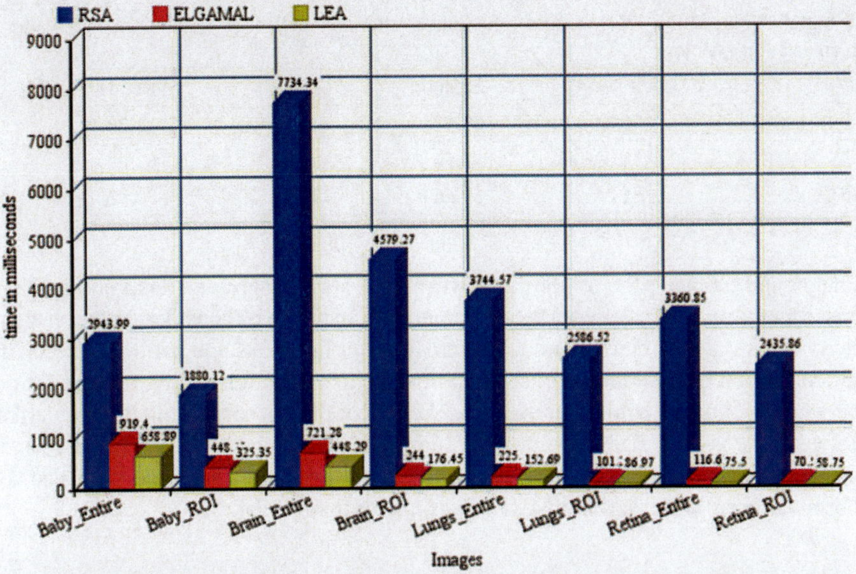

Fig. 4 Comparative analysis of computational times of respective algorithms for the entire image and ROI of respective brain, lungs, retina images

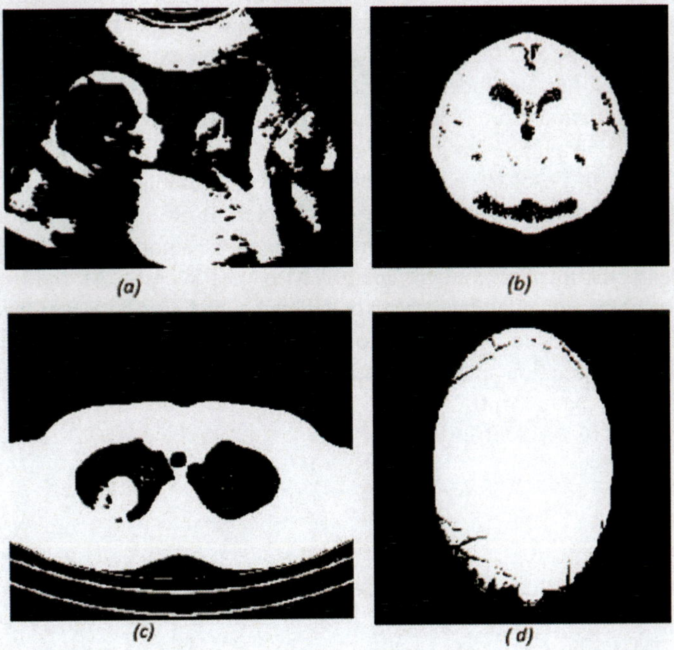

Fig. 5 Result of Otsu thresholding (**a**). **a** Image of baby of 259 × 188 pixels, **b** image of brain of 442 × 442 pixels, **c** image of lungs of dimensions of 256 × 256 pixels, **d** image of retina of 300 × 200 pixels

Table 2 PSNR, SSIM, NCC values of the images after the application of the Otsu thresholding segmentation technique

	Baby	Brain	Lungs	Retina
PSNR (dB)	29.34	33.12	28.50	31.28
SSIM (%)	35.03	72.96	33.97	67.02
NCC (%)	84.24	89.66	97.78	80.52

images and estimates the similarity between them accordingly, and it is a perception-based method which calculates the degradation of an image by considering the change in its structural information and takes into account luminous and contrast effects of the images. When coming to NCC technique is also used for finding the similarity between the images but here it uses correlation coefficient for finding the similarity between the images rather than taking luminous and contrast effects of the image and considered effective in finding the similarity between images, which can also be observed from Table 2 where NCC values higher than SSIM.

5 Conclusion

Here, extraction of the region of interest (ROI) of the required biomedical image of a baby, brain, lungs, and retina is performed. After the extraction, encryption and decryption operations have been performed on the actual image and respected ROI, and their computational times are calculated and tabulated. Comparative study of those times is shown in the bar graph. From those experimental results and comparative study, it can be observed that there is a considerable amount of reduction of computational times when operations performed on ROI and entire image, i.e., computational times of ROI are lesser. Also among three algorithms chosen, the LEA takes the least amount of computational time. Also, the PSNR, SSIM, and NCC values of otsu threshold segmented images are calculated, and discussions have been made. On the whole importance of otsu threshold segmentation and contour detection techniques for ROI extraction and optimization of computational times, achievement is discussed in this manuscript.

References

1. Siogkas GK, Dermatas ES (2006) Detection, tracking and classification of road signs in adverse conditions. In: MELECON 2006–2006 IEEE Mediterranean electrotechnical conference. IEEE, pp 537–540

2. Han X, Gao Y, Lu Z, Zhang Z, Niu D (2015) Research on moving object detection algorithm based on improved three frame difference method and optical flow. In: 2015 Fifth international conference on instrumentation and measurement, computer, communication and control (IMCCC). IEEE, pp 580–584
3. Huang W, Liu Z, Pan W (2007) The precise recognition of moving object in complex background. In: Third international conference on natural computation (ICNC 2007), vol 2. IEEE, pp 246–252
4. Ansari MA, Mahraj SK (2018) A robust method for identification of paper currency using otsu's thresholding. In: 2018 International conference on smart computing and electronic enterprise (ICSCEE). IEEE, pp 1–5
5. Xu X, Xu S, Jin L, Song E (2011) Characteristic analysis of otsu threshold and its applications. Pattern Recogn Lett 32(7):956–961
6. Sargin ME, Altinok A, Manjunath BS, Rose K (2010) Variable length open contour tracking using a deformable trellis. IEEE Trans Image Process 20(4):1023–1035
7. Seitz P, Ruegsegger P (1983) Fast contour detection algorithm for high precision quantitative CT. IEEE Trans Med Imaging 2(3):136–141
8. Dragusu M, Mihalache AN, Solea R (2012) Practical applications for robotic arms using image processing. In: 2012 16th International conference on system theory, control and computing (ICSTCC). IEEE, pp 1–6
9. Gurav RM, Kadbe PK (2015) Real time finger tracking and contour detection for gesture recognition using openCV. In: 2015 International conference on industrial instrumentation and control (ICIC). IEEE, pp 974–977
10. Schindler K, Suter D (2008) Object detection by global contour shape. Pattern Recogn 41(12):3736–3748
11. Mclvor C, McLoone M, McCanny JV (2003) Fast Montgomery modular multiplication and RSA cryptographic processor architectures. In: The thirty-seventh Asilomar conference on signals, systems & computers, vol 1. IEEE, pp 379–384
12. Kiltz E, Pietrzak K (2010) Leakage resilient Elgamal encryption. In: International conference on the theory and application of cryptology and information security. Springer, Berlin, pp 595–612
13. Chevallier-Mames B, Paillier P, Pointcheval D (2006) Encoding-free Elgamal encryption without random oracles. In: International workshop on public key cryptography. Springer, Berlin, pp 91–104
14. Bhardwaj I, Kumar A, Bansal M (2017) A review on lightweight cryptography algorithms for data security and authentication in IoTs. In: 2017 4th International conference on signal processing, computing and control (ISPCC). IEEE, pp 504–509
15. Fujisaki E, Okamoto T (2013) Secure integration of asymmetric and symmetric encryption schemes. J Cryptol 26(1):80–101

Chapter 68
Implementation of Production Planning Using Overall Equipment Efficiency (OEE) for IIoT

K. Amar Prem

1 Introduction

The biggest challenge for enterprises to retain in the competitive market is to improve productivity. Overall equipment efficiency (OEE) index is a metric to determine the effectiveness of a planned production with complete utilization in the scheduled running periods. OEE was developed in Japan for Japan Institute of Plant Maintenance (JIPM) association. Overall equipment efficiency is a key element for improvising quality in total quality management (TQM) and lean production. OEE plays a major role in key performance indication (KPI) with three core elements—availability, performance, and quality [1].

1.1 Availability

Availability is an estimation of the time period; an equipment is available for scheduled production comparing to the actual time of earlier productions. The factors like scheduled equipment maintenance, fixed downtime, and trials are not considered as available time [2]. The actual estimation of available time allows the plant manager to readily identify the definite factors of the equipment in the timeline [3]. Availability (A) is calculated using operating time(OT) and planned production time (PPT).

$$A = \frac{\text{OT}}{\text{PPT}} \tag{1}$$

K. Amar Prem (✉)
IIoT Consultant, Thoucentric Technology Pvt Ltd, Bengaluru, Karnataka, India
e-mail: k.amarprem@gmail.com

I. Jeena Jacob et al. (eds.), *Data Intelligence and Cognitive Informatics*,
Algorithms for Intelligent Systems,
https://doi.org/10.1007/978-981-15-8530-2_68

1.2 Performance

Performance is a comparison between an actual production of equipment with a theoretical estimation in a specified operating time. Performance also compares a facility to the available downtime of availability along with efficiency [4]. The analysis points out any issues of equipment like less output, excess downtime. Additionally, the problems like recurring efficiency, performance measurement portray the hiccups of the machinery instead of the issue of the operator. Performance (P) is estimated with ideal cycle time (ICT) and operating time(OT) [5].

$$P = \frac{\text{ICT}}{\text{OT}} \tag{2}$$

Ideal cycle time (ICT) is the minimum cycle time expected to achieve ideal situations. ICT can also be called as design cycle time or theoretical cycle time. The time is the reciprocal of run rate; performance (P) can also be calculated with total count (TC), run time (RT), and ideal run rate (IRR).

$$P = \frac{\text{TC/RT}}{\text{IRR}} \tag{3}$$

Performance should not be exceed 100%, or it indicates that the ICT is incorrect or high.

1.3 Quality

Quality is assessed with the percentage of products qualifies the quality inspection [6]. Quality allows the plant manager to assess the product consistency between different equipment and manufactures with specifications including operators. Quality (Q) is estimated with good count (GC) and total count (TC).

$$Q = \frac{\text{GC}}{\text{TC}} \tag{4}$$

Quality also counts the products that do not meet the manufacturing standards along with rework. OEE quality is based on the first pass yield that defines the products qualifying the manufacturing standards without involving any rework [7]. Equation 4 is similar to consider the ratio of full production time (FPT) without any stop time (ST) to no stop time indicating the Net Runtime (NRT).

Table 1 Comparison of OEE factors with availability, performance, and quality

OEE factor	Shift 1 (%)	Shift 2 (%)
Availability	90.0	93.0
Performance	95.0	95.4
Quality	99.5	96.6
OEE	**85.072**	**85.7**

1.4 Overall Equipment Efficiency

Overall equipment efficiency considers the three core elements—availability, performance, quality and is computed as

$$\text{OEE} = A * P * Q * 100 \tag{5}$$

The improvement of OEE is not the only objective. Table 1 shows sample data of two shifts compared based on the OEE factors. The data of two shifts based on availability, performance, and quality are given below:

The achieved data of OEE in Shift 1 and Shift 2 is highlighted in the Table 1. The data in Table 1 might appear that shift-2 is performing ahead of shift-1 as the OEE is higher. However, some enterprises might focus on the availability with 6% increase and decline of 3% in quality. OEE produces the factual about three core elements which can provide provisions to decide on daily bases and assists in visualizing for future needs.

2 Purpose of OEE

The purpose of OEE is to implement a system to analyze and measure of overall equipment efficiency. The results assist the manufactures to analyze the performance of the equipment, operational procedures, and improve the maintenance process. The six categories of productivity losses which are globally experienced by manufactures are

- Equipment Failures
- Setup & Adjustments
- Minor Stoppages
- Reduced Speed
- Process Defects
- Reduced Yield.

Table 2 Mapping of six major losses to the core elements in OEE

Availability	Performance	Quality
Equipment failures	Minor stoppages	Defects in process
Setup and adjustments	Reduced speed	Reduced yield

The six major equipment losses involve sub-loss category involving the ten states of equipment (E-10) state and OEE core element effect by the major loss [8]. Availability is the core element which is affected by the equipment failure and setup and adjustments. Availability is the core element which is affected by the equipment failure and setup and adjustments. The core element and performance are affected by the minor stoppages and reduced speed. Process defects and reduced yield affects the final core element—quality [9]. Table 2 portrays the mapping of six major losses to the core elements in OEE.

Equipment Failures The sub-loss categories like equipment failure, facility failure, and process failure affect the major loss category like equipment failures. The unscheduled downtime of E-10 state is affected by the equipment downtime due to the equipment failures. The tooling failures, unplanned maintenance, unexpected breakdowns, and failures are the general reasons for the equipment failures [9]. Flexibility is as to be set on the threshold between equipment loss and minor stoppages.

Setup and Adjustments The setup and adjustments influence the availability element of the OEE. The major loss of setup and adjustments is affected by the changeovers, planned maintenance, raw material or operator shortages and warm-up time of the equipment [10]. The planned downtime management, workplace organization, pre-planning are the countermeasures for the setup and adjustments major loss. Setup time reduction programs address the major loss of the equipment failures.

Minor Stoppages Obstructed product flow, component jams, misfeeds, delivery or sensor failure, cleaning, and checking are the major losses of minor stoppages. The small stops affect the performance element of the OEE. Alternate arrangement, reserved equipment, line minor stop audits are the alternative solutions for the minor stoppages. Nominal break periods that do not involve maintenance personnel only should be programmed.

Reduced Speed The operation running lower than the expected or specified speed, operator inefficiency, under nameplate, and design capacity are the losses of reduced speed. Line balance adjustments and line balance optimization are the countermeasures for reduced speed losses. The core element and performance are influenced by the reduced speed. The alternates for reduced speed major losses are ideal run rate or nameplate capacity which will keep the process running in regulated speed.

Defects in Process The defects in process are occurred by the product misconfigurations, damages, scraps, expiration, incorrect assembly. The countermeasures for defects in process are six sigmas, error proofing, root cause analysis. Quality is impacted by the defects in process. The rejects obtained during the warm-up, startup or other early production may be due to improper setup, warm-up periods.

Reduced Yield The reduced yield is intervened by the product out of specification, scrap produced after changeover before nominal running, damaged produced after planned maintenance activity. The countermeasures are planned downtime activity, workplace organization, standard operational procedures, precision settings. The core element, quality is influenced by the reduced yield. The rejects are accumulated during steady-state productions.

3 Six Major Equipment Losses Calculation

The calculation of six major equipment losses is represented with an example of a 480-minute shift with a machine expected to produce 100 products/minute.

Maximum output = 480 mins * 100 units = 48000 units
Shift info: Output (Good Production) = 32000 units
Speed = 98 units/minute
Planned downtime = 82 mins
Bottlenecks due to equipment failures = 30 mins
Defects in process = 1255 in 8hr shift
Output of OEE = 32000/48000 =66%
480 * 66.7% = 320 mins
Total Loss = 160 mins

Six Loss Calculation

Max theoretical units possible at actual speed = 98 *480 = 47040
= (32000/47070) - (32000/48000) = 68.03% - 66.67% = 1.36%

480 * 1.36%	= 6.53 mins/80	= 1.36%
Planned Downtime	= 82 mins/480	= 17.08%
Equipment Failures	= 30 mins/80	= 6.25%
Defects in process = 1255/98 (Actual running speed)	= 12.81 mins/480	= 2.67%
Minor stoppages = 480-320-6.53-82-30-12.81	= 28.66 mins/480	= 5.97%
Total Loss	**160 mins**	**33.33%**

OEE Calculations (Time in Mins)

Production time	= 480	Time less in availability loss	= 368	Time less in performance loss	= 333
<u>Availability Loss</u>		<u>Performance Loss</u>		<u>Quality Loss</u>	
Setup & Adjustments	= 82	Minor Stoppages (<5Mins)	= 28.66	Defects in Process	= 0
Equipment Failures	= 30	Reduced Speed	= 6.53	Reduced Yield	= 12.81
Total	= 112	Total	= 35.19	Total	= 12.81
Availability (368/480)	= 77%	*Performance (333/368)*	= 90%	*Quality (320/333)*	= 96%

OEE = 0.7*0.9 *0.96 = 66.7%

4 Enterprise Benefits

The implementation of OEE in an adequate system obtains substantial financial growth in manufacturing operations. Some of the benefits are

- Reduction in downtime costs
- Reduction in repair costs
- Increase in labor efficiencies
- Reduction in quality costs
- Increase in personnel production
- Increase in production capability.

The benefits of enterprise benefits are elaborated as follows

4.1 Reduced Downtime Costs

As a critical machine turns into inoperable, it moves major operations to standstill. The impact could affect or postpone the delivery commitments projecting a negative opinion on the manufacturer. The downsides can alter the cash flow and revenue of the manufacturer. Using the example of a semiconductor, the downtime of every year estimates to $1,00,000 of revenue loss for a process equipment in a critical unit. Alternatively, reducing the downtime to 1% on half of the critical equipment could gain a revenue and cost profit of %100,000,000 annually.

4.2 Reduced Repair Costs

The predictive maintenance which substantial reduces the repair costs is enabled using OEE. With the sequential growth in the database of issues related to downtime raises, the maintenance department will predict the impeding reasons for failure. The maintenance department will purview the predictive maintenance measures as the OEE system can be interfaced with a computer or android-based maintenance management system.

The maintenance department could keep track of raw materials and pre-order the required items in case of shortage predictions. The mandatory checks for equipment and timeline-based servicing can also reduce the repair costs. Savings the high repair costs would benefit the manufacturer in greater scales instead of repairing the machine after the equipment failures.

4.3 Increased Labor Efficiencies

With the raise in economic conditions, several manufactures have downsized suitably. Subsequently, the manufacturers are keen to optimize the production using the existing workforce. OEE system captures operator downtime periods along with the productivity data [11]. The manufacturer utilizes the information obtained by the OEE system to allocate proper resources inline with the productivity of a particular personnel. As the business gains profit, OEE systems predict managers to identify supplemental capacity inside the existing workforce rather than the new labor.

4.4 Reduced Quality Costs

Quality is a mark between the produced good yield compared to the total production. OEE is responsible to capture the quantity of total parts produced, the count of scraps and manufacturing defects and the cause of the latter. The information is obtained at the particular machine or line level. The ability to capture the quality in context of a particular product is attained by OEE [12]. The production managers will be able to identify the reasons and provide solutions to eradicate further costs involved in reworking the scrap by tracking the context-rich quality data using OEE and with enhanced focus on quality at every stage of production will reduce quality costs.

4.5 Increased Personnel Productivity

OEE systems reduce the paperwork involving the automation with computers or android-enabled devices. The floor managers and plant managers spend enormous time in calculating, analyzing, and recording the downtime and reasons in an spreadsheet with reasons for a particular issue [13]. OEE system captures and reports the downtime efficiently and saves time and additional workloads. OEE system alerts the respective worker, plant manager to the floor manager if the issues immediately.

4.6 Increased Production Capability

The reduced machine downtime due to the equipment failures require additional workforce to increase the productivity and decrease the defects. Increased production capability is an ability to achieve higher production levels with the available resources [14].

4.7 Conclusion

OEE can be increased by minimizing the breakdowns and changeovers losses which are associated with availability and by minimizing the defects and setup scraps losses which are associated with quality. The benefit of OEE is to establish benchmarks for success; it divides improvement opportunities into measurable categories; and it provides a measuring mechanism that allows for production improvement. Therefore, it is a goal to implement production planning using overall equipment efficiency (OEE) for IIoT to achieve maximum profit. However, the runway exists to raise goals as the OEE process is in place. Ultimately, the OEE measurement mechanism encourages implementable actions.

References

1. Zhu X (2011) Analysis and improvement of enterprise's equipment effectivenessbased on OEE. In: 2011 international conference on electronics, communications and control (ICECC), Ningbo, pp 4167–4171. https://doi.org/10.1109/ICECC.2011.6067706
2. Pomorski T (1997) Managing overall equipment effectiveness [OEE] to optimize factory performance. In: 1997 IEEE international symposium on semiconductor manufacturing conference proceedings (Cat. No.97CH36023). San Francisco, CA, USA, pp A33–A36. https://doi.org/10.1109/ISSM.1997.664488
3. Bai ZQ, Dai M, Wei QY, Zhang ZS (2018) An OEE improvement method based on TOC. In: 2018 25th international conference on mechatronics and machine vision in practice (M2VIP), Stuttgart, pp 1–6. https://doi.org/10.1109/M2VIP.2018.8600875
4. Yu C, Kuo C, Chiu C, Wen W, Zhang M (2018) Unveil the black box for performance efficiency of OEE for semiconductor wafer fabrication. In: 2018 international symposium on semiconductor manufacturing (ISSM). Tokyo, Japan, pp 1–4. https://doi.org/10.1109/ISSM.2018.8651146
5. Herng LK, Hee QC (2018) New benchmarked standard in yield and OEE improvement—SENS PSSO STS Gen2 taper performance improvement with breakthrough innovative solutions. In 2018 IEEE 38th international electronics manufacturing technology conference (IEMT). Melaka, pp 1–5. https://doi.org/10.1109/IEMT.2018.8511730
6. Chong KE, Ng KC (2016) Relationship between overall equipment effectiveness, throughput and production part cost in semiconductor manufacturing industry. In: 2016 IEEE international conference on industrial engineering and engineering management (IEEM), Bali, pp 75–79. https://doi.org/10.1109/IEEM.2016.7797839
7. de Ron AJ, Rooda JE (2005) Equipment effectiveness: OEE revisited. IEEE Trans Semicond Manuf 18(1):190–196. https://doi.org/10.1109/TSM.2004.836657
8. Krachangchan K, Thawesaengskulthai N (2018) Loss time reduction for improve overall equipment effectiveness (OEE). 2018 5th international conference on industrial engineering and applications (ICIEA), Singapore, pp 396–400. https://doi.org/10.1109/IEA.2018.8387132
9. Kao Y, Chang S, Dauzere-Peres S, Blue J (2016) Opportunity for improving fab effectiveness by predictive overall equipment effectiveness (POEE) In: 2016 e-Manufacturing and design collaboration symposium (eMDC). Hsinchu, pp 1–4
10. Wu Y, Wang S, Chen L, Yu C (2017) Streaming analytics processing in manufacturing performance monitoring and prediction. In: IEEE international conference on big data (big data), Boston, MA, pp 3285–289. https://doi.org/10.1109/BigData.2017.8258312

11. Kuan CW, Onn LS, Hanif M (2007) Improving direct labour productivity through minimizing time wastage approach. In: 2007 international symposium on semiconductor manufacturing. Santa Clara, CA, pp 1–4. https://doi.org/10.1109/ISSM.2007.4446822
12. Joochim O, Meekaew J (2016) Application of TPM in production process of aluminium stranded conductors. In: 2016 International conference on industrial engineering, management science and application (ICIMSA), Jeju, pp 1–5. https://doi.org/10.1109/ICIMSA.2016.7503996
13. Sanderson M, Shelton M, Mulligan S (1993) IBM Austin industrial business center total productive maintenance—the beginning. In: Proceedings of 15th IEEE/CHMT international electronic manufacturing technology symposium, Santa Clara, CA, USA, pp 211–214. https://doi.org/10.1109/IEMT.1993.398201
14. da Costa SEG, de Lima EP (2002) Uses and misuses of the 'overall equipment effectiveness' for production management. In: IEEE international engineering management conference, vol 2. Cambridge, UK, pp 816–820. https://doi.org/10.1109/IEMC.2002.1038543

Chapter 69
Counting and Tracking of Vehicles and Pedestrians in Real Time Using You Only Look Once V3

O. Swetha and C. Ramachandran

1 Introduction

In applications such as self-driving cars, intelligent security systems, machine learning, artificial intelligence, and many other similar areas, object detection, counting, tracking, etc. are currently important concerns. In a real-world scenario, object detection and tracking are challenging tasks. This paper under computer vision aims at detection, tracking, and counting objects over images and sequences of images called video. It helps the machine to understand, describe the object's direction of movement all by itself, once trained efficiently.

Our main concern in this paper will be to track moving traffic participants, such as people and vehicles. The human and vehicle counter was developed with the aid of deep neural network and tracking algorithms. The human counter was also used in situations where the number of students entering and leaving a class, college, or store to be counted. The algorithm used for tracking is the correlation tracking algorithm. Using the open-source dlib (digital library) toolkit, trackers can be applied to the bounding box obtained after the object detection stage. There are other applications where you need to count items from single pictures, such as taking attendance from classroom photos.

A large dataset is a pre-requisite for accurate results in image processing. The accuracy of a system lies in the dataset used in its training stages. A good dataset contributes good precision and recall. In the realm of object detection datasets like Pascal, ImageNet, and COCO are highly recommended by researchers. In this paper, the COCO dataset [1] has been used. COCO stands for (Common Objects in Context).

O. Swetha (✉) · C. Ramachandran
College of Engineering, Thalassery, Kannur, Kerala, India
e-mail: swetha.othayoth@gmail.com

C. Ramachandran
e-mail: crcsagar@gmail.com

I. Jeena Jacob et al. (eds.), *Data Intelligence and Cognitive Informatics*, Algorithms for Intelligent Systems,
https://doi.org/10.1007/978-981-15-8530-2_69

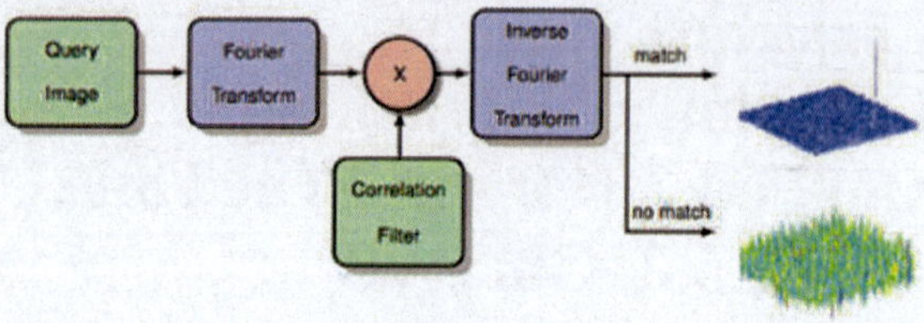

Fig. 1 Standard correlation tracker process [2]

It helped to increase the mAP (mean average precision) to 57.9 with YOLOv3. As the name suggests, images in the COCO dataset are taken from everyday scenes. The COCO contains more than 200,000 images and 80 object categories. As on track, moving objects are being concentrated, the dataset with is the person and different vehicles will be worked.

The basic block diagram of the proposed system is shown in Fig. 1. Section 2 does the literature review. Section 3 describes the various stages of the proposed system. Section 4 explains the software requirements, Sect. 5 covers the experimental system used for the implementation of the project. Section 6 shows the experimental results.

2 Literature Review

Just a CNN (convolution neural network) based frameworks for solving physical problems has been going down gradually. Several methods that made improvements to CNN based methods are discussed below.

2.1 Object Detection Algorithms

In 2014, an object detection technique called RCNN Region-based Convolutional Neural network [3] was introduced. Different region proposals were provided to the network. SVM classifier is used here. RCNN was incompatible to be used for real-time applications as the average time prediction taken on a single image is 40–50 s. Its successors, Fast RCNN [4] and Faster RCNN [5] tackle some of the performance and efficiency problems prevalent in RCNN. Here softmax classifiers were used. Fast RCNN added a bounding box regression head and took 2 s on each image. Faster RCNN added a Region Proposal Network and took just 0.2 s on each test image.

On the other hand, YOLO (You Look Only Once) method works by, passing the entire image through the network once. The first version of YOLO was introduced in 2016 [6] and was the basis for the new versions that followed. The architecture used 24 convolutional layers for feature extraction and 2 dense layers for doing the predictions. The architecture used is Darknet, which is an open-source neural network framework. The algorithm divides an image into a grid of cells, with each cell predicting bounding boxes. From the infinite bounding box predicted, the confidence scores help choose the credible one. The confidence is higher when the detected object overlaps with the ground truth. The version was not giving accurate results for objects that were near YOLOv2 [7] introduced anchor boxes. These anchor boxes are pre-determined sets of boxes on which corrections are done in the form of offset to find bounding box, instead of finding a bounding box directly. More fine-grained features helped in estimating smaller objects better. YOLOv3 [8] also uses the concept of anchor boxes with generating up to 3 bounding boxes from each cell. The darknet is updated to have 53 convolutional layers.

Review on some of the tracking algorithm is done below:

Kumar et al. proposed background modeling using Gaussian Mixture Model (GMM) for blob extraction of multiple human motions and multiple people tracking using Kalman filter for tracking multiple objects for real-world surveillance scenarios [9]. However, there can be a possibility where one object will be hidden by another object in the field of view of any one of the cameras. This problem is referred to as occlusion. Thus, this proposed system has a problem to track humans in the video sequences under occlusion in the dynamic background. Rao et al. suggested the detection using a median approximation technique and done tracking based on template matching using the Kalman filter estimation technique for a single object in a sequence of frames [10]. Nevertheless, there also remains a problem to track an object the video sequences when happening to done occlusion. Prabhakar et al. presented a concept of frame differencing for detection and done technique of object template matching for tracking [11]. Mishra et al. proposed a technique for multiple objects tracking from a single frame by calculating the centroid of objects using some structure information [12].

2.2 Correlation Filter Tracker

To create a fast tracker, correlation is computed in the Fourier domain Fast Fourier Transform (FFT). First, the 2D Fourier transform of the input image: $F = F(f)$, and of the filter: $H = F(h)$ are computed. The Convolution Theorem states that correlation becomes an element-wise multiplication in the Fourier domain. Using the $\odot$ symbol to explicitly denote element-wise multiplication and $*$ to indicate the complex conjugate, correlation takes the form:

$$G = F \odot H* \tag{1}$$

The correlation output is transformed back into the spatial domain using the inverse FFT. The bottleneck in this process is computing the forward and inverse FFTs so that the entire process has an upper bound time of $O\ (P \log P)$ where P is the number of pixels in the tracking window [13].

3 Proposed System

A video is a series of frames. All input video frames undergo the detection of targets using Yolo v3. After this stage, the output obtained will be the bounding box around the detected object. Obtaining the bounding box implies that from the input frame the object has been found. Only the target object should be identified by the detector (that is, People in People counter application). After the bounding box is obtained the next stage would be to locate the bounding box center. Every centroid identified is given a unique ID. The centroid of each new object that enters the frame is determined and a new ID is given for each of the new objects. The bounding box coordinates are passed on to the tracking algorithm for objects, which will track each object throughout the video. Section 3.1 discusses more about the tracking algorithm (Fig. 2).

3.1 Object Detection

YOLO V3 The object detection algorithm used is the latest version of YOLO (You Look Only Once) which is Yolov3. The input image is looked upon only once and then the images are divided into $S * S$ grids. Each grid is checked for the presence of objects and a bounding box is drawn. Each grid can predict up to three bounding boxes. There will be a large number of prior boxes called anchors boxes are present with fixed aspect ratio. These aspect ratios are determined before training by running K-means on the entire dataset. The ones which have high objectness score are kept. That is, a score of less than 0.5 are discarded. If the value is 1, it means the bounding box overlaps the ground truth object than any other bounding boxes.

Darknet-53 is the architecture of yolov3. 53 refer to the number of layers in the network [8]. For the task of detection, another 53 layers are added, giving it a 106 fully convolutional underlying architecture for Yolo v3. The previous version of YOLO failed in the detection of smaller objects. This network performs multiple object detection from a single image. Features are learned during the network training process. This network performed better and achieved higher precision results comparing to the methods which use the sliding window approach like R-CNN, Faster R-CNN (Table 1).

Each block contains convolution layer 1×1 followed by 3×3 plus a skip connection. A 2 stride after each block reduces the dimension. A feature map is

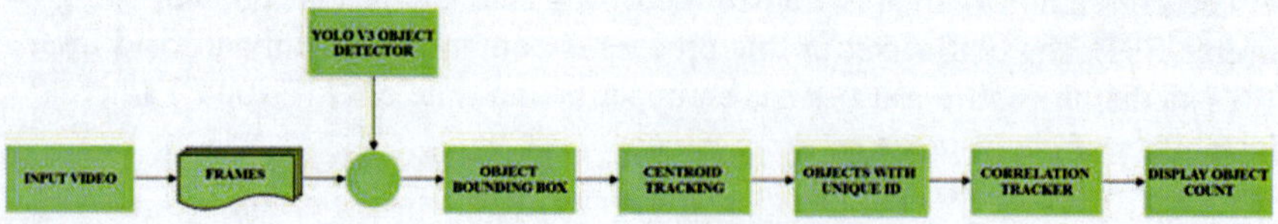

Fig. 2 Proposed architecture

Table 1 Darknet 53 [8]

	Type	Filters	Size	Output
	Convolutional	32	3 × 3	256 × 256
	Convolutional	64	3 × 3 / 2	128 × 128
1x	Convolutional	32	1 × 1	
	Convolutional	64	3 × 3	
	Residual			128 × 128
	Convolutional	128	3 × 3 / 2	64 × 64
2x	Convolutional	64	1 × 1	
	Convolutional	128	3 × 3	
	Residual			64 × 64
	Convolutional	256	3 × 3 / 2	32 × 32
8x	Convolutional	128	1 × 1	
	Convolutional	256	3 × 3	
	Residual			32 × 32
	Convolutional	512	3 × 3 / 2	16 × 16
8x	Convolutional	256	1 × 1	
	Convolutional	512	3 × 3	
	Residual			16 × 16
	Convolutional	1024	3 × 3 / 2	8 × 8
4x	Convolutional	512	1 × 1	
	Convolutional	1024	3 × 3	
	Residual			8 × 8
	Avgpool		Global	
	Connected		1000	
	Softmax			

obtained finally after passing through all those layers. Softmax classifiers are not used here instead depends on logical regression to calculate how likely the object belongs to a specific label. The last two layers Avgpool and connected layers are for classification.

Training Download darknet, an open-source neural network framework to train the detector for the training.

First, collect the images of objects for the dataset. A file with an extension of.names will store the object name. For YOLOv3, each image should have a corresponding text file in the same directory, with the same file name as the image (Fig. 3).

Labellmg is the graphical image annotation tool used to create this .txt file. The 5 numbers shown in the .txt file above corresponds to class id, x and y center coordinates of the bounding box and finally height and width of the bounding box. Txt files are created using software called Labellmg (Fig. 4).

A file with an extension of.data includes train and test data sets and backup file information. The yolov3.cfg file needs to be modified according to our custom dataset. Changes are made to parameters like classes and filters.

Fig. 3 .txt file format

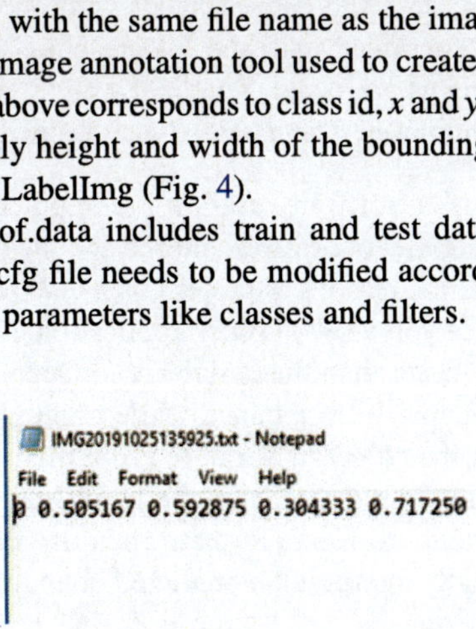

Fig. 4 Labeling using labeling

$$\text{Here, classes} = 4 \tag{2}$$

$$\text{Filter} = (\text{classes} + 5) * 3. \tag{3}$$

Since classes = 4 (here, person, car, bus, motorcycle), the value of the filter is 27.

Pre-trained YOLO weights are available to perform detection. Though these pre-trained weights may be trained on datasets that don't be required, still using these weights for training our datasets is called Transfer learning. Transfer learning helps our network learn faster. The dataset needs to be trained for higher epochs. The training process is continued until the loss reaches a certain threshold. For every 1000 iterations, weights are saved. The final weight generated after training can be used for detection purposes. The datasets were trained for 50 k epochs.

Testing OpenCV is a software used for computer vision applications. Using OpenCV the image or the frames are captured and the bounding box is drawn around the objects. Some of the reasons to prefer Open CV over darknet for testing are that it is easy to integrate yolov3 with other applications that use OpenCV, OpenCV in CPU version is 9× faster than the darknet, and OpenCV supports python. Keras is used to call the models which obtained while training and also to use the model to find the position of the object in a frame. From the models, the object coordinates are obtained and then implement it on the input image and video. The pre-trained weights are converted into Keras format. Then use the pre-trained model to make predictions, and finally interpret the predicted bounding boxes.

3.2 *Video Analysis*

Centroid Tracking
The centroid tracking will find the center (*x*, *y*) of the bounding box, which was obtained from the object detection algorithm. The bounding box will be provided to the centroid tracker through the YOLO object detector. A new ID is given to the objects in a frame.

In a video, there will be movement of the object in the consecutive frame. Any new objects entering the frame have to be correctly identified and assign them a new ID number.

The Euclidean distance for a pair of existing centroids and new centroids is determined. It is with the assumption that the Euclidean distance between a new object entering the frame and the already existing object will be more. Then the next step will be to register the new object with a new ID number.

Correlation Tracker In python, the correlation filter tracker is implemented using an open-source library called Dlib (Digital Library), which deals with various machine learning problems. Dlib's implementation of the correlation tracker is used to track the object as it moves in the subsequent frames. The location of the object bounding box in the previous frame and from the current frame is collected to infer where the new location of the object is. The working of the dlib correlation filter is explained in the flowchart in Figs. 5 and 7 .

- Python can use dlib tools. One such tool used here is dlib.correlation_tracker. It is a tool to track moving objects in a video frame.
- The default format in OpenCV is BGR which is converted to RGB format.

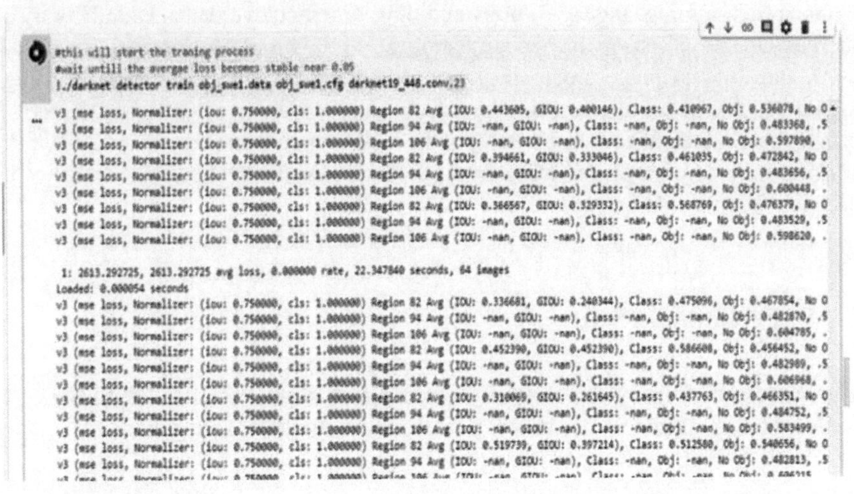

Fig. 5 Training process

- The object detector should only detect the object wanted to track. For that, the class id should match. For instance, for a person counter, the object id should be 1.
- Once the object is detected, it is tracked in further frames.
- Tracker.update (image) gives the correlation tracker the latest frame and tracker.get position() returns the object coordinates are being tracked.
- In applications that require counting the number of people leaving and entering a class or store, needed to figure out whether the target is moving upwards or downwards. The direction is found by the following method:

  (X,Y) coordinate of current centroid—mean of the previous centroids

$$=> 0, \text{Moving outside college}$$
$$; < 0. \text{Moving inside college}$$

Improvements

The correlation filter previously monitors the moving object by continuously measuring the highest value of the correlation between the input image and a filter, also the detection algorithm had to be performed separately for each frame, to detect an object. The use of the YOLO object detection algorithm which performs object detection at once for all frames rectifies this problem.

4 Dependencies

Python programming language is used including its associative packages and libraries such as OpenCV, Tensor flow (deep learning), dlib, NumPy (numerical computation), and Pillow (image processing), imutils libraries. Dlib is a machine learning toolkit capable of solving real-time problems. Anaconda navigator assists to recognize objects in the image dataset. To build the YOLO open-source darknet framework and pre-trained weights for training the model need to be downloaded. Other software used are Google colab, pycharm, and visual studio.

5 Experimental Setup

The processor of the system is Intel(R) CoreTM i5-5200U CPU @ 22.20 GHz. A webcam for recording scenes in real time can be mounted or handheld.

6 Results

The first process is object detection. Figure 6 demonstrates object detection outcomes using YOLOv3. Depending on the item in view, the bounding box varied in shape and scale.

6.1 *YOLO Object Detection*

Table 2 summarizes the features of different object detection techniques.

6.2 *Real-Time People Counter*

People counter has been implemented two ways

- First, counting people randomly moving left, right in a frame. The screenshots of the results are shown in Fig. 8.
- Secondly, counting the number of people entering and exiting college. The result is shown in Figs. 9 and 10.

To count students from both directions, the camera should be placed such that the top view of the path is visible. Accurate results are seen when the camera is placed vertically. All the students entering and exiting has been counted successfully (Fig. 11).

Vehicle Counter
See Fig. 12.

The accuracy is computed as:

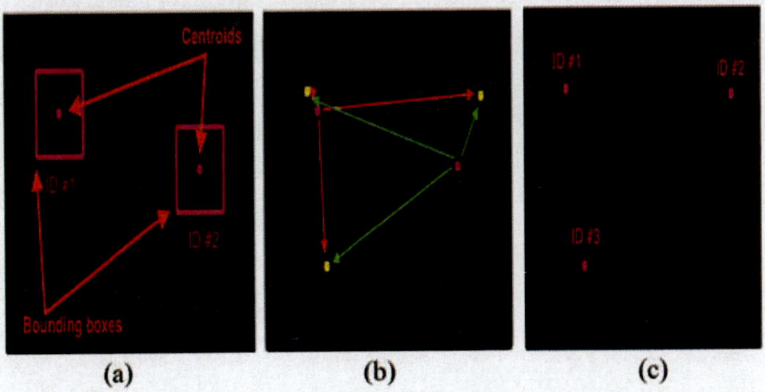

Fig. 6 **a** Computing centroids of the bounding box. **b** Calculating the Euclidean distance between each pair of centroids. **c** Registering the new object with a new ID number

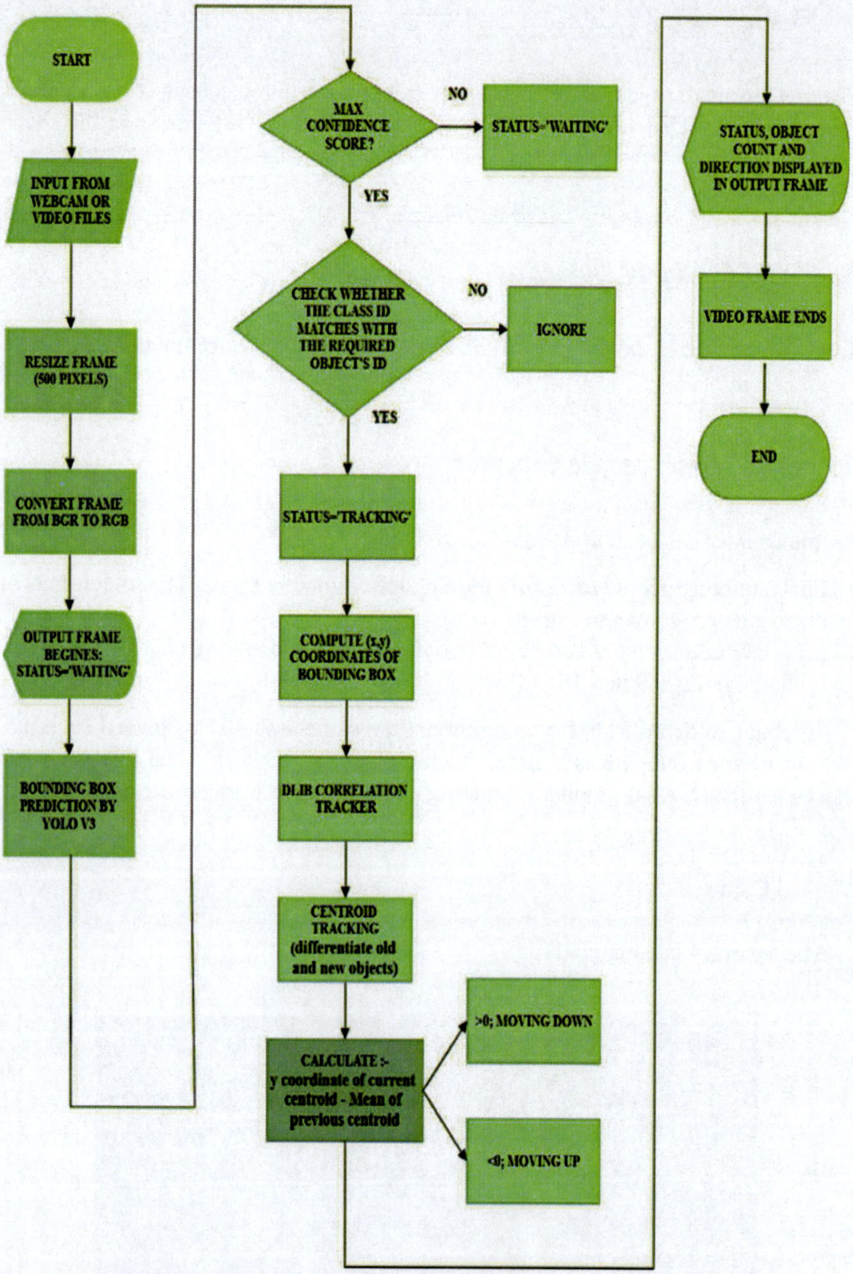

Fig. 7 Flowchart of the proposed scheme

Table 2 Comparison of techniques [3–8]

Model	Architecture	mAP	FPS
Fast RCNN	CNN	70	0.5
Faster RCNN	VGG 16	73.2	7
Faster RCNN	ResNet	76.4	5
YOLO (448 × 448)	VGG-16	63.4	45
YOLO v2 (544 × 544)	Darknet-19	78.6	40
YOLO v2 (416 × 416)	Darknet-19	76.8	67
YOLO V3	Darknet 53	57.9	50

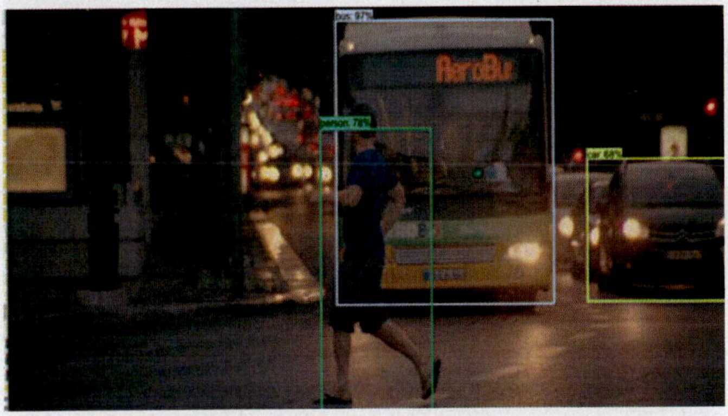

Fig. 8 Object classification and localization using YOLOV3

(a) (b)

Fig. 9 **a** Video 1, **b** video 2

$$\text{Accuracy} = \frac{\text{Number of correct detection}}{\text{Number of ground truth detection}}$$

With reference to the calculations of Tables 3 and 4, the accuracy of the tracker is found to be around 95%.

(c) (d)

Fig. 10 **c** Real time capture, **d** occlusion scene

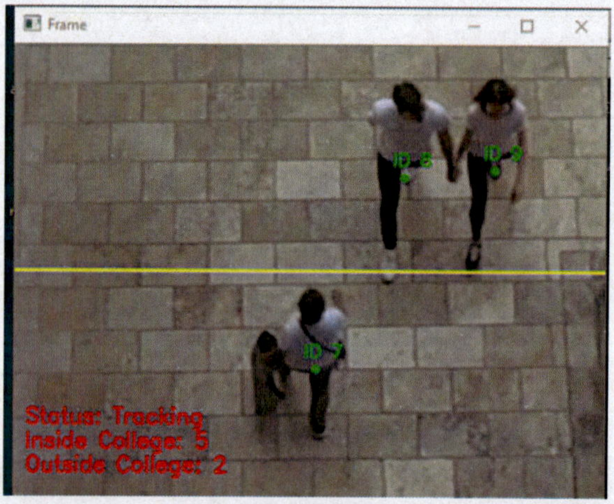

Fig. 11 Students tracking video

(e) (f)

Fig. 12 Vehicle Counter, **e** video 1, **f** video 2

Table 3 Output analysis of videos (a–d)

Inputs	Predicted count	Actual count	FPS
Video 1	3	4	19.22
Video 2	2	2	45.60
Real time	1	1	20.70
Occluded scene	5	5	14.37

Table 4 Output analysis of videos (e) and (f)

Inputs	Predicted count	Actual count	FPS
Video 1	3	3	51.20
Video 2	2	2	14.37

7 Conclusion

Different video inputs and real-time capture were used to analyze the human and vehicle counter. OpenCV made video processing in CPU much faster. The processing speed can be increased with the use of GPU. The detection of objects finds its application in a self-driving vehicle. It is a massive project which offers endless possibilities. Image analysis can be used in crowd research applications, to analyze how many children enter college within the time given, or can be used in shops to help the shopkeeper take the required steps for increased or decreased traffic to improve their profits. There are endless applications where computer vision is the core.

References

1. Lin TY et al (2014) Microsoft COCO: common objects in context. In: Fleet D, Pajdla T, Schiele B, Tuytelaars T (eds) Computer vision—ECCV 2014. Lecture notes in computer science, vol 8693. Springer, Cham
2. Kumar BVKV, Mahalanobis A, Juday RD (2005) Correlation pattern recognition, Cambridge University Press
3. Girshick R, Donahue J, Darrell T, Malik J (2014) Rich feature hierarchies for accurate object detection and semantic segmentation. In: 2014 IEEE Conference on computer vision and pattern recognition, Columbus, OH
4. Girshick R (2015) Fast R-CNN. In: 2015 IEEE International conference on computer vision (ICCV), Santiago
5. Ren S et al (2015) Faster R-CNN: towards real-time object detection with region proposal networks. In: Proceedins of NIPS
6. Redmon J, Divvala S, Girshick R, Farhadi A (2016) You only look once: unified, real-time object detection. In: 2016 IEEE Conference on computer vision and pattern recognition (CVPR), Las Vegas, NV
7. Redmon J, Farhadi A (2017) YOLO9000: better, faster, stronger. In: 2017 IEEE Conference on computer vision and pattern recognition (CVPR), Honolulu, HI

8. Redmon J, Farhadi A (2018) YOLOv3: an incremental improvement. C. J. Kaufman, Rocky Mountain Research Lab., Boulder, CO, private communication, May 1995
9. Kumar A Sureshkumar C (2014) Multiple human tracking in surveillance videos. SIJ Trans Comput Sci Eng Appl (CSEA) 2(6):202–207
10. Rao G, Satyanarayana C (2014) Object tracking system using approximate median filter, Kalman filter and dynamic template matching. Int J Intell Syst Appl 6(5):83–89
11. Prabhakar N, Vaithiyanathan V, Sharma A (2012) Object tracking using frame differencing and template matching. Res J Appl Sci Eng Technol 4(24):5497–5501
12. Mishra R, Chouhan M, Nitnawwre D (2012) Multiple object tracking by kernel based centroid method for improve localization. Int J Adv Res Comput Sci Softw Eng 2(7):137–140
13. Bolme DS, Beveridge JR, Draper BA, Lui YM (2010) Visual object tracking using adaptive correlation filters. In: Proceedings of the IEEE conference on computer vision and pattern recognition, San Francisco, CA, USA, 13–18 June 2010, pp 2544–2550

Chapter 70
Comprehensive Analysis of Multimodal Recommender Systems

Viomesh Kumar Singh, Sangeeta Sabharwal, and Goldie Gabrani

1 Introduction

This is the era of the Internet, mobile phones, smart devices and social network where information is growing explosively. Available information can be used for decision making, but one of the major problems is of information volume being huge which requires so much time to analyze the information and most of the time it creates confusion as well. Information growth is explosive after the introduction of mobile phones, smart devices and social networks. Users might think that more choices/information is better for decision making. However, in 2000, psychologists Iyengar and Lepper [1] published a remarkable study which says, "More is not always better". More choices require time, and it creates confusion and sometimes even anxiety.

Recommender systems are designed to solve the immense issue of information overload by supporting customers in decision making with personalized and more relevant information [2]. Since the introduction of mobile phones, smart devices and social networks, information growth is explosive. Information is not only huge but is also very diversified. This information is in various forms such as text, social tags, images, location, etc. Social networking is encouraging people to communicate and

V. K. Singh (✉)
Netaji Subhas Institute of Technology, Delhi, India
e-mail: singh.viomesh@gmail.com

S. Sabharwal
Netaji Subhas University of Technology, Delhi, India
e-mail: ssab63@gmail.com

G. Gabrani
BML Munjal University, Gurgaon, India
e-mail: goldie.gabrani@bmu.edu.in

I. Jeena Jacob et al. (eds.), *Data Intelligence and Cognitive Informatics*,
Algorithms for Intelligent Systems,
https://doi.org/10.1007/978-981-15-8530-2_70

provide more and more information, but the modality of information remains very diversified.

The choice of the user depends on many factors like his mood, contextual environment, system's interface, location, etc. Normally, user interacts with the system with more than one mode such as facial expression, speech and gesture. All these factors affect users in their decision-making process. So, multimodal analysis of user-generated content [3] is required to understand user preferences better and to perform better user personalization. Recommender systems based on multimodality are best suitable to analyze user behavior. Multimodal recommender systems can enhance the accuracy as different modalities can support an object more accurately than the single modality.

In the past two decades, the tremendous successes of recommender systems have been witnessed, multimedia information and multimodal interaction between the user and the system. Multimodality is revolutionizing the architectures of recommender systems dramatically and is also enhancing opportunities by reinventing the techniques used for the analysis of both users and items for better customer satisfaction. Recent advances in multimodal recommender systems have gained significant attention by overcoming obstacles of conventional unimodal recommender systems and achieving high recommendation quality. Furthermore, it extracts relationships within the data itself, such as textual, visual, audio and contextual information.

In the last few years, several surveys on traditional recommender systems have been published. For example, Adomavicius and Tuzhilin [4] presented a state-of-the-art of current recommender systems with limitations and discussed possible extensions to enhance the capabilities of recommender systems by incorporating contextual information and multi-criteria ratings; Burke [5] published a comprehensive survey on hybrid recommender system; Yang et al. [6] and Milicevic et al. [7] presented a survey on social networks and social tagging-based recommender systems. Zhang et al. [8] analyzed recent research on deep learning-based recommender systems. Cantador et al. [9] and Zhang et al. [10] analyzed the cross-domain recommendation. However, there is a no extensive analysis of multimodal-based recommender systems. To the extent of our best knowledge, no survey paper has been published on multimodal recommender systems. Given the rising popularity and potential of multimedia information, multimodality applied in recommender systems to enhance the accuracy, a systematic analysis is required. Various multimodal recommender systems are also analyzed and presented some insights. In the end, this paper provides a critical conclusion, pointing out weak and strong issues of recommendation techniques based on multimodality.

2 Recommender Systems

Recommender systems are destined to solve the immense issue of information overload by supporting customers in decision making with personalized more relevant information [2]. Recommender systems are shown in Fig. 1 suggest the most relevant

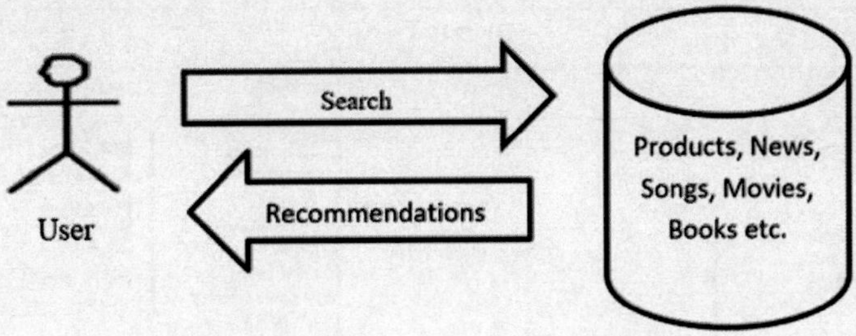

Fig. 1 Overview of a recommender system

products or services based on the user's profile. Most of the recommender systems mainly consider the content of items and previous choices/feedback of users and estimate users' preferences on items they have not seen. Recommender systems are broadly classified into three main categories content-based, collaborative filtering and hybrid.

Content-based recommender systems establish relationships among items based on their attributes and properties. In collaborative filtering approach, the system computes correlation among users by examining the user-item matrix based on users' previous choices and their feedbacks. The hybrid recommender system is a combination of both collaborative filtering and contentbased systems. Besides these new recommender systems can also utilize other available information parameters like time, location, the mood of the user, context, etc., to provide a more accurate recommendation. Moreover, with the growth of smartphones, social networks and IoT devices can provide more and accurate information about users, which can enhance the quality of recommendation systems.

3 Multimodality

User and system both require some sort of interface to communicate. An interface, on the other hand, depends on number and variety of its inputs and outputs which are known as communication channels that enable users to interact with a system. Each distinct independent single channel is called a modality. If a system allows only one modality as an interface between system and user, then it is known as a unimodal system, and if the system allows interface with more than one modality (shown in Fig. 2), then it is called multimodal system.

Naturally, humans interact with more than one mode like facial expression, gesture and speech at the same time. To provide the best possible recommendations, the system needs to know human interaction which is generally not unimodal. This is similar to a situation when someone is watching a movie, and his reaction might be smiling or feeling thrilled (Gesture). Further, he might talk positive about the movie or humming song/music of the movie (Audio) or may write comments and provide

Multimodal				
Linguistic	**Aural**	**Visual**	**Spatial**	**Gestural**
• spoken and written language • text • blog • metaphore • information structure	• music • rhythm • sound effects • noises • pitch etc.	• Images (still/moving), • vectors • viewpoint • symbols • video	• environmental spaces and architectural spaces, • Geographical direction, layout, position • organization of objects in space	• movement of the body • facial expressions • demeanors through the use of rhythm • speed, and stillness

Fig. 2 Key modalities used in multimodal recommender systems

movie rating (text). Usually, text modality in the form of ratings or comments is only considered, but very few people bother to provide ratings or comments. This is the reason a very sparse matrix of a user is got and item which is typically error-prone in nature.

Recommender systems which consider multimodal systems are best suited to analyze the behavior of a wide range of clients, tasks, situations and context. Due to multimodal feedback, can get redundancy that increases the likelihood of user preferences because there are many simultaneous references to the same interest. Recommender systems based on multimodality can also enhance accuracy. This is because usage of more than one modality can indicate an object more authoritatively than the case when a single modality is used. One mode of communication can resolve ambiguities or refine imprecision, provided by another mode to enhance synergy. For example, in the case of merchandise, textual description and images of clothes are both available. A text description provides valuable information like size material, etc., whereas images provide information like shades, color, style, etc. Images of packing and branding also matter in the case of most of the products. In the case of restaurants, similarities can be found based on images providing information like ambience, garden, rooftop seating, lighting, view, etc.

4 Literature Review

Kaklauskas et al. [11] proposed an affect-based multimodal video recommendation system (ARTIST) by analyzing limbic reactions such as emotions, instinctive and instantaneous human reactions conveyed by facial expressions and movements of legs, arms and/or other body parts. ARTIST is capable of analyzing movements of legs, arms and/or other body parts, facial expressions and physiological parameters of a video's viewer. ARTIST recommends next video based on analysis of user's limbic reactions, and if nothing suitable is found for any user, then it recommends video according to viewer's mood (neutral, scared, angry, happy, sad, disgusted and surprised). This recommender system works in a real-time environment.

Oramas et al. [12] developed a recommender system by applying deep learning techniques on multimode (text and audio) information. This system learns about an artist by applying feature embedding on biographies with semantic information, metadata and analyzes songs and music by applying track embedding techniques on an audio signal. The author also exploited the usage of data and available feedbacks. Then, they built a deep learning-based recommender system which combines text and audio features in the multimodal network. They successfully overcome the cold start problem.

Yukawa et al. [13] proposed a group recommender system with the multimodal user interface by enabling users to recognize, coordinate and reconsider the diverse information aspects from visual and haptic information. They performed two tasks: First, they provide a recommendation for a temporary group based on the user's social network. Second, they computed the candidate groups of candidate members to provide trustful group recommendation.

Kuo et al. [14] proposed a framework for automatic background music recommendation for video. They applied multimodal latent semantic analysis audio and video techniques on YouTube videos. The framework extract features from music and videos. It utilizes those features to find-out the co-occurrence relationship between background music and videos. They also proposed an algorithm to compute alienability of music and video for music beats and specific video shot.

Domingues et al. [15] built a multimodal online hybrid recommender system for music based on usage, audio and social tags. The author in his work generated three different recommendation list based on usage, audio and social tags and exploited user's playlist. They applied cosine angle method on the usage data to compute usage-based similarity among music tracks. To calculate the relationship among social tags assigned to music tracks, latent semantic analysis has been applied. Authors used MARSYAS framework [16] to extract audio features because it has a low dimensional matrix and requires low computational time, and similarity among audio features is calculated by Euclidian distance. The author built three different recommendation list (rank wise) based on audio, social tags and usage then combine rank based on each list and compute an average of them and produce a final recommendation.

Da Costa and Manzato [17] proposed a recommender system based on three different ensemble approaches to exploit multimodal user interaction in order to integrate different types of individually processed feedbacks. Authors applied three ensemble algorithms, two of them are heuristic-based methods: tag-based heuristic technique, average-based heuristic technique, and the third one is ensemble Bayesian personalized ranking (BPR) learning which they created to analyze how much a certain interaction contributes. Finally, authors ensemble rankings produced by different unimodal-based aforementioned approaches.

Da Costa et al. [18] proposed a robust framework to provide more accurate personalized ranking in recommendation based on multimodal interactions. Authors applied SVD++ and BPR MF (Bayesian personalized ranking) to produce a ranking for unimodal recommender system on different kinds of feedbacks. SVD++ is specifically used to provide ranking on explicit feedbacks, and BPR MF is used for implicit feedback. Authors also proposed a post-processing step to ensemble rankings generated by unimodal recommender systems.

Arapakis et al. [19] proposed a multimodal recommender system for videos which integrates real-time facial expression into a user profile to generate a better recommendation. Authors created user profile of 24 participants of diverse ethnicity and educational background based on query submission at the time of registration. This user profile is updated based on implicit feedback such as user's search query and click-through action. Then, user captures user facial expression (chuckles, frowns, smiles, smirks, etc.) and analyzes their emotions (anger, disgust, happiness, sadness and surprise). Further user's emotions were integrated with the user's profile data and used an support vector machine model (SVM) to find out relevant videos.

Cho et al. [20] proposed a recommendation approach by exploiting contextual information to access much real-time and plentiful information like location, time schedule, event, weather condition, etc. under omnipresent environment. Based on the aforementioned information, the Authors added a new dimension 'Time Schedule' to the existing User-Item matrix. Then the new 'User-Item-Time Schedule' matrix was utilized to make context-aware recommendations. Authors also computed similarity between items and contextual information manually, which is not feasible for a real scenario as the number of items is huge and it is not possible to compute similarity manually.

Peska [21] developed a multimodal recommender for small scale e-commerce by utilizing heterogeneous implicit feedback to address the lack of relevant feedback and data sparsity problems. Authors concentrate on heterogeneous implicit feedback and proposed relevant context of multimodal feedback and derived models of preferential relations and negative implicit feedback.

Da Costa and Manzato [22] presented an ensembling technique for movie recommendation that consolidates rankings produced by different unimodal recommenders based on specific interaction types. This method enables users to interact with the system in various ways and collect implicit and explicit feedbacks based on different interactions paradigms to know more information about the user's preferences to provide better recommendations.

Jia et al. [23] proposed an innovative approach for video recommendation on mobile phones by exploiting mobile data. Authors developed an Android-based application to analyze user's preferences based on their Android application usage. This model incorporates text information and video features with the user's short-term and long-term preferences. This multimodal heterogeneous information is utilized to provide recommendations for short-term relevance and the user's long-term preferences.

Laurier et al. [24] presented multimodal music mood classification approach based on lyrics and audio. Latent semantic analysis and distance-based methods were applied to classify the lyrics in an unsupervised way, and music information retrieval evaluation exchange (MIREX) was applied to classify audio mood in a supervised way. Authors combined audio and lyrics features in the same space within one classifier. Integration of music and audio (multimodal: audio and text) provided better performance.

Yang et al. [25] proposed a novel approach for online video recommender system based on multimodal relevance and feedback. Authors first computed multimodal relevance between two videos as the combination of visual, textual and aural relevance by automatic adjustment of intra weights of aforementioned three modalities. Finally, they applied attention fusion function to combine individual visual, aural and textual modalities.

Campigotto et al. [26] proposed a three-step framework for personalized and situation-aware route recommendation by utilizing multimodal information. In the first step, authors collected users' information like name, age, vehicle type, home location, profession, workplace and other important places and classify them. In the second step, mass preference prior is applied to encode the knowledge collected at the first stage, user is provided with two realistic routes, and his preference between them was asked. In the final step, the Bayesian learning technique is applied to update the user's profile based on his choice of route. This framework recommends route based on the user's profile as well as actually chosen route, and the user's profile is regularly updated based on the choices he makes.

Rafailidis et al. [27] proposed a temporal collective matrix factorization (TCMF) model to understand users' preference evolution by considering multimodal information and social relationship over time. Authors explore multimodal user-item interaction to address the data sparsity problem and exploit social relationship over time to compute users' preference evolution by extracting the users' temporal pattern based on changes in their friend list time to time.

Wang et al. [28] presented a location recommender system based on semantic similarity among venues by user-generated multimodal information from different social networks. Authors exploited user's generated multimodal information like comments, description, photos, venue context (geographical information) and check-in behavior. The integration of these user's generated multimodal information is used to generate location ontology.

Zhao et al. [29] built a movie recommender system by utilization of multimodal heterogeneous information: text, poster, rating of the movies and social relationship. The presented framework learns movie content by the multimodal neural network in two phases: The relationship among text information of movies is calculated by recurrent neural network, and relationship among images of movies is computed by a convolutional neural network. These two neural networks are then integrated, and multimodal fusion layer is formed to represent movies. Authors applied random walk-based learning technique on the multimodal representation of the movie, user feedback and social relationship in a social network to create ranking matrices.

Alhamid et al. [30] presented a framework that utilizes the real-time physiological and environmental information with user-item ratings, user's purchase history and social network information. They developed a multimodal interface to collect the physiological parameters and ambient intelligent (AmI) environment information such as ECG signal light intensity, contextual information to provide Android-based smart biosensor recommendation and interactive TV recommendations.

Liu and Gan [31] presented a recommendation system for tourism products based on multimodal topic modeling. They explained various multimodal topic modeling methods based on latent Dirichlet allocation (LDA) and probabilistic latent semantic analysis (PLSA). However, they did not elaborate cross-lingual text mining and effect of multilingual setting on multimodal topic modeling.

Huang et al. [32] built a recommender system based on multimodal learning for Internet of things (IoT). They applied four parallel five-layer CNN-based triplet network on global features of users and two parallel three-layer CNN-based neural networks on multimodal features of items to provide a better recommendation.

Salah et al. [33] developed a framework "Cornac" to handle text, image and graph modality to build, access and evaluate recommender systems. Cornac allows researchers to read, format, parse and transform three different modalities: text, image and graph. It also allows researchers to utilize various recommendation models to implement, evaluate and compare on abovementioned three modalities.

4.1 Modality-Based Analysis

Most of the unimodal-based recommender systems are based on text-based modality like feedbacks, ratings and purchase history. To exploit most of the available information, researchers should utilize images, audio, videos, locations, etc., with text-based information. The knowledge gained from these different modalities is combined in the form of multimodal information to provide better recommendations. In our survey, most of the multimodal recommender systems have been found to utilize text-based modality (86%). Apart from text, researchers also used temporal parameters (43%), audio (33%), images (19%), video (29%), location and social parameters as shown in Fig. 3. Temporal and demography parameters play an important part in decision making as preferences change with time, culture and further language changes with

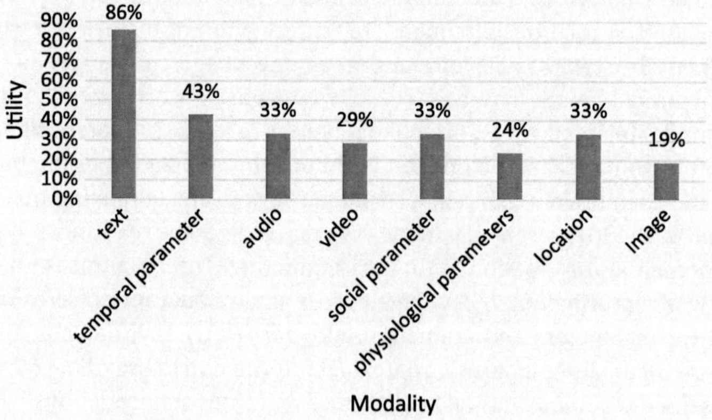

Fig. 3 Utilization of each modality in multimodal recommender systems

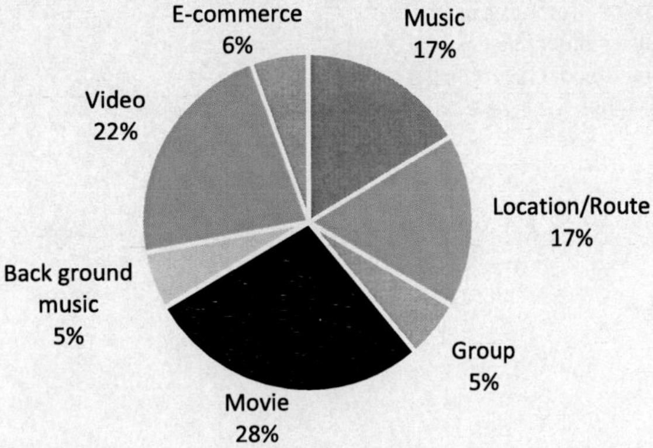

Fig. 4 Illustration of multimodal recommender system for different application

location. Physiological parameters are exploited to get knowledge about users' mood in real time to provide recommendation according to their mood.

4.2 Application-Based Analysis

In our analysis, multimodal analysis is mainly used to recommend application like movies, videos and music have been found as shown in Fig. 4. Researchers exploit multimodal information according to the application. For example, video recommendation researches exploit the mood of users by taking into account various psychological parameters and for music recommendations, researchers exploit text, audio, time and location.

Multimodal information is not much utilized in the recommendation of many areas like e-commerce, tourist, news, restaurants, etc. But multimodal recommender system can provide better recommendation in these areas as it utilizes additional alternative information which can support to make a better decision. For example, in the case of merchandise, textual description and images of clothes both are available. A text description provides valuable information like material, size, etc., whereas images provide information like shades, color, of restaurants, apart from rating similarities can be calculated based on locations, images providing information like ambience, garden, rooftop seating, lighting, view, etc.

4.3 Technique-Based Analysis

In our analysis, there is no single technique can be utilized for handling multimodal information have been found. Most of the researchers applied different techniques on

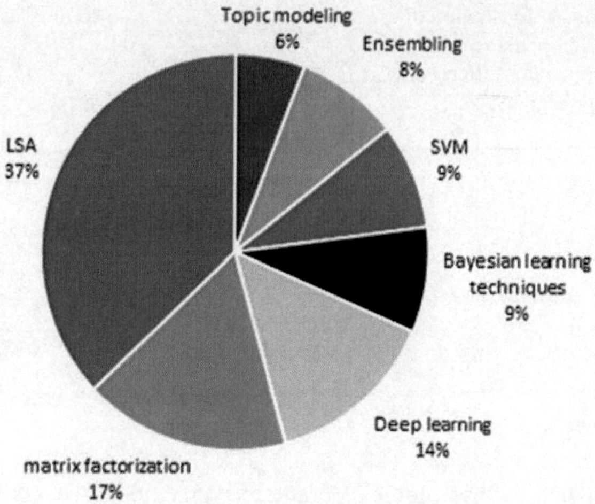

Fig. 5 Application of different techniques in multimodal recommender systems

different modes based on the suitability of a technique for particular unimodal information. For example, all machine learning techniques like SVM, matrix factorization, KNN, topic modeling, etc., are utilized for users' similarity according to rating, word embedding, TF-IDF for content similarity, audio embedding to compute audio-based similarity as shown in Fig. 5. LSA is mostly applied to compute the similarity between different modalities.

4.4 Modality- and Application-Based Analysis

In our paper, music recommendations have found that, researchers combine text like lyrics, artist, metadata with audio. Oramas et al. [12] utilized lyrics (text), audio and feedbacks. Domingues et al. [15] exploited usage and content data (audio and textual tags). Laurier et al. [24] exploited lyrics, audio and tags. Kuo et al. [14] utilized video, music,audio word, visual word to provide automatic background music recommendation.

Whereas for location/route recommender systems, researchers utilize multimodal information like location coordinates, mobility option, socio-demographics, text descriptions, photos, user check-in patterns, venue context, etc. Campigotto et al. [26] utilized location, mobility option and socio-demographics. Wang et al. [28] utilize text descriptions, photos, check-in patterns and location. Liu and Gan [31] exploited multimodal data for tourist recommendation.

For movie recommender systems, researchers mainly exploited history, implicit and explicit rating, tags, poster image and social relationship. Da Costa et al. [17, 18, 22] utilized implicit interaction (browsing history, rental history), explicit rating (feedback, tags). Rafailidis et al. [27] utilized user preference with temporal and

Table 1 Different modalities in multimodal recommender system for each application

Paper	Modality	Application
[12, 15, 24]	Text (artist, metadata, name, lyrics), audio, feedback, social tags, usage, biographies	Music
[14]	Video, music	Background music recommendation for automatic video editing
[26, 28, 31]	Location, mobility option, socio-demographics, text descriptions, photos, user check-in patterns, venue context	Location/route recommendation
[17, 18, 22, 27, 29]	User preference with temporal and social media relationship, browsing history, tags, rating, textual description, movie-poster image	Movie
[11, 19, 23, 25]	Human emotions, real-time facial expression, video content, mobile data, textual (video query, title, tags)	Video
[21]	Implicit feedback, comments, rating, image of product, location	E-commerce

social media relationship. Zhao et al. [29] utilized user ratings, social relationships, textual description and poster image of the movie.

For video recommender systems, researchers exploited user behavior, real-time facial expression/emotions, visual aural and textual description. Kaklauskas et al. [11] analyzed human emotions and video content. Whereas Arapakis et al. utilized user feedback and real-time facial expression. Jia et al. [23] exploited user behavior on Android. Yang et al. [25] analyzed visual, aural and textual content of the video (Table 1).

Authors utilized text-based comments, feedbacks, rating with poster images, music, human emotions, real-time facial expression to provide video/movie recommendation. In the case of e-commerce recommender system, authors exploit text information like implicit feedback, comments, rating with an image of a product, packing and location. Peska [21] exploited volumes of text, images, page dimensions, browser's visible area dimensions, visible area ratio, handheld device indicator to build a recommender system for small e-commerce.

4.5 Application- and Technique-Based Analysis

In our analysis, matrix factorization and latent semantic analysis have been found are most prominently applied techniques for multimodal recommender system.

For music recommender system, Oramas et al. [12] exploits deep learning techniques—text and audio embedding to find out similarities between lyrics and music. Domingues et al. [15] applied three different techniques: first: cosine angle to compute usage-based similarity among music tracks, second: latent semantic analysis to compute similarity among social tags assigned to music tracks and third Euclidian distance to calculate similarity among audio features. Laurier et al. [24] Latent semantic analysis and distance-based methods were applied to classify the lyrics in an unsupervised way, and music information retrieval evaluation exchange (MIREX) was applied to classify audio mood in a supervised way (Table 2).

For background music recommender systems, Kuo et al. [14] proposed a framework for automatic background music recommendation for video. They applied multimodal latent semantic analysis to find out the co-occurrence relationship between background music and video.

For location/route recommender system, Campigotto et al. [26] exploit mass preference prior to encode the knowledge of vehicle type, home location, profession, workplace and provide two realistic routes. Finally, Bayesian learning techniques applied on the user's choice, to update the user's profile based on his choices of route. Wang et al. [28] proposed location recommendation with venue semantics based on the latent semantic representation of venues and matrix factorization. Liu and Gan [31] presented a recommendation system using multimodal topic modeling based on latent Dirichlet allocation (LDA) and probabilistic latent semantic analysis (PLSA).

Table 2 Different techniques applied for development of multimodal recommender systems for each application

Papers	Application	Technique
[12, 15, 24]	Music	Text and audio embedding, latent semantic analysis, ensemble rankings on SVD++, matrix factorization, support vector machine (SVM), deep learning
[14]	Background music	Multimodal latent semantic analysis
[26, 28, 31]	Location/route recommendation	Bayesian techniques, matrix factorization, multimodal topic modeling
[13]	A group recommendation	Multi-agent system
[17, 18, 22, 27, 29]	Movie	Matrix factorization (MF), SVD, SVD++, random walk learning, Bayesian personalized ranking (BPR), recurrent neural network (RNN), convolutional neural network
[11, 19, 23, 25]	Video	Support vector machine (SVM), latent semantic analysis, attention fusion function (AFF)
[21]	E-commerce	Vector space model (VSM)

Yukawa et al. [13] proposed a group recommender system based on a multi-agent system using a user's social network to provide trustful group recommendation.

For movie recommender systems, Da Costa et al. [17, 18, 22] published three papers on multimodal recommender system. Authors mainly applied Bayesian personalized ranking (BPR), matrix factorization, SVD, SVD++ on the unimodal system and uses an ensemble technique as a post-processing step that consolidates rankings produced by different unimodal recommenders. Rafailidis et al. [27] proposed a temporal collective matrix factorization (TCMF) model to understand users' preference evolution by considering multimodal information and social relationship over time. Zhao et al. [29] compute relationship among text information of movie by recurrent neural network and relationship among images of movie convolutional neural network. These two neural networks were integrated. They also utilize random walk learning technique on the multimodal representation of the movie.

For video recommender systems, Arapakis et al. [19] integrate user's emotions with a user's profile data and applied the support vector machine model (SVM) to find out relevant videos. Jia et al. [23] proposed multimodal learning for video recommendation (MVR), an Android background application to analyze user's preferences based on their Android application usage. Authors proposed graphical models where they computed topic distribution by two different Dirichlet distributions, and classification is performed mainly by two-layer learning model, in stage 1, restricted Boltzmann machine (RBM) is applied, then in stage 2, supervised training is conducted on learned parameters from the stage 1. Yang et al. [25] compute multimodal relevance between two videos as the combination of visual, textual and aural relevance by attention fusion function (AFF). Peska [21] developed a multimodal recommender for small-scale e-commerce by utilizing a vector space model (VSM).

5 Conclusions

Since last two decades, it has been observed that recommender systems play a significant role in the decision-making process in various applications like e-commerce, movie, video, music, IPTV, restaurant, news, etc. However, recommender systems still require many more improvements. Most recommender systems suffer from limited information problem as users do not provide rating/feedback very often. Multimodal recommender system exploits most of the available information by investigating different information modes. In our paper, a multimodal recommender system is founded to provide better recommendation by utilizing most of the available information, but some key limitations of using multimodal information have been found as well. (i) There is no specific technique to process multimodal information. (ii) Dataset of multimodal information is also not available. (iii) The computational cost of processing of multimodal information is huge. (iv) Due to huge computational cost, it cannot be used by most of the industry, and it is not suitable for a real-time recommendation as well.

Hence, this paper has made clear more research needs to be done in the multimodal recommendation to solve the limited information problem. Researchers are expected to provide more public dataset supporting multimodal information to analyze users and item. Researchers need to find solutions to the stated problems by developing a specific technique/framework to process multimodal information. These techniques should be cost-effective. The researcher can utilize cloud computing to solve the computational cost problem. Researchers are required to explore further to achieve complete coherency to resolve challenges of a multimodal information system to provide an optimal solution efficiently.

A well-tailored framework for multimodal recommender system is required, which should be capable of handling multimodal data have also been observed. The framework should enable researchers to develop, access, evaluate and compare unimodal and multimodal recommender system algorithms.

References

1. Iyengar SS, Lepper MR (2000) When choice is demotivating: can one desire too much of a good thing? J Pers Soc Psychol 79(6):995
2. Ricci F, Rokach L, Shapira B (2015) Recommender systems: introduction and challenges. In: Recommender systems handbook. Springer, Boston, pp 1–34
3. Shah RR (2016) Multimodal analysis of user-generated content in support of social media applications. In: Proceedings of the ACM on international conference on multimedia retrieval, pp 423–426
4. Adomavicius G, Tuzhilin A (2005) Toward the next generation of recommender systems: a survey of the state-of-the-art and possible extensions. IEEE Trans Knowl Data Eng 17(6):734–749
5. Burke R (2002) Hybrid recommender systems: survey and experiments. User Model User-Adap Inter 12(4):331–370
6. Yang X, Guo Y, Liu Y, Steck H (2014) A survey of collaborative filtering based social recommender systems. Comput Commun 41:1–10
7. Milicevic AK, Nanopoulos A, Ivanovic M (2010) Social tagging in recommender systems: a survey of the state-of-the-art and possible extensions. Artif Intell Rev 33(3):187–209
8. Zhang S, Yao L, Sun A, Tay Y (2019) Deep learning based recommender system: a survey and new perspectives. ACM Comput Surv (CSUR) 52(1):1–38
9. Cantador I, Fernández-Tobías I, Berkovsky S, Cremonesi, P (2015) Cross-domain recommender systems. In: Recommender systems handbook. Springer, Boston, pp 119–159
10. Zhang Q, Wu D, Lu J, Liu F, Zhang G (2017) A cross-domain recommender system with consistent information transfer. Decis Support Syst 104:49–63
11. Kaklauskas A, Gudauskas R, Kozlovas M, Peciure L, Lepkova N, Cerkauskas J, Banaitis A (2016) An affect-based multimodal video recommendation system. Stud Inf Control 25(1):6
12. Oramas S, Nieto O, Sordo M, Serra X (2017) A deep multimodal approach for cold-start music recommendation. In: Proceedings of the 2nd workshop on deep learning for recommender systems, pp 32–37
13. Yukawa M, Hayashi Y, Ogawa H, Kryssanov VV (2012) A group recommendation system with a multi-modal user interface. In: The 6th international conference on soft computing and ıntelligent systems, and The 13th international symposium on advanced ıntelligence systems. IEEE, pp 2158–2163

14. Kuo FF, Shan MK, Lee SY (2013) Background music recommendation for video based on multimodal latent semantic analysis. In: 2013 IEEE international conference on multimedia and expo (ICME). IEEE, pp 1–6
15. Domingues MA, Gouyon F, Jorge AM, Leal JP, Vinagre J, Lemos L, Sordo M (2013) Combining usage and content in an online recommendation system for music in the long tail. Int J Multimedia Inf Retrieval 2(1):3–13
16. Tzanetakis G, Cook P (2000) Marsyas: a framework for audio analysis. Organised Sound 4(3):169–175
17. Da Costa AF, Manzato MG (2016) Exploiting multimodal interactions in recommender systems with ensemble algorithms. Inf Syst 56:120–132
18. Da Costa AF, Domingues MA, Rezende SO, Manzato MG (2014, Aug) Improving personalized ranking in recommender systems with multimodal interactions. In: 2014 IEEE/WIC/ACM International joint conferences on web intelligence (WI) and intelligent agent technologies (IAT), vol 1. IEEE, pp 198–204
19. Arapakis I, Moshfeghi Y, Joho H, Ren R, Hannah D, Jose JM (2009) Integrating facial expressions into user profiling for the improvement of a multimodal recommender system. In: IEEE International conference on multimedia and expo, pp 1440–1443
20. Cho S, Lee M, Jang C, Choi E (2006) Multidimensional filtering approach based on contextual information. In: International conference on hybrid ınformation technology, vol 2. IEEE, pp 497–504
21. Peska L (2017) Multimodal ımplicit feedback for recommender systems. In: ITAT, pp 240–245
22. Da Costa AF, Manzato MG (2014) Multimodal interactions in recommender systems: an ensembling approach. In: Brazilian conference on ıntelligent systems. IEEE, pp 67–72
23. Jia X, Wang A, Li X, Xun G, Xu W, Zhang A (2015) Multi-modal learning for video recommendation based on mobile application usage. In: International conference on big data (big data). IEEE, pp 837–842
24. Laurier C, Grivolla J, Herrera P (2008) Multimodal music mood classification using audio and lyrics. In: Seventh international conference on machine learning and applications. IEEE, pp 688–693
25. Yang B, Mei T, Hua XS, Yang L, Yang SQ, Li M (2007) Online video recommendation based on multimodal fusion and relevance feedback. In: Proceedings of the 6th ACM international conference on image and video retrieval, pp 73–80
26. Campigotto P, Rudloff C, Leodolter M, Bauer D (2016) Personalized and situation-aware multimodal route recommendations: the FAVOUR algorithm. IEEE Trans Intell Transp Syst 18(1):92–102
27. Rafailidis D, Kefalas P, Manolopoulos Y (2017) Preference dynamics with multimodal user-item interactions in social media recommendation. Expert Syst Appl 74:11–18
28. Wang X, Zhao YL, Nie L, Gao Y, Nie W, Zha ZJ, Chua TS (2014) Semantic-based location recommendation with multimodal venue semantics. IEEE Trans Multimedia 17(3):409–419
29. Zhao Z, Yang Q, Lu H, Weninger T, Cai D, He X, Zhuang Y (2017) Social-aware movie recommendation via multimodal network learning. IEEE Trans Multimedia 20(2):430–440
30. Alhamid MF, Rawashdeh M, El Saddik A (2013) Towards context-aware recommendations of multimedia in an ambient intelligence environment. In: IEEE International symposium on multimedia. IEEE, pp 409–414
31. Liu L, Gan J (2017) Using multi-modal topic modeling in national culture resources: methods and applications. In: 9th International conference on ıntelligent human-machine systems and cybernetics (IHMSC), vol 2. IEEE, pp 298–301
32. Huang Z, Xu X, Ni J, Zhu H, Wang C (2009) Multimodal representation learning for recommendation in Internet of Things. IEEE Internet Things J 6(6):10675–10685
33. Salah A, Truong QT, Lauw HW (2020) Cornac a comparative framework for multimodal recommender systems. J Mach Learn Res 21(95):1–5

Chapter 71
All-Zero Pre-processing Over ISI Channels: A Methodology for TCM Schemes

Vanaja Shivakumar

1 Introduction

Communication channels are subjected to noise and interference effects which are unavoidable. It is mainly due to the natural phenomena occurring in the atmosphere, and some of them are due to man-made systems. Various types of interference that appear in communication systems are: co-channel interference, adjacent-channel interference, self-interference, multiple access interference and intersymbol interference. The co-channel interference can be defined as the disturbance to the transmitted signal because of the reuse of the frequency band for transmission. This interference will corrupt the original message or information and results in crosstalk. In cellular communication systems, to overcome this situation, no two adjacent cells are operated on the same frequency range. Now, coming to the adjacent-channel interference, it can be said that it is the disturbance in the transmission to the information signal because multiple channels in the same geographical region use neighboring ranges of frequencies. It can be minimized by managing the division of the frequency spectrum by introducing guard bands between two adjacent channels. Another category is the self-interference, which is due to the interference induced in the signals which share a common transmitter. Also, the nonlinearities in the electronic devices and circuits, namely amplifier nonlinearities, etc., also result in interference. Generally, filters and self-interference cancelation techniques are employed to minimize this interference. The multiple access interference in another type comes across in the communication systems, and it is mainly due to the use of the same frequency channel by multiple users. One good example is the code division multiple access systems. Here, the same frequency range is used for multiple transmissions. In cellular communication

V. Shivakumar (✉)
CMR Institute of Technology, Bengaluru, India
e-mail: vanaja.s@cmrit.ac.in

I. Jeena Jacob et al. (eds.), *Data Intelligence and Cognitive Informatics*,
Algorithms for Intelligent Systems,
https://doi.org/10.1007/978-981-15-8530-2_71

systems, this type of interference could be caused either by the user within the cell interference or caused by the user in some other cell. To overcome this, one of the methodologies implemented is spread spectrum technology. Coding techniques will also be considered to minimize this interference effect. Another very well-known type of interference which is taken into consideration is the intersymbol interference (ISI) as a result of the time spread phenomena. The spreading of the transmission pulses beyond the allotted time interval causes a previous interval symbol to interfere with the present interval pulses. Due to the time spreading phenomena, past interval transmissions will have its effect on more than one neighboring intervals' symbols. In cellular communication systems, methods implemented to overcome this effect are: refining the impulse response for a shorter duration, developing efficient sequence detection algorithms and use of equalizers.

2 Review on the Previous Work

In the paper [1], a block diagram of the encoder–modulator is shown. Trellis encoder structure possessing the linearity and nonlinearity has been explained, and the classification of TCM schemes has been explained. Various parameters which are treated as important to study and analyze the performance of various TCM schemes on AWGN channel have been explained. How these parameters are related to each other and how they are used in the performance evaluation have been explained. As given in [1], the parameters such as the free distance of the code, number of nearest neighbors in the sense of the free distance of the symbol sequences, the distance spectrum and the probability measure of the occurrence of the error events are all depending only on the free distance of the code sequences, and for the parameter bit error probability, it requires the complete knowledge of the mapping that takes place in the encoder–modulator unit. Signal partitioning for various TCM schemes has been explained. Considering all possibilities of information transmission, trellis transition in the trellis diagram of the encoder and subsets associated with each transmission have been explained.

The paper [2] explains the TCM scheme with a punctured convolutional code, and the code generation has been demonstrated. In this case, the authors have come across the mapping process which leads to a time-variant trellis structure, and the other way it can be stated that due to the puncturing process, the resulting trellis code will be a code with a variable rate. For this variable rate code, it has been shown that a reduced complexity decoding in a modified version of conventional VA would serve the purpose better. For this approach, hyper states are defined in this paper as applicable to the code generated.

In paper [3], by R. Raheli, A. Polydoros and Ching-Kae Tzou, the authors have presented various algorithms with an objective of pre-estimation of the communication channel parameters as required by the VA. Here, the channel parameters under consideration are unknown. Additionally, the data sequence which is transmitted through the noisy communication channel needs to be estimated. In this paper, various methods are described for the estimation of channel information and the data sequence transmitted simultaneously. Many algorithms are developed for the purpose. Improved performance has been demonstrated over conventional methodologies. The channel information estimation and the estimation of the transmitted data sequence have been explained considering the fact that the information is available in the conventional VA structure itself. He has demonstrated that the code sequences associated with each of the survivor paths of the trellis structure of the decoder will aid the estimation process. This information has been used by the authors in the pre-survivor estimation process. According to this paper, the pre-survivor estimation module should be such that in the absence of the channel noise, the estimator should be able to produce the correct estimation of the channel parameters when the correct data-Aided sequence is supplied. In the view of the authors' approach, for MLSE in the suboptimal sense, based on the least mean square estimation, the authors have followed pre-survivor processing (PSP) method to know the unknown parameters, and PSP-based suboptimal algorithms are presented. In this approach, conventional MLSE receivers' structure has been considered for the channel parameters estimation. During the channel information estimation, the data decision process has not been considered, instead it has been assumed that the correct decision is available at the receiver, and also, at this stage, it has been assumed that the zero delays have been introduced in the data received at the receiver.

The paper [4] explains a two-stage decoding procedure followed for pragmatic TCM. In this approach, the conventional VA, as well as symbol-by-symbol decoding methods, is implemented. Conventional VA is implemented in the first stage of decoding, and followed by a recoding process, the second stage of decoding is done. Improved performance of this decoding system and reduction in the memory requirement have been demonstrated in this paper. In [5], a modified Viterbi decoder has been developed for the TCM schemes through a technique called windowing mechanism. And improved error performance has been demonstrated.

The paper [6] is focussed on developing reduced state decoding algorithms. The approach here is that in reduced state trellis processing two iterations survivor paths mapping is considered together for a posteriori analysis schemes. The paper [7] is about the research for space-time coded modulation, and fading channel effects are considered. Iterative detection methods are developed for MIMO applications. The paper [8] is focussed on the investigation of a subclass of adaptive iterative receivers in which the parameter estimates are not exchanged as part of the iterative procedure, rather they are generated and are confined inside the adaptive module. Coded linear modulations over AWGN channel have been considered in paper [9], and noncoherent methodologies are developed for demodulation and decoding as a combined approach. Bit errors performance has been analyzed. Improved bit error performance is shown over coherent detection. Reduced state sequence detection is also considered. In the decoding process of transmitted TCM signals, for both branch metric and path metric computations, T-algorithm has been implemented, and with this methodology, a new hybrid algorithm has been developed in [10] and has implemented it on 8 PSK TCM. About 50% reduction in complexity over conventional method has been shown. In paper [11], a modified approach for the decoding of trellis coded modulation scheme with multidimensional structure has been discussed, and the paper [12] discussed modified branch metric computations of Viterbi decoder.

3 Transmission on AWGN Channel

Assuming that the AWGN channel is under consideration for transmission, the equivalent representation of the channel is as shown in Fig. 1. Any signal when transmitted on the AWGN channel will be corrupted by the AWGN. The received noisy signal $r(t)$ is given by Eq. (1).

$$r(t) = s_{\mathrm{m}}(t) + w(t); \quad 0 \leq t \leq T \tag{1}$$

where T is the symbol interval duration, $w(t)$ is complex AWGN samples with power spectral density $N_0/2$ W/Hz and $s_{\mathrm{m}}(t)$ is the signal under the transmission. The set of signals is given by Eq. (2),

$$\{s_{\mathrm{m}}(t), \quad m = 1, 2, \ldots, M\} \tag{2}$$

The optimum detector produces the received signal vector r, given by Eq. (3),

$$r = [r_1, r_2, \ldots, r_N] \tag{3}$$

The Maximum Likelihood (ML) sequence detection algorithm selects the sequence that minimizes the Euclidean distance metric D, given by Eq. (4),

$$D\left(r, s^{(m)}\right) = \sum_{k=1}^{K} (r_k - s_k^{(m)})^2 \tag{4}$$

where K elements are under consideration.

Fig. 1 AWGN channel equivalent representation

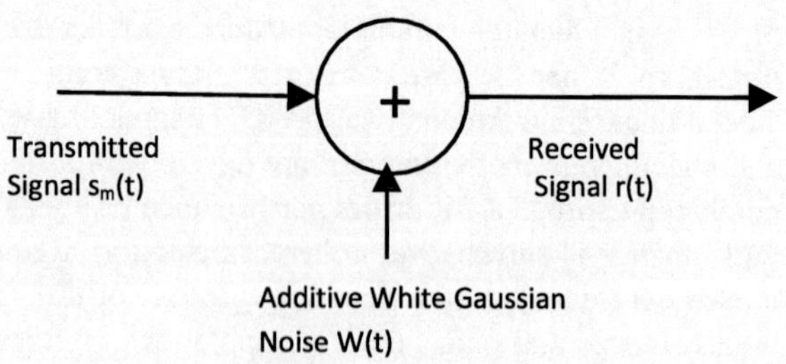

4 Bandlimited ISI Channel

The channel is assumed to be linear and time-invariant. Assuming that the symbols transmitted are $a(n)$ and the channel impulse response is $g(n)$ which is complex-valued, and the channel output is corrupted by the complex AWGN $w(t)$; considering tapped delay filter model, the received waveform is

$$r(t) = \sum_{k} a(n)g(t - kT) + w(t) \tag{5}$$

In a discrete time model, it is

$$r(n) = \sum_{i=0}^{L} g_i.a(n - i) + w(n) \tag{6}$$

where $\{g_i\}$ is the tap gain that corresponds to the sampled channel impulse response, and $\{w_n\}$ is the sample of complex-valued Gaussian noise with zero mean and unit variance. The noise and the data sequences are assumed to be uncorrelated, and the number of taps is $(L + 1)$, where L represents the channel memory length.

5 Description of TCM Schemes

The state-of-the-art description of TCM schemes of Ungerboeck can be found in the paper [1]. A generic TCM encoder and modulator unit as explained by Ungerboeck is shown in Fig. 2. It comprises serial to parallel bit converter, a convolutional encoder and a mapper. Internally, the convolutional encoder comprises delay unit and modulo-2 adder. The delay unit basically contains serial shift registers assembly. This unit

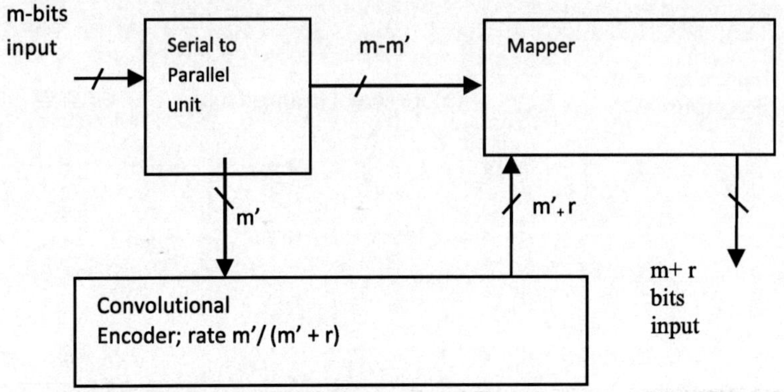

Fig. 2 General TCM encoder/modulator unit

serves the purpose of storage of the information bits which are required by the encoding process, and the required delay will be introduced among the information bits. Also, from this delay unit, the encoder state information can be extracted. The convolutional encoding unit is an integral part of the trellis encoder. So, all the inherent properties of the convolutional codes are carried by the trellis code. The well-known fact of the convolutional encoder to add extra bits for redundancy into the code generated will carry the redundancy into the trellis codes also. The output of the convolutional code generated possesses a structural property. Each of the code generated has a dependency on the codes generated in the past transmission intervals. This, in turn, creates dependency among the information bits transmitted during each symbol interval. The definite patterns of the dependency can be expressed in the form of the graphical representation, called a trellis diagram. This diagram relates the possible states of the convolution encoder and corresponding transitions considering all possible combinations of the binary data inputted. In the trellis encoding process, as shown in Fig. 2, the convolutional codes generated constitute one part of the trellis code. The remaining segment of the trellis code is the input binary information which is left uncoded by the convolutional encoder. If the convolutional encoder is of rate n/r, the number of information bits to be transmitted is m, and the bits to be encoded by the convolutional encoder is $n = m'$, then the uncoded bits available in the trellis encoder is $m - m'$, and the convolutional encoder output will be of $m' + r$ bits code. Here, r is the redundancy added by the convolutional encoder. Now, the rate of the encoder will be $m'/(m' + r)$.

Figure 3 shows the convolutional encoder used in systematic TCM encoder, for the case of 4-state design. This encoder comprises of two delay elements and a modulo-2 adder. There is no feedback from the code generated.

Fig. 3 Convolutional encoder for 4-state systematic TCM

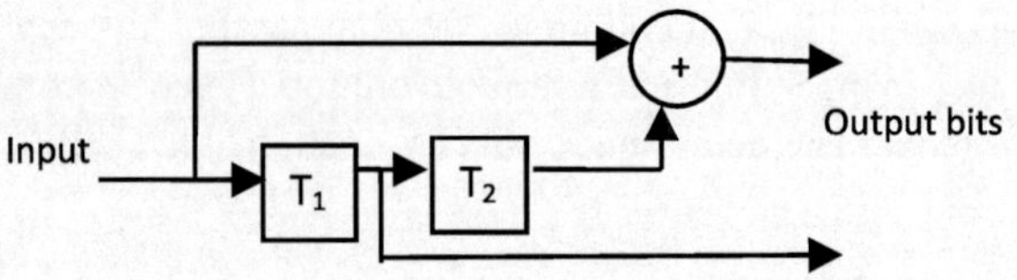

Fig. 4 Trellis diagram of 4-state encoder; 16 QAM

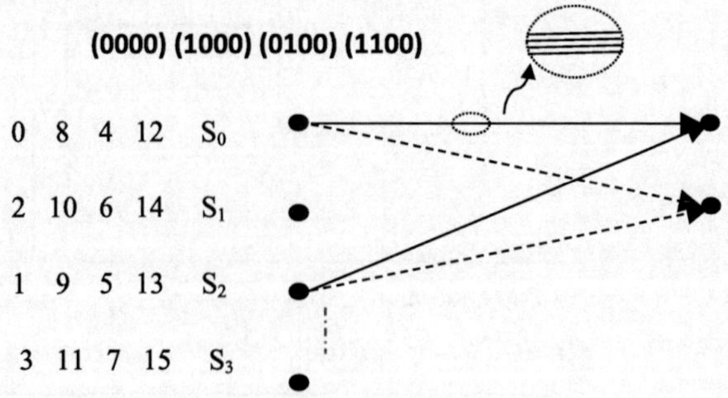

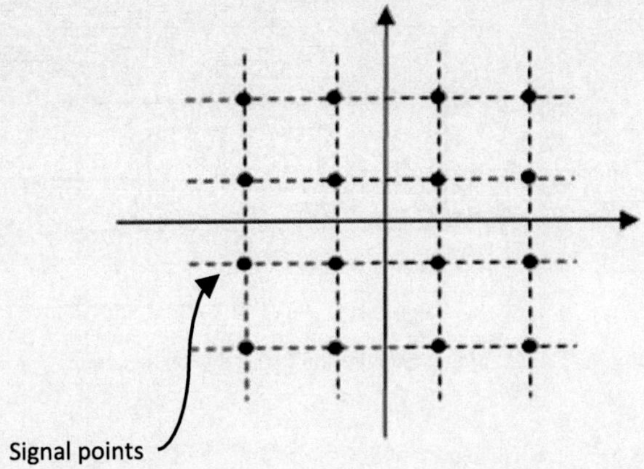

Fig. 5 Sixteen QAM constellation

In Fig. 4, the state diagram of the 4-state TCM scheme has been shown for QAM signal constellation of size 16 points which are shown in Fig. 5. The TCM encoder state transitions and the associated subsets are shown for each bit of information transmission in the trellis diagram. The number of input bits transmitted is $m = 3$, the number bits encoded by the convolutional encoder is $m' = 1$, and hence, there are 2 transitions from each state. The number of uncoded bits is two, which indicates that there are four parallel paths in each transition. The trellis has four node points each representing one of the states of the encoder. The straight-line transitions in the trellis diagram representations indicate the next state attained by the encoder. The direction of the arrow is an indication of the state transitions after a new bits-set is received. In response to the information, inputs fed to the convolutional encoder observed the generation of the trellis code and the state change of the encoder. The transitions in the trellis diagram are associated with the corresponding output symbols generated, similar to the symbolic representations in the signal constellation; for example, the encoder is in the state s_0; if the information is 1, the state transition is from state s_0 to the state s_1, and the associated symbol group, also called the subset, is {2, 10, 6, 14}. In the diagram of Fig. 4 shown, the labeling to the left of each state node is for parallel transitions.

In Fig. 6, the convolutional encoder for 8-state design of TCM scheme has been shown, for the QAM signal constellation size 16. The representation of QAM constellation of 16 signal points, shown in Fig. 5, has been considered. In Fig. 7, the state diagram of the 8-state TCM scheme has been shown for QAM signal constellation of size 16 points which are shown in Fig. 5. The TCM encoder state transitions and the associated subsets are shown in the trellis diagram for each bit of information transmission. The number of input bits transmitted is $m = 3$, the number bits encoded by the convolutional encoder is $m' = 2$, and hence, there are four transitions from each state. The number of uncoded bits 1 indicates that there are two parallel paths in each transition.

In Fig. 7, trellis structure of 8 state TCM encoder for sixteen QAM constellation has been given. The trellis has eight node points each representing one of the states

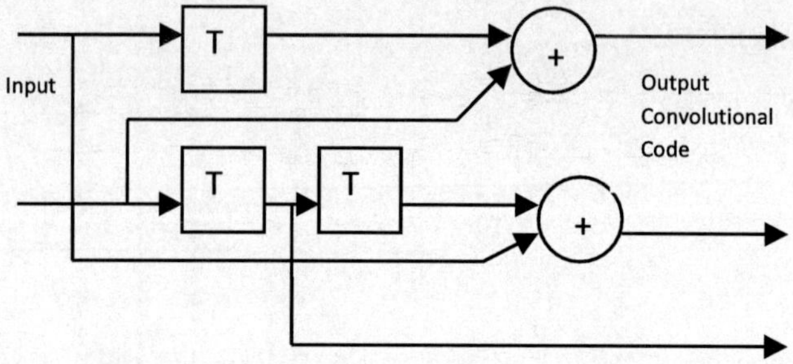

Fig. 6 Convolutional encoder for 8 state systematic code

Fig. 7 Trellis diagram of 8 state encoder; 16 QAM

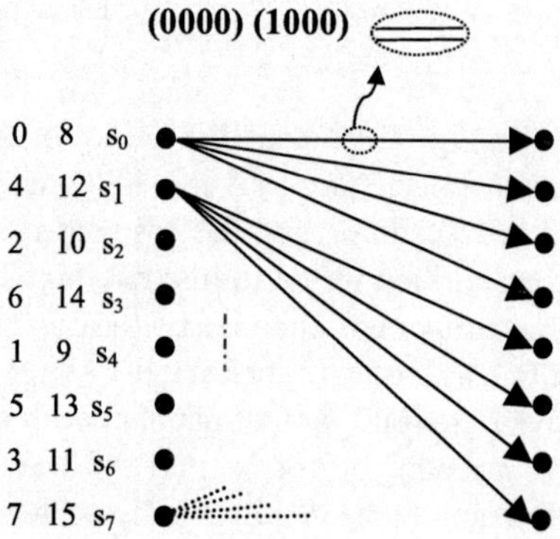

of the encoder, and straight-line transitions in the trellis diagram representations indicate the state attained by the encoder. In response to the information, inputs fed to the convolutional encoder observed the generation of the trellis code and the state change of the encoder. The encoder being in the state s_0, if the information is 001, the state transition is from state s_0 to the state s_1, and the associated symbol group, also called the subset, is {4, 12}. In the diagram shown in Fig. 7, the labeling from the left corresponds to the transitions from the top in the trellis diagram.

6 Selection of Subset

The selection of a subset from the signal constellation of a particular TCM scheme is done by the signal set partitioning, an inherent concept in Ungerboeck's TCM scheme. In the process of trellis code generation and modulation, once the convolutional code is generated by the encoder and its output is fed to a mapper, which is another essential component of the TCM encoder/modulator. The mapper will map the convolutional codes to select a group of signal points from the signal constellation, and this group is called a subset. Each subset will have more than one signal point. One of these signal points will be selected for transmission by the uncoded information bits. In Fig. 8, the set portioning of 8-state TCM scheme for sixteen

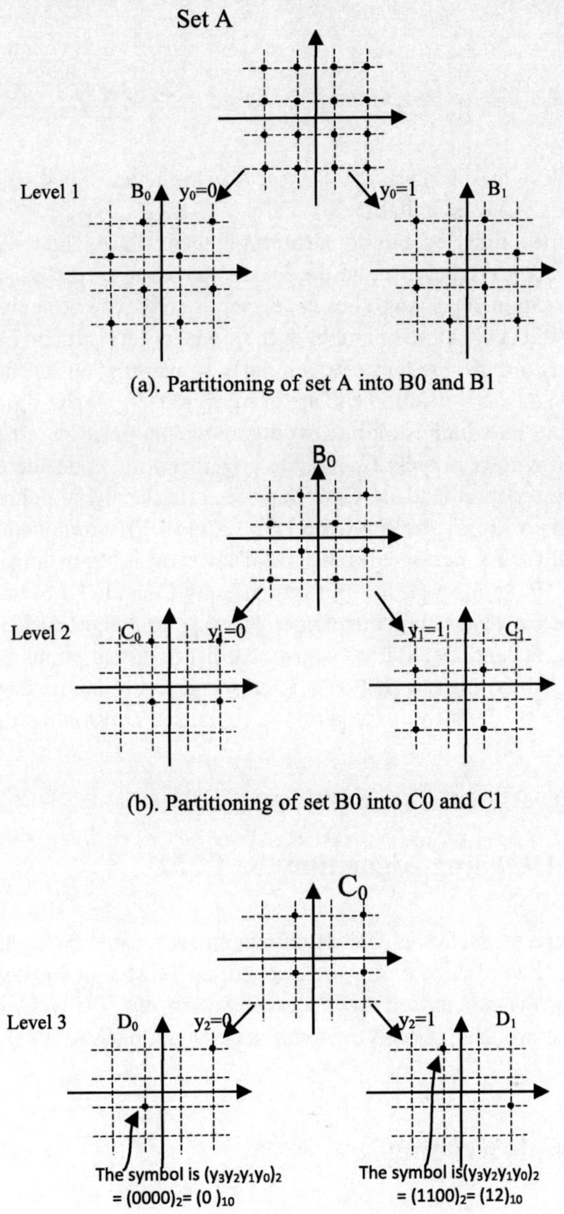

(a). Partitioning of set A into B0 and B1

(b). Partitioning of set B0 into C0 and C1

(c). Partitioning of set C0 into D0, D1

Fig. 8 Set partitioning of 16 QAM at different levels

Table 1 Set partitioning table for 8-state TCM scheme

*i*th level of set partitioning	Resulting subsets (for 8 state and sixteen QAM TCM scheme)							
Level 0	A							
Level 1	B_0	B_1						
Level 2	C_0	C_1	C_2	C_3				
Level 0	D_0	D_1	D_2	D_3	D_4	D_5	D_6	D_7

QAM constellation has been shown. The set A is the actual signal set defined by the 16 points of the QAM constellation.

In the set partitioning level 0, no partitioning takes place, and the corresponding signal set is A. At level 1, based on the least significant bit (LSB) y_0 of the trellis code, set A is partitioned into two subsets, B_0 and B_1. This can be analyzed in another way that as if y_0 bit is 0, half the number of the signal points is found having 0 at LSB position and they are grouped into B_0; similarly, remaining half of the signal points will have 1 bit at the LSB position and are grouped into B_1. At the signal partitioning level 2, it is the y_1 bit which is in the next higher weight position of LSB, and it will split B_0 into two groups, namely C_0 and C_1. Together with if considered y_1y_0 at LSB position, 00 combination leads to C_0 group; here, all the signal points will have 00 at lower weight positions, starting from LSB. And the 01 combination leads to C_1 group where all the elements will have 10 at lower weight positions starting from LSB. Similarly, B_1 group will be split into subsets C_2 and C_3. In the next level of partitioning, the bit y3 will be considered. The resulting eight subsets are: D_0, D_1, D_2, D_3, D_4, D_5, D_6 and D_7. Table 1 shows the list of signal points present in each of the subsets. All the D_i, $i = 0, 1, \ldots, 7$, subsets contain only two symbols which are differentiated by the bit y_3. Based on the signal set partitioning, the code design rules are followed [4].

7 Optimal Decoding Algorithm for TCM

The ML sequence estimation is the optimum approach in estimating the transmitted sequence at the time of decoding if the transmitted TCM signal is corrupted by the AWGN. During the process of sequence estimation, the VA with the soft output computational approach has been treated as a conventional VA in this paper.

8 Suboptimal Decoding

The optimal ML sequence estimation procedure becomes highly complicated if the channel has ISI effect. For transmission on bandlimited channels subjected to ISI,

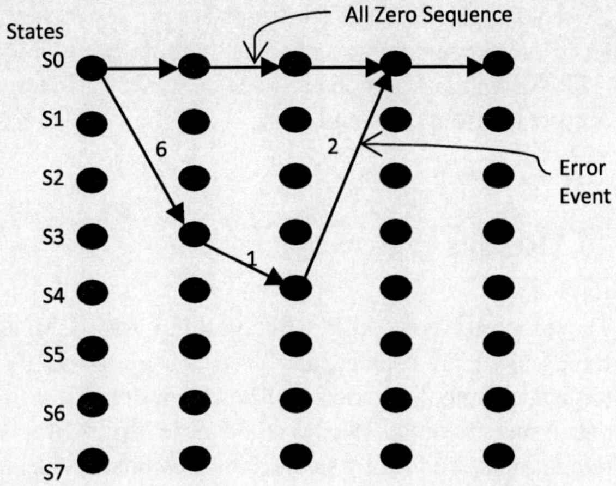

Fig. 9 Error event in 8 state 16 QAM TCM scheme

suboptimal approaches will be the preferred technologies. By reducing the computational complexity through near-optimal solutions, sequence estimation and symbol detection will be done which will reduce the error involved in the detection. In the presence of ISI, by merging the state of the ISI channel with the encoder states, resulting reduced complexity sequence estimation will provide a better solution to decoding strategies to handle complicated situations. Also, data-aided pre-survivor processing methodology can also be considered in addition to reduced states' trellis structures of the combined ISI encoder trellis for the improvement in the sequence estimation and detection process. Data-aided pre-survivor processing will minimize the occurrence of error events and bit errors. Figure 9 is illustrating the occurrence of an error event in an 8 state 16 QAM TCM scheme. Referring to the subset assignments of Fig. 7, the detected sequence path is the symbol sequence 6-1-2, deviated from the all-zero sequences 0-0-0 path, and hence, an event of error occurred. Here, the all-zero sequence path is the reference sequence, and since the deviation is for 3 symbols duration, the error event is of length 3.

9 All-Zero Pre-processing Sequence Estimation

A new All-zero Pre-processing Sequence Estimation (AZPPSE) methodology is proposed as a near-optimal solution to reduce the complexity of decoding of the TCM schemes. Complexity reduction indirectly expects some means for better estimation of the communication channel parameters and the transmitted data sequence estimation either with optimal or near-optimal approaches. Here, a reduced complexity decoding system is being proposed with integrated processing for the TCM signals which are noisy. In our approach, the information associated with the trellis structure of the receiver has been used efficiently, and the transmission of all-zero sequence to the requirement has been considered as the key factor in order to strengthen the data-aided pre-processing sequence estimation. Improved performance of the proposed AZPPSE implementation has been documented in terms of the improved error event occurrence from the simulation.

10 Results

The proposed new AZPPSE algorithm for TCM schemes has been simulated. Bandlimited ISI channel has been considered as the transmission channel. Zero sequence symbol transmission has been done. It will have an effect on the actual data transmission by 1%. It can be varied from 1 to 4% efficiently according to the requirement. Reduced state sequence estimation has been implemented along with the AZPPSE. ISI effect for one symbol interval has been assumed. Conventional VA has been modified to implement the proposed AZPPSE methodology. Improved performance has been documented. Results are obtained for 4 state TCM scheme with 16 QAM constellation. Table 2 contains the simulation results. Column 6 contains a total number of error event occurred when the simplest RSSE is implemented. Column 7 contains the number of error events occurred when AZPPSE has been implemented. Reduced number of error events are observed in column 8 over the tabulation in column 7. Column 1 says a number of states of the TCM scheme, column 2 says about the constellation size, column 3 contains encoder states, and column 4 tells about the number parallel paths of the trellis structure.

Table 2 Error event probability comparison

TCM scheme				Error event (EE) (symbols transmitted 10^4)		
No. of states	QAM size	Encoder states	Parallel paths	SNR in dB	EE in simplest RSSE	EE in AZPPSE (all-zero 25 symbols)
4	16	4	4	15	03	2
				14	16	10
				13	77	61
				12	200	160
				11	337	288
				10	471	457

11 Conclusion

The simulation results show that the new proposed AZPPSE technique of suboptimal decoding approach for TCM scheme performs better over the simplex reduced state sequence estimation (RSSE) conventional technique. The computational complexity of the AZPPSE is less over conventional simplex RSSE. AZPPSE technique can be extended to TCM schemes with a greater number of states. As the number of states, the computational complexity of AZPPSE decreases because of all-zero sequence transmission. With the transmission of additional data of 1–4% depending upon the application, better error improvement will be achieved. The AZPPSE can be implemented in any application because the zero symbol sequence transmission would be an integral part of the data transmission. Only the decoder differentiates these symbols.

Acknowledgements The author would like to thank all the reviewers of this paper.

References

1. Benedetto S, Mondin M, Montorsi G (1993) Performance evaluation of trellis-coded modulation schemes. In: Proceedings of the IEEE—PIEEE. 82, vol 3, pp 1785–1790. https://doi.org/10.1109/GLOCOM.1992.276691
2. Schuh F, Huber JB (2013, Dec) Low complexity decoding for punctured trellis-coded modulation over ıntersymbol ınterference channels. arXiv:1312.1482. v1, [cs.IT]
3. Raheli R, Polydoros A, Tzou C-K (1995, Feb) Per-survivor processing: a general approach to MLSE ın uncertain environments. IEEE Trans Commun 43(2/3/4):354–364. https://doi.org/10.1109/26.380054
4. Morelos-Zaragoza RH, Mogre A (2001, Sept) A two-stage decoder for pragmatic trellis-coded M-PSK modulation using a symbol transformation. IEEE Trans Commun 49(9)
5. Kahina R, Bassou A (2018, Feb) Improving the performance of viterbi decoder using window system. Int J Electr Comput Eng (IJECE) 8(1):611–621. ISSN: 2088-8708, https://doi.org/10.11591/ijece.v8i1.pp611-621
6. Colavolpe G, Ferrari G, Raheli R (2001, May) Reduced-state BCJR-type algorithms. IEEE J Sel Areas Commun 19(5)
7. Visoz R, Berthet AO (2003, Aug) Iterative decoding and channel estimation for space-time BICM over MIMO block fading multipath AWGN channel. IEEE Trans Commun 51(8):1358–1367
8. Anastasopoulos A, Chugg KM (2000, Oct) Adaptive soft-ınput soft-output algorithms for ıterative detection with parametric uncertainty. IEEE Trans Commun 48(10):1638–1649
9. Colavolpe G, Raheli R (1998, Aug) On noncoherent sequence detection of coded QAM. IEEE Commun Lett 2(8):211
10. He J, Wang Z, Liu H (2010, May) An efficient 4-D 8PSK TCM decoder architecture. IEEE Trans Very Large Scale Integr (VLSI) Syst 18(5)
11. Wang Q, Wei L (2002) Iterative viterbi algorithm for concatenated multidimensional TCM. IEEE Trans Commun 50:12–15
12. Chandel S, Mathur M (2019, Aug) Viterbi decoder plain sailing design for TCM decoders. Int J Trend Sci Res Dev (IJTSRD) 3(5). e-ISSN: 2456-6470

Chapter 72
New Technical Using Nano in Medical Field to Determine Medications that are Suitable Activities for COVID-19

Shuker Mahmood Khalil and Nadia M. Ali Abbas

1 Introduction

In the recent years, there are many studies are discussed the epidemics which are cause diseases for a large number of people within a short period of time. The COVID-19 is the one of these epidemics. However, there are many clinical trials which are now underway to choose the best medication, but until now the worldwide response to this epidemic has been only limited to monitoring and containment see [1–4]. Therefore, in this paper, we will focus on the best technical that may help coronavirus patients. This technical using the lower approximation and upper approximation in (NTS). The notion of (NTS) with respect to a subset η of a universe set Π is introduced by *L.* Thivagar and Riclord [5]. There are many applications on classical and non-classical sets like nano sets (NS), fuzzy sets (FS), soft sets (SS), and others in mathematical field, see [6–19], and in medical field, see [20, 21]. In this work, we will use (NS) in medical field to select the best medications that are necessary for coronavirus patients. Sometimes the patients may take more than one medication. However, they do not heal within a same fixed time limit. Therefore, in this research, the technical using concepts in (NTS) are studied. This technical help us to choose some of these medications that are necessary to health rate recovery for coronavirus patients.

S. M. Khalil (✉)
Department of Mathematics, College of Science, Basrah University, Basrah 61004, Iraq
e-mail: shuker.alsalem@gmail.com

N. M. Ali Abbas
Ministry of Education, Directorate General of Education, Al-Kark/3, Baghdad, Iraq
e-mail: mali.nadia@yahoo.com

I. Jeena Jacob et al. (eds.), *Data Intelligence and Cognitive Informatics*,
Algorithms for Intelligent Systems,
https://doi.org/10.1007/978-981-15-8530-2_72

2 Definitions and Notations

The following definitions have been used to consider our results.

Definition Assume that $\Im$ is an equivalence relation (ER) on finite set of objects Π (universe set) and $\eta \subseteq \Pi$ [22]. Then,

1. We say $L_{\Im}(\eta) = \bigcup_{x\in\Pi} \{\Im(x) : \Im(x) \subset \eta\}$ is a lower approximation of η with respect to $\Im$ and is referred by $L_{\Im}(\eta)$, where $\Im(x)$ denotes the equivalence class (EC) contains x.
2. We say $H_{\Im}(\eta) = \bigcup_{x\in\Pi} \{\Im(x) : \Im(x) \cap \eta \neq \varphi\}$ is an upper approximation of η with respect to $\Im$ and is referred by $H_{\Im}(\eta)$.
3. We say $B_{\Im}(\eta) = H_{\Im}(\eta) - L_{\Im}(\eta)$ is boundary region of η with respect to $\Im$ and is referred by $B_{\Im}(\eta)$. In other side, η is said to be a rough set if $H_{\Im}(\eta) \neq L_{\Im}(\eta)$.

Definition Suppose that $\Im$ is an (ER) on Π and $\eta \subseteq \Pi$ [5]. We say $\tau_{\Im}(\eta) = \{\Pi, \varphi, L_{\Im}(\eta), H_{\Im}(\eta), B_{\Im}(\eta)\}$ is nano topology (NT) on Π and $(\Pi, \tau_{\Im}(\eta))$ is a nano topological space (NTS). Also, each member in $\tau_{\Im}(\eta)$ is said to be a nano open set (NOS), and its complement is said to be a nano closed set (NCS).

Nano Topology with Its Application for COVID-19

First of all, the coronavirus (COVID-19) is an infectious disease caused by a new virus. This disease spreads by touching an infected person, or when they cough or sneeze. Moreover, it spreads when a person touches something like surface which has the virus on it. There are some medications for (COVID-19), like Chloroquine, Arbidol, Tamiflu, Kaletra, Remdesivir, and other medications. However, these medications may be harmful or useful if the coronavirus patients take some of these together with different rates. The next table gives data about eight patients, and it show that each one of them what is the using rate of the set $\psi = \{C, A, T, K, R\}$ which contains five medications, see Table 1. The columns clarify the attributes (Health rate recovery), and the rows clarify the thingummies (the patients) in Table 1, put $\Pi = \{\text{Cor}_1, \text{Cor}_2, \ldots, \text{Cor}_8\}$, then:

Table 1 Health rate recovery for patients

Patients	C (%)	A (%)	T (%)	K (%)	R (%)	Health rate recovery (%)
Cor_1	100	100	75	75	50	75
Cor_2	100	100	75	75	90	100
Cor_3	100	75	75	75	80	100
Cor_4	75	75	75	75	90	75
Cor_5	75	100	100	100	80	75
Cor_6	100	75	75	75	80	75
Cor_7	100	100	100	100	80	100
Cor_8	100	100	75	75	90	100

Case I Let $\eta = \{Cor_2, Cor_3, Cor_7, Cor_8\}$ be the set of coronavirus patients having (100%) health rate recovery. Assume that $\Im$ is an (ER) on Π with respect to ψ the set of medications. The (EC) corresponding to $\Im$ is considered as $\Pi/I(\Im) = \{\{Cor_1\}, \{Cor_4\}, \{Cor_5\}, \{Cor_7\}, \{Cor_2, Cor_8\}, \{Cor_3, Cor_6\}\}$. This implies that $\tau_{\Im}(\eta) = \{\Pi, \varphi, \{Cor_3, Cor_6\}, \{Cor_2, Cor_7, Cor_8\}, \{Cor_2, Cor_3, Cor_6, Cor_7, Cor_8\}\}$ is (NT) on Π with respect to η. If we ignore the medication "*C*," then we consider that $\Pi/I(\Im - (C)) = \{\{Cor_1\}, \{Cor_4\}, \{Cor_3, Cor_6\}, \{Cor_2, Cor_8\}, \{Cor_5, Cor_7\}\}$. Hence, $\tau_{\Im-(C)}(\eta) = \{\Pi, \{Cor_2, Cor_8\}, \{Cor_3, Cor_5, Cor_6, Cor_7\}, \{Cor_2, Cor_3, Cor_5, Cor_6, Cor_7, Cor_8\}\} \neq \tau_{\Im}(\eta)$. If we ignore the medication "*A*," then we obtain $\Pi/I(\Im - (A)) = \Pi/I(\Im)$, and we fined $\tau_{\Im-(A)}(\eta) = \tau_{\Im}(\eta)$. If we ignore the medication "*T*," then we obtain $\Pi/I(\Im - (T)) = \Pi/I(\Im)$, and we fined $\tau_{\Im-(T)}(\eta) = \tau_{\Im}(\eta)$. If we ignore the medication "*K*," then we obtain $\Pi/I(\Im - (K)) = \Pi/I(\Im)$, and we fined $\tau_{\Im-(K)}(\eta) = \tau_{\Im}(\eta)$. If we ignore the medication "*R*," then we obtain, $\Pi/I(\Im - (R)) = \{\{Cor_4\}, \{Cor_5\}, \{Cor_7\}, \{Cor_3, Cor_6\}, \{Cor_1, Cor_2, Cor_8\}\}$. Therefore, we have $\tau_{\Im-(R)}(\eta) = \{\Pi, \varphi, \{Cor_7\}, \{Cor_1, Cor_2, Cor_3, Cor_6, Cor_8\}, \{Cor_1, Cor_2, Cor_3, Cor_6, Cor_7, Cor_8\}\} \neq \tau_{\Im}(\eta)$. From Case (I), we obtain $Core(\Im) = \{C, R\}$.

Case II Let $\eta = \{Cor_1, Cor_4, Cor_5, Cor_6\}$ be the set of coronavirus patients not having (100%) health rate recovery. Hence, $\prod/I(\Im) = \{\{Cor_1\}, \{Cor_4\}, \{Cor_5\}, \{Cor_7\}, \{Cor_3, Cor_6\}, \{Cor_2, Cor_8\}\}$, and this implies that $\tau_{\Im}(\eta) = \{\Pi, \varphi, \{Cor_3, Cor_6\}, \{Cor_1, Cor_4, Cor_5\}, \{Cor_1, Cor_3, Cor_4, Cor_5, Cor_6\}\}$. If we ignore the medication "*C*," then we get $\Pi/I(\Im) = \{\{Cor_1\}, \{Cor_4\}, \{Cor_2, Cor_8\}, \{Cor_3, Cor_6\}, \{Cor_5, Cor_7\}\}$, and we consider $\tau_{\Im-(C)}(\eta) = \{\prod, \varphi, \{Cor_1, Cor_4\}, \{Cor_3, Cor_5, Cor_4, Cor_5, Cor_6, Cor_7\}\{Cor_1, Cor_3, Cor_4, Cor_5, Cor_6, Cor_7\}\} \neq \tau_{\Im}(\eta)$. If the medication "*A*" is ignored that means we will consider that

$$\Pi/I(\Im - (A)) = \{\{Cor_1\}, \{Cor_4\}, \{Cor_5\}, \{Cor_7\}, \{Cor_2, Cor_8\}, \{Cor_3, Cor_6\}\}$$

we obtain, which is the same as $\Pi/I(\Im)$, and we fined $\tau_{\Im-(A)}(\eta) = \tau_{\Im}(\eta)$. If the medication "*T*" is ignored we obtain, $\Pi/I(\Im - (T)) = \{\{Cor_1\}, \{Cor_4\}, \{Cor_5\}, \{Cor_7\}, \{Cor_2, Cor_8\}, \{Cor_3, Cor_6\}\}$, which is the same as $\Pi/I(\Im)$, and we fined $\tau_{\Im-(T)}(\eta) = \tau_{\Im}(\eta)$. When the medication "*K*" is ignored that means we will consider that $\Pi/I(\Im - (K)) = \{\{Cor_1\}, \{Cor_4\}, \{Cor_5\}, \{Cor_7\}, \{Cor_2, Cor_8\}, \{Cor_3, Cor_6\}\}$ which is the same as $\Pi/I(\Im)$, and we fined $\tau_{\Im-(K)}(\eta) = \tau_{\Im}(\eta)$. If the medication "*R*" is ignored we obtain, $\Pi/I(\Im - (R)) = \{\{Cor_4\}, \{Cor_5\}, \{Cor_7\}, \{Cor_3, Cor_6\}, \{Cor_1, Cor_2, Cor_8\}\}$. Therefore, $\tau_{\Im-(R)}(\eta) = \{\Pi, \varphi, \{Cor_4, Cor_5\}, \{Cor_1, Cor_2, Cor_3, Cor_6, Cor_8\}, \{Cor_1, Cor_2, Cor_3, Cor_4, Cor_5, Cor_6, Cor_8\}\} \neq \tau_{\Im}(\eta)$. From Case II, we obtain $Core(\Im) = \{C, R\}$.

Observation From the two cases above, we investigate that "*C*" and "*R*" with their using rates are the necessary to say that a patient has approximation (100%) health rate recovery. So, we recommend that the best medications for (COVID-19) are

Chloroquine and Remdesivir. Therefore, this technical is so useful to help us to determine the medications that suitable activities for coronavirus patients.

References

1. Elfiky AA (2020) Anti-HCV, nucleotide inhibitors, repurposing against COVID-19. Life Sci 248:117477
2. Dong L, Hu S, Gao J (2020) Discovering drugs to treat coronavirus disease 2019 (COVID-19). Drug Discov Ther 14(1):58–60
3. Lai CC, Shih TP, Ko WC, Tang HJ, Hsueh PR (2020) Severe acute respiratory syndrome coronavirus 2 (SARS-CoV-2) and corona virus disease-2019 (COVID-19): the epidemic and the challenges. Int J Antimicrob Agents 105924
4. Li G, De Clercq E (2020) Therapeutic options for the 2019 novel coronavirus (2019-nCoV). Nat Rev Drug Discov 19(3):149–150
5. Thivagar ML, Riclord C (2013) On nano forms of weakly open sets. Int J Math Stat Invention 1(1):31–37
6. Jehad RK, Ahmed HA (2019) Properties of a complete fuzzy normed algebra. BSJ 16(2):382–388
7. Samira NK, Muna JM (2019) On fully stable Banach algebra modules relative to an ideal. BSJ 14(4):813–815
8. Khalil SM, Hameed F (2020) Applications of fuzzy ρ-ideals in ρ-algebras. Soft Comput. https://doi.org/10.1007/s00500-020-04773-3
9. Kadhim SN (2018) On fully stable Banach algebra modules and fully pseudo stable Banach algebra modules. BSJ 15(1):102–105
10. Abdullah HK, Jawad HK (2019) New types of pseudo ideals in pseudo Q-algebra. BSJ 16(2):403–408
11. Al-Musawi AM, Mahmood S, Ulrazaq MA (2019) Soft (1,2)-strongly open maps in bi-topological spaces. IOP Conf Ser Mater Sci Eng 571:012002. https://doi.org/10.1088/1757-899X/571/1/012002
12. Mahmood S, Hasab MH (2020) Decision making using new distances of intuitionistic fuzzy sets and study their application in the universities, INFUS (to appear)
13. Mahmood S, Hameed F (2018) An algorithm for generating permutations in symmetric groups using soft spaces with general study and basic properties of permutations spaces. J Theor Appl Inform Technol 96(9):2445–2457
14. Mahmood S, Hameed F (2018) An algorithm for generating permutation algebras using soft spaces. J Taibah Univ Sci 12(3):299–308
15. Mahmood S, Ulrazaq MA, Abdul-Ghani S, Al-Musawi AM (2018) σ-algebra and σ-Baire in fuzzy soft setting. Adv Fuzzy Syst 2018:10 p, Article ID 5731682
16. Mahmood S, Abdul-Ghani S (2019) Soft M-ideals and soft S-ideals in soft S-algebras. J Phys Conf Ser 1234:012100. https://doi.org/10.1088/1742-6596/1234/1/012100
17. Mahmood S (2019) Decision making using algebraic operations on soft effect matrix as new category of similarity measures and study their application in medical diagnosis problems J. Intell Fuzzy Syst 37(2):1865–1877. https://doi.org/10.3233/JIFS-179249
18. Mahmood S, Hameed F (2020) Applications on cyclic soft symmetric. IOP Conf Ser J Phys 1530:012046. https://doi.org/10.1088/1742-6596/1530/1/012046
19. Mahmood S (2016) Soft dissimilarity fuzzy soft points and their applications. J Fuzzy Inf Eng 8:281–294
20. Ashraf SN, El-Atik AA (2019) A model of a human heart via graph nano topological spaces. Int J Biomath 12(1):1950006

21. Thivagar ML, Priyalatha SPR (2017) Medical diagnosis in a indiscernibility matrix based on nano topology. Cogent Math 4(1):1–9
22. Pawlak Z (1982) Rough sets. Int J Comput Inf Sci 11(5):341–356

Chapter 73
Usability Evaluation of Virtual Reality-Based Fire Training Simulator Using a Combined AHP and Fuzzy Comprehensive Evaluation Approach

El Mostafa Bourhim and Abdelghani Cherkaoui

1 Introduction

The high demand and rapid advance of VR training simulators have attracted many study interests. Since the event of VR-based fire training simulators with an easy-to-use interface is essential for successful adoption and use of the applications, one of the main concerns of the research is how to assess the acceptable usability of a VE. Usability is an evaluation method used to measure the extent to which users can use a specific software system. An effective usability assessment should be able to obtain feedback from users on their use of a VE simulator and how they wish to use it. Traditional evaluation techniques [1] apply to VEs. However, they may not be comprehensive enough to explain the specific usability features of interactive and immersive 3D environments such as VEs. Larger VE interfaces are intrinsically dissimilar from conventional graphical user interfaces (GUIs), with separate input/output (I/O) devices, physiological interactions, and perspectives. For example, when the usability specialist wants to adopt the usability approaches to evaluating VEs, they find few, if any. Therefore, it is essential to develop specific guidelines for assessing the usability of a VE. During this study, a new model supported the combination of the AHP weighting approach and the FCE method was used to propose a systematic satisfaction assessment model to provide guidance to fire researchers on usability assessment and techniques to improve VR fire simulation.

The remainder of the article is organized into five sections: Sect. 2 briefly presents the literature related to the topic. Methodological details are then given in Sect. 3.

E. M. Bourhim (✉) · A. Cherkaoui
Research Team EMISYS: Energetic, Mechanic & Industrial Systems, Engineering 3S Research Center, Industrial Engineering Department, Mohammadia School of Engineers, Mohammed V University, Rabat, Morocco
e-mail: elmostafabourhim@research.emi.ac.ma

I. Jeena Jacob et al. (eds.), *Data Intelligence and Cognitive Informatics*,
Algorithms for Intelligent Systems,
https://doi.org/10.1007/978-981-15-8530-2_73

In Sect. 4, the results and discussion of the results are provided. The final section presents the conclusions, limitations, and directions for future research.

2 Literature Survey

Customary appraisal techniques [1] apply to VEs. Be that as it may, they may not be adequately nitty gritty to depict the particular ease of use properties of VEs. Actually, studies [2–4] have demonstrated that VR framework originators cannot utilize just the techniques that have been produced for 2D graphical interfaces (gis) on the grounds that their collaboration standards are altogether particular from standard gis. Albeit inadequate work on the ease of use of VEs has been done to date, there are spearheading considers that plan to improve VEs from the client's perspective by utilizing an efficient way to deal with ease of use appraisal [5–7]. In talking about the characteristics of VEs, their exploration featured the deficiency of existing techniques for assessing virtual environment frameworks. Various investigations [6–10] on the ease of use of VEs are situated toward the arrangement of two fundamental classes for the ease of use of VR-based frameworks. These are both the virtual environment framework interface and the UI. Despite the fact that these two allotments and their subclasses may cover as parts of virtual environment framework structure, qualifications can be made on how every model adds to ease of use. Subsequently, these differentiations can encourage productive and successful assessment of the ease of use of the virtual environment framework. Although there are several methodologies for estimating the usability of human–computer applications, these methodologies have some limits, particularly for evaluation VEs. For example, the larger usability assessment procedures are suited for a limited variety of interfaces and have few innovative features such as those found in VEs. Although there are several methodologies for estimating the usability of human–computer applications, these methodologies have some limits, particularly for evaluation VEs. For example, the larger usability assessment procedures are suited for a limited variety of interfaces and have few innovative features such as those found in VEs.

Assessing the usability of VEs is a multi-criteria decision problem (MCDM), hence the interest in creating a set of indexing system to evaluate and determine the weight of each criterion. Methods cited in the literature focus on the subjective dimension which involves the user's perception and the cognitive method to relate what the user feels when interacting with the system such as AHP, judgment matrix, point estimation, fuzzy AHP, etc. As these techniques are human dependent, people have complex behavior and perpetually embedded uncertainty, which makes it impossible to take into account precise values in these cases. Consequently, fuzzy approaches are widely selected for assessing usability [11]. Expert knowledge is seen as a contribution to the assignment of subjective value to each criterion.

Fig. 1 Experiment procedure

3 Research Methodology

3.1 Participants

In this experiment, which simulates the effectiveness of VR in the case of fires, a random sample of 200 participants was taken for this experiment. Of these 200 subjects, 19 people were unable to complete the experiment due to fear or stress caused by the virtual simulation, so the rate of participants who completed the experiment is 90.5%.

3.2 Experiment Description

Our virtual fire experiment has been carried out entirely in a VE. The experiment is constituted by five rounds, namely (a) a brief video tutorial at the commencement of the test, (b) the warm-up training (c) the interaction of the participants with the VR situation, and (d) the accomplishment of an administered survey. After having finished the questionnaire, participants were entirely notified about the goals of the test, and they were given the opportunity to request additional questions (see Fig. 1).

3.3 Framework for VE Usability Evaluation

Estimating the usability of VEs is an MCDM subject, in that case, it is vital to structure a hierarchy system to evaluate and conclude the weight for each criterion. A range of approaches has been discussed in the literature to quantify the weights of the MCDM subject. These approaches highlight the subjective dimension which involves the user's opinion and the cognitive approach to construct a relationship with how the user feels while interacting with the system such as AHP, judgment matrix, point estimation, fuzzy AHP, etc. As these techniques are humanly dependent and individuals have a distinct point of view and uncertainty always built-in, there is no justification for considering specific values in these circumstances. Thus, fuzzy approaches are widely adopted for assessing usability [11]. The strong personal knowledge of experts is seen as a contribution to the subjective value assigned to each criterion. In this research, an AHP weighting procedure and a fuzzy evaluation approach were applied to provide a metric that combines both the design dimension and human input. The suggested method for estimating the usability of VEs is shown in Fig. 2.

3.4 Evaluation Index System

Structuring an effective and usable intelligent VR reproduction framework is a test for human variables experts and framework engineers. Shockingly, regular ease of use rules does not mirror the particular highlights of VR frameworks, for example, object determination, control, wayfinding, and route, notwithstanding the combination of tactile, haptic, and visual framework yields. VR makers must upgrade nearness, immersion, and solace in the VE while diminishing both the ailment and VR side effects simultaneously. The structure presented by Stanney et al. [4] was a reasonable technique for this exploration to surveying VE convenience in the created VR fire simulation framework. It gives a deliberate methodology that licenses VE makers to perform usability in VE framework structure and assessment. The created VE secured haptic, yet for the nonexistence of material impact in the created VE, the standard concerning contact feeling was dropped.

The motivation behind the model that is utilized the ease of use of the VE is situated at the top. The essential level incorporates the models that ought to be fulfilled to accomplish the worldwide target. The essential level gave two primary models: the VE framework interface and the UI. Every one of these two guidelines should likewise be delegated a particular component at the subsequent level. A significant distinction must be given regarding input gadgets contrasted with collaboration aptitudes, as info gadgets are just the physical equipment used to perform complex cooperation abilities. In light of this distinction, the VE framework interface can be arranged into collaboration strategies and multimodal framework yields. Accordingly, the cooperation is separated into three sub-factors: route, object control, and direction. Multimodal yield incorporates: tangible, visual demeanors, and haptic input. These models are applied to include observation, cognizance, and relational abilities to comprehend the capacity of VEs. Hence, the yield of the multimodal framework is deteriorated into two sub-factors: visual and sound-related. The UI is additionally separated into two sub-factors: commitment and reactions. Hence, commitment concerns both framework-driven inclusion (for example submersion) and client-driven contribution (e.g., nearness) [11]. These standards are proposed here as the two principle sub-components of commitment. Additionally, solace, disorder, and eventual outcomes are picked as a third symptom standard.

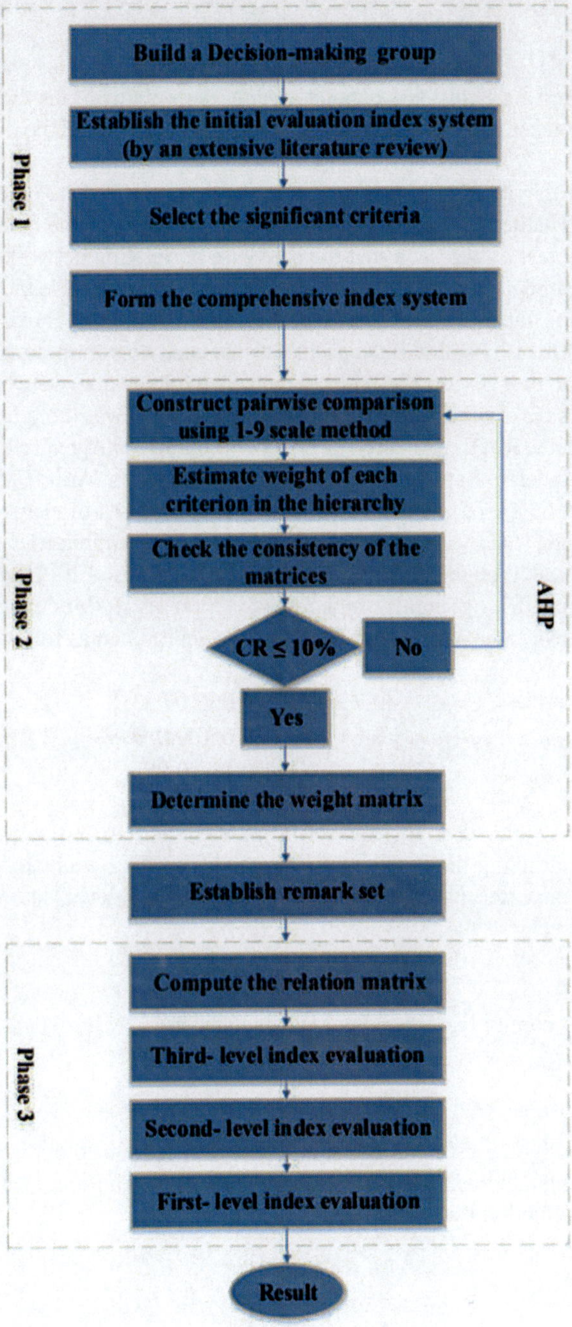

Fig. 2 Flowchart of the evaluation procedure

4 Results

When developing a usability assessment system, the weights of the factors and sub-factors must be taken into account in a meaningful way. To obtain the factor and sub-factor weights, a group of decision-makers was created. Surveys were distributed to stakeholders to gather their views. Characteristic matrices were obtained. At that point, the C.R. values for all grids were resolved. In light of the consequences of the consistency test, it was seen that the pair-wise correlation networks gave by the leaders were completely steady. The AHP was then used to perceive the weighting of elements and sub-factors (see Table 1). In view of the rating list framework and the arrangement of decisions plot in the past areas, the members' impressions of the VR preparing reenactment is got. The record assessment frameworks were then prepared. Let us take the case of the framework R_11, when "Route" was thought of, 73.5% of members evaluated it "High," 16.6% of members appraised it "generally high," 6.6% appraised it "Moderate," and 3.3% evaluated it "Low"; when "Article Manipulation" was considered, 65.7% of members evaluated it "High," 25.4% of members evaluated it "generally high," 5% evaluated it "Moderate," and 3.9% appraised it "Low"; when "Wayfinding" was thought of, 66.3% of members appraised it "High," 17.1% of members evaluated it "generally high," 11% appraised it "Moderate," and 5.5% appraised it "Low." Thus, the lattice R_11 can be determined as follows:

$$R_{11} = \begin{bmatrix} 0.735\ 0.166\ 0.066\ 0.033 \\ 0.657\ 0.254\ 0.050\ 0.039 \\ 0.663\ 0.171\ 0.110\ 0.055 \end{bmatrix}$$

Likewise, the framework R_12,R_21, R_22 were gotten, and they are appeared in Table 1. By utilizing the fluffy administrator (•, +), FCE sets can be gotten. At that point got, the extensive assessment set of the principal level [1].

$$\mathrm{B} = \begin{bmatrix} 0.643\ 0.149\ 0.112\ 0.097 \end{bmatrix}$$

The overall score of the experiment is calculated as:

$$N_{score} = B \times V = 90.038$$

As indicated by the estimation result, the overall score of the investigation was 90.038, implying that members have an elevated level of fulfillment with the VR fire preparing reproduction framework.

Table 1 Evaluation results for each grade percentage [1]

First-level index	Weight	Second-level index	Weight	Third-level index	Weight	Relation matrix	The proportion of each evaluation level			
							High	Relatively high	Moderate	Low
VE System Interface (U_1)	0.250	Interaction (U_{11})	0.167	Navigation (U_{111})	0.172	R_{11}	0.735	0.166	0.066	0.033
				Object Manipulation (U_{112})	0.726		0.657	0.254	0.050	0.039
				Wayfinding (U_{113})	0.102		0.663	0.171	0.110	0.055
		Multimodal System Output (U_{12})	0.833	Visual (U_{121})	0.800	R_{12}	0.624	0.160	0.166	0.050
				Auditory (U_{122})	0.200		0.613	0.210	0.138	0.039
VE User Interface (U_2)	0.750	Engagement (U_{21})	0.667	Presence (U_{211})	0.500	R_{21}	0.718	0.160	0.099	0.022
				Immersion (U_{212})	0.500		0.652	0.166	0.144	0.039
		Side Effects (U_{22})	0.333	Comfort (U_{221})	0.701	R_{22}	0.807	0.110	0.028	0.055
				Sickness (U_{222})	0.193		0.028	0.033	0.166	0.773
				Aftereffects (U_{223})	0.106		0.011	0.039	0.072	0.878

5 Conclusions

To accurately evaluate the VR-based fire training simulator, the above research builds based on fuzzy model for fire researchers and helps to improve VR fire simulation. The overall rate of the trial was 90.038, which explains why the subjects were very satisfied with the VR-based fire training simulator. The application of a VR fire simulation for training situations has several advantages over real-world studies. The most significant advantage is the safety of a VR performed in an unrestricted laboratory, which gives the subjects minimal risk. Also, the simulator used in this study is a highly controlled environment that allows the analysis of scenarios that cannot be tested in a real environment.

References

1. El Mostafa Bourhim, Research Team, and Research Team Abdelghani Cherkaoui (2020) Efficacy of virtual reality for studying people's pre-evacuation behavior under fire. Int J Hum-Comput Stud 102484
2. Nielsen J (1993) Usability engineering. Academic Press, Boston
3. Bowman DA, Hodges LF (1997) An evaluation of techniques for grabbing and manipulating remote objects in immersive virtual environments. In: Proceedings of the 1997 symposium on interactive 3D graphics, pp 35–38. ACM Press, New York
4. Poupyrev I, Ichikawa T (1999) Manipulating objects in virtual worlds: categorization and empirical evaluation of interaction techniques. J. Vis Lang Comput 10(1):19–35
5. Stanney KM, Mollaghasemi M, Reeves L, Breaux R, Graeber DA (2003) Usability engineering of virtual environment (VEs): identifying multiple criteria that drive effective VE system design. Int J Hum-Comput Stud 58:447–481
6. Hix D, Gabbard JL (2002) Usability engineering of virtual environments. In: Stanney KM (ed) Handbook of Virtual Environments: Design, Implementation, and Applications. Lawrence Erlbaum Associates, Mahwah, NJ, pp 681–699
7. Kalawsky RS (1999) VRUSE—a computerized diagnostic tool: for usability evaluation of virtual/synthetic environments systems. Appl Ergon 30:11–25
8. Bowman D (1999) Interaction techniques for common tasks in immersive virtual environments: Design, evaluation, and application. Unpublished doctoral dissertation, Georgia University of Technology, Georgia
9. McCauley-Bell P (2002) Ergonomics in virtual environments. handbook of virtual environments: design, implementation, and applications. Lawrence Erlbaum Associates, Mahwah, pp 807–826
10. Sadowski W, Stanney KM (2002) Measuring and managing presence in virtual environments. Handbook of virtual environments: design, implementation, and applications, pp 791–806. Lawrence Erlbaum Associates, Mahwah
11. Turk M, Robertson G (2000) Perceptual user interfaces. Commun ACM 43(3):33–34
12. Chang E, Dillon,Mar TS (2006) A usability-evaluation metric based on a soft-computing approach. IEEE Trans Syst Man Cybern. A, Syst Hum 36(2):356–372
13. Bourhim EM, Cherkaoui A (2020) Selection of optimal game engine by using ahp approach for virtual reality fire safety training. In: Abraham A, Cherukuri A, Melin P, Gandhi N (eds.) Intelligent Systems Design and Applications. ISDA 2018. Advances in Intelligent Systems and Computing, vol 940. Springer, Cham

14. Bourhim EM, Cherkaoui A (2020) Exploring the potential of virtual reality in fire training research using A'WOT hybrid method. In: Thampi S et al. (eds.) Intelligent Systems, Technologies and Applications. Advances in Intelligent Systems and Computing, vol 1148. Springer, Singapore

Chapter 74
Study the Basic Survey of Food Quality Parameter Analysis Technique Using Image Development Methodology

Sanket Mungale and Deepak S. Khot

1 Introduction

Food quality is the most crucial task in our life because proper food quality has to be maintained in our day to day life. In the Sanskrit language, one word is vital "anna-data sukhi bhav" is the best approach in our life. Many countries like China, India, Malaysia, Bhutan, America and UAE created sensible food quality of human life. In India, due to vast geographical regions and culture provides a variety of food products and hence necessary to improve the quality and standards of the manufacturing process. To test the food products many quality institutes are available. International Monde selection is one of the oldest in food quality evaluation. The Indian government improves the methods and standards of food production to produce healthy and hygienic food for the consumers and supports the life of farmers and improve their living standards.

2 Objective

The objective of the study as follows:

- Time and energy saving of food grading.
- Increase reliability when compared to Manual Grading.

S. Mungale (✉) · D. S. Khot
Department of ENTC, AMGOI, Kolhapur, Maharashtra, India
e-mail: smungale044@gmail.com

D. S. Khot
e-mail: dskhot0012@gmail.com

I. Jeena Jacob et al. (eds.), *Data Intelligence and Cognitive Informatics*,
Algorithms for Intelligent Systems,
https://doi.org/10.1007/978-981-15-8530-2_74

- Grading supported External Quality issue.
- Higher Accuracy than manual Sorting

3 Need of Project

The prime need of the project is to analyze the quality of the food products and the measures for checking the parameters of food quality. Artificial Intelligent plays a vital role in the evaluation and analyses of food characteristics. Automated food processing technology also helps to improve the production and quality standards. This paper surveys all the latest image processing techniques.

4 Food Quality Management

Food quality management system helps the food industry to improve the efficiency and quality of food production by managing safety risks. Many systems and tools are necessary to assist in the implementation of quality standards in the food manufacturing process to improve security and competitiveness. Many quality assurance systems are available to achieve the expected quality level. To meet the objective, appropriate tools and systems are necessary to improve the quality of the product. The food industry needs to be organized systematically for improving quality and standards and customer satisfaction.

5 Proposed Work

The segmented sample should not exceed twenty, in the course of future food storage in the box. The common half location, and seven of common half perimeter length whereas running on 940 nm key wavelength diode illumination. Proxy Mate is also a durable, compact and easy to use at-line NIR instrument for the food and feed business.

It reduces the amount in production and provides fast control of batch samples. The intuitive software permits operation by everybody and an honest alternative of pre-calibration packages produce it applicable for all applications.

The NIR Master skilled IP65 is the world's first dedicated FT-NIR standalone prism. Spectroscope with IP65 protection class. Its hygienic vogue is certified to be proof against water jet cleaning procedures. It entirely meets the requirements for the safety of control instruments used in farm and meat food production.

NIR Cal is also a robust chemometric code package to develop durable NIR ways in which lead to fast and reliable NIR prediction results.

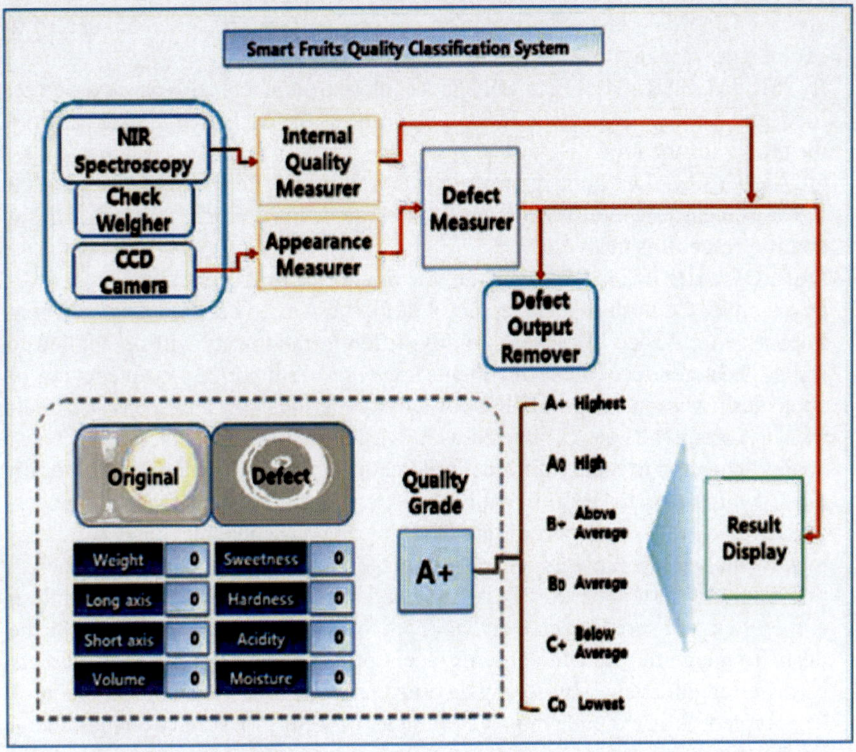

Fig. 1 System block diagram

6 Proposed System

(Figure 1).

7 Description of Components

1. NIR METER: It helps to determine the moisture level at the surface and to evaluate and determine the substances like fats, oils and sugars. In food quality, also amendment their work Holik and paralysis work to internal and external food quality grading some minimum changes measure acquired. Ferment grades, food (human grade) and non-food (industrial grade).
2. CHECK WEIGHER: It is the automatic machine for weight checking of packed products and is additionally developed the parameter result and can be effective of automatic grading machine. It ensures the proper weight of the package.
3. CCD CAMERA: It is coupling camera device which might additionally rotate and inspect food containers for stains and other defects. The photographs can

check of internal content and use for analysis purpose. It can be connected to external devices like laptops and other vision systems.

4. INTERNAL PARAMETER CHECK: the internal parameter also check the color, food-grade, sweet or left tend food. It also checks the content in food to measure the quality of the food.
5. QUALITY GRADE: Food quality grade is additionally very important as a result of the standard of food for human consumption. Food grading sorts foods for quality, value, freshness etc.
6. DEFECT CHECKER: Over storage, the quality of fruits typically changes. As an example, the flesh of some apples will flip brown. To cut back this browning impact and enhance the vision, totally different treatments will be applied to fruits. The potential of those treatments is evaluated supported a visible review of apple flesh, associate degree of correct activity is critical to match the varied treatments. A study has been conducted with IRIS visual instrument on ten batches of apples subjected to totally different treatments aimed toward deceleration down the browning of their flesh. To gauge the extent of this vision defect, the analysis with IRIS visual instrument consists of taking an image of apples cut in two items then to analyze the pictures. Thanks to image pre-processing, the world cherish the vision defect will be exactly isolated and also the proportion of broken flesh is therefore measured objectively, as a proportion of the overall surface of the flesh. To gauge the visual quality of recent of apple, a top-quality management card was engineered up by setting the most proportion of vision defect that may be accepted, as an example most 1 Chronicles of the brown surface on the general flesh surface.
7. DEFECT REMOVER SYSTEM: Food article defect removal equipment includes a manifold that defines one or additional chambers for holding controlled fluid. Initial channels extend from a primary facet of the manifold into fluid communication with the chambers. Second channels extend from the primary facet of the manifold to a second facet of the manifold. Valves by selection connect corresponding initial channels and second channels along to dispense the controlled fluid from the manifold. The manifold will have AN exterior wall that a minimum of part defines a primary chamber and a second chamber, with an inside wall, disposed between the primary chamber and also the second chamber. the inside wall will outline the second channels. The valves are often enclosed with a valve assembly, which incorporates a driver operably as well as the valves. The valves are often pluggable as well as the driving force.
8. RESULT DISPLAY: Food quality result which can be also displayed on some customer point of view on internal food taste, whether this food is good or not also any chemical content is involved or not and what type food quality is provided and this barrier can be removed from the result for a minimum and maximum content analysis in the display.

RESULT: Sample of orange fruit.

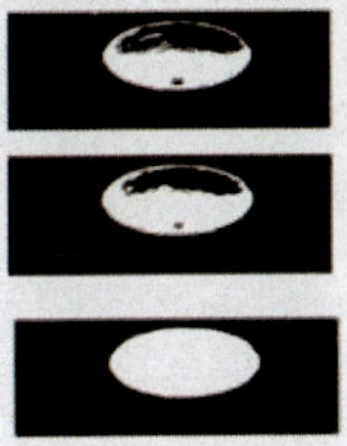

8 Conclusion

This project analyses the quality and standards of food production and proposed a system using Artificial Intelligence in the evaluation of food characteristics using image processing techniques.

9 Future Scope

This work can also be extended in future using IoT based system and Artificial neural network methods to compete for the challenges in the food industry.

References

1. Ali MAH (2017) Automated fruit grading system. In: IEEE 3rd international symposium on robotics and manufacturing automation (ROMA)
2. Choi HS, Cho JB, Kim SG, Choi HS (2018) A real-time smart fruit quality grading system classifying by external appearance and internal flavor factors
3. Tumas P, Serackis A (2017) Effective background subtraction algorithm for food inspection using a low-cost near infrared camera. Department of Electronic Systems Vilnius Gediminas Technical University, Naugarduko
4. Singhal P, Dubey AK, Goyal A (2017) A comparative approach for image segmentation to identify the defected portion of Apple. In: International Conference on Intelligent. Amity University Uttar Pradesh, Noida, India, vol 4, Texas A&M University, Kingsville
6. Bulanon M, Kataoka T (2010) Fruit detection system associate degreed an finish effectorfor robotic gather of Fuji apples. Agric Eng Int CICR J 12
7. Schertz E, Brown GK (1968) Basic issues in mechanizingcitrus harvest. Trans ASAE II(3):343–346
8. Parrish E, Goksel K Pictorial pattern recognition applied to fruit harvesting. Trans ASAE 20
9. Hayashi S, Ota T, Kubota K, Ganno K, Kondo N (2005) Robotic harvesting technology for fruit vegetables in protected agriculture production.In: Data and technology for property fruit and vegetable production, Frutic, Motpellier, France, vol 05
10. Chinchuluun R, Lee WS, Burks TF (2006) Machine vision-based Citrus yield mapping system. In: Proceedings of the Florida State Horticultural Society, vol 119, pp 142–147

11. Bulanon DM, Kataoka T, Okamoto H, Hata S (2004) Development of areal-time machine vision system for the Apple gather mechanism. In: SiCE annual conference in sapporo, pp 4–6, August 2004
12. Plebe A, Grasso G (2001) Localization of spherical fruits for roboticharvesting. Mach Vis Appl 13:70–79
13. Edan Y, Flash T, Miles GE (2000) Robotic Melon gather. IEEET Robot 16

Author Index

I. Jeena Jacob et al. (eds.), *Data Intelligence and Cognitive Informatics*, Algorithms for Intelligent Systems,
https://doi.org/10.1007/978-981-15-8530-2